Particulate Plastics in Terrestrial and Aquatic Environments

Particulate Plastics in Terrestrial and Aquatic Environments

Edited by
Nanthi S. Bolan
M.B. Kirkham
Claudia Halsband
Dayanthi Nugegoda
Yong Sik Ok

CRC Press is an imprint of the
Taylor & Francis Group, an **informa** business

First edition published 2020
by CRC Press
6000 Broken Sound Parkway NW, Suite 300, Boca Raton, FL 33487-2742

and by CRC Press
2 Park Square, Milton Park, Abingdon, Oxon, OX14 4RN

Library of Congress Cataloging-in-Publication Data

Names: Bolan, Nanthi, S., editor. | Kirkham, M. B., editor. | Halsband, Claudia, editor. | Nugegoda, Dayanthi, editor. | Ok, Yŏng-sik, 1944- editor.
Title: Particulate plastics in terrestrial and aquatic environments / edited by Nanthi S. Bolan, M.B. Kirkham, Claudia Halsband, Dayanthi Nugegoda, Yong Sik Ok.
Description: First edition. | Boca Raton : CRC Press, 2020. | Includes bibliographical references and index. | Summary: "Particulate Plastics in Terrestrial and Aquatic Environments provides a fundamental understanding of the sources of these plastics and the threats they pose to the environment. It demonstrates the ecotoxicity of particulate plastics using case studies and offers management practices to mitigate environmental contamination"-- Provided by publisher.
Identifiers: LCCN 2020013284 (print) | LCCN 2020013285 (ebook) | ISBN 9781138543928 (hardback) | ISBN 9780367511401 (paperback) | ISBN 9781003053071 (ebook)
Subjects: LCSH: Microplastics--Environmental aspects. | Pollution prevention.
Classification: LCC TD196.P5 P37 2020 (print) | LCC TD196.P5 (ebook) | DDC 628.5/2--dc23
LC record available at https://lccn.loc.gov/2020013284
LC ebook record available at https://lccn.loc.gov/2020013285

ISBN: 978-1-138-54392-8 (hbk)
ISBN: 978-0-367-51140-1 (pbk)
ISBN: 978-1-003-05307-1 (ebk)

Typeset in Times
by Lumina Datamatics Limited

Contents

Foreword ix
Preface xiii
Acknowledgments xv
Editors xvii
Contributors xix

SECTION I Sources of Particulate Plastics in the Environment

Chapter 1 Sources of Particulate Plastics in Terrestrial Ecosystems 3

Lauren Bradney, Hasintha Wijesekara, Nanthi S. Bolan, and M.B. Kirkham

Chapter 2 Particulate Plastics from Agriculture 19

M.B. Kirkham, Reshma M. Antony, and Nanthi S. Bolan

Chapter 3 Polyacrylamide (PAM) as a Source of Particulate Plastics in the Terrestrial Environment 39

Franklin M. Chen

Chapter 4 Analytical Methods for Particulate Plastics in Soil and Water 51

Upekshya Welikala, Chanaka M. Navarathna, Samadhi Nawalage, Binoy Sarkar, Todd E. Mlsna, and Sameera R. Gunatilake

SECTION II Distribution and Characteristics of Particulate Plastics

Chapter 5 An Introduction to the Chemistry and Manufacture of Plastics 85

Nanthi S. Bolan, Kandaswamy Karthikeyan, Shiv Shankar Bolan, M.B. Kirkham, Ki-Hyun Kim, and Deyi Hou

Chapter 6 Interaction of Dissolved Organic Matter with Particulate Plastics 95

Yubo Yan, Qiao Li, Shiv Shankar Bolan, Nanthi S. Bolan, Yong Sik Ok, M.B. Kirkham, and Eilhann E. Kwon

Chapter 7 Characteristics of Particulate Plastics in Terrestrial Ecosystems 107

Kumuduni Niroshika Palansooriya, Hasintha Wijesekara, Lauren Bradney, Prasanna Kumarathilaka, Jochen Bundschuh, Nanthi S. Bolan, Teresa Rocha-Santos, Cheng Gu, and Yong Sik Ok

Chapter 8 Facilitated Transport of Zinc on Plastic Colloids through Soil Columns 125

Christopher Barton

Chapter 9 Microbial Plastisphere: Microbial Habitation of Particulate Plastics in Terrestrial and Aquatic Environments 135

Nanthi S. Bolan, M.B. Kirkham, B. Ravindran, Anu Kumar, and Weixin Ding

Chapter 10 Aggregation Behavior of Particulate Plastics and Its Implications 147

Xinjie Wang, Yang Li, Jiajun Duan, Yuan Liu, Shengdong Liu, Enxiang Shang, Nanthi S. Bolan, and Yining Wang

SECTION III Ecotoxicity of Particulate Plastics

Chapter 11 Environmentally Toxic Components of Particulate Plastics 165

Sanchita Mandal, Nanthi S. Bolan, Binoy Sarkar, Hasintha Wijesekara, Lauren Bradney, and M.B. Kirkham

Chapter 12 Particulate Plastics as Vectors of Heavy Metal(loid)s 181

Hasintha Wijesekara, Lauren Bradney, Sanchita Mandal, Binoy Sarkar, Hocheol Song, Nanthi S. Bolan, and M.B. Kirkham

Chapter 13 Water Relations and Cadmium Uptake of Wheat Grown in Soil with Particulate Plastics 193

M.B. Kirkham

Chapter 14 Microplastics as Vectors of Chemicals and Microorganisms in the Environment ... 209

Yini Ma, Lin Wang, Ting Wang, Qianqian Chen, and Rong Ji

Chapter 15 Ecological Impacts of Particulate Plastics in Marine Ecosystems 231

Claudia Halsband and Andy M. Booth

Chapter 16 Sub-Lethal Responses to Microplastic Ingestion in Invertebrates: Toward a Mechanistic Understanding Using Energy Flux 247

Charlene Trestrail, Jeff Shimeta, and Dayanthi Nugegoda

Chapter 17 Particulate Plastics and Human Health 285

Nanthi S. Bolan, M.B. Kirkham, Shiv Shankar Bolan, Daniel C.W. Tsang, Yiu Fai Tsang, and Hailong Wang

SECTION IV Case Studies of Particulate Plastics in the Environment

Chapter 18 Status of Particulate Marine Plastics in Sri Lanka: Research Gaps and Policy Needs.......297

T.W.G.F. Mafaziya Nijamdeen, Thilakshani Atugoda, P.B. Terney Pradeep Kumara, A.J.M. Gunasekara, and Meththika Vithanage

Chapter 19 Case Studies of Particulate Plastic Distribution and Ecotoxicity in Japan.......327

Shosaku Kashiwada

Chapter 20 Particulate Plastic Distribution and Ecotoxicity in Marine Ecosystems and a Case Study in Thailand.......355

Suchana Chavanich, Voranop Viyakarn, Somkiat Khokiattiwong, and Wenxi Zhu

Chapter 21 The Current Status of Plastics: A New Zealand Perspective.......363

Louis A. Tremblay, Xavier Pochon, Olivier Champeau, Virginia Baker, and Grant L. Northcott

Chapter 22 Plastic Food for Fledgling Short-Tailed Shearwaters (*Ardenna tenuirostris*): A Case Study.......377

Jacinta Colvin, Peter Dann, and Dayanthi Nugegoda

SECTION V Management of Particulate Plastics

Chapter 23 Management of Particulate Plastic Waste Input to Terrestrial and Aquatic Environments.......397

Binoy Sarkar, Nanthi S. Bolan, Raj Mukhopadhyay, Shiv Shankar Bolan, M.B. Kirkham, and Jörg Rinklebe

Chapter 24 Evaluation and Mitigation of the Environmental Impact of Synthetic Microfibers.......413

Francesca De Falco, Mariacristina Cocca, Emilia Di Pace, Maria Emanuela Errico, Gennaro Gentile, and Maurizio Avella

Chapter 25 Biodegradable Bioplastics: A Silver Bullet to Plastic Pollution?.......425

Steven Pratt, Nanthi S. Bolan, Bronwyn Laycock, Paul Lant, Emily Bryson, and Leela Dilkes-Hoffman

Index.......435

Foreword

It is my great pleasure to pen a few thoughts for this wonderful compilation of expert views on the plastic problem – a problem that has resulted from us as humans being too clever in our quest to develop materials that offer benefits not found in nature. We have developed synthetic replacements for ivory, for pearl shell buttons, for rubber, for timber, but we have not developed a way to dispose of our creations. The durability of plastics means that they accumulate in nature and will remain there at least for the next four centuries. This book highlights the evidence and environmental consequences of the mechanical evolution of the plastic product into particulate forms. We are on the cusp of the health consequences from our ecosystems, foods, and own bodies being swamped with these indigestible microplastics and nanoplastics. The outcomes will be chronicled in future history books.

Plastics offer benefits not found in natural materials, not the least of which is that they are resistant to the enzymes of living organisms. First, we had celluloid, and then ivory was replaced by a cheap chemical produced from cellulose and camphor, a boon for the elephants, but the start of our plastic pollution. In 1907, Bakelite became the first real synthetic mass-produced plastic. Plastic soon became the name for this category of materials made from natural and man-made polymers. Imagine the power of being able to make from these cheap plastics any product that was previously made of wood, bone, tusk, shell, and stone. These new materials did indeed have great environmental and societal benefits, as they replaced the harvesting of raw materials from the geosphere and the biosphere. A golden age was on hand, with durable, cheap plastic items being available to all levels of society and changing the way we live. Silk stockings were one such luxury item replaced by nylon. World War II saw an exponential growth in the type and range of plastics, and in the 1950s there was an almost Utopian view of a future where cheap, safe plastics would be universally available for every human need. It is only now we are regretting the durability of our modern plastics, as they accumulate in the environment. Indeed, plastic debris in the oceans was being reported as early as the 1960s, but little did we understand the compounding enormity of the problem, nor the physical degradation of these man-made discarded objects into particulate forms that could then invade the food chain. Let us hope that this book will shake into wakefulness, awareness, and action both society and our elected officials in all levels of government.

Globally, plastic production has continued to rise for more than 50 years, and it is expected that by 2020, annual production will reach 335,000,000 tons. Every year, over 6 million tons of garbage end up in the world's oceans, of which, 80% are plastics. We have all seen the heart-wrenching pictures of marine birds with stomachs full of indigestible plastic, the turtle head shrouded in plastic, and reports of the tiny microbeads – used in toothpastes and facial scrubs – passing through the filters of water treatment plants and filling the ocean, ready indigestible targets for marine animals.

We can but applaud the initiatives that ban the use and distribution of plastic bags, and the container deposits that have encouraged recycling rather than throwing out drink containers. However, an individual can also contribute to preserving and protecting our ocean through using the free plastic bags recycling facilities and the purchase of biodegradable bags and reusable shopping bags, natural cleansing products, and reusable coffee cups. Legendary local hero Tim Silverwood has been the champion of "Take 3," an Australian not-for-profit organization formed in 2009 that aims to raise awareness of marine debris by encouraging visitors to the beach, or waterway, to simply take three pieces of rubbish with them when they leave.

Sometimes we are not spurred into action until a chance physical encounter. At a recent scientific conference, the poster of a research student graphically detailing his project on microbeads in biowastes caught my eye because it was accompanied by a practical demonstration of the huge number of tiny plastic beads that are in use in household products, such as toothpastes, body scrubs, shampoos, and many others. On display was a small vial with tens of thousands of tiny beads extracted from one bottle of facial cleanser. Since their introduction into our households about 20 years ago, microbeads have become one of the biggest threats to marine wildlife because they are so tiny that they are too small to be captured and filtered out at the water treatment plant, eventually ending up in the ocean as indigestible targets for marine animals. Further, the plastic beads can accumulate dangerous toxins and heavy metals on their trip through our waste disposal network, which are released back into the environment with the death of the organism. A typical usage of a facial scrub might flush 100,000 of these microbeads into sewers and eventually into the marine environment. Along with plastic bags, microbeads represent another major threat to our marine environment. These particles can be transferred to higher levels in the food chain causing adverse effects and may serve as a global transport mechanism for accumulated contaminants, such as persistent organic pollutants.

My personal hope is for a biological solution to the plastic waste problem of the world. I believe that soon we might be able to grow up cultures of specific plastic-eating bacteria and let them loose into landfills. The question is not whether such bacteria exist, but rather how does one find them, isolate them, and then culture them to mega quantities for use on a commercial scale. The very process of evolution tells us that there will be such microbes living in plastic environments that have been naturally selected for their ability to survive on plastics as food. Indeed, it is well known that insect pests found in warehouses where grains are stored in plastic packages are able to chew and eat through the plastic. A recent study from China of mealworms that were able to eat Styrofoam has shown that in the gut of these larvae there are bacteria that can actually breakdown and digest this polystyrene that we once thought was non-biodegradable. Isolation of this *Exiguobacterium* gives hope for the future. Similarly, in Japan, it was reported in 2016 that the soil of a PET bottle recycling plant contained a bacterium *Ideonella sakaiensis* that was capable of digesting this plastic.

Our own biodegradation studies of polystyrene have resulted in the isolation of a number of bacteria from landfills that can break down polystyrene in the laboratory. This opens the possibility of industrial scale cultures of the relevant organisms that might be applied commercially to landfills to specifically breakdown such plastics and contribute significantly to biogas production.

Rachel Carson raised awareness in the 1950s of the devastation caused by our unfettered chemical industrialization. Sharon Bader's book *Toxic Fish and Sewer Surfing* published in 1989 set down in print the facts that were all too apparent to the swimmers and surfers of the day that the beaches around Sydney, Australia, had become open sewers. She detailed how mega liters of nearly raw sewage mixed with toxic industrial waste were being dumped daily into the sea off Sydney right next to the iconic beaches of Bondi and Manly, fouling bathing waters and causing fish contamination. The intervening decades have seen a dramatic improvement in our ocean waters, but even today, we are warned to avoid swimming for at least one day after heavy rain at ocean beaches, and at least three days at harbor beaches, and warned off eating fish and other seafood caught west of Sydney Harbour Bridge because of their contamination with industrial chemicals including dioxins! Ironically, and perhaps illogically, it is a slightly better story if you fish east of the Sydney Harbour Bridge where, depending on the species of fish you catch, your monthly consumption per month will range from 50 g for sea mullet to 150 g for bream and tailor to over a kilogram for flounder. But do not ask the fish if it has crossed under the bridge! And now, with particulate plastics, we leave a further legacy to our grandchildren and their children.

The solutions to the plastic problem must look both forward to biodegradable plastics and the circular economy, and backward to finding ways of dealing with the existing accumulated particulate plastics in our food webs. Sadly, only about 12% of the 3 million tons of plastic waste produced each year in Australia is recycled. Australia has a National Waste Policy that clearly shows us the pathway to a sustainable plastic future. The United Nations has given us the tools to make this happen through the 17 transformative Sustainable Development Goals (SDGs), adopted by all 193 UN Member States in 2015. It is time for us all to step up efforts to deal with the health and environmental challenges of past and future plastic, in this time of the GREAT ACCELERATION of humankind's effects on every one of our Earth's systems.

Emeritus Professor Tim Roberts
School of Environmental & Life Sciences
The University of Newcastle
Australia

Preface

More than 8.3 billion tons of plastic are estimated to exist in the world, the majority of which exists in the form of new products, some as products currently in use, and the rest in landfills, in the wider environment, or in stockpiles waiting for new plastic manufacturing markets. Of the 8.3 billion tons produced worldwide, approximately 79% is estimated to be in landfills or in the broader environment, resulting in the contamination of terrestrial and aquatic environments (Geyer et al. 2017).

The manufacture of plastic, as well as its indiscriminate disposal and destruction by incineration, pollutes atmospheric, terrestrial, and aquatic ecosystems. Synthetic plastics do not biodegrade and tend to accumulate in the environment. Plastics in the environment can occur in a range of physical size fractions covering macro, micro, and nanoplastics. Although the term "microplastics" is used extensively to represent plastic contamination, we deliberately choose the term "particulate plastics" to represent the range of plastic contamination in both terrestrial and aquatic ecosystems. These particulate plastics present in terrestrial and aquatic ecosystems are becoming a major source of pollutants.

There are two major sources of particulate plastics reaching terrestrial and aquatic environments: (i) primary particulate plastics are manufactured and are a direct result of anthropogenic use of plastic-based materials (e.g., microbeads in cosmetics) and (ii) secondary particulate plastics are plastic fragments derived from the breakdown of larger plastic debris. Both types persist in terrestrial (i.e., soil) and aquatic (i.e., marine) ecosystems. Because plastics do not break down readily, they can be ingested and incorporated into the tissues of some terrestrial and aquatic organisms. Large quantities of particulate plastics retained within the marine environment have originated from land-based sources, having been transported through processes, such as sediment transfer or soil erosion. Despite this link to land-based sources, the majority of scientific research on particulate plastics has focused on their effects in aquatic environments. The results indicate that particulate plastics can have a direct impact on biota ingesting them, and they also can act as a vector for pollutants and impact aquatic environments through long-range transport of these pollutants.

Particulate plastics reach terrestrial ecosystems through their indiscriminate disposal in landfills and through compost and application of biosolids. There has been renewed interest in the large-scale application of composts and biosolids to soil, mainly to increase soil health and also to enhance carbon sequestration in soil. Although medium- and large-sized plastic materials are generally segregated during the composting and wastewater treatment processes through sieving, a significant portion of small-sized plastics reach the end products, such as composts and biosolids. Because of the subsequent milling of these products, most plastics end up as micro/nanoplastics.

This book covers five major themes in particulate plastics that include: (i) the sources of particulate plastics in the environment, (ii) distribution and characteristics of particulate plastics, (iii) ecotoxicity of particulate plastics, (iv) case studies of particulate plastics in the environment, and (v) management of particulate plastics. The chapters in these five themes describe various components of particulate plastics; particulate plastic inputs to terrestrial and aquatic environments through wastewater discharge and application of biosolids and compost and through the use of polythene mulching; characteristics of particulate plastics in the environment as impacted by weathering processes; distribution and ecotoxicity of particulate plastics in both terrestrial and aquatic ecosystems; and management practices to mitigate particulate plastic contamination in the environment.

The key features of this book include:

- A fundamental understanding of the various sources of particulate plastic inputs to terrestrial and aquatic ecosystems.
- The various physical and chemical properties of particulate plastics in terrestrial and aquatic ecosystems.

- The ecotoxicity of particulate plastics to terrestrial and aquatic organisms.
- Case studies of particulate plastic contamination in terrestrial and aquatic ecosystems in various regions.
- Various approaches to managing particulate plastic contamination in terrestrial and aquatic ecosystems.

Although a large number of books have been published covering particulate plastic contamination in the environment, these books tend to focus on aquatic ecosystems, whereas this book covers both terrestrial and aquatic ecosystems. This book is suitable for undergraduate and postgraduate students majoring in marine, environmental, earth, and soil sciences, environmental and marine scientists, and environmental regulators.

Nanthi S. Bolan
M.B. Kirkham
Claudia Halsband
Dayanthi Nugegoda
Yong Sik Ok

REFERENCE

Geyer, R., Jambeck, J. R. and Law, K. L. (2017) Production, use, and fate of all plastics ever made. *Science Advances*, 3 (7), e1700782. doi:10.1126/sciadv.1700782.

Acknowledgments

Professor Nanthi S. Bolan acknowledges the University of Newcastle (www.newcastle.edu.au), Australia and the Cooperative Research Center for High Performance Soils (Soil CRC; www.soil-crc.com.au), Callaghan, Australia.

Professor M.B. Kirkham acknowledges Kansas State University, Manhattan, USA.

Dr. Claudia Halsband was supported by Akvaplan-niva and the Fram Centre program "Plastic in the Arctic."

Professor Dayanthi Nugegoda acknowledges RMIT University, Melbourne Australia and AQUEST (Aquatic Environmental Stress Research Group) at RMIT.

Professor Yong Sik Ok was supported by the Cooperative Research Program for Agriculture Science and Technology Development (Effects of Plastic Mulch Wastes on Crop Productivity and Agro-Environment, Project No. PJ014758), Rural Development Administration, Republic of Korea.

Editors

Nanthi S. Bolan, PhD, completed his PhD in soil science and plant nutrition at the University of Western Australia and is currently working as a professor of environmental science at the University of Newcastle. His teaching and research interests include agronomic value of manures, fertilizers, soil amendments, soil acidification, nutrient and carbon cycling, interactions of pesticides and metal pollutants in soils, greenhouse gas emission, soil remediation, and waste and wastewater management. Dr. Bolan is a fellow of the American Soil Science Society, American Society of Agronomy, and New Zealand Soil Science Society and was awarded the Communicator of the Year award by the New Zealand Institute of Agricultural Sciences. He has supervised more than 50 postgraduate students and was awarded the Massey University Research Medal for excellence in postgraduate student supervision. Dr. Bolan has published more than 400 book chapters and journal papers and was awarded the M.L. Leamy Award by the New Zealand Soil Science Society in recognition of his meritorious contribution to soil science. He has served as the associate editor of *Critical Reviews in Environmental Science and Technology* and technical editor of the *Journal of Environmental Quality*. Dr. Bolan is one of the Web of Science Globally Highly Cited Researchers for 2018 and 2019.

M.B. Kirkham, PhD, is university distinguished professor in the Department of Agronomy at Kansas State University and a graduate of Wellesley College (BA) and the University of Wisconsin, Madison (MS and PhD). Dr. Kirkham's research deals with water movement in the soil-plant atmosphere continuum, uptake of heavy metals by plants grown on contaminated soil, and effects of particulate plastics on soil-plant-water relations. Dr. Kirkham has written three textbooks dealing with soil-plant-water relations, is the author or coauthor of more than 300 contributions to scientific publications, and is on the editorial boards of 16 journals. Every year, Dr. Kirkham teaches at Kansas State University a graduate course entitled Plant-Water Relations. Dr. Kirkham is a fellow of the American Society of Agronomy, the Crop Science Society of America, the Soil Science Society of America, and the American Association for the Advancement of Science and has been elected honorary member of the International Union of Soil Sciences. Dr. Kirkham has received several awards, including, most recently, the International Soil Science Award from the Soil Science Society of America.

Claudia Halsband, PhD, a marine ecologist with a joint PhD from the Universities of Paris (France) and Oldenburg (Germany), is currently working as a senior scientist at Akvaplanniva, a private research institute in Tromsø (Norway), where she leads the section Ecosystem Understanding with a focus on Arctic environmental issues. Her research interests include marine ecology, life histories, and reproductive biology of zooplankton, in addition to impacts of environmental change on ecosystem function, including global warming, ocean acidification, and plastic pollution. Dr. Halsband studies microplastic-biota interactions with a focus on experimental toxicity studies with microplastic spheres, fragments, and fibers, as well as crumb rubber and associated contaminants. Dr. Halsband has published more than 40 papers, co-supervised 5 PhD students, and is a former EU Marie-Curie postdoctoral fellow. She is a member of ARCTOS (Arctic Marine Ecosystem Research Network) and the World Association of Copepodologists (WAC) and a former founding member of APECS (the Association of Polar Early Career Scientists).

Dayanthi Nugegoda, PhD, is a professor of ecotoxicology at RMIT University and a lead researcher in the Aquatic Environmental Stress (AQUEST) Research Group. Her group has developed novel methods to assess, monitor, and evaluate the effect of toxicants and environmental stressors on aquatic organisms and ecosystems. She has published more than 160 peer-reviewed journal articles, 3 books, 5 book chapters, and over 30 reports to industry. She has graduated 24 PhD and 5 MSc research students and currently supervises 7 PhD candidates. She served (2012–2014) on the Independent Expert Scientific Panel on Coal Seam Gas and Large Coal Mining and the Forest Stewardship Council (FSC) of Australia. She advises the New South Wales Environment Trust; the Australian Shipowners Association; the Research Councils of Hong Kong, South Africa, Norway, France, Croatia, and the Netherlands; and the OECD Validation Management Group for Ecotoxicity Tests (VMG-eco). In 2015, she served on the Scientific Reference Panel for Onshore Natural Gas and Water, Victoria, and the Science Panel for the Victorian Coastal Council. In 2017, she was invited by the lead scientist of Victoria to serve with the Scientific Reference Group for Onshore Conventional Gas until 2020. Dr. Nugegoda was elected president of the Society for Environmental Toxicology and Chemistry (SETAC) Australasia for the period 2011to 2013. In 2014, she was awarded a visiting professorship at Tianjin, Chengxian University, China; in 2009, she was awarded a visiting professorship at the Helmholtz Research Centre for Environment and Health, Munich, Germany, and a visiting research fellowship at the Flinders University Research Centre for Coastal and Catchment Environments. In 2019, she was awarded the RMIT Vice-Chancellors Award for Research Impact.

Yong Sik Ok, PhD, is a full professor and global research director of Korea University in Seoul, Korea. His academic background covers waste management, the bioavailability of emerging contaminants, and bioenergy and value-added products (such as biochar). Professor Ok also has experience in fundamental soil science and the remediation of various contaminants in soils and sediments. Together with graduate students and colleagues, Professor Ok has published over 600 research papers, 77 of which have been ranked as Web of Science ESI top papers (71 nominated as highly cited papers [HCPs] and 6 nominated as hot papers) since 2009. He has been a Web of Science Highly Cited Researcher (HCR) since 2018. In 2019, he became the first Korean to be selected as an HCR in the field of environment and ecology. He maintains a worldwide professional network through his service as an editor (former co-editor in chief) of the *Journal of Hazardous Materials*, an editor (former co-editor) of the *Critical Reviews in Environmental Science and Technology*, an editor of the *Environmental Pollution*, and a member of the editorial boards of *Renewable and Sustainable Energy Reviews*, *Chemical Engineering Journal*, *Chemosphere*, and *Journal of Analytical and Applied Pyrolysis*, along with several other top journals. Professor Ok has served in a number of positions worldwide, including as honorary professor at the University of Queensland (Australia), visiting professor at Tsinghua University (China), adjunct professor at the University of Wuppertal (Germany), and guest professor at Ghent University (Belgium). He currently serves as director of the Sustainable Waste Management Program for the Association of Pacific Rim Universities (APRU). He has served as chairman of numerous major conferences such as Engineering Sustainable Development 2019, organized by the APRU, and the Institute for Sustainability of the American Institute of Chemical Engineers (AIChE).

Contributors

Reshma M. Antony
Department of Agronomy
2004 Throckmorton Plant Sciences Center
Kansas State University
and
Ruminant Nutrition Laboratory
Department of Animal Sciences and Industry
Kansas State University
Manhattan, Kansas

Thilakshani Atugoda
Ecosphere Resilience Research Center
Faculty of Applied Sciences
University of Sri Jayewardenepura
Nugegoda, Sri Lanka

Maurizio Avella
Institute for Polymers
Composites and Biomaterials of Italian
Research Council (IPCB-CNR)
Pozzuoli, Italy

Virginia Baker
Institute of Environmental Science and
Research Limited (ESR)
Porirua, New Zealand

Christopher Barton
Department of Forestry and Natural Resources
University of Kentucky
Lexington, Kentucky

Nanthi S. Bolan
Global Centre for Environmental Remediation
(GCER)
Advanced Technology Centre
Faculty of Science
The University of Newcastle
and
Cooperative Research Centre for High
Performance Soil (Soil CRC)
The University of Newcastle
Callaghan, New South Wales, Australia

Shiv Shankar Bolan
Global Centre for Environmental Remediation
(GCER)
Advanced Technology Centre
Faculty of Science
The University of Newcastle
Callaghan, New South Wales, Australia

Andy M. Booth
SINTEF Ocean
Trondheim, Norway

Lauren Bradney
Global Centre for Environmental Remediation
(GCER)
Advanced Technology Centre
Faculty of Science
The University of Newcastle
Callaghan, New South Wales, Australia

Emily Bryson
School of Health Medical and Applied
Sciences
Central Queensland University
Rockhampton, Victoria, Australia

Jochen Bundschuh
Faculty of Health, Engineering, and Sciences
School of Civil Engineering and Surveying
University of Southern Queensland
and
UNESCO Chair on Groundwater Arsenic
within the 2030 Agenda for Sustainable
Development
University of Southern Queensland
Toowoomba, Queensland, Australia

Olivier Champeau
Cawthron Institute
Nelson, New Zealand

Suchana Chavanich
Faculty of Science
Department of Marine Science
Chulalongkorn University
Bangkok, Thailand

Franklin M. Chen
University of Wisconsin–Green Bay
Green Bay, Wisconsin

Qianqian Chen
State Key Laboratory of Pollution Control and Resource Reuse
School of the Environment
Nanjing University
Nanjing, People's Republic of China

Mariacristina Cocca
Institute for Polymers, Composites, and Biomaterials of Italian Research Council (IPCB-CNR)
Pozzuoli, Italy

Jacinta Colvin
Ecotoxicology Research Group
School of Science
RMIT University
Bundoora, Victoria, Australia

Peter Dann
Phillip Island Nature Parks
Cowes, Victoria, Australia

Francesca De Falco
Institute for Polymers, Composites, and Biomaterials of Italian Research Council (IPCB-CNR)
Pozzuoli, Italy

Emilia Di Pace
Institute for Polymers, Composites, and Biomaterials of Italian Research Council (IPCB-CNR)
Pozzuoli, Italy

Leela Dilkes-Hoffman
Faculty of Engineering, Architecture, and Information Technology
School of Chemical Engineering
The University of Queensland
Brisbane, Queensland, Australia

Weixin Ding
State Key Laboratory of Soil and Sustainable Agriculture
Institute of Soil Science
Chinese Academy of Sciences
Nanjing, People's Republic of China

Jiajun Duan
State Key Laboratory of Water Environment Simulation
School of Environment, Beijing Normal University
Beijing, People's Republic of China

Maria Emanuela Errico
Institute for Polymers, Composites, and Biomaterials of Italian Research Council (IPCB-CNR)
Pozzuoli, Italy

Gennaro Gentile
Institute for Polymers, Composites, and Biomaterials of Italian Research Council (IPCB-CNR)
Pozzuoli, Italy

Cheng Gu
State Key Laboratory of Pollution Control and Resource Reuse
School of the Environment
Nanjing University
Nanjing, People's Republic of China

A. J. M. Gunasekara
Marine Environment Protection Authority (MEPA)
Colombo, Sri Lanka

Sameera R. Gunatilake
College of Chemical Sciences
Institute of Chemistry Ceylon
Rajagiriya, Sri Lanka

Claudia Halsband
Akvaplan-niva
Fram Centre
Tromsø, Norway

Deyi Hou
School of Environment
Tsinghua University
Beijing, People's Republic of China

Rong Ji
State Key Laboratory of Pollution Control and Resource Reuse
School of the Environment
Nanjing University
Nanjing, People's Republic of China

Kandaswamy Karthikeyan
Faculty of Agriculture
Department of Soil Science and Agricultural Chemistry
Annamalai University
Chidambaram, India

Shosaku Kashiwada
Department of Life Sciences
Research Centre for Life and Environmental Sciences
Toyo University
Itakura, Gunma, Japan

Somkiat Khokiattiwong
Department of Marine and Coastal Resources and
Phuket Marine Biological Center
Phuket, Thailand
UNESCO-IOC/WESTPAC
Bangkok, Thailand

Ki-Hyun Kim
Department of Civil and Environmental Engineering
Hanyang University
Seoul, South Korea

M.B. Kirkham
Department of Agronomy
Throckmorton Plant Sciences Center
Kansas State University
Manhattan, Kansas

Anu Kumar
CSIRO Land and Water
PMB 2, Glen Osmond
Adelaide, South Australia, Australia

Prasanna Kumarathilaka
School of Civil Engineering and Surveying
Faculty of Health, Engineering, and Sciences
University of Southern Queensland
Toowoomba, Queensland, Australia

Eilhann E. Kwon
Department of Environment and Energy
Sejong University, Korea

Paul Lant
School of Chemical Engineering
Faculty of Engineering, Architecture, and Information Technology
The University of Queensland
Brisbane, Queensland, Australia

Bronwyn Laycock
Faculty of Engineering, Architecture, and Information Technology
School of Chemical Engineering
The University of Queensland
Brisbane, Queensland, Australia

Qiao Li
School of Environmental and Biological Engineering
Nanjing University of Science and Technology
Nanjing, People's Republic of China

Yang Li
State Key Laboratory of Water Environment Simulation
School of Environment, Beijing Normal University
Beijing, People's Republic of China

Shengdong Liu
State Key Laboratory of Water Environment Simulation
School of Environment, Beijing Normal University
Beijing, People's Republic of China

Yuan Liu
State Key Laboratory of Water Environment Simulation
School of Environment, Beijing Normal University
Beijing, People's Republic of China

Yini Ma
State Key Laboratory of Pollution Control and Resource Reuse
School of the Environment
Nanjing University
Nanjing, People's Republic of China

Sanchita Mandal
Department of Animal and Plant Sciences
The University of Sheffield
Sheffield, United Kingdom

Todd E. Mlsna
Department of Chemistry
Mississippi State University
Starkville, Mississippi

Raj Mukhopadhyay
ICAR-Central Soil Salinity Research Institute
Karnal, India

Chanaka M. Navarathna
Department of Chemistry
Mississippi State University
Starkville, Mississippi

Samadhi Nawalage
College of Chemical Sciences
Institute of Chemistry Ceylon
Rajagiriya, Sri Lanka

T.W.G.F. Mafaziya Nijamdeen
Faculty of Applied Sciences
Department of Biological Sciences
South Eastern University of Sri Lanka
Sammanthurei, Sri Lanka

Grant L. Northcott
Northcott Research Consultants
Hamilton, New Zealand

Dayanthi Nugegoda
Ecotoxicology Research Group
School of Science RMIT University
and
AQuatic Environmental STress research group (AQUEST)
RMIT University
Bundoora, Victoria, Australia

Yong Sik Ok
Division of Environmental Science and Ecological Engineering
Korea Biochar Research Center
O-Jeong Eco-Resilience Institute (OJERI)
Korea University
Seoul, South Korea

Kumuduni Niroshika Palansooriya
Division of Environmental Science and Ecological Engineering
Korea Biochar Research Center
Korea University
Seoul, South Korea

Xavier Pochon
Cawthron Institute
Nelson, New Zealand
and
Institute of Marine Science
University of Auckland
Auckland, New Zealand

Steven Pratt
Faculty of Engineering, Architecture, and Information Technology
School of Chemical Engineering
The University of Queensland
Brisbane, Queensland, Australia

B. Ravindran
Department of Environmental Energy Engineering
Kyonggi University
Suwon, Republic of Korea

Teresa Rocha-Santos
Centre for Environmental and Marine Studies (CESAM)
Department of Chemistry
University of Aveiro
Aveiro, Portugal

Jörg Rinklebe
Laboratory of Soil- and Groundwater-Management
School of Architecture and Civil Engineering
Institute of Foundation Engineering, Water- and Waste-Management
University of Wuppertal
Wuppertal, Germany

Binoy Sarkar
Lancaster Environment Centre
Lancaster University
Lancaster, United Kingdom

Enxiang Shang
College of Science and Technology
Hebei Agricultural University
Huanghua, People's Republic of China

Jeff Shimeta
Centre for Environmental Sustainability and Remediation
RMIT University
Bundoora, Victoria, Australia

Hocheol Song
Department of Environment and Energy
Sejong University
Seoul, South Korea

P.B. Terney Pradeep Kumara
Marine Environment Protection Authority (MEPA)
Colombo, Sri Lanka
and
Department of Oceanography and Marine Geology
University of Ruhuna
Matara, Sri Lanka

Louis A. Tremblay
Cawthron Institute
Nelson, New Zealand
and
School of Biological Sciences
University of Auckland
Auckland, New Zealand

Charlene Trestrail
Ecotoxicology Research Group
School of Science RMIT University
and
Centre for Environmental Sustainability and Remediation
RMIT University
Bundoora, Victoria, Australia

Daniel C.W. Tsang
Department of Civil and Environmental Engineering
The Hong Kong Polytechnic University
Hong Kong, China

Yiu Fai Tsang
Department of Science and Environmental Studies
The Education University of Hong Kong
Tai Po, Hong Kong

Meththika Vithanage
Ecosphere Resilience Research Center
Faculty of Applied Sciences
University of Sri Jayewardenepura
Nugegoda, Sri Lanka

Voranop Viyakarn
Faculty of Science
Department of Marine Science
Chulalongkorn University
Bangkok, Thailand

Hailong Wang
Biochar Engineering Technology Research Center of Guangdong Province
School of Environment and Chemical Engineering
Foshan University
Foshan, People's Republic of China

Lin Wang
State Key Laboratory of Pollution Control and Resource Reuse
School of the Environment
Nanjing University
Nanjing, People's Republic of China

Ting Wang
State Key Laboratory of Pollution Control and Resource Reuse
School of the Environment
Nanjing University
Nanjing, People's Republic of China

Yining Wang
State Key Laboratory of Water Environment Simulation
School of Environment, Beijing Normal University
Beijing, People's Republic of China

Xinjie Wang
State Key Laboratory of Water Environment Simulation
School of Environment, Beijing Normal University
Beijing, People's Republic of China

Upekshya Welikala
College of Chemical Sciences
Institute of Chemistry Ceylon
Rajagiriya, Sri Lanka

Hasintha Wijesekara
Department of Natural Resources
Faculty of Applied Sciences
Sabaragamuwa University of Sri Lanka
Belihuloya, Sri Lanka

Yubo Yan
School of Chemistry and Chemical Engineering
Huaiyin Normal University
Jiangsu, People's Republic of China

Wenxi Zhu
UNESCO-IOC/WESTPAC
Bangkok, Thailand

Section I

Sources of Particulate Plastics in the Environment

1 Sources of Particulate Plastics in Terrestrial Ecosystems

Lauren Bradney, Hasintha Wijesekara, Nanthi S. Bolan, and M.B. Kirkham

CONTENTS

1.1 Introduction 3
1.2 Sources of Particulate Plastics 5
  1.2.1 Primary Sources 5
    1.2.1.1 Microbeads 5
    1.2.1.2 Polyacrylamide 8
  1.2.2 Secondary Sources 9
    1.2.2.1 Plastic Mulch 9
    1.2.2.2 Biosolids and Composts 11
    1.2.2.3 Wastewater for Irrigation 12
    1.2.2.4 Indiscriminate Disposal in Landfills and Littering 13
    1.2.2.5 Atmospheric Input 13
1.3 The Future 13
1.4 Conclusions 14
References 14

1.1 INTRODUCTION

Particulate plastics are synthetic polymers no greater than 5 mm in diameter. They encompass both "microplastics" (100 nm to 5 mm) and "nanoplastics" (<100 nm) (Ng et al. 2018). Particulate plastics can be further classified based on their origin, as follows: primary particulate plastics or secondary particulate plastics. Primary particulate plastics are manufactured for their inclusion in industrial products. A typical example of this is the manufacture of microbeads for their use in the cosmetic industry. Another soil input is polyacrylamide (PAM), which is used to promote flocculation and aid soil stability (Li et al. 2011; Sojka et al. 2007). Alternatively, secondary microplastics arise from the breakdown of macroplastics. Photodegradation, hydrolysis, and oxidation are just some of the environmental processes which cause secondary particulate plastics (Andrady 2011; Fok and Cheung 2015; Murphy et al. 2016). The weathering and breakdown of agricultural plastic mulch, or the presence of plastics within compost, are both examples of secondary particulate plastics within soils (Liu et al. 2014; Weithmann et al. 2018).

The widespread inclusion and use of plastics within industries have led to their presence within the marine, freshwater, terrestrial, and atmospheric environments. As a result, concerns surrounding particulate plastics have risen in recent years. In 2016, the United Nations Environmental Programme (UNEP) included microplastics as one of their six emerging environmental concerns in their annual edition of the UNEP Frontiers. UNEP's inclusion of particulate plastics highlights both their widespread presence and their potential threat to the environment (UNEP 2016). Furthermore, many studies have not only demonstrated their pervasiveness, but also their impact on animals, humans, and ecosystems (Akhbarizadeh et al. 2018; Besseling et al. 2013; Kedzierski et al. 2018; Khan et al. 2015; Liebezeit and Liebezeit 2014; Rochman et al. 2013; Sleight et al. 2017;

Song et al. 2014). Early research focused primarily on the presence of particulate plastics within marine environments and the resulting impacts on marine organisms (Andrady 2011; Brennecke et al. 2016; Browne et al. 2011; Cole et al. 2011; Thompson et al. 2009). Now, research is also examining terrestrial and atmospheric environments (Bläsing and Amelung 2018; Dris et al. 2016; Duis and Coors 2016; Horton et al. 2017; Ng et al. 2018; Prata 2018; Rillig 2012). This chapter will focus predominately on the sources and pathways of particulate plastics within the terrestrial environment.

Particulate plastics within soils can arise from several sources. Some sources (e.g., biosolids, composts, plastic mulch, and PAM application) are used in the agricultural and horticulture industry to improve soil health. Plastic mulch can aid plant growth, suppress weeds, and conserve water. Unfortunately, ineffective disposal of plastic mulch can lead to its degradation and accumulation within the soil and result in crop yield reduction. Biosolids and compost are another source of particulate plastics and are applied to the land to help improve the soil's physical and chemical qualities (e.g., as a soil conditioner). However, composts and biosolids can often contain small amounts of particulate plastics that have bypassed treatment and remain within the final product. Therefore, an increased application onto the land may result in an increased concentration of particulate plastics becoming exposed to the soil. Most research has focused primarily on particulate plastics located in biosolids or treated sewage (Mahon et al. 2016; Wijesekara et al. 2018a; Zubris and Richards 2005). Despite the presence of particulate plastics in composts, there are limited studies regarding their occurrence and fate (Bläsing and Amelung 2018).

While some sources of particulate plastics are beneficial to soil, other sources arise from human carelessness (e.g., indiscriminate plastic waste disposal, poorly managed landfills, and littering) and are not beneficial to soil health. The location of poorly managed landfills and littering can determine how much plastic enters each environment. For example, if mismanaged waste occurs within coastal populations, this further increases the chance of secondary particulate plastics entering rivers and oceans (Duis and Coors 2016; Jambeck et al. 2015).

Research has suggested that a large percentage of marine plastics has originated from land-based sources (Andrady 2015; Duis and Coors 2016; UNEP 2009). Sediment transfer (i.e., soil erosion) is one pathway for the transportation of particulate plastics from terrestrial to aquatic environments (Nizzetto et al. 2016a). Several factors influence the extent of soil erosion. These factors include, but are not limited to, the surrounding land use (e.g., agricultural and farming practices) or climatic factors (e.g., the intensity and frequency of rainfall or wind velocity). Therefore, particulate plastics in soils that are more exposed to the natural elements and close to bodies of water (e.g., rivers or oceans) have a higher chance of being transported from terrestrial to aquatic environments. It is, therefore, vital for research to focus on the presence of particulate plastics within multiple environments, as they are often interlinked.

Particulate plastics can also facilitate the long-range transport of contaminants to aquatic environments. Contaminants can adsorb onto, or desorb from, the surface of particulate plastics, and it is this process that enables transportation. The concentration and type of contaminant are dependent on both environmental factors and contaminate levels within the given location, as well as the characteristics of the plastic themselves (GESAMP 2015; Holmes et al. 2012, 2014; Turner and Holmes 2015). Virgin particulate plastics have a high surface area and polarity, and these factors aid the sorption of inorganic contaminants. Additionally, the surface of aged particulate plastics can often contain organic matter, and this can increase metal sorption (Wijesekara et al. 2018a). Increased metal sorption can also occur on aged particulate plastics due to surface degradation and modification. Changes to the surface may arise from processes including, but not limited to, photodegradation, photo-oxidation, or thermal degradation (Brennecke et al. 2016; Turner and Holmes 2015). However, a vast majority of research surrounding the fate, transport, and effects of particulate plastics has focused on aquatic environments. Therefore, the overall fate of particulate plastics within soils is less clear. Rillig (2012) suggested that this lack of research is partly due to the difficulties surrounding the extraction of particulate plastics from the soil matrix.

This chapter will discuss both the various sources of terrestrial-based primary and secondary particulate plastics, as well as the pathways that they take before reaching the terrestrial ecosystem.

1.2 SOURCES OF PARTICULATE PLASTICS

Particulate plastics can enter the terrestrial and aquatic environment through a variety of sources. As mentioned earlier, particulate plastics are not often static; instead, they can move between the atmospheric, aquatic, and terrestrial environments (Figure 1.1). Unfortunately, research is so far unable to compare accurately among the three different environments due to many complex factors, including differences in the types of impact and target organisms (Rillig 2012).

1.2.1 Primary Sources

Primary particulate plastics are synthetic polymers that are manufactured for industrial use, such as those used within the cosmetic or agricultural industry. For many years, the cosmetic industry included plastic microbeads in their products to prolong shelf life, improve durability, increase chemical resistance, as an exfoliant, or to improve aesthetics (GESAMP 2015; Murphy et al. 2016). The agricultural industry also uses particulate plastics, such as the application of PAM. When applied to the soil, PAM helps to improve the stability and physical characteristics of the soil.

1.2.1.1 Microbeads

In this chapter, microbeads are defined as spherically manufactured plastics, no greater than 1 mm in size (Figure 1.2) (Wardrop et al. 2016). Microbeads have become an increasing concern in recent years, due to their persistence in the environment. Wastewater treatment plants (WWTPs) constitute a significant source of microbeads entering the aquatic environment (i.e., treated effluent) or soils (i.e., treated biosolids). Additionally, accidental spills during manufacture or transportation can also result in the release of microbeads. Once present in the environment, microbeads can cause direct

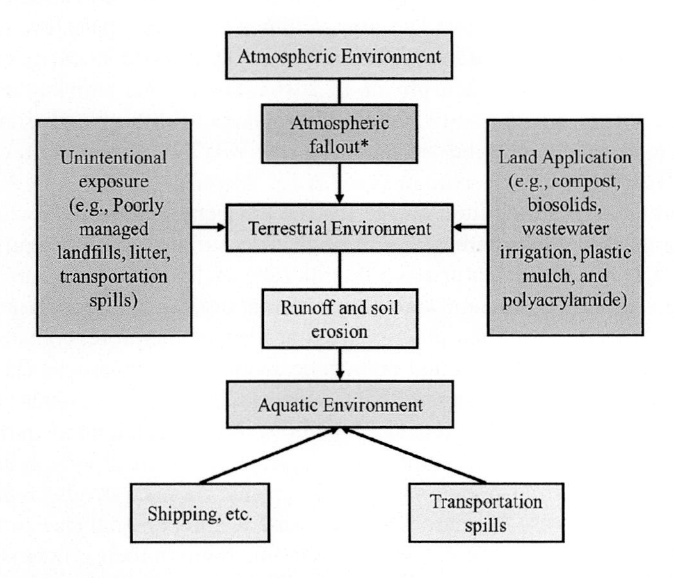

FIGURE 1.1 Conceptual diagram demonstrating the interactions between the sources of particulate plastics within the atmospheric, terrestrial, and aquatic environments. *Atmospheric fallout is also a source for aquatic environments.

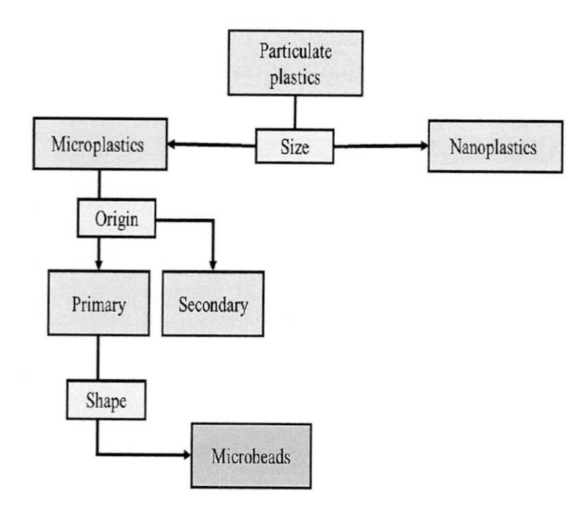

FIGURE 1.2 Conceptual diagram demonstrating the categorization (i.e., size, shape, and origin) of microbeads within particulate plastic.

contamination (i.e., the microbeads themselves) and indirect contamination (i.e., as a vector for other contaminants) to organisms.

Natural materials (e.g., ground almond or apricot pits) were often added to cosmetic and personal care products as exfoliants. Over the years, these natural products have been substituted with synthetic materials (i.e., plastics), commonly referred to as microbeads (Rochman et al. 2015). Facial scrubs, toothpaste, and soaps are all products which have been identified as containing microbeads. Unfortunately, these products are designed to be rinsed off once used. Once rinsed off, the particulate plastics are then transported into WWTPs through raw effluent (Murphy et al. 2016). WWTPs release microbeads into the environment one of two ways. First, they can be released through the discharge of treated effluent into rivers or oceans. Murphy et al. (2016) found that, although particulate plastics, including microbeads, were efficiently removed from WWTPs, some still managed to bypass the treatment process (Table 1.1). Large volumes of effluent pass through WWTPs every year; thus, even a small percentage of microplastics bypassing the system can result in large quantities entering aquatic environments (Murphy et al. 2016). The second pathway involves their presence within treated sludge, known as biosolids. Microbeads' diminutive size, non-biodegradable nature, and ability to associate with the sewage mean that WWTPs often fail to remove, effectively, microbeads from their biosolids (Rochman et al. 2015; Wardrop et al. 2016). As a result, significant volumes of microbeads are emitted into terrestrial environments from biosolids and composts (UNEP 2015). Another potential pathway is through direct routes, such as spills that may occur during the transportation or manufacturing of the microbeads and associated products.

Plastics are produced from petroleum, and many different organic compounds are added during the manufacturing process to improve their quality. Microbeads can, therefore, contain organic contaminants, such as petroleum hydrocarbons and polycyclic aromatic hydrocarbons (Horton et al. 2017). The microplastics then transport these organic contaminants to other environments through adsorption, desorption, degradation, and leaching (Teuten et al. 2009). It has been demonstrated that polychlorinated biphenyls, polycyclic aromatic hydrocarbons, petroleum hydrocarbons, organochlorine pesticides, and bisphenol A can be transported into marine systems via microbeads (Teuten et al. 2009).

Concerns surrounding the use of microbeads in cosmetic and personal care products have seen a shift in their use, with many companies no longer including them in their products. Unfortunately, it will still be some time before microbeads are universally banned within commercial products, and even longer before the effects of banning particulate plastics are seen in the environment.

TABLE 1.1
Selected References on the Sources and Quantity of Particulate Plastics That Enter Terrestrial Environments (i.e., Soils)

Source	Country	Plastic Type	Quantity of Particulate Plastic			Reference
			Mean	Range	Units	
WWTPs (effluent)	Germany	PE	N.D.	81–257	mg per m^3	Majewsky et al. (2016)
WWTPs (influent)	Scotland	Acrylic, alkyde, PET, PA, polyester, PE, PP, PS, PUR, PVF, PS acrylic, PV acrylate, PVA, PVC, PVE	15.70	N.D.	MP per liter	Murphy et al. (2016)
WWTPs (effluent)	Scotland	Acrylic, alkyde, PET, PA, polyaryl ether, polyester, PE, PP, PS, PVA	0.25	N.D.	MP per liter	Murphy et al. (2016)
WTP 1 (raw water)	Czech Republic	PET, PP, PS, PAM, PVC	1473	1383–1575	Particles per liter	Pivokonsky et al. (2018)
WTP 2 (raw water)	Czech Republic	PET, PP, PS, PAM, PBA, PVC	1812	1648–2040	Particles per liter	Pivokonsky et al. (2018)
WTP 3 (raw water)	Czech Republic	PET, PP, PE, PS, PAM, PBA, Bakelite, PMMA, PPTA, PTT	3605	3123–4464	Particles per liter	Pivokonsky et al. (2018)
WTP 1 (treated water)	Czech Republic	PET, PP, PAM, PVC	443	369–485	Particles per liter	Pivokonsky et al. (2018)
WTP 2 (treated water)	Czech Republic	PET, PP, PAM, PVC, PBA	338	243–466	Particles per liter	Pivokonsky et al. (2018)
WTP 3 (treated water)	Czech Republic	PET, PP, PAM, PBA, PE, PPTA, PTT	628	562–684	Particles per liter	Pivokonsky et al. (2018)
Mismanaged plastic waste[a]	China[b]	N.D.	8.82	N.D.	Million metric tons per year	Jambeck et al. (2015)
Mismanaged plastic waste[a]	India[b]	N.D.	0.60	N.D.	Million metric tons per year	Jambeck et al. (2015)
Mismanaged plastic waste[a]	United States[b]	N.D.	0.28	N.D.	Million metric tons per year	Jambeck et al. (2015)
Biosolids-agroecosystems[a]	Australia	N.D.	N.D.	2800–19,000	Metric tons	Ng et al. (2018)
Biosolids-agroecosystems[a]	Europe	N.D.	N.D.	63,000–430,000	Tons	Nizzetto et al. (2016b)
Biosolids-agroecosystems[a]	North America	N.D.	N.D.	44,000–300,000	Tons	Nizzetto et al. (2016b)

(*Continued*)

TABLE 1.1 (*Continued*)
Selected References on the Sources and Quantity of Particulate Plastics that Enter Terrestrial Environments (i.e., Soils)

Source	Country	Plastic Type	Quantity of Particulate Plastic			Reference
			Mean	Range	Units	
Compost	Germany	N.D.	N.D.	2.38–180	mg per kg	Bläsing and Amelung (2018)
Atmospheric fallout	France	N.D.	118	29–280	Particles per m per day	Dris et al. (2016)

Abbreviations: N.D. = no data available; WWTPs = wastewater treatment plants; WTP = Water treatment plant; MP = microplastic; PET = polyethylene terephthalate; PA = polyamide; PE = polyethylene; PS = polystyrene; PUR = polyurethane; PVF = polyvinyl fluoride; PVA = polyvinyl alcohol; PVC = polyvinyl chloride; PVE = polyvinyl ethers; PAM = polyacrylamide; PMMA = poly(methyl methacrylate); PPTA = poly-p-phenylene terephthalamide; PTT = polytrimethylene terephthalate.

[a] Estimated values only.

[b] Populations located within 50 km of the coast.

1.2.1.2 Polyacrylamide

PAM [poly(1-carbamoylethylene)] is a polymer formed from acrylamide base units. Cross-linked PAMs readily absorb water, becoming a soft gel when hydrated (i.e., hydrogel). Alternatively, water-soluble PAMs are linear, containing little to no cross-linking. It is the anionic water-soluble form of PAM which is used for agricultural purposes and other environmental activities (Sojka et al. 2007). Anionic-PAM is used due to its lower aquatic toxicity when compared to its cationic and non-ionic forms. The lower aquatic toxicity of anionic-PAM helps to reduce toxicity within the surrounding waters (e.g., rivers), if surface runoff should occur (Sojka et al. 2007).

Agricultural and horticulture industries use PAM due to its ability to flocculate and stabilize soil (Li et al. 2011; Sojka et al. 2007). Further, the large-scale application of PAM onto soil helps reduce soil erosion. Depending on the concentrations used (i.e., <10 ppm in water), PAM has also been reported to increase the infiltration of silt and clay soils, limit soil sealing, and reduce nutrient desorption (Sojka et al. 2007).

The application of PAM onto soil can improve soil health by increasing pore space and water infiltration in soils containing clay, preventing soil crusting, reducing soil erosion and water runoff, helping make friable soil easier to cultivate, and allowing soils to dry faster and be worked sooner (Sojka et al. 2007). Consequently, these translate into significant benefits in agricultural production, particularly for plant growth. PAM application helps plants have more extensive root systems, earlier seedling emergence and crop maturity, more efficient water utilization, better response to fertilizers, fewer root diseases related to poor soil aeration, and decreased energy requirement for tillage (Sojka et al. 2007).

The ionic form of PAM is used in the potable water treatment industry as an effective flocculating agent. The addition of PAM helps facilitate the removal of particulate organic matter in potable water sources. A study conducted by Pivokonsky et al. (2018) analyzed particulate plastics found in raw and treated drinking water at three WTPs (Table 1.1). PAM was detected in the raw water of all three WTPs, although it only accounted for <10% of all plastic found. Furthermore, PAM was also present in the treated water of all three WTPs at higher concentrations of around 10%–25%. One WTP contained a significant increase in the level of PAM, and it was speculated that this was due to the use of PAM as a coagulant during the treatment process (Pivokonsky et al. 2018).

Despite the many benefits of using PAM, there are negative aspects surrounding its use, particularly within the agricultural and horticultural industries. However, concerns surrounding PAM (which is relatively non-toxic) are primarily due to the potential presence of acrylamide. Acrylamide is a known neurotoxin and potential cariogenic for animals and humans (Sojka et al. 2007). Residual acrylamide may remain within PAM products after production. Alternatively, chemical or thermal degradation of PAM may result in the formation of acrylamide through de-polymerization. Chemical degradation is initiated by a chemical change in the surroundings, resulting in free radical polymerization (Xiong et al. 2018). A common example is when the amide radical hydrolyzes (i.e., base hydrolysis) at an elevated temperature or pH, and this results in the evolution of ammonia and a remaining carboxyl group. Thermal degradation occurs at high temperatures. During the thermal degradation of PAM, ammonia is released to form imides (at around 200°C–300°C) and nitriles (above 300°C).

Primary particulate plastics have long been used within industries. However, with the growing concern surrounding the presence of non-biodegradable particulate plastics within aquatic, terrestrial, and atmospheric environments, this has seen a shift in industries using plastics. Preferably, plastic alternatives (e.g., biodegradable plastics or plant-based exfoliants instead of plastic microbeads) are now being implemented. Unfortunately, it will be many years before non-biodegradable particulate plastics are entirely replaced within industries worldwide.

1.2.2 Secondary Sources

Secondary sources of particulate plastics occur when macroplastics are broken down by natural processes, such as photodegradation, hydrolysis, or oxidation. Examples of secondary sources include the breakdown and weathering of plastic mulch and inputs through compost and biosolid application.

1.2.2.1 Plastic Mulch

The agricultural industry uses plastic material for a variety of applications, with this reliance on plastic commonly referred to as "plasticulture." The 1990s saw the transition from paper to plastic mulch as a way to improve agricultural practices (Kasirajan and Ngouajio 2012). Many countries have seen the rise of plastic mulch, although some countries have experienced this more keenly than others. China's plastic mulch usage has significantly increased over the years, rising by approximately 926,000 tons between 1991 and 2011 (Liu et al. 2014). A review conducted by Gao et al. (2019) used a meta-analysis to compare 266 publications on Chinese agriculture. The study found that the short-term use of plastic mulching significantly increased both the crop yield and water use efficiency at approximately 24% and 28%, respectively. However, Gao et al. (2019) also found that the crop yield decreased significantly (i.e., approximately 11%), when exposed to residual plastic over 240 kg/ha and decreased by approximately 24% when over 480 kg/ha.

Plastic mulch is commonly made from low-density polyethylene and is used to change the crop microclimate, aid plant growth, and suppress weeds (Kasirajan and Ngouajio 2012). The use of plastic mulch helps plant growth (i.e., earlier planting and faster growth) through the increase of soil temperatures (Liu et al. 2014). By preventing sunlight from reaching the soils, it also helps reduce weed growth.

Furthermore, plastic mulch is used by the agricultural industry for water conservation and is commonly used in conjunction with drip irrigation (Liu et al. 2014). Plastic mulch helps to reduce water usage by lessening evaporation. Plastic mulch coupled with drip irrigation can also help to distribute the water evenly, again decreasing the water volume required (Liu et al. 2014). Soil erosion is also reduced by using plastic mulch. The mulch covers the soil, thereby limiting the effects of rain and sunlight, which can cause crusting. Crusting not only increases erosion, but also decreases water infiltration, which can lead to the delay of seedling emergence (Borselli et al. 1996).

Despite the many advantages of using plastic mulches for crop production, there are also disadvantages, and these mainly center on their disposal. Disposal of conventional plastic mulch is both technically and economically burdensome (Hayes et al. 2017). One of the more sustainable options for plastic mulch disposal is recycling. However, soil and plant debris often remain attached to the plastic mulch after removal, meaning many companies are unable or unwilling to recycle the plastic (Saglam et al. 2017). Other standard disposal methods involve incineration

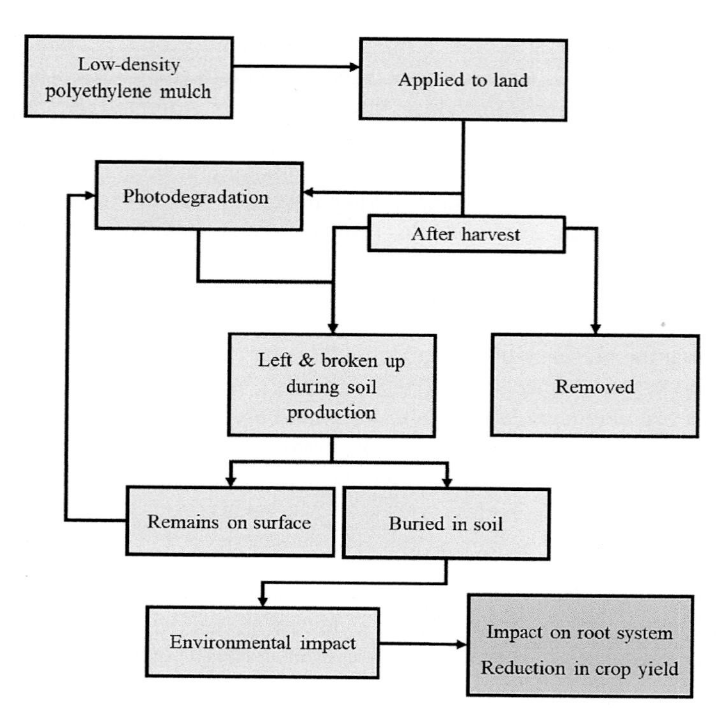

FIGURE 1.3 Conceptual diagram demonstrating the input of secondary particulate plastics to soil due to the breakdown and weathering of plastic mulch used on agricultural and horticultural crops.

or burying them in landfills. Unfortunately, because the removal and disposal of plastic mulch are burdensome, they are often incorrectly discarded (e.g., on-farm burning). This can result in the accumulation of plastic and particulate plastics within the soil and ultimately lead to unsustainable farmland use and can cause crop yield reduction and other environmental impacts (Figure 1.3) (Hayes et al. 2017). A perspective written by Liu et al. (2014) has suggested that the use of plastic mulch in China's agricultural industry is shifting from "white revolution" and moving toward "white pollution."

Unfortunately, the short-term benefits of using plastic mulch may be outweighed by the long-term impacts on soil health (Steinmetz et al. 2016). A review by Steinmetz et al. (2016) examined several studies and concluded that the use of plastic mulch could impair soil functions. This review found that, despite initial applications of plastic mulch aiding water conservation, long-term exposure of plastic mulch could cause water repellency within the soil. Other negative impacts found for long-term use included the potential accelerated turnover of carbon and nitrogen, as well as the depletion of essential soil nutrients (Steinmetz et al. 2016). Other concerns regarding plastic mulch are the inclusion of several additives, such as bis(2-ethylhexyl) phthalate. Bis(2-ethylhexyl) phthalate is a potential endocrine-disrupting chemical as well as carcinogenetic. Plastic mulch could introduce higher concentrations of bis(2-ethylhexyl) phthalate into the soil. This could result in a decrease of soil organisms, as well as reducing crop quality if it is taken up by plants (Steinmetz et al. 2016).

Overall, the initial benefits of using plastic mulch are outweighed by the long-term environmental impacts and the labor and disposal costs surrounding its use. The environmental impacts of non-biodegradable plastic mulch have resulted in an increased interest in using biodegradable plastic mulch instead. However, the long-term effects on soil health are still widely unknown (Sintim et al. 2019).

1.2.2.2 Biosolids and Composts

Composts and biosolids contain high levels of nutrients and organic matter, and when applied to the land can increase the health of the soil (Wijesekara et al. 2017). The benefits of composts and biosolids for land application are that, not only do they provide the soils with essential nutrients, but they also help increase moisture retention, reduce soil erosion, and increase plant growth and crop production. Despite these benefits, composts and biosolids also can contain contaminants (i.e., particulate plastics and toxic metals).

Although composts are a potential source of particulate plastic pollution, concentrations in composts are not widely reported. Instead, research tends to focus on the type of particulate plastics found within sewage sludge or biosolids. One of the few studies conducted on plastics within compost reported a concentration of 2.38–180 mg per kg (Table 1.1) (Bläsing and Amelung 2018).

Composts undergo a range of processes before they are applied to the land. Two conventional methods include sieving and milling. Sieving, whether by mechanical processes or handpicking, allows for the removal of large and medium plastic fragments. Unfortunately, particulate plastics are often smaller than the sieve mesh size and are not removed. Furthermore, milling or grinding of the compost can result in the creation of smaller secondary particulate plastics. The smaller the particulate plastics becomes (e.g., down to nano-range), the less chance it will have of being removed by the sieve. Consequently, sieve size is a crucial factor in determining the concentration and quantity of particulate plastics that bypass the removal process (Weithmann et al. 2018).

Many countries regulate the concentrations of plastic allowed to remain in composts for land application. For example, the regulation in New South Wales (Australia) states that composts supplied for land application must contain less than 0.5% (dry weight) of glass, metal, and rigid plastics (>2 mm), or 0.05% (dry weight) for flexible plastics (>5 mm) (NSW Environmental Protection Agency 2016). Unfortunately, if large amounts of compost are continually applied to the land, then even small quantities of particulate plastics will eventually become concentrated within the soil over time.

WWTPs have been reported as being a major pathway of particulate plastics reaching aquatic environments through the discharge of particulate plastic within treated effluent. However, WWTPs are also a significant pathway of particulate plastics entering soils through biosolid application (Figure 1.4). Particulate plastics enter WWTP from several sources: domestic (i.e., sewage), industrial (i.e., wastewater), and urban (i.e., stormwater and landfill leachate). The origin of the influent can cause variations in the types (e.g., shape and base material) and concentrations of particulate plastics. For example, clothing fibers are commonly found within WWTPs, and these are most

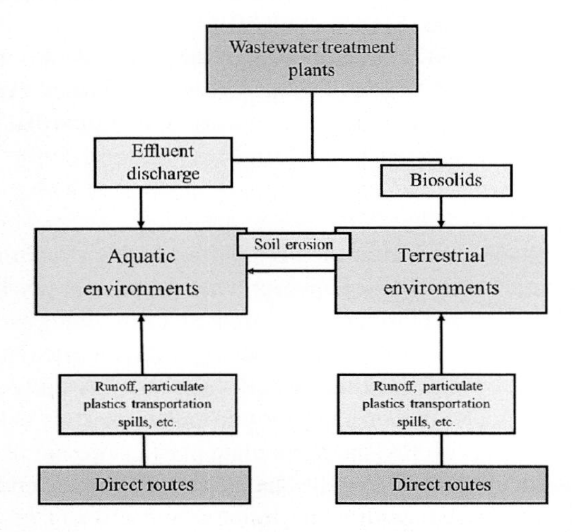

FIGURE 1.4 Conceptual diagram demonstrating wastewater treatment plants as a source of particulate plastics in aquatic and terrestrial environments. (Adapted from Bradney, L. et al., *Environ. Int.*, 131, 104937, 2019.)

likely the result of the machine washing of clothing (i.e., domestic). Alternatively, a study by Li et al. (2018) found that, on average, particulate plastic concentrations increased when the influent came from industrial sources.

Treated biosolids from WWTPs also can contain particulate plastics, which bypass the removal stage due to their minute size. Fragments, fibers (e.g., clothing fibers), and microbeads are just some of the particulate plastics identified within WWTPs (Carr et al. 2016; Mahon et al. 2016; Ziajahromi et al. 2017). Treated biosolids can be used on the land to improve soil health and carbon sequestration (Bolan et al. 2013). The benefits regarding biosolid application have seen the renewed and increased interest in the large-scale application of composts and biosolids. Unfortunately, increased biosolid usage has inadvertently caused a greater chance of soil becoming exposed to particulate plastics.

A study conducted by Wijesekara et al. (2018a) extracted particulate plastics from biosolids collected from a wastewater treatment plant in Sydney. Extracted particulate plastics were examined by size, ranging from less than 50–1000 μm. In terms of size, the highest concentration of particulate particles (i.e., 352 particles per kilogram of biosolids) was less than 50 μm. The other concentrations were 146, 324, and 174 particles per kilogram of biosolids for 50–100, 100–250, and 250–1000 μm size fractions, respectively.

Biosolids are often used in agriculture as nutrient-rich fertilizers. However, specific guidelines and regulations must be met before land application. These regulations require the pre-treatment of the biosolids to remove any contaminants (e.g., heavy metals) that may be present (Mahon et al. 2016). These treatments can include, but are not limited to, lime stabilization, anaerobic digestion, and thermal drying. Lime stabilization involves the addition of lime to help breakdown organic compounds and reduce pathogen content. Anaerobic digestion is the process by which organic waste is degraded in an oxygen-free environment by microbial processes resulting in biogas and the treated biosolids (Zhen et al. 2017).

A study conducted by Mahon et al. (2016) found that differences in the treatment process influenced the concentration of particulate plastic remaining in the treated biosolids. They found that lime stabilization contained a higher abundance of particulate plastics within the smaller size range of 250–400 μm. This study suggests that when lime stabilization is used as the treatment, it may result in the increased sheering of particulate plastics. This decrease in size makes the removal of plastic more difficult and could cause increased exposure of plastic to the soil. Alternatively, one sample site used anerobic digestion and had a significantly lower concentration of particulate plastics than the other sites. This decrease in concentration suggests that this treatment process may reduce particulate plastics abundance (Mahon et al. 2016).

It has been estimated that in 2017, biowaste (e.g., compost and biosolids) application caused Australian soils to become exposed to 280 billion particulate plastics (Wijesekara et al. 2018b). It is therefore crucial for tighter regulation to be put in place to ensure that soils are not exposed to particulate plastics through biowaste.

1.2.2.3 Wastewater for Irrigation

Irrigation is a common method used within the agricultural and horticulture industries and is particularly crucial in areas with low rainfall. However, with water scarcity becoming an increasing concern, there has been a shift toward farmers using treated, or in some cases untreated, wastewater for irrigation (Bläsing and Amelung 2018). Unfortunately, just as particulate plastics are present in treated biosolids, so too are they present within treated wastewater. A study conducted by Majewsky et al. (2016) analyzed two WWTPs in Germany for particulate plastics within wastewater effluent (Table 1.1). The concentration of polyethylene particulate plastics present in the wastewater samples was 81 mg per m^3 (Majewsky et al. 2016). Additionally, a study conducted by Murphy et al. (2016) examined particulate plastics present within the influent and effluent of a wastewater treatment plant in Scotland. The influent and effluent were found to contain 15.70 and 0.25 microplastics particles per liter, respectively. This study demonstrates that both treated and untreated wastewater can contain particulate plastics and cause increased concentrations of particulate plastics within soil if used for irrigation. However, exposure is significantly less if treated wastewater is used.

1.2.2.4 Indiscriminate Disposal in Landfills and Littering

Indiscriminate disposal in landfills and littering are sources of secondary particulate plastics that arise from human carelessness. Disposal of plastics at landfills can cause the plastic to leach from the site and travel to soils or rivers. Furthermore, if poorly disposed of plastics occur within coastal populations, this further increases the chance of particulate plastics entering rivers and oceans (Jambeck et al. 2015). A study conducted by Jambeck et al. (2015) estimated the annual quantity of mismanaged plastic waste in coastal locations (i.e., <50 km from the coast). It was calculated that 31.9 million metric tons of plastic were mismanaged in coastal regions, with oceans receiving between 4.8 and 12.7 million metric tons of poorly managed plastic waste (Table 1.1). Furthermore, poorly discarded plastics have the potential to break down, resulting in secondary particulate plastics (Duis and Coors 2016). Stormwater overflow and runoff are other routes by which particulate plastics can enter aquatic environments and soil, particularly in highly developed landscapes (Duis and Coors 2016).

1.2.2.5 Atmospheric Input

Research has suggested the occurrence of particulate plastics within the atmospheric environment. Although there is limited research surrounding airborne particles as a source or pathway to other environments, it is reasonable to assume that atmospheric fallout is a source for terrestrial environments. Wind and rain are two key features in dispersing or removing the particles from an area, and this dispersal could cause plastic and particulate plastics to be transported to soils that would otherwise not be exposed (Bläsing and Amelung 2018; Prata 2018).

A study by Dris et al. (2016) examined the particulate plastic fallout that occurred in both urban and sub-urban sites of France (Table 1.1). It was found that between 2 and 355 particles per m^2 per day of atmospheric fallout occurred, with 29% of the fibers analyzed found to contain plastic. Furthermore, these results were extrapolated by Dris et al. (2016), and it was estimated that the annual atmospheric fallout of synthetic fibers could be anywhere between 3 and 10 tons. Overall, there was less fallout in the sub-urban site when compared to the urban site. It was suggested that this difference in fallout might have been due to the urban location containing a larger human population. Furthermore, the weather conditions influenced the amount of atmospheric fallout that occurred each day. Fewer particles were observed on dry days, while the highest level of atmospheric fallout was recorded on a wet day. Interestingly, there was no correlation between the amount of rainfall and the atmospheric fallout, signifying that it is a combination of factors which affect particle fallout. This research highlights the need for a better understanding of the factors which contribute to increased atmospheric fallout and how these individual factors affect one another.

1.3 THE FUTURE

Particulate plastics are now prevalent in terrestrial, aquatic, and atmospheric environments. In terrestrial environments, particulate plastics become incorporated into the soil either through the usage and application of products (e.g., compost, biosolids, and plastic mulch) or through the poor management and disposal of these plastics. Although the use of plastics and particulate plastics can be beneficial within terrestrial environments, they can also be detrimental to some aspects of soil health and terrestrial organisms.

Therefore, it is crucial that humans limit their use of particulate plastics. There have been increasing efforts to ban particulate plastics in personal care products; however, it will be many years before this is adopted worldwide. The United States was one of the first countries to ban the use of microbeads within personal care products (Jiang 2018). From July 1, 2017, the U.S. House of Representatives legislation, Microbead-Free Waters Act of 2015, H.R. 1321, placed a ban on the manufacture and intentional addition of any rinse-off particulate plastics used in the cosmetic industry (U.S. House of

Representatives 2015). However, this act only bans the presence of microbeads in "rinse-off" products, leaving a loophole that still allows microbeads to enter the environment. In 2017, the Canadian Environmental Protection Act, 1999, saw the inclusion of the Microbeads in Toiletries Regulation, which bans the manufacture, importation, or sale of toiletries containing microbeads as of July 1, 2018 (Canadian Environmental Protection Act 2017). The Australian government chose a less forceful approach, calling for industries to voluntarily phase-out microbeads by July 2017 (Australian Government 2016). Therefore, although plans to stop further contamination of microbeads are in progress, it could be many years before they are wholly implemented. Further, as particulate plastics can persist in the environment for many hundreds of years, problems arising from their contamination will continue even after governmental bans. This increases the need for better understanding of the effect of particulate plastic contamination on different environments.

Furthermore, industries are now focusing on the use of biodegradable plastics to replace the currently used non-biodegradable plastics. An example of this is the use of biodegradable plastic mulch instead of the more traditional, low-density polyethylene. Biodegradable plastics can be degraded by microorganisms and include plastics such as poly(butylene succinate), poly(ε-caprolactone), and poly(lactic acid) (Urbanek et al. 2018).

A recent study by Dr. Melik Demirel, who has been leading a project at Pennsylvania State University, aims to address the problem surrounding the synthetics used within clothing (Gabbatiss 2019). Clothing fibers or microfibers have been identified by research as one of the primary sources of particulate plastics. Dr. Demirel aims to use biosynthetic materials to create microfibers. This project looks at growing clothing fibers in a fermentation tank using proteins present in squid tentacle suction cups.

Unfortunately, some commercial products claiming to contain biodegradable plastic are not entirely transparent. A study conducted by Nazareth et al. (2019) examined plastic products which claimed to be biodegradable and found that greenwashing occurred. Greenwashing is a term used in marketing to make a product appear more environmentally friendly than it actually is. Nazareth et al. (2019) analyzed "biodegradable" plastics from three different countries (i.e., Brazil, Canada, and the United States), with two samples per country. The results showed that out of the six products sampled, four of them contained elements of greenwashing. Aside from greenwashing, the degradation time for some "biodegradable" plastics [e.g., poly(lactic acid)] are still unknown in the marine environment (Nazareth et al. 2019). This study demonstrates the need for tighter controls in ensuring consumers receive the correct information surrounding the type of plastic materials they are purchasing.

1.4 CONCLUSIONS

Particulate plastics enter our environments through a range of different pathways and a variety of sources. In the terrestrial environment, particulate plastics are often introduced into the soil through the application and use of plastic materials (i.e., plastic mulch or PAM) or inadvertently via composts and biosolids. Particulate plastics can have both beneficial and detrimental effects on soil health. However, there is limited research on the overall and long-term fate of particulate plastics in terrestrial environments. Furthermore, particulate plastics can move between environments. Therefore, a better understanding of the linkage and effects of particulate plastics between environments is required.

REFERENCES

Akhbarizadeh, R., Moore, F., Keshavarzi, B. 2018. Investigating a probable relationship between microplastics and potentially toxic elements in fish muscles from northeast of Persian Gulf. *Environmental Pollution*, **232**, 154–163.

Andrady, A.L. 2011. Microplastics in the marine environment. *Marine Pollution Bulletin*, **62**(8), 1596–1605.

Andrady, A.L. 2015. *Plastics and Environmental Sustainability*. John Wiley & Sons, Hoboken, NJ.

Australian Government. 2016. *Voluntary Industry Phase-Out of Solid Plastic Microbeads from 'Rinse-Off' Personal Care, Cosmetic and Cleaning Products*. Department of Environment and Energy, Canberra, ACT.

Besseling, E., Wegner, A., Foekema, E.M., van den Heuvel-Greve, M.J., Koelmans, A.A. 2013. Effects of microplastic on fitness and PCB bioaccumulation by the *Lugworm arenicola marina* (L.). *Environmental Science & Technology*, **47**(1), 593–600.

Bläsing, M., Amelung, W. 2018. Plastics in soil: Analytical methods and possible sources. *Science of the Total Environment*, **612**, 422–435.

Bolan, N.S., Kunhikrishnan, A., Naidu, R. 2013. Carbon storage in a heavy clay soil landfill site after biosolid application. *Science of the Total Environment*, **465**, 216–225.

Borselli, L., Biancalani, R., Giordani, C., Carnicelli, S., Ferrari, G.A. 1996. Effect of gypsum on seedling emergence in a kaolinitic crusting soil. *Soil Technology*, **9**(1), 71–81.

Bradney, L., Wijesekara, H., Palansooriya, K.N., Obadamudalige, N., Bolan, N.S., Ok, Y.S., Rinklebe, J., Kim, K.-H., Kirkham, M.B. 2019. Particulate plastics as a vector for toxic trace-element uptake by aquatic and terrestrial organisms and human health risk. *Environment International*, **131**, 104937.

Brennecke, D., Duarte, B., Paiva, F., Caçador, I., Canning-Clode, J. 2016. Microplastics as vector for heavy metal contamination from the marine environment. *Estuarine, Coastal and Shelf Science*, **178**, 189–195.

Browne, M.A., Crump, P., Niven, S.J., Teuten, E., Tonkin, A., Galloway, T., Thompson, R. 2011. Accumulation of microplastic on shorelines worldwide: Sources and sinks. *Environmental Science & Technology*, **45**(21), 9175–9179.

Canadian Environmental Protection Act. 2017. Microbeads in toiletries regulations. *Canada Gazette*, 151(12), 14 June 2017. Registration SOR/2017-111 June 2, http://www.gazette.gc.ca/rp-pr/p2/2017/2017-06-14/html/sor-dors111-eng.html.

Carr, S.A., Liu, J., Tesoro, A.G. 2016. Transport and fate of microplastic particles in wastewater treatment plants. *Water Research*, **91**, 174–182.

Cole, M., Lindeque, P., Halsband, C., Galloway, T.S. 2011. Microplastics as contaminants in the marine environment: A review. *Marine Pollution Bulletin*, **62**(12), 2588–2597.

Dris, R., Gasperi, J., Saad, M., Mirande, C., Tassin, B. 2016. Synthetic fibers in atmospheric fallout: A source of microplastics in the environment? *Marine Pollution Bulletin*, **104**(1), 290–293.

Duis, K., Coors, A. 2016. Microplastics in the aquatic and terrestrial environment: Sources (with a specific focus on personal care products), fate and effects. *Environmental Sciences Europe*, **28**(1), 2.

Fok, L., Cheung, P.K. 2015. Hong Kong at the Pearl River Estuary: A hotspot of microplastic pollution. *Marine Pollution Bulletin*, **99**(1), 112–118.

Gabbatiss, J. 2019. Fabrics made from squid-based material could cut ocean plastic pollution. *Independent* (Online).

Gao, H., Yan, C., Liu, Q., Ding, W., Chen, B., Li, Z. 2019. Effects of plastic mulching and plastic residue on agricultural production: A meta-analysis. *Science of the Total Environment*, **651**, 484–492.

GESAMP. 2015. Sources, fate and effects of microplastics in the marine environment: A global assessment. IMO/FAO/UNESCO-IOC/UNIDO/WMO/IAEA/UN/UNEP/UNDP Joint Group of Experts on the Scientific Aspects of Marine Environmental Protection.

Hayes, D.G., Wadsworth, L.C., Sintim, H.Y., Flury, M., English, M., Schaeffer, S., Saxton, A.M. 2017. Effect of diverse weathering conditions on the physicochemical properties of biodegradable plastic mulches. *Polymer Testing*, **62**, 454–467.

Holmes, L.A., Turner, A., Thompson, R.C. 2012. Adsorption of trace metals to plastic resin pellets in the marine environment. *Environmental Pollution*, **160**, 42–48.

Holmes, L.A., Turner, A., Thompson, R.C. 2014. Interactions between trace metals and plastic production pellets under estuarine conditions. *Marine Chemistry*, **167**, 25–32.

Horton, A.A., Walton, A., Spurgeon, D.J., Lahive, E., Svendsen, C. 2017. Microplastics in freshwater and terrestrial environments: Evaluating the current understanding to identify the knowledge gaps and future research priorities. *Science of the Total Environment*, **586**, 127–141.

Jambeck, J.R., Geyer, R., Wilcox, C., Siegler, T.R., Perryman, M., Andrady, A., Narayan, R., Law, K.L. 2015. Plastic waste inputs from land into the ocean. *Science*, **347**(6223), 768.

Jiang, J.-Q. 2018. Occurrence of microplastics and its pollution in the environment: A review. *Sustainable Production and Consumption*, **13**, 16–23.

Kasirajan, S., Ngouajio, M. 2012. Polyethylene and biodegradable mulches for agricultural applications: A review. *Agronomy for Sustainable Development*, **32**(2), 501–529.

Kedzierski, M., D'Almeida, M., Magueresse, A., Le Grand, A., Duval, H., César, G., Sire, O., Bruzaud, S., Le Tilly, V. 2018. Threat of plastic ageing in marine environment. Adsorption/desorption of micropollutants. *Marine Pollution Bulletin*, **127**, 684–694.

Khan, F.R., Syberg, K., Shashoua, Y., Bury, N.R. 2015. Influence of polyethylene microplastic beads on the uptake and localization of silver in zebrafish (*Danio rerio*). *Environmental Pollution*, **206**, 73–79.

Li, X., Chen, L., Mei, Q., Dong, B., Dai, X., Ding, G., Zeng, E.Y. 2018. Microplastics in sewage sludge from the wastewater treatment plants in China. *Water Research*, **142**, 75–85.

Li, Y., Shao, M., Horton, R. 2011. Effect of polyacrylamide applications on soil hydraulic characteristics and sediment yield of sloping land. *Procedia Environmental Sciences*, **11**, 763–773.

Liebezeit, G., Liebezeit, E. 2014. Synthetic particles as contaminants in German beers. *Food Additives & Contaminants: Part A*, **31**(9), 1574–1578.

Liu, E.K., He, W.Q., Yan, C.R. 2014. "White revolution" to "white pollution" – Agricultural plastic film mulch in China. *Environmental Research Letters*, **9**(9), 091001.

Mahon, A.M., O'Connell, B., Healy, M.G., O'Connor, I., Officer, R., Nash, R., Morrison, L. 2016. Microplastics in sewage sludge: Effects of treatment. *Environmental Science & Technology*, **51**(2), 810–818.

Majewsky, M., Bitter, H., Eiche, E., Horn, H. 2016. Determination of microplastic polyethylene (PE) and polypropylene (PP) in environmental samples using thermal analysis (TGA-DSC). *Science of the Total Environment*, **568**, 507–511.

Murphy, F., Ewins, C., Carbonnier, F., Quinn, B. 2016. Wastewater treatment works (WwTW) as a source of microplastics in the aquatic environment. *Environmental Science & Technology*, **50**(11), 5800–5808.

Nazareth, M., Marques, M.R.C., Leite, M.C.A., Castro, Í.B. 2019. Commercial plastics claiming biodegradable status: Is this also accurate for marine environments? *Journal of Hazardous Materials*, **366**, 714–722.

Ng, E.-L., Huerta Lwanga, E., Eldridge, S.M., Johnston, P., Hu, H.-W., Geissen, V., Chen, D. 2018. An overview of microplastic and nanoplastic pollution in agroecosystems. *Science of the Total Environment*, **627**, 1377–1388.

Nizzetto, L., Bussi, G., Futter, M.N., Butterfield, D., Whitehead, P.G. 2016a. A theoretical assessment of microplastic transport in river catchments and their retention by soils and river sediments. *Environmental Science: Processes & Impacts*, **18**(8), 1050–1059.

Nizzetto, L., Futter, M., Langaas, S. 2016b. Are agricultural soils dumps for microplastics of urban origin? *Environmental Science & Technology*, **50**(20), 10777–10779.

NSW Environmental Protection Agency. 2016. Resource Recovery Order under Part 9, Clause 93 of the Protection of the Environment Operations (Waste) Regulation 2014: The compost order 2016.

Pivokonsky, M., Cermakova, L., Novotna, K., Peer, P., Cajthaml, T., Janda, V. 2018. Occurrence of microplastics in raw and treated drinking water. *Science of the Total Environment*, **643**, 1644–1651.

Prata, J.C. 2018. Airborne microplastics: Consequences to human health? *Environmental Pollution*, **234**, 115–126.

Rillig, M.C. 2012. Microplastic in terrestrial ecosystems and the soil? *Environmental Science & Technology*, **46**(12), 6453–6454.

Rochman, C.M., Hoh, E., Kurobe, T., Teh, S.J. 2013. Ingested plastic transfers hazardous chemicals to fish and induces hepatic stress. *Scientific Reports*, **3**, 3263.

Rochman, C.M., Kross, S.M., Armstrong, J.B., Bogan, M.T., Darling, E.S., Green, S.J., Smyth, A.R., Veríssimo, D. 2015. Scientific evidence supports a ban on microbeads. *Environmental Science & Technology*, **49**(18), 10759–10761.

Saglam, M., Sintim, H.Y., Bary, A.I., Miles, C.A., Ghimire, S., Inglis, D.A., Flury, M. 2017. Modeling the effect of biodegradable paper and plastic mulch on soil moisture dynamics. *Agricultural Water Management*, **193**, 240–250.

Sintim, H.Y., Bandopadhyay, S., English, M.E., Bary, A.I., DeBruyn, J.M., Schaeffer, S.M., Miles, C.A., Reganold, J.P., Flury, M. 2019. Impacts of biodegradable plastic mulches on soil health. *Agriculture, Ecosystems & Environment*, **273**, 36–49.

Sleight, V.A., Bakir, A., Thompson, R.C., Henry, T.B. 2017. Assessment of microplastic-sorbed contaminant bioavailability through analysis of biomarker gene expression in larval zebrafish. *Marine Pollution Bulletin*, **116**(1), 291–297.

Sojka, R.E., Bjorneberg, D.L., Entry, J.A., Lentz, R.D., Orts, W.J. 2007. Polyacrylamide in agriculture and environmental land management. *Advances in Agronomy*, **92**, 75–162.

Song, Y.K., Hong, S.H., Jang, M., Kang, J.-H., Kwon, O.Y., Han, G.M., Shim, W.J. 2014. Large accumulation of micro-sized synthetic polymer particles in the sea surface microlayer. *Environmental Science & Technology*, **48**(16), 9014–9021.

Steinmetz, Z., Wollmann, C., Schaefer, M., Buchmann, C., David, J., Tröger, J., Muñoz, K., Frör, O., Schaumann, G.E. 2016. Plastic mulching in agriculture: Trading short-term agronomic benefits for long-term soil degradation? *Science of the Total Environment*, **550**, 690–705.

Teuten, E.L., Saquing, J.M., Knappe, D.R.U., Barlaz, M.A., Jonsson, S., Björn, A., Rowland, S.J., et al. 2009. Transport and release of chemicals from plastics to the environment and to wildlife. *Philosophical Transactions of the Royal Society B: Biological Sciences*, **364**(1526), 2027–2045.

Thompson, R.C., Swan, S.H., Moore, C.J., vom Saal, F.S. 2009. Our plastic age. *Philosophical Transactions of the Royal Society B: Biological Sciences*, **364**(1526), 1973–1976.

Turner, A., Holmes, L.A. 2015. Adsorption of trace metals by microplastic pellets in fresh water. *Environmental Chemistry*, **12**(5), 600–610.

UNEP. 2009. *Marine Litter – A Global Challenge*. United Nations Environment Programme, Nairobi, Kenya, p. 232.

UNEP. 2015. *Plastic Cosmetics: Are We Polluting the Environment Through Our Personal Care?* United Nations Environment Programme, Nairobi, Kenya.

UNEP. 2016. *Marine Plastic Debris and Microplastics – Global Lessons and Research to Inspire Action and Guide Policy Change*. United Nations Environment Programme, Nairobi, Kenya.

Urbanek, A.K., Rymowicz, W., Mirończuk, A.M. 2018. Degradation of plastics and plastic-degrading bacteria in cold marine habitats. *Applied Microbiology and Biotechnology*, **102**(18), 7669–7678.

U.S. House of Representatives. 2015. Microbead-Free Waters Act of 2015, Report. pp. 114–371.

Wardrop, P., Shimeta, J., Nugegoda, D., Morrison, P.D., Miranda, A., Tang, M., Clarke, B.O. 2016. Chemical pollutants sorbed to ingested microbeads from personal care products accumulate in fish. *Environmental Science & Technology*, **50**(7), 4037–4044.

Weithmann, N., Möller, J.N., Löder, M.G.J., Piehl, S., Laforsch, C., Freitag, R. 2018. Organic fertilizer as a vehicle for the entry of microplastic into the environment. *Science Advances*, **4**(4), eaap8060.

Wijesekara, H., Bolan, N.S., Bradney, L., Obadamudalige, N., Seshadri, B., Kunhikrishnan, A., Dharmarajan, R., et al. 2018a. Trace element dynamics of biosolids-derived microbeads. *Chemosphere*, **199**, 331–339.

Wijesekara, H., Bolan, N.S., Vithanage, M., Xu, Y., Mandal, S., Brown, S., Hettiarachchi, G., et al. 2017. *Utilization of Biowaste for Mine Spoil Rehabilitation*. Advances in Agronomy, London, UK.

Wijesekara, H., Bradney, L., Kunhikrishnan, A., Bolan, N. 2018b. Particulate plastics in soil: Friend or foe? *Remediation Australasia* (20), 20–22. https://www.remediationaustralasia.com.au/articles/particulate-plastics-soil-friend-or-foe

Xiong, B., Loss, R.D., Shields, D., Pawlik, T., Hochreiter, R., Zydney, A.L., Kumar, M. 2018. Polyacrylamide degradation and its implications in environmental systems. *NPJ Clean Water*, **1**(1), 17.

Zhen, G., Lu, X., Kato, H., Zhao, Y., Li, Y.-Y. 2017. Overview of pretreatment strategies for enhancing sewage sludge disintegration and subsequent anaerobic digestion: Current advances, full-scale application and future perspectives. *Renewable and Sustainable Energy Reviews*, **69**, 559–577.

Ziajahromi, S., Neale, P.A., Rintoul, L., Leusch, F.D.L. 2017. Wastewater treatment plants as a pathway for microplastics: Development of a new approach to sample wastewater-based microplastics. *Water Research*, **112**, 93–99.

Zubris, K.A.V., Richards, B.K. 2005. Synthetic fibers as an indicator of land application of sludge. *Environmental Pollution*, **138**(2), 201–211.

2 Particulate Plastics from Agriculture

M.B. Kirkham, Reshma M. Antony, and Nanthi S. Bolan

CONTENTS

2.1 Introduction 19
2.2 Uses of Plastics in Agriculture 20
 2.2.1 Greenhouses 20
 2.2.2 High Tunnels 21
 2.2.3 Low Tunnels 21
 2.2.4 Plastic Mulches 22
 2.2.5 Bagging of Fruit 23
 2.2.6 Windbreaks 23
 2.2.7 Seed Coatings and Seed Priming 23
 2.2.8 Other Uses 23
2.3 Effects of Particulate Plastics on Soil Organisms 24
 2.3.1 Bacteria and Fungi 24
 2.3.2 Worms 25
 2.3.3 Nematodes 26
 2.3.4 Collembola 26
 2.3.5 Isopods 26
2.4 Effects of Particulate Plastics on Plants 26
2.5 Disposal of Agricultural Plastics 27
 2.5.1 Incineration 27
 2.5.2 Landfilling 28
 2.5.3 Recycling 28
 2.5.4 Photodegradable Plastics 29
 2.5.5 Biodegradable Plastics 30
2.6 Future 33
Acknowledgments 34
References 34

2.1 INTRODUCTION

Since the middle of the 1950s, plastic polymers, such as polyethylene or polypropylene, have been widely used in agriculture, and they have allowed farmers to increase crop production (Kyrikou and Briassoulis 2007; Kasirajan and Ngouajio 2012; Rodríguez-Seijo et al. 2019). Currently, billions of pounds of agricultural plastics are used around the world each year (Grossman 2015). It is estimated that U.S. agriculture alone uses about a billion pounds (454,000 metric tons) annually (Grossman 2015). To face the food needs of the growing population, the use of plastics will increase by 50% (9.5 million metric tons) by 2030 (Le Moine and Ferry 2018). By then, the waste from agricultural plastics will grow as well, up to 17 million metric tons, which will increase the plasticultors' responsibility to remove the used plastic from the environment (Le Moine and Ferry 2018). The plastics break down in the soil into microplastics, which are synthetic organic polymer particles with a size between 100 nm and 5 mm (Duis and Coors 2016). In this review, the more general term, particulate plastics, will be used, which include both microplastics and nanoplastics with sizes from 5 mm down to the nano-meter range (Bradney et al. 2019). However, the terms microplastics and nanoplastics will be used, when authors of the references cited herewith use them in their papers.

This chapter will review the use of plastics in agriculture, the consequences of particulate plastics on soil organisms and plants, and then describe methods that are being developed to reduce plastic contamination due to agriculture. The review will not consider particulate plastics in sewage sludge added to the soil. This topic has been reviewed (Carr et al. 2016; Mahon et al. 2017; Wijesekara et al. 2018).

2.2 USES OF PLASTICS IN AGRICULTURE

Plastics are used in agriculture in many different ways, including for manufacture of greenhouses, high tunnels, low tunnels, plastic mulches, bags for fruits, windbreaks, and seed coatings (Figure 2.1).

Each of these will be considered in the following sections.

2.2.1 Greenhouses

For centuries, horticulturists and agronomists have attempted to modify the environment in which crops are grown to extend the season for increased production (Lamont 2005). Early structures were glass greenhouses. Commercial production of the polyethylene polymer began in the late 1930s, and by the 1950s, plastic films were available for construction of greenhouses (Lamont 2005).

Emery Emmert at the University of Kentucky, known as the "Father of Plasticulture," was a pioneer in the development of plastic-covered greenhouses (Lamont 2005). Crops had been successfully grown in polyethylene houses in Lexington, KY, for 8 years before Emmert started his studies in 1953 to give proof that crops grew well in them (Emmert 1955). His main goal was to show that it was possible to grow-out-of-season crops much cheaper in them than in glasshouses.

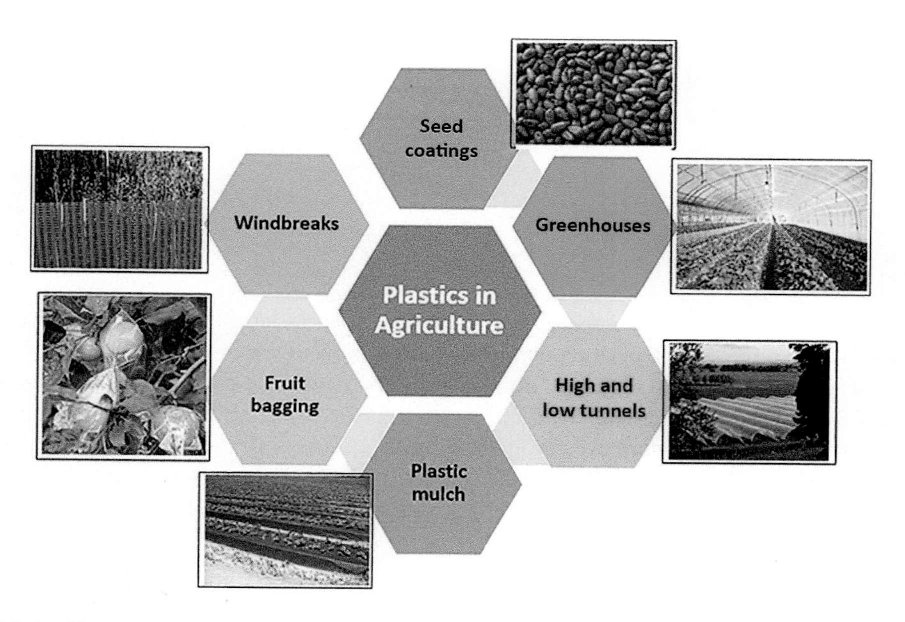

FIGURE 2.1 Uses of plastics in agriculture.

The global use of greenhouses for food production has increased six-fold over the past 20 years to more than 9,000,000 acres (3,600,000 ha) today (McNutty 2017). Film plastic rather than glass is the leading greenhouse covering for two reasons (Nelson 2012, p. 45). First, film plastic greenhouses with permanent metal frames cost less than glass greenhouses. Second, film plastic greenhouses are popular because the cost of heating them is approximately 40% lower compared to glass (Nelson 2012, p. 45). The use of polyethylene for greenhouses has increased rapidly and continues to do so. Film plastic greenhouses constitute the largest portion (71%) of greenhouses in the United States today (Nelson 2012, p. 46).

As a percent of current global greenhouse or protected agriculture production, the United States has only a small fraction of the total area (Janke et al. 2017). The leader is the People's Republic of China, with an estimated area for production in greenhouses at 2.6 million ha, or about 90% of estimated global area under greenhouses. In 1999, the country's greenhouse production area was about 1.4 million ha, up from 9180 ha in 1981, with 460,000 ha of the total greenhouse area in high tunnels (see next section). Other major producers of horticultural crops in greenhouses include South Korea, Spain, Japan, Turkey, and Italy. The current area in the United States is estimated at only 0.3% of global area under greenhouses (Janke et al. 2017).

One disadvantage of using film plastics for greenhouses is that these covering materials are short-lived compared to glass. The highest-quality, ultraviolet light-resistant, 6-mil thick (0.006 in., 0.15 mm) polyethylene films last up to four years. Ultraviolet light from the sun causes the plastic to darken, thereby lowering light transmission, and it also causes the plastic to become brittle, which subjects it to breakage in the wind (Nelson 2012, p. 46).

2.2.2 High Tunnels

A high tunnel is clear plastic covering a frame high enough to walk inside, heated by solar radiation, and cooled by passive ventilation (Knewtson et al. 2010b). High tunnels also are called "hoop houses" because the plastic is stretched over hoops of metal or polyvinyl chloride (Janke et al. 2017). Producers use high tunnels to elevate temperatures to allow earlier planting in the spring, earlier ripening, and extension of fall harvests. Compared with the open field, other benefits of high tunnels include wind and rain protection; enhanced crop quality and yield; and exclusion of pests, such as animals, diseases, and insects (Knewtson et al. 2010a,b).

High tunnels have been used in the United States for more than 50 years (Janke et al. 2017). In the Central Great Plains, the number of vegetable, fruit, and flower growers using high tunnels has increased steadily, and, a survey, conducted between 2005 and 2007 of 81 growers managing 185 high tunnels in the region, found that the oldest one was 15 years old (Knewtson et al. 2010a). Growers reported satisfaction with them; those with more than one high tunnel had often added tunnels following the success of crop production in an initial tunnel.

Since 2009, conservation programs of the United States Department of Agriculture has been providing funding for high tunnel construction, resulting in over $61 million for more than 13,000 high tunnels by the end of 2014 (Janke et al. 2017).

2.2.3 Low Tunnels

Low tunnels or row covers (Lamont 2005) are temporary structures that are easy to assemble and disassemble with each crop (Arancibia 2018). This mobility offers an advantage over high tunnels because it allows for rotations with cover crops or other field crops to improve and maintain productivity (Arancibia 2018). With low tunnels, early production of vegetables is possible by creating a mini-greenhouse effect. Low tunnels increase air and soil temperatures, protect crops against cold temperatures and mild freezes, and promote vegetable growth early in the spring and late fall (Arancibia 2018). In the northern states of the United States, farmers use low tunnels inside high tunnels as an additional protection layer against extremely cold temperatures to promote growth of cool-season crops (Arancibia 2018). Low tunnels are more affordable than high tunnels, so more farmers have access to this technology, and, thereby, they may be able to maintain the sustainability of their operations (Arancibia 2018).

The first row covers used for high-value field-grown vegetable crops were not plastic, but were parchment paper used for covering early spring plantings of celery (*Apium graveolens* L.) for protection against wind, cold rains, and frost in the Grand Rapids and Kalamazoo areas of Michigan (Lamont 2005; Lucas and Wittwer 1956). The use of paper preceded the use of plastic by more than 50 years (Lamont 2005). As for plastic greenhouses, Emmert (1957) was a pioneer in the research on using plastics (polyethylene) for row covers. Now the use of row covers for protection of early warm-season crops is widespread throughout the United States. (Lamont 2005).

Low tunnels can be of various heights and are covered mainly with three types of plastic materials: perforated or slit plastic film, spun-bonded fabrics, and insect nets (Arancibia 2018). Plastic film increases the temperature, but its use in low tunnels is limited to small plants, because of water condensation that promotes decay of the foliage in contact with the film. This problem is rare with spun-bonded fabrics, because they are permeable and allow airflow with little condensation inside the tunnel. Insect nets made of plastic material exclude insects of different sizes, depending upon the width of the holes in the netting.

2.2.4 Plastic Mulches

The dictionary defines mulch as "leaves, straw, or other loose material spread on the ground around plants to prevent evaporation of water from soil, freezing of roots, etc." (Webster's New World Dictionary of the American Language 1959). When one speaks of "plastic mulch," the mulch is plastic, not organic material. Plastic mulches are used when crops are planted in rows and the plastic is put over the rows where the seeds are planted. Soil is exposed between the rows covered with the plastic. Plastic mulches are commercially used for both vegetables and small fruit crops (Maughan and Drost 2016). They are employed to conserve moisture, suppress growth of weeds, prevent packing of the soil by rain (Emmert 1957), and to increase the soil temperature in the spring, which extends the season of growth. Warmer soil temperatures allow for early planting dates, and they speed up plant growth. Plastic mulches are used in the summer to reduce the soil temperature, because they can reflect heat (Maughan and Drost 2016).

Plastic mulches also are used to act as a barrier to keep methyl bromide, a fumigant against insects, weeds, and soil-borne diseases, in the soil. However, the use of methyl bromide has been phased out due to its potential for depleting stratospheric ozone (Chellemi et al. 2011; Yates et al. 2011). Alternatives to methyl bromide, such as the use of other fumigants (Chellemi et al. 2011) or solarization (Yates et al. 2011), have been developed, but the methods still use soil beds that are covered with plastic film after fumigation (Yates et al. 2011).

Plastic mulches have been used commercially on vegetables since the early 1960s. The first mulches were made of paper or aluminum. Paper was too costly (Lucas and Wittwer 1956) and deteriorated before the season was over, and aluminum did not decompose and also was costly. Black polyethylene was found to be the cheapest and best mulch (Emmert 1957). Much of the early research was conducted on the impact of color (black and clear) on soil and air temperatures, moisture retention, and vegetable yields (Lamont 2005).

The use of plastic mulch in agriculture has increased dramatically throughout the world due to the large increases in yields of both horticultural and field crops obtained with them (Kasirajan and Ngouajio 2012; Steinmetz et al. 2016). In 1999, over 30 million acres (12,000,000 ha) of agricultural land (over 185,000 acres or 74,000 ha in the United States) were covered with plastic mulch, and the figures have increased since then (Kasirajan and Ngouajio 2012). It is estimated that 1 million metric tons of mulch film are used worldwide every year in agriculture. Today, production of fresh-market vegetables using raised beds covered with plastic mulch and drip irrigated is a standard method (Kasirajan and Ngouajio 2012).

2.2.5 Bagging of Fruit

The covering of bunches of bananas with plastic has become a common practice in commercial production. In 1955, Australia led the use of plastic covers (Berrill 1956; Santosh et al. 2017). Bunch covers are typically made of thin plastic, usually low-density polyethylene with thickness of 5–40 µm (Santosh et al. 2017). Bananas on a tree are bagged to protect the fruit against damage caused by insects, diseases, birds, mechanical damage, or by the application of chemical products. Bagging also improves the visual quality of the fruit by promoting skin coloration and reducing blemishes (Santosh et al. 2017). In hot and humid climates, the bunch covers are perforated to prevent overheating, rotting, or premature ripening. Bunch covering is an ancient practice, and the first covers were old banana leaves or paper bags (Santosh et al. 2017).

2.2.6 Windbreaks

Windbreaks are made out of plastic, webbed materials (Lamont 2005). Plastics mulches can also be used with windbreaks that are not plastic, but are trees or annual, small grains. If small grains are used as windbreaks, the grain is planted in the fall in strips that are 10–12 feet (3.0–3.7 m) in width and placed far enough apart to accommodate five or six mulched raised beds, when the crop to be protected is planted in the spring. The grain then provides a protection from cold spring winds. Another option for an annual, small grain windbreak is to plant the entire field as a solid grain cover crop in the fall. In the spring, the area where the plastic mulch and drip irrigation beds are to be made is tilled, leaving strips of small grain between five to six plastic-covered raised beds. Once wind protection is no longer required, the grain strips are mowed and used as rows to drive equipment for spraying for insect and disease control and for harvesting (Lamont 2005).

2.2.7 Seed Coatings and Seed Priming

Coatings have been applied to seeds before planting for many years. The coating agents are pesticides, bactericides, fertilizers, plant growth regulators (Wang et al. 2016), or gelatin (Taylor et al. 1982). When gelatin is used, seeds are first germinated in controlled conditions, and then suspended in a fluid gelatin, which is extruded behind the furrow opener or a conventional planter. The major advantages of sowing germinated seeds, compared to dry seeds, are earlier and more uniform emergence. Another major advantage is the capacity of a germinated seed to continue growth under environmental conditions sub-optimal for normal germination to occur (Taylor et al. 1982).

Treating the seeds with moisture, such as coating the seeds with fluid gelatin, is called "seed priming." In recent years, the coatings have been covered with a plastic. Encapsulating fungicides and insecticides in film coatings applied to agronomic seeds has become a widely accepted method for enhancing seed germination by protecting against diseases and early season insect pests (Accinelli et al. 2019).

A specific type of seed priming is called "osmotic priming" or "osmopriming," in which seeds are soaked in chemical solutions of varying osmotic potentials. The chemicals are often polyethylene glycol, such as polyethylene glycol with the molecular weight of 8000. For example, Zhang et al. (2015) found that soaking sorghum (*Sorghum bicolor* L. Moench) seeds in a 20% (w/v) polyethylene glycol 8000 solution for 48 hours before planting improved seed germination and seedling establishment under drought conditions.

2.2.8 Other Uses

Plastics are used in drip-irrigation tubing, tape, pipes, filters, fittings and connectors, containers for growing plants, and packaging materials (Lamont 2005) including bags for grain. Some of these materials last a long time and do not break down as readily as other plastics used in agriculture.

Subsurface drip irrigation, developed when plastics were available after World War II, is of particular interest due to the pressure to conserve water (Camp et al. 2000), especially in arid and semiarid regions. The reliability and longevity of modern materials has increased (Camp et al. 2000). Subsurface drip irrigation systems can last 15 years or more (Alam and Rogers 2005).

A subsurface water retention technology, which uses plastic films placed at various depths below a plant's root zone to retain soil water, has been developed at Michigan State University (Smucker et al. 2018). The film is made of an impermeable, low-density polyethylene. The technique is called the "soil water retention technology." The soil water retention technology plastic membranes are U-shaped troughs that are installed at two depths below the soil surface (40 and 55 cm), using patented mechanical equipment, and then rows of seeds are planted. Smucker et al. (2018) found that the membranes increased volumetric water content in the plant root zone and increased yields of green bell pepper (*Capsicum frutescems* L.) by 20% and cucumber (*Cucumis sativus* L.) by 24%.

2.3 EFFECTS OF PARTICULATE PLASTICS ON SOIL ORGANISMS

Most studies with particulate plastics have focused on the marine environment rather than the terrestrial environment (de Lorenzo 2017; Hurley and Nizzetto 2018; Ren et al. 2018; Rillig 2012), even though the topic now is getting more attention (Zhu, Y.-G. et al. 2019). The research that has been done relates mainly to soil organisms. Nevertheless, information about the bioavailability and bioaccumulation of microplastics in soil organisms is generally lacking (Ng et al. 2018).

2.3.1 BACTERIA AND FUNGI

Esan et al. (2019) used DNA extraction and sequencing techniques to analyze the structure of bacterial and fungal communities in both mature bulk compost piles and compost piles associated with partially degraded low-density polyethylene plastics from four compost facilities across Nova Scotia, Canada. The compost piles had ages ranging between 2 and 10 years. They found no effect of the plastics on the structure of the microbial communities in the compost piles. Esan et al. (2019) concluded that the presence of low-density polyethylene plastics is not harmful for compost ecosystems. However, the lack of evidence of ecological impact from microplastics and nanoplastics in agroecosystems does not equate to the evidence of absence (Ng et al. 2018). Further studies are needed.

Kathiresan (2003) studied the biodegradation of polythene bags and plastics cups, buried at the 5 cm depth, in soil with mangroves (*Rhizophora* sp. and *Avicennia* sp.) along the southeast coast of India. The bags and cups were sampled at intervals of 2, 4, 6, and 9 months and were weighed for percent weight loss. Another set of samples was analyzed for the bacteria and fungi associated with the plastics. The microorganisms were isolated and grown in shaker cultures. The biodegradation of the polythene bags was higher (up to 4.21% in 9 months) than that of the plastic cups (up to 0.25% in 9 months). The microbial species found associated with the degrading materials were identified as five Gram-positive and two Gram-negative bacteria and eight fungal species of *Aspergillus*. The species that were predominant were *Streptococcus*, *Staphylococcus*, *Micrococcus* (these three are Gram-positive), *Moraxella*, and *Pseudomonas* (these two are Gram-negative), and two species of fungi (*Aspergillus glaucus* and *A. niger*). Among the bacteria grown in shaker cultures, the *Pseudomonas* species degraded 20.54% of the polythene and 8.16% of the plastic cups in a one-month period; among the fungal species, *Aspergillus glaucus* degraded 28.80% of the polythene and 7.26% of the plastic cups in a one-month period. The results indicate that it is going to take a long time for plastics to degrade in this mangrove soil.

The spread of antibiotic-resistant genes by soil microorganisms is an increasing concern, due to the potential adverse effect on human health (Zhu, F. et al. 2019). Studies based on aquatic ecosystems reveal that microplastics can serve as points of gene exchange between phylogenetically different microorganisms by introducing additional surfaces. This could result in increased spread of antibiotic-resistant genes and antibiotic-resistant pathogens in water and sediments (Zhu, F. et al. 2019). In soil ecosystems, the presence of polystyrene microplastics has been shown to increase the retention time of antibiotics and antibiotic-resistant genes (Zhu, F. et al. 2019).

Sun et al. (2018) demonstrated that the presence of microplastics in a greenhouse soil inhibited the dissipation of soil antibiotics and antibiotic-resistant genes. This may arise from the porous structure and high specific surface area of microplastics, which allow these materials to adsorb pollutants, leading to a reduced bioaccessiblity of the antibiotic (da Costa et al. 2019).

2.3.2 Worms

Several studies have been done to determine the effects of particulate plastics on earthworms. Huerta Lwanga et al. (2016) showed that earthworms (*Lumbricus terrestris*) can ingest microplastic particles and egest them in casts. Huerta Lwanga et al. (2017) demonstrated that earthworms can incorporate microplastic particles in surface soil and carry them into their burrow walls. The earthworm holes cause a risk of leaching of the microplastics through preferential flow into groundwater. In addition, microplastic ingestion may represent a removal from soils, when this results in systemic translocation or trophic transfer (Hurley and Nizzetto 2018), such as when birds eat earthworms.

Yu et al. (2019) studied how the activities of earthworms (*Lumbricus terrestris*) affected the distribution of microplastics in soil. They used 50-cm long soil columns, divided into 10-cm layers, which had a dry litter of black poplar (*Populus nigra* L.) added to the top 10 cm. A low-density microplastic at a concentration of 7% was mixed into the surface soil of the columns that received microplastics. The particle distribution of microplastics was 50% 250 μm–1 mm; 30% 150–250 μm; and 20% less than 150 μm. They then added two adult earthworms to each column. Results showed that low-density polyethylene particles could be introduced into the soil by the earthworms. Microplastic particles were detected in each soil sample and within different soil layers for the earthworm treatments. Earthworms showed a tendency to transport the smaller microplastic particles. After leaching, microplastics only were detected in the leachate from the treatments with the earthworms. The results showed that earthworms can mobilize microplastic transport from the surface into the soil and into drainage water. Thus, the activities of earthworms constitute a potential pathway for microplastics to be transported into groundwater. Huerta Lwanga et al. (2016) reached a similar conclusion.

Rodríguez-Seijo et al. (2019) wanted to know if microplastics can transport contaminants like pesticides to the soil matrix. They exposed earthworms (*Eisenia fetida*) for 14 days to soil containing two differently sized microplastics (5 mm and 0.25 μm–1 mm) with or without chlorpyrifos, a pesticide. In the tests with the chlorpyrifos, the microplastics were sprayed with the pesticide before being added to the soil. More chlorpyrifos was in the soil at the end of the experiment with the small microplastics compared to the large microplastics. Despite the ability of the microplastics to release chlorpyrifos to the soil, the earthworms avoided the contaminated microplastics. No evidence of microplastic uptake was observed, and they concluded that microplastics are not carriers of pesticides to earthworms. However, more studies with other pesticides and earthworms are needed.

Microplastics derived from starch-based biodegradable films had more effect on earthworm growth than conventional low-density polyethylene films. This was possibly because the biodegradable plastics were mainly composed of polyethylene terephthalate and polybutylene terephthalate, which might be more toxic than polyethylene (Zhu, F. et al. 2019).

Studies with worms show that the effect of particulate plastics is dependent on the amount of the plastics present. There was a concentration-dependent effect of polystyrene nanoplastics on the weight of the soil oligochaete *Enchytraeus crypticus*, an annelid worm. When incorporated into oatmeal at 0.5%, there was no effect, but at 10%, there was a negative effect. A shift in the gut microbiota was only observed under the highest exposure (10%) (Zhu, F. et al. 2019). Huerta Lwanga et al. (2016) also found that mortality of *Lumbricus terrestris* was concentration dependent.

Ng et al. (2018) said that water infiltration appears to be correlated to earthworm biomass and burrow length. That is, soil porosity is linked to earthworm presence. They hypothesized that when earthworm mortality is high as a result of high microplastic contamination, such as Huerta Lwanga et al. (2016) found, soil porosity would be impacted, but they did not say what the impact would be. If the worms die, there will be no burrows, and infiltration probably will decrease.

2.3.3 Nematodes

When assessing the toxicological effects of microplastics on nematodes, the size of the microplastic is an important factor to be considered (Zhu, F. et al. 2019). *Caenorhabditis elegans*, a nematode about 1 mm long that lives in temperate soil environments, was exposed to polystyrene particles of size 0.1, 0.5, 1.0, 2.0, and 5.0 μm for 3 days. The nematodes placed with the 1.0 μm group had the lowest survival rate. Apparently, the 1.0 μm-sized polystyrene particles were more readily taken up by the nematodes than the other sizes.

2.3.4 Collembola

The class Collembola includes springtails and allies. Soil Collembola have been shown to be sensitive to microplastic pollution. Exposure to 0.1% polyvinyl chloride microplastics for 56 days inhibited the growth and reproduction in soil of *Folsomia candida*, a species of springtail. Reproduction of *Folsomia* was inhibited by polyethylene, too (Zhu, F. et al. 2019). The gut bacterial community due to microplastic exposure was altered. These results suggest the Collembola may be used as a bioindicator of microplastic pollution in soil (Zhu, F. et al. 2019).

In laboratory studies, Collembola have been shown to move microplastic beads. Such active, incidental small-scale transport could spread microplastic particles horizontally, which may facilitate their subsequent entry into the soil (Rillig et al. 2017).

2.3.5 Isopods

An isopod is a crustacean with a flat, oval body and seven pairs of legs of similar size and form, each pair attached to a segment of the thorax (Webster's New World Dictionary of the American Language 1959). Isopods are commonly used as test species in ecotoxicity studies, due to their important role in plant litter decomposition (Zhu, F. et al. 2019). When the isopod *Porcellio scaber* was fed food pellets with 0.4% polyethylene microplastics for 14 days, there were no effects on food ingestion or body mass or mortality. This suggested little hazardous effects of polyethylene microplastics to this isopod. (Zhu, F. et al. 2019). However, more work is needed to investigate long-term effects of such exposure as well as the effects of other microplastics in the soil (Zhu, F. et al. 2019).

2.4 EFFECTS OF PARTICULATE PLASTICS ON PLANTS

When it comes to plants, people ask two questions: Can plants absorb and accumulate microplastics? How do microplastics affect plant growth and food quality? Currently, such information is scarce for two reasons: it is difficult to identify microplastics in plant tissues and the effect on crops has not attracted enough attention (Zhu, F. et al. 2019).

It is generally felt that the uptake of microplastics by plants is unlikely (Hurley and Nizzetta 2018; Ng et al. 2018). The high molecular weight or the large size of the plastic particles prevents their penetration through the cellulose-rich plant cell wall (Ng et al. 2018). However, if roots are damaged, particulate plastics can be taken up by plants (Kirkham 2020).

Currently, apparently only two studies have been carried out to investigate the impacts of microplastics on plants. By adding 1% biodegradable or low-density polyethylene microplastic particles in soil, Qi et al. (2018) found that both types of microplastics reduced root biomass after 2 months of growth, but only the biodegradable plastic reduced shoot growth after 2 months of growth. The low-density polyethylene did not affect the shoot biomass.

Kirkham (2020) studied the effect of particulate plastics (polyethylene glycol of molecular weight 8000) on plant-water relations and growth of wheat and showed that they increased stomatal resistance and reduced evapotranspiration and height. The particulate plastics also were a potent vector for the uptake of cadmium by the wheat. However, in neither study (Kirkham 2020; Qi et al. 2018) was the uptake of the particulate plastics by the wheat determined.

In contrast to microplastics, nanoplastics have been shown to enter plant cells. Uptake of 20 and 40 nm nano-polystyrene beads by tobacco (*Nicotiana tabacum* L.) cells in cell culture via endocytosis has been demonstrated, while 100 nm beads were excluded. However, apparently no studies have investigated whole-plant, instead of plant cell culture, uptake of nanoplastics (Ng et al. 2018). As a rule of thumb, particles less than 6 nm in one dimension may be able to permeate the cell wall (Ng et al. 2018). Studies on plant uptake of engineered carbonaceous nanoparticles may provide information about nanoplastic interactions with plants and bioavailability (Ng et al. 2018). The proposed pathways for entry of carbonaceous nanoparticles into plants include endocytosis through the plasmodesmata; passage via ion transport channels, carrier proteins, or aquaporins; and soil-carbon or root-exudate-mediated entry.

As noted, microplastics can move through macropores in the soil made by earthworms. However, there are no data on roots. In agricultural soils after harvest, macropores from roots could be a major transport pathway, as roots decompose and leave biopores (Rillig et al. 2017).

Because of the lack of studies focusing on the effects of microplastics in soils, more research is needed concerning the risks that particulate plastics pose to soil biota (da Costa et al. 2019). Studies are needed even more for plants. The uptake of particulate plastics by plants is essentially unknown, so the danger that particulate plastics in food poses to human health is not known.

2.5 DISPOSAL OF AGRICULTURAL PLASTICS

Disposal of plastics from agriculture has long been recognized as a problem (Wittwer 1993). The one question that is always asked by people who are using plastics to modify the cropping environment concerns how to dispose of them (Lamont 2005). Lamont (2005) listed several methods of disposal: incineration; recycling; and use of photodegradable or biodegradable materials. Another method is landfilling of the plastics. Each method is being used for disposal.

2.5.1 Incineration

In 2005, when Lamont (2005) published his article, incineration was an option for disposal of agricultural plastics. The possibility of incinerating plastics to produce usable heat and electricity has been long recognized (Hemphill 1993). The high heating value of plastics is comparable to that of heating oil, which can be recovered if plastics are burned (Lamont 2005). Used plastics have been burned in waste-to-energy plants, but one problem is that plastics burn hotter than the surrounding waste; thus, "hot spots" are created in the waste stream. Lamont (2005) reported that a team of researchers at Pennsylvania State University had developed "Plastofuel" made from used agricultural plastics, such as mulch film, landscape pots and flats, and fertilizer containers (PennState 2004). The Plastofuel can be burned as a supplement to coal or other waste products or by itself in a specially designed burner unit that can handle very high temperature combustion (PennState 2004).

However, since Lamont (2005) wrote his article, most states have enacted rules against outdoor burning of plastics (Grossman 2015). For example, in New York a statewide ban on backyard or farm burning of plastics was passed in 2009 (Grossman 2015). Burning of plastics is illegal in Wisconsin (Wisconsin Department of Natural Resources 2019). Other states in the United States that prohibit open burning of plastics are Idaho, Michigan, Oregon, and Wyoming (Kasirajan and Ngouajin 2012). But Florida does allow burning, and the bulk of agricultural film mulch in Florida is managed by open burning at the end of the growing season (Jones 2018). The allowance of open burning in the field has led to a disincentive of the recycling of the material (Jones 2018). Canada also permits burning, and in that country, every year tons of plastic wastes are burned on farms (Weber 2018).

The burning of plastics contaminated with fertilizers and pesticides releases potent environmental toxins, such as dioxins and furans into the air (Kasirajan and Ngouajin 2012; Mayer 2016). Dioxins are known endocrine disruptors and carcinogens (Kasirajan and Ngouajin 2012). These toxins can be inhaled by humans and animals and redeposited in soil and surface water (Mayer 2016). Even the developers of Plastofuel recognized that some plastics could not be burned due to the production of dioxins (PennState 2004). Burning also releases greenhouse gases to the air (personal communication, Keith L. Bristow, senior principal research scientist, CSIRO, Townsville, Australia, April 11, 2018). Residue from burning also can contaminate the soil and groundwater and enter the human food chain through crops and livestock (Mayer 2016). Moreover, unburned portions of plastic become litter on the ground, with larger piles creating breeding grounds for mosquitoes (Mayer 2016). Exposure to fine particles (diameter less than 2.5 mm) from open burning has been associated with many health effects, such as increased risk of stroke, asthmatic attacks, decreased lung function, respiratory diseases, and premature death (Kasirajan and Ngouajin 2012).

Because of high transportation cost and landfill tipping fees, farmers consider on-site burning to be economically more favorable (Kasirajan and Ngouajin 2012), if they are allowed to do it.

2.5.2 Landfilling

In Canada, tons of plastic wastes are buried or dumped in municipal landfills (Weber 2018). In Alberta, Canada, a 2012 governmental study found that 3000 metric tons of plastic wastes were being generated every year and burning in an open fire or burying the plastic were common ways to dispose of it. Much of it was dumped in the nearest landfill, but buried plastic does not biodegrade (Weber 2018). Only 17% of farmers said they had sent plastic for recycling. The Alberta farmers would like to recycle, but there was no place where they could take their plastics for recycling. This issue occurs across Canada (Weber 2018). It is estimated that 40,000 metric tons of plastic wastes are created on farms in the country, and the one group in Canada that specifically collects agricultural plastic, Cleanfarms, Inc., a non-profit organization funded by the plastics industry, collects less than 10% of it. Recycling is unlikely to get the economies of scale that it needs to be viable without some form of legislation (Weber 2018). The federal government is only responsible for hazardous wastes, so rules for plastic must come from the provinces (Weber 2018).

Disposal of agricultural plastics by on-site dumping is not recommended since seepage of water that has been in contact with buried agricultural plastics, caused by irrigation or rainfall, can contaminate groundwater with various agrochemicals (Kasirajan and Ngouajin 2012).

2.5.3 Recycling

The hardest problem to deal with in recycling is the debris, dirt, and moisture associated with plastics, such as mulch films and drip irrigation tapes, which makes it difficult to recycle them (Lamont 2005). Most agricultural film mulches are manufactured with a micro pattern for rigidity purposes. The micro pattern acts like a suction cup causing moisture and associated soil to adhere to the film (Jones 2018). Therefore, recycling is a relatively minor practice among farmers because of the difficulty in removing the dirt and moisture remaining on the plastics (Rivin 2015). Sending plastics to landfills is the option used by many farmers (Rivin 2015). The requirement of cleaning the plastic before being sent to a recycling facility is put on the farmer, and farmers do not want to pay for the labor to clean the plastic before it goes to recycling. It is economically not possible for them to do this (Rivin 2015). Consequently, farmers pile the plastic in huge heaps or send it to landfills (Rivin 2015).

Soil retained in the recovered film damages shredding equipment that is used before the use of the plastics for fuel. In one recycling trial in Florida with Suwannee American Cement, the amount of soil contained on the film forced the company to change its shredder blades at the end of every week at a cost of $16,000. After the trial was over, Suwannee American Cement decided not to recycle the film (Jones 2018).

In 2019 in Wales in the United Kingdom, the only company dedicated to collecting agricultural plastic from farms suspended its services for a year, because the place where the company took the plastics for recycling started to charge fees instead of paying for farm plastics (*BBC News* 2019). Rules state that farmers are allowed to store plastic wastes for 12 months, but then must dispose of them correctly (*BBC News* 2019). However, since the only company dedicated to collecting it from farms across Wales has suspended its services, farmers will need to resort to burning or burying plastic waste, because there is no other way to dispose of it, even though farmers would rather recycle plastic wastes (*BBC News* 2019).

Wisconsin recommends sending any plastics that one cannot recycle to the landfill, rather than burning or burying them (Wisconsin Department of Natural Resources 2019). If plastics are put in a landfill or buried, they are not recycled, and their degradation will lead to particulate plastics. A statewide survey conducted by the University of Wisconsin–Extension and the Wisconsin Department of Natural Resources in 2015 of 1500 farms and greenhouses in the state reported that about two-thirds of farmers use landfills for disposing of their used agricultural plastics (Mayer 2016). The second most common method of disposal was burning, at 30% (even though it is illegal in Wisconsin), with 10% of respondents reporting that they recycled the plastic. If the plastics are recycled, they can be processed and used in products, such as plastic lumber and garbage bags (Wisconsin Department of Natural Resources 2019).

Another problem with recycling is the cost of transport to the recycling facility. The survey done in Wisconsin (Mayer 2016) reported that 85% of farmers are willing to haul their plastics to a collection site for recycling. Farmers will drive up to 30 miles (47 km) to recycle their plastics if there is no disposal charge. The farmers in Wisconsin reported that they are willing to pay up to $1500 per year to dispose of the plastics in landfills, if it cannot be recycled. Green County in Wisconsin started recycling agricultural plastics in 2014. It has spring and fall collections, during which time farmers can dispose of their used agricultural plastics free of charge by hauling it to a local landfill. Green County's landfill department then bales the used agricultural plastics into one English ton (907 kg) parcels and sends them to a facility in Arkansas, where the material is cleaned, processed, and recycled into trash can liners and other products (Mayer 2016). Updated technology has solved some of the cleanliness issues associated with agricultural plastics processing. However, farmers are asked to shake any excess mud, gravel, dirt, stones, and feed off the plastic before storing it. Farmers are encouraged to landfill plastics with large amounts of dirt.

For plastics recovered from the field, the majority used to be sent overseas to the People's Republic of China, Vietnam, and Malaysia. However, all three of these countries have put a stop to taking agricultural plastics (Jones 2018). As a result, the United States has an oversupply of agricultural plastics with no viable market. Most of the material is now ending up in landfills, or even worse, being burned in fields (Jones 2018).

2.5.4 Photodegradable Plastics

In the 1960s and 1970s, scientists started to investigate the possibility of using photodegradation as a self-destructive disposal technique for plastic film (Kasirajan and Ngouajio 2012). According to Lamont (2005), photodegradable plastic mulches were developed by Ennis (1987). Photodegradable plastic looks much like other plastic mulches, but it breaks down by exposure to ultraviolet sunlight (Lamont 2005).

Pure polyethylene does not absorb in the ultraviolet range, but, during polymerization and film manufacture, heat causes chemical reactions that increase sensitivity to ultraviolet radiation (Hemphill 1993). Hydroperoxides are produced; these functional groups are stable at room temperature, but act as absorbers of ultraviolet light. Transition metal chelates, most often nickel or cobalt dithiocarbamates, are added to the films and act as heat stabilizers and decomposers of hydroperoxides. In contrast to nickel and cobalt compounds, iron and copper chelates become active radiation sensitizers after an induction period. Free metal ions catalyze non-reversible and self-accelerating chain reactions that degrade the polymers. The selection and concentration of photoactivators control the length of the induction period (Hemphill 1993). For example, photodegradable polyethylene films can contain a ferric ion complex that accelerates the rate of embrittlement (Brown et al. 1991). When exposed to sunlight, the material can disintegrate within 1–3 months, and the residual inorganic materials disperse in the soil (Brown et al. 1991). The relative proportion of a second nickel or cobalt dithiocarbamate makes it possible to achieve a graduated concentration vs. time with the embrittlement relationship (Brown et al. 1991).

A photodegradable mulch film must retain its physical properties until the end of the crop cycle, and then break down rapidly and completely, and it must not leave a toxic residue; the degradative system must not cause instability during manufacture; and degradation must not start until the film is exposed to sunlight (Hemphill 1993). However, the development of such desirable photodegradable materials seems to be limited. Photodegradable plastic mulches have proven to be unreliable, as well as expensive to use (Kasirajan and Ngouajio 2012). There have been many reports on preparation, properties, and application of degradable polymeric materials, but few reports are related to the agricultural application of degradable mulching films (Kasirajan and Ngouajio 2012). The ability of photodegradable mulches, which are manufactured with petroleum-based ingredients, to degrade into carbon dioxide and water has been questioned (Kasirajan and Ngouajio 2012).

The actual rate of breakdown depends on several factors, including temperature, the proportion of the plastic shaded by the crop, and the amount of sunlight received during the growing season (Lamont 2005). Degradation is inhibited by crops that cover the mulch as they grow, because exposure to ultraviolet light is reduced or prevented (Kasirajan and Ngouajio 2012). When using photodegradable plastic mulch, decomposition of the buried edges (commonly referred to as the tuck) is initiated only after lifting them out of the soil and exposing them to sunlight (Lamont 2005).

Photobiodegradable polyethylene films containing starch have been developed and used in agriculture (Kasirajan and Ngouajio 2012). Tests show that they are better able to raise temperature, preserve moisture, and raise yield than common polyethylene films, and they can be degraded after use. The photobiodegradation induction periods of photobiodegradable polyethylene films tested have ranged from 46 to 64 days, which satisfies the needs of agricultural cultivation. Almost no film exists on the surface of the ridges 2–3 months after the induction periods. The photobiodegradable polyethylene films buried in soil also have good degradability (Kasirajan and Ngouajio 2012).

2.5.5 Biodegradable Plastics

Plastics are resistant against microbial attack, because during their short time of presence in nature, evolution has not yet designed new enzymes capable of degrading them (Kasirajan and Ngouajio 2012). Alternative methods for the disposal of plastic films include the use of biodegradable materials that can be integrated into the soil, where microflora supposedly transforms them into carbon dioxide or methane and water (Kasirajan and Ngouajio 2012; Moore and Wszelaki 2016). An advantage of biodegradable plastics is that bio-based plastics generally show lower life-cycle greenhouse gas emissions than their fossil fuel-based counterparts (Zheng and Suh 2019).

Biodegradable polymers may be naturally occurring or may be synthesized by chemical means. Biodegradable polymers can be divided, in general, into three groups: (i) natural polymers, such as starch, cellulose, proteins, poly-b-hydroxybutyrte; (ii) natural polymers biologically or chemically modified (e.g., cellulose acetate, lignocellulose esters, polyalkanoate copolymers); and (iii) readily biodegradable synthetic polymers modified with added natural biodegradable components (starch, reclaimed cellulose, natural rubber) (Kyrikou and Briassoulis 2007).

Biodegradation of polymers requires microorganisms to metabolize all organic components of the polymer. Biodegradation in soil involves several key steps: (i) colonization of the polymer surface by microorganisms, (ii) secretion of extracellular microbial enzymes that de-polymerize the polymer into low molecular weight compounds, and (iii) microbial uptake and utilization of these compounds, incorporation of the polymer carbon into biomass, or releasing it as carbon dioxide (Zumstein et al. 2018).

Biodegradable mulch is made from about 20% bio-based feedstocks, such as cornstarch, and 80% from fossil fuels (Dentzman and Goldberger 2019). A small percentage of additives, such as colorants and processing aids are also included. Instead of being removed from the field at the end of the season, biodegradable mulch can be tilled directly into soil for degradation by microbes. Standards set by the American Society for Testing and Materials International and the European Committee for Standardization required that biodegradable mulch biodegrade 90% in soil within 2 years (Dentzman and Goldberger 2019). Biodegradable mulch's impact on soil quality depends on the season and geographic location, and it has been shown to both improve and degrade different soil quality parameters (Dentzman and Goldberger 2019). Wang et al. (2019) reported that the plastic mulches break into small segments that reduce soil permeability and impede the absorption of water and nutrients by crop roots, which result in reduced agricultural production.

In recent years, large-scale studies have been undertaken to try and increase the share of starch in starch-plastic composites to the maximum possible level. Biodegradable composites consist of biodegradable polymers as a matrix material and biodegradable fibers, usually biofibers. Jute (*Corchorus* sp.), hemp (*Cannabis sativa* L.) (Bergmeier 2019), lignocellulose, flax (*Linum* sp.) fiber, coir [fiber from the husks of coconuts (*Cocos nucifera* L.)], cotton (*Gossypium* sp.), and oriental plant fibers are the most commonly applied materials (Oniszczuk et al. 2016). Various fibers can be added to raw-material blends in amounts ranging from 1% to 50%. The main disadvantage of these fibers is that their mechanical properties depend on several factors, such as moisture content (Chocyk et al. 2015).

Kapanen et al. (2008) evaluated the degradability in soil of biodegradable starch-based soil mulches and biodegradable low-tunnel films. The lifetimes of the biodegradable mulches and biodegradable low-tunnel films were 9 and 6 months, respectively. Air temperature under the biodegradable low-tunnel films was 2°C higher than under the low-density polyethylene films, resulting in an up to 20% higher yield of strawberries (*Fragaria* sp.). At the end of the cultivation period, the biodegradable mulches were broken up and buried in the field soil together with the plant residues. One year after burial, less than 4% of the initial weight of the biodegradable film was found in the soil. There was no evidence of toxicity in the soil during the biodegradation process.

Wang et al. (2019) compared plastic mulch made from polyethylene and four types of biodegradable mulch under drip irrigation when cotton (*Gossypium hirsutum* L.) was grown. All were made from polybutyrate adipate terephthalate with different levels of thickness (0.01–0.012 mm) and induction periods (45–60 days). The polyethylene mulch was the most efficient in water and heat conservation in the soil. Although polyethylene mulch resulted in the highest yield, there was no difference in water-use efficiency between polyethylene mulch and two of the biodegradable mulches. Wang et al. (2019) concluded that some of the commercially available biodegradable mulches can be used as an alternative to plastic mulch to control soil pollution with plastics.

Moore-Kucera et al. (2014) studied four biodegradable plastic mulches tilled into the soil at the end of a growing season. The mulches were used to grow tomatoes (*Lycopersicon esculentum* Mill.) for a growing season at three locations in the United States (Knoxville, Tennessee, Lubbock, Texas, and Mount Vernon, Washington), and then they were buried in field soil. Two of the mulches were made from substances obtained from plants. One mulch contained 100% polylactic acid, a commonly used polymer that is biodegradable by microorganisms. The fourth mulch was a paper-like material made from cellulose fiber. Degradation after 6 months in soil was minimal for all, but the cellulosic mulch. After removal of mulches from soil, bacterial isolates were nearly undetectable, so only fungal isolates were studied. The majority of culturable soil fungi that colonized biodegradable mulches were within the family Trichocomaceae, which includes beneficial and pathogenic species of *Aspergillus* and *Penicillium*. No fungal isolate substantially degraded any mulch. While Moore-Kucera et al. (2014) said the fungal genera isolated in this study are common in the environment and potentially useful for degradation of biodegradable mulches, there are drawbacks to enhancing artificially their populations in agricultural environments, because some fungal species, including *Aspergillus* and *Penicillium*, are plant pathogens, as well as human pathogens. *Aspergillus* causes aspergillosis, an infection usually of the lungs. *Penicillium* causes superficial infections and allergic pulmonary disease. Dehghani et al. (2017) showed that inhalation of street dust contaminated with microplastics is a potential health risk. In Tehran, Iran, they found that a mean of 3223 and 1063 microplastic particles per year is ingested by children and adults, respectively. If the microplastics are biodegradable ones with fungi attached, the seriousness of the contamination could be compounded by the presence of the microorganisms.

A sprayable biodegradable polymer membrane technology for use in agricultural crop production is being developed in Australia (CSIRO 2019). It is a patented process, but it is known that an enzyme is put in the polymer to break it down and 90% of the carbon is gone in 2 years (Keith L. Bristow, personal communication, April 11, 2018). Crops are planted on top of beds, and then nozzles are used to spray the polymer on the plant row. The film redirects rain water into the plant row to enhance yield. The polymer membrane can be applied using existing hand sprayers or large agricultural spray machines, making it accessible to small scale farmers in developing countries and large-scale, highly mechanized farmers and agribusinesses in developed countries (CSIRO 2019). The sprayable biodegradable polymer membrane technology will help farmers eliminate the costs associated with retrieval and disposal of plastics, and it will reduce the pollution in soil and water due to petroleum-based products (CSIRO 2019). The sprayable polymer is white, but eventually it turns translucent and takes on the color of soil (Keith L. Bristow, personal communication, April 11, 2018). The material has been tested successfully with tomatoes (*Lycopersicon esculentum* Mill.) and melons (*Cucumis melo* L.) (CSIRO 2017), and with the membrane, there was a 16%–18% increase in yield, which was more than the average with other plastic films, where a 10% yield increase is good (Keith L. Bristow, personal communication April 11, 2018). Another advantage of the sprayable membrane is that it remains intact and on the soil during windstorms, while the biodegradable films are blown away and destroyed by strong winds (Keith L. Bristow, personal communication, April 11, 2018).

Biodegradable mulch has drawbacks. It is not certified for use in organic agriculture in the United States due to utilizing fossil fuel resources and genetically modified bacteria in manufacturing, which is an ironic situation given that polyethylene mulch is approved for organic use despite being made from 100% fossil fuel resources (Dentzman and Goldberger 2019). Biodegradable mulch also has drawbacks, such as high cost, variable rates of biodegradation, and limited research on the long-term impacts on soil (Dentzman and Goldberger 2019). Additionally, the aesthetics of these mulches may present a challenge. Given that the rate of biodegradation is variable depending on weather, soil quality, and other factors, small visible scraps may remain in the field for as long as 2 years, preventing biodegradable mulch from being aesthetically appreciated by farmers (Dentzman and Goldberger 2019).

The literature shows that the materials and technology to develop biodegradable mulch films for agricultural application exist. The major limitation remains the high cost of these materials that prevents their adoption by farmers (Kasirajan and Ngouajio 2012). However, the added cost is offset by the costs to remove and dispose of the standard plastic mulches (Kasirajan and Ngouajio 2012).

2.6 FUTURE

To avoid pollution from particulate plastics on agricultural land, the use of plastics will need to be stopped. Perhaps agriculture can return to technologies that were in use before plastics were developed. Plastic mulches, plastic irrigation tubing, and plastic pots might be replaced using these methods.

Mulches used to be made of paper (Emmert 1957). Durable and cost-effective paper mulches might be developed, just as boxed water is now being sold to replace the use of plastic bottles (https://boxedwaterisbetter.com). Thick paper and fiber mats are being used as alternatives to plastic mulches, but they are expensive (Kasirajan and Ngouajio 2012). Shredded newspaper has been reported to be used in organic high-tunnel cucumber production (Kasirajan and Ngouajio 2012). Or other old-fashioned mulches could be used like straw, leaves, woodchips, and other types of organic matter (Barth 2015). Cover crops also might be used in place of plastic mulches (Barth 2015). There are many examples of financially viable modern-day farms that use organic materials rather than plastic mulches (Barth 2015).

Greenhouses could be made of glass again, which would help revive the glass industry that has suffered closures, bankruptcy, mergers, and acquisitions in the past 35 years. The glass could be safety glass, so it would be less likely to break or to pose a threat when broken. Great Britain is encouraging the return to glass by launching a three-year project entitled "Glass against Plastic," costing 750 thousand pounds. Glass bottles will be filled at 20 different locations in London, reducing the environmental impact and costs associated with the purchase of water in plastic bottles in supermarkets (Misso and Varlese 2018).

Before plastic tubing was developed, farms used to be drained with earthen ditches or tiles (earthenware pipes) (Kirkham and Gaskell 1950). The use of ditches or tiles might be considered again. In dry farming, crops depend upon rainfall, and plastic irrigation supplies are irrelevant (Barth 2015). Farmers in dry regions have mastered the craft of trapping moisture in the soil so irrigation is unnecessary. Dry farming was practiced successfully by early settlers in the western United States (Widtsoe 1912). Modern dry-farming techniques might be developed to avoid the use of plastics (Barth 2015).

Instead of using plastic pots in vegetable production, which last two to three years before they are dumped in a landfill (Barth 2015), farmers might use biodegradable paper pots. They have been shown to be economically viable for growers with both small and large operations (Barth 2015). Pots used to be made of clay. Perhaps clay pots could be reintroduced into horticultural production.

The understanding of the effects of particulate plastics in the terrestrial environment is limited by the lack of universally accepted methods for their sampling, identification, and quantification in soil (da Costa et al. 2019). Astner et al. (2019) noted that most studies with particulate plastics employ idealized microplastic materials as models, such a monodisperse polystyrene spheres. In contrast, plastics that reside in agricultural soils consist of polydisperse fragments resulting from degraded films. Astner et al. (2019) developed a method to make representative microplastics that would occur from the cutting of plastic films during tillage into soil. More studies, such as those by Astner et al. (2019) are needed to implement methodologies from which environmentally relevant data may be used to gain knowledge about particulate plastics in soil (da Costa et al. 2019).

As Barth (2015) said, the use of alternatives to plastics depends upon the economics. However, it also depends upon the environmental consequences of using plastics. If it is shown that particulate plastics in soil are causing contamination of food crops that is detrimental to human health, then the public will demand the elimination of plastics from the environment.

It is estimated that it takes about 300 years to entirely degrade films of low-density polyethylene (a non-degradable polymer) with a thickness of 60 μm (Kyrikou and Briassoulis 2007). However, this estimation implies an unrealistically high rate of biodegradation for the low-density polyethylene films, and data in many published reports show that it takes several thousand years for the low-density polyethylene to be completely degraded (Kyrikou and Briassoulis 2007). But even with this overestimated rate (300 years), if the accumulation rate in the soil is considered, the end result is that the low-density polyethylene remains will continue to increase with time, instead of decreasing, with irreversible contamination of the soil (Kyrikou and Briassoulis 2007).

Due to these long times, the degradation of plastics applied to land is expected to be very small within the time scale of a human lifetime (Ng et al. 2018). Measures to counter plastic pollution should be enacted immediately. Actions taken today to mitigate emissions of particulate plastics on land will benefit the wider environment (Ng et al. 2018) and future generations.

ACKNOWLEDGMENTS

The research was funded by the State of Kansas Organized Research Grant No. 381041, OR-19 and by Hatch Grant No. 371047, H-08. This is contribution No. 20-076-B from the Kansas Agricultural Experiment Station.

REFERENCES

Accinelli, C., H.K. Abbas, W.T. Shier, A. Vicari, N.S. Little, M.R. Aloise, and S. Giacomini. 2019. Degradation of microplastic seed film-coating fragments in soil. *Chemosphere* 226:645–650. https://doi.org/10.1016/j.chemosphere.2019.03.161.

Alam, M., and D.H. Rogers. 2005. Field performance of subsurface drip irrigation (SDI) in Kansas. In: *Technical Session Proceedings: Irrigation Show 2005.* Irrigation Association, Fairfax, VA, pp. 1–5. https://www.irrigation.org/IA/Resources/Technical-Paper-Library.aspx.

Arancibia, R.A. 2018. Low tunnels in vegetable crops: Beyond season extension. *Virginia Cooperative Extension Publication HORT-291.* Virginia Cooperative Extension, Virginia Tech, Blacksburg, and Virginia State University, Petersburg, VA, 6 p.

Astner, A.F., D.G. Hayes, H. O'Neill, B.R. Evans, S.V. Pingali, V.S. Urban, and T.M. Young. 2019. Mechanical formation of micro- and nano-plastic materials for environmental studies in agricultural ecosystems. *Sci. Total Environ.* 685:1097–1106.

Barth, B. 2015. 3 ways farmers are kicking the plastic habit. *Modern Farmer.* https://modernfarmer.com/2015/agriculture-plastic-waste/.

BBC News. 2019. Farmers consider 'burning or burying' plastic waste. https://www.bbc.com/news/uk-wales-48246060.

Bergmeier, O. 2019. Industrial hemp in Kansas: A sustainable crop for the future. *The Collegian.* 124(60):1, 6.

Berrill, F.W. 1956. Bunch covers for bananas. *Queensland Agr. J.* 82(8):435–439.

Bradney, L., H. Wijesekara, K.N. Palansooriya, N. Obadamudalige, N.S. Bolan, Y.S. Ok, J. Rinklebe, K.-H. Kim, and M.B. Kirkham. 2019. Particulate plastics as a vector for toxic trace-element uptake by aquatic and terrestrial organisms and human health risk. *Environ. Int.* 131:104937. https://doi.org/10.1016/j.envint.2019.104937.

Brown, J.E., C. Stevens, V.A. Khan, G.J. Hochmuth, W.E. Splittstoesser, D.M. Granberry, and B.C. Early. 1991. Development in plastics for soil solarization. Chapter 8 (13 pages). In: DeVay, J.E., J.J. Stapleton, and C.L. Elmore (Editors). Soil Solarization. FAO Plant Production and Protection Paper 109. *Proceedings of the First International Conference on Soil Solarization*, Amman, Jordan, February 19–25, 1990. Food and Agriculture Organization of the United Nations, Rome, Italy.

Camp, C.R., F.R. Lamm, R.G. Evans, and C.J. Phene. 2000. Subsurface drip irrigation – Past, present and future. In: *Proceedings of the 4th Decennial National Irrigation Symposium*, Phoenix, AZ, November 14–16, 2000, pp. 363–372. https://www.ksre.k-state.edu/sdi/reports/2000/campis.html.

Carr, S.A., J. Liu, and A.G. Tesoro. 2016. Transport and fate of microplastic particles in wastewater treatment plants. *Water Res.* 91:174–182.

Chellemi, D.O., H.A. Ajwa, D.A. Sullivan, R. Alessandro, J.P. Gilreath, and S.R. Yates. 2011. Soil fate of agricultural fumigants in raised-bed, plasticulture systems in the Southeastern United States. *J. Environ. Quality* 40:1204–1214.

Chocyk, D., B. Gładyszewska, A. Ciupak, T. Oniszczuk, L. Mościcki, and A. Rejak. 2015. Influence of water addition on mechanical properties of thermoplastic starch foils. *Int. Agrophys.* 29:267–275.

CSIRO. 2017. TranspiratiONal: Sprayable biodegradable polymer membrane technology for agricultural crop production systems. Agriculture & Food Brochure. Commonwealth Scientific and Industrial Organisation, Canberra, Australia, 2 p.

CSIRO. 2019. TranspiratioONal. Case study. Commonwealth Scientific and Industrial Organisation. https://www.csiro.au/en/Research/AF/Areas/Sustainable-farming-systems/Sprayable-biodegradable-polymer-membrane.

da Costa, J.P., A. Paço, P.S.M. Santos, A.C. Duarte, and T. Rocha-Santos. 2019. Microplastics in soils: Assessment, analytics and risks. *Environ. Chem.* 16:18–30.

de Lorenzo, V. 2017. Seven microbial bio-processes to help the planet. *Microbial Biotechnol.* 10:995–998.

Dehghani, S., F. Moore, and R. Akhbarizadeh. 2017. Microplastic pollutants in deposited urban dust, Tehran metropolis. *Iran. Environ. Sci. Pollution Res.* 24:20360–20371.

Dentzman, K., and J.R. Goldberger. 2019. Plastic scraps: Biodegradable mulch films and the aesthetics of 'good farming' in US specialty crop production. *Agr. Human Values.* https://doi.org/10.1007/s10460-019-09970-x (published online July 5, 2019) 14 p.

Duis, K., and A. Coors. 2016. Microplastics in the aquatic and terrestrial environment: Sources (with a specific focus on personal care products), fate and effects. *Environ. Sci. Europe.* 28(2):unpaged (25 pages). https://doi.org/10.1186/s12302-015-0069-y.

Emmert, E.M. 1955. Progress report on low-cost plastic greenhouses, pp. 3–7. In: *Agricultural Experiment Station Progress Report 28.* University of Kentucky, Lexington, KY.

Emmert, E.M. 1957. Black polyethylene for mulching vegetables. *Proc. Am. Soc. Hort. Sci.* 69:464–469.

Ennis, R.W. 1987. Plastigone®: A new, time-controlled photodegradable plastic mulch film. *Proc. 20th Nat. Agr. Plastics Congr.* 20:83–90. (Cited by Lamont, 2005).

Esan, E.O., L. Abbey, and S. Yurgel. 2019. Exploring the long-term effect of plastic on compost microbiome. *PLoS ONE* 14(3):e0214376. https://doi.org/10.1371/journal.pone.0214376.

Grossman, E. 2015. How can agriculture solve its $5.87 billion plastic problem? https://www.greenbiz.com/article/how-can-agriculture-solve-its-1-billion-plastic-problem.

Hemphill, D.D., Jr. 1993. Agricultural plastics as solid waste: What are the options for disposal? *HortTechnology* 3:70–73.

Huerta Lwanga, E., H. Gertsen, H. Gooren, P. Peters, T. Salánki, M. van der Ploeg, E. Besseling, A.A. Koelmans, and V. Geissen. 2016. Microplastics in the terrestrial ecosystem: Implications for *Lumbricus terrestris* (Oligochaeta, Lumbricidae). *Environ. Sci. Technol.* 50:2685–2691.

Huerta Lwanga, E., H. Gertsen, H. Gooren, P. Peters, T. Salánki, M. van der Ploeg, E. Besseling, A. Koelmans, and V. Geissen. 2017. Incorporation of microplastics from litter into burrows of *Lumbricus terrestris. Environ. Pollution* 220:523–531.

Hurley, R.R., and L. Nizzetto. 2018. Fate and occurrence of micro(nano)plastics in soils: Knowledge gaps and possible risks. *Curr. Opin. Environ. Sci. Health* 1:6–11.

Janke, R.R., M.E. Altamimi, and M. Khan. 2017. The use of high tunnels to produce fruit and vegetable crops in North America. *Agr. Sci.* 8:692–715.

Jones, G. 2018. Recovering agricultural plastics: Obstacles and opportunities. https://wasteadvantagemag.com/recovering-agricultural-plastics-obstacles-and-opportunities.

Kapanen, A., E. Schettini, G. Vox, and M. Itävaara. 2008. Performance and environmental impact of biodegradable films in agriculture: A field study on protected cultivation. *J. Polymer Environ.* 16:109–122.

Kasirajan, S., and M. Ngouajin. 2012. Polyethylene and biodegradable mulches for agricultural applications: A review. *Agron. Sustain. Dev.* 32:501–529.

Kathiresan, K. 2003. Polythene and plastic-degrading microbes in an Indian mangrove soil. *Rev. Biol. Trop.* 51:629–633.

Kirkham, D., and R.E. Gaskell. 1950. The falling water table in tile and ditch drainage. *Soil Sci. Soc. Amer. Proc.* 15:37–48.

Kirkham, M.B. 2020. Water relations and cadmium uptake of wheat grown in soils with particulate plastics. In: N.S. Bolan, M.B. Kirkham, C. Halsband, and Y.S. Ok (Editors). *Particulate Plastics in Terrestrial and Aquatic Environments.* CRC Press, Taylor & Francis Group, Boca Raton, FL.

Knewtson, S.J.B., E.E. Carey, and M.B. Kirkhm. 2010a. Management practices of growers using high tunnels in the Central Great Plains of the United States. *HortTechnology* 20:639–645.

Knewtson, S.J.B., R. Janke, M.B. Kirkham, K.A. Williams, and E.E. Carey. 2010b. Trends in soil quality under high tunnels. *HortScience* 45:1534–1538.

Kyrikou, I., and D. Briassoulis. 2007. Biodegradation of agricultural plastic films: A critical review. *J. Polymer Environ.* 15:125–150.

Lamont, W.J., Jr. 2005. Plastics: Modifying the microclimate for the production of vegetable crops. *HortTechnology* 15:477–481.

Le Moine, B., and X. Ferry. 2018. XXI International congress on plastics in agriculture. *Chron. Hort.* 58(3):33–34.

Lucas, R.E., and S.H. Wittwer. 1956. Celery production in Michigan. Extension Bulletin 339. Cooperative Extension Service, Michigan State University, East Lansing, MI. 28 p.

Mahon, A.M., B. O'Connell, M.G. Healy, I. O'Connor, R. Officer, R. Nash, and L. Morrison. 2017. Microplastics in sewage sludge: Effects of treatment. *Environ. Sci. Technol.* 51:810–818.

Maughan, T., and D. Drost. 2016. Use of plastic mulch for vegetable production. Utah State University Extension, Horticulture. Horticulture/Vegetables/2016-01. Utah State University, Logan, UT. 6 p.

Mayer, M. 2016. Tackling agriculture's plastic waste problem. https://www.wiscontext.org/tackling-agricultures -plastic-waste-problem.

McNutty, J. 2017. Solar greenhouses generate electricity and grow crops at the same time, UC Santa Cruz study reveals. USC Newscenter. University of California, Santa Cruz. https://news.uscs.edu/2017/11/loik-greenhouse.html.

Misso, R., and M. Varlese. 2018. Agri-food, plastic and sustainability. *Calitatea-Acces La Succes* (Quality-Access to Success) 19(Suppl. S1):324–330.

Moore, J., and A. Wszelaki. 2016. Plastic mulch in fruit and vegetable production: Challenges for disposal. Report No. FA-2016-02. Montana State University, Bozeman; Washington State University, Pullman; Institute of Agriculture, University of Tennessee, Knoxville. 4 p. https://ag.tennessee.edu/biodegradablemulch/Documents/Plastic%20Mulch%20in%20Fruit%20and%20Vegetable%20Production_12_20factsheet.pdf.

Moore-Kucera, J., S.B. Cox, M. Peyron, G. Bailes, K. Kinloch, K. Karich, C. Miles, D.A. Inglis, and M. Brodhagen. 2014. Native soil fungi associated with compostable plastics in three contrasting agricultural settings. *Appl. Microbiol. Biotechnol.* 98:6467–6485.

Nelson, P.V. 2012. *Greenhouse Operation and Management.* Seventh edition. Prentice Hall, Boston, MA. 607 p.

Ng, E.-L., E. Huerta Lwanga, S.M. Eldridge, P. Johnston, H.-W. Hu, V. Geissen, and D. Chen. 2018. An overview of microplastic and nanoplastic pollution in agroecosystems. *Sci. Total Environ.* 627:1377–1388.

Oniszczuk, T., A. Wójtowicz, L. Mościcki, M. Mitrus, K. Kupryaniuk, A. Kusz, and G. Bartnik. 2016. Effect of natural fibres on the mechanical properties of thermoplastic starch. *Int. Agrophys.* 30:211–218.

PennState. 2004. Penn State researchers receive grant for turning waste plastics into fuel. https://news.psu.edu/story/213715/2004/11/03/penn-state-researchers-receive-grant-turning-waste-plastics-fuel.

Qi, Y., X. Yang, A.M. Pelaez, E.H. Lwanga, N. Beriot, H. Gertsen, P. Garbeva, and V. Geissen. 2018. Macro- and micro-plastics in soil-plant system: Effects of plastic mulch film residues on wheat (*Triticum aestivum*) growth. *Sci. Total Environ.* 645:1048–1056.

Ren, X.-W., J.-C. Tang, C. Yu, and J. He. 2018. Advances in research on the ecological effects of microplastic pollution on soil ecosystems. *J. Agro-Environmental Sci.* 37:1045–1058. (In Chinese with English abstract).

Rillig, M.C. 2012. Microplastic in terrestrial ecosystems and the soil? *Environ. Sci. Technol.* 46:6453–6454.

Rillig, M.C., R. Ingraffia, and A.A. de Souza Machado. 2017. Microplastic incorporation into soil in agroecosystems. *Frontiers Plant Sci.* 8:1805. https://doi.org/10.3389/fpls.2017.01805.

Rivin, G. 2015. Bill allows farmers to burn plastics, DENR offers support. https://www.northcarolinahealthnews.org/2015/06/01/bill-allows-farmers-to-burn-plastics-denr-offers-approval/.

Rodríguez-Seijo, A., B. Santos, E. Ferreira da Silva, A. Cachada, and R. Pereira. 2019. Low-density polyethylene microplastics as a source and carriers of agrochemicals to soil and earthworms. *Environ. Chem.* 16:8–17.

Santosh, D.T., K.N. Tiwari, and R. Gopala Reddy. 2017. Banana bunch covers for quality banana production: A review. *Int. J. Current Microbiol. Appl. Sci.* 6(7):1275–1291.

Smucker, A.J.M., B.C. Levene, and M. Ngouajio. 2018. Increasing vegetable production on transformed sand to retain twice the soil water holding capacity in plant root zone. *J. Hort.* 5(4):unpaged (7 pages). https://doi.org/10.4172/2376-0354.1000246.

Steinmetz, Z., C. Wollmann, M. Schaefer, C. Buchmann, J. David, J. Tröger, K. Muñoz, O. Frör, and G.E. Schaumann. 2016. Plastic mulching in agriculture: Trading short-term agronomic benefits for long-term soil degradation? *Sci. Total Environ.* 550:690–705.

Sun, M., M. Ye, W. Jiao, Y. Feng, P. Yu, M. Liu, J. Jiao et al. 2018. Changes in tetracycline partitioning and bacteria/phage-comediated ARGs in microplastic-contaminated greenhouse soil facilitated by sophorolipid. *J. Hazard. Mater.* 345:131–139.

Taylor, A.G., J.E. Motes, and M.B. Kirkham. 1982. Germination and seedling growth characteristics of three tomato species affected by water deficits. *J. Am. Soc. Hort. Sci.* 107:282–285.

Wang, W., Q. Chen, S. Hussain, J. Mei, H. Dong, S. Peng, J. Huang, K. Cui, and L. Nie. 2016. Pre-sowing seed treatments in direct-seeded early rice: Consequences for emergence, seedling growth and associated metabolic events under chilling stress. *Sci. Rep.* 6:19637. https://doi.org/10.1038/srep19637.

Wang, Z., Q. Wu, B. Fan, X. Zheng, J. Zhang, W. Li, and L. Guo. 2019. Effects of mulching biodegradable films under drip irrigation on soil hydrothermal conditions and cotton (*Gossypium hirsutum* L.) yield. *Agr. Water Manage.* 213:477–485.

Weber, B. 2018. 'Lack of options': Farm groups seek to recycle plastic instead of burning it. The Canadian Press, Toronto, ON. https://www.citynews1130.com/2018/03/02/lack-of-options-farm-groups-seek-to-recycle-plastic-instead-of-burning-it/.

Webster's New World Dictionary of the American Language, College Edition. 1959. World Publishing Company, Cleveland, ON. 1724 p.

Widtsoe, J.A. 1912. *Dry-Farming: A System of Agriculture for Countries under a Low Rainfall.* The Macmillan Co., New York. 445 p.

Wijesekara, H., N.S. Bolan, L. Bradney, N. Obadamudalige, B. Seshadri, A. Kunhikrishnan, R. Dharmarajan et al. 2018. Trace element dynamics of biosolids-derived microbeads. *Chemosphere* 199:331–339.

Wisconsin Department of Natural Resources. 2019. Managing agricultural plastics. https://dnr.wi.gov/topic/Recylcing/agplastics.html.

Wittwer, S.H. 1993. World-wide use of plastics in horticultural production. *HortTechnology* 3:6–19.

Yates, S.R., D.J. Ashworth, M.D. Yates, and L. Luo. 2011. Active solarization as a nonchemical alternative to soil fumigation for controlling pests. *Soil Sci. Soc. Amer. J.* 75:9–16.

Yu, M., M. van der Ploeg, E. Huerta Lwanga, X. Yang, S. Zhang, X. Ma, C.J. Ritsema, and V. Geissen. 2019. Leaching of microplastics by preferential flow in earthworm (*Lumbricus terrestris*) burrows. *Environ. Chem.* 16:31–40.

Zhang, F., J. Yu, C.R. Johnston, Y. Wang, K. Zhu, F. Lu, Z. Zhang, and J. Zou. 2015. Seed priming with polyethylene glycol induces physiological changes in sorghum (*Sorghum bicolor* L. Moench) seedlings under suboptimal soil moisture environments. *PLoS ONE* 10(10): e0140620. (no page numbers) https://doi.org/10.1371/journal.pone.0140620.

Zheng, J., and S. Suh. 2019. Strategies to reduce the global carbon footprint of plastics. *Nat. Clim. Chang.* 9:374–378.

Zhu, F., C. Zhu, C. Wang, and C. Gu. 2019. Occurrence and ecological impacts of microplastics in soils systems: A review. *Bull. Environ. Contamination Toxicol.* 102:741–749.

Zhu, Y.-G., D. Zhu, T. Xu, and J. Ma. 2019. Impacts of (micro) plastics on soil ecosystem: Progress and perspective. *J. Agro-Environ. Sci.* 38:1–6. (In Chinese with English abstract.)

Zumstein, M.T., A. Schintimeister, T.F. Nelson, R. Baumgartner, D. Woebken, M. Wagner, H.-P. E. Kohler, K. McNeill, and M. Sander. 2018. Biodegradation of synthetic polymers in soils: Tracking carbon into CO_2 and microbial biomass. *Sci. Adv.* 4:eaas9024. 8. https://doi.org/10.1126/sciadv.as9024.

3 Polyacrylamide (PAM) as a Source of Particulate Plastics in the Terrestrial Environment

Franklin M. Chen

CONTENTS

3.1 Sources and Distribution of Polyacrylamide (PAM) Particulate Plastics in the Environment 39
3.2 Ecotoxicity of PAM Particulate Plastics in the Environment 40
3.2.1 Transport of PAM and HPAM 40
3.2.2 Transport of the Acrylamide Monomer (AMD) 41
3.2.3 Degradation of PAM and HPAM 41
3.2.4 Release of AMD in the Environment 43
3.2.5 Degradation of AMD 44
3.2.6 Aquatic Toxicity of PAM and HPAM 45
3.2.7 Aquatic Toxicity of AMD 45
3.2.8 Terrestrial Toxicity of AMD 46
3.3 Conclusions 46
References 47

3.1 SOURCES AND DISTRIBUTION OF POLYACRYLAMIDE (PAM) PARTICULATE PLASTICS IN THE ENVIRONMENT

Polyacrylamide (PAM), as well as hydrolyzed polyacrylamide (HPAM), can be synthesized through various polymerization techniques (Caulfield et al. 2003) to achieve a polymer of ultra-high molecular weight (~20 million). These high molecular weight polymers find many applications that are highly related to terrestrial (soil) and aquatic (i.e., marine and freshwater) environments. When used as flocculants, they clarify drinking water, as well as domestic wastewater (Zhu 1996). In mining management, they are used as an additive to the drilling mud to remove cuttings and solids from the shaft of a well being drilled (Keas 1974). They are also used in agriculture to facilitate tillage and to promote soil aeration (Smith et al. 1996).

Many studies have shown that PAM has a significant role in controlling soil erosion induced by irrigation water and increased infiltration (Kornecki et al. 2005; Smith et al. 1996; Sojka et al. 2007; Szögi et al. 2007). Polyacrylamides are also mixed with various herbicides to aid in reducing spray drift and increasing the herbicides' surfactant capabilities (Bouse et al. 1986). These polymers are formulated into solutions referred to as thickening agents and contain between 25% and 30% of the polymer. PAMs are permitted to be used in food application, including production of sugar and corn starch hydrolysate, preparation of food-contact paper, and as additives in animal feed (Barvenik et al. 1996). When PAM or HPAM are grafted to polyethylene plastics, they can be used for biomedical applications (Bamford and Al-Lamee 1996; Chabrecek and Lohmann 2001; Marshall and Williams 1991; Pavlyk 1996).

PAM and HPAM have been widely used to enhance oil recovery from depleted oil fields (Bao et al. 2010; Kolya and Tripathy 2014; Prabu and Thatheyus 2007). HPAM is dissolved in flooding water to accelerate oil mobility during the flooding process (Deng et al. 2002; Sang et al. 2015). Similarly, polyacrylamide is used in hydraulic fracturing (Jenning and Sprunt 1995), or fracking, a relatively new method of hydrocarbon extraction also practiced mostly in the United States.

The global polyacrylamide market is expected to reach $ 6.91 billion by 2019 (Transparency Market Research 2013). In terms of volume, the demand for polyacrylamide was 1,337,500 tons in 2012 and is expected to cross 2,205,000 tons by 2019, growing at a compound annual growth rate of 7.4% from 2013 to 2019 (Transparency Market Research 2013). Approximately 6% of total polyacrylamide is in particulate form (Guezennec et al. 2015). It is estimated that 132,000 tons of PAM in particulate form annually are used in flocculation, soil stabilization, and various other applications.

In applications as a flocculent, soil stabilizing agent, or in chemicals to enhance oil recovery, PAM or HPAM are often mixed with water to form solutions. When used in oil recovery, large quantities of PAM and HPAM in water solution are discarded into the environment along with the recovered crude oil in the post oil recovery phase. These discarded solutions contain polyacrylamide particulates that are often the cross-linked type of the polymer, or the unused residual solid residues of the initial PAM (or HPAM) particulate plastics. The PAM (or HPAM) particulate plastics also exist when they are grafted to other plastics for medical devices, or when they are formulated into herbicides.

3.2 ECOTOXICITY OF PAM PARTICULATE PLASTICS IN THE ENVIRONMENT

When PAM is grafted to other plastics, such as polyethylene (Bamford and Al-Lamee 1996; Barvenik et al. 1996), the parent particulate plastics will account for the ecotoxicity of the grafted material. The effects of its ecotoxicity depend on three factors (Lambert et al. 2017) of parent plastics: (1) physical, which include particle size, shapes, degree of crystallinity, and surface areas; (2) chemical, which include polymer type, additives, and surface chemistry; and (3) biological (Harrison et al. 2017), which include propensity for biofilm formations.

When PAM or HPAM are used without being grafted, their ecotoxicity depends on their transport (from land to water), toxicity of PAM or HPAM, toxicity of residual monomers (mainly the acrylamide monomer), and toxicity of their degraded products with degradation resulting from heat, moisture, pH changes, light (photons), and biological factors. Other environmental considerations may also include the formation of a stagnant emulsion with the oil (in oil recovery applications), so that it is difficult to separate oil and water. As a consequence, the water produced during oil recovery will contain a high oil content, which poses concerns with regard to local discharge limits.

3.2.1 Transport of PAM and HPAM

The transport of PAM and HPAM from land to water depends on their ability to be adsorbed to the soil and clay mineral surfaces, because more adsorption implies less transport from land to water. Many studies (Deng et al. 2006; Tekin et al. 2005, 2006, 2010) have demonstrated that the adsorption of PAM (or HPAM) to soil and clay mineral surfaces is rapid and irreversible. The dissolved high molecular weight polymers are, in fact, readily adsorbed to soil particles via electrostatic, hydrogen, and chemical bonding, and by displacement of inner solvation-sphere water molecules (Jin et al. 1987; Laird 1997; LaMer and Healvy 1963; Malik et al. 1991; Mortland 1970). The degree of adsorption (Lambert et al. 2017) is dependent on polymer conformation, soil and mineral properties, and solution characteristics. Sorption isotherms of PAM generally fit with the form of the Langmuir equation rather than the Freundlich isotherm, and they are usually linear (Harrison et al. 2017; Lu et al. 2002). Results of sorption experiments and parameters using the Langmuir isotherm have indicated a high affinity of PAM to soil minerals. For anionic PAMs, adsorption on kaolinite, illite, and quartz is enhanced in the presence of divalent cations through a cationic bridge process (Taylor et al. 2002). Cationic bridges take place between divalent cations and negative groups of an anionic polymer, on one hand, and divalent cations and negatives sites of mineral surfaces, on the other hand. This binding mode occurs at high concentrations of divalent cations (Ca^{2+} and Mg^{2+}) and enhances the sorption of anionic polymers on negative mineral surfaces (Chiappa et al. 1999). Clays with a high specific outer-surface area, such as montmorillonite and smectite, are better PAM adsorbents than clays with low specific outer-surface areas, such as kaolinite and illite (Harrison et al. 2017; Lambert et al. 2017). Dry soil adsorbs more polymers than wet soils, because adsorbed water reduces the number of potential soil-binding sites (Chang et al. 1991). According to Lu et al. (2002), organic matter decreases the amount of anionic PAM adsorption on soil. In summary, due to the ability of PAM or HPAM to adsorb onto mineral particles, their transport in surface water, groundwater, and soils is rather limited and restricted to specific conditions.

3.2.2 Transport of the Acrylamide Monomer (AMD)

Polyacrylamide is synthesized from acrylamide (AMD) and acrylic acid or acrylate. There are several approaches employed by the industry to eliminate or reduce AMD from the final products. These approaches include adding sodium sulfite (Kurenkov and Abramova 1992); treatment with ammonium or amine (Jones 1958); or adding amidase to assist bacterial removal of AMD (Farrar et al. 1989) using methanol-water mixtures (Yamauchi and Nishihata 1978). Nevertheless, it is necessary to discuss the transport and other ecotoxicity issues associated with AMD, because it is difficult to completely remove AMD when PAM or HPAM applications are concerned.

Adsorption of AMD to mineral and clay surfaces has been little studied, but it is expected to be insignificant. AMD is a polar molecule with a molecular weight much smaller than that of polyacrylamide, with a molecular weight of 20 million. AMD lacks key adsorption advantages (to mineral and clay surfaces) as a polymer. Key disadvantages for AMD adsorption include not being able to form loops, tails, and trains on mineral or clay surfaces to facilitate an effective bridging mechanism (Sharma et al. 2006), and being easily carried away by water. Brown et al. (1980a) reported that, at AMD concentrations between 0.5 and 10 mg L^{-1}, no AMD was lost from sterile river water samples in contact with kaolinite or montmorillonite or anionic, cationic, and hydrophobic resins. These authors reported that AMD was lost from solutions in contact with peat. However, there was a noticeable lag period, inferring a bacteriological breakdown rather than adsorption. Additional AMD survey studies using samples from the environment (DeArmond and DiGoregorio 2013), water-recycling systems, and sludge in aggregate industries (Junqua et al. 2015) corroborate the conclusions of Brown et al. (1980a).

3.2.3 Degradation of PAM and HPAM

Thermal degradation of PAM has been reviewed (Caulfield et al. 2003). No degradation takes place when samples are below 200°C. Between 200° and 300°C, the polymer undergoes both intra- and intermolecular imidization reactions that occur via the pendant amide groups. At this stage, H_2O, NH_3, and minor quantities of CO_2 are released as by-products of the imide formation and degradation. It has also been proposed that some breakdown of the cyclic imide groups occurs at this stage. As the temperatures increase above 300°C, the reactions are characterized by the decomposition of imides to form nitriles and the release of volatiles, such as CO_2 and H_2O. Further degradation occurs, and there is evidence for some main-chain scission affording various substituted glutarimides. At higher temperatures, the predominant reactions are random bond scissions of the polymeric main chain backbone forming long chain hydrocarbons. Barvenik et al. (1996) argued that there is no regeneration of AMD from the thermal degradation of PAM, because forming a double bond in AMD is not thermodynamically feasible.

PAM photodegradation is reported to be a free radical process that can lead to a cleavage of the polymer backbone (bond scission) and the formation of lower molecular weight products (LaMer and Healvy 1963). The presence of oxygen in the system has a strong impact on the photodegradation of PAM. In addition, impurities contained in the initial polymer influence its photodegradation. A study led by Caulfield et al. (2003) reported that strong UV radiation at 254 nm released AMD from solutions of a non-ionic PAM. However, the release was very small. A solution of commercial anionic PAM with different concentrations of Fe^{3+} (from 0.02 to 4.4 ppm) was submitted to UV irradiation in a simulator able to closely approximate solar irradiance in the troposphere (equivalent to late summer noontime sunlight at 40° latitude, 300 nm) (Woodrow et al. 2008). In the presence of Fe^{3+}, AMD release was observed at acid/neutral pH, whereas, at alkaline pH (~8.0), PAM/Fe^{3+} remained stable under irradiation (Farrar et al. 1989).

Extensive studies (Grula et al. 1994; Joshi and Abed 2017; Senft 1993; Yu et al. 2015; Zhao et al. 2016) in the last two decades have shown that microorganisms utilize not only acrylamide, but also PAM and its derivatives as the sole source of nitrogen and/or carbon under aerobic, as well as anaerobic conditions. Microbial degradation lowers the molecular weight of the polymer and its viscosity, and the amide nitrogen is degraded to ammonia. Two PAM degrading bacterial strains, *Enterobacter agglomerans* and *Azomonas macrocytogenes*, isolated from soil samples, grew on media containing PAM as the sole source for both carbon and nitrogen. After 27 hours of growth, almost 20% of the total organic carbon in the initial medium had been consumed and the average molecular weight of the PAM was reduced from approximately 2,000,000–500,000 g/mol (Nakamiya and Kinoshita 1995). After 72 hours of incubation at 30°C under aerobic conditions, two bacterial strains, *Bacillus cereus* and *Bacillus flexus*, isolated from activated sludge and oil-contaminated soil, also were reported to consume >70% of the PAM as their sole carbon and nitrogen source (Wen et al. 2010). The bacterial strains *Bacillus sphaericus* no. 2 and *Acinetobacter sp.* no. 11 were isolated from soil and could degrade 16%–19% of PAM under aerobic conditions at 37°C (Matsuoka et al. 2002). PAM was also shown to be degraded to polyacrylic acids using mixed microbial communities under aerobic conditions (Liu et al. 2012).

Under anaerobic conditions and in environments rich in carbon, but lacking nitrogen sources, PAM can act as a nitrogen source and stimulate methanogenesis. *Clostridium bifermentans* H1 was isolated from a curing pot at an HPAM distribution center. The bacteria removed 30.8% of HPAM by using it as the sole carbon source, hydrolyzing the side chain, and changing some functional groups (Ma et al. 2008). The HPAM containing wastewater was treated by Fenton oxidation followed by anaerobic biological treatment, and the observed HPAM removal was 91.06% (Yongrui et al. 2015). Hence, at present, it is concluded that PAM or HPAM could be utilized by bacteria for carbon and/or nitrogen sources and degraded under aerobic or anaerobic conditions.

The pathway of biodegradation of PAM has been summarized by Joshi and Abed (2017). The microbes obtain nitrogen by hydrolyzing amide groups from PAM or HPAM using the amidase enzyme in aqueous solutions. During the reaction, $-NH_3$ is released and the –COOH group is introduced. After many cycles, subsequent catalysis by monooxygenase oxidation on the main PAM carbon chain leads to oxidation of α-[–CH2–]–COH–, which is transferred to –CHO, and then oxidized to –COOH (Eubeler et al. 2009, 2010; Li et al. 2006). During this process, the main carbon backbone is cleaved and the PAM or HPAM molecules are transformed into smaller molecules, such as acrylamide and/or acrylic acid, which can be further degraded. Wampler and Ensign (2005) suggested that acrylate can be degraded to acryl CoA and to β-hydroxypropionate via a hydroxylation reaction, then oxidized to CO_2 or reduced to propionic acid.

3.2.4 Release of AMD in the Environment

Acrylamide is a hazardous substance with irritant, neurotoxic, carcinogenic, and mutagenic properties; it may impair fertility (Bonnard et al. 2007; USEPA 2010). The maximum value of acrylamide permitted in drinking water, proposed by the World Health Organization, is 0.5 $\mu g\ L^{-1}$ (or 0.5 ppb) (WHO 1993), while the European directive 98/83 for drinking water fixes this limit to 0.1 $\mu g\ L^{-1}$ (or 0.1 ppb) (EC 1998).

Because of the toxicity of acrylamide, it is important to know how it is released to the environment.

There are several sources of AMD that are released to the environment. The first is the discharge from manufacturing sites. Residual acrylamide concentrations in 32 polyacrylamide flocculants approved for water treatment plants ranged from 0.5 to 600 ppm (Michalanko et al. 1989). Acrylamide may remain in water after treatment (Croll et al. 1974) and after flocculation with polyacrylamides due to its high solubility and because it is not readily adsorbed by sediments (Rogacheva and Ignatov 2001). The second is due to photodegradation or chemical degradation in the presence of Fe^{3+}, which could also release AMD (Caulfield et al. 2003; Woodrow et al. 2008). The third source of AMD release comes from soil grouting (for fracking to extract oil) (Howard 1989).

Acrylamide released to land and water from 1987 to 1993 totaled over 18.16 tons, of which, about 85 percent was to water, according to the Toxic Chemical Release Inventory of the United States (U.S.) Environmental Protection Agency (EPA 1994). These releases were primarily from the plastics industries, which use acrylamide as a monomer. In 1992, discharges of acrylamide, reported to the Toxic Chemical Release Inventory by certain U.S. industries, included 12.71 tons to the water acrylamide based sewer grouting and wastepaper recycling; 4.54 tons to surface water; 1906.8 tons to underground injection sites; and 0.44 tons to land (Exon 2006; Howard 1989).

Concentrations of 0.3 ppb to 5 ppm acrylamide have been detected in terrestrial and aquatic ecosystems near industrial areas that use acrylamide and/or polyacrylamides (Chang et al. 2002; Weideborg et al. 2001). In another EPA study of five industrial sites that produce acrylamide and polyacrylamide, acrylamide (1.5 ppm) was found in only one sample downstream from a polyacrylamide producer and no acrylamide was detected in soil or air samples (Brown et al. 1980b; Wen et al. 2010). Cases of human poisoning have been documented from water contaminated with acrylamide from sewer grouting. The acrylamide monomer was found to remain stable for more than 2 months in tap water (Croll et al. 1974). Atmospheric levels around six U.S. plants were found on average to be $<0.2\ \mu g/m^3$ (0.007 ppb) in either vapor or particulate form (Croll et al. 1974). The vapor phase chemical could react with photochemically produced hydroxyl radicals (half-life 6.6 hours) and be washed out by rain (Eubeler et al. 2009; European Union Risk Assessment Report 2002).

A separate study of the European Chemical Bureau on Risk Assessment Report (Igisu et al. 1975) indicates that a total continental acrylamide of 280 kg/day is released to water and 0.38 kg/day to air and a total regional acrylamide of 30.4 kg/day is released to water and 0.22 kg/day to air. Acrylamide has been measured in a number of water systems in the European Union, and the background levels of acrylamide are generally low (3.4 $\mu g/L$ or 0.0034 ppm) and in most cases below the level of detection. Higher values have been observed in water systems near production sites.

Very few measured exposure data are available relating to the use of acrylamide-based grouts in sewer and pipeline repairs within the European Union. In the United States, most monitoring data are based upon occupational exposure, not environmental exposure. In Japan in 1975, there was an incidence of a drinking water sample taken 2.5 m from a pipeline grouted with acrylamide, and it contained acrylamide at a concentration of 400 mg/L or 400 ppm (Igisu et al. 1975).

Acrylamide-based grouts are known to have been used in large tunneling projects in Norway and Sweden, and these have led to high levels of acrylamide being detected in water courses downstream of the construction operations. For example, in southern Sweden at the Hallandsås Ridge, an 8.6 km tunnel was driven through bedrock. The ridge has a high water content, and the chosen grout contained residual acrylamide. The large-scale use of the product started in August 1997. River water samples taken from the Vadbäken Creek immediately downstream from the construction site at the end of September 1997 had acrylamide concentrations of 92 ppm (European Union Risk Assessment Report 2002; Wampler and Ensign 2005). At the same time, samples taken in fish ponds that were connected to the Vadbäcken Creek contained 2 ppm acrylamide (European Union Risk Assessment Report 2002; Wampler and Ensign 2005).

Releases to air during production and polymer production are reported by NIOSH (1976) in the United States The releases ranged from 100 to 900 μg/m³ for weekly monitoring.

3.2.5 Degradation of AMD

AMD is quite reactive. There are two reactive sites on the molecule: the amide group and the double bond group. The amide group readily hydrolyzes in the presence of base or acid to yield acrylic acid. The rate and degree of completion of this hydrolysis is dependent on the pH conditions and temperature. The hydrolysis rate is somewhat higher under basic than acidic conditions, and it is quite slow at pH close to neutral (Haberman 1991; Jones 1958; Kurenkov and Abramova 1992; Smith and Oehme 1991). AMD will polymerize with other molecules of AMD or other vinyl monomers, such as acrylic acid, in the presence of free radicals and the absence of oxygen. This non-reversible polymerization can yield polymers with molecular weights ranging from 10^3 to $>10^7$. In addition, under moderate pH and temperature conditions, a variety of compounds will react with AMD at the double bond, to yield less toxic end products. These chemicals include ammonia, amines, alcohols, cellulose, starch, mercaptans, sulfite and bisulfite, sulfides, and strong oxidizing agents, such as chlorine and hypochlorite, bromine, permanganate, and ozone. It is noted that sulfite-bisulfite addition (Buranasilp and Charoenpanich 2011; LaMer and Healvy 1963), and ammonia-amine treatment (Guezennec et al. 2015; Mortland 1970), in post PAM synthesis were used to remove AMD from the final product.

Acrylamide, because of its small molecular size, can easily pass through biological membranes, and affect bacterial growth (Shukor et al. 2009), despite inhibitory effects on sulfhydryl proteins of the biological membranes. Bacterial strains affected by AMD are species within the genera *Bacillus* (Shukor et al. 2009), *Pseudomonas* (Labahn et al. 2010), and *Rhodococcus* (Nawaz et al. 1998). Other strains documented by Joshi and Abed (2017), such as *Enterobacter aerogenes* (Buranasilp and Charoenpanich 2011; Nakamiya and Kinoshita 1995), *Enterococcus faecalis* (Buranasilp and Charoenpanich 2011; Nakamiya and Kinoshita 1995), and *Klebsiella pneumonia* (Nawaz et al. 1993), are able to grow in the presence of acrylamide and degrade it partially or completely. The majority of the strains isolated from various ecosystems, such as soils (Lakshmikandan et al. 2014), water and sediments (Labahn et al. 2010; Yongrui et al. 2015), and waste water (Jebasingh et al. 2013) degrade acrylamide to acrylic acid and other derivatives including ammonia, carboxylic acids, and carbon dioxide, while others showed a complete acrylamide degradation to CO_2 and H_2O. The biodegradation of AMD was first shown to be aerobic, but recent studies have demonstrated the involvement of some bacteria in the anaerobic biodegradation of AMD (Labahn et al. 2010; Yongrui et al. 2015). The microbial degradation of AMD is due to the presence of an enzyme named acrylamide amidohydrolase (EC 3.5.1.4) or amidase. Amidases catalyze the hydrolysis of an amide to free carboxylic acids and free ammonium (Sharma et al. 2009).

Besides bacteria, fungi that degrade acrylamide have been documented (Joshi and Abed 2017; Woodrow et al. 2008). *Phanerochaete chrysosporium BKMF-1767*, a white rot fungus, was able to degrade more than 80% acrylamide to carbon dioxide (Sutherland et al. 1997). Filamentous fungi, routinely used in food and beverage industries, such as *Aspergillus oryzae KBN1010*, *A. terreus NRRL1960*, and 20 other strains showed acrylamide degradation ability (Wakaizumi et al. 2009). The pathway for fungi degradation of AMD is still not clear, and it is in an area under intense investigation.

3.2.6 Aquatic Toxicity of PAM and HPAM

Cationic polyelectrolytes including PAMs can be toxic to fish and other aquatic animals. Tests have resulted in low LC_{50} values (0.271 to 1.733 ppm.) (Buchholz 1992; Goodrich et al. 1991). The toxicity of cationic polymers does not result from residual monomers, but rather from mechanical suffocation caused by binding of the cationic polymer with anionic sites on fish gills. However, these polymers also have a high affinity for suspended solids and dissolved anionic solutes. Their aquatic toxicity is substantially reduced in the presence of these suspended solids. Goodrich et al. (1991) have shown that, in the presence of humic acid at 5 mg/L, the toxicity of the polymers was reduced 7- to 16-fold. At higher humic acid concentrations (50 mg/L), cationic polymer toxicity was reduced 33- to 75-fold.

Anionic PAMs exhibit low toxicity with high LC_{50} values (>100 ppm) in fish (Barvenik 1994). Data summarized by Buchholz (1992) indicate that some species of fish appear to be unaffected by anionic PAM at concentrations of 100 ppm in 90 day tests. However, minnows were killed by 2500 ppm solutions of anionic PAM due to the extreme viscosity of the fluid. This is more than two orders of magnitude higher than the recommended concentration for addition to irrigation water.

3.2.7 Aquatic Toxicity of AMD

The acute toxicity of acrylamide measured as a 96-hour LC_{50} on bluegill sunfish (*Lepomis macrochirus*), the most sensitive freshwater species, is 100 mg/L (ABC Labs 1982) (or 100 ppm). The sensitivity of the other species appears to be in a similar range (96-hour LC_{50} between 100 and 180 mg/L).

Petersen et al. (1987) studied the behavioral and histological effect of acrylamide on rainbow trout (*Oncorhynchus mykiss*). The fish were exposed under static conditions to various concentrations of acrylamide for 15 days, followed by a 7-day depuration period. Histological lesions were observed in the gills and liver in fish exposed to 25 mg/L for 15 days. Fish exposed to 50 mg/L developed lesions in the cephalic lateral line and peripheral lateral line, in addition to the gills and liver. After the depuration period, additional lesions were observed in the sagittal and proximal nerve plexus (25 mg/L exposure) and in the optic nerve (50 mg/L exposure only). Swimming behavior of the fish was unaffected at exposure concentrations below 25 mg/L. At 50 mg/L, fish had difficulty in orientating themselves when swimming.

In fish exposed to acrylamide following grout application incidents in Scandinavia, researchers (EC 2002) have observed gill alterations with a thickening of the epithelial cells, hyperplasia, and fusion of the secondary lamellae. Between the covering of epithelial cells and the underlying blood vessels, edema and eosinophilic granular cells were seen. In the liver, alterations were observed resulting in necrotic liver cells. An increase in hemoglobin adduct levels also was observed.

Acrylamide is toxic to aquatic invertebrate. In acute tests, the water flea *Daphnia magna* is the most sensitive species with a 48-hour LC_{50} of 98 mg/L (ABC Labs 1983). Tests with the saltwater shrimp *Mysidopsis bahia* showed a similar sensitivity (48-hour LC_{50} of 109 mg/L) (Spingborn Bionomics 1985). Long-term toxicity data are only available for saltwater species, with a 28 day no observed effect concentration of 2.04 mg/L being reported (EG&G Bionomics 1986) for *Mysidopsis bahia*.

Edwards (1975) studied the effects of acrylamide on frogs (*Rana temporaria*). The frogs were given acrylamide either by injection of saline solution into the dorsal sac or by exposing them to a solution containing acrylamide. Three doses of 50 μg/g in 7 days killed three out of five frogs, and a 2-hour exposure to a 2% (w/v) solution of acrylamide killed two out of three frogs. No adverse effects were observed in the surviving frogs.

The EU (2002) used PNEC (predicted no effect concentration) as a measure of risk assessment. The PNEC is the level below which the probabilities suggest that an adverse environmental effect will not occur. For acrylamide toxicity risk assessment, a PNEC of 20 μg/L is obtained for aquatic species exposed to acrylamide.

3.2.8 Terrestrial Toxicity of AMD

Bilderback (1981) studied the effect of toxic substances on pollen germination and pollen tube growth using the pollen of *Impatiens sultanii*. The study indicated that, when acrylamide was added to the basal medium at concentrations ranging from 10 to 2000 ppm, there was no significant effect upon germination, tube formation, or tube growth.

Kuboi and Fujii (1984) studied the toxicity of acrylamide and cationic polymer flocculants to higher plants. About 50 seeds of turnip (*Brassica rapa* L. cv. Chuusei-kanamachi), rapa (*Brassica rapa* L. cv. Tokiwa-jibai), Chinese cabbage (*Brassica pekinensis*), sesame (*Sesamum indicum*), cucumber (*Cucumis sativus*), upland rice (*Oryza sativa*), and wheat (*Triticum aestivum*) were planted. The EC50 was calculated as the concentration of flocculant or monomer, where the root elongation rate was equal to 50% of the control. The study indicated that 100 mg/L acrylamide was found to retard root elongation by 39% compared to the control, and the EC50 was calculated as 220 mg/L. No significant effect on seed germination was observed.

Sonoda et al. (1977) studied the behavior of polyacrylamide as a cohesive agent in soil-plant systems. Seeds of Chinese cabbage (*Brassica pekinensis*) were allowed to germinate in soil treated with 5–100 mg acrylamide/kg. Germination was inhibited with concentrations of more than 50 mg acrylamide/kg soil, and plants did not grow normally thereafter. Concentrations of less than 10 mg acrylamide/kg soil affected the growth of the plants, because growth was delayed.

The potential for the uptake and accumulation of acrylamide into plant tissue has been examined using lettuce plants (*Lactuca saliva* L) (Hazelton Laboratories America 1987). C^{14} labeled acrylamide monomer was added to 100 ML of nutrient solution and mixed with 4000 g of air-dried soil to obtain a uniform concentration of 5.0 ppm. The shoots of the plants were analyzed for C^{14} after 18 days. The roots, soil, leachate, and shoots were analyzed separately. They found that in soils treated with acrylamide, germination and growth were slow and the plants showed signs of necrosis. C^{14} was detected in the shoots and roots of treated plants, and it was also present in the soil and leachate. The C^{14} in the leachate and plant tissue did not appear to be acrylamide.

Castle et al. (1991) grew tomato plants hydroponically using commercial polyacrylamide gel bags containing 0.018% acrylamide. Acrylamide was not detected in the tomato fruit. The limit of detection for the tomato tissue, using their gas chromatography-mass spectrometry method, was 1 ppb. The same method was used by Castle (1993), who found no evidence of acrylamide uptake by edible mushrooms grown in a casing mix containing polyacrylamide gel, with a detection limit of 0.5 ppb in the mushroom tissue.

Nishikawa et al. (1983) demonstrated migration of the acrylamide monomer into the roots (1.7 ppb) and stalks (41 ppb) of rice grown in a polyacrylamide hydroponics medium containing 50 ppm acrylamide. The study also presented evidence of growth rate inhibition of Chinese cabbage in the presence of >5 ppm acrylamide.

In conclusion, acrylamide shows a slight, toxic effect on plant growth at concentrations of 10 mg/kg soil. No effect on seed germination has been observed. The EC (2002) indicated PNEC for acrylamide for terrestrial organisms is 220 μg/L.

3.3 CONCLUSIONS

PAM and HPAM are both water soluble polymers. Even if they are supplied as powders with size distribution generally between 150 and 2000 microns (or 0.15–2 mm), most of the powders will dissolve and will not be considered as microplastics as other polymers are.

As far as the ecotoxicity of PAM or HPAM is concerned, we should first address the presence and the toxicity of the monomer acrylamide from which the polymers are made. A substantial amount of acrylamide can be released when polymers are used in grouting operations. When acrylamide grouts are used in large-scale operations, the estimated exposures for reasonable worst-case scenarios are high. Under such conditions, one may be concerned about the ecotoxicity of the polymers in aquatic and terrestrial environments.

Under special circumstances in which either PAM or HPAM is grafted to other water insoluble plastics, then the general principles of microplastics ecotoxicity of the host water insoluble polymers can be applied.

REFERENCES

ABC Labs. 1982. Dynamic 96 hours acute toxicity of acrylamide monomer to bluegill sunfish (*Lepomis macrochirus*). Test Report # 29736.

ABC Labs. 1983. Dynamic 96 hour acute toxicity to acrylamide monomer to water fleas (*Daphnia magna*). Test Report # 29736.

Bamford, C.H., Al-Lamee, K.G. 1996. Studies in polymer surface modification and grafting for biomedical uses: 2. Application to arterial blood filters and oxygenators. *Polymer* 37:4885–4889.

Bao, M., Chen, Q., Li, Y., Jiang, G. 2010. Biodegradation of partially hydrolysed polyacrylamide by bacteria isolated from production water after polymer flooding in an oil field. *J Hazard Mater* 184:105–110.

Barvenik, F.W. 1994. Polyacrylamide characteristics related to soil applications. *Soil Sci* 158:235–243.

Barvenik F.W., Sojka, R.E, Lentz, R.D, Andrewes, F.F, Meissner, L.S. 1996. Fate of acrylamide monomer following application of polyacrylamide to cropland. *Proceedings from conference held at College of Southern Idaho Twin Falls*, Twin Falls, ID, May 6–8: pp. 103–110. https://eprints.nwisrl.ars.usda.gov/1192/1/912.pdf.

Bilderback, D.E. 1981. Impatiens pollen germination and tube growth as a bioassay for toxic substances. *Environ. Health Perspect* 37:95–103.

Bonnard, N., Jargot, D., Miraval, S., Pillière, F., Schneider, O. 2007. Fiche toxicologique FT 119 Acrylamide, INRS (in French).

Bouse, L.F., Carlton, J.B., Jank, P.C. 1986. Use of polymers for control of spray droplet size. *Am Soc Agric Eng* AA-86-005:1–18.

Brown, L., Bancroft, K.C.C., Rhead, M.M. 1980a. Laboratory studies on the adsorption of acrylamide monomer by sludge, sediments, clays, peat and synthetic resins. *Water Res* 14:779–781.

Brown, L., Rhead, M.M., Bancroft, K.C.C., Allen, N. 1980b. Case studies of acrylamide pollution resulting from industrial use of acrylamides. *Water Pollut Control* 79:507–510.

Buchholz, F.L. 1992. Polyacrylamides and poly(acrylic acids). pp. 143–156. In: B. Elvers, S. Hawkins and G. Shulz (eds.), *Ullmann's Encyclopedia of Industrial Chemistry*, Vol. A21, pp. 143–146, Wiley-VCH, Weinheim, Germany.

Buranasilp, K., Charoenpanich, J. 2011. Biodegradation of acrylamide by Enterobacter aerogenes isolated from wastewater in Thailand. *J Environ Sci* 23:396–403.

Castle, L. 1993. Determination of acrylamide monomer in mushrooms grown on polyacrylamide gel. *J Agric Food Chem* 41:1261–1263.

Castle, L., Campos, M.J., Gilbert, J. 1991. Determination of acrylamide monomer in hydroponically grown tomato fruits by capillary gas chromatography-mass spectrometry. *J Sci Food Agric* 54:549–555.

Caulfield, M.J., Hao, X, Qiao, G.G., Solomon, D.H. 2003. Degradation on polyacrylamides. Part I Linear Polyacrylamide. *Polymer* 44:1331–1337.

Chabrecek, P., Lohmann, D. 2001. Method for modifying the surface of biomedical articles. European Patent Application EP 1,095,711.

Chang, L., Bruch, M.D., Griskowitz, N.J., Dentel, S.K. 2002. NMR spectroscopy for determination of cationic polymer concentrations. *Water Res* 36:2255–2264.

Chang, S.H., Ryan, M.E., Gupta, R.K., Swiatkiewicz, B. 1991. The adsorption of water-soluble polymers on mica, talc limestone, and various clay minerals. *Coll Surf* 59:59–70.

Chiappa, L., Mennella, A., Lockhart, T.P., Burrafato, G. 1999. Polymer adsorption at the brine/rock interface: The role of electrostatic interactions and wettability. *J Pet Sci Eng* 24:113–122.

Croll, B.T., Arkell, G.H., Hodge, R.P.J. 1974. Residues of acrylamide in water. *Water Resour* 8:989–993.

DeArmond, P.D., DiGoregorio, A.L. 2013. Characterization of liquid chromatography-tandem mass spectrometry method for the determination of acrylamide in complex environmental samples. *Anal Bioanal Chem* 405:4159–4166.

Deng, S., Bai, R., Chen, J.P., Yu, G., Jiang, Z., Zhou, F. 2002. Effects of alkaline/surfactant/polymer on stability of oil droplets in produced water from ASP flooding. *Colloids Surf A Physicochem Eng Asp* 211:275–284.

Deng, Y.J., Dixon, J.B., White, G.N., Loeppert, R.H., Juo, A.S.R. 2006. Bonding between polyacrylamide and smectite. *Coll Surf Physicochem Eng Asp* 281:82–91.

EC (European Commission). 1998. European Council Directive 98/83/ EC of November 3, 1998 on the Quality of Water Intended for Human Consumption. European Union, Brussels, Belgium.

EC (European Commission). 2002. European Union Risk Assessment Report Acrylamide, EUR19835EN. Office for Official Publications of the European Communities, Luxembourg.

Edwards, P.M. 1975. Neurotoxicity of acrylamide and its analogues and effects of these analogues and other agents on acrylamide neuropathy. *Br J Med*, 32:31–38.

EG&G Bionomics. 1986. Acute toxicity of acrylamide to mysid shrimp (*Mysidopsis bahia*). Report No. BP-83-5-58-R.

EPA. 1994. Part V 40 CFR Part 372 addition of certain chemicals; toxic chemical release reporting; community right-to-know; final rule, Federal Register, Washington D.C.

Eubeler, J.P., Bernhard, M., Knepper, T.P. 2010. Environmental biodegradation of synthetic polymers II. Biodegradation of different polymer groups. *TrAC Trends Anal Chem* 29:84–100.

Eubeler, J.P., Bernhard, M., Zok, S., Knepper, T.P. 2009. Environmental biodegradation of synthetic polymers I: Test methodologies and procedures. *TrAC Trends Anal Chem* 28:1057–1072.

European Union Risk Assessment Report. 2002. CAS no: 79-06-1, EINECS no: 201-173-7, European Chemical Bureau, 24, Acrylamide.

Exon, J.H. 2006. A review of the toxicology of acrylamide. *J Toxicol Env Heal*, Part B, 9:397–412. doi: 10.1080/10937400600681430.

Farrar, D., Flesher, P., Lawrence, P.R.B. 1989. Polymeric compositions and their production. European Patent Application EP 329,324.

Goodrich, M.S., Dulak, L.H., Friedman, M.A., Lech, J.J. 1991. Acute and long-term toxicity cationic polymers to rainbow trout (*Oncorhynchus mykiss*) and the modification of toxicity by humic acid. *Environ Toxicol Chem*, 10:509–515.

Grula, M.M., Huang, M.L., Sewell, G. 1994. Interactions of certain polyacrylamides with soil bacteria. *Soil Sci* 158:291–300.

Guezennec, A.G., Michel, C., Bru, K., Touze, S., Desroche, N., Mnif, I., Motelica-Heino, M. 2015. Transfer and degradation of polyacrylamide-based flocculants in hydrosystems: A review. *Environ Sci Pollut Res* 22:6390–6406.

Haberman, C.E. 1991. Acrylamide. pp. 251–266. In: Kroschwitz J.L. and Howe-Grant M. (eds.), *Kirk-Othmer Encyclopedia of Chemical Technology*, 4th ed., Vol. 1., John Wiley & Sons, New York.

Harrison, J.P., Hoellein, T.J., Sapp, M., Tagg, A.S., Ju-Nam, Y., Ojeda, J.J. 2017. Microplastic-associated biofilms: A comparison of freshwater and marine environments. In: Wagner M., Lambert S. (eds.), *Freshwater Microplastics: Emerging Environmental Contaminants*, Springer, Heidelberg, Germany. doi:10.1007/978-3-319-61615-5_9.

Hazelton Laboratories America. 1987. Plant growth study to estimate the potential for the uptake and accumulation of residual acrylamide monomer in plant tissue. HLA Study No. 6015-310.

Igisu, H., Goto, I., Kawamura, Y., Kato, M., Izumi, K., Kuroiura, Y. 1975. Acrylamide encephaloneuropathy due to well water pollution. *J Neurol, Neurosurg Psychiatry* 38:581–584.

Jebasingh, S.E.J., Lakshmikandan, M., Rajesh, R.P., Raja, P. 2013. Biodegradation of acrylamide and purification of acrylamidase from newly isolated bacterium *Moraxella osloensis* MSU11. *Int Biodeter Biodegr* 85:120–125.

Jennings, A.R., Sprunt, E.S. 1995. Unique method for hydraulic fracturing. U.S. Patent 5,402,846.

Jin, R.R., Hu, W.B., Hou, X.J. 1987. Mechanism of selective flocculation of hematite from quartz with hydrolyzed polyacrylamide. *Coll Surf* 26:317–331.

Jones, G.D. 1958. Treatment of acrylamide polymers. U.S. Patent 2,831,841.

Joshi, S.J., Abed, R.M.M. 2017. Biodegradation of polyacrylamide and its derivatives. *Environ Process* 4:463–476.

Junqua, G., Spinelli, S., Gonzalez, C. 2015. Occurrence and fate of acrylamide in water-recycling systems and sludge in aggregate industries. *Environ Sci Pollut Res* 22:6452–6460. doi:10.1007/s11356-014-3022-5.

Keas, W. 1974. Liquid flocculent additive from polyacrylamide. U.S. Patent 3,817,891.

Kolya, H., Tripathy, T. 2014. Biodegradable flocculants based on polyacrylamide and poly(*N*, N-dimethylacrylamide) grafted amylopectin. *Int J Biol Macromol* 70:26–36.

Kornecki, T.S., Grigg, B.C., Fouss, J.L., Southwick, L.M. 2005. Polyacryalmide (PAM) application effectiveness in reducing soil erosion from sugarcane fields in southern Louisiana. *App Eng Agric* 21:189–196.

Kuboi, T., Fujii, K. 1984. Toxicity of cationic polymer flocculants to higher plants: I. Seedling assays. *J Soil Sci Plant Nut*, 30:311–320.

Kurenkov, V.F., Abramova, L.I. 1992. Acrylamide inverse polymerization. *Polym Plast Technol Eng* 31:659–704.

Labahn, S.K., Fisher, J.C., Robleto, E.A., Young, M.H., Moser, D.P. 2010. Microbially mediated aerobic and anaerobic degradation of acrylamide in a western United States irrigation canal. *J Environ Qual* 39:1563–1569.

Laird, D.A. 1997. Bonding between polyacrylamide and clay mineral surfaces. *Soil Sci* 162:826–832.

Lakshmikandan, M., Sivaraman, K., Raja, S.E., Vasanthakumar, P., Rajesh, R.P., Sowparthani, K., Jebasingh, S.E.J. 2014. Biodegradation of acrylamide by acrylamidase from *Stenotrophomonas acidaminiphila* MSU12 and analysis of degradation products by MALDI-TOF and HPLC. *Int Biodeter Biodegr* 94:214–221.

Lambert, S., Scherer, C., Wagnery, M. 2017. Ecotoxicity testing of microplastics: Considering the heterogeneity of physical chemical properties. *Integr Environ Asses Manag* 13:470–475.

LaMer, V.K., Healvy, T.W. 1963. Adsorption-flocculation reactions of macromolecules at the solid–liquid interface. *Rev Pure Appl Chem* 13:112–132.

Li, Y.Q., Shen, C.H., Jing, G.C., Hu, J.Q. 2006. Mechanism of HPAM-biodegradation and its application. *Pet Explor Dev* 33:738–743 (in Chinese).

Liu, L., Wang, Z., Lin, K., Cai, W. 2012. Microbial degradation of polyacrylamide by aerobic granules. *Environ Technol* 33:1049–1054.

Lu, J.H., Wu, L., Letey, J. 2002. Effects of soil and water properties on anionic polyacrylamide sorption. *Soil Sci Soc Am J* 66:578–584.

Ma, F., Wei, L., Wang, L., Chang, C.C. 2008. Isolation and identification of the sulphate reducing bacteria strain H1 and its function for hydrolyzed polyacrylamide degradation. *Int J Biotechnol* 10:55–63.

Malik, M., Nadler, A., Letey, J. 1991. Mobility of polyacrylamide and polysaccharide polymer through soil materials. *Soil Technol* 4:255–263.

Marshall, T., Williams, K.M. 1991. The simplified technique of high-resolution two-dimensional polyacrylamide gel electrophoresis: Biomedical applications in health and disease. *Electrophoresis* 12:461–471.

Matsuoka, H., Ishimura, F., Takeda, T., Hikuma, M. 2002. Isolation of polyacrylamide degrading microorganisms from soil. *Biotechnol Bioprocess Eng* 7:327–330.

Michalanko, E.M., Gray, D.A., Sage, G.W., Jarvis, W.F. 1989. Large production and priority pollutants. In: P.H. Howard (Ed.), *Handbook of Environmental Fate and Exposure Data for Organic Chemicals*. Lewis Publishers, London, UK.

Mortland, M.M. 1970. Clay-organic complexes and interactions. *Adv Agron* 22:75–117.

Nakamiya, K., Kinoshita, S. 1995. Isolation of polyacrylamide-degrading bacteria. *J Ferment Bioeng* 80:418–420.

National Institute for Occupational Safety and Health (NIOSH). 1976. Criteria for a recommended standard-occupational exposure to acrylamide. NTIS Order No. PB-273871.

Nawaz, M.S., Billedeau, S.M., Cerniglia, C.E. 1998. Influence of selected physical parameters on the biodegradation of acrylamide by immobilized cells of *Rhodococcus* sp. *Biodegradation* 9:381–387.

Nawaz, M.S., Franklin, W., Cerniglia, C.E. 1993. Degradation of acrylamide by immobilized cells of a Pseudomonas sp. and Xanthomonas maltophilia. *Can J Microbiol* 39:207–212.

Nishikawa, J., Hara, T., Sonoda, Y. 1983. Absorption of acrylamide by plants. *Nippon Dojyou Hiryou Gaku Zasshi* 54:55–57.

Pavlyk, B.I. 1996. Biologically compatible hydrogel world. Patent WO 9,604,943.

Petersen, D.W., Cooper, K.R., Friedman, M.A., Lech, J.J. 1987. Behavioral and histological effects of acrylamide in rainbow trout. *Toxicol Appl Pharm* 87:177–184.

Prabu, C.S., Thatheyus A.J. 2007. Biodegradation of acrylamide employing free and immobilized cells of *Pseudomonas aeruginosa*. *Int Biodeter Biodegr* 60:69–73.

Rogacheva, S.M., Ignatov, O.V. 2001. The respiratory activity of *Rhodococcus rhodochrous* M8 cells producing nitrile-hydrolyzing enzymes. *Appl Biochem Microbiol* 37:282–286.

Sang, G., Pi, Y., Bao, M., Li, Y., Lu, J. 2015. Biodegradation for hydrolysed polyacrylamide in the anaerobic baffled reactor combined aeration tank. *Ecol Eng* 84:121–127.

Senft, R. 1993. Erosion takes a powder. *Agric Res* 41:16–17.

Sharma, B.R., Dhuldhoya, N.C., Merchant, U.C. 2006. Flocculant: An ecofriendly approach. *J. Polym. Environm* 14:195–202.

Sharma, M., Sharma, N., Bhalla, T. 2009. Amidases: Versatile enzymes in nature. *Rev Environ Sci Biotechnol* 8:343–366.

Shukor, M.Y., Gusmanizar, N., Azmi, N.A., Hamid, M., Ramli, J., Shamaan, N.A., Shed, M.A. 2009. Isolation and characterization of an acrylamide-degrading *Bacillus cereus*. *J Environ Biol* 30:57–64.

Smith, E.A., Oehme, F.W. 1991. Acrylamide and polyacrylamide: A review of production, use, environmental fate and neurotoxicity. *Rev Environ Health* 9:215–228.

Smith, E.A., Prues, S.L., Oehme, F.W. 1996. Environmental degradation of polyacrylamides: 1. Effects of artificial environmental conditions: Temperature, light, and pH. *Ecotoxicol Environ Saf* 35:121–135.

Sojka, R.E., Bjoneberg, D.L., Entry, J.A., Lentz, R.D., Orts, W.J. 2007. Polyacrylamide in agriculture and environmental land management. *Adv. Agron* 92:75–162.

Sonoda, Y., Kano, K., Hara, T. 1977. The behavior of polyacrylamides as cohesive agent in soil-plant system. *Daigaku Nogakubu Kenkyu Hokuku* 40:61–69.

Spingborn Bionomics. 1985. Toxicity test report: Chronic toxicity of acrylamide monomer to mysid. NTIS/OTS Order No. 0510508. Doc # 40-8631565.

Sutherland, G.R., Haselbach, J., Aust, S.D. 1997. Biodegradation of crosslinked acrylic polymers by a white-rot fungus. *Environ Sci Pollut Res* 4:16–20.

Szögi, A.A., Leib, B.G., Redulla, C.A., Stevens, R.G., Mathews, G.R., Strausz, D.A. 2007. Erosion control practices integrated with polyacrylamide for nutrient reduction in rill irrigation runoff. *Agric Water Manag* 91:43–50.

Taylor, M.L., Morris, G.E., Self, P.G., Smart, R.S. 2002. Kinetics of adsorption of high molecular weight anionic polyacrylamide onto kaolinite: The flocculation process. *J Colloid Interface Sci* 250:28–36.

Tekin, N., Demirbas, O., Alkan, M. 2005. Adsorption of cationic polyacrylamide onto kaolinite. *Micropor Mesopor Mat* 85:340–350.

Tekin, N., Dincer, A., Demirbas, O., Alkan, M. 2006. Adsorption of cationic polyacrylamide onto sepiolite. *J Hazard Mater* 134:211–219.

Tekin, N., Dincer, A., Demirbas, O., Alkan, M. 2010. Adsorption of cationic polyacrylamide (C-PAM) on expanded perlite. *Appl Clay Sci* 50:125–129.

Transparency Market Research Publication info: PR Newswire; New York [New York] December 23, 2013.

U.S. Environmental Protection Agency (U.S. EPA). 1994. Integrated Risk Information System (IRIS) Online. Coversheet for Acrylamide. In: Office of Health and Environmental Assessment, U.S. EPA, Cincinnati, OH.

USEPA. 2010. *Toxicological Review of Acrylamide, EPA/635/R-07/009F.* U.S. Environmental Protection Agency, Washington, DC.

Wakaizumi, M., Yamamoto, H., Fujimoto, N., Ozeki, K. 2009. Acrylamide degradation by filamentous fungi used in food and beverage industries. *J Biosci Bioeng* 108:391–393.

Wampler, D.A., Ensign, S.A. 2005. Photoheterotrophic metabolism of acrylamide by a 2005 newly isolated strain of *Rhodopseudomonas palustris. Appl Environ Microbiol* 7:5850–5857.

Weideborg, M., Källqvist, T., Ødegård, K.E., Sverdrup, L.E, Vik, E.A. 2001. Environmental risk assessment of acrylamide and methyloacrylamide from a grouting agent used in the tunnel construction of Romeriksporten, Norway. *Water Res* 35:2645–2652.

Wen, Q., Chen, Z., Zhao, Y., Zhang, H., Feng, Y. 2010. Biodegradation of polyacrylamide by bacteria isolated from activated sludge and oil-contaminated soil. *J Hazard Mater* 175:955–959.

WHO (World Health Organization). 1993. *Guidelines for Drinking Water Quality.* vol. 1, 2nd ed., Recommendations, vol. 1. WHO, Geneva, Switzerland.

Woodrow, J.E., Seiber, J.N., Miller, G.C. 2008. Acrylamide release resulting from sunlight irradiation of aqueous polyacrylamide/iron mixtures. *J Agric Food Chem* 56:2773–2779.

Yamauchi, T., Nishihata, T. 1978. Japanese Patent Application J P. 53051289.

Yongrui, P., Zheng, Z., Bao, M., Li, Y., Zhou, Y., Sang, G. 2015. Treatment of partially hydrolyzed polyacrylamide wastewater by combined Fenton oxidation and anaerobic biological processes. *Chem Eng J* 273:1–6.

Yu, F., Fu, R., Xie, Y., Chen, W. 2015. Isolation and characterization of polyacrylamide degrading bacteria from dewatered sludge. *Int J Environ Res Public Health* 12:4214–4230.

Zhao, L., Bao, M,, Yan, M., Lu, J. 2016. Kinetics and thermodynamics of biodegradation of hydrolyzed polyacrylamide under anaerobic and aerobic conditions. *Bioresour Technol* 216:95–104.

Zhu, H., Smith, D.W., Zhou, H., Stanley, S.J. 1996. Improving removal of turbidity causing material by using polymers as a filter aid. *Water Res* 30:103–114.

4 Analytical Methods for Particulate Plastics in Soil and Water

Upekshya Welikala, Chanaka M. Navarathna, Samadhi Nawalage, Binoy Sarkar, Todd E. Mlsna, and Sameera R. Gunatilake

CONTENTS

4.1 Introduction 52
4.2 Sampling and Sample Preparation 53
4.2.1 Sampling 53
4.2.2 Extraction and Separation 54
4.2.2.1 Visual Examination 54
4.2.2.2 Sieving 55
4.2.2.3 Density Separation 55
4.2.2.4 Froth Flotation 56
4.2.2.5 Munich Plastic Sediment Separator 57
4.2.2.6 Pressurized Fluid Extraction 58
4.2.2.7 Chromatographic Extraction Methods 58
4.2.2.8 Other Techniques 59
4.2.3 Purification 60
4.2.4 Storage and Preservation 61
4.3 Identification 61
4.3.1 Challenges of Detecting Small MPs 61
4.3.2 Classification 61
4.3.3 Visual Sorting 62
4.3.4 Microscopic Methods 63
4.3.5 Fourier-Transform Infrared Spectroscopy and Raman Spectroscopy Analyses 65
4.3.5.1 Raman Imaging 65
4.3.5.2 Raman Libraries 66
4.3.5.3 Micro-Raman Spectroscopy and Other Raman Techniques 66
4.3.6 Infrared Spectroscopy 67
4.3.6.1 Micro-FTIR Techniques 67
4.3.6.2 Near-Infrared Reflectance Spectroscopy 68
4.3.7 Atomic Force Microscopy 68
4.3.8 Chromatographic, Mass Spectrometric, and Thermometric Techniques 68
4.3.8.1 Chromatography 69
4.3.8.2 Time-of-Flight–Secondary Ion Mass Spectrometry 69
4.3.8.3 Matrix-Assisted Laser Desorption Ionization–Time-of-Flight–Mass Spectrometry 69
4.3.8.4 Pyro–Gas Chromatographic–Mass Spectrometry 69

4.3.8.5 Thermal Extraction and Desorption–Gas Chromatography–Mass Spectrometry 70
4.3.8.6 Thermal Desorption–Gas Chromatography–Mass Spectrometry 71
4.3.8.7 Thermo-Gravimetric–Mass Spectrometry 71
4.3.8.8 Differential Scanning Calorimetry 71
4.3.9 Other Techniques 71
4.3.9.1 Nuclear Magnetic Resonance Spectroscopy 71
4.3.9.2 X-ray Photoelectron Spectroscopy 71
4.3.9.3 Miscellaneous Techniques 72
4.4 Quantification 72
4.4.1 Microscopic Techniques 73
4.4.1.1 Stereoscopic Microscopy 73
4.4.1.2 Confocal Microscopy and Image Processing Counting 73
4.4.2 Thermal Extraction Desorption–Gas Chromatography–Mass Spectrometry and Size Exclusion Chromatography 73
4.4.3 High-Temperature–Gel Permeation–Chromatography 74
4.4.4 Reversed-Phase Liquid Chromatography 74
4.4.5 Quantitative Nuclear Magnetic Resonance 74
4.4.5.1 Detection Limits 75
4.4.6 Fluorescence Staining 75
4.4.7 Reporting Units 76
4.5 Conclusions 77
References 78

4.1 INTRODUCTION

Over the past years, plastics have become a quotidian term due to the tremendous impact they have had on human lifestyle (Thompson et al. 2009). Plastic production required to meet unabating demands was reported to increase from 0.5 million tons per year in 1960 to as high as 300 million tons in 2013 (Rocha-Santos and Duarte 2015). Plastics are polymers comprised of many monomeric subunits and are modified with either organic or inorganic additives (Coppock et al. 2017). Their high disposability, high persistence, and low regeneration, and so on, has made waste plastics a worldwide research focus as they accumulate uncontrollably in the environment.

Plastics were discerned as problematic compounds by the United Nations Environment Program (Lusher et al. 2017). Although their breakdown in the environment was thought to be advantageous, it was later found that the resulting smaller fragments, called microplastics (MPs), were as hazardous as their precursors (Hale 2017). As was defined by the National Oceanic and Atmospheric Administration International research workshop, MPs are plastic particles having a size less than 5 mm (Hanvey et al. 2017). Microplastics can be categorized into primary and secondary types, with the former being nurdles and microbeads and the latter being the chemically or mechanically fragmented products of larger plastic material (Hanvey et al. 2017).

Although MPs are emerging contaminants, their presence in the environment has been unsuccessfully screened, which has resulted in their global dispersion (Hale 2017). Microplastic debris have been recovered from distant isolated areas such as the poles, benthic zones, on mountain peaks, and open oceans and shorelines, which are regions where direct human impact is minimal. Various contaminants have been reported to concentrate more on MP fragments than in the surroundings where they are found. Thereby, MPs have also been referred to as "vectors" or carriers of contaminants in ecosystems (Coppock et al. 2017).

Microplastics are hazardous because they prompt detrimental health consequences via multiple mechanisms, and elicit chemical, physical, and biochemical toxicity on various systems of an organism, especially systems involved in breathing and digestion. The threat posed by MPs varies as a function of their size and weight, where the smaller particle size favors penetration through cell membranes upon ingestion due to its high mobility (Hale 2017). Microplastics are solid, non-volatile materials, and if left in contact with the environment for a significant time, they would form associations with other fragments like themselves and other materials in the environment due to their high surface area (Hale 2017). The smaller particles are not as prominent as the larger fragments, making the characterization of MPs difficult in ecosystems.

In order to gain a comprehensive understanding on how the availability of MPs in the environment affects life, robust and optimal analytical methods to quantify the abundance of plastic fragments in systems must be developed. This requires that sampling techniques, sample preparation, and detection methods all be compatible. Microplastics are chemically heterogeneous and their characterization must take into consideration particle size, composition, shape, and concentration of available MPs (Hidalgo-Ruz et al. 2012; Song et al. 2015). Additionally, there can be a significant contribution from the sample matrix to the MP analysis, for example, a sludge sample or wastewater sample must undergo significant purification before analysis (Hale 2017). Plastic interactions with other environmental chemicals and natural phenomena, such as weathering can impact analysis (Hale 2017). Sampling and preparation methods should be employed after careful experimentation using different matrices and sample sizes because techniques used for a laboratory scale determination might not yield concordant results for a field-scale operation. Therefore, it is a requirement that consistent analytical techniques exist for these analyses (Hanvey et al. 2017).

The techniques currently used for detection, such as naked-eye detection, microscopy, optical spectroscopy, chromatography, mass spectrometry, and thermal analysis are a few such analytical methods, although all are with their associated advantages and disadvantages (Wang and Wang 2018). Traditional liquid and gas chromatographic techniques used to determine soluble and semi-volatile substances cannot be incorporated directly for MP quantification due to their non-volatility and low solubility. Sometimes a single technique would not suffice in an instance, where the matrix is highly complex. In such cases, hyphenated techniques are used to improve the reliability of the result. The hyphenated techniques, in which separational methods are coupled with multidimensional detectors, have become useful alternatives. Hyphenated methods couple two analytical techniques by using an appropriate interface (e.g., gas chromatography–mass spectrometry [GC-MS], thermo-gravimetric analysis coupled to differential scanning calorimetry [TGA-DSC], thermal desorption GC-MS [TDS-GC-MS], etc.). Such hyphenated techniques, although accurate, tend to be convoluted (Song et al. 2015; Zhu 2015).

Pollution caused by MPs span the globe, and the true extent of the dangers posed to the environment is a growing concern. Therefore, increasingly, research is being carried out to reveal the fate and ecological risks of MPs in soil and water. Knowledge on a wide spectrum of analytical techniques used for sampling, identification, and quantification of MPs is dispersed over various sources. The chapter at hand is an attempt to centralize and unify information on a plethora of techniques used for the analysis of particulate plastics in soil and water. The comprehensive information presented in this chapter is intended for an avid reader or a researcher keen in understanding MPs.

4.2 SAMPLING AND SAMPLE PREPARATION

4.2.1 Sampling

Sampling and analysis of MPs are becoming increasingly tedious and demanding, as interest expands to include smaller particle sizes. The intentions of new method development are as follows: (1) to obtain high accuracy, precision, and certainty of MP identification by avoiding false-positive and/or false-negative misidentification; (2) to improve analytical results with an indication of polymer types; and (3) to transfer labor intensive steps from manual to automated routines (Bläsing and Amelung 2018).

Three main strategies are used when sampling MPs: (1) sediment selection, (2) volume-reduction, and (3) bulk sampling methods (Frias, Sobral, and Ferreira 2010; Hidalgo-Ruz et al. 2012). In sediment-selective sampling, plastic pellets and fragments are collected using tweezers, spoons, or manually by hand (Hidalgo-Ruz et al. 2012). During volume-reduced sampling, the volume of the sample (either sediment of seawater) is reduced greatly during the sampling period, while preserving only that portion of the sample of interest for further processing (Hidalgo-Ruz et al. 2012). Also, in volume-reduced sampling, samples are usually obtained by filtering large volumes of water using nets (Hidalgo-Ruz et al. 2012).

Neuston plankton and manta trawl nets are frequently used for collecting plastics from surface water samples. The Neuston net design features a large, rectangular net frame and a relatively long net for sampling considerable water volumes. A manta trawl is a net system which is used for sampling the surface of the ocean. Bulk sampling is more suitable for MPs that cannot be easily identified visually when: (1) they are covered by sediment particles, (2) their abundance is small, requiring sorting and filtering of large volumes of sediment and water, or (3) they are too small to be identified with the naked eye. In contrast, during bulk sampling, the volume of water is not reduced, and only a few studies have been reported using this method (Law and Thompson 2014; Li, Liu, and Chen 2018).

The process of sampling for plastic pellets is often selective, and this is due to their size range (1–6 mm diameter), which makes them easily recognizable in the flotsam deposits of sandy beaches. However, when MPs are mixed with other debris or have no characteristic shapes (e.g., irregular, rough, or angular), there is a greater probability of overlooking them and particular care needs to be taken during selective sampling in the field (Hidalgo-Ruz et al. 2012).

4.2.2 Extraction and Separation

Microplastics can directly enter the environment from large plastic items that are continuously fragmenting until they attain micrometric dimensions. Microplastics tend to sediment after reaching soil and aquatic systems via effluents from wastewater treatment plants. Their analysis is a tedious task, because their separation must be performed prior to analysis. Some of the commonly used methods for separating MPs from soil and sediments in practice include visual examination and selection, filtration (size fractionation or size exclusion sieving), density separation, floatation, centrifuging organic matter, oxidation, and digestion (Erni-Cassola et al. 2017; Zhang, Yang et al. 2018; Zobkov and Esiukova 2017).

4.2.2.1 Visual Examination

Visual examination of the concentrated sample remnants is a common step in sampling procedures (Silva et al. 2018). To improve the accuracy of identification results, a series of selection criteria are recommended to be strictly followed: (1) suspected particles or fibers should not possess visible organic or cellular structures, (2) fibers should have consistent thickness along the entire length, (3) particles must be clear and uniformly colored, and (4) transparent and white particles should be further confirmed under a microscope having a magnification within the range of 10× to 40× (Zobkov and Esiukova 2017). Visual sorting of residues is vital in separating the plastics from other materials, such as organic debris (e.g., shell fragments, animal parts, dried algae, and seagrasses) and other foreign substances (e.g., metal paint coatings, tar, and glass) (Zobkov and Esiukova 2017). This is carried out by examining the sample with the naked eye or by using a dissecting microscope (Hidalgo-Ruz et al. 2012). If it is a marine sample, a drop of olive oil can be added to the salt solution before stirring as a means of improving the percentage recovery. The percentage recovery (R%) of MPs takes the following formula (Quinn, Murphy, and Ewins 2017),

$$\%\,\text{Recovery}\,(\text{R}\%) = \left(\frac{\text{Final weight of plastic}}{\text{Initial weight of plastic}}\right) \times 100\%$$

This causes plastic particles to assemble on the oil droplets and collect on the filter instead of sticking to glass walls (Silva et al. 2018). The size of the MPs used for recovery tests can also affect the recovery percentage (Quinn, Murphy, and Ewins 2017).

4.2.2.2 Sieving

Sieving is a commonly used method for isolating MPs from water and sediment matrices (Hidalgo-Ruz et al. 2012). The sieves can be made of stainless steel or copper. Solid materials that are larger than the mesh sizes are retained during sieving, which permits the removal of water and smaller particles from the sample. Multi-tier sieving has been conducted in numerous studies by using a series of sieves with a decreasing mesh size which allows the separation of MPs into differently sized categories. The mesh size of sieves for MPs typically varies from 0.035 to 4.75 mm, and it depends on the desired size range of MPs to be collected. For sediment samples, sieving reduces the sample volume for subsequent extraction (Wang and Wang 2018).

4.2.2.3 Density Separation

Density separation using various brine solutions has been commonly used for the extraction of MPs via solid-liquid separation (Sánchez-Nieva et al. 2017). In seawater, polymers with lower densities than seawater have a higher mobility and buoyancy when compared to high-density polymers that tend to settle out (Erni-Cassola et al. 2017). Bottom sediments have a noticeably higher specific density (≥2.65 g/mL) than most plastics (0.05–1.70 g/mL), and this difference is employed to separate comparatively light plastics from heavy sediments (Zobkov and Esiukova 2017). Numerous salts are reported to have been used in density separation, and they cover a wide spectrum of density values when dissolved in water, which enables one to match the salt solution with the density of plastics. Densities of some common plastics are listed in Table 4.1, and the density of some common salts and their extractable plastic densities are listed in Table 4.2 (Käppler et al. 2016; Qiu et al. 2016; Quinn, Murphy, and Ewins 2017; Sánchez-Nieva et al. 2017; Silva et al. 2018; Zhang, Shi et al. 2018; Zobkov and Esiukova 2017).

TABLE 4.1
Densities of Some Common Plastics

Plastic Type	Density (g/mL)	Reference
Polyvinyl chloride (PVC)	1.40	Imhof et al. (2012)
Polyoxy methylene (POM)	1.43	Imhof et al. (2012)
Polypropylene (PP)	0.85–0.94	Hidalgo-Ruz et al. (2012)
Polyethylene (PE)	0.92–0.97	Hidalgo-Ruz et al. (2012)
Polystyrene (PS)	1.04–1.1	Wang and Wang (2018)
Polyethylene terephthalate (PET)	1.4–1.6	Wang and Wang (2018)
Low-density polyethylene (LDPE)	0.917–0.930	Zhang, Yang et al. (2018)
High-density polyethylene (HDPE)	0.93–0.97	Hidalgo-Ruz et al. (2012)
Expanded polystyrene (EPS)	1.04	Babu, Babu, and Wee (2005)
Polyamide (PA)	1.14	Utracki, Dumoulin, and Toma (1986)
Polycarbonate (PC)	1.20–1.22	Golden, Hammant, and Hazell (1967)
Polyester (PES)	1.39	Zeronian and Collins (1989)
Poly(ethylene-vinyl acetate)/ethylene vinyl acetate (PEVA/EVA)	0.936	Takidis et al. (2003)
Polyurethane (PUR)	>0.27	Saint-Michel, Chazeau, and Cavaillé (2006)
Cross-linked polystyrene (PSXL)	0.1	Williams and Wrobleski (1988)
Poly methyl methacrylate (PMMA)	1.18	Zhang et al. (1998)

TABLE 4.2
Density of Common Salts and the Extraction Plastic Density

Salt	Density (g/mL)	Extractable Plastic Density	Reference
Tap water	~1	Low	Qiu et al. (2016)
Seawater	1.03	Low	Qiu et al. (2016)
Sodium chloride	1.2	Medium	Sánchez-Nieva et al. (2017)
Calcium chloride	1.30–1.35	Medium	Zobkov and Esiukova (2017), Quinn, Murphy, and Ewins (2017)
Sodium bromide	1.5	Medium	Zhang, Shi et al. (2018), Quinn, Murphy, and Ewins (2017)
Potassium formate	1.5	Medium	Zhang, Shi et al. (2018), Quinn, Murphy, and Ewins (2017)
Zinc bromide	1.5	Medium	Zhang, Shi et al. (2018), Quinn, Murphy, and Ewins (2017)
Zinc chloride	1.5–1.7	Higher	Silva et al. (2018), Käppler et al. (2016), Quinn, Murphy, and Ewins (2017)
Sodium iodide	1.6–1.8	Higher	Silva et al. (2018), Sánchez-Nieva et al. (2017), Käppler et al. (2016)
Lithium metatungstate	1.62	Higher	Zobkov and Esiukova (2017), Quinn, Murphy, and Ewins (2017)
Sodium polytungstate	1.8	Higher	Käppler et al. (2016), Quinn, Murphy, and Ewins (2017)

Sequential density extractions have been applied to shed light on the potential environmental behavior of particles (Hurley et al. 2018). Food-grade table salt and reagent-grade high purity (>99.5%) NaCl has been studied for density separation procedures, and the food-grade table salt has been reported to show poor recovery percentages due to the presence of impurities (Sánchez-Nieva et al. 2017). Costs and the environmental impact are important factors to be considered when using these salts. Both NaCl and NaI are relatively inexpensive, and they have more environmental compatibility when compared to other salts (Sánchez-Nieva et al. 2017). Solutions with higher densities are often expensive and toxic to the environment (except NaI). Their usage for MP recovery has not been widely studied (Quinn, Murphy, and Ewins 2017).

The use of either NaI or $ZnBr_2$ is particularly advantageous during density separation, as they only require a single cycle sediment wash for MP removal. In contrast, NaCl requires at least three wash cycles (Quinn, Murphy, and Ewins 2017). Both NaI and $ZnBr_2$ have exhibited similar percentages of recovery and are more environmentally friendly. However, when NaI is used on sediment samples, it turns the filter paper black, making it difficult to isolate the MPs. This was thought to occur due to the iodine generated from excess NaI, which develops the dark purple starch-iodine complex (Quinn, Murphy, and Ewins 2017). For density separation of the larger MP size class (800–1000 μm), a pronounced trend of increasing percentage recoveries with increasing brine solution density was observed. The recovery percentages for both water and NaCl are lower for the larger MP size when compared to the smaller particles of the same polymer (Quinn, Murphy, and Ewins 2017).

4.2.2.4 Froth Flotation

Froth flotation is analogous to classical density separation, with the only difference being the absence of the skimming chamber (Imhof et al. 2012). It is used for the selective separation of hydrophobic materials from hydrophilic and in the recycling industry (mineral processing, paper recycling, and wastewater treatment) to separate plastic particles from waste and mixtures of different plastics

(Maes et al. 2017). This technique depends largely on the physical properties of MPs, such as bulk density, particle size, shape, surface energy, and surface roughness (Imhof et al. 2012). Froth flotation involves the introduction of air bubbles into a mixture of finely divided ore or another material with water and another chemical that helps the attachment of the bubbles to the particles of the desired material. Thereafter, the mixture is recovered as a froth (Imhof et al. 2012). However, the technique has several limitations including complexity, high consumable cost, low extraction efficiencies, incompatibility with very fine sediments, and particle degradation from flotation media (Coppock et al. 2017). In addition, extraction by flotation can be difficult when MPs are adsorbed onto soil particles (Zhang, Yang et al. 2018).

In froth flotation, the most important feature is the wettability of the plastic. It is possible to modulate the wettability by changing the surface tension of the medium or by chemical conditioning of the particles with hydrophobic modifications or by the adsorption of chemicals (Imhof et al. 2012; Maes et al. 2017). Multiple experiments using different surfactants and foam generators can be performed initially to improve the wetting of all plastic particles (Imhof et al. 2012). Surfactants, in addition to reducing the surface tension, degrade organic material and aid the detachment of plastic particles from other fragments.

Examples of surfactants include detergents, bleachers, and enzymes. Imhof et al. (2012) report of an effective method of plastic flotation, which incorporates a combination of a wetting agent and froth conditioner. The foam-reducing properties of the wetting agent are stabilized and strengthened by the froth conditioner to collect even greater fragments (Imhof et al. 2012). A density of ≤1.35 g/mL in MP-dispersed solutions allow floatation of polymers, such as polyethylene (PE) and polypropylene (PP) and small amounts of sediment material, while a larger portion of the sediment material settled to the bottom of the centrifuge tube. A considerable amount of sediment material floats on the surface under these conditions. The density could be reduced to decrease the floating fraction, but with the possibility that some denser plastic particles have a chance to settle to the bottom (Zhang, Yang et al. 2018).

4.2.2.5 Munich Plastic Sediment Separator

A most commonly used device for MP separation is the Munich Plastic Sediment Separator (MPSS). Density-based separation involves the selection of a suitable salt and oxidative digestion material. The mixture is then continuously stirred for few minutes and left to settle for several hours. The stirring rate of the salt solution, shaking, settling down, drying times, and sinking rate depends on the type of salt used, nature of the plastic, and size of the sample (Hidalgo-Ruz et al. 2012). Afterward, the supernatant which contains the floating plastic particles is collected and filtered through a Buchner funnel apparatus (Dehaut et al. 2016; Quinn, Murphy, and Ewins 2017). Subsequently, the filter containing the floating solids is rolled up, tied, placed on a petri dish, and dried under ambient conditions for 24 hours (Zobkov and Esiukova 2017).

The MPSS consists of a sediment container, a standpipe, and a dividing chamber with a ball valve and a filter holder. The sample chamber is between the filter holder and ball valve (Imhof et al. 2012). The sample chamber and filter holder can be turned upside down to change from extraction mode when they are on top of the MPSS to filtering mode when a vacuum is attached.

The MPSS provides a successive separation of MP particles (20 to <1 mm). It analyzes large MP particles that range in size from ~ 1 to 5 mm), and small MP particles that are ~<1 mm (Imhof et al. 2012). The MPSS can hold large quantities of sediment (6 kg). It is robust, is manufactured from stainless steel, and stands approximately 1.75 m tall. However, the manufacturing costs are high; it has limited portability and feasibility when processing replicates of small samples. The MPSS can be used for the extraction of MPs from beach sediment samples (Käppler et al. 2016).

The density separator in MPSS is made using a glass funnel with latex tubing attached to the bottom of the stem and a pinch clamp attached to control liquid flow from the funnel (Zobkov and Esiukova 2017). Filters can be made from glass fibers, nitrocellulose, polycarbonate membranes, or they may be zooplankton filters or isopore filters. The pore size of filters generally ranges from

0.45–20 μm (Hidalgo-Ruz et al. 2012; Imhof et al. 2016; Wang and Wang 2018). The time for analysis depends on the pore size of the filter and is inversely proportional to the pore size (Frère et al. 2016). Validation of the MPSS has been performed using common plastic particles, including polyvinyl chloride (PVC) and polyoxymethylene that have a relatively high specific density (Imhof et al. 2012).

Although the MPSS process is a simple technique, there are several associated limitations due to the presence of microscopic particulates or debris. These materials can rapidly clog the filter media and lower its effectiveness. These issues can be mostly overcome by reducing the solution volume, settling liquids for a longer time to facilitate the separation of heavier solid particles from the supernatant, performing a pre-filtration step using a filter with a larger pore size, or by adding auxiliary chemicals, such as ferrous sulfate ($FeSO_4$) to the liquid in order to flocculate the solid fraction. The glassware walls should be thoroughly rinsed to minimize the adherence of MPs to the glass wall (Wang and Wang 2018).

4.2.2.6 Pressurized Fluid Extraction

Pressurized fluid extraction (PFE) is another method used for extraction of MPs, which involves a sample preparation technique that employs elevated temperatures and pressure with liquid solvents to achieve rapid and efficient extraction of the analytes from the solid matrix. Plastics can be physically separated from waste and soil samples by optimizing the PFE conditions (Wagner et al. 2017). This method utilizes two extractions, as shown in Figure 4.1. In the initial extraction, methanol at 100°C is used to remove semi-volatile organic compounds, such as fats and oils, and in the second extraction, PFE dichloromethane is used to recover MPs. The collected dichloromethane extracts are evaporated to dryness and subsequently measured gravimetrically (Silva et al. 2018). Pressurized fluid extraction is only recommended for extractions of MPs that are less than <30 μm in soil (Fuller and Gautam 2016). Although plastic particles as small as 30 μm can be extracted efficiently, the challenges of using the method include: (1) inability of assessing the size distribution and (2) morphological changes of MPs after the extraction (Li, Liu, and Chen 2018). Pressurized fluid extraction is a technique that utilizes solvents at sub-critical temperature and pressure for the recovery of semi-volatile organics from solid materials. The technique has achieved status as a standard extraction technique and is commonly used in environmental laboratories for the extraction of organic pollutants from soils, sediments, and wastes (Fuller and Gautam 2016). Preliminary results indicated that polycarbonate (PC) and polyurethane (PUR) foam can be extracted by using PFE (Fuller and Gautam 2016).

The PFE technique may also be applicable to liquid matrices by means of an initial filtering step through glass fiber filters with subsequent PFE of the filters. However, this approach is yet to be validated. Some benefits of using this technique include simplicity, cost, speed, and reproducibility of results. The extraction component of the method can be fully automated and thereby disregards the need of an operator. The sensitivity of the method depends on the precision and accuracy of gravimetric techniques (Fuller and Gautam 2016). This method can address certain limitations of the current MP methods and provide laboratories with a simple, analytical method for quantifying common MPs in a variety of environmental samples (Fuller and Gautam 2016).

4.2.2.7 Chromatographic Extraction Methods

Two chromatographic extraction methods, namely, thermal extraction desorption-GC-MS (TED-GC-MS) and liquid extraction with subsequent size exclusion chromatography (SEC) for soil samples with known reference amounts of PE, PP, polystyrene (PS), and PE terephthalate (PET) were reported for MP separation. The results obtained were comparable to those found by other techniques in terms of measurement time, technique handling, detection limits, and requirements for sample preparation.

The liquid-liquid extraction method is a rapid technique that permits the extraction of relatively large sample amounts (>500 mg), which is advantageous. Some of the commonly used SEC solvents include hexafluoro-2-propanol, tetrahydrofuran, dimethylacetamide, or dimethylformamide (except

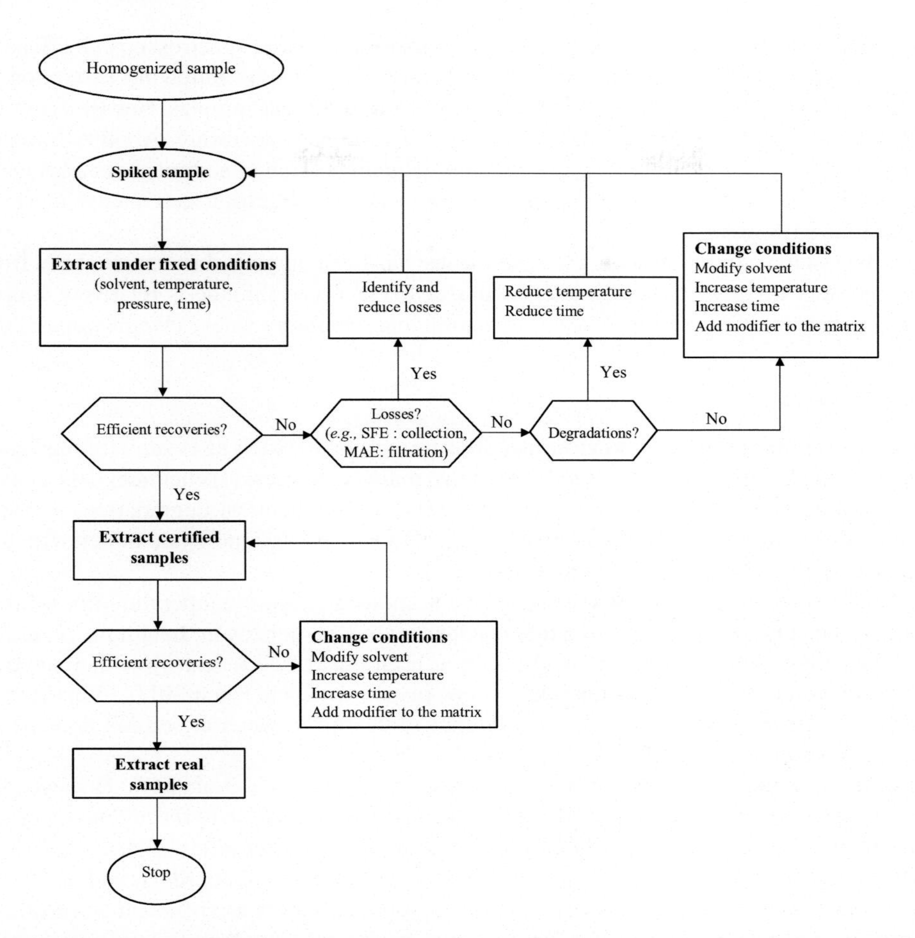

FIGURE 4.1 General strategy for optimizing extraction conditions for SFE (supercritical fluid extraction), PFE, and MAE (microwave-assisted extraction). (Camel, V., *Analyst*, 126, 1182–1193, 2001. Reproduced by permission of the Royal Society of Chemistry.)

for PE, PP, or their networks). The specific characteristics exhibited by the solution are an indication of the existence of different polymer classes (Elert et al. 2017). These chromatographic extraction methods can handle considerably higher sample amounts (20 and 500 mg used for TED-GC-MS and SEC, respectively), and this could be further scaled up. The main limitation involved in scaling up is the high cost of extraction solvents (Elert et al. 2017).

4.2.2.8 Other Techniques

Elutriation and two-step air-induced overflow followed by NaI density separation produces higher MP recovery percentages because the method: (1) is relatively fast, (2) is reproducible, (3) involves minimum technology, and (4) is relatively cheap (Quinn, Murphy, and Ewins 2017). The principle of separating particles is based on their size, shape, and density. A stream of gas or liquid flows in a direction usually opposite to the direction of sedimentation. This method is mainly employed for particles smaller than 1 μm. The smaller or lighter particles rise to the top because their terminal sedimentation velocities are lower than the velocity of the rising fluid.

A new and improved, cost-effective oil extraction protocol (OEP) was demonstrated as an alternative to density-based approaches by taking advantage of the oleophilic properties of MPs. The OEP is a cheap method, requiring minimal amounts of reagents and basic laboratory equipment

(Crichton et al. 2017). Further, the OEP is Fourier-transform infrared spectroscopy (FTIR) compatible, allowing the possibility of successful identification of polymers following extraction. It can also be used to extract degraded environmental MPs from sediment samples. However, there is the possibility that biofouling can degrade samples stored at room temperature. Another limitation of the OEP pertains to the organic content of samples. The processing of sediment samples with high organic content is advantageous and may benefit from an added digestion step to enable visual enumeration of the MPs and FTIR.

Pulsed ultrasonic extraction with ultra-pure water has been employed to remove MPs from fish stomachs without dissolving the stomach tissues or MPs. The technique is relatively simple and eliminates issues associated with hazardous disposal and laboratory safety (Wagner et al. 2017).

4.2.3 Purification

Purification is an important step to eliminate interfering matter, such as organic tissues and inorganic dusts, and this process can be divided into two main categories: (1) chemical degradation and (2) enzymatic degradation (Li, Liu, and Chen 2018). During chemical degradation, the MPs are treated with different chemicals, mainly 10% or 30% H_2O_2 solution or peroxide mixed with mineral acids such as H_2SO_4 (Li, Liu, and Chen 2018).

Non-oxidizing acids, such as HCl at low concentrations and room temperature are insufficient, because they yield large amounts of organic residues after digestion, even though the damage done to MPs may be less. Strong oxidizing acids, such as H_2SO_4 and HNO_3 destroy or damage the MPs made from polymers that have low tolerance at low pH (Li, Liu, and Chen 2018). Other techniques such as ultra-sonication can be used in the presence of deionized water or sodium dodecyl sulfate solution (Li, Liu, and Chen 2018).

Digestion of whole organisms or excised tissues is widely used to extract MPs. However, special attention must be paid when selecting an appropriate digestive agent due to the potential destruction of contaminants (Lusher et al. 2017). Digestive agents at high concentrations, such as 50% HCOOH, >35% $HClO_4$, >40% HF, >80% H_2O_2, >50% HNO_3, >70% $HClO_4$, >50% KOH, and >95% H_2SO_4, can cause damage to polymers (Lusher et al. 2017). Some polymers are resistant to certain digestive agents. For instance, PET and PVC are resistant to 50% HNO_3, albeit PE and PP are partially resistant and polyamide, PC, and PS are non-resistant to 50% HNO_3.

Efficiency of digestion increases with the increase in the molarity and the temperature. A study on digestion of MPs using 1 M NaOH displayed high effectiveness, where 90% of the MPs digested were recovered (Li, Liu, and Chen 2018). Plankton samples are reported to be digested in the presence of 10 M NaOH at 60°C. However, such harsh conditions would damage the MPs (Li, Liu, and Chen 2018). A concentration of 10% KOH (at 60°C and left overnight) and enzymatic digestion protocols are reported to be effective digestion methods when compared to oxidative acid digestion (Lusher et al. 2017).

Density separation cannot stand alone for sewage or soil samples, because they contain high contents of organic matter and aggregates. For example, soil organic matter usually exhibits a density of 1.0–1.4 g/mL and, therefore, will not be separated effectively from MPs during density extraction. Therefore, additional pre-treatment steps are required (Hurley et al. 2018). One approach is photodegradation of the organic matter with Fenton's reagent (H_2O_2 with ferrous ion, typically $FeSO_4$), which produces reactive oxygen species, such as hydroxyl, peroxyl, and superoxide radicals that can oxidize the organic matter. The reactive oxygen species generation is triggered in the presence of light. The technique has no effect or minimal effect on plastic degradation and, therefore, does not negatively affect the recovery (Hurley et al. 2018). The sequence of the analytical procedure (organic matter removal followed by density separation, and vice versa) had no noticeable effect on the recovery of the different MP particles. Hence, the step where organic matter is removed can be performed within existing protocols for MP isolation through density separation based on preference or convenience (Hurley et al. 2018).

4.2.4 Storage and Preservation

Plastic fragments that have been previously isolated also can be further washed to remove any adsorbed impurities. This can be attained by ultrasonic cleaning in a liquid medium or deionized water. Samples can be preserved in their original form without initial sorting, or they can be immediately sorted and stored. Plastics separated from the sample should be dried completely and kept in a dark and temperature-controlled environment (stable room temperature) in order to prevent or reduce degradation during storage (Hidalgo-Ruz et al. 2012).

4.3 IDENTIFICATION

Microplastics can be identified based on their specific functional groups and chemical composition. This section focuses on MP identification based on their size, shape, color, and surface properties (Rocha-Santos and Duarte 2015). Identification techniques for MPs can be broadly introduced as visual sorting, microscopic methods, chromatography and mass spectrometry, spectroscopy, densitometry, and thermal analysis techniques (Huppertsberg and Knepper 2018). The microscopic techniques include optical, light, binocular, dissecting, stereo, and scanning electron microscopy (SEM) (Hanvey et al. 2017; Rocha-Santos and Duarte 2015; Silva et al. 2018). The chromatographic and mass spectrometric techniques in use are liquid chromatography (LC), pyro–gas chromatography coupled with mass spectrometry (Pyr-GC-MS, SEC) (Elert et al. 2017), TED-GC-MS (Hanvey et al. 2017), time-of-flight–secondary ion mass spectrometry (ToF-SIMS), and matrix-assisted laser desorption ionization-ToF-MS (MALDI-ToF-MS) (Elert et al. 2017; Hanvey et al. 2017; Huppertsberg and Knepper 2018; Rocha-Santos and Duarte 2015; Silva et al. 2018).

Fourier-transform infrared spectroscopy (Silva et al. 2018), Raman spectroscopy (Frère et al. 2016), SEM/energy-dispersive X-ray spectroscopy (SEM/EDS) (Silva et al. 2018), environmental SEM/EDS (Rocha-Santos and Duarte 2015), and X-ray photoelectron spectroscopy (Herrera et al. 2018) are some of the spectroscopic methods. Thermal analysis techniques include DSC and TGA (Shim, Hong, and Eo 2017).

4.3.1 Challenges of Detecting Small MPs

The sensitivity of the sampling and extraction techniques can be considered as the factor that governs the detection of lower size limits of MPs (Hanvey et al. 2017; Silva et al. 2018). Lower size limits of MPs and the techniques by which they are sampled and extracted could lead to underestimation of their concentrations (Hanvey et al. 2017). The smaller the sizes (<0.1 mm) of MPs, the more tedious the analysis. Challenges that can be encountered in such cases include low percentages in confirming the particle size (Lenz et al. 2015), longer time for analysis, and poor recovery percentages (Fischer and Scholz-Böttcher 2017; Zhao et al. 2017). Natural degradative processes, such as UV-induced photodegradation, thermal degradation, and biodegradation, can significantly alter the composition of polymers, which makes the identification process increasingly difficult (Lenz et al. 2015). This problem becomes more pronounced for smaller particles, where the high surface to volume ratio makes the signals from surface material more significant (Maes et al. 2017).

4.3.2 Classification

Microplastics can be classified according to their particle shape and size. The shape varies from irregular to spherical, and they can be found in fibrous, granular, and film type forms (Herrera et al. 2018; Hidalgo-Ruz et al. 2012). They can be grouped into 10 size classes based on the longest dimension of individual MPs (Figure 4.2). A suitable analytical technique should be selected based on their size and physical appearance (Nor and Obbard 2014). The size distribution (Figure 4.2) of MPs depends on the sieve and filter pore sizes or the measurements of length that are assisted by optical microscopy (Rocha-Santos and Duarte 2015).

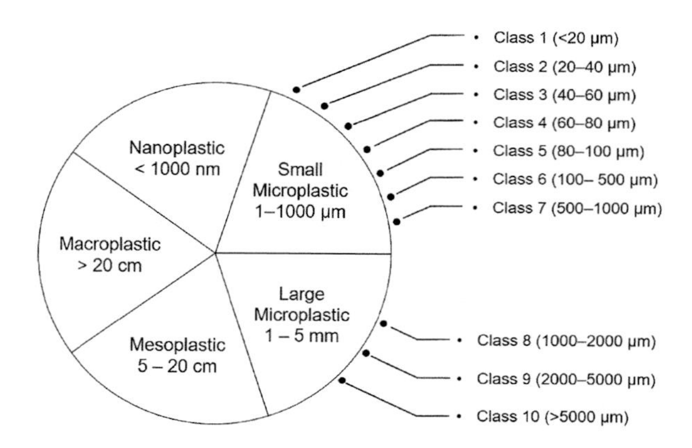

FIGURE 4.2 Plastics are categorized into different size classes and MP are further grouped into different classes. (From Hanvey, J.S. et al., *Anal. Methods*, 9, 1369–1383, 2017; Nor, N.H.M. and Obbard, J.P. *Mar. Pollut. Bull.*, 79, 278–283, 2014.)

Knowledge of plastic concentrations, spatial and temporal changes, size and polymer distribution, and fragmentation dynamics is vital for understanding the fate and impact of MPs (Lenz et al. 2015). Environmental MPs mostly result from the fragmentation of commercial polymers that exist in different morphologies (foam, sheet, fiber, and particles) (Lenz et al. 2015) and contain different compositional additives, fillers, and coloring agents (Araujo et al. 2018).

4.3.3 Visual Sorting

Pellets are commonly cylindrical, disk, flat, ovoid, spheroid, tablet-like, oblong, and disk shaped. They mostly exist as spheroid to ovoid with rounded ends (Hidalgo-Ruz et al. 2012). In fragments, rounded, sub-rounded, sub-angular, angular, irregular, elongated, degraded, rough, and broken edge shapes are observed (Hidalgo-Ruz et al. 2012). The shape of plastic fragments depends on the fragmentation process, as well as the residence time in the environment. Sharp edges might indicate either recent introduction to the sea or the recent breakup of larger pieces, while smooth edges are often associated with older fragments that have been polished continuously by other particles or sediment. Circularity varies inversely with particle size. Larger particles have more elongated shapes or irregular surfaces, while progressively smaller particles are consistently more circular. Particles continue to fragment and degrade to even smaller particles with time (Hidalgo-Ruz et al. 2012). Irregularly shaped MPs produce non-interpretable spectra due to refractive errors (Li, Liu, and Chen 2018).

Each particle can be classified based on its color as follows: light (white, yellow, and yellow brown), medium (brown, blue, green, and red), dark-hued (dark blue, dark green, dark red, and gray-black), transparent, crystalline, white, clear-white-cream, red, orange, blue, opaque, black, gray, brown, green, pink, tan, yellow, and pigmented. Low-density polyethylene (LDPE) has opaque colors, while poly(ethylene-vinyl acetate) (PEVA) is contained in clear and pseudo-transparent pellets (Hidalgo-Ruz et al. 2012). Color has also been used as an index of photodegradation and residence time on the sea surface and the degree of tarring or weathering. It has been suggested that discolored PE pellets may contain higher amounts of polychlorinated biphenyls than non-discolored pellets because the discoloration process (yellowing) is indicative of longer exposure time to seawater, which increases the chances of the polymers becoming oxidized. Black and aged pellets composed of PS and PP present the highest diversity of adsorbed pollutants for both polyaromatic hydrocarbons and polychlorinated biphenyls (Hidalgo-Ruz et al. 2012).

Polymer types retrieved in recovery experiments include PE, PP, PVC, PET, PS, expanded polystyrene, and PUR, and their colors were red, blue, orange-brown, green, yellow-white, white, and yellow, respectively (Nuelle et al. 2014). Color can facilitate separation, where MPs are scattered among large quantities of other debris. Particles with eye-catching colors have a high probability of being isolated when compared to those possessing dull colors (Hidalgo-Ruz et al. 2012).

Visual sorting is one of the most important methods used in the identification of MPs and is more suited for colored MPs in the size range of 2–5 mm (Shim, Hong, and Eo 2017), and higher success rates have been observed for blue color (86%), green (54%), and red (52%). In order for MPs to be determined visually, certain criteria should be met, which include: (1) visual absence of cellular or organic structures in the MPs, (2) uniformly thick fibers throughout their entire length, (3) presence of clear and homogeneous colors, and (4) the ability to observe the presence of transparent or white particles under high magnification. These particles can also be observed under a fluorescence microscope to confirm whether they are of organic origin (Qiu et al. 2016).

The drawbacks of visual sorting include plastic overestimation by misidentification, underestimation of concentrations because of barely visible or transparent plastic items, and the difficulties in identifying the type of plastic (Bläsing and Amelung 2018). Visual sorting sometimes requires a combination with different analytical methods (e.g., FTIR and Raman spectroscopy) for identification of MPs, and it is integrated with microscopic and spectroscopic techniques during the analysis of multiple samples (Renner, Schmidt, and Schram 2017; Song et al. 2015).

4.3.4 Microscopic Methods

Microscopy provides detailed information on surface morphology and can be used as a pre-screening technique to reduce the number of particles that needs to be analyzed by other spectroscopic methods (Wang et al. 2017). As the size of MPs increases, microscopic analysis is combined with spectroscopic or thermal analysis techniques (Shim, Hong, and Eo 2017). Occasionally, suspect particles are pre-selected by light microscopy followed by FTIR. However, transparent and small particles could be overlooked during the pre-selection process (Primpke et al. 2017). Stereo-(or dissecting) microscopy is a commonly used identification method. For fibers and dark particles, visual counting is more suitable, while for fragments and light particles, Raman spectroscopy works best (Zhao et al. 2017). Polarized light microscopy has been applied successfully to identify synthetic fibers in soil (Allen, Kalivas, and Rodriguez 1999).

Microscopy is an essential tool for measuring the physical characteristics of MPs. When only large MPs are the target, microscopy can be used on its own to analyze the physical characteristics, or together with an additional test (e.g., prodding with a needle) to identify the plastics (Shim, Hong, and Eo 2017). Optical microscopy provides information on shape, color, and morphological characteristics (Rocha-Santos and Duarte 2015; Wang et al. 2017). Wang et al. (2017) used optical microscopy as a pre-screening technique to reduce the number of particles that were needed to be analyzed by SEM. Particles such as organic, plant, or animal residues, and even some shells, were easily identified as non-plastic through optical microscopy due to their texture and color and were excluded from further analysis (Wang et al. 2017). Different microscopic techniques and their features are given in Table 4.3 (Dehaut et al. 2016; Qiu et al. 2016; Sánchez-Nieva et al. 2017).

A commonly used microscopic technique for MP determination uses a scanning electron microscope, which provides high magnification using a backscattered-electron detector (Wang et al. 2017). Vivid structural images of MPs that were less than 1 mm (Eriksen et al. 2013; Wagner et al. 2017) were effectively identified, and they were distinguished from organic impurities (Shim, Hong, and Eo 2017; Wang and Wang 2018). The technique gives comprehensive information on size, shape, and additives (Hanvey et al. 2017). It also provides information about cracks that result from weathering (mechanical or oxidative), polymer degradation, and pigment-particle additives (Wang et al. 2017). Mechanical weathering shows surface features, such as grooves and gouges, while deep cracks and fractures indicate degradation (Wang et al. 2017).

The SEM/EDS is used for chemical-composition analysis of MPs (Reddy et al. 2006; Rocha-Santos and Duarte 2015; Silva et al. 2018; Vianello et al. 2013). The elemental composition of

TABLE 4.3
Different Types of Microscopes Used to Observe MPs

Microscope	Illumination Source	Color	Resolution	Advantages/Limitations	Reference
Ordinary microscope	Light	Dark view	Poor	Lowest cost with the worst observation	Qiu et al. (2016)
Confocal microscope	Light	Dark view	Poor	Confocal images are free of defocus blur, but had to be edited concerning optical parameters, such as contrast, light, and depth of focus in order to optimize the quantification	Sánchez-Nieva et al. (2017)
Stereomicroscope/ dissecting microscope	Light	Clear view	Moderate	Stereoscopy with clear discriminability	Qiu et al. (2016)
Fluorescent microscope	Light	Bright colors	Good	Fluorescence with best observation and accurate counting	Qiu et al. (2016)
Binocular microscope	Light	Dark view	Good	Allows good observation and detection of microplastics	Dehaut et al. (2016)
Scanning electron microscope	Electron beam	Only black and white	Great	High resolution to gain the chemical and morphological characterization of MPs with highest cost	Qiu et al. (2016)

particles is useful for identifying carbon-dominant plastics from inorganic particles (titanium dioxide (TiO_2)-nanoparticles, barium, sulfur, and zinc) (Fries et al. 2013; Rocha-Santos and Duarte 2015; Shim, Hong, and Eo 2017; Wagner et al. 2017). However, EDS spectra cannot identify the elemental signatures for additives and adsorbed contaminants (Huppertsberg and Knepper 2018; Silva et al. 2018; Wang et al. 2017). The unique chlorine signals of PVC enable its easy identification using this technique (Wang et al. 2017). Other MPs, such as PP and PE, have no distinct EDS peaks other than a strong carbon peak (Wang et al. 2017).

Analysis with both SEM or SEM/EDS is time consuming (Prata et al. 2018; Silva et al. 2018), relatively expensive, tedious, and requires coatings of Pt and Au to avoid charging interferences, which may result in inaccuracies when identifying the surface texture and color of MPs (Karunanayake et al. 2018; Rocha-Santos and Duarte 2015). The colors of the particles cannot be used as identifiers in SEM analysis. Therefore, this technique is only suitable for specific plastic particles (Shim, Hong, and Eo 2017), and this limitation may result in inaccuracies (Silva et al. 2018). However, SEM/EDS combined with optical microscopy can screen rapidly the MP particles prior to FTIR or Raman spectroscopy (Wang et al. 2017), but it involves a high capital cost (Wagner et al. 2017).

Environmental scanning electron microscope/energy dispersive spectroscopic analysis is used both for characterizing the surface morphology of MPs and determining the elemental composition of polymers. It is also used to determine whether each particle is potentially plastic, based on elemental composition of MPs and the inorganic additives they contain (Rocha-Santos and Duarte 2015; Wagner et al. 2017; Wang et al. 2017; Wang and Wang 2018). It can be used to validate the main atomic composition of polymers previously identified by micro-FTIR (μ-FTIR) analysis, but it cannot be used for semi-quantitative measurements. The environmental SEM/EDS analyses were reported for material contained on filters, which revealed various polymer types (Vianello et al. 2013).

4.3.5 Fourier-Transform Infrared Spectroscopy and Raman Spectroscopy Analyses

Spectra from absorption or scattering phenomena are analyzed for the presence or absence of characteristic frequency bands that arise from excited vibrational states in the sample material. The energy difference between an excited and ground state is characteristic for a molecular structure and gives rise to a characteristic spectral signal. Micro–Fourier-transform infrared spectroscopy has been used in this context and can allow polymer identification of single particles (Lenz et al. 2015).

For transparent and white MPs with fragment morphology, FTIR is a better detection method than microscopy (Silva et al. 2018). Pre-programmed, semi-automatic mapping without the need for microscopic pre-selection of particles for analysis can be used to minimize effort in the FTIR process (Crichton et al. 2017). While mapping methods are not feasible for visualizing large surface areas, focal plane array (FPA)-based imaging is likely to improve the ability to identify plastics within sediment retentates. This technique captures spectra from the entire sample surface instead of small and spatially separated locations. When high spatial resolution is required, Raman microspectroscopy may also provide an alternative to FTIR analyses of MPs. Since measurements used for the generation of molecular micrographs consist of full-range IR spectra (4000–700 cm^{-1}), several types of plastics may be imaged within a single micrograph by selecting different combinations of absorbance band characteristics to a given type of plastic, followed by mapping or imaging (Harrison, Ojeda, and Romero-González 2012).

In Raman spectroscopy, information on surface morphology can be well recognized and detected. This method could be evaluated, because the technique is most sensitive to the surface of the particle and, as such, could provide detailed characterization of the chemistry and the stage of degradation of the tested particles. By applying the FTIR technique, the oxidation of the sample can also be detected (it is highly sensitive to the C=O bond) (Elert et al. 2017).

However, the signals from the surface of MPs cannot be well differentiated, especially when transmission FTIR is applied. Additionally, the liquid-extraction method could be applied when looking at molar mass changes in the measured sample, because it shows promising results for evaluating the degradation status of the polymer (Elert et al. 2017). A semi-automated Raman microspectroscopic method can be used additionally for surface characterization of MPs. Some advantages include fast analysis time (<3 hours), reproducibility, the requirement of minimum operator intervention, and high throughput (Frère et al. 2016).

4.3.5.1 Raman Imaging

Raman spectroscopy is a scattering method exclusive for the molecules that can undergo a change in the polarizability of a chemical bond (Zhao et al. 2017). Raman imaging is commonly used to identify the type of polymer and can determine particle sizes and distributions in MPs. Raman microspectroscopy reveals vibrational information of the molecular structure of samples; high resolutions are attained with modern equipment, with horizontal and vertical resolutions below 1 μm (Lenz et al. 2015).

The effect of Raman scattering increases with increasing excitation energy, and, therefore, short laser wavelengths can be used to improve the detection of weakly Raman active compounds. Improved detection can be achieved by using techniques, such as resonance Raman, surface enhanced Raman, coherent anti-Stokes Raman, or Fourier-transformed Raman. This increases the signal to noise (S/N) ratio, but is comparatively expensive and needs advanced equipment or complex sample preparation steps (Frère et al. 2016). The integration of more peaks, known as mapping, allows more accurate and quicker detection of different particles (Elert et al. 2017). In contrast to FTIR, a noteworthy advantage of Raman spectroscopy is the lack of limitations on the part of the thickness or shape of the sample. The complete wavelength region can be used for identification because the signal is not affected by water and atmospheric CO_2 (Elert et al. 2017). Fluorescence interferences that lead to low S/N ratios (Imhof et al. 2016) usually arise due to the presence of microbiological, organic (humic substance), or inorganic (clay minerals and unperfected crystals structures) entities. The choice of appropriate acquisition parameters (laser wavelength and power, measurement time, and photobleaching of the sample prior to measurements) could overcome this problem (Elert et al. 2017). The imaging technique would benefit from collecting spectra at each pixel and would allow surface contamination of the particle to be estimated (Elert et al. 2017).

4.3.5.2 Raman Libraries

A good Raman library includes several polymer samples with different characteristics and features, such as flexibility, color, and form, such as fibrous, solid, film, and foamed, and spectra of degraded polymer stages. To improve the exclusion of misleading non-particles, expected non-plastic contaminants and organic constituents present in the marine environment, i.e., cellulose, methylcellulose, cellulose acetate, viscose rayon, keratin, and aragonite, are included (Lenz et al. 2015). The chemical identification of particles can be performed using commercial Raman libraries (KnowItAll Informatics Systems, Bio-Rad, and Raman ID Expert) and the home database (Zhao et al. 2017). Polymers are identified by fingerprints and characteristic peaks by comparing the pyrograms obtained from samples to those from the analysis of standard polymer materials (Dekiff et al. 2014). A well established polymer spectrum library enables the confirmation of plastics and specific polymer types (Shim, Hong, and Eo 2017).

Limitations of Raman spectroscopy include: (1) destruction of small, dark, or fibrous particles, as they are prone to destruction; (2) spectra with low S/N ratios, which require lower laser power to avoid sample destruction, resulting in spectra with weak signals that, at times, are unsuitable for interpretation; and (3) Raman spectra of a particular polymer, which can be overlaid by signals (foreign bands, fluorescence, and absorbance; Lenz et al. 2015) due to additives (fillers, pigments, and dyes), resulting in a considerable modification of the spectra.

Raman measurements preferably should take place on a non-Raman active carrier, such as microscopic glass slides, to avoid background signals (Lenz et al. 2015). Laser-induced sample heating can cause background emission, occasionally followed by polymer degradation (Araujo et al. 2018). Colored particles are rather difficult to identify using Raman spectroscopy, because they have high fluorescence background emission and will, therefore, lead to misidentification (Imhof et al. 2016). In order to increase the S/N ratio, an electron-multiplying charged coupled device detector is used. Compared to conventional charged coupled device detectors, electron-multiplying charged coupled device detectors are equipped with a multiplication register that amplifies the gain up to 1000 times (Araujo et al. 2018).

4.3.5.3 Micro-Raman Spectroscopy and Other Raman Techniques

μ-Raman spectroscopy only identifies the presence of MPs with pigments because it overlaps with polymer signals (Hermabessiere et al. 2018). A chemometric approach has been applied in order to separate the spectral sources in a mixture (e.g., polymers and additives), but it is not possible to reveal the polymer spectrum by merely subtracting the pigment spectrum due to its more intense Raman signal (Renner, Schmidt, and Schram 2017).

A semi-automated Raman microspectroscopic method coupled to static image analysis allows the screening of a large quantity of MPs (Frère et al. 2016). This method is reproducible and time effective, with minimal machine operator intervention for faster and thorough morphological and chemical characterization of MPs (Frère et al. 2016). In order to minimize laser-induced damage to a sample, laser power and acquisition times can be varied in Raman spectroscopy depending on how prone the sample is to thermal damage (Wagner et al. 2017). The time needed to record the Raman image strongly depends on the integration time and the number of accumulated scans per spectrum (Elert et al. 2017).

4.3.6 Infrared Spectroscopy

Infrared spectroscopy is one of the most widely available techniques for relatively quick and reliable detection of polymers (Elert et al. 2017). There are three different operating modes available for FTIR, namely, (1) transmission, (2) reflection, and (3) attenuated total-reflectance (ATR) mode (Li, Liu, and Chen 2018). Contrary to Raman spectroscopy, which depends on a change in the polarizability of a chemical bond, IR absorption depends on the change in the permanent dipole moment of a chemical bond. Therefore, polar functional groups (e.g., carbonyl groups) are easily detectable (Elert et al. 2017). For FTIR analysis, samples must be dry to avoid interferences from water (Shim, Hong, and Eo 2017). Polar oxidative functional groups can be observed easily by IR measurements (Lenz et al. 2015).

In addition to chemical differentiation of MPs, FTIR in the transmission mode provides visual characterization of the analyzed material. Consequently, FTIR equipped with the FPA detector technique is an important tool for the detection and identification of MPs in environmental samples. Performing FTIR in the transmission mode requires additional caution because of the thickness of the measured particles (Elert et al. 2017). Furthermore, the sample must be placed on the IR transparent substrate. This is not problematic when loose samples are analyzed; but often when a water sample is to be measured, the water must be filtered by aluminum oxide filters. Aluminum oxide is water resistant and IR transparent over a wide range of IR light. However, IR permeability by applying such filters drops significantly below 1300 cm^{-1}. Special etched Si filters can extend the spectral range so that fingerprint regions (below 1300 cm^{-1}) can also be detected (Elert et al. 2017). If a fluorescent filter is installed on a FTIR microscope, spectroscopic identification can be performed for the same particles immediately after fluorescence microscopy. A combination of fluorescence microscopy using Nile Red (NR) fluorescent dye staining and FTIR confirmation would reduce the likelihood of missing MPs in the identification of field samples (Araujo et al. 2018).

4.3.6.1 Micro-FTIR Techniques

Micro-FTIR spectroscopy is a useful tool for simultaneous visualization, mapping, and spectral collection of smaller particles (Li, Liu, and Chen 2018). The abundance of MP fragments of all size classes is high in FTIR. A clear trend of increasing abundance with decreasing size is observed in FTIR (Renner, Schmidt, and Schram 2017). The MPs from each sample must be categorized by size, shape, and color. The length and mass of each particle must be recorded prior to subjecting particles to ATR-FTIR in order to determine the MP chemical composition (Wessel et al. 2016). Two feasible measuring modes in ATR-FTIR are reflectance and transmittance modes. The transmission mode gives a high-quality spectrum, but is limited to IR transparent substrates. Analysis done using the reflectance mode is limited to thick samples with a regular shape. If otherwise, the signal will be disturbed or distorted by reflection errors caused by light scattering (Li, Liu, and Chen 2018). The μ-FTIR method proves to be time consuming when finding suitable MP particles for analysis. Refractive errors that arise when analyzing irregularly shaped MPs in the reflectance mode show spectral distortion. Infrared transparent filters would be required; therefore, the transmittance mode is limited by thick MPs due to the total absorption patterns (Qiu et al. 2016).

The ATR-FTIR technique is an extremely sensitive method (Primpke et al. 2017) that improves the information on irregular MPs, which, in contrast to transmission FTIR, is also applicable to thick and opaque samples (Dehaut et al. 2016). In contrast to the transmission mode, the reflectance and ATR modes do not require the sample preparation step (Shim, Hong, and Eo 2017). The ATR and FTIR measurement is a form of surface contact analysis (Shim, Hong, and Eo 2017). When the ATR contacts the surface of the sample, a beam of IR light is reflected at least once by passing through the ATR crystal. The relatively high pressure produced by the ATR probe may damage fragile MPs, and tiny plastic particles can be pulled from the filter paper by adhesion to, or electrostatic interaction with, the probe tip.

An ATR probe made of germanium is prone to damage in contact analysis of hard and sharp inorganic particle remnants on a filter paper from sand samples (Araujo et al. 2018). The μ-FTIR mapping, with the sequential measurement of IR spectra at spatially separated and manually selected points on the sample surface, can only analyze small areas of the filter paper, and this process is very time consuming. The screening and analysis for the whole sample filter paper may become impossible with this technique (Li, Liu, and Chen 2018). The commercial availability of software programs using μ-FTIR systems for MP analysis is relatively limited. For analysis, the μ-FTIR data are usually transformed to false color images on the basis of integration on specific spectral regions (Primpke et al. 2017). Using μ-ATR-FTIR, pseudo-plastic particles can be identified by microscopy, and afterward, chemical confirmation can be carried out by spectroscopy. However, MPs with a maximum length <50 μm are needed. A trial and error approach is usually required to obtain a clear spectra that enables accurate identification (Shim, Hong, and Eo 2017). The many advantages of μ-ATR-FTIR spectroscopy are ease of handling, short analytical time, and analysis of a higher number of polymers (Shim, Hong, and Eo 2017).

Fourier-transform infrared spectroscopy enables a faster screening of larger surfaces using an FPA detector. However, this method is limited to particles with a minimum size of ~20 mm. When compared to manual analysis, a seven-fold increase in the number of polymer particles was found with the automated analysis (Silva et al. 2018). A FPA-based FTIR imaging with several detectors placed in a grid pattern was applied for MP analysis, and it allowed the detailed and high-throughput screening of total MPs on the whole filter paper. It can simultaneously record several thousand spectra in a targeted area within a single measurement run, generating chemical images for the whole filter paper (Li, Liu, and Chen 2018). Fourier-transform IR imaging with FPA is not necessarily more time efficient than Raman mapping (Araujo et al. 2018). Even by using FPA-based μ-FTIR, the determination and quantification of MPs is still a time consuming and tedious process, because several manual data management and analyses steps are required (Primpke et al. 2017). But FPA-based FTIR reduces the analysis time significantly (Silva et al. 2018). Although FPA-based μ-FTIR generally improves the accuracy of MP analyses, the final analysis is still affected by human bias (Primpke et al. 2017).

4.3.6.2 Near-Infrared Reflectance Spectroscopy

The use of near-infrared reflectance spectroscopy has received attention for plastic identification (Allen, Kalivas, and Rodriguez, 1999). It gives rapid measurement of chemical composition of MPs without requiring any chemical pre-treatment. However, Raman spectroscopy provides improved results for identification of high-density polyethylene (HDPE) and LDPE plastics compared to near-infrared reflectance spectroscopy (Allen, Kalivas, and Rodriguez, 1999). The Pyr-GC-MS and μ-Raman spectrometric methods allowed the identification of the native molecule for PE and PS, but it failed to differentiate between the subtype of polymers, i.e. LDPE vs. HDPE or PS vs. cross-linked polystyrene (Dehaut et al. 2016).

4.3.7 Atomic Force Microscopy

Atomic force microscopy (AFM) used in combination with either IR or Raman spectroscopy is a potential candidate for nano- or sub-microplastic analysis and determination of the composition of the plastic (Shim, Hong, and Eo 2017). The AFM probes can be used in both contact and non-contact modes with objects. Raman analysis is a combination of two instruments using the simultaneous or independent scanning of the same object, while AFM-IR is the actual merging and integration of two instruments to a single instrument (Shim, Hong, and Eo 2017).

4.3.8 Chromatographic, Mass Spectrometric, and Thermometric Techniques

Several chromatographic, mass spectrometric, and thermometric techniques continue to be updated as identification methods for MPs.

4.3.8.1 Chromatography

Liquid chromatography is a method that enables the detection of only PS and PET particles. For PP and PE, high-temperature SEC needs to be applied. However, the main limitation of the liquid extraction technique is the setup of the equipment used. By using specific calibrations, it is possible to quantify the polymers and mass concentrations that are found in the samples. Spectroscopic methods, as alternatives, could be used for numerical quantification (particle count) (Elert et al. 2017).

Size exclusion chromatography gives information on the relative molar mass of the existing polymer content, and, when coupled with FTIR, leads to the identification of chemical heterogeneity. High-performance liquid chromatography (HPLC) methods for further separation of different polymers are yet to be developed. When HPLC is coupled to SEC in 2D, it would allow the estimation of molecular weights of different polymers.

Furthermore, for receiving information on chemical heterogeneity, HPLC fractions should be collected and measured using other techniques, such as FTIR, MS, and nuclear magnetic resonance (NMR). By applying HPLC, quantitative results also could be obtained (Elert et al. 2017). Laser diffraction can be used as a method to confirm the median diameter obtained by optical analysis for PS MPs (Rocha-Santos and Duarte 2015).

4.3.8.2 Time-of-Flight–Secondary Ion Mass Spectrometry

Tof-SIMS is commonly employed to characterize fragments emerging from plastics for composition and size analysis (Jungnickel et al. 2016; Schirinzi et al. 2016). The mass spectrum is produced by summing the detected secondary ion intensities and graphing them against the mass channels (Jungnickel et al. 2016; Schirinzi et al. 2016). Time-of-flight–SIMS imaging can be used without initial cleanup steps (Jungnickel et al. 2016), and it provides a detailed mass spectral pattern of the polymer for a specific ion fragment of the particle surface. By utilizing the recent generation of SIMS instruments equipped with the latest ion cluster sources, it is possible to use them even in the presence of SiO_2 interferences (Jungnickel et al. 2016).

4.3.8.3 Matrix-Assisted Laser Desorption Ionization–Time-of-Flight–Mass Spectrometry

The MALDI-Tof-MS spectra can provide detailed morphological information (size and shape) on imaging. It can be used to examine modifications on polymer surfaces during the degradative process. In this method, samples are placed on indium-tin oxide glass slides, and the matrix is sublimated and deposited onto the target through a special coating (Huppertsberg and Knepper 2018). A limitation of MALDI-ToF-MS is that different polymers require different cationizing agents for their ionization. The potential to identify different polymer materials with a high degree of selectivity makes the application of MALDI-ToF-MS analysis interesting. Possible applications include the direct identification of additives or sorbed pollutants on particles (Huppertsberg and Knepper 2018). The use of MALDI-ToF-MS and TGA-DSC require further research and development in order to realize their full potentials (Huppertsberg and Knepper 2018). Mass spectrometry-based methods are not constrained by particle sizes and may, therefore, be more versatile and applicable to plastic particle determinations. Spectroscopic methods provide more accurate information for large particles (Huppertsberg and Knepper 2018).

4.3.8.4 Pyro–Gas Chromatographic–Mass Spectrometry

Pyro–gas chromatography–MS is a widely used method for the identification of single particles by analyzing their thermal degradation products (Bläsing and Amelung 2018; Ceccarini et al. 2018; Silva et al. 2018). It is a destructive method (Qiu et al. 2016). In addition to the polymer types, organic plastic additives, which cannot be dissolved, extracted, or hydrolyzed easily could be analyzed by applying sequential Pyr-GC-MS (Qiu et al. 2016). This method does not provide information regarding size and morphology (Silva et al. 2018), and it cannot be carried out for MPs below 50 μm (Hermabessiere et al. 2018). The GC-MS methodology is a time-consuming process, and the pre-selection of a single particle is a necessary step. It has a lower size limitation than FTIR or Raman spectroscopy (Bläsing and Amelung 2018). The Pyr-GC-MS method is applied to simultaneously identify polymer types of MP particles and associated organic plastic additives that are found in MPs (Fries et al. 2013). It is usually preferred over methods based on solvent extraction that use large amounts of solvents and which also lead to background effects (Dekiff et al. 2014). The Pyr-GC-MS method relies more on particle weight than on size (Hermabessiere et al. 2018). The optimized Pyr-GC-MS method requires a pyrolysis temperature of 700°C with a split ratio of 5 and an injector temperature of 300°C (A split ratio is a term used in gas chromatography and determines the amount of a sample entering the column.).

The main advantage of using sequential Pyr-GC-MS over commonly applied FTIR spectroscopy is that polymer types and organic plastic additives can be analyzed in a single run. Such a sequential procedure can be a useful tool for extracting organic plastic additives at lower temperatures before determining the pyrolysis products of polymers in a single run. The disadvantage of Pyr-GS-MS is the manual placement of the particle in the instrument, which can result in size limitations and only one particle can be run per sample (Lusher et al. 2017). The technique of Pyr-GC-MS is not adequate for use in analysis of MPs in the environment (Fries et al. 2013). The maximum size of plastic particles that can be analyzed using the Pyr-GC-MS method is governed by the diameter of the thermal desorption tubes. Larger particles can be cut into smaller pieces to overcome this size limitation (Fries et al. 2013).

The method allows the full identification of pigments and certain fibers. It could be combined with improved separation methods to retrieve smaller particles (Hermabessiere et al. 2018). The Pyr-GC-MS method could be used as a complementary identification method followed by μ-Raman spectroscopy (Hermabessiere et al. 2018).

The use of Pyr-GC-MS by itself does not permit the determination of the number, type, or morphology of MPs, because it only provides the mass of the polymer per sample, and, therefore, it requires pre-selection of MPs by optical techniques (Huppertsberg and Knepper 2018). The Pyr-GC-MS is used to obtain structural information of macromolecules by studying their thermal degradation products. Sequential pyrolysis is performed by analyzing identical samples under varied conditions (e.g., by increasing the pyrolysis temperature) (Fries et al. 2013).

4.3.8.5 Thermal Extraction and Desorption–Gas Chromatography–Mass Spectrometry

The TED-GC-MS method has a limited scope when detecting MPs, because it only allows for the identification of PE, PP, and PS in complex environmental matrices (sample matrices like soil), however, it requires no time-consuming pre-selection of particles (Bläsing and Amelung 2018). It fails to provide an exact quantification of PE in standardized laboratory experiments (Bläsing and Amelung 2018). This method is not constrained to size limits, but has limits of quantification (Bläsing and Amelung 2018). However, the TED-GC-MS method produces valuable information rapidly on polymer mass fraction in environmental samples, and it does not require special pre-treatment. However, direct information on particle sizes and their distribution cannot be obtained unless an additional pre-sieving or pre-filtration is done (Elert et al. 2017). Single particles and bulk samples can also be analyzed with TED-GC-MS, which provides summed MP concentration data on a weight basis (w/w) (Shim, Hong, and Eo 2017).

Larger sample sizes can be used in the TED-GC-MS method compared to the Pyr-GC-MS method. Samples can range from 0.5 to 100 mg. Increased sensitivities are obtained for larger sample sizes (Huppertsberg and Knepper 2018). The TED-GC-MS method can analyze simultaneously chemicals added to MPs. However, when bulk analysis is done, the method fails to provide information related to the number, size, and shape of analyzed MPs. The method is time consuming and requires a well trained and experienced operator (Shim, Hong, and Eo 2017).

4.3.8.6 Thermal Desorption–Gas Chromatography–Mass Spectrometry

TED-GC-MS is an advantageous identification method, which can be employed for both qualitative and quantitative determinations. Sample preparation is not required, and it is a fast method. However, the sample is destroyed, resulting in loss of crucial particle size information (Peez, Janiska, and Imhof 2019). The TED-GC-MS method has been able to provide identification for four polymers in a single measurement (Elert et al. 2017).

4.3.8.7 Thermo-Gravimetric–Mass Spectrometry

In TGA-MS, samples are initially thermally degraded, and the resultant products are subsequently sent to a mass spectrometer for analysis. The collected data are compared with reference data to obtain sample information, such as identity and concentration. The TGA-MS method can only handle samples less than 500 μm. The method is usually not used for mixtures with high concentrations of impurities (Li, Liu, and Chen 2018). The TGA method is also combined with TGA-solid-phase extraction and TDS-GC-MS. That is, thermo-gravimetric analysis is often combined with solid-phase extraction, and subsequently with TDS-GC-MS, which enables the use of larger sample sizes when compared to Pyr-GC-MS and provides better resolution versus GC-MS and DSC (Shim, Hong, and Eo 2017).

4.3.8.8 Differential Scanning Calorimetry

The DSC and TGA-DSC techniques are based on the principle that elevated temperatures can cause a phase transition in a polymer sample. Differential scanning calorimetry can be used in combination with TGA to analyze PE and PP, but it fails to identify PVC, polyamide, polyester, PET, and PUR due to their overlapping phase transition signals (Shim, Hong, and Eo 2017). Preparation of TGA-DSC samples involves filtration and separation from sediments with an oxidative treatment afterward. Sample preparation for Pyr-GC-MS samples only requires grinding via cryomilling, which results in representing the samples better. Identification is based on GC-MS analysis of products after pyrolysis, and it is done under oxygen-free conditions. Elucidation of specific polymer pyrolyzation products needs to be performed prior to analysis. This allows simultaneous determination of sorbed substances if their markers are known. Instrumental limitations, such as blockages of capillaries by products of pyrolysis, can be avoided through addition of polymer tweezers that sorb emerging pyrolysis products. Upon complete pyrolyzation, the tweezers are transferred into the GC inlet and the sorbates are thermally desorbed (Huppertsberg and Knepper 2018).

4.3.9 Other Techniques

4.3.9.1 Nuclear Magnetic Resonance Spectroscopy

NMR allows the molecular structure of a material to be analyzed by observing and measuring the interaction of nuclear spins when placed in a powerful magnetic field. Nuclear magnetic resonance is a simple identification method that offers a big advantage compared to other detection methods, such as Pyr-GC-MS, FTIR, or Raman spectroscopy. When considering the instrument, detection limits of the sample are not dependent on the particle size, because any sample that can dissolve in a solvent can be analyzed easily by the NMR. The application of NMR techniques in environmental samples can be challenging, as additional procedures are needed for sample preparation including digestion of the biological matrix, as well as the separation of MPs from additional constituents of the sample (Peez, Janiska, and Imhof 2019).

4.3.9.2 X-ray Photoelectron Spectroscopy

Surface-sensitive X-ray photoelectron spectroscopy is a useful tool in elemental identification (Hernandez, Yousefi, and Tufenkji 2017). In X-ray photoelectron spectroscopy, the instrument probes the elemental composition and chemical oxidation states at the surface and near the surface (~10–100 Å). Low-resolution, wide-scan survey spectra depict the surface elemental composition, and high-resolution spectra provide information on functional groups and their quantities.

4.3.9.3 Miscellaneous Techniques

Curie-point Pyr-GC-MS and thermo-chemolysis are methods where molecules produced by the anaerobic decomposition of organic matter are separated by GC and subsequently detected by mass spectrometry. A method based on this technique, which allows the simultaneous identification and quantification of six types of common polymers (PE, PP, PS, PET, PVC, and poly(methyl methacrylate)) has been reported (Silva et al. 2018). Techniques, such as inductively coupled plasma-MS, GC-ion trap-MS, LC-MS, and X-ray fluorescence, are used for analysis of chemicals accumulated in MPs as well (Silva et al. 2018).

Unidentified particles could be MPs whose spectra failed to match pure materials due to degradation of the constituent polymers, which highlights the need to develop specific dedicated libraries of spectra of such materials by subjecting the particles to degradation (Huppertsberg and Knepper 2018). Degradation and erosion of the particle surface are caused by biological breakdown, photodegradation, chemical weathering, and physical forces, such as wave action, wind, or sandblasting. These processes can result in visible cracks on the plastic surface, producing a wide variety of different particle shapes. Images from SEM revealed that angular and sub-angular particles produced after weathering featured conchoidal fractures, while rounded particles displayed linear fractures and adhering particles, which are attributes useful for characterizing different MP particles (Hidalgo-Ruz et al. 2012). Several surface scratches on predominantly eroded angular plastic fragments (<1 cm^2) may be caused by continuous particle–particle collision (Hidalgo-Ruz et al. 2012).

Contrary to spectroscopic methods, thermo-analytical methods are destructive, only capable of chemical characterization, and are unable to determine morphological properties, such as particle size and size distributions (Huppertsberg and Knepper 2018). Automated particle tracking and image analysis is another possible analytical combination that may be applicable for identification of MPs (Shim, Hong, and Eo 2017). Although the signal-to-noise ratios of the spectra improved with increasing aperture size, large apertures may limit the detection of small plastic particles in heterogeneous matrices. Small aperture sizes of automated particle tracking and image analysis give rise to high spatial resolution and are likely to aid molecular mapping of MPs, favoring the use of small apertures during spectral collection (Harrison, Ojeda, and Romero-González 2012).

Although there are analytical methods like FTIR and Raman spectroscopy, there is no gold standard method that is capable of characterizing and quantifying the chemical compounds derived from MPs in real environmental samples and, consequently, to assess their concentration (Rocha-Santos and Duarte 2015). Only a handful of studies report the identification of particles with sizes below 500 µm and even fewer below 50 µm (Imhof et al. 2016). Furthermore, there is also a need for future studies to evaluate rates of degradation of different types of MPs and the related leachability of pollutants and the fate and effects of MP-related compounds (e.g., phthalates and bisphenol A) in marine organisms (Rocha-Santos and Duarte 2015). Organic plastic additives in polymers have been analyzed using extraction techniques, such as supercritical fluid extraction and Soxhlet extraction (Fries et al. 2013). Thermo-analytical techniques have been used to analyze additives contained in polymers that are insoluble and those that are difficult to extract or hydrolyze easily (Fries et al. 2013).

4.4 QUANTIFICATION

Quantification of MPs can be carried out by analytical methods, such as manual counting under a microscope (Li, Liu, and Chen 2018), TED-GC-MS (Elert et al. 2017), SEC (Elert et al. 2017), reversed phase-LC (Elert et al. 2017), high temperature gel permeation chromatography (Elert et al. 2017), dynamic light scattering (Islam et al. 2019), Py-GC-MS (Huppertsberg and Knepper 2018), TGA-DSC (Huppertsberg and Knepper 2018), PFE (Camel 2001), qualitative NMR spectroscopy (Peez, Janiska, and Imhof 2019), TDS-GC-MS (Huppertsberg and Knepper 2018), and fluorescence tagging (Maes et al. 2017). An appropriate method must be chosen based on whether quantification of MPs is needed or a rough estimate of the extent of contamination by plastics is required. A combination of methods can be employed to obtain information about MPs in environmental samples (Elert et al. 2017). Quantification of polymer materials is determined by comparison of peak areas with isotope-labeled internal standards (markers). There is a demand for databases for polymer and additive markers. Each marker is required to be validated in every examined matrix to exclude interferences caused by matrix compounds (Huppertsberg and Knepper 2018).

4.4.1 Microscopic Techniques

The two main techniques used for quantification are: (1) stereoscopic microscopy and visual sorting and (2) confocal microscopy and image processing and counting (Sánchez-Nieva et al. 2017).

4.4.1.1 Stereoscopic Microscopy

Manual counting under a stereomicroscope is relatively easy and, therefore, widely used if the following criteria has been met by samples of MPs or fibers: (1) free from biological organisms; (2) has a three-dimensional structure, (3) homogeneously colored, and (4) transparent or whitish in color. The sample can be studied under high magnification with the aid of fluorescence microscopy (Li, Liu, and Chen 2018).

4.4.1.2 Confocal Microscopy and Image Processing Counting

Confocal microscopy (direct observation) and image processing and counting are two techniques that make use of object detection algorithms, which not only allow quantification of plastic particles, but also allow classifying them into size groups. Magnified images of different sections of the filter can be recorded until the entire filter area is covered, which thereby allows the detection of particles down to a few micrometers if combined with FTIR and Raman spectroscopy (Sánchez-Nieva et al. 2017). Special attention must be paid in controlling the illumination during the image capture by the confocal microscope. The recorded images have to be edited with concern about optical parameters, such as contrast, light, and depth of focus in order to optimize quantification (Sánchez-Nieva et al. 2017).

Particularly for the smaller MPs, a slightly higher count has been observed by image processing compared to direct counting (Sánchez-Nieva et al. 2017). Results obtained by visual sorting depend strongly on: (a) the user, (b) the technical characteristics of the microscope, and (c) the sample matrix. Therefore, it is strongly suggested to use image processing when counting MPs in order to avoid any bias occurring during visual counting, because counts based on image processing are less time consuming (Sánchez-Nieva et al. 2017). In addition, visual counting suffers from the drawback of size limitation due to the resolution of the microscopes (Li, Liu, and Chen 2018).

4.4.2 Thermal Extraction Desorption–Gas Chromatography–Mass Spectrometry and Size Exclusion Chromatography

As noted before, the analytical techniques of TED-GC-MS and SEC permit the testing of larger sample amounts compared to Pyr-GC-MS. Optical methods could be applied for more detailed characterization of the collected MPs (Elert et al. 2017). The aforementioned two methods include a calibration step and allow the quantification of the polymers, but do not deliver direct information regarding the size (Elert et al. 2017). For qualitative determination, no sample preparation is needed, and both methods produce high recovery percentages. However, the sample is destroyed, and this results in loss of particle size information (Huppertsberg and Knepper 2018; Peez, Janiska, and Imhof 2019).

4.4.3 High-Temperature–Gel Permeation–Chromatography

High temperature gel permeation chromatography is a technique to determine the amount of polyolefin MPs. In order to enable the analysis of other types of polymers with higher densities than polyolefins, the extraction procedure needs to be modified based on the salt solution used for the density separation (Hintersteiner, Himmelsbach, and Buchberger 2015). This includes a fast sample cleanup involving density separation (twice) and subsequent quantitation (Hintersteiner, Himmelsbach, and Buchberger 2015). This method minimizes the errors caused by weighing after insufficient separation of plastics and matrix. Therefore, time-consuming visual sorting is reduced (Hintersteiner, Himmelsbach, and Buchberger 2015). High-temperature gel permeation chromatography does not require any visual sorting, and it gives detailed characterization of the particles' size and shape and polymer distribution, as well as a qualitative analysis regarding stabilizers (Hintersteiner, Himmelsbach, and Buchberger 2015). This method is coupled with infrared detection in order to obtain accurate results for the determination of MPs. By calibration of the IR detector with a known amount of PE, the determination of the amount of PE in a sample using the peak area is possible (Hintersteiner, Himmelsbach, and Buchberger 2015); however, particle sizes cannot be directly determined (Elert et al. 2017).

4.4.4 Reversed-Phase Liquid Chromatography

In reversed-phase liquid chromatography, the process of quantification is based on HPLC analysis (Elert et al. 2017; Li, Liu, and Chen 2018). Liquid chromatography cannot be used to determine the physical characteristics, such as size, but it can be used only for limited polymer types, such as PS and PET. Additionally, only a few samples can be assessed per run (Li, Liu, and Chen 2018).

4.4.5 Quantitative Nuclear Magnetic Resonance

Quantitative determination of MPs by quantitative nuclear magnetic resonance is centered on the proportional relationship of integrated signal area and number of resonant nuclei. The method is cost effective, rapid (about 1 minute per measurement), non-destructive, size-independent, and simple, and it can be described as a quantification method with a high quantitative accuracy (>98%) (Peez, Janiska, and Imhof 2019). Table 4.4 shows several polymer types and their respective ^{1}H NMR ranges. A drawback of quantitative ^{1}H NMR spectroscopy is the dissolution of the analytes in deuterated solvents, which leads to a loss of size information of MPs (Peez, Janiska, and Imhof 2019). This technique is limited to solution NMR only, as the solid state is not sensitive enough for accurate quantification (Peez, Janiska, and Imhof 2019). Therefore, it is important to find appropriate conditions for the analysis of MP particles by means of qualitative nuclear magnetic resonance in order to dissolve different types of polymers (PE, PP, PET, PS, and PVC). This is challenging due to the inherent chemical and physicochemical properties of polymers compared with other organic substances (Peez, Janiska, and Imhof 2019). The solvents used and their residual proton signals have the possibility of overlapping with polymer signals, and measurement temperatures can be above the manageable range for quantitative analysis. Sometimes extreme temperatures (>100°C) are applied, which probably leads to strong peak broadening that makes quantitative determination difficult. Therefore, the main task is to find solvents and to validate the calibration curve (Peez, Janiska, and Imhof 2019).

TABLE 4.4
Comparison of the MP Polymer Types

MP	^{1}H NMR Signals in ppm	Commonly Used Solvents	Reference
LDPE	1.29–1.33; 0.89–0.93	ODCB-d^4/TCB-d^3 3:1 (130°C), Toluene-d^8 (60°C)	Peez, Janiska, and Imhof (2019), Pauli, Jaki, and Lankin (2005), Brandolini and Hills (2000)
HDPE	0.84–1.87	ODCB-d^4/TCB-d^3 (130°C)	Brandolini and Hills (2000)
PET	7.6–8.1; 4.11–4.77	$CDCl_3$/TFA 8:1 (25°C), $CDCl_3$/TFA 4:1 (25°C)	Peez, Janiska, and Imhof (2019, de Ilarduya, and Muñoz-Guerra (2014); Barding, Salditos, and Larive (2012)
PS	1.51–2.3; 6.56–7.5	$CDCl_3$ (25°C–50°C),	Peez, Janiska, and Imhof (2019), Pauli, Jaki, and Lankin (2005), Brandolini and Hills (2000)

ODCB, 1,2-dichlorobenzene; TCB, trichlorobenzene.
The ^{1}H signals in ppm are indicated and the associated solvents and measurement temperatures (T) according to the literature (Peez, Janiska, and Imhof 2019) are given.

4.4.5.1 Detection Limits

Hydrogen NMR is a method that has a major advantage compared to other detection methods, such as pyrolysis GC-MS, FTIR, or Raman spectroscopy. The detection limit does not depend on the size of the MPs, and there is no upper or lower limit, because the MP particles are dissolved. However, application of the method for environmental samples needs additional experiments. The main problems are the digestion of the biological matrix and the separation of MPs from additional constituents of the sample (Peez, Janiska, and Imhof 2019). The analytical techniques currently in use for MP detection and their detection limits are shown in Figure 4.3.

4.4.6 Fluorescence Staining

The method of fluorescence staining using NR has proven to be highly effective in the quantification of small PE, PP, PS, PC, PUR, and PEVA, but not for PVC, polyamide, polyester, and nylon 6 (Greven et al. 2016; Shim et al. 2016). Nile red is a lipid soluble fluorescent dye that permits the in situ staining of lipids (Maes et al. 2017). Lipophilic dyes can be employed to visualize MPs under

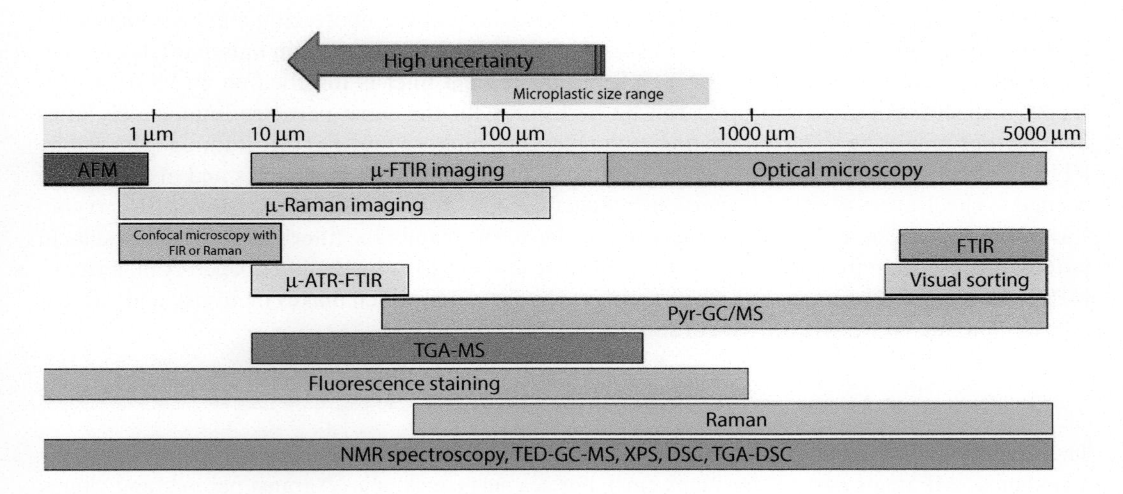

FIGURE 4.3 Detection limits of various MPs based on their size for different analytical techniques.

a fluorescence microscope (Shim et al. 2016). The dye adsorbs onto plastic surfaces and leads to their fluorescence when irradiated with blue light. Fluorescence emission is detected by means of simple photography through an orange filter, and image analysis allows fluorescent particles to be identified and counted.

The solvatochromic nature of NR offers the advantage of plastic categorization based on surface polarity characteristics of identified particles. This behavior might permit MPs to be categorized into types based on their general hydrophobicity (e.g., polyolefin, polyaromatic, polar polyesters/nylons), or it could provide useful information to evaluate the residence time through temporal changes in surface properties due to oxidation or biofouling in the environment. Microplastics of a range of sizes (e.g., small size ranges [<1 mm]) with low densities could be detected and counted in marine sediment samples. The method of fluorescence staining has identified the same particles as those found by scanning a filter area with IR-microscopy (Maes et al. 2017).

This method presents a way to detect small MPs and is an alternative to human visual sorting; it is a sensitive and semi-automated procedure (Erni-Cassola et al. 2017). The main limitation is that NR can also stain natural organic matter and, therefore, pre-purification is required. The staining method cannot be used alone without removing the interfering organic matter (Li, Liu, and Chen 2018). This method is sensitive, cheap, uses readily available equipment, and can be semi-automated for high throughput sample analysis. This method requires a sample purification step, fluorescence microscopy (green fluorescence protein settings), and free image analysis software to quantify MPs (Erni-Cassola et al. 2017). The NR staining method does not provide information about the chemical identity of the detected plastic particles (as in FTIR and Raman spectroscopy), but it does enable quantifying and measuring small particles in environmental samples (Erni-Cassola et al. 2017). This fluorescence staining method, in combination with density separation, provides a simple and sensitive approach for detecting the most common polymer fragments in marine sediments (Maes et al. 2017). Due to the relatively polar nature of NR molecules when compared to plastics, the partitioning of NR molecules from the carrier solvent to plastics can be made more feasible in non-polar carrier solvents, such as n-hexane than in polar solvents (Shim et al. 2016).

Using the method, a simple fluorescence index can be calculated as (R + G)/R ("R" and "G" are the 8-bit color intensity values of red and green, respectively). This equation is normalized, and the overall intensity of the fluorescence and maximized differences in color are determined. The equation produces a single value that can be used to represent the "polarity" of the polymer surface (Maes et al. 2017). The minute amounts of NR adsorbed on the particles have not proven to interfere with IR or Raman spectroscopy so far (Maes et al. 2017).

Micro-Raman spectroscopy is used to verify the identity of the fluorescing and non-fluorescing particles found on the filters in order to assess the specificity of NR to stain only particles of synthetic origin (Erni-Cassola et al. 2017). When a fluorescent filter is installed on an FTIR microscope, a spectroscopic identification can be performed for the same particles immediately after fluorescence microscopy. A combination of fluorescence microscopy after NR staining followed by FTIR confirmation would minimize the likelihood of missing MPs in samples and also the time needed to check every plastic-like particle by spectroscopy (Shim et al. 2016). Organic NR-carrying solvents may cause de-coloration or physical damage of the membrane filter papers, which results in failure of subsequent microscopic and FTIR identification of MPs on the filter paper. A majority of MPs from water and sediment are transparent or white in color, which makes their visual identification challenging on white filter paper (Shim et al. 2016).

4.4.7 Reporting Units

The most commonly used units for mass in sediment samples are "grams of MP per m^2" and for abundance "MP items per m^2" (or items cm^2). For sea surface samples, "grams per m^2" and "items per m^2" are the most commonly used values for mass and abundance, respectively. However, a considerable number of studies also have reported "items per volume" (items m^3). Only one study for water column samples quantified mass values in "milligrams per m^3;" abundance usually was reported as "items per m^3" (Hidalgo-Ruz et al. 2012).

4.5 CONCLUSIONS

Microplastic analysis is a rapidly growing field. In the process, methodologies and techniques have evolved to allow classification, identification, and quantification. However, there are numerous challenges to be addressed with respect to sampling, identification, and quantification in different environmental matrices. Also, there is a high demand for simple, cost-effective, comparable, and robust methods.

The concentrations of MPs simulated in most experiments are several orders of magnitude higher than current environmental concentrations, and experiments are short term. In addition, the MPs possess uniformly shaped particles in experiments. Microplastics are not prone to degradation within a short time frame. Testing at high, environmentally unrealistic concentrations and using uniformly shaped microplastics do not provide any information on the current adverse effects or risks to marine ecosystems. More specifically, long-term exposure assessments of environmentally relevant concentrations of naturally occurring assemblages of microplastics (different sizes, shapes, and types) are recommended. Also, standard protocols or methods for MP analyses, including thresholds for the World Health Organization or Environmental Protection Agency regulatory guidelines are lacking. Standards, certified reference materials, and internal standards are also unavailable to validate analytical methodologies, sensitivity, recovery, and interferences. The National Institute of Standards and Technology and other standard developing research institutes should be encouraged to develop measurement standards that can fill these gaps. Few polymeric materials are being studied, and they do not provide a representation of the whole family of MPs that exist in the environment. Such standards also will enable the expansion of current MP reference libraries required for some analytical techniques.

There is also a need to develop efficient and detailed sampling strategies, because sampling is crucial to obtain a representative sample containing plastic particles. In general, depending on the research question addressed, sampling strategies will differ. The complete set of sampling details, sampling depth, sediment weight or volume, sediment density, and water content must be taken into consideration. Extraction techniques, with respect to their efficiency and cost dimension need to be categorized. A standard extraction or separation technique, especially for regular monitoring purposes, should be implemented.

A multitude of MP analytical techniques are in use today, including naked-eye detection and microscopy. Spectroscopic methods used with microscopy are time consuming and cumbersome because of sample preparation difficulties and extended measurement times. Thermo-mass spectrometric analytical methods are rapid, but they suffer from not being able to determine the size distribution in samples. Some powerful techniques also have particle and size limitations. However, most of the described methods in this chapter are applicable for MP analysis, depending on the defined analytical question. Promising methods for MP detection in environmental samples have benefits and limitations. For routine analysis, spectroscopic (Raman and FTIR) and thermo-analytical (Pyr-GC-MS and TED-GC-MS) are the ideal, whereas techniques, such as MALDI-ToF-MS and TGA-DSC need further development to be considered for such use. Some less studied techniques, such as AFM, should be explored further to push down the size limitations. Mass spectrometry-based methods are not limited to particle sizes, and they are more versatile and applicable to plastic particle-size determinations. Spectroscopic methods are suitable for large particles, and they will find applications for particle analysis at micro- and macro-levels. Furthermore, methods such as SEM, fluorescence microscopy, and dynamic light scattering can be applied, even though they have special experimental requirements. Protocols for simultaneous multi-MP analysis are still lacking.

Since MPs occur in different sizes, there is no definite size limit for their classification or its nomenclature. There should be definite lower and upper size limits defined for MPs. Particles in nm ranges or sub-μm ranges have received little attention in studies. Secondary pollutants originating from MP degradation, including additives, such as coloring agents, anti-dispersing agents, and any plastic-associated chemicals, must be taken into consideration. The wavelength of electromagnetic radiation is a limiting parameter for nano-sized plastic analysis, which is the reason that small MPs are not detectable by all single-particle analytical techniques.

Ultimately, these considerations could lead to implementation of standardized methodologies for sampling, identification, and quantification of MPs in the environment. Only then can collected data allow a thorough assessment of the potential eco-toxicological effects of these materials. The information will contribute toward bridging knowledge gaps and toward seeing the possibility of developing miniaturized, portable, and less sophisticated instrumentation and sensors for regular environmental monitoring.

REFERENCES

Allen, Valerie, John H. Kalivas, and Rene G. Rodriguez. 1999. Post-consumer plastic identification using Raman spectroscopy. *Applied Spectroscopy* 53 (6):672–681.

Araujo, Catarina F., Mariela M. Nolasco, Antonio M.P. Ribeiro, and Paulo J.A. Ribeiro-Claro. 2018. Identification of microplastics using Raman spectroscopy: Latest developments and future prospects. *Water Research* 142:426–440.

Babu, D., Saradhi, K. Ganesh Babu, and T.H. Wee. 2005. Properties of lightweight expanded polystyrene aggregate concretes containing fly ash. *Cement and Concrete Research* 35 (6):1218–1223.

Barding, Gregory A., Ryan Salditos, and Cynthia K. Larive. 2012. Quantitative NMR for bioanalysis and metabolomics. *Analytical and Bioanalytical Chemistry* 404 (4):1165–1179.

Bläsing, Melanie, and Wulf Amelung. 2018. Plastics in soil: Analytical methods and possible sources. *Science of the Total Environment* 612:422–435.

Brandolini, Anita J., and Deborah D. Hills. 2000. *NMR Spectra of Polymers and Polymer Additives*. New York: CRC Press.

Camel, Valérie. 2001. Recent extraction techniques for solid matrices—Supercritical fluid extraction, pressurized fluid extraction and microwave-assisted extraction: Their potential and pitfalls. *Analyst* 126 (7):1182–1193.

Ceccarini, Alessio, Andrea Corti, Francesca Erba, et al. 2018. The hidden microplastics: New Insights and figures from the thorough separation and characterization of microplastics and of their degradation byproducts in coastal sediments. *Environmental Science & Technology* 52 (10):5634–5643.

Coppock, Rachel L., Matthew Cole, Penelope K. Lindeque, Ana M. Queirós, and Tamara S. Galloway. 2017. A small-scale, portable method for extracting microplastics from marine sediments. *Environmental Pollution* 230:829–837.

Crichton, Ellika M., Marie Noël, Esther A. Gies, and Peter S. Ross. 2017. A novel, density-independent and FTIR-compatible approach for the rapid extraction of microplastics from aquatic sediments. *Analytical Methods* 9 (9):1419–1428.

de Ilarduya, Antxon Martínez, and Sebastián Muñoz-Guerra. 2014. Chemical structure and microstructure of poly (alkylene terephthalate) s, their copolyesters, and their blends as studied by NMR. *Macromolecular Chemistry and Physics* 215 (22):2138–2160.

Dehaut, Alexandre, Anne-Laure Cassone, Laura Frère, et al. 2016. Microplastics in seafood: Benchmark protocol for their extraction and characterization. *Environmental Pollution* 215:223–233.

Dekiff, Jens H., Dominique Remy, Jörg Klasmeier, and Elke Fries. 2014. Occurrence and spatial distribution of microplastics in sediments from Norderney. *Environmental Pollution* 186:248–256.

Elert, Anna M., Roland Becker, Erik Duemichen, et al. 2017. Comparison of different methods for MP detection: What can we learn from them, and why asking the right question before measurements matters? *Environmental Pollution* 231:1256–1264.

Eriksen, Marcus, Sherri Mason, Stiv Wilson, et al. 2013. Microplastic pollution in the surface waters of the Laurentian Great Lakes. *Marine Pollution Bulletin* 77 (1–2):177–182.

Erni-Cassola, Gabriel, Matthew I. Gibson, Richard C. Thompson, and Joseph A. Christie-Oleza. 2017. Lost, but found with Nile red: A novel method for detecting and quantifying small microplastics (1 mm to 20 μm) in environmental samples. *Environmental Science & Technology* 51 (23):13641–13648.

Fischer, Marten, and Barbara M. Scholz-Böttcher. 2017. Simultaneous trace identification and quantification of common types of microplastics in environmental samples by pyrolysis-gas chromatography–mass spectrometry. *Environmental Science & Technology* 51 (9):5052–5060.

Frère, Laura, Ika Paul-Pont, J. Moreau, et al. 2016. A semi-automated Raman micro-spectroscopy method for morphological and chemical characterizations of microplastic litter. *Marine Pollution Bulletin* 113 (1–2):461–468.

Frias, J.P.G.L., P. Sobral, and Ana Maria Ferreira. 2010. Organic pollutants in microplastics from two beaches of the Portuguese coast. *Marine Pollution Bulletin* 60 (11):1988–1992.

Fries, Elke, Jens H. Dekiff, Jana Willmeyer, Marie-Theres Nuelle, Martin Ebert, and Dominique Remy. 2013. Identification of polymer types and additives in marine microplastic particles using pyrolysis-GC/MS and scanning electron microscopy. *Environmental Science: Processes & Impacts* 15 (10):1949–1956.

Fuller, Stephen, and Anil Gautam. 2016. A procedure for measuring microplastics using pressurized fluid extraction. *Environmental Science & Technology* 50 (11):5774–5780.

Golden, J.H., B.L. Hammant, and E.A. Hazell. 1967. The effect of thermal pretreatment on the strength of polycarbonate. *Journal of Applied Polymer Science* 11 (8):1571–1579.

Greven, Anne-Catherine, Teresa Merk, Filiz Karagöz, et al. 2016. Polycarbonate and polystyrene nanoplastic particles act as stressors to the innate immune system of fathead minnow (*Pimephales promelas*). *Environmental Toxicology and Chemistry* 35 (12):3093–3100.

Hale, Robert C. 2017. Analytical challenges associated with the determination of microplastics in the environment. *Analytical Methods* 9 (9):1326–1327.

Hanvey, Joanne S., Phoebe J. Lewis, Jennifer L. Lavers, Nicholas D. Crosbie, Karla Pozo, and Bradley O. Clarke. 2017. A review of analytical techniques for quantifying microplastics in sediments. *Analytical Methods* 9 (9):1369–1383.

Harrison, Jesse P., Jesús J. Ojeda, and María E. Romero-González. 2012. The applicability of reflectance micro-Fourier-transform infrared spectroscopy for the detection of synthetic microplastics in marine sediments. *Science of the Total Environment* 416:455–463.

Hermabessiere, Ludovic, Charlotte Himber, Béatrice Boricaud, et al. 2018. Optimization, performance, and application of a pyrolysis-GC/MS method for the identification of microplastics. *Analytical and Bioanalytical Chemistry* 410 (25):6663–6676.

Hernandez, Laura M., Nariman Yousefi, and Nathalie Tufenkji. 2017. Are there nanoplastics in your personal care products? *Environmental Science & Technology Letters* 4 (7):280–285.

Herrera, Alicia, Paloma Garrido-Amador, Ico Martínez, et al. 2018. Novel methodology to isolate microplastics from vegetal-rich samples. *Marine Pollution Bulletin* 129 (1):61–69.

Hidalgo-Ruz, Valeria, Lars Gutow, Richard C. Thompson, and Martin Thiel. 2012. Microplastics in the marine environment: A review of the methods used for identification and quantification. *Environmental Science & Technology* 46 (6):3060–3075.

Hintersteiner, Ingrid, Markus Himmelsbach, and Wolfgang W. Buchberger. 2015. Characterization and quantitation of polyolefin microplastics in personal-care products using high-temperature gel-permeation chromatography. *Analytical and Bioanalytical Chemistry* 407 (4):1253–1259.

Huppertsberg, Sven, and Thomas P. Knepper. 2018. Instrumental analysis of microplastics—Benefits and challenges. *Analytical and Bioanalytical Chemistry* 410 (25):6343–6352.

Hurley, Rachel R., Amy L. Lusher, Marianne Olsen, and Luca Nizzetto. 2018. Validation of a method for extracting microplastics from complex, organic-rich, environmental matrices. *Environmental Science & Technology* 52 (13):7409–7417.

Imhof, Hannes K., Christian Laforsch, Alexandra C. Wiesheu, et al. 2016. Pigments and plastic in limnetic ecosystems: A qualitative and quantitative study on microparticles of different size classes. *Water Research* 98:64–74.

Imhof, Hannes K., Johannes Schmid, Reinhard Niessner, Natalia P. Ivleva, and Christian Laforsch. 2012. A novel, highly efficient method for the separation and quantification of plastic particles in sediments of aquatic environments. *Limnology and Oceanography: Methods* 10 (7):524–537.

Islam, Shohana, Lina Apitius, Felix Jakob, and Ulrich Schwaneberg. 2019. Targeting microplastic particles in the void of diluted suspensions. *Environment International* 123:428–435.

Jungnickel, H., R. Pund, J. Tentschert, et al. 2016. Time-of-flight secondary ion mass spectrometry (ToF-SIMS)-based analysis and imaging of polyethylene microplastics formation during sea surf simulation. *Science of the Total Environment* 563:261–266.

Käppler, Andrea, Dieter Fischer, Sonja Oberbeckmann, et al. 2016. Analysis of environmental microplastics by vibrational microspectroscopy: FTIR, Raman or both? *Analytical and Bioanalytical Chemistry* 408 (29):8377–8391.

Karunanayake, Akila G., Olivia Adele Todd, Morgan Crowley, et al. 2018. Lead and cadmium remediation using magnetized and nonmagnetized biochar from Douglas fir. *Chemical Engineering Journal* 331:480–491.

Law, Kara Lavender, and Richard C. Thompson. 2014. Microplastics in the seas. *Science* 345 (6193):144–145.

Lenz, Robin, Kristina Enders, Colin A. Stedmon, David M.A. Mackenzie, and Torkel Gissel Nielsen. 2015. A critical assessment of visual identification of marine microplastic using Raman spectroscopy for analysis improvement. *Marine Pollution Bulletin* 100 (1):82–91.

Li, Jingyi, Huihui Liu, and J. Paul Chen. 2018. Microplastics in freshwater systems: A review on occurrence, environmental effects, and methods for microplastics detection. *Water Research* 137:362–374.

Lusher, A.L., N.A. Welden, P. Sobral, and M. Cole. 2017. Sampling, isolating and identifying microplastics ingested by fish and invertebrates. *Analytical Methods* 9 (9):1346–1360.

Maes, Thomas, Rebecca Jessop, Nikolaus Wellner, Karsten Haupt, and Andrew G. Mayes. 2017. A rapid-screening approach to detect and quantify microplastics based on fluorescent tagging with Nile Red. *Scientific Reports* 7:44501.

Nor, Nur Hazimah Mohamed, and Jeffrey Philip Obbard. 2014. Microplastics in Singapore's coastal mangrove ecosystems. *Marine Pollution Bulletin* 79 (1–2):278–283.

Nuelle, Marie-Theres, Jens H. Dekiff, Dominique Remy, and Elke Fries. 2014. A new analytical approach for monitoring microplastics in marine sediments. *Environmental Pollution* 184:161–169.

Pauli, Guido F., Birgit U. Jaki, and David C. Lankin. 2005. Quantitative 1H NMR: Development and potential of a method for natural products analysis. *Journal of Natural Products* 68 (1):133–149.

Peez, Nadine, Marie-Christine Janiska, and Wolfgang Imhof. 2019. The first application of quantitative 1 H NMR spectroscopy as a simple and fast method of identification and quantification of microplastic particles (PE, PET, and PS). *Analytical and Bioanalytical Chemistry* 411 (4):823–833.

Prata, Joana Correia, João P. da Costa, Armando C. Duarte, and Teresa Rocha-Santos. 2018. Methods for sampling and detection of microplastics in water and sediment: A critical review. *TrAC Trends in Analytical Chemistry* 110:150–159.

Primpke, Sebastian, Claudia Lorenz, Richard Rascher-Friesenhausen, and Gunnar Gerdts. 2017. An automated approach for microplastics analysis using focal plane array (FPA) FTIR microscopy and image analysis. *Analytical Methods* 9 (9):1499–1511.

Qiu, Qiongxuan, Zhi Tan, Jundong Wang, Jinping Peng, Meimin Li, and Zhiwei Zhan. 2016. Extraction, enumeration and identification methods for monitoring microplastics in the environment. *Estuarine, Coastal and Shelf Science* 176:102–109.

Quinn, Brian, Fionn Murphy, and Ciaran Ewins. 2017. Validation of density separation for the rapid recovery of microplastics from sediment. *Analytical Methods* 9 (9):1491–1498.

Reddy, M. Srinivasa, Shaik Basha, S. Adimurthy, and G. Ramachandraiah. 2006. Description of the small plastics fragments in marine sediments along the Alang-Sosiya ship-breaking yard, India. *Estuarine, Coastal and Shelf Science* 68 (3–4):656–660.

Renner, Gerrit, Torsten C. Schmidt, and Jürgen Schram. 2017. A new chemometric approach for automatic identification of microplastics from environmental compartments based on FT-IR spectroscopy. *Analytical Chemistry* 89 (22):12045–12053.

Rocha-Santos, Teresa, and Armando C. Duarte. 2015. A critical overview of the analytical approaches to the occurrence, the fate and the behavior of microplastics in the environment. *TrAC Trends in Analytical Chemistry* 65:47–53.

Saint-Michel, Fabrice, Laurent Chazeau, and Jean-Yves Cavaillé. 2006. Mechanical properties of high density polyurethane foams: II Effect of the filler size. *Composites Science and Technology* 66 (15):2709–2718.

Sánchez-Nieva, Julio, José Antonio Perales, Juan Maria González-Leal, and Elisa Rojo-Nieto. 2017. A new analytical technique for the extraction and quantification of microplastics in marine sediments focused on easy implementation and repeatability. *Analytical Methods* 9 (45):6371–6378.

Schirinzi, Gabriella, Marta Llorca, Josep Sanchís, Marinella Farré, and Damià Barceló. 2016. Analytical characterization of microplastics and assessment of their adsorption capacity for organic contaminants. *Monitoring for a Sustainable Management of Marine Resources*. G. Sacchettini, and M. Giulivo (eds.), Fidenza, Italy: Mattioli 1885 S.p.A., pp. 15–19.

Shim, Won Joon, Sang Hee Hong, and Soeun Eo Eo. 2017. Identification methods in microplastic analysis: A review. *Analytical Methods* 9 (9):1384–1391.

Shim, Won Joon, Young Kyoung Song, Sang Hee Hong, and Mi Jang. 2016. Identification and quantification of microplastics using Nile Red staining. *Marine Pollution Bulletin* 113 (1–2):469–476.

Silva, Ana B., Ana S. Bastos, Celine I.L. Justino, João P. da Costa, Armando C. Duarte, and Teresa A.P. Rocha-Santos. 2018. Microplastics in the environment: Challenges in analytical chemistry—A review. *Analytica Chimica Acta* 1017:1–19.

Song, Young Kyoung, Sang Hee Hong, Mi Jang, et al. 2015. A comparison of microscopic and spectroscopic identification methods for analysis of microplastics in environmental samples. *Marine Pollution Bulletin* 93 (1–2):202–209.

Takidis, G, D.N. Bikiaris, G.Z. Papageorgiou, D.S. Achilias, and I. Sideridou. 2003. Compatibility of low-density polyethylene/poly (ethylene-co-vinyl acetate) binary blends prepared by melt mixing. *Journal of Applied Polymer Science* 90 (3):841–852.

Thompson, Richard C., Shanna H. Swan, Charles J. Moore, and Frederick S. Vom Saal. 2009. Our plastic age. *Philosophical Transactions of the Royal Society B: Biological Sciences* 364 (1526):1973–1976.

Utracki, L.A., Michel M. Dumoulin, and P. Toma. 1986. Melt rheology of high density polyethylene/polyamide-6 blends. *Polymer Engineering & Science* 26 (1):34–44.

Vianello, A., A. Boldrin, P. Guerriero, et al. 2013. Microplastic particles in sediments of Lagoon of Venice, Italy: First observations on occurrence, spatial patterns and identification. *Estuarine, Coastal and Shelf Science* 130:54–61.

Wagner, Jeff, Zhong-Min Wang, Sutapa Ghosal, Chelsea Rochman, Margy Gassel, and Stephen Wall. 2017. Novel method for the extraction and identification of microplastics in ocean trawl and fish gut matrices. *Analytical Methods* 9 (9):1479–1490.

Wang, Wenfeng, and Jun Wang. 2018. Investigation of microplastics in aquatic environments: An overview of the methods used, from field sampling to laboratory analysis. *TrAC Trends in Analytical Chemistry* 108:195–202.

Wang, Zhong-Min, Jeff Wagner, Sutapa Ghosal, Gagandeep Bedi, and Stephen Wall. 2017. SEM/EDS and optical microscopy analyses of microplastics in ocean trawl and fish guts. *Science of the Total Environment* 603:616–626.

Wessel, Caitlin C., Grant R. Lockridge, David Battiste, and Just Cebrian. 2016. Abundance and characteristics of microplastics in beach sediments: Insights into microplastic accumulation in northern Gulf of Mexico estuaries. *Marine Pollution Bulletin* 109 (1):178–183.

Williams, Joel M., and Debra A. Wrobleski. 1988. Spatial distribution of the phases in water-in-oil emulsions: Open and closed microcellular foams from cross-linked polystyrene. *Langmuir* 4 (3):656–662.

Zeronian, S. Haig, and Martha J. Collins. 1989. Surface modification of polyester by alkaline treatments. *Textile Progress* 20 (2):1–26.

Zhang, Cheng, Xiao-Su Yi, Hiroshi Yui, Shigeo Asai, and Masao Sumita. 1998. Morphology and electrical properties of short carbon fiber-filled polymer blends: High-density polyethylene/poly (methyl methacrylate). *Journal of Applied Polymer Science* 69 (9):1813–1819.

Zhang, Kai, Huahong Shi, Jinping Peng, et al. 2018. Microplastic pollution in China's inland water systems: A review of findings, methods, characteristics, effects, and management. *Science of the Total Environment* 630:1641–1653.

Zhang, Shaoliang, Xiaomei Yang, Hennie Gertsen, Piet Peters, Tamás Salánki, and Violette Geissen. 2018. A simple method for the extraction and identification of light density microplastics from soil. *Science of the Total Environment* 616:1056–1065.

Zhao, Shiye, Meghan Danley, J. Evan Ward, Daoji Li, and Tracy J. Mincer. 2017. An approach for extraction, characterization and quantitation of microplastic in natural marine snow using Raman microscopy. *Analytical Methods* 9 (9):1470–1478.

Zhu, X. 2015. Optimization of elutriation device for filtration of microplastic particles from sediment. *Marine Pollution Bulletin* 92 (1–2):69–72.

Zobkov, M., and E. Esiukova. 2017. Microplastics in Baltic bottom sediments: Quantification procedures and first results. *Marine Pollution Bulletin* 114 (2):724–732.

Section II

Distribution and Characteristics of Particulate Plastics

5 An Introduction to the Chemistry and Manufacture of Plastics

Nanthi S. Bolan, Kandaswamy Karthikeyan, Shiv Shankar Bolan, M.B. Kirkham, Ki-Hyun Kim, and Deyi Hou

CONTENTS

5.1 Polymer Sources and Polymerization Reactions 85
5.2 Types of Plastics 87
5.2.1 Grouping of Plastics Based on the Stability during Heating 87
5.2.2 Grouping of Plastics Based on the Sources of Polymers 88
5.3 Processing of Plastics 89
5.4 Conclusions 93
References 93

5.1 POLYMER SOURCES AND POLYMERIZATION REACTIONS

There are two major sources of polymers used in plastic manufacture, which are commonly defined as non-renewable (i.e., synthetic) and renewable (i.e., natural) sources (Vroman and Thghzert 2009; Brazel and Rosen 2012). Synthetic polymers, derived from crude oil, natural gas, and coal, are broad enough to include Teflon, polyethylene, polyester, and nylon. Natural polymers, derived from plants and microorganisms also include diverse forms of products such as rayon and polyester (Saito et al. 2012; Babu et al. 2013). It should also be noted that there are some polymer products (e.g., polyester) derived from both natural and synthetic sources, while being used similarly for plastic manufacturing (Dijkmans et al. 2013).

The first step in the processing of plastic from non-renewable sources is cracking, which converts the crude oil-based naphtha and the natural gas-based ethane into ethylene, a starting point for a variety of polymer products. There are two major origins of renewable sources of plastics, such as polylactic acid (PLA) derived from plant-based polysaccharides (i.e., starch and sugars) and polyhydroxyalkanoate (PHA) derived from microorganisms (Majid et al. 2010). Polymers, which are the primary components of plastics, are formed through polymerization reactions of monomers based on the processes of either addition or condensation through three basic steps: initiation, propagation, and termination (Hiorns et al. 2012) (Figure 5.1). In the process of addition polymerization, the last "mer" in the chain sequence becomes the binding site for the monomers.

The addition polymerization process can result in the formation of polyethylene, polystyrene, and acrylic plastics. The thermoplastic properties of polymers derived from the addition polymerization process involve a softening stage through the heating process, and then the hardening process takes place as they are cooled down (Seymour and Carraher 1990). During condensation polymerization, a small molecule, such as water or alcohol is discharged as the monomers adhere. Condensation polymers can be thermoplastic or thermosetting (Figure 5.2) to include nylons, polyesters, and urethanes. During condensation polymerization, the production of useful products is predicated by the removal of unwanted by-products so that the chemical reactions can continue. The major difference

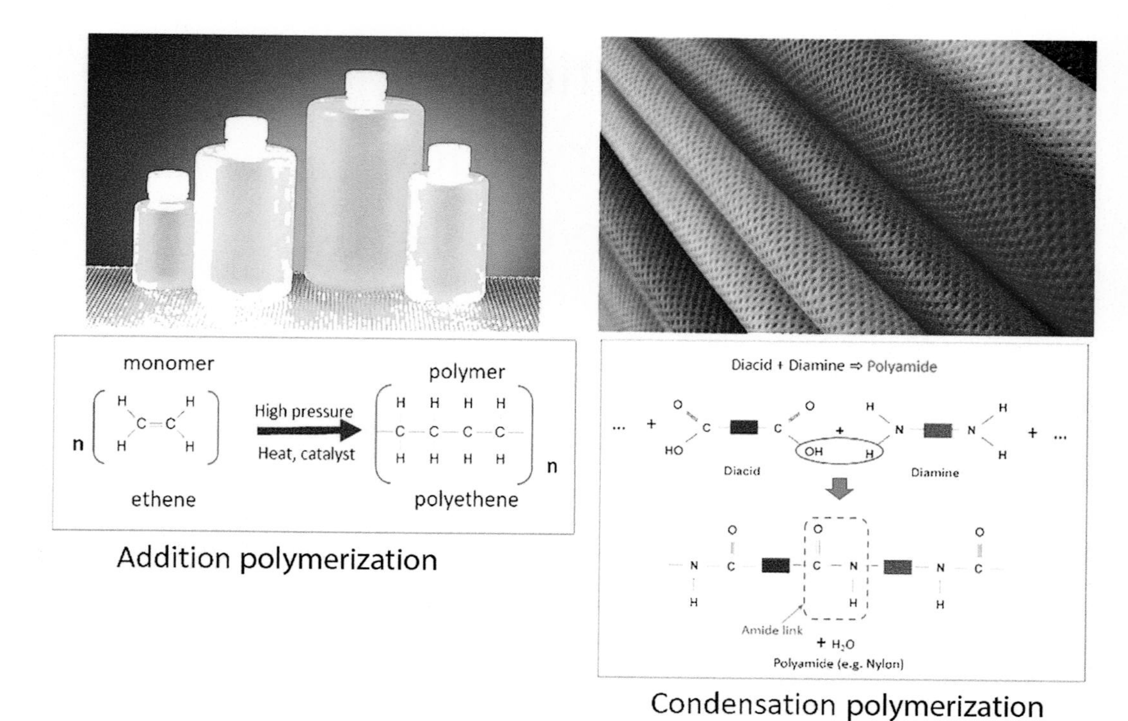

FIGURE 5.1 Polymerization reactions.

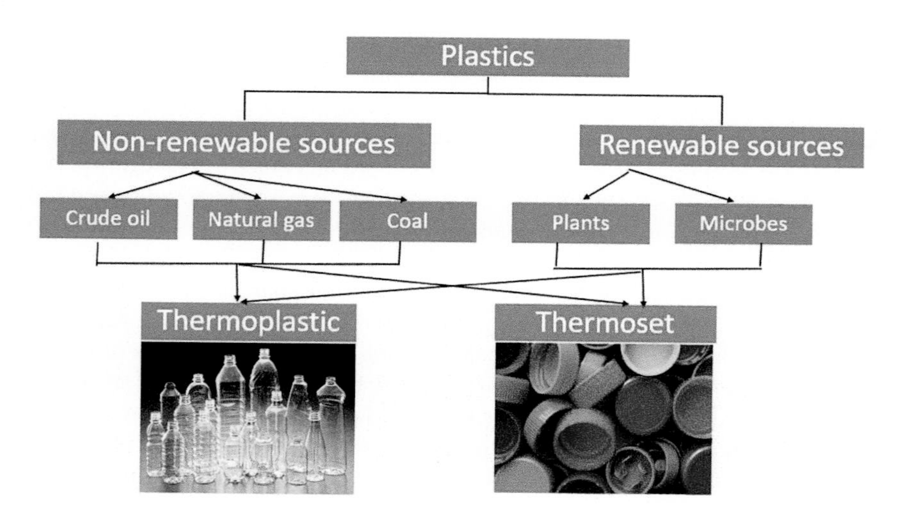

FIGURE 5.2 Major grouping of plastics.

between addition and condensation polymerization reactions is the quality and durability of the plastic. Generally, condensation polymers are more readily degradable than addition polymers.

Monomers can be combined in different configurations to produce plastic resins with variable properties and characteristics. A homogenous polymer, referred to as a homopolymer, can be produced through the combination of the same monomers together. Copolymers are, on the other hand, referred to as plastics, where multiple monomers are used. Monomers confer specific properties and characteristics to the plastic resins, with combinations of monomers producing various copolymers with further variations in their property.

The first step in the processing of plastic from non-renewable sources is cracking, which converts the crude oil-based naptha and the natural gas-based ethane into ethylene, a starting point for a variety of polymer products. There are two major origins of renewable sources of plastics that include polylactic acid (PLA) derived from plant based polysaccharides (i.e., starch and sugars), and polyhydroxyalkanoate (PHA) derived from microorganisms (Majid et al. 2010).

Polymers, which are the primary components of plastics, are formed through polymerization reactions of monomers. Polymerization requires either the processes of addition or condensation reactions to occur (Hiorns et al. 2012), which involve three basic steps: initiation, propagation, and termination (Figure 5.1). For the process of addition polymerization, the last “mer” in the chain-sequence becomes the binding site for the monomers.

The addition polymerization process can result in the formation of polyethylene, polystyrene, and acrylic plastics. The thermoplastic properties of polymers derived from the addition polymerization process allow the heating to soften them, and then the process of hardening takes place when they are cooled (Seymour and Carraher 1990). During condensation polymerization, a small molecule such as water or alcohol is discharged as the monomers adhere. Condensation polymers can be thermoplastic or thermosetting (Figure 5.2) and include nylons, polyesters, and urethanes. During condensation polymerization, the production of useful products is predicated by the removal of unwanted byproducts for the chemical reactions to continue. The major difference between addition and condensation polymerization reactions is the quality and durability of the plastic. Generally, condensation polymers are relatively more readily degradable compared to addition polymers.

Monomers can be combined in different configurations to produce plastic resins with variable properties and characteristics. A homogenous polymer referred to as a homopolymer can be produced through the combination of the same monomers together. Copolymers are on the other hand referred to as plastics where multiple monomers are used. Monomers confer specific properties and characteristics to the plastic resins with combinations of monomers to produce various copolymers with further variations in their property.

5.2 TYPES OF PLASTICS

Plastics are grouped on the basis of their origin (natural and synthetic), structure (thermoplastic and thermoset), and the source of polymer (petroleum and plants (or microbes)). On the basis of the origin of polymers, plastics can be grouped into natural (derived from animals and plants) and synthetic (artificially synthesized by chemical processes). For example, cellulose polymer used for making sticky tape occurs naturally, whereas nylon is a synthetic polymer made in a factory. Because the majority of plastics are synthetic, the groupings based on stability during heating, and the sources of polymers, are more frequently used in the applications of plastics.

5.2.1 Grouping of Plastics Based on the Stability during Heating

The behavior of plastics when treated thermally allows the differentiation into two broad categories, namely: thermoplastics and thermosets. Thermoplastics make up the majority of plastics, with approximately 92% of plastics falling into this category. Thermoplastic polymers when subjected to heat are softened due to weak secondary bonding forces that hold together the molecules of the polymer plastic. These polymers return to their original orientation when reverted to ambient temperatures; thus, thermoplastic, when softened, can be shaped for a vast array of applications through extrusion, molding, or pressing processes due to enhanced adaptability. Examples of thermoplastics include polyethylene or polythene (plastic bottles and sheets), polystyrene (packaging material), polypropylene (plastic ropes), polyvinylchloride (toys and drainage pipes), polycarbonate (plastic windows and car headlamps), and polyamide (nylon purposed for stockings and swimming wear).

A thermoset is a polymer that solidifies or "sets" irreversibly when heated or cured. Larger polymer chains are used to make thermoset plastics rather than thermoplastics. During the initial manufacturing process, dense structures with strong cross-links bind long molecular chains together via heating and compression. These connections between carbon atoms lattice together and are involved in the formation of two- and three-dimensional frames as opposed to one-dimensional chains. These thermoset plastics are not meltable and hence have specific values for their durability and strength. Common examples of thermosets include polyurethane (insulating material in buildings), polytetrafluoroethylene (non-stick coatings), melamine (hard plastic crockery), and epoxy resin.

5.2.2 Grouping of Plastics Based on the Sources of Polymers

Polymers used for the manufacture of plastics are derived either from non-renewable or renewable sources. Non-renewable sources for plastic manufacture include crude oil, natural gas, and coal. More than 90% of the global plastic production is derived from these non-renewable sources, and some of the common polymers used in plastic production from non-renewable sources are given in Table 5.1. The use of renewable sources is generally confined to the manufacture of bioplastics. The term "bioplastic" broadly represents plastic substances that are derived wholly or in part from biomass-based feedstock rather than petroleum-based feedstock material.

Most of the bioplastics are biodegradable, and there are three broad descriptors of bioplastics: (i) bio-based plastics, which include all plastic materials, derived wholly or in part from plant-based material, and starch and cellulose are two of the most common renewable feedstocks used to synthesize bioplastics; (ii) degradable plastics, which include only those that degrade within a relatively short period of time, and bioplastics that do not degrade within a short period are sometimes called "durable plastic;" and (iii) compostable plastics, which include those that undergo the process of biological decomposition in a specialized compost site and break down to carbon dioxide, water, inorganic compounds, and biomass, at a rate similar to other known compostables (e.g., cellulose).

There are two main sources of bioplastics that include polylactic acid, derived from plant-based polysaccharides, and polyhydroxyalkanoate, derived from microorganisms (Hartmann 1998). Polylactic acid is typically synthesized from starch and sugars derived from various crops that include corn, cassava, sugarcane, and sugarbeet. Biodegradable polymers synthesized from renewable materials can be classified into three categories according to their raw material and synthetic method (Tschan et al. 2012): (i) polymers extracted from agricultural products, such as cellulose; (ii) polymers made by microorganisms, such as poly(hydroxyalkanoate); and (iii) synthetic polymers using monosaccharides that can be obtained from agricultural products, such as PLA. For example, starch includes long chains of carbon molecules, similar to the carbon chains in plastic from fossil fuels. Starch is converted into PLA in stepwise changes. PLA is produced from LA, and the current industrial scale manufacture of LA is established through microbial carbohydrate fermentation.

PHA is synthesized by microorganisms when they are exposed to high carbon environments, while being deprived of other nutrients, such as nitrogen and phosphorus (López et al. 2015). This causes the production of PHA as carbon reserves, which are stored in granules until they have more of the other limiting nutrients that they need to grow and reproduce. Various organic sources can be used to feed microbes the carbon source that they need to produce PHA (Jiang et al. 2016). These carbon sources include wastewater and solid waste (such as biosolids, food waste, and crop

TABLE 5.1
Various Monomers Derived from Non-Renewable Sources Used in Plastic Manufacture

Monomer	Formula	Polymer	Trivial name	Structure
Ethene	$H_2C{=}CH_2$	LDPE Low-density poly(ethene)	low-density polythene	$-CH_2-CH_2-CH_2-CH_2-$
Chloroethene	$H_2C{=}CHCl$	Poly(chloroethene)	polyvinyl chlorine, PVC	$-CH_2-CH(Cl)-CH_2-CH(Cl)-$
Propene	$H_2C{=}CH-CH_3$	Poly(propene)	polypropylene	$-CH_2-CH(CH_3)-CH_2-CH(CH_3)-$
Propenonitrile	$H_2C{=}CH-CN$	Poly(propenonitrile)	polyacrylonitrile	$-CH_2-CH(CN)-CH_2-CH(CN)-$
Methyl 2-methylpropenoate	$H_2C{=}C(CO_2CH_3)-CH_3$	Poly(methyl 2-methylpropenoate)	polymethyl methacrylate	$-CH_2-C(CO_2CH_3)(CH_3)-CH_2-C(CO_2CH_3)(CH_3)-$
Phenylethene	$H_2C{=}CH-C_6H_5$	Poly(phenylethene)	polystyrene	$-CH_2-CH(C_6H_5)-CH_2-CH(C_6H_5)-$
Tetrafluoroethene	$F_2C{=}CF_2$	Poly(tetrafluoroethene) (PTFE)	polytetrafluoroethylene PTFE	$-CF_2-CF_2-CF_2-CF_2-$

Source: http://www.essentialchemicalindustry.org/polymers/polymers-an-overview.html

residues). Microorganisms first convert the waste's organic carbon into volatile fatty acids. Then, the plastic-producing microbes feed on the volatile fatty acids. These microbes are continually subjected to alternate "feast" and "famine" phases leading to the synthesis of PHA. Cyanobacteria, which use sunlight to produce carbohydrates through photosynthesis, have been used as a source of carbon for PHA synthesis (Kamravamanesh et al. 2018). Similarly, methane is fed to plastic-producing bacteria that transform it into PHA (Muhammadi et al. 2015).

5.3 PROCESSING OF PLASTICS

Plastic polymers known as resins are utilized in the manufacturing of various plastic products. Some of the major plastic processing and shaping methods include extrusion, injection, blow molding, calendering, and spinnerating (Rosato et al. 2001) (Table 5.2). The continuous extrusion process uses plastic granules, pellets, or powder that are first loaded into a hopper, and then fed into a heated extruder, through which the plastic is moved by the mechanism of a continuously spinning screw (Hanson 2005). The plastic is melted by the mechanical work of the screw and the heat from the extruder wall. After this process, the molten plastic is extruded through a small opening called a die to form the shape of the finished product. In injection molding, which is not a continuous process, the plastic material is fed into a hopper, and an extruder screw advances the plastic through the heating chamber, which melts the material (Grelle 2005). At the end of the extruder, the molten plastic is forced at high pressure into a closed cold mold. The plastic then cools to a solid state, the mold opens, and the finished product is ejected. Blow molding is a technique by which hollow

TABLE 5.2
General Characteristics of Plastic Processing Methods

Process	Characteristics	Example Products
Injection	Complex shapes of various sizes; thin walls; very high production rates; costly tooling; and good dimensional accuracy	
Extrusion	Continuous, uniformly solid or hollow, and complex cross sections; high production rates; relatively low tooling costs; and wide tolerance	
Calendering	Continuous process; high output and the ability to deal with low melt strength; the thickness is well maintained; and surface made smooth	
Spinnerate	Fine fibers; high precision and high production cost; and relatively high tooling and maintenance costs	
Blow molding	Hollow, thin-walled parts and bottles of various sizes; high production rates; and relatively low tooling costs	

Source: https://pritamashutosh.wordpress.com/2014/02/11/forming-and-shaping-of-plastics/

plastic parts are made (Lee 2005). In general, there are three main types of blow molding: extrusion blow molding, injection blow molding, and injection stretch blow molding. The blow molding process begins with the melting of the plastic and forming a parison. The parison is a tube-like piece of plastic with a hole in one end through which compressed air can pass. The air pressure pushes the plastic out to match the mold. Once the plastic has cooled and hardened, the part is separated from the mold. Calendering is a continuous process used for high output and the ability to deal with low melt strength. The thickness is well maintained, and the surface is made smoothly by the polished rollers. Spinnerating is an extruding process in which the melted plastic is forced through a microscopically small sieve, called a spinneret, to make thin fibers used for the production of various types of end-products, including toothbrushes and nylon stockings.

A number of additives are added either during the manufacture of plastics or during the processing stages of plastics into consumer products (Table 5.3) (Hahladakis et al. 2018). Improvements to the basic chemical, physical, and mechanical properties of plastics can be done through the incorporation of additives. These also have protective effects by reducing degradation of the polymer from light, heat, or bacteria. They affect polymer-processing properties, such as melt flow and viscosity. Additives also provide product color and other special characteristics, such as improved surface appearance, reduced friction, and flame retardancy.

TABLE 5.3
Additives Used during Plastic Manufacture and Processing

Types of Additives	Function	Example
Antioxidants	For plastic processing and outside application, where weathering resistance is needed	 Pentaerythritol tetrakis (3,5-di-tert-butyl-4-hydroxyhydrocinnamate), Tris(2,4-di-tert-butylphenyl)phosphite
Colorants	For coloring plastic products	 Diarylide pigment, Sudan stain
Foaming agents	For expanding plastic products, as in the case of polystyrene cups, building boards, and polyurethane carpet underlayment	 Chlorofluorocarbons, Isocyanate
Plasticizers	To slow down decomposition from light; used in wire insulation, flooring, gutters, and some films	 Bis(2-ethylhexyl) phthalate

(Continued)

TABLE 5.3 (*Continued*)
Additives Used during Plastic Manufacture and Processing

Types of Additives	Function	Example
Lubricants	To produce flexible plastics; used for making fibers and squeeze bottles	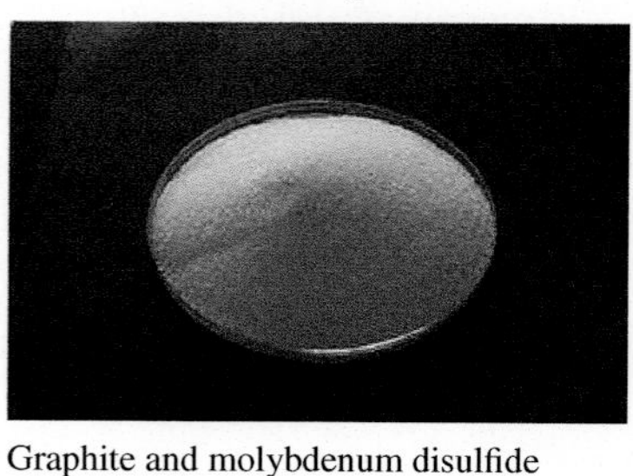 Graphite and molybdenum disulfide
Anti-statics	To reduce dust collection by static electricity attraction	 Indium tin oxide, glycerol monostearate
Anti-microbials	To control biofilm formation; used for shower curtains and wall coverings (Bithionol)	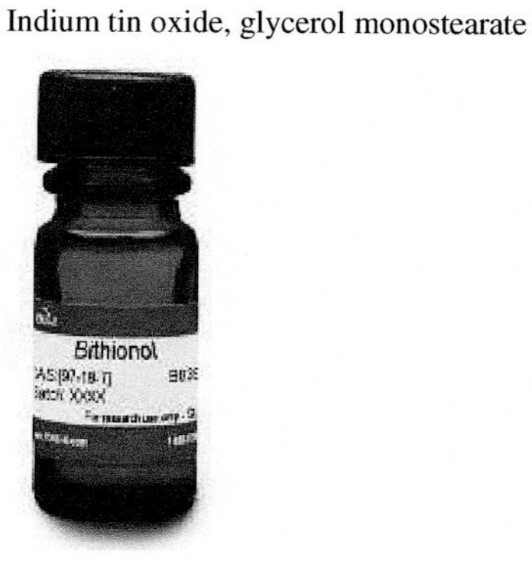 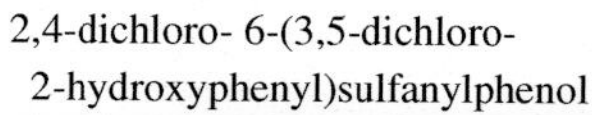 2,4-dichloro- 6-(3,5-dichloro-2-hydroxyphenyl)sulfanylphenol
Flame retardants	To improve the safety of wire and cable coverings and cultured marble	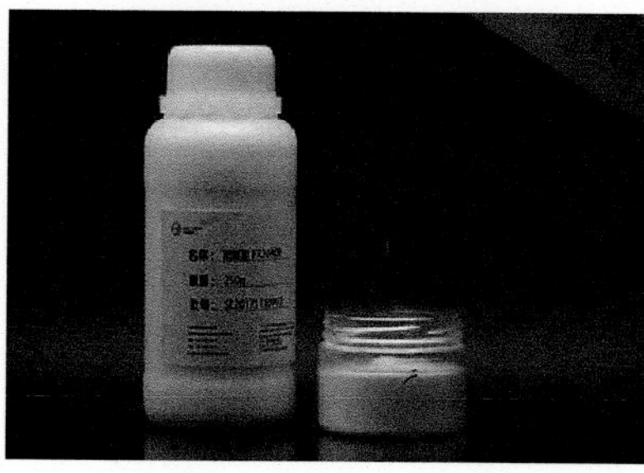 *Decabromodiphenyl* ether

5.4 CONCLUSIONS

Polymers used for plastics are derived from both non-renewable and renewable sources. The non-renewable sources of polymers include crude oil, natural gas, and coal, while the renewable sources include plant- and microbial-derived polysaccharides. There are two types of polymerization reactions, which include addition and condensation polymerization reactions. The nature and stability of plastics depend on the polymerization reactions. Plastics are grouped mainly on the basis of the stability of plastics during heating (thermoplastics and thermosets) and the origin of polymers (natural and synthetic plastics). A number of techniques are used for processing plastics into usable commodity products, during which a range of additives are added to enhance the durability and mechanical properties of plastic products.

REFERENCES

Babu, R.P., O'Connor, K. and Seeram, R. (2013). Current progress on bio-based polymers and their future trends. *Prog Biomater.*, 2: 8. https://doi.org/10.1186/2194-0517-2-8.

Brazel, C.S. and Rosen, S.L. (2012). *Fundamental Principles of Polymeric Materials* (3rd ed.) p. 404. John Wiley & Sons, Hoboken, NJ.

Dijkmans, T., Pyl, S.P., Reyniers, M.F., Abhari, R., van Greem, K.M. and Martin, G.B. (2013). Production of bio-ethene and propene: Alternatives for bulk chemicals and polymers. *Green Chem.*, 15, 3064–3076.

Grelle, P.F. (2005). Injection molding. In: *Handbook of Plastic Processes* (Ed. Harper, C.A.) pp. 1–123. John Wiley & Sons, Hoboken, NJ.

Hahladakis, J.N., Velis, C.E., Weber, R., EleniIacovidou. J. and Purnell, P. (2018). An overview of chemical additives present in plastics: Migration, release, fate and environmental impact during their use, disposal and recycling. *J. Hazard. Mater.*, 344, 179–199.

Hanson, D.R. (2005). Sheet extrusion. In: *Handbook of Plastic Processes* (Ed. Harper, C.A.) pp. 189–289. John Wiley & Sons, Hoboken, NJ.

Hartmann, M.H. (1998). High molecular weight polylactic acid polymers. In: *Biopolymers from Renewable Resources; Macromolecular Systems-Materials Approach* Chapter 15 (Ed. Kaplan, D.L.) pp. 367–411. Springer-Verlag, Berlin, Germany.

Hiorns, R.C., Boucher, R.J., Duhlev, R., Hellwich, K.H., Hodge, P., Jenkins, A.D. and Jones, R.G., et al. (2012). Brief guide to polymer nomenclature. *Pure Appl. Chem.* 84, 2167–2169.

Jiang, G., Hill, D.J., Kowalczuk, M., Johnston, B., Adamus, G., Irorere, V. and Radecka, I. (2016). Carbon sources for polyhydroxyalkanoates and an integrated biorefinery. *Int. J. Mol. Sci.*, 17(7), 1157.

Kamravamanesh, D., Lackner, M. and Herwig, C. (2018). Bioprocess engineering aspects of sustainable polyhydroxyalkanoate production in cyanobacteria. *Bioengineering (Basel).* 5(4), 111.

Lee, N.C. (2005). Blow molding. In: *Handbook of Plastic Processes* (Ed. Harper, C.A.) pp. 189–289, John Wiley & Sons, New York.

López, N.I., Pettinari, M.J., Nikel, P.I. and Méndez, B.S. (2015). Polyhydroxyalkanoates: Much more than biodegradable plastics. *Adv. Appl. Microbiol.*, 93:73–106.

Majid, J., Elmira, A.T., Muhammad, I., Muriel, J. and St'ephane D. (2010). Poly-lactic acid: Production, applications, nanocomposites, and release studies. *Comprehensive Rev. Food Sci. Safety.* 9(5), 552–571.

Muhammadi, S., Afzal, M. and Hameed, S. (2015). Bacterial polyhydroxyalkanoates-eco-friendly next generation plastic: Production, biocompatibility, biodegradation, physical properties and applications. *Green Chem. Lett. Rev.*, 8:3–4, 56–77.

Rosato, D.V., Rosato, D.V. and Rosato, M.G. (2001). Plastic processing. In: *Plastics Design Handbook* (Eds. Rosato, M.G. and Rosato, D.V.) pp. 433–566. Springer, Boston, MA.

Saito, T., Brown, R.H., Hunt, M.A., Pickel, D.L., Messman, J.M., Baker, F.S., Keller, M. and Naskar, A.K. (2012). Turning renewable resources into value-added polymer: Development of lignin-based thermoplastic. *Green Chem.*, 14, 3295–3303.

Seymour, B. and Carraher, C.E. (1990). *Giant Molecules: Essential Materials for Everyday Living and Problem Solving*, 2nd ed. Wiley-Interscience, New York.

Tschan, M.J.L., Brulé, E., Haquette, P. and Thomas, C.M. (2012). Synthesis of biodegradable polymers from renewable resources. *Polym. Chem.*, 3, 836–851.

Vroman, I. and Thghzert, L. (2009). Biodegradable polymers. *Materials* 2(2): 307–344.

6 Interaction of Dissolved Organic Matter with Particulate Plastics

Yubo Yan, Qiao Li, Shiv Shankar Bolan, Nanthi S. Bolan, Yong Sik Ok, M.B. Kirkham, and Eilhann E. Kwon

CONTENTS

6.1 Sources of Particulate Plastics 95
6.2 Sources of Dissolved Organic Matter in Terrestrial and Aquatic Ecosystems 96
6.3 Particulate Plastics and Dissolved Organic Carbon Assemblages 97
6.4 Implications of PPs and DOM Assemblages 100
6.4.1 Biofilm Formation 100
6.4.2 Aggregation of Microplastics 101
6.4.3 Retention and Transport of Contaminants 101
6.5 Conclusions 103
References 103

6.1 SOURCES OF PARTICULATE PLASTICS

The sources of particulate input to terrestrial/aquatic ecosystems are described in Chapter 1. Particulate plastics (PPs) include a subset of polymers (Chapter 1, Table 1.1), which can be abandoned into ecosystems and are smaller than 1 mm, down to the μm range (Browne et al. 2011). Two groups of particulate plastics can enter the environment, namely, primary and secondary PPs (Duis and Coors 2016; Wijesekara et al. 2018). Primary PPs are a direct result of anthropogenic adoption of plastic-based materials (microbeads in cosmetics), whereas plastic fragments that form due to the breakdown of larger plastics are called secondary PPs. While wastewater discharge is a major source of PP input to aquatic ecosystems, biowastes including biosolids and composts are a major source of PP input to soil (Rillig 2012; Mahon et al. 2016). Both primary and secondary PPs persist in terrestrial (soil) and aquatic (marine and freshwater) ecosystems.

The United Nations Environment Programme has identified land-based sources of PPs, which make up the majority of them found within the marine environments (UNEP 2016). Soil erosion is an important process in PP transport, as it allows for sediment transfer that transports PPs from the terrestrial to the aquatic ecosystems (Duis and Coors 2016; Rillig et al. 2017). Currently, the greater part of research on PPs has been focused on elucidating their detrimental effects in aquatic marine environments, despite this link to land-based origins (Thompson et al. 2009; Browne et al. 2011; Cole et al. 2011). Biowastes, such as biosolids/composts have been increasingly thought to be a productive avenue for large-scale applications for nutrient improvement and soil health, but they have posed the unwanted risk of plastic exposure in soil (Wijesekara et al. 2016). During the composting process, medium- and large-sized plastics could be segregated and removed. Unfortunately, a large portion of smaller-sized PPs remain in the final composting and biosolid products. These sources, along with weathering issues with plastic film mulch used over agricultural fields, offer unwanted pathways of PP input into soils (Rochman 2018).

6.2 SOURCES OF DISSOLVED ORGANIC MATTER IN TERRESTRIAL AND AQUATIC ECOSYSTEMS

Dissolved organic matter (DOM) can be seen both as a link and bottleneck among various ecological compartments. It also can be seen as a warning indicator in ecological processes, notably, in relation to aquatic systems. Thus, there has been an increasing trend in studies exploring the processes of DOM and its distribution in aquatic systems (Dafner and Wangersky 2002). However, terrestrial environments and aquatic ecosystems offer the different DOM dynamics. In aquatic ecosystems, primary producers have a relatively small biomass, allochthonous sources of DOM are dominant (i.e., sources found in a place remote from the site of formation), the surface area of reactive solid particles (i.e., sediments) is small, and the fate and dynamics of DOM are strongly influenced by photolysis and other light-mediated spontaneous reactions. However, PPs in aquatic ecosystems provide an important "semi-permanent" solid substrate for interaction with DOM, thereby impacting the chemo/biodynamics of PPs in this ecosystem. In the case of the terrestrial environment, DOM dynamics are mostly influenced through interactions with abiotically and biotically reactive solid components including soil components and PPs.

The major source of organic matters into aquatic ecosystems, which includes rivers and oceans, is through primary production. Although biomass from living organisms forms less than 1% of the total organic carbon in seawater, non-living DOM accounts for more than 90% of the organic carbon (Bolan et al. 2011). The DOM that is produced biologically at the ocean surface is consumed by microbial respiration; the refractory component resistant to microbial degradation remains in deep water. In an aquatic ecosystem, DOM is derived from photosynthetic aquatic higher plants and algae, hydrolysis, and transport from terrestrial ecosystems. Aquatic organic matter, in a large portion, comes from the movement of carbon that is produced in terrestrial environments.

In terrestrial ecosystems, as far as forest ecosystems are concerned, the bulk of the DOM comes from throughfall and stemflow, root exudates and decaying roots, as well as decomposition and metabolic by-products and leachates of microbiologically processed soil organic matter (Figure 6.1)

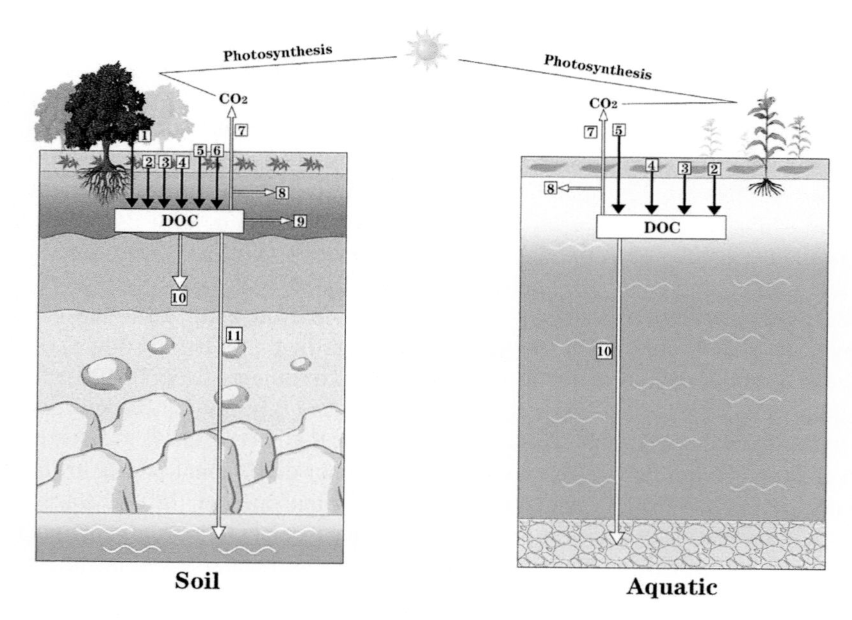

FIGURE 6.1 Pathways of inputs and outputs of dissolved organic matter in forest and agricultural soils (Inputs: 1. Throughfall and stem flow; 2. Root exudates; 3. Microbial lysis; 4. Humification; 5. Litter/and crop residue decomposition; 6. Organic amendments; Outputs: 7. Microbial degradation; 8. Microbial assimilation; 9. Lateral flow; 10. Sorption; and 11. Leaching).

(McDowell 2003; Bolan et al. 2011). Exogenous biological waste materials are added to soils and contain DOM, enhancing the solubilization of natural soil organic matter. Examples include livestock manures. Addition of specific manures, for example, poultry manure and lime-stabilized biosolids, causes the pH to increase and, therefore, improves the solubilization of soil organic matter (Chowdhury et al. 2015). Soils and aquifers contain DOM that can be influenced by cultivation, fire, clear cutting, wetland drainage, acidic precipitation, eutrophication, and climate change (Yallop and Clutterbuck 2009; Ramesh et al. 2018).

Forest ecosystems differ from cultivated and pastoral soils in terms of sources of DOM. In forest environments, the canopy and forest floor layers are the primary sources of DOM, which represent a significant proportion of the total carbon allotment (Kalbitz et al. 2007). Plant remains and residues are the major sources of DOM in cultivated and pastoral soils, while litter and throughfall are the major sources in forest soils (Ghani et al. 2007; Laik et al. 2009). The rhizosphere is commonly associated with a large carbon flux due to root decay and exudation (Muller et al. 2009; Ramesh et al. 2018). Microbial activity in the rhizosphere is enhanced by readily available organic substances that serve as an energy source for these organisms (Paterson et al. 2007; Phillips et al. 2008). Because of its rapid turnover, soil microbial biomass is also considered as an important source of DOM in soils (Steenwerth and Belina 2008).

6.3 PARTICULATE PLASTICS AND DISSOLVED ORGANIC CARBON ASSEMBLAGES

DOM interacts with organic/inorganic moiety in clay minerals, sediments, and PPs, and the nature of the interaction depends on the characteristics of both the DOM and colloidal particles. In the aquatic nature, PPs offer a major solid surface for interactions with DOM, while in a terrestrial ecosystem, soil mineral particles, such as silt and clay provide the major solid surface for interactions with DOM.

DOM constituents can be grouped into the labile/recalcitrant parts (Marschner and Kalbitz 2003). DOM that is labile is made up of mostly carbohydrate compounds (glucose and fructose), low molecular weight organic acids, amino sugars, and low molecular weight proteins (Kaiser et al. 2001). Recalcitrant DOM consists of polysaccharides (breakdown products of cellulose and hemicellulose) and other plant compounds, and/or biogenic degradation products (Marschner and Kalbitz 2003) (Table 6.1). DOM is also grouped into various fractions based on solubility, molecular weight, and sorption chromatography (Table 6.2). Molecular size distribution is one of the important characteristics of DOM that controls the interactions of it with colloidal particles including PPs. In detail, Ogawa and Tanoue (2003) indicated that the low molecular weight fractions (less than 1 kDa) account for 65%–80% of the bulk DOM, while the high molecular weight fractions are minor portions of DOM (ca. 20%–35% for >1 kDa and 2%–7% for >10 kDa). Under environmental pH values, DOM carries a negative net charge due to the presence of carboxylic/phenolic functional groups (Kinniburgh et al. 1999).

PPs in terrestrial and aquatic ecosystems vary in size, shape, and surface characteristics including functional groups and surface charge that play a critical role in the interactions of PPs with DOM. Macroplastic (>5 mm), microplastic (1–5 mm), and nanoplastic (1–1000 nm) particles have been reported in terrestrial and aquatic ecosystems. PPs carry various functional groups that are involved in the adsorption of DOM and contaminants (Bradney et al. 2019). Ibrahim et al. (2017) showed that the functional groups that make up microplastic composition show a strong peak at 3342 cm^{-1} for N-H stretching vibrations from the terminal amine group, 1510 cm^{-1} for the N-H bending, 2852 cm^{-1} for C-H aliphatic stretching modes, 1428 cm^{-1} for bending of C-H_2 vibrations, showing the existence of alkyl chains, and 3342 cm^{-1}, which corresponds to the C=O carbonyl stretching group vibrations. These functional groups are involved in the adsorption of DOM and contaminants. For example, Kim et al. (2017) observed that the immobilization of *Daphnia magna* exposed to Ni combined with PPs in the presence of the -COOH functional group was higher than

TABLE 6.1
Components Identified in Specific Fractions of Dissolved Organic Matter

Fraction	Compounds
Hydrophobic neutrals	Hydrocarbons
	Chlorophyll
	Carotenoids
	Phospholipids
Weak (phenolic) hydrophobic acids	Tannins
	Flavonoids
	Other polyphenols
	Vanillin
Strong (carboxylic) hydrophobic acids	Fulvic acid and humic acid
	Humic-bound amino acids and peptides
	Humic-bound carbohydrates
	Aromatic acids (including phenolic carboxylic acids)
	Oxidized polyphenols
	Long-chain fatty acids
Hydrophilic acids	Humic-like substances with lower molecular size and higher COOH/C ratios
	Oxidized carbohydrates with COOH groups
	Small carboxylic acids
	Inositol and sugar phosphates
Hydrophilic neutrals	Simple neutral sugars
	Non-humic-bound polysaccharides
	Alcohols
Bases	Proteins
	Free amino acids and peptides
	Aromatic amines
	Amino-sugar polymers (such as from microbial cell walls)

Source: Bolan, N.S. et al., *Adv. Agron.*, 110, 1–75, 2011.

TABLE 6.2
Molecular Parameters of Dissolved Organic Matter

DOM-Type	Molecular Weight [kDa]	Most Common Functional Groups	Charge (4 < pH <10)	Solubility in Water
Humic acids	2–5	Aliphatic and aromatic COOH, OH, and OCH_3 aliphatic CO	Negative	Well soluble at high pH
Fulvic acids	0.5–2	Aliphatic and aromatic COOH, OH, and OCH_3 aliphatic CO	Negative	Well soluble
Carbohydrates	0.18–3000	OH, CO, and COOH	Side-group dependent	Side-group dependent
Proteins	10–a few 1000	NH_2, COOH, OH, and SH	Side-group dependent	Side-group dependent
Fatty acids	0.25–0.85	COOH	Negative	Chain-length dependent
Amino acids	<0.2	CNH_2 and COOH	Side-group dependent	Well soluble

Source: Philippe, A., and Schaumann, G.E., *Environ. Sci. Technol.*, 48, 8946–8962, 2014.

that of *D. magna* exposed to Ni combined with PPs alone, indicating that -COOH functional groups enhanced the adsorption of Ni to PPs and the subsequent toxicity to aquatic creatures. The surface area of PPs varies with their size and weathering stage. The specific surface area (surface area per unit mass) increases with decreasing particle size and increases with weathering (Bradney et al. 2019).

An important part in the interaction process of organic/inorganic colloids including PPs is related to the adsorption of DOM onto them. Various mechanisms have been proposed for the adsorption reactions of DOM with colloids including soil particles, nano-materials, and microplastics (Figure 6.2; Philippe and Schaumann 2014). Ligand exchange reactions; Coulomb, *van der Waals*, and hydrophobic forces; hydrogen bonding; cation bridging; and surface ion chelation are the main routes in the DOM adsorption process. A combination of the noted adsorption mechanisms is involved in the interaction of DOM with colloidal particles including PPs (Philippe and Schaumann 2014). Negatively charged DOM adsorbs unselectively onto both net negatively and net positively charged colloidal surfaces including PPs, indicating both hydrophobic partition forces and electrostatic binding reactions are involved in the adsorption process (Bolan et al. 2011).

Adsorption is readily affected by sorbent (particulate plastics), sorbate (DOM), and environmental medium characteristics. These aspects that determine the sorbent properties include reactive surface area, surface charge, crystalline phase, and aggregation state. In the case of PPs, the nature of them and their weathering stage in the environment are important factors affecting the interactions with DOM (Bradney et al. 2019). Factors affecting sorbate (DOM) characteristics are the fractions

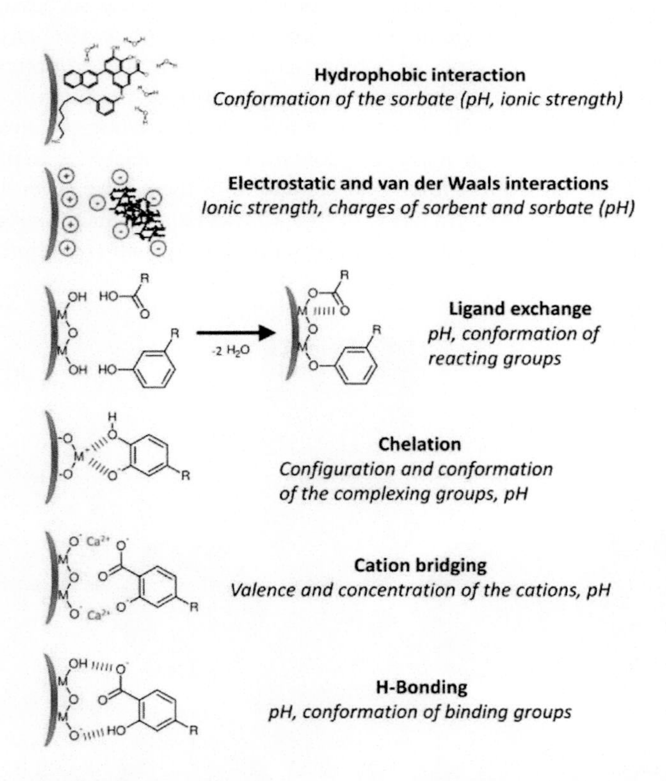

FIGURE 6.2 Schematic description of the diverse sorption mechanisms of DOM on the surfaces of colloid particles including particulate plastics reported in the literature. The most important parameters influencing the interaction are written in italics. (From Philippe, A., and Schaumann, G.E., *Environ. Sci. Technol.*, 48, 8946–8962, 2014.)

of DOM components and the charge, hydrophobicity, and molecular weight distribution. As such, Zhang et al. (2018) noticed that adsorption of oxytetracycline antibiotics by PPs was enhanced more in the presence of humic acid than of fulvic acid, probably because of π-π conjugations between the humic acid and the surface of plastic, which offer strong electrostatic attraction for oxytetracycline. Solution properties such as pH, ionic strength, valence of solute ions, and temperature influence both DOM and particulate plastics characteristics, thereby impacting their interactions.

6.4 IMPLICATIONS OF PPs AND DOM ASSEMBLAGES

PPs and DOM assemblages are important in biofilm formation and subsequent habitation by microbes, especially in the aquatic ecosystems. Figure 6.3 shows particulate plastic-dissolved organic matter assemblages in terrestrial and aquatic ecosystems.

6.4.1 Biofilm Formation

PP fragments in the terrestrial/aquatic environments have been known to host a plethora of microorganisms. In terrestrial ecosystems, soil colloidal particles provide a major substrate for biofilm formation, resulting in microbial habitation. In aquatic ecosystems, PPs provide an important solid surface for biofilm formation. The hydrophobic nature of plastic surfaces facilitates the adsorption of dissolved organic carbon in soil and aquatic ecosystems. This stimulates the rapid growth of microorganisms, thereby resulting in the subsequent creation and accumulation of biofilms. Such facts influence a diverse range of metabolic processes and cause succession of other micro and macro-organisms. The assemblage of ecosystems colonizing the plastic environment is often referred to as the plastisphere. The types of polymers, environmental conditions including nutrient status and salinity, and season affect the microbial community of the biofilm. Microbial habitation of PPs provides a favorable condition for the movement of microorganisms, especially in the aquatic environments, and impacts the toxicity of contaminants associated with these PP fragments. Biofilm formation and the subsequent microbial habitation of them have several implications that include (Figure 9.5, Chapter 9): (i) degradation of PPs; (ii) horizontal gene transfer between microorganisms; (iii) toxicity of contaminants associated with plastics; and (iv) migration or spreading

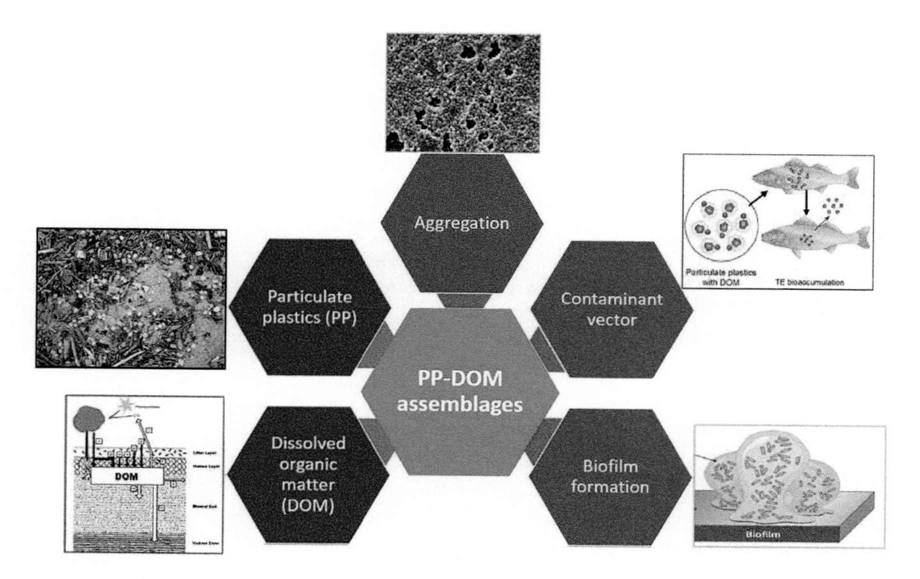

FIGURE 6.3 Implications of PP-DOM assemblages.

of microorganisms especially in aquatic ecosystems, such as oceans and rivers. The formation of biofilms on PPs and the subsequent environmental implications of biofilm formation are described in Chapter 9.

6.4.2 Aggregation of Microplastics

Interactions of DOM with colloids including PPs impacts aggregation properties of DOM-PP assemblages. For example, Michels et al. (2018) noticed that DOM interactions with PPs lead to biofilm formation, and biofilm-covered microplastics have a greater adhesion potential than pristine microplastics, leading to aggregate formation. They demonstrated that the aggregation of microplastics and biogenic particles was enhanced by microbial biofilm formation on the plastic surfaces. The aggregation of microplastics enhances the movement of microplastics from surface layers of oceans to deeper depths, and thus is an important factor in the movement and redistribution of microplastics in oceans.

DOM, for the most part, has an intrinsic negative charge, and hence its adsorption onto colloidal surfaces carries with it a greater negative charge to the surface. Surface charges that are originally positive can be rendered neutral with the additional inclusion of small amounts of DOM. These have the function of reducing the intensity of the Coulomb forces. Consequently, the adsorption of DOM can induce aggregation, resulting from electrostatic attractions between DOM-PP assemblages.

The connection of colloids, such as clays through cation bridging, especially with Ca^{2+}, was observed for humic acids, fulvic acids, and polysaccharides (Philippe and Schaumann 2014). Subsequent aggregation through cation bridging is faster than that through electrostatic destabilization because the macromolecular coating enlarges the diffuse layer (collision radius) of the particles. DOM chains adsorbed onto different colloidal particles can also bridge together through H-bonding, as was observed for a polysaccharide-rich humic acid fraction (Philippe and Schaumann 2014). If the surface charge is initially negative or if the amount of adsorbed DOM is large enough for completely reversing the original surface charge, colloids will be electrostatically stabilized. Furthermore, DOM can complex multivalent cations, which neutralize a part of the negative charges of DOM adsorbed on the surface, thereby modifying the conformation of DOM and the subsequent stabilization of DOM-associated colloidal particles including particulate plastics.

Aggregation of PPs through their interactions with DOM leads to depth-wise redistribution of them in aquatic ecosystems since it offers the density variations. Thus, apart from the accumulation and transport of plastics via food matrices, aggregation and accumulation of plastics and their subsequent weighted sinking are one of the main transport pathways of microplastics in deep-sea sediments. The ramifications of this accumulation process are an increased availability of microplastics to benthic organisms and the accumulation of microplastics over longer time frames within marine sediments. Biofilms can form on surfaces of microplastics that subsequently can reduce the hydrophobic nature of the plastics, making smaller plastic particles and other plastics neutrally buoyant by reducing their density in relation to seawater. Thus, plastics, such as polyethylene, which normally floats at the surface when it enters the ocean in a fresh and clean state, can be brought into suspension. This effect is likely to be intensified by the aggregation of the PPs with denser biogenic particles.

6.4.3 Retention and Transport of Contaminants

A wide variety of pollutants, organic and inorganic, has an interactive role with microplastics (Wijesekara et al. 2016; Wang et al. 2018). Certain environmental conditions, such as weathering and surface area, interaction with DOM, and microbial activity (i.e., biofilm formation) (Jiang et al. 2018) play critical roles in the adsorption and desorption processes of pollutants in PPs in soil and aquatic environments. PPs have a high surface area, which can be increased via exposure to the environmental elements, such as weathering. This increased surface area increases the adsorption of contaminants and also increases the transport of chemical pollutants via leaching (Teuten et al. 2009; Holmes et al. 2014).

Concentrations of pollutants can have an effect on the adsorption-desorption processes of pollutants with particulate plastics, which can then produce long-range transport issues and impacts in aquatic environments (GESAMP 2010). For example, research related to PP interactions with organic contaminants has reported that polychlorinated biphenyls, polycyclic aromatic hydrocarbons, petroleum hydrocarbons, and organochlorine pesticides have all been transported by PPs into marine systems (Teuten et al. 2009), and they can be transferred from PPs into biota (Chua et al. 2014; Wardrop et al. 2016). PPs also interact with and transport inorganic contaminants, such as heavy metals, with this area coming into sharp focus in contemporary research (Brennecke et al. 2016; Wang et al. 2017; Wijesekara et al. 2018) (Figure 6.4).

Pristine PPs are generally hydrophobic in nature and do not interact readily with aqueous solutes including, especially, organic contaminants (hydrophobic pollutants) (Wijesekara et al. 2018). However, DOM, in terrestrial and aquatic ecosystems, interact readily with PPs and forms particulate plastic-DOM assemblages (Cai et al. 2018; Bradney et al. 2019). These PP-DOM assemblages promote PPs as a vector for the transport of contaminants, and its adverse impacts have not been fully elucidated. The adsorption of contaminants to PP-DOM assemblages (biofilms) can cause an accumulation in soil and aquatic ecosystems, providing a further ongoing source of contaminants (Jiang et al. 2018).

Pristine PPs have a lower likelihood of interacting with trace elements when compared to PP-DOM assemblages. In the latter case, there is strong interaction and increased retention of trace elements (Figure 6.4). In detail, Wijesekara et al. (2018) reported the adsorption of trace elements (i.e., Cu) onto PPs that had modified surfaces due to DOM adsorption. The findings implied that modified PPs adsorbed significantly greater concentrations of Cu than pristine PPs (Figure 6.4). Furthermore, long-term pre-modification (e.g., photo-oxidation and attrition of charged materials) that contributes to natural degradation of plastics imparts a metal sorption capability. In sum, DOM attachment contributes to greater metal sorption (Turner and Holmes 2015).

PPs have been shown to adsorb and transport contaminants in aquatic environments (Holmes et al. 2014; Turner and Holmes 2015). Hence, microplastics can be a pathway for exposure to chemicals (e.g., polychlorinated biphenyls and polycyclic aromatic hydrocarbons) and heavy metals that

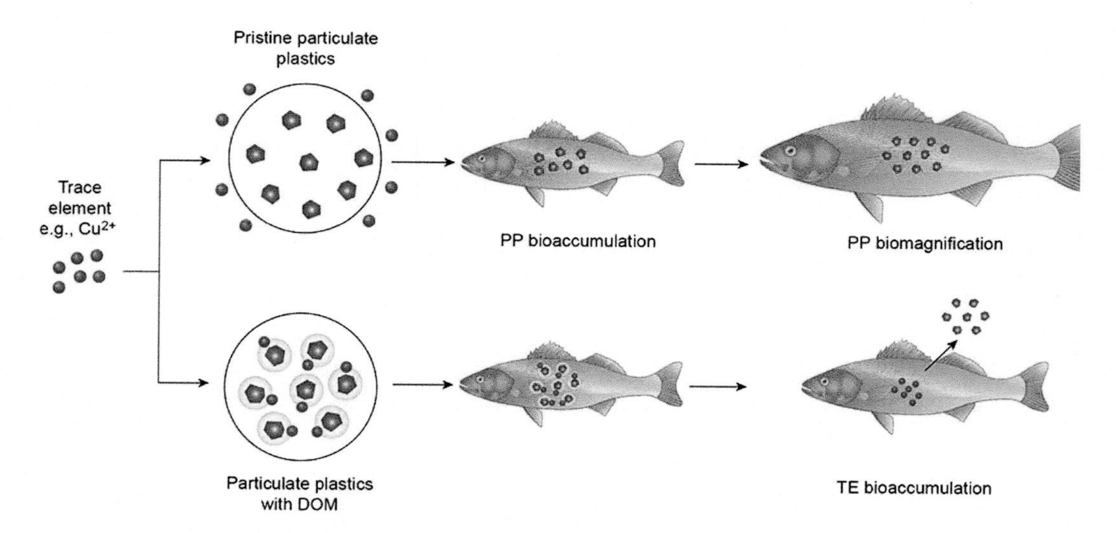

FIGURE 6.4 Conceptual diagram illustrating the impact of PPs derived trace elements (TE) to aquatic organisms (Cu = copper). (From Bradney, L. et al., *Environ. Int.*, 131, 104937, 2019.)

are widely distributed by microplastics. This can then cause toxicity and lead to increasing bioaccumulation in aquatic and terrestrial environments (Chua et al. 2014; Wardrop et al. 2016). Also, PPs in the environment cause detrimental effects on earthworms (Huerta Lwanga et al. 2016), aquatic birds (Holland et al. 2016; Zhao et al. 2016), freshwater invertebrates (Blarer and Burkhardt-Holm 2016), fish (Lusher et al. 2017), and oysters (Vegter et al. 2014; Cole and Galloway 2015).

The environmental effects of plastics are a cause for concern, and various regulatory bodies have called for the prohibition of consumer grade PPs including microbeads (https://www.beatthemicrobead.org/). However, even if concerned parties move to ban the use of plastics including PPs, currently, there are millions of tons of plastics that have already abandoned environmental ecosystems, and these plastics are persistent and could detrimentally remain for long durations without cleanup and interventions (Rillig et al. 2017).

6.5 CONCLUSIONS

Pristine PPs are generally hydrophobic in nature and do not interact readily with aqueous solutes including hydrophobic organic pollutants, such as polycyclic aromatic hydrocarbons. However, DOM, in terrestrial and aquatic ecosystems, interact with PPs and forms PM-DOM assemblages. The association of DOM with PPs in terrestrial and aquatic ecosystems impacts the chemo and biodynamics of them, and plays a crucial role in the environmental implications of them in these ecosystems. PPs and DOM assemblages are important in the aggregation of microplastics, which facilitates the redistribution of the assemblages in the aquatic environment; in the retention and transport of contaminants; and in biofilm formation and habitation by microbes. Association of DOM with PPs expedites the aggregation of particulate plastics, especially in aquatic ecosystems. Similarly, DOM and PPs enhance the adsorption of both heavy metals and organic contaminants, thereby serving as a vector for these pollutants. Nonetheless, the quantitative analysis in line with their detrimental environmental consequences has not been fully elucidated.

REFERENCES

Blarer, P., Burkhardt-Holm, P. 2016. Microplastics affect assimilation efficiency in the freshwater amphipod *Gammarus fossarum*. *Environmental Science and Pollution Research* 23: 23522–23532.

Bolan, N.S., Adriano, D.C., Kunhikrishnan, A., James, T., McDowell, R., Senesi, N. 2011. Dissolved organic matter: Biogeochemistry, dynamics, and environmental significance in soils. *Advances in Agronomy* *110*(C), 1–75.

Bradney, L., Wijesekara, H., Palansooriya, K.N., Obadamudalige, N., Bolan, N.S., Ok, Y.S., Rinklebe, J., Kim, K.H., Kirkham, M.B. 2019. Particulate plastics as a vector for toxic trace-element uptake by aquatic and terrestrial organisms and human health risk. *Environment International* 131: 104937.

Brennecke, D., Duarte, B., Paiva, F., Caçador, I., Canning-Clode, J. 2016. Microplastics as vector for heavy metal contamination from the marine environment. *Estuarine, Coastal and Shelf Science* 178: 189–195.

Browne, M.A., Crump, P., Niven, S.J., Teuten, E., Tonkin, A., Galloway, T., Thompson, R. 2011. Accumulation of microplastic on shorelines worldwide: Sources and sinks. *Environmental Science & Technology* 45: 9175–9179.

Cai, L., Hu, L., Shi, H., Ye, J., Zhang, Y., Kim, H. 2018. Effects of inorganic ions and natural organic matter on the aggregation of nanoplastics. *Chemosphere* 197: 142–151.

Chowdhury, S., Farrell, M., Butler, G., Bolan, N. 2015. Assessing the effect of crop residue removal on soil organic carbon storage and microbial activity in a no-till cropping system. *Soil Use & Management* 31: 450–460.

Chua, E.M., Shimeta, J., Nugegoda, D., Morrison, P.D., Clarke, B.O. 2014. Assimilation of polybrominated diphenyl ethers from microplastics by the marine amphipod, *Allorchestes compressa*. *Environmental Science & Technology* 48: 8127–8134.

Cole, M., Galloway, T.S. 2015. Ingestion of nanoplastics and microplastics by pacific oyster larvae. *Environmental Science & Technology* 49: 14625–14632.

Cole, M., Lindeque, P., Halsband, C., Galloway, T.S. 2011. Microplastics as contaminants in the marine environment: A review. *Marine Pollution Bulletin* 62: 2588–2597.

Dafner, E.V, Wangersky, P.J. 2002. A brief overview of modern directions in marine DOC studies. Part I: Methodological aspects. *Journal of Environmental Monitoring* 4: 48–54.

Duis, K., Coors, A. 2016. Microplastics in the aquatic and terrestrial environment: Sources (with a specific focus on personal care products), fate and effects. *Environmental Sciences Europe* 28: 1–25.

GESAMP. 2010. GESAMP Reports and Studies. *Proceedings of the GESAMP International Workshop on Plastic Particles as a Vector in Transporting Persistent, Bio-Accumulating and Toxic Substances in the Oceans* 82: 68.

Ghani, A., Dexter, M., Carran, R.A., Theobald, P.W. 2007. Dissolved organic nitrogen and carbon in pastoral soils: The New Zealand experience. *European Journal of Soil Science* 58: 832–843.

Holland, E.R., Mallory, M.L., Shutler, D. 2016. Plastics and other anthropogenic debris in freshwater birds from Canada. *Science of The Total Environment* 571: 251–258.

Holmes, L.A., Turner, A., Thompson, R.C. 2014. Interactions between trace metals and plastic production pellets under estuarine conditions. *Marine Chemistry* 167: 25–32.

Huerta Lwanga, E., Gertsen, H., Gooren, H., Peters, P., Salanki, T., van der Ploeg, M., Besseling, E., Koelmans, A.A., Geissen, V. 2016. Microplastics in the terrestrial ecosystem: Implications for *Lumbricus terrestris* (Oligochaeta, Lumbricidae). *Environmental Science & Technology* 50: 2685–2691.

Ibrahim, Y.S., Rathnam, R., Anuar, S.T., Khalik, W.M.A.W.M. 2017. Isolation and characterisation of microplastic abundance in Lates calcarifer from Setiu Wetlands, Malaysia. *Malaysian Journal of Analytical Sciences* 21: 1054–1064.

Jiang, P., Zhao, S., Zhu, L., Li, D. 2018. Microplastic-associated bacterial assemblages in the intertidal zone of the Yangtze Estuary. *Science of the Total Environment* 624: 48–54.

Kaiser, K., Guggenberger, G., Haumaier, L., Zech, W. 2001. Seasonal variations in the chemical composition of dissolved organic matter in organic forest floor layer leachates of old-growth Scots pine (*Pinus sylvestris* L.) and European beech (*Fagus sylvatica* L.) stands in northeastern Bavaria, Germany. *Biogeochemistry* 55: 103–143.

Kalbitz, K., Meyer, A., Yang, R., Gerstberger, P. 2007. Response of dissolved organic matter in the forest floor to long-term manipulation of litter and throughfall inputs. *Biogeochemistry* 86: 301–318.

Kim, D., Chae, Y., An, Y.J. 2017. Mixture toxicity of nickel and microplastics with different functional groups on Daphnia magna. *Environmental Science & Technology* 51: 12852–12858.

Kinniburgh, D.G., van Riemsdijk, W.H., Koopal, L.K., Borkovec, M., Benedetti, M.F., Avena, M.J. 1999. Ion binding to natural organic matter: Competition, heterogeneity, stoichiometry and thermodynamic consistency. *Colloids and Surfaces A: Physicochemical and Engineering Aspects* 151: 147–166.

Laik, R., Kumar, K., Das, D.K., Chaturvedi, O.P. 2009. Labile soil organic matter pools in a calciorthent after 18 years of afforestation by different plantations. *Applied Soil Ecology* 42: 71–78.

Lusher, A.L., Welden, N.A., Sobral, P., Cole, M. 2017. Sampling, isolating and identifying microplastics ingested by fish and invertebrates. *Analytical Methods* 9: 1346–1360.

Mahon, A.M., O'Connell, B., Healy, M.G., O'Connor, I., Officer, R., Nash, R., Morrison, L. 2016. Microplastics in sewage sludge: Effects of treatment. *Environmental Science & Technology* 51: 810–818.

Marschner, B., Kalbitz, K. 2003. Controls of bioavailability and biodegradability of dissolved organic matter in soils. *Geoderma* 113: 211–236.

McDowell, W.H. 2003. Dissolved organic matter in soils: Future directions and unanswered questions. *Geoderma* 113: 179–186.

Michels, J., Stippkugel, A., Lenz, M., Wirtz, K., Engel, A. 2018. Rapid aggregation of biofilm-covered microplastics with marine biogenic particles. *Proceedings of the Royal Society B: Biological Sciences* 285: 20181203.

Muller, M., Alewell, C., Hagedorn, F. 2009. Effective retention of litter-derived dissolved organic carbon in organic layers. *Soil Biology and Biochemistry* 41: 1066–1074.

Ogawa, H., Tanoue, E. 2003. Dissolved organic matter in oceanic waters. *Journal of Oceanography* 59: 129–147.

Paterson, E., Gebbing, T., Abel, C., Sim, A., Telfer, G. 2007. Rhizodeposition shapes rhizosphere microbial community structure in organic soil. *New Phytologist* 173: 600–610.

Philippe, A., Schaumann, G.E. 2014. Interactions of dissolved organic matter with natural and engineered inorganic colloids: A review. *Environmental Science & Technology* 48: 8946–8962.

Phillips, R.P., Erlitz, Y., Bier, R., Bernhardt, E.S. 2008. New approach for capturing soluble root exudates in forest soils. *Functional Ecology* 22: 990–999.

Ramesh, T., Manjaiah, K., Abinandan, S., Arunachalam, A., Rajasekar, K., Nishant, A.D., Ngachan, S.V. 2018. Evaluating organic carbon fractions, temperature sensitivity and artificial neural network modeling of CO_2 efflux in soils: Impact of land use change in subtropical India (Meghalaya). *Ecological Indicators* 93: 129–141.

Rillig, M.C. 2012. Microplastic in terrestrial ecosystems and the soil? *Environmental Science & Technology* 46: 6453–6454.

Rillig, M.C., Ziersch, L., Hempel, S. 2017. Microplastic transport in soil by earthworms. *Scientific Reports* 7: 1–6.

Rochman, C.M. 2018. Microplastics research: From sink to source. *Science* 360: 28–29.

Steenwerth, K., Belina, K.M. 2008. Cover crops enhance soil organic matter, carbon dynamics and microbiological function in a vineyard agroecosystem. *Applied Soil Ecology* 40: 359–369.

Teuten, E.L., Saquing, J.M., Knappe, D.R.U., Barlaz, M.A., Jonsson, S., Björn, A., Rowland, S.J., et al. 2009. Transport and release of chemicals from plastics to the environment and to wildlife. *Philosophical Transactions of the Royal Society B: Biological Sciences* 364: 2027–2045.

Thompson, R.C., Swan, S.H., Moore, C.J., vom Saal, F.S. 2009. Our plastic age. *Philosophical Transactions of the Royal Society B: Biological Sciences* 364: 1973–1976.

Turner, A., Holmes, L.A. 2015. Adsorption of trace metals by microplastic pellets in fresh water. *Environmental Chemistry* 12: 600–610.

UNEP. (United Nations Environmental Programme). 2016. *Marine Plastic Debris and Microplastics: Global Lessons and Research to Inspire Action and Guide Policy Change.* https://wedocs.unep.org/handle/20.500.11822/7720 (Accessed September 16, 2018.)

Vegter, A.C., Barletta, M., Beck, C., Borrero, J., Burton, H., Campbell, M.L., Costa, M.F., et al. 2014. Global research priorities to mitigate plastic pollution impacts on marine wildlife. *Endangered Species Research* 25: 225–247.

Wang, F., Wong, C.S., Chen, D., Lu, X., Wang, F., Zeng, E.Y. 2018. Interaction of toxic chemicals with microplastics: A critical review. *Water Research* 139: 208–219.

Wang, J., Peng, J., Tan, Z., Gao, Y., Zhan, Z., Chen, Q., Cai, L. 2017. Microplastics in the surface sediments from the Beijiang River littoral zone: Composition, abundance, surface textures and interaction with heavy metals. *Chemosphere* 171: 248–258.

Wardrop, P., Shimeta, J., Nugegoda, D., Morrison, P.D., Miranda, A., Tang, M., Clarke, B.O. 2016. Chemical pollutants sorbed to ingested microbeads from personal care products accumulate in fish. *Environmental Science & Technology* 50: 4037–4044.

Wijesekara, H., Bolan, N.S., Bradney, L., Obadamudalige, N., Seshadri, B., Kunhikrishnan, A., Dharmarajan, R., et al. 2018. Trace element dynamics of biosolids-derived microbeads. *Chemosphere* 199: 331–339.

Wijesekara, H., Bolan, N.S., Vithanage, M., Xu, Y., Mandal, S., Brown, S.L., Hettiarachchi, G.M., et al. 2016. Utilization of biowaste for mine spoil rehabilitation. *Advances in Agronomy* 138: 97–173.

Yallop, A.R., Clutterbuck, B. 2009. Land management as a factor controlling dissolved organic carbon release from upland peat soils 1: Spatial variation in DOC productivity. *Science of the Total Environment* 407: 3803–3813.

Zhang, H., Wang, J., Zhou, B., Zhou, Y., Dai, Z., Zhou, Q., Chriestie, P., Luo, Y. 2018. Enhanced adsorption of oxytetracycline to weathered microplastic polystyrene: Kinetics, isotherms and influencing factors. *Environmental Pollution* 243: 1550–1557.

Zhao, S., Zhu, L., Li, D. 2016. Microscopic anthropogenic litter in terrestrial birds from Shanghai, China: Not only plastics but also natural fibers. *Science of the Total Environment* 550: 1110–1115.

7 Characteristics of Particulate Plastics in Terrestrial Ecosystems

Kumuduni Niroshika Palansooriya, Hasintha Wijesekara, Lauren Bradney, Prasanna Kumarathilaka, Jochen Bundschuh, Nanthi S. Bolan, Teresa Rocha-Santos, Cheng Gu, and Yong Sik Ok

CONTENTS

7.1 Introduction 107
7.2 Characteristics of Particulate Plastics 109
7.2.1 Physical Properties 109
7.2.1.1 Size 109
7.2.1.2 Color 110
7.2.1.3 Shape 110
7.2.2 Chemical Properties 111
7.2.2.1 Polymer Type 111
7.2.2.2 Associated Chemical Bonds 111
7.2.2.3 Chemical Additives 112
7.2.3 Advanced Characteristics 112
7.2.3.1 Trace Metals 112
7.2.3.2 Biodegradable and Biosynthetic Particulate Plastics 113
7.3 Identification and Quantification of Particulate Plastics 114
7.3.1 Extraction 115
7.3.2 Removal of Organic Matter 116
7.3.3 Identification and Characterization of Particulate Plastics 116
7.3.3.1 Morphological Characterization 116
7.3.3.2 Chemical Characterization 117
7.4 Conclusions 119
Acknowledgment 120
References 120

7.1 INTRODUCTION

Rapid development in industrial and agricultural activities and overconsumption by humans have sped up the manufacturing of plastic wastes and their consequent disposal into terrestrial and aquatic ecosystems. Recent statistics showed that worldwide production of plastics has increased at an unprecedented rate (from 1.5×10^6 tons in 1950 to 335×10^6 tons in 2016)

(Liu et al. 2018). Even though different methods (e.g., recycling, reuse, and reduction) have been introduced to minimize the disposal of plastics, most of the plastic debris is still released into the environment.

Contamination of terrestrial environments with particulate plastics (PPs) is continuously increasing and is considered to be one of the major threats to human and animal health (Horton et al. 2017). Depending upon the characteristics of the PPs in terrestrial environments, their physical and biogeochemical behavior and risks to humans and animals can show large variation (Bradney et al. 2019). For instance, PP size is an important physical characteristic that determines the occurrence and movement of PPs in soil. After being disposed of in the environment, PPs gradually break down into small fragments, due to the action of physical, chemical, and biological drivers. Liu et al. (2018) demonstrated that two different types of PPs, namely, microplastics (sizes of 20 μm–5 mm) and mesoplastics (5 mm–2 cm), occurred in farmland soils of Shanghai, China. Their study also found that the abundance of microplastics and mesoplastics were 78.0 and 6.75 items/kg in shallow soils, respectively, whereas the abundance of microplastics and mesoplastics were 62.5 and 3.3 items/kg in deep soils, respectively. The results show that the distribution of PPs varies in the soil profile due to their different sizes. These PPs can be subjected to migration and transformation through various means, such as soil erosion, mechanical abrasion, and photodegradation (Song et al. 2017; Hurley and Nizzetto 2018); however, the degree of migration and transformation depends on where the PPs are present in the soil profile.

The size, color, shape, and abundance of PPs in terrestrial environments are the key factors that determine their bioavailability to soil organisms. Due to distinguishable sizes, shapes, or colors, PPs can be readily ingested by terrestrial animals, especially birds (Zhao et al. 2016; Auta et al. 2017). For example, seabirds, such as little auks are found to be color selective on their prey. These little auks generally consume light-colored PPs rather than dark-colored PPs, thereby signifying active contamination of birds that mistake PPs for their natural prey (Amélineau et al. 2016). Furthermore, Huerta Lwanga et al. (2016) reported that earthworms (*Lumbricus terrestris*) preferentially take in small-sized (<50 μm) PPs, exhibiting a size selective ingestion. These types of feeding behaviors of terrestrial organisms must be studied to understand the fate and risk of PPs in terrestrial ecosystems.

The chemical properties of PPs also play an important role in terrestrial ecosystems. The chemical composition and the availability of various chemical additives in PPs may determine their degradability and associated risk to terrestrial ecosystems. Various additives, such as stabilizers, plasticizers, monomers, antioxidants, clarifiers, colorants, and flame retardants are used during manufacturing of plastics to improve their properties (de Souza Machado et al. 2018). Therefore, PP products have embedded within their morphological structure a complex chemical formation. In the soil, these additives can leach out and may affect soil fauna. In addition, soil PPs can also adsorb contaminants, such as polycyclic aromatic hydrocarbons, polychlorinated biphenyls, and organochlorine pesticides because of their hydrophobic surface (Horton et al. 2017). Subsequently, these toxicants can be released into terrestrial environments depending upon environmental conditions. Previous studies observed that organic and inorganic contaminants released from PPs accumulate and adversely affect terrestrial organisms, such as soil microorganisms (Wijesekara et al. 2018), earthworms (Hodson et al. 2017), and seabirds (Fife et al. 2015).

With the increasing concern regarding the potential impacts of PPs in terrestrial environments, there is a growing need to develop techniques to extract, identify, characterize, and quantify these impacts. The methods for analysis of PPs in terrestrial environments follow the same ones used for sediments and water columns, since no standardized methods have been developed for extracting and quantifying PPs in terrestrial environments. Techniques, such as scanning electron microscopy–energy dispersive X-ray spectroscopy, environmental scanning microscopy–energy dispersive X-ray spectroscopy, Fourier-transform infrared spectroscopy (FTIR), micro-FTIR (μ-FTIR) and micro-Raman (μ-Raman) spectrometry, thermal desorption gas chromatography mass spectrometry (TED-GCMS), and pyrolysis gas chromatography mass spectrometry (Py-GCMS) are used for morphological and chemical characterizations of PPs.

In this chapter, the characteristics of PPs in terrestrial ecosystems are discussed, so as to thoroughly outline their behavioral impacts on terrestrial environments and potential risks. Summaries of physical, chemical, and advanced characteristics of PPs, and different extraction, identification, and quantification techniques, are also presented, and research gaps are highlighted.

7.2 CHARACTERISTICS OF PARTICULATE PLASTICS

Particulate plastics are broadly defined as synthetic polymers less than 5 mm in size, and they can be of a primary or secondary origin (Frias and Nash 2019). However, this broad definition means that PPs can encompass a diverse range of physical and chemical properties or characteristics. Polyethylene, polypropylene, polystyrene, polyester, polyamide, and polyacrylonitrile are a few examples of the many synthetic polymers used to manufacture PPs. Additionally, chemicals such as phthalates, bisphenol A (BPA), or trace metals are incorporated with the polymers as additives. Furthermore, the negative environmental impacts of PPs have resulted in renewed interest in biodegradable plastics and environmentally sustainable alternatives.

The physical and chemical properties of PPs play an important role in terrestrial ecosystems. For instance, properties, such as surface roughness, hydrophilic/hydrophobic properties, chemical structure, and molecular weight have an impact on PP vulnerability to biodegradation. Table 7.1 shows the impact of PP properties on their susceptibility to biodegradation.

7.2.1 Physical Properties

7.2.1.1 Size

Particulate plastic is a broad term encompassing micro and nanoplastics. The literature, however, is often divided when it comes to defining their size, thereby making comparisons of results between papers difficult. Arthur et al. (2009) first defined microplastics as <5 mm. Although generally accepted (Ng et al. 2018; Ferreira et al. 2019; Schwaferts et al. 2019), researchers have argued that the upper size limit of microplastics should be more "intuitive," that is, <1 mm (Browne et al. 2011; Van Cauwenberghe et al. 2013).

TABLE 7.1
General Impact of Certain Particulate Plastic Properties on Susceptibility to Biodegradation

Property	Impact on Biodegradation	Sample Format
Molecular weight	Only low molecular weight compounds can be assimilated by microbial cells and enzymatically degraded. Carbon-chain backbones do not biodegrade until the molecular weight is <1000 g/mol	Molecular
Chemical structure and morphology	Certain functional groups provide sites for enzymatic cleavage (ester, ether, amide, urethane)	Molecular
	Branched structures are more difficult for microbes to assimilate	Molecular
	Amorphous materials biodegrade faster than crystalline ones	Macro
Surface hydrophobicity	Hydrophobic surfaces inhibit biofilm formation, hydrophilic surfaces (water contact angle 40°–70°) promote it	Surfaces of thin films
Water absorption	Bulk hydrophilicity and water absorption give microbes access throughout the bulk material	Macro
	Water absorption softens polymers, and softer materials biodegrade faster than harder ones	Macro
Surface roughness	Microbes adhere to rougher surfaces more easily than smooth ones	Surfaces of thin films

Source: Reproduced from Ng, E.-L. et al., *Sci. Total. Environ.*, 627, 1377–1388, 2018. With the permission from publisher.

The lower size limit of microplastics is far more varied, and, consequently, there is often an uncertainty between the upper limits of nanoplastics and the lower limits of microplastics. While nanoplastics typically begin at 100–1000 nm (i.e., 0.1–1 μm) (Gigault et al. 2018; Ng et al. 2018; Ferreira et al. 2019), a variety of values has been reported for the lower size limit of microplastics, e.g., 0.1 μm, 1 μm, 20 μm, or 333 μm (Arthur et al. 2009; Wagner et al. 2014; Ng et al. 2018; Schwaferts et al. 2019). Other papers also provided sub-categories to further classify plastics. For example, Wagner et al. (2014) specified "large" microplastics as being between 1 and 5 mm and "small" microplastics as 20 μm–1 mm. Additionally, a paper from Schwaferts et al. (2019) used the term "sub-μ-plastic" for plastics between 100 nm and 1 μm.

Alternatively, some papers do not use the term nanoplastics, instead choosing to leave the lower limits of microplastics undefined. However, Arthur et al. (2009) did acknowledge that sampling restrictions (i.e., net size) meant that recovering particulates smaller than 333 μm from the marine environments would be a challenge. Although the use of 333 μm mesh nets is standard sampling practice, mesh sizes can be smaller, for example, 80 μm plankton nets (Dris et al. 2015). Therefore, sampling restrictions are influential in determining the size range of microplastics. However, these restrictions may differ between aquatic and terrestrial environments, thereby hindering comparisons between them.

This chapter uses the term PPs, which encompasses both micro and nanoplastics, and is therefore defined as ranging from 5 mm down to the nano-meter range.

7.2.1.2 Color

Particulate plastics are manufactured in a wide array of colors. Commonly reported colors include black (Gewert et al. 2017; Anderson et al. 2018; Nelms et al. 2018), clear (Nelms et al. 2018), red (Gewert et al. 2017; Anderson et al. 2018; Nelms et al. 2018), blue (Eriksson and Burton 2003; Gewert et al. 2017; Anderson et al. 2018), white (Eriksson and Burton 2003; Gewert et al. 2017), green (Eriksson and Burton 2003; Gewert et al. 2017), yellow (Eriksson and Burton 2003), and brown (Eriksson and Burton 2003). Furthermore, the color of the PP is often influenced by its shape. A substantial percentage of fibers were reported as red, blue, and black (Gewert et al. 2017; Anderson et al. 2018; Nelms et al. 2018), whereas pellets were often reported as white (pristine) or yellow and off-white (aged) (Karkanorachaki et al. 2018; Wang et al. 2018; Acosta-Coley et al. 2019).

The particulate plastic's colors also influence how easily they are observed, and therefore can impact their sampling likelihood. PPs that are brighter in color, for example, red, are more likely to be observed and extracted than those that blend into the substrate (Nelms et al. 2018). PP color may also influence the concentrations that are ingested by organisms. An organism may mistakenly ingest PP if the plastic's color closely resembles its normal food. Although papers readily report color as a characteristic, it is subjective, and should not be relied on to accurately determine and differentiate PPs (Lusher et al. 2017).

7.2.1.3 Shape

Particulate plastics can be manufactured into a range of different shapes (Figure 7.1), e.g., fibers, beads, hexagonal patterns, fragments, foams (i.e., polystyrene), and pellets (i.e., nurdles) (Lusher et al. 2017). Fibers are one of the more common shapes and are used in the clothing industry to make synthetic textiles (Frias and Nash 2019). Unfortunately, clothes that are machine washed will shed fibers. This results in the release of PPs into the environment. One study found that about 100 fibers were released into washing machine effluent per polyester fabric, with one garment shedding about 1900 fibers (Browne et al. 2011). Furthermore, polyester fleece sheds 180% more fibers when compared to polyester blankets and shirts (Browne et al. 2011). PPs can also be spherical (i.e., beads), and these are commonly referred to as microbeads. Microbeads have received considerable attention in recent years and are now banned from many commercial cosmetic products (U.S. House of Representatives 2015; Canadian Environmental Protection Act 2017). Fragments are also prevalent in the environment, but are not usually manufactured into a microbead shape. Instead, they occur as secondary PPs (i.e., the breakdown of larger plastic) (Tanaka and Takada 2016).

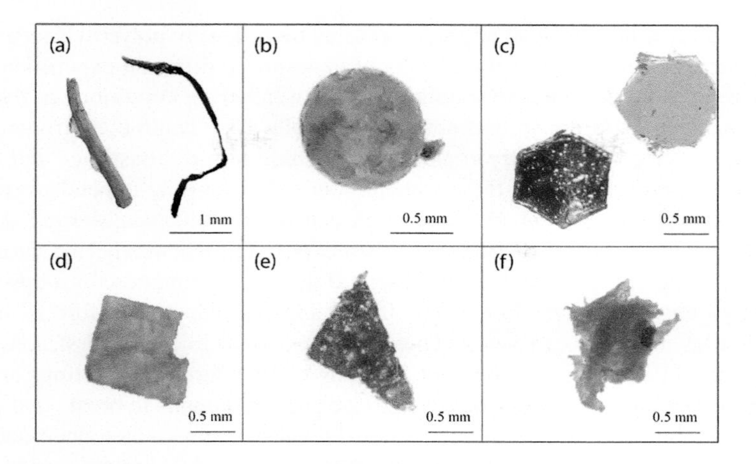

FIGURE 7.1 Selected examples of differently shaped particulate plastics found in Australian biosolids and compost under stereomicroscopy: (a) fibers, (b) spherical, (c) hexagonal, (d) square, (e) triangle, and (f) irregular. (Reproduced from Bradney, L. et al., *Environ. Int.*, 131, 104937, 2019. With the permission from publisher.)

7.2.2 Chemical Properties

7.2.2.1 Polymer Type

Particulate plastics are manufactured using a range of base polymers, e.g., polyethylene, polypropylene, and polystyrene. The choice in polymer is dependent on how a product will be used and its desired properties and characteristics. For example, low-density polyethylene is a flexible polymer and is used for plastic bags and film wrap, whereas polystyrene is an excellent insulator, making it ideal for roofing, refrigerators, and disposable food containers (Wijesekara et al. 2018).

Polyester, polyamide, and polyacrylonitrile are frequently used as synthetic fibers (Schwarz et al. 2019). However, one study found that a large proportion of fibers released into the environment are made from "natural" sources, for example, wool and cotton, rather than PPs (Stanton et al. 2019). Another study found that two-thirds of sampled fibers were made from synthetic rubber (i.e., neoprene, nitrile rubber, styrene butadiene rubber, and ethylene propylene), with nearly half the fibers made from ethylene propylene (Nelms et al. 2018). The remaining third were identified as synthetic plastics (i.e., polyacrylamide (PAM), polyethylene terephthalate, poly butylene terephthalate, polyethylene, and polypropylene) (Nelms et al. 2018). Most of the fibers found in the environment are made from plastic-based polymers, although this is not always the case. Therefore, fibers must be properly identified before being classified as PPs.

Microbeads replaced natural products (e.g., ground almonds and apricot pits) to provide exfoliation from facial cleansers and other personal care products. Although many countries are now banning the use of microbeads, some products still contain them. A 2018 report by O'Farrell (2018) analyzed personal care and cosmetic products sold commercially in Australia. Polyethylene appeared on the ingredients list of 138 products. This was followed by nylon-12 contained within 92 products, polymethyl methacrylate within 33 products, and polyethylene terephthalate within 15 products. All other PPs were found in less than 10 products and included polylactic acid (PLA), polypropylene, polytetrafluoroethylene, and several other nylon polymers.

7.2.2.2 Associated Chemical Bonds

Fourier-transform infrared spectroscopy is commonly used to identify PP polymers based on their chemical bonds. However, chemical bonds may be affected by factors such as weathering. A study conducted by Brandon et al. (2016) examined changes in the chemical bonds (i.e., hydroxyl, carbonyl groups, and carbon-oxygen) of polypropylene, high-density polyethylene, and low-density polyethylene pellets before and after three years of exposure to different experimental treatments. The four treatments were designed to simulate natural weathering conditions in the marine environments and included exposure to darkness and seawater (i.e., benthic environments), seawater and sunlight (i.e., marine surface), dry conditions and sunlight (i.e., coastline), and dry conditions and darkness (i.e., control). Overall, the chemical bonds showed time-dependent (i.e., non-linear) changes in their bond indices. For example, high-density polyethylene showed decreased peak heights for all three chemical bond structures when exposed to the weathering treatment of darkness and seawater. Ioakeimidis et al. (2016) observed that native functional groups decreased and new functional groups (i.e., alkyne bonds at 620 cm^{-1} and 1435 cm^{-1}) were formed in polyethylene terephthalate bottles aged for <15 years. These changes in chemical bond structures hinder the identification of aged PPs in the environment, as many FTIR reference spectrums are designed for pristine polymers. Degradation by UV light can cause chemical bonds to break, and weathered PPs may therefore have slight variations in their chemical bond structures. Photodegradable plastics are specifically designed to undergo UV degradation, as they contain UV-sensitive additives that cause a weakening of bonds when exposed to sunlight (Fotopoulou and Karapanagioti 2019).

7.2.2.3 Chemical Additives

Particulate plastics not only consist of synthetic polymers, but also contain several chemical additives. These additives are used to help improve the physical and chemical properties of the plastics. For example, additives may be included to improve the mechanical strength of the plastics (e.g., fillers and reinforcements), increase resistance to degradation from heat and light (e.g., stabilizers), improve the flexibility of the plastic (e.g., plasticizers), or function as a flame retardant (Fries et al. 2013; Avio et al. 2017; Hahladakis et al. 2018; Godoy et al. 2019).

Many additives are included within the manufacturing of PPs, e.g., nonylphenol, phthalates, BPA, and polybrominated diphenyl ethers (Talsness et al. 2009; Koelmans et al. 2014). Phthalates are used as plasticizers and are commonly added to rigid PPs, such as polyvinyl chloride (Fries et al. 2013; Hahladakis et al. 2018). Furthermore, phthalates are susceptible to leaching, thereby adding another source of potential contamination caused by PPs; this contamination has been shown to affect marine organisms (Halden 2010; Fries et al. 2013). BPA is another well-known plastic additive. It is used as a flame retardant and stabilizer or antioxidant, where it is commonly used in polycarbonate plastics (Hahladakis et al. 2018). BPA has received considerable negative attention surrounding its use in food containers due to leaching and its subsequent implications on human health (e.g., endocrine disruptor). Other studies have also demonstrated that BPA can impact marine organisms (Anderson et al. 2016). Polybrominated diphenyl ethers is included in PPs as a flame retardant (Hahladakis et al. 2018). Unfortunately, polybrominated diphenyl ethers is a persistent organic pollutant, a known endocrine disruptor, and can cause neurotoxicity in certain organisms. Nonylphenol acts as an antioxidant or plasticizer and can be added to polyvinyl chloride and high-density polyethylene plastic bottles (Loyo-Rosales et al. 2004; Koelmans et al. 2014). These chemical additives are transported by the PPs and leach into the environment over time. Therefore, organisms not normally exposed to the chemicals may ingest them, thereby causing adverse effects.

7.2.3 Advanced Characteristics

7.2.3.1 Trace Metals

Trace metals may be introduced to the PPs either as additives or as a by-product of the manufacturing process. A previous study suggested that a majority of trace elements found on PPs are derived from the manufacturing process (Wang et al. 2017). A study conducted by Godoy et al. (2019) identified trace metal additives (i.e., Al and Mg) within cosmetic PPs in Spain, and another study, conducted by Fries et al. (2013), found Al, Ti, Ba, and Zn. Furthermore, titanium dioxide nanoparticles (added as white pigments or UV blockers) were also identified on the surface of PPs (Fries et al. 2013). Pb and Cd were often included in plastics as inorganic pigments. However, environmental laws and regulations have facilitated a reduction in the use of these additives (Hahladakis et al. 2018). Pb and Cd are still used as heat stabilizers in plastics, such as polyvinyl chloride (e.g., medical grade), along with Ba, Zn, and Sn (Sastri 2010). However, they are strongly bound to the polymer matrix and the likelihood of migration and leaching from the polymer is small.

Alternatively, trace metals can also sorb to PPs from the environment, and this process is influenced by factors, such as weathering and aging of the plastic's surface (Bradney et al. 2019). Particulate plastics can act as a vector for trace metal contamination. Furthermore, PPs and associated trace metals can be ingested by organisms, leading to bioaccumulation and food web contamination.

7.2.3.2 Biodegradable and Biosynthetic Particulate Plastics

In recent years, there has been a shift toward using biodegradable polymers due to concerns surrounding the environmental persistence of non-biodegradable PPs. However, not all the additives in biodegradable plastics completely degrade.

Plastic fibers have received considerable attention regarding biodegradable and biosynthetic alternatives. A study recently reported that clothing fibers can be made from proteins that are commonly found in squid tentacles. These "biosynthetic" fibers would break down naturally and thus have no impact on organisms or the environment (Gabbatiss 2019). PLA is another example of a biodegradable polymer that can be used in the textile industry. However, this polymer can also be used in soil erosion control or for films and coatings (e.g., paper) (Lunt and Shafer 2000). PLA is produced from the fermentation of foods, such as corn or potatoes and can be produced one of two ways. The first method involves removing water through the use of a solvent (e.g., acetone or dimethylformamide), with a strong vacuum and high temperature (Lunt and Shafer 2000; Jahangir et al. 2017). The second method uses a vacuum distillation (with no solvent) to produce a cyclic intermediate dimer known as lactide. One advantage of PLA is that it does not produce toxic products when it degrades (Gupta et al. 2007). Unfortunately, PLA can be highly flammable, thereby restricting its application. However, there has been an increase in research surrounding environmentally sustainable flame retardants for PLA over the last decade; for example, one study addressed phosphorus-based flame retardants, such as melamine polyphosphate and aluminum phosphinate (Tawiah et al. 2018). There are also many alternatives to non-biodegradable plastic microbeads, including plant-based substitutes (e.g., cellulose, nuts, seeds, and grains) and mineral substitutes (e.g., silica, mica, sea salt, and quartz sand) (Coombs Obrien et al. 2017; Scudo 2017). Calcium alginate (derived from seaweed) microbeads have also been suggested as a potential alternative to plastic microbeads and will rapidly degrade in seawater (Bae et al. 2019).

However, biodegradable PPs may also act as vectors for contaminants. Zuo et al. (2019) reported that the biodegradable PPs of poly butylene adipate co-terephthalate had a higher sorption capacity for phenanthrene when compared to the non-biodegradable plastics of polyethylene and polystyrene. Additionally, Balestri et al. (2017) examined biodegradable plastic bags made from Mater-Bi (vegetable oils and corn starch) and raised concerns about its involvement in an increased intensity for intra- and interspecific competition within marine plants. Therefore, although biodegradable PPs may be more sustainable than their non-biodegradable counterpart, there are still concerns surrounding their use.

On the other hand, oxo-plastic is another type of biodegradable plastic, which is considered as a potential solution to reduce the accumulation of plastic waste and litter in the soil (Chiellini and Corti 2016). The presence of an additive (pro-oxidant) facilitates the molecular structure of oxo-plastic to break down when exposed to heat or sunlight (Napper and Thompson 2019). Eventually, the degraded plastics are digested by microorganisms until they are fully degraded. Hence, oxo-plastics could be an alternative for minimizing plastic waste and thereby reducing PP accumulation in terrestrial ecosystems.

7.3 IDENTIFICATION AND QUANTIFICATION OF PARTICULATE PLASTICS

Particulate plastics can enter the terrestrial ecosystems through different anthropogenic activities and natural processes. For instance, PPs can reach agricultural lands and aquatic systems due to application of biosolids. Dry and wet depositions of airborne PPs also lead to their accumulation in terrestrial ecosystems (de Souza Machado et al. 2018; Ng et al. 2018). However, our understanding of the impact of PPs in terrestrial environments, as mentioned in the introduction section, is limited. Therefore, there is an increasing demand to develop analytical methods to extract, identify, characterize, and quantify PPs in terrestrial environments. At present, there is a gap in research on PPs in these environments, as no standardized methods have been developed for extracting and quantifying PPs in soils.

Most often, the procedures for analyzing PPs in soils follow similar methods as those used in sediments and water columns. Methods used in laboratories to quantify PPs in waters and sediments are described by Masura et al. (2015). Figure 7.2 shows a schematic diagram for analytical procedures used for PP particles in terrestrial environments. First, soil samples are properly collected for the analysis of PPs. Soils from the topsoil layer and from different depths in the soil profile are usually collected. Second, the collected soil samples are dried (i.e., air dried or mechanically dried) and sieved to desired fractions. Then the soil samples are floated, filtered, and separated by density. Following this stage, there is an oxidation or digestion step to remove organic matter from PP particles. Finally, morphological and physical characterization and a quantification of PPs are carried out using optical microscopy techniques and other analytical instruments. The following sections provide detailed information on the extraction process, removal of organic matter, and morphological and chemical characterization of PPs particles in terrestrial environments, thereby highlighting the efficiencies, advantages, and challenges of the different methods and analytical instruments.

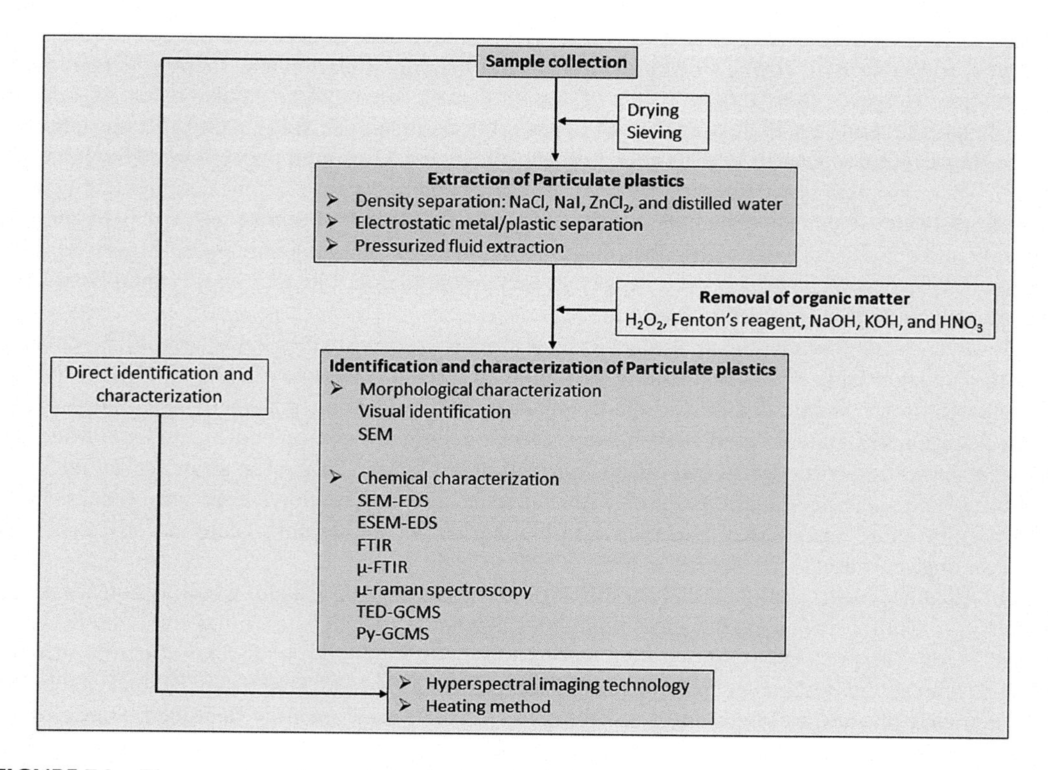

FIGURE 7.2 The procedures for analysis of microplastics in terrestrial environments.

Quality control and quality assurance are important requirements in PP analysis to ensure the quality of the research results. Since there are no well-established QC and QA systems for PP investigations, some existing methods can be adopted to ensure the quality of the results.

Some studies report that analytical errors, such as misidentification and background contamination, can occur during PP analysis. For instance, insufficient spectroscopic measurements may lead to misidentification of synthetic fibers with artificial cellulose or lignin fibers (Remy et al. 2015; Wesch et al. 2016). Moreover, analysis of PPs may also be confounded by post sampling contaminants during laboratory processing. Thus, some precautionary measures, such as use of tools without any particle contamination (Lusher et al. 2013), avoiding air circulation in the working area, and wearing clothes made of cotton (Masura et al. 2015; Frias et al. 2016), would be helpful to avoid contamination by aerial PPs during analysis. Nevertheless, determination of airborne contamination levels in laboratories is an important step prior to the analysis (Hidalgo-Ruz et al. 2012). To avoid PP contamination in laboratories, isolation of the working space (such as by using pyramid glove boxes and hermetic enclosure devices) (Torre et al. 2016), use of clean-air devices (such as laminar flow cabinets) (Van Cauwenberghe and Janssen 2014; Wiesheu et al. 2016), and use of fume hoods (De Witte et al. 2014, 2015) have been used in some studies.

7.3.1 Extraction

The initial PP extraction process involves sieving of dry soils to remove large particles (i.e., macroscopic debris) (Figure 7.2). In general, soil samples are initially passed through a 2 mm sieve. The densities of sand and sediments are both approximately 2.65 g cm^{-3}. Thereafter, density flotation can be used to isolate lighter PPs from the heavier sand and sediments (Figure 7.2). Salt-saturated solutions of known densities are used to separate PPs from the soil matrix. When the soils are mixed with high-density salt solutions, PP particles float to the surface of the solution or remain in suspension; however, the densest sand or sediment particles settle at the bottom of the solution. The supernatant is then collected for further analysis of the PPs.

Different types of salt-saturated solutions have been examined for their extraction efficiencies for PPs. The salt NaCl has been used to extract PPs from terrestrial environments. However, the density of the NaCl (1.2 g cm^{-3}) is too low to enable the floatation of some polymers, such as those containing additives (Coppock et al. 2017). High-density solutions, such as NaI and $ZnCl_2$ (1.6–1.8 g cm^{-3}) are preferred to separate the PP polymers containing additives (Coppock et al. 2017). Those alternatives are relatively cost-expensive compared to NaCl. Therefore, developing cost-effective methods with an excellent recovery rate is essential. Zhang et al. (2018) demonstrated that light-density PPs (i.e., polyethylene and polypropylene) in agricultural soils can be extracted using distilled water, which is a simple and cost-effective method. Since the densities of both polyethylene and polypropylene are <1 g cm^{-3}, saturated salt solutions can be replaced by distilled water (density of 1 g cm^{-3}) for agricultural soils. Liu et al. (2018) developed a method that has a number of extraction and ultrasonic treatments with NaCl for separating PPs. They found that their method can successfully extract seven out of nine tested PPs (polyethylene, polypropylene, polycarbonate, polymethyl methacrylate, polyamide, polystyrene, and acrylonitrile butadiene styrene). The use of a $CaCl_2$ solution (density = 1.5 g cm^{-3}) also shows a greater extraction recovery of PPs in floodplain soils compared to a solution of NaCl (Scheurer and Bigalke 2018). To avoid the negative impacts of salt-saturated solutions, Felsing et al. (2018) investigated the electrostatic behavior of PPs to facilitate their separation from a sample matrix using a modified electrostatic metal/plastic separator. The recovery rate for PPs under a modified electrostatic metal/plastic separator was approximately 100% (Felsing et al. 2018). Nevertheless, still there is uncertainty regarding the efficacy of this method for large-scale extractions of PPs from terrestrial environments. A pressurized fluid extraction technique has also been developed as a promising alternative for assessing the concentration and identification of PPs in terrestrial environments (Fuller and Gautam 2016). This technique utilizes solvents at subcritical temperature and pressure conditions to recover semi-volatile organics from solid materials.

The method was developed and recovered 85%–94% of the spiked PPs in a municipal waste sample (Fuller and Gautam 2016). The method efficiently extracted PP particles at <30 μm. Limitations of this method include the inability to measure the fractional sizes of PP particles and the change in morphology of the PPs (Fuller and Gautam 2016).

7.3.2 Removal of Organic Matter

Another challenge involved in analyzing PPs in soils is the presence of soil organic matter. Most commonly, PPs particles are extracted based on their densities. However, this approach is not effective for analyzing PPs in soil samples that may contain organic matter up to 99%. The densities of soil organic matter typically range between 1.0 and 1.4 g cm^{-3}. These density values are mostly similar to several other PPs, such as nylon and polyethylene terephthalate. Visual examination and sorting of the concentrated samples constitute the initial steps separating PPs from organic matter (i.e., animal and plant residues) and other non-plastics (i.e., tar and glass). Obviously, visual inspection has a minimal effect on distinguishing organic debris from PPs. Therefore, additional procedural steps are needed to remove organic matter from PP samples (Figure 7.2). Various methods, including acidic, alkaline, or oxidizing treatments, as well as enzymatic digestions, have been investigated for removing organic matter from PP samples. For example, Hurley et al. (2018) found that Fenton's reagent effectively removed organic matter from PP samples, when compared with alkaline digestion with NaOH and KOH and oxidation with H_2O_2. Scheurer and Bigalke (2018) demonstrated that an HNO_3 solution can remove organic matter in a short period of time; however, the morphology of some PPs can be affected due to the acidity of HNO_3. Alkaline solutions may also cause surface degradation of PPs (Hurley et al. 2018). Currently, H_2O_2 is the most commonly used chemical for removing organic matter from the PP samples, even though H_2O_2 can slightly change the shape of the PPs. Recent studies have focused on the development of efficient methods to remove organic matter, while maintaining the shape of the PPs. The heat bleach method has been used to reduce the H_2O_2 volume needed for removing organic matter from PP samples. For instance, Sujathan et al. (2017) demonstrated that the H_2O_2 volume required for organic matter removal from return activated sludge is approximately 6% of the sample volume. Elevated temperatures (~ 70°C) may also enhance the degradation of organic matter, thereby minimizing the time required for purification steps to 24 hours or less, and leaving the shape of the PPs unaffected (Sujathan et al. 2017). Overall, the establishment of a reliable and standardized method for the precise extraction of PPs is necessary, since this precision is a crucial pre-requisite to avoid misidentifications or underestimations of PPs in terrestrial environments.

7.3.3 Identification and Characterization of Particulate Plastics

Particulate plastics extracted from terrestrial environments must be accurately identified and quantified. Analysis of PPs in terrestrial environments can be categorized into two main categories: (1) morphological characterization and (2) chemical characterization. The following sections describe these two categories in detail.

7.3.3.1 Morphological Characterization

PPs have unique properties, such as shape, color, size, and density. Visual identification is an important step in identifying and classifying different PPs in a given sample. However, there are several limitations of visual identification, such as overestimations or misidentifications of PPs. Lenz et al. (2015) confirmed through an FTIR analysis that 70% of PP particles were erroneously identified by visual observation. Eriksen et al. (2013) also demonstrated that aluminum silicate and coal ash (approximately 20%) were misidentified as PPs through visual identification. Therefore, visual identification must be combined with other analytical instrumentation methods to avoid inaccurate identifications and under/overestimations.

Morphological characteristics of PPs can be confirmed by different analytical techniques (Figure 7.2). Optical microscopy (i.e., stereomicroscopy) has been used to assess the shape and color of PP particles (Claessens et al. 2011; Eriksen et al. 2013). Scanning electron microscopy can also be used to characterize the surface morphology of PP particles, as it provides high-magnified, high resolution structural images of the particles. Prior to an analysis, scanning electron microscopy requires coating over the sample (i.e., gold coating), which may change the actual color of PPs, and particles might change the texture as well. Both scanning electron microscopy-energy dispersive X-ray spectroscopy and environmental scanning microscopy-energy dispersive X-ray spectroscopy can provide data on the surface morphology, as well as the elemental composition of PP particles (Vianello et al. 2013).

SEM images of different PPs are shown in Figure 7.3. Disintegration features of PPs can be seen on PP surfaces; for instance, patterns such as pits, fractures, flakes, and adhering particles have been observed in PPs due to changes in the surrounding environment.

Sampling and extraction methods also affect the morphological characteristics of PP particles (i.e., size distribution). For instance, using different sieve and filter-pore sizes can lead to various size categories in extracted PPs. Therefore, the establishment of a standardized method for the sampling and extraction of PPs would help to ensure precise morphological identification of PP particles.

7.3.3.2 Chemical Characterization

Chemical characterization is an important step for assessing the composition and the quantification of PPs. Chemical characterization also confirms the results obtained through visual inspection and optical analysis. A variety of analytical instrumentation methods have been used for the chemical characterization of PP particles (Figure 7.2). Infrared spectroscopy, which uses non-destructive vibrational technology, is the most commonly used analytical method for the chemical characterization of PPs. Infrared spectra of unknown PP samples can be compared with the infrared spectra of known PP samples or with the known PP polymers stored in infrared spectra libraries. μ-FTIR spectrometry and μ-Raman spectrometry have been extensively used in infrared spectroscopies. These instruments possess automated scanning coupled with microspectrometry (He et al. 2018).

μ-FTIR spectroscopy can collect spectrums in different modes, such as attenuated total reflectance (ATR), transmittance, and reflectance. In addition to the collecting of spectra, μ-FTIR spectroscopy provides simultaneous visualization and mapping of PPs particles. The ATR mode of μ-FTIR spectroscopy facilitates the chemical characterization of irregularly shaped PPs particles. There is a different spatial distinction between μ-FTIR and μ-Raman spectrometry. The assay size limit of μ-FTIR is approximately 10–20 μm, while μ-Raman spectrometry can detect PPs particles as low as 1 μm. The key disadvantage of both μ-FTIR and μ-Raman spectrometry is that organic matter in an analyzed sample may interfere with the signal of the spectrometer. Therefore, organic matter must be completely removed from the extracted PPs samples prior to μ-FTIR or μ-Raman spectrometry analysis. In addition, another major disadvantage of μ-FTIR and μ-Raman spectrometry is their time-consuming scanning procedures. Therefore, μ-FTIR or μ-Raman spectrometry is difficult to perform for routine analyses of PP particles.

LUMOS II FT-IR spectrometer introduced by the Bruker corporation is an easy-to-use technique with automated measurements and a high precision sample stage (Bruker Corporation 2020). This technique enables analyzing PPs, which are embedded in a sediment/soil matrix, without sample preparation or special filters. Moreover, ATR transmission and reflection modes can be switched instantly, and ATR measurement is further enhanced by focal-plane array imaging (Bruker Corporation 2020).

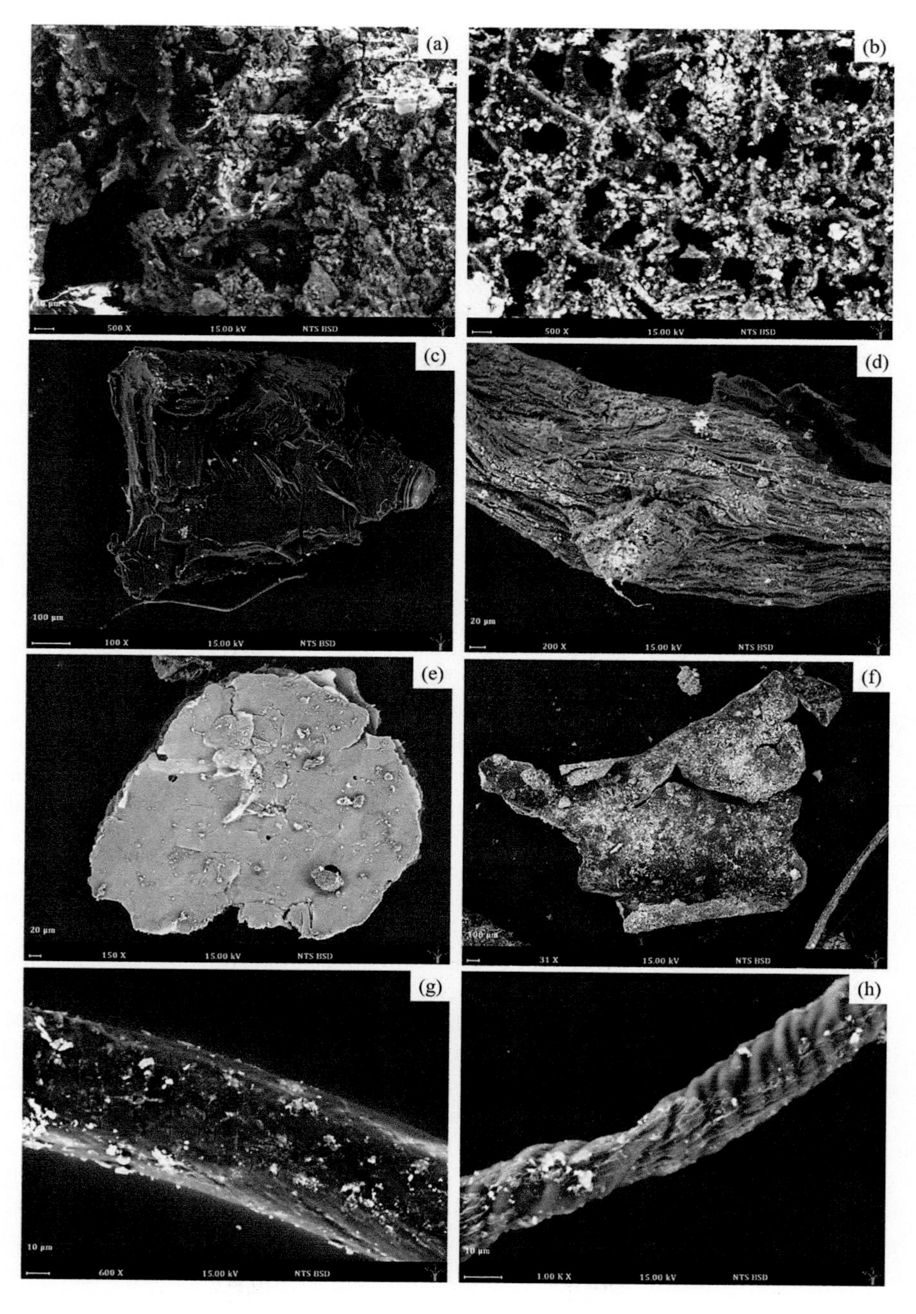

FIGURE 7.3 SEM images displaying different disintegration features in particulate plastics/microplastics (a,b) pits; (c,d) fractures; (e,f) flakes; and (g,h) adhering particles. (Reproduced from Shruti, V. et al., *Sci. Total Environ.*, 654, 154–163, 2019. With the permission from publisher.)

Recent studies have focused on developing reliable and time-saving methods to determine PPs particles in terrestrial environments. Shan et al. (2018) developed a hyperspectral imaging technology as a potential method to determine and visualize the PPs with particle sizes ranging between 0.5 and 5 mm in soils. Dümichen et al. (2017) also developed a thermo-analytical method to identify PP particles in terrestrial environments. A homogenized sample of 20 mg is subjected to complete thermal decomposition. Specific degradation products are then adsorbed onto solid-phase absorbers, followed by subsequent analysis through TED-GCMS. This thermo-analytical method enabled researchers to identify polypropylene, polyethylene, and polystyrene in a biogas plant in terms of mass (Dümichen et al. 2017). Since the TED-GCMS is a quantifiable technique, it can measure the amount of PPs (micrograms) present in one liter of water in addition to the large amounts of natural particles (BAM Institute 2020). The other advantage of TED-GCMS is that the measurements can be completed within a few hours.

In general, Py-GCMS may be used to analyze polymers, as well as PP particles. In this method, single polymer particles are pyrolyzed under inert environments; consequently, thermal degraded products are cryo-trapped, separated, and quantified through the GCMS. This method works perfectly for single particles; therefore, a time-consuming preselection of a single particle is mandatory. The small amount of the sample (approximately 0.1–0.5 mg) used in Py-GCMS may not represent complex sample mixtures, such as environmental samples. Although the Py-GCMS method provides rapid measurements, information regarding the size distribution might be lost since the polymeric particles can be melted prior to degradation. A simple and cost-effective method was also developed to identify and quantify light-density PPs, such as polyethylene and polypropylene (Zhang et al. 2018). In this method, PPs and impurities are identified under a heating protocol (3–5 s at 130°C). When the sample is exposed to heat, PP particles melt and are transformed into circular and transparent particles, whereas impurities (i.e., silicates and organic matter) are not changed by the heat. The size and the number of the PP particles are evaluated using a camera connected to a microscope. An empirical model was developed to quantify PPs in the sample (Zhang et al. 2018).

In summary, analyzing PPs in terrestrial environments is still highly challenging due to the presence of complex and organic-rich substrates. There are inconsistencies of units (i.e., particles m^{-2}, particles m^{-3}, and kg of dry sediment) in different studies when the concentrations of PPs are reported. As mentioned previously, scientifically based improvements are required to standardize the methods for effective extraction and characterization of PP particles in terrestrial environments.

7.4 CONCLUSIONS

Particulate plastics are ubiquitous and persistent contaminants of increasing concern. Based on the properties of particulate plastics, their physical and biogeochemical behavior and associated risks to humans and animal health can differ. The physical and chemical properties of PPs and their abundance in terrestrial environments are the key factors that determine their bioavailability to soil organisms. The chemical composition and the availability of various chemical additives (such as trace elements, stabilizers, plasticizers, monomers, antioxidants, clarifiers, colorants, and flame retardants used during plastic manufacturing) in particulate plastics may determine their degradability and associated risks to terrestrial ecosystems. At present, various analytical techniques are used to extract, identify, characterize, and quantify particulate plastics in the environment. However, no standardized methods have been yet developed for the extraction and quantification of particulate plastics in soils. Generally, the procedures for analyzing particulate plastics in soils are highly similar to those used in sediments and water columns. After preparing the samples, morphological and chemical characterization can be performed with techniques, such as scanning electron microscopy-energy dispersive X-ray spectroscopy, environmental scanning microscopy-energy dispersive X-ray spectroscopy, FTIR, μ-FTIR and micro-Raman (μ-Raman) spectrometry, TED-GCMS, and Py-GCMS. There is urgent need to prioritize scientific-based research on the environmental behavior and ecotoxicity of PPs in terrestrial environments, with a particular focus on the development of standard methods for extraction, identification, and quantification of PPs. This information would be useful for policy planning and to develop strategic management of PP pollution in terrestrial environments.

ACKNOWLEDGMENT

This work was carried out with the support of the Cooperative Research Program for Agriculture Science and Technology Development (Effects of Plastic Mulch Wastes on Crop Productivity and Agro-Environment, Project No. PJ014758), Rural Development Administration, Republic of Korea.

REFERENCES

Acosta-Coley, I., Mendez-Cuadro, D., Rodriguez-Cavallo, E., de la Rosa, J., Olivero-Verbel, J. 2019. Trace elements in microplastics in Cartagena: A hotspot for plastic pollution at the Caribbean. *Marine Pollution Bulletin*, **139**, 402–411.

Amélineau, F., Bonnet, D., Heitz, O., Mortreux, V., Harding, A.M.A., Karnovsky, N., Walkusz, W., Fort, J., Grémillet, D. 2016. Microplastic pollution in the Greenland Sea: Background levels and selective contamination of planktivorous diving seabirds. *Environmental Pollution*, **219**, 1131–1139.

Anderson, J.C., Park, B.J., Palace, V.P. 2016. Microplastics in aquatic environments: Implications for Canadian ecosystems. *Environmental Pollution*, **218**, 269–280.

Anderson, Z.T., Cundy, A.B., Croudace, I.W., Warwick, P.E., Celis-Hernandez, O., Stead, J.L. 2018. A rapid method for assessing the accumulation of microplastics in the sea surface microlayer (SML) of estuarine systems. *Scientific Reports*, **8**(1), 9428.

Arthur, C., Baker, J.E., Bamford, H.A. 2009. *Proceedings of the International Research Workshop on the Occurrence, Effects, and Fate of Microplastic Marine Debris*, September 9–11, 2008, University of Washington Tacoma, Tacoma, WA.

Auta, H., Emenike, C., Fauziah, S. 2017. Distribution and importance of microplastics in the marine environment: A review of the sources, fate, effects, and potential solutions. *Environment International*, **102**, 165–176.

Avio, C.G., Gorbi, S., Regoli, F. 2017. Plastics and microplastics in the oceans: From emerging pollutants to emerged threat. *Marine Environmental Research*, **128**, 2–11.

Bae, S.B., Nam, H.C., Park, W.H. 2019. Electrospraying of environmentally sustainable alginate microbeads for cosmetic additives. *International Journal of Biological Macromolecules*, **133**, 278–283.

Balestri, E., Menicagli, V., Vallerini, F., Lardicci, C. 2017. Biodegradable plastic bags on the seafloor: A future threat for seagrass meadows? *Science of the Total Environment*, **605**, 755–763.

BAM institute. 2020. Bundesanstalt für Materialforschung und -prüfung (BAM). Environment: Tracing microplastics (Accessed on 13 January 2020). Available at https://www.bam.de/Content/EN/Standard-Articles/Topics/Environment/article-measuring-microplastics.html

Bradney, L., Wijesekara, H., Palansooriya, K.N., Obadamudalige, N., Bolan, N.S., Ok, Y.S., Rinklebe, J., Kim, K.-H., Kirkham, M. 2019. Particulate plastics as a vector for toxic trace-element uptake by aquatic and terrestrial organisms and human health risk. *Environment International*, **131**, 104937.

Brandon, J., Goldstein, M., Ohman, M.D. 2016. Long-term aging and degradation of microplastic particles: Comparing in situ oceanic and experimental weathering patterns. *Marine Pollution Bulletin*, **110**(1), 299–308.

Browne, M.A., Crump, P., Niven, S.J., Teuten, E., Tonkin, A., Galloway, T., Thompson, R. 2011. Accumulation of microplastic on shorelines worldwide: Sources and Sinks. *Environmental Science & Technology*, **45**(21), 9175–9179.

Bruker Corporation. 2020. FT-IR Spectrometer LUMOS II (Accessed on 12 January 2020). Available at https://www.bruker.com/products/infrared-near-infrared-and-raman-spectroscopy/ft-ir-routine-spectrometers/ftir-microscope-lumos-ii.html

Canadian Environmental Protection Act. 2017. Microbeads in toiletries regulations. *Canada Gazette*, 151(12), 14 June 2017. Registration SOR/2017-111 June 2. http://www.gazette.gc.ca/rp-pr/p2/2017/2017-06-14/html/sor-dors111-eng.html

Chiellini, E., Corti, A. 2016. Oxo-biodegradable plastics: Who they are and to what they serve—present status and future perspectives. In: S. Kalia (ed.), *Polyolefin Compounds and Materials*, Switzerland, Springer, pp. 341–354.

Claessens, M., De Meester, S., Van Landuyt, L., De Clerck, K., Janssen, C.R. 2011. Occurrence and distribution of microplastics in marine sediments along the Belgian coast. *Marine Pollution Bulletin*, **62**(10), 2199–2204.

Coombs Obrien, J., Torrente-Murciano, L., Mattia, D., Scott, J.L. 2017. Continuous production of cellulose microbeads via membrane emulsification. *ACS Sustainable Chemistry & Engineering*, **5**(7), 5931–5939.

Coppock, R.L., Cole, M., Lindeque, P.K., Queirós, A.M., Galloway, T.S. 2017. A small-scale, portable method for extracting microplastics from marine sediments. *Environmental Pollution*, **230**, 829–837.

de Souza Machado, A.A., Kloas, W., Zarfl, C., Hempel, S., Rillig, M.C. 2018. Microplastics as an emerging threat to terrestrial ecosystems. *Global Change Biology*, **24**(4), 1405–1416.

De Witte, B., Devriese, L., Bekaert, K., Hoffman, S., Vandermeersch, G., Cooreman, K., Robbens, J. 2014. Quality assessment of the blue mussel (*Mytilus edulis*): Comparison between commercial and wild types. *Marine Pollution Bulletin*, **85**(1), 146–155.

Devriese, L.I., van der Meulen, M.D., Maes, T., Bekaert, K., Paul-Pont, I., Frère, L., Robbens, J., Vethaak, A.D. 2015. Microplastic contamination in brown shrimp (*Crangon crangon*, Linnaeus 1758) from coastal waters of the Southern North Sea and Channel area. *Marine Pollution Bulletin*, **98**(1–2), 179–187.

Dris, R., Gasperi, J., Rocher, V., Saad, M., Renault, N., Tassin, B. 2015. Microplastic contamination in an urban area: A case study in Greater Paris. *Environmental Chemistry*, **12**(5), 592–599.

Dümichen, E., Eisentraut, P., Bannick, C.G., Barthel, A.-K., Senz, R., Braun, U. 2017. Fast identification of microplastics in complex environmental samples by a thermal degradation method. *Chemosphere*, **174**, 572–584.

Eriksen, M., Mason, S., Wilson, S., Box, C., Zellers, A., Edwards, W., Farley, H., Amato, S. 2013. Microplastic pollution in the surface waters of the Laurentian Great Lakes. *Marine Pollution Bulletin*, **77**(1–2), 177–182.

Eriksson, C., Burton, H. 2003. *Origins and Biological Accumulation of Small Plastic Particles in Fur Seals from Macquarie Island.* AMBIO: *A Journal of the Human Environment*, **32**(6), 380–384.

Felsing, S., Kochleus, C., Buchinger, S., Brennholt, N., Stock, F., Reifferscheid, G. 2018. A new approach in separating microplastics from environmental samples based on their electrostatic behavior. *Environmental Pollution*, **234**, 20–28.

Ferreira, I., Venâncio, C., Lopes, I., Oliveira, M. 2019. Nanoplastics and marine organisms: What has been studied? *Environmental Toxicology and Pharmacology*, **67**, 1–7.

Fife, D.T., Robertson, G.J., Shutler, D., Braune, B.M., Mallory, M.L. 2015. Trace elements and ingested plastic debris in wintering dovekies (Alle alle). *Marine Pollution Bulletin*, **91**(1), 368–371.

Fotopoulou, K.N., Karapanagioti, H.K. 2019. Degradation of various plastics in the environment. In: *Hazardous Chemicals Associated with Plastics in the Marine Environment.* (Eds.) H. Takada, H.K. Karapanagioti, Springer International Publishing, Cham, Switzerland, pp. 71–92.

Frias, J., Gago, J., Otero, V., Sobral, P. 2016. Microplastics in coastal sediments from Southern Portuguese shelf waters. *Marine Environmental Research*, **114**, 24–30.

Frias, J.P.G.L., Nash, R. 2019. Microplastics: Finding a consensus on the definition. *Marine Pollution Bulletin*, **138**, 145–147.

Fries, E., Dekiff, J.H., Willmeyer, J., Nuelle, M.-T., Ebert, M., Remy, D. 2013. Identification of polymer types and additives in marine microplastic particles using pyrolysis-GC/MS and scanning electron microscopy. *Environmental Science: Processes & Impacts*, **15**(10), 1949–1956.

Fuller, S., Gautam, A. 2016. A procedure for measuring microplastics using pressurized fluid extraction. *Environmental Science & Technology*, **50**(11), 5774–5780.

Gabbatiss, J. 2019. Fabrics made from squid-based material could cut ocean plastic pollution. In: *Independent* https://www.independent.co.uk/news/science/new-fabric-plastic-pollution-microplastics-microfibres-biosynthetic-a8782606.html.

Gewert, B., Ogonowski, M., Barth, A., MacLeod, M. 2017. Abundance and composition of near surface microplastics and plastic debris in the Stockholm Archipelago, Baltic Sea. *Marine Pollution Bulletin*, **120**(1), 292–302.

Gigault, J., Halle, A.T., Baudrimont, M., Pascal, P.-Y., Gauffre, F., Phi, T.-L., El Hadri, H., Grassl, B., Reynaud, S. 2018. Current opinion: What is a nanoplastic? *Environmental Pollution*, **235**, 1030–1034.

Godoy, V., Martín-Lara, M.A., Calero, M., Blázquez, G. 2019. Physical-chemical characterization of microplastics present in some exfoliating products from Spain. *Marine Pollution Bulletin*, **139**, 91–99.

Gupta, B., Revagade, N., Hilborn, J. 2007. Poly(lactic acid) fiber: An overview. *Progress in Polymer Science*, **32**(4), 455–482.

Hahladakis, J.N., Velis, C.A., Weber, R., Iacovidou, E., Purnell, P. 2018. An overview of chemical additives present in plastics: Migration, release, fate and environmental impact during their use, disposal and recycling. *Journal of Hazardous Materials*, **344**, 179–199.

Halden, R.U. 2010. Plastics and health risks. *Annual Review of Public Health*, **31**(1), 179–194.

He, D., Luo, Y., Lu, S., Liu, M., Song, Y., Lei, L. 2018. Microplastics in soils: Analytical methods, pollution characteristics and ecological risks. *TrAC Trends in Analytical Chemistry.*

Hidalgo-Ruz, V., Gutow, L., Thompson, R.C., Thiel, M. 2012. Microplastics in the marine environment: A review of the methods used for identification and quantification. *Environmental Science & Technology*, **46**(6), 3060–3075.

Hodson, M.E., Duffus-Hodson, C.A., Clark, A., Prendergast-Miller, M.T., Thorpe, K.L. 2017. Plastic bag derived-microplastics as a vector for metal exposure in terrestrial invertebrates. *Environmental Science & Technology*, **51**(8), 4714–4721.

Horton, A.A., Walton, A., Spurgeon, D.J., Lahive, E., Svendsen, C. 2017. Microplastics in freshwater and terrestrial environments: Evaluating the current understanding to identify the knowledge gaps and future research priorities. *Science of the Total Environment*, **586**, 127–141.

Huerta Lwanga, E., Gertsen, H., Gooren, H., Peters, P., Salánki, T.S., van der Ploeg, M., Besseling, E., Koelmans, A.A., Geissen, V. 2016. Microplastics in the terrestrial ecosystem: Implications for *Lumbricus terrestris* (Oligochaeta, Lumbricidae). *Environmental Science & Technology*, **50**(5), 2685–2691.

Hurley, R.R., Lusher, A.L., Olsen, M., Nizzetto, L. 2018. Validation of a method for extracting microplastics from complex, organic-rich, environmental matrices. *Environmental Science & Technology*, **52**(13), 7409–7417.

Hurley, R.R., Nizzetto, L. 2018. Fate and occurrence of micro(nano)plastics in soils: Knowledge gaps and possible risks. *Current Opinion in Environmental Science & Health*, **1**, 6–11.

Ioakeimidis, C., Fotopoulou, K.N., Karapanagioti, H.K., Geraga, M., Zeri, C., Papathanassiou, E., Galgani, F., Papatheodorou, G. 2016. The degradation potential of PET bottles in the marine environment: An ATR-FTIR based approach. *Scientific Reports*, **6**, 23501.

Jahangir, M.A., Rumi, T.M., Wahab, M.A., Khan, M.I., Rahman, M.A., Sayed, Z.B. 2017. Poly Lactic Acid (PLA) fibres: Different solvent systems and their effect on fibre morphology and diameter. *American Journal of Chemistry*, **7**(6), 177–186.

Karkanorachaki, K., Kiparissis, S., Kalogerakis, G.C., Yiantzi, E., Psillakis, E., Kalogerakis, N. 2018. Plastic pellets, meso- and microplastics on the coastline of Northern Crete: Distribution and organic pollution. *Marine Pollution Bulletin*, **133**, 578–589.

Koelmans, A.A., Besseling, E., Foekema, E.M. 2014. Leaching of plastic additives to marine organisms. *Environmental Pollution*, **187**, 49–54.

Lenz, R., Enders, K., Stedmon, C.A., Mackenzie, D.M., Nielsen, T.G. 2015. A critical assessment of visual identification of marine microplastic using Raman spectroscopy for analysis improvement. *Marine Pollution Bulletin*, **100**(1), 82–91.

Liu, M., Lu, S., Song, Y., Lei, L., Hu, J., Lv, W., Zhou, W., Cao, C., Shi, H., Yang, X. 2018. Microplastic and mesoplastic pollution in farmland soils in suburbs of Shanghai, China. *Environmental Pollution*, **242**, 855–862.

Loyo-Rosales, J.E., Rosales-Rivera, G.C., Lynch, A.M., Rice, C.P., Torrents, A. 2004. Migration of nonylphenol from plastic containers to water and a milk surrogate. *Journal of Agricultural and Food Chemistry*, **52**(7), 2016–2020.

Lunt, J., Shafer, A.L. 2000. Polylactic acid polymers from com. applications in the textiles industry. *Journal of Industrial Textiles*, **29**(3), 191–205.

Lusher, A., Mchugh, M., Thompson, R. 2013. Occurrence of microplastics in the gastrointestinal tract of pelagic and demersal fish from the English Channel. *Marine Pollution Bulletin*, **67**(1–2), 94–99.

Lusher, A.L., Welden, N.A., Sobral, P., Cole, M. 2017. Sampling, isolating and identifying microplastics ingested by fish and invertebrates. *Analytical Methods*, **9**(9), 1346–1360.

Masura, J., Baker, J.E., Foster, G.D., Arthur, C., Herring, C. 2015. Laboratory methods for the analysis of microplastics in the marine environment: Recommendations for quantifying synthetic particles in waters and sediments: NOAA Technical Memorandum NOS-OR&R-48; National Oceanic and Atmospheric Administration: Silver Spring, MD.

Napper, I.E., Thompson, R.C. 2019. Environmental deterioration of biodegradable, oxo-biodegradable, compostable, and conventional plastic carrier bags in the sea, soil, and open-air over a 3-year period. *Environmental Science & Technology*, **53**(9), 4775–4783.

Nelms, S.E., Galloway, T.S., Godley, B.J., Jarvis, D.S., Lindeque, P.K. 2018. Investigating microplastic trophic transfer in marine top predators. *Environmental Pollution*, **238**, 999–1007.

Ng, E.-L., Lwanga, E.H., Eldridge, S.M., Johnston, P., Hu, H.-W., Geissen, V., Chen, D. 2018. An overview of microplastic and nanoplastic pollution in agroecosystems. *Science of the Total Environment*, **627**, 1377–1388.

O'Farrell, K. 2018. An assessment of the sale of microbeads and other nonsoluble plastic polymers in personal care and cosmetic products currently available within the Australian retail (in store) market. Project report. Envisage Works. Available at https://www.environment.gov.au/protection/waste-resource-recovery/publications/assessment-sale-microbeads-within-retail-market.

Remy, F.O., Collard, F., Gilbert, B., Compère, P., Eppe, G., Lepoint, G. 2015. When microplastic is not plastic: The ingestion of artificial cellulose fibers by macrofauna living in seagrass macrophytodetritus. *Environmental Science & Technology*, **49**(18), 11158–11166.

Sastri, V.R. 2010. Chapter 6—Commodity thermoplastics: Polyvinyl chloride, polyolefins, and polystyrene. In: *Plastics in Medical Devices.* (Ed.) V.R. Sastri, William Andrew Publishing, Boston, MA, pp. 73–119.

Scheurer, M., Bigalke, M. 2018. Microplastics in Swiss floodplain soils. *Environmental Science & Technology*, **52**(6), 3591–3598.

Schwaferts, C., Niessner, R., Elsner, M., Ivleva, N.P. 2019. Methods for the analysis of submicrometer- and nanoplastic particles in the environment. *TrAC Trends in Analytical Chemistry*, **112**, 52–65.

Schwarz, A.E., Ligthart, T.N., Boukris, E., van Harmelen, T. 2019. Sources, transport, and accumulation of different types of plastic litter in aquatic environments: A review study. *Marine Pollution Bulletin*, **143**, 92–100.

Scudo, A., Liebmann, B., Corden, C., Tyrer, D., Kreissig, J., Warwick, O. 2017. Intentionally added microplastics in products. European Commission (DG Environment). Report. Amec Foster Wheeler Environment & Infrastructure UK Limited. Available at: https://ec.europa.eu/environment/chemicals/reach/pdf/39168%20Intentionally%20added%20microplastics%20-%20Final%20report%2020171020.pdf.

Shan, J., Zhao, J., Liu, L., Zhang, Y., Wang, X., Wu, F. 2018. A novel way to rapidly monitor microplastics in soil by hyperspectral imaging technology and chemometrics. *Environmental Pollution*, **238**, 121–129.

Shruti, V., Jonathan, M., Rodriguez-Espinosa, P., Rodríguez-González, F. 2019. Microplastics in freshwater sediments of Atoyac River Basin, Puebla City, Mexico. *Science of the Total Environment*, **654**, 154–163.

Song, Y.K., Hong, S.H., Jang, M., Han, G.M., Jung, S.W., Shim, W.J. 2017. Combined effects of UV exposure duration and mechanical abrasion on microplastic fragmentation by polymer type. *Environmental Science & Technology*, **51**(8), 4368–4376.

Stanton, T., Johnson, M., Nathanail, P., MacNaughtan, W., Gomes, R.L. 2019. Freshwater and airborne textile fibre populations are dominated by 'natural', not microplastic, fibres. *Science of the Total Environment*, **666**, 377–389.

Sujathan, S., Kniggendorf, A.-K., Kumar, A., Roth, B., Rosenwinkel, K.-H., Nogueira, R. 2017. Heat and bleach: A cost-efficient method for extracting microplastics from return activated sludge. *Archives of Environmental Contamination and Toxicology*, **73**(4), 641–648.

Talsness, C.E., Andrade, A.J., Kuriyama, S.N., Taylor, J.A., Vom Saal, F.S. 2009. Components of plastic: Experimental studies in animals and relevance for human health. *Philosophical Transactions of the Royal Society B: Biological Sciences*, **364**(1526), 2079–2096.

Tanaka, K., Takada, H. 2016. Microplastic fragments and microbeads in digestive tracts of planktivorous fish from urban coastal waters. *Scientific Reports*, **6**, 34351.

Tawiah, B., Yu, B., Fei, B. 2018. Advances in flame retardant poly(lactic acid). *Polymers*, **10**(8), 876.

Torre, M., Digka, N., Anastasopoulou, A., Tsangaris, C., Mytilineou, C. 2016. Anthropogenic microfibres pollution in marine biota: A new and simple methodology to minimize airborne contamination. *Marine Pollution Bulletin*, **113**(1–2), 55–61.

USA House of Representatives. 2015. Microbead-Free Waters Act of 2015. HR 1321, Report 114–371, December 7, 2015.

Van Cauwenberghe, L., Claessens, M., Vandegehuchte, M.B., Mees, J., Janssen, C.R. 2013. Assessment of marine debris on the Belgian Continental Shelf. *Marine Pollution Bulletin*, **73**(1), 161–169.

Van Cauwenberghe, L., Janssen, C.R. 2014. Microplastics in bivalves cultured for human consumption. *Environmental Pollution*, **193**, 65–70.

Vianello, A., Boldrin, A., Guerriero, P., Moschino, V., Rella, R., Sturaro, A., Da Ros, L. 2013. Microplastic particles in sediments of Lagoon of Venice, Italy: First observations on occurrence, spatial patterns and identification. *Estuarine, Coastal and Shelf Science*, **130**, 54–61.

Wagner, M., Scherer, C., Alvarez-Muñoz, D., Brennholt, N., Bourrain, X., Buchinger, S., Fries, E., et al. 2014. Microplastics in freshwater ecosystems: What we know and what we need to know. *Environmental Sciences Europe*, **26**(1), 12.

Wang, F., Wong, C.S., Chen, D., Lu, X., Wang, F., Zeng, E.Y. 2018. Interaction of toxic chemicals with microplastics: A critical review. *Water Research*, **139**, 208–219.

Wang, J., Peng, J., Tan, Z., Gao, Y., Zhan, Z., Chen, Q., Cai, L. 2017. Microplastics in the surface sediments from the Beijiang River littoral zone: Composition, abundance, surface textures and interaction with heavy metals. *Chemosphere*, **171**, 248–258.

Wesch, C., Barthel, A.-K., Braun, U., Klein, R., Paulus, M. 2016. No microplastics in benthic eelpout (*Zoarces viviparus*): An urgent need for spectroscopic analyses in microplastic detection. *Environmental Research*, **148**, 36–38.

Wiesheu, A.C., Anger, P.M., Baumann, T., Niessner, R., Ivleva, N.P. 2016. Raman microspectroscopic analysis of fibers in beverages. *Analytical Methods*, **8**(28), 5722–5725.

Wijesekara, H., Bolan, N.S., Bradney, L., Obadamudalige, N., Seshadri, B., Kunhikrishnan, A., Dharmarajan, R., Ok, Y.S., Rinklebe, J., Kirkham, M. 2018. Trace element dynamics of biosolids-derived microbeads. *Chemosphere*, **199**, 331–339.

Zhang, S., Yang, X., Gertsen, H., Peters, P., Salánki, T., Geissen, V. 2018. A simple method for the extraction and identification of light density microplastics from soil. *Science of the Total Environment*, **616**, 1056–1065.

Zhao, S., Zhu, L., Li, D. 2016. Microscopic anthropogenic litter in terrestrial birds from Shanghai, China: Not only plastics but also natural fibers. *Science of the Total Environment*, **550**, 1110–1115.

Zuo, L.-Z., Li, H.-X., Lin, L., Sun, Y.-X., Diao, Z.-H., Liu, S., Zhang, Z.-Y., Xu, X.-R. 2019. Sorption and desorption of phenanthrene on biodegradable poly(butylene adipate co-terephtalate) microplastics. *Chemosphere*, **215**, 25–32.

8 Facilitated Transport of Zinc on Plastic Colloids through Soil Columns

Christopher Barton

CONTENTS

8.1 Introduction ... 125
8.2 Methods ... 126
    8.2.1 Soil Columns ... 126
    8.2.2 Leaching Experiment ... 126
    8.2.3 Adsorption Isotherms ... 127
    8.2.4 Plastic Colloid Stability ... 127
8.3 Results and Discussion ... 127
    8.3.1 Plastic Colloid Stability ... 127
    8.3.2 Adsorption Isotherms ... 129
    8.3.3 Plastic Colloid Transport ... 129
    8.3.4 Zinc Transport ... 130
8.4 Conclusions ... 132
References ... 132

8.1 INTRODUCTION

The mobility of colloids, dissolved organic matter, and engineered nanoparticles through soil has been widely documented (McCarthy and Zachara 1989; Mills et al. 1991; Sen and Khilar 2006; Tian et al. 2010). Environmental contaminants, such as heavy metals, strongly bind to solid-phase constituents in the soil and are generally considered to have limited mobility. However, colloid-sized particles can move through the porous media of soils and can carry contaminants bound to them and, thus, facilitate contaminant transport. Several researchers have documented the co-transport of radionuclides (Penrose et al. 1990), metals (Barton and Karathanasis 2003a; Kaplan et al. 1995; Karathanasis 1999; Xie et al. 2018), hydrophobic organic compounds (Johnson and Amy, 1995), and herbicides (Barton and Karathanasis 2003a; Seta and Karathanasis 1997a) with colloids in subsurface environments. For metals, mobility is controlled by several chemical and physical reactions. Cation exchange (weak outer sphere complexation) and specific adsorption (strong inner sphere complexation) are the primary mechanisms for metals to bind to charged colloids (Sposito 1984). Soil organic matter can also play an important role in metal retention due to its high charge and ability to form stable complexes with organic ligands (Elliott et al. 1986).

Particulate plastics have been identified as a vector for toxic trace elements in the environment and are increasingly becoming a contaminant of concern (Bradney et al. 2019). Much research has focused on the fate of microplastics in aquatic environments (Holmes et al. 2012; Turner and Holmes 2015), but the fate of microplastics in terrestrial environments has also gained recent attention (Horton et al. 2017; Rillig 2012). Plastics, either disposed of or surface applied, on the soil surface will break down and can be translocated from the site by wind or water (Nizzetto et al. 2016), but some will remain in place, decompose, and potentially be incorporated in the soil. As plastics in the soil continue to breakdown and become smaller, their surface area increases, and their ability to interact with contaminants and become a mechanism for contaminant transport increases (Holmes et al. 2014; Teuten et al. 2009).

Heavy metal contamination of soils from anthropogenic activities is of global concern (Li et al. 2019). Barton and Karathanasis (2003a and 2003b) showed that the metal zinc (Zn) could bind to clay colloids and migrate through soil macropores and potentially become a source of groundwater contamination. Given that microplastics can have a similar size and surface chemistry as clay colloids, an experiment was undertaken to determine if colloidal-sized plastics exhibited sufficient stability to remain dispersed in water, such that they could migrate through the porous media of soils. Moreover, the ability of plastics to serve as a vector for contaminant transport via facilitated transport was also examined.

8.2 METHODS

8.2.1 Soil Columns

Intact soil columns from the Maury silt loam series (fine, mixed, mesic Typic Paleudalf) were prepared in the field at the University of Kentucky Agricultural Experiment Station in Lexington, Kentucky. Four columns were prepared by carving cylindrical pedestals of 13-cm diameter and 20-cm length. A 16-cm diameter by 20-cm length of polyvinyl chloride (PVC) pipe was placed over each pipe and the annulus between the pipe and soil pedestal was filled with expandable polyurethane foam. After the foam had hardened, the columns were cut from their base, wrapped in plastic to retain moisture, and transported to the laboratory.

Soil samples from the same depth as the columns were taken adjacent to the area where columns were created. Soils were analyzed for pH, organic carbon, soluble salts, cation exchange capacity, base saturation, texture, and bulk density. Soil pH was measured in a 1:1 soil:water paste, and soluble salts (soil electrical conductivity) was measured in a 1:3 soil:water solution with a conductivity bridge (Soil and Plant Analysis Council 2000). Concentrations of exchangeable Na, K, Ca, and Mg were measured by Mehlich III extraction and analyzed by inductively coupled plasma optical emission spectrometer (ICP-OES) (Soil and Plant Analysis Council 2000). Organic C was quantified using a CHN Analyzer (LECO Corporation, St. Joseph, MI, USA) (Nelson and Sommers 1982). Particle size distribution was evaluated by the micropipette method (Miller and Miller 1987). Cation exchange capacity and base saturation were assessed using the ammonium acetate method at pH 3 (Soil and Plant Analysis Council 2000). Bulk density was determined using the core method (NRCS 1996).

8.2.2 Leaching Experiment

Colloidal carboxylated polystyrene nano-spheres were purchased from Alpha Nanotech Inc. www.alphananotechne.com (Vancouver, Canada). The standard colloid solution contained 50 mg mL^{-1} of 300 nm diameter particles. Two mL of the standard solution was diluted to 1 L with ultrapure water (Barnstead Easypure II, Thermo Scientific, Langenselbold, Germany), resulting in an input solution that contained 100 mg of plastic colloids per liter. This concentration was chosen because it produced a strong linear calibration curve ($r^2 = 0.998$) for colloid detection using a turbidity meter (Turner Designs, Sunnyvale, CA). The colloid solution was spiked with 100 mg L^{-1} Zn (from zinc chloride, anhydrous). The mixture was allowed to equilibrate overnight before application to the monoliths. A solution containing 100 mg L^{-1} Zn with no colloid was prepared for the control treatment.

Prior to the leaching experiment, the columns were wetted with 1 L of d-H_2O to moisten the soil. The experiment was conducted under unsaturated, gravitational flow conditions. Under these conditions, a plastic colloid-Zn mixture or control Zn solution was applied onto the soil columns in 50 mL pulses, at eight-hour intervals. Column set I was leached for nearly 40 days, while column set II was leached for nearly 24 days.

Eluents from the columns were collected in 50 mL increments and analyzed for colloid and Zn concentrations. Zn was then determined using a Varian Vista Pro (Varian Inc., Palo Alto, CA) ICP-OES. Because of the low colloid concentration in the input solution and small size, separating the solid from solution phase in the eluents was problematic. As such, we spiked a range of colloid solutions from 1 to 100 mg colloids per liter under acidified conditions (pH < 2 with HNO_3) and tested whether the colloid solution could be analyzed without filtration or centrifugation. Zn recovery in the spiked samples was high (93%–100%), and the plastic colloids did not appear to interfere with the operation of the ICP-OES operation. As such, Zn concentrations in the eluents are expressed as total Zn. Plastic colloid concentrations were measured as NTUs (Nephelometric Turbidity Unit) using the turbidity meter and converted to colloid concentration (mg L^{-1}) using the calibration curve equation (Colloid (mg L^{-1}) = 0.4978 × NTU + 1.6739).

8.2.3 Adsorption Isotherms

Batch equilibrium experiments were performed using 50 mL Teflon test tubes. A 50 mg sample of plastic colloid was added to each test tube along with 30 mL of adsorbate solution containing 0–10.0 mg L^{-1} Zn. Samples were shaken on a reciprocating shaker for 24 hours and centrifuged for 30 minutes at 10,200 × g (10,000 rpm). Because of the hydrophobic nature of the plastic colloids, supernatants were filtered through a 0.2 μm membrane filter to ensure the solution was free of any solids prior to analysis by ICP-OES. A similar procedure was followed using 250 mg of air-dried Maury silt loam soil instead of the colloid. Adsorption isotherms were constructed by plotting the concentration of Zn retained by the plastic colloids or soil against the equilibrium Zn concentration in solution. The Freundlich isotherm equation was used to describe the adsorption data in the soil sample. An adsorption maximum for Zn on the plastic colloids was observed, so a Langmuir adsorption isotherm was utilized for this association.

8.2.4 Plastic Colloid Stability

The stability of plastic colloids in ultrapure water was evaluated under varying pH and electrical conductivity (EC) conditions. A series of 50 mL test tubes were filled with 40 mL aliquots of the 100 mg L^{-1} colloid solution with a variety of pH and EC conditions. In the pH stability experiment, suspensions were adjusted in pH increments ranging from 2.0 to 10.0 with HNO_3 or NaOH. A 200 μL suspension aliquot was pipetted from the 5 cm depth of each tube after 0, 1, 6, 24, and 48 hours of settling time. The concentration of the colloid in the pipetted suspension was evaluated using the turbidity meter. Colloid suspensions used in the EC stability experiment were adjusted in EC increments ranging from 50 to 1000 μS with KCl and were pipetted at the 5 cm depth and analyzed after 0, 1, 2, 6, 24, 48, and 72 hours of settling time. After 72 hours, the tubes from the EC experiment were placed in an incubator at 36°C for 24 hours and analyzed. The colloid stability at each pH or EC level was expressed in terms of percent of colloid concentration remaining in suspension vs. pH or EC, respectively. The higher the percentage of colloid remaining in suspension, the higher its colloid stability.

8.3 RESULTS AND DISCUSSION

8.3.1 Plastic Colloid Stability

Settling-rate experiments were performed to determine if the plastic colloids will remain in suspension over time under different pH and ionic strength, as measured by EC, conditions. Soil colloid stability has been shown to be greatly influenced by time due to gravity and Stokes's law (Hillel 1998). Changes in the pH and ionic strength can also influence stability in soil colloids (Barton and Karathanasis 2003a; Ryan and Geschwend 1998; Seta and Karathanasis 1997b). For instance, Barton and Karathanasis (2003b) found that decreasing the pH below 4.0 resulted in a sharp decline in clay colloid suspension and attributed this to flocculation promoted by a reduction in the net surface potential of the colloid. Similarly, increases in ionic strength in clay colloids can suppress the thickness of the double layer, which allows colloids to physically approach each other and flocculate (Jekel 1986).

Unlike soil colloids, plastic colloids were shown to be highly stable regardless of the pH or ionic strength. Colloid dispersal remained high in solutions ranging from pH 2 to 10 over a 48-hour period. The highest reduction in colloid stability over initial time zero rates was only 7% (93% dispersed) after 24 hours in a pH 4 solution. Likewise, dispersal remained high in solutions with EC values ranging from 50 to 1000 μS over a 72-hour period. Here, the highest reduction in colloid stability over initial time zero rates was a mere 9% (91% dispersed) after 72 hours in a 750 μS solution. In an effort to see if heating influenced stability, tubes from the EC experiment were placed in an incubator at 36°C for 24 hours. Results indicated that heating had little effect on stability, as dispersal was reduced by only 3% over a 24-hour heating period.

Unlike other natural, charged colloids (clays; iron and aluminum oxides), plastic colloids showed high stability under a variety of chemical conditions. The unaltered polystyrene colloids exhibited an initial zeta potential of −40.9 mV (Table 8.1), which would support high colloid stability (Wang and Keller 2009), but dispersal levels >90% for the duration of the settling experiment and differing pH and EC conditions were somewhat unexpected. Lu et al. (2018) measured the zeta potential of polystyrene microspheres under similar pH and ionic strength conditions and also noted high dispersion (zeta potential of −40 to −75 mV) from pH 2 to 10. Substitution of K^+ with Mg^{2+} tended to slightly decrease colloid stability in their ionic strength experiments, and the addition of humic acid decreased zeta potential under alkaline (pH >8) conditions (Lu et al. 2018). Regardless, dispersion of the plastic colloids under the conditions of the leaching procedure in this experiment (pH 6.3, EC 410 μS) should be maintained, thus allowing for colloid migration through soil macropores.

TABLE 8.1
Physicochemical Characteristics of Soil, Colloid, and Leaching Solution

Properties[a]	Soil[b]	Colloid	Colloid-Zn Solution
Clay (%)	16.6 ± 1.8		
Silt (%)	74.4 ± 2.0		
Sand (%)	9.6 ± 0.1		
pH	5.5 ± 0.1		6.3
EC (μS)	12.0 ± 0.4	19.9	410
CEC (cmol kg^{-1})	17.2 ± 2.4		
TEB (cmol kg^{-1})	9.6 ± 1.0		
OC (%)	1.8 ± 0.2		
BD (g cm^3)	1.3 ± 0.1		
Colloid diameter (nm)[c]		300	
Zeta potential (mV)[c]		−40.9 ± 3.6	

[a] EC = electrical conductivity; CEC = cation exchange capacity; TEB = total exchangeable bases; OC = organic carbon; BD = bulk density.
[b] Mean ± standard deviation (n = 3).
[c] Values provided by manufacturer.

8.3.2 Adsorption Isotherms

Equilibrium adsorption isotherms of Zn on the Maury silt loam soil and the plastic colloids were prepared to assess sorption affinities. Unfortunately, sorption of the metal to the solid phase appeared to interact differently between the plastic and soil and different sorption models had to be used. In a study that examined the use of isotherm models for sorption of hydrophobic organic contaminants on particulate plastics, Velez et al. (2018) noted that interactions between contaminants and particulate plastics are complex and many factors need to be considered when choosing an appropriate model. Zn sorption to the soil enabled the calculation of a partition coefficient ($K_f = 2.3$, $N = 0.2$, $r^2 = 0.99$, where K_f is the sorption capacity and N is the intensity factor) through the use of the Freundlich equation. Barton and Karathanasis (2003a) and Karathanasis (1999) noted a similar sorption affinity for Zn to Maury soil and suggested that the bonding was primarily with outer hydration sphere complexes in the double layer. The plastic colloid, on the other hand, exhibited an adsorption maximum (Qm = 11.56 mg g^{-1}), so a Langmuir model was utilized. Knowing the adsorption maximum and assumptions of the Langmuir model that suggest adsorption will occur until a monolayer is formed on the surface of the colloid, an estimation of the maximum amount of the Zn sorbed to the plastic colloid in our Zn-colloid leachate was determined to be about 2.5 mg L^{-1}.

8.3.3 Plastic Colloid Transport

Data from the leaching experiment were transformed into breakthrough curves based on reduced concentration (ratio of effluent concentration to influent concentration = C/Co) versus pore volume of colloid suspension passed through the columns. The leaching experiment for column I produced 7.75 pore volumes of effluent, whereas column II had 4.74 pore volumes eluted (Figures 8.1a and 8.2a, respectively). In column I, the plastic colloid breakthrough gradually increased to 0.15 over 4.5 pore volumes, and then exhibited a spike in breakthrough to a maximum of 0.59 (Figure 8.1a). Subsequently, colloid breakthrough showed a fairly rapid decline to <0.2 at 7.75 pore volumes. Column II showed a much more rapid breakthrough of the plastic colloids to a maximum of 0.76 at 1.5 pore volumes (Figure 8.2a). Subsequently, the breakthrough suddenly dropped to nearly 0.1, and then quickly rebounded to 0.7 at 2.5 pore volumes, after which, the breakthrough steadily declined to 0.05 by 4.75 pore volumes.

The gradual breakthrough of colloids in column I suggests that adsorption and/or filtration processes were responsible for some colloid retention within the soil during the early leaching stages (Seta and Karathanasis 1997b). As micropores become clogged, flow becomes more predominant in macropores, and the spike in breakthrough was observed. The ability of transported colloids to block micropore entrances and alter the pathway of chemicals through macropores, where the velocity of flow is higher, has been documented elsewhere (Barton and Karathanasis 2003a; Kretzschmar et al. 1995; Seta and Karathanasis 1997a). Over time, macropores also become clogged, or pore diameters are reduced due to colloid deposition on pore walls, and the breakthrough diminishes. The colloid sorption in the column can also be attributed to the presence of Fe and Al oxides and hydroxides in the soil, where the negatively charged particle is attracted to the positively charged metal (Goldberg et al. 1990; McBride 1989). Barton and Karathanasis (2003a) noted a very similar breakthrough pattern for clay colloids in Maury soils. Both columns in this study exhibited a similar maximum colloid breakthrough concentration (mean = 67.7%), but the column II breakthrough occurred much faster (1.5 versus 6.5 pore volumes). When carving the soil pedestals, insect activity and burrowing were noted more prevalently where the column II and its control column were excavated. The large macropores created by the burrowing may have contributed to the faster colloid breakthrough exhibited in column II.

In a study that characterized soil macro-porosity, Barton and Karathanasis (2002) found that the Maury silt loam soil (0–15 cm depth) had a pore area of 23.9 mm^2 100 mm^{-2} and a mean pore diameter of 2.18 mm. Given that the plastic colloid had a mean diameter of 300 μm and exhibited

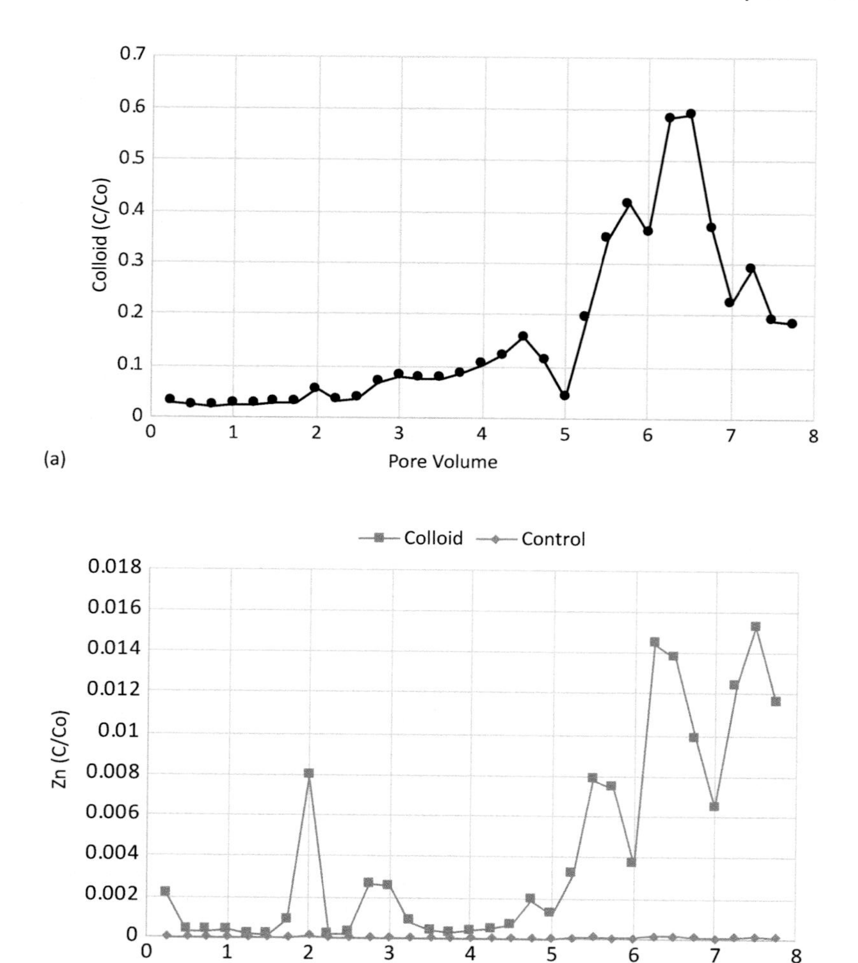

FIGURE 8.1 Breakthrough curves of plastic colloid (a) and Zn (b) for the column I set (leached for 40 days) with Maury silt loam soil. The set had two columns. One was leached with a plastic colloid-Zn mixture and one was leached with a Zn solution (the control).

high stability in solution, the potential for plastic colloid movement through an environment with this porous structure is clearly feasible and results support that it can occur. The fate of the colloid migration through in situ soils is difficult to predict, but there is the potential for microplastics to continue migration through the soil and into groundwater, migrate through interflow and into surface waters, or the plastic colloids could be filtered in the porous media. Regardless, they pose an environmental contamination concern.

8.3.4 Zinc Transport

In the absence of colloids (control), zinc exhibited essentially no breakthrough by either control column, suggesting nearly complete sorption to the soil matrix (Figures 8.1b and 8.2b). Given that the soil had a modest cation exchange capacity (Table 8.1) and K_f, Zn sorption to the soil was anticipated. As such, a high concentration of Zn in the leaching solution (100 mg L^{-1}) was utilized in hopes that some Zn breakthrough in the control would be observed. After 7.75 pore volumes in column I,

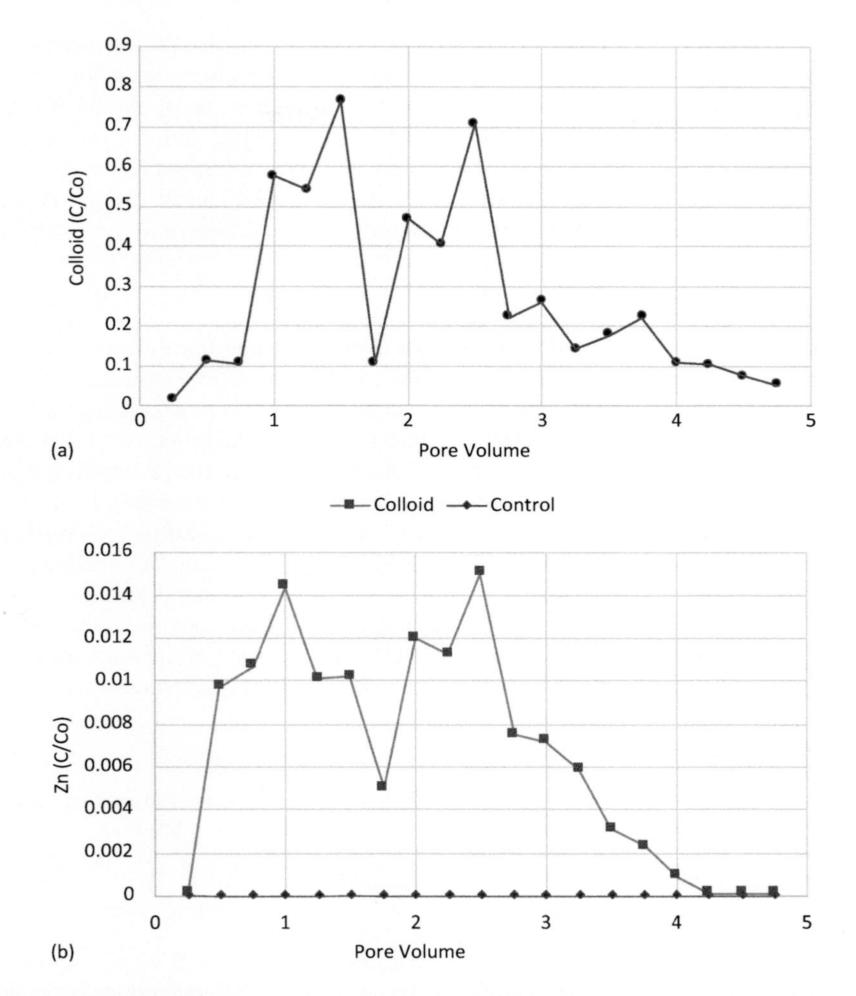

FIGURE 8.2 Breakthrough curves of plastic colloid (a) and Zn (b) for the column II set (leached for 24 days) with Maury silt loam soil. The set had two columns. One was leached with a plastic colloid-Zn mixture and one was leached with a Zn solution (the control).

however, no breakthrough had occurred. Using a Loradale soil column, which is found in the same association as Maury, over 65 pore volumes of a 10 mg L^{-1} Zn solution were eluted before a breakthrough was detected (Barton and Karathanasis 2003b). When plastic colloid elution diminished after 7 and 3 pore volumes in columns I and II, respectively, the leaching procedure was ended.

In the presence of the plastic colloid, Zn elution was observed in both soil columns (Figures 8.1b and 8.2b). In column I, little Zn breakthrough was observed in the first 4.5 pore volumes. Subsequently, as colloids were eluted through the column, Zn elution was also observed. The goodness of fit between eluted colloids and eluted Zn yielded an r^2 of 0.59 for column I. At 6.25 pore volumes, the Zn breakthrough in column I peaked at 0.014 C/Co, then declined sharply and peaked again to a maximum breakthrough of 0.015 at 7.5 pore volumes (Figure 8.2b). Column II also showed a strong correlation between eluted plastic colloids and Zn (r^2 = 0.60). As with the colloid elution, the Zn breakthrough occurred rapidly with an initial breakthrough at 1 pore volume of 0.014 and a maximum breakthrough of 0.015 at 2.5 pore volumes (Figure 8.2b). As the colloid breakthrough diminished in column II, so did the Zn elution.

A strong sorption affinity for Zn to the plastic colloid was responsible for the migration of Zn in both soil columns. As noted previously, the Qm developed from the adsorption isotherm experiment suggested that the plastic colloids could adsorb 2.5 mg L^{-1} from the eluent solution. The maximum concentration measured from the eluents was 1.93 and 1.50 mg L^{-1} in columns I and II, respectively. Given that the mean maximum colloid breakthrough was 0.67, the observed values conform to what the Langmuir isotherm predicted. Since the plastic colloids had an adsorption maximum, the remaining Zn in solution likely sorbed to the soil column, as was observed in the control columns.

8.4 CONCLUSIONS

As has been reported previously, particulate plastics can accumulate trace elements and become a vector for their transport in aquatic environments (Brennecke et al. 2016; Holmes et al. 2012; Turner and Holmes 2015). This study showed that plastic colloids were able to accumulate a heavy metal (Zn) and facilitate its migration in a terrestrial system through soil macropores. Given their dispersive nature and high stability in water, migration of plastic colloids in the pedosphere may serve as another mechanism for contamination of groundwater and surface water sources. In locations where plastic wastes and other contamination sources are co-mingled (i.e., landfills), microplastics could be a major vector for contaminant transport through the soil environment. The facilitated transport of contaminants by colloids and nanoparticles in soils has been well demonstrated, but the role of colloidal plastics in this process has not been well documented. Additional research utilizing other metals and differing soils will be needed to gain a better appreciation of the magnitude of concern of plastic-metal soil migration and its importance in environmental risk management.

REFERENCES

Barton, C.D. and A.D. Karathanasis. 2002. A novel method for measurement and characterization of soil macroporosity. *Communications in Soil Science and Plant Analysis* 33: 1305–1322.

Barton, C.D. and A.D. Karathanasis. 2003a. Influence of soil colloids on the migration of atrazine and zinc through large soil monoliths. *Water, Air and Soil Pollution* 143: 3–21.

Barton, C.D. and A.D. Karathanasis. 2003b. Colloid-enhanced desorption of zinc in soil monoliths. *International Journal of Environmental Studies* 60: 395–409.

Bradney, L., H. Wijesekara, K.N. Palansooriya, N. Obadamudalige, N.S. Bolan, Y.S. Ok, J. Rinklebe, K. Kim and M.B. Kirkham. 2019. Particular plastics as a vector for toxic trace-element uptake by aquatic and terrestrial organisms and human health risk. *Environment International* 131: 104937.

Brennecke, D., B. Duarte, F. Paiva, I. Cacador and J. Canning-Clode. 2016. Microplastics as a vector for heavy metal contamination from the marine environment. *Estuary and Coastal Shelf Science* 178: 189–195.

Elliott, H.A., M.R. Liberati and C.P. Huang. 1986. Competitive adsorption of heavy metals by soils. *Journal of Environmental Quality* 15: 214–219.

Goldberg, S., B.S. Kapoor and J.D. Rhoades. 1990. Effect of aluminum and iron oxides and organic matter on flocculation and dispersion of arid zone soils. *Soil Science* 150: 588–593.

Hillel, D. 1998. *Environmental Soil Physics*. Academic Press: San Diego, CA. 769 p.

Holmes, L.A., A. Turner and R.C. Thompson. 2012. Adsorption of trace metals to plastic resin pellets in the marine environment. *Environmental Pollution* 160: 42–48.

Holmes, L.A., A. Turner and R.C. Thompson. 2014. Interactions between trace metals and plastic production pellets under estuarine environments. *Marine Chemistry* 167: 25–32.

Horton, A.A., A. Walton, D.J. Spurgeon, E. Lahive and C. Svendsen. 2017. Microplastics in freshwater and terrestrial environments: Evaluating the current understanding to identify knowledge gaps and future research opportunities. *Science of the Total Environment* 586: 127–141.

Jekel, M.R. 1986. The stabilization of dispersed mineral particles by adsorption of humic substances. *Water Resources Research* 20: 1543–1554.

Johnson, W.P. and G.L. Amy. 1995. Facilitated transport of and enhanced desorption of polycyclic aromatic hydrocarbons by natural organic matter in aquifer sediments. *Environmental Science and Technology* 29: 807–817.

Kaplan, D.I., P.M. Bertsch and D.C. Adriano. 1995. Facilitated transport of contaminant metals through an acidified aquifer. *Ground* Water 33: 708–717.

Karathanasis, A.D. 1999. Subsurface migration of Cu and Zn mediated by soil colloids. *Soil Science Society of America Journal* 63: 830–838.

Kretzschmar, R., W.P. Robarge and A. Amoozegar. 1995. Influence of natural organic matter on colloid transport through saprolite. *Water Resources Research* 31: 435–445.

Li, C., K. Zhou, W. Qin, C. Tian, M. Qi, X. Yan and W. Han. 2019. A review on heavy metals contamination in soil: Effects, sources and remediation techniques. *Soil and Sediment Contamination: An International Journal* 28 (4): 380–394.

Lu, S., K. Zhu, W. Song, G. Song, D. Chen, T. Hayat, N. S. Alharbi, C. Chen and Y. Sun. 2018. Impact of water chemistry on surface charge and aggregation of polystyrene microspheres suspensions. *Science of the Total Environment* 630: 951–959.

McBride, M.B. 1989. Surface chemistry of soil minerals, In *Minerals in Soil Environments*, J.B. Dixon and S.B. Weed (Eds.), Soil Science Society of America, Madison, WI, pp. 35–88.

McCarthy, J.F. and J.M. Zachara. 1989. Subsurface transport of contaminants. *Environmental Science and Technology* 23: 496–503.

Miller, W. and D.A. Miller. 1987. A micro-pipette method for soil mechanical analysis. *Communications in Soil Science and Plant Analysis* 18: 1–15.

Mills, W.B., S. Liu and F.K. Fong. 1991. Literature review and model(COMET) for colloid/metal transport in porous media. *Ground Water* 29: 199–208.

Natural Resources Conservation Service (NRCS). 1996. Soil survey laboratory methods manual. *Soil Survey Investigations Report Number 42* USDA, Washington DC.

Nelson, D.W. and L.E. Sommers. 1982. Total carbon, organic carbon, and organic matter. In *Methods of Soil Analysis, Part 2: Chemical and Microbiological Properties*, 2nd ed., A.L., Miller, R.H., Keeney, D.R. (Eds.), ASA-SSSA: Madison, WI.

Nizzetto, L., G. Bussi, M.N. Futter, D. Butterfield, and P.G. Whitehead. 2016. A theoretical assessment of microplastic transport in river catchments and their retention by soils and river sediments. *Environment Science: Processes & Impacts* 18(8): 1050–1059.

Penrose, W.R., W.L. Polzer, E.H. Essington, D.M. Nelson and K.A. Orlandini. 1990. Mobility of plutonium and americium through a shallow aquifer in a semiarid region. *Environmental Science and Technology* 24: 228–234.

Rillig, M.C. 2012. Microplastic in terrestrial ecosystems and soil? *Environmental Science and Technology* 46: 6453–6454.

Ryan, J.N. and P.M. Gschwend. 1998. Effect of solution chemistry on clay colloid release from an iron oxide-coated aquifer sand. *Environmental Science and Technology* 28: 1717–1785.

Sen, T.K. and K.C. Khilar. 2006. Review on subsurface colloids and colloid-associated contaminant transport in saturated porous media. *Advances in Colloid and Interface Science* 119: 71–96.

Seta, A.K. and A.D. Karathanasis. 1997a. Atrazine adsorption by soil colloids and co-transport through subsurface environments. *Soil Science Society of America Journal* 61: 612–617.

Seta, A.K., and A.D. Karathanasis. 1997b. Stability and transportability of water-dispersible soil colloids. *Soil Science Society of America Journal* 61: 604–611.

Soil and Plant Analysis Council. 2000. *Soil Analysis Handbook of Reference Methods*. CRC Press: Boca Raton, FL.

Sposito, G. 1984. *The Surface Chemistry of Soils*. Oxford University Press: New York. 277 p.

Teuten, E.L., J.M. Saquing, D.R.U. Knappe, M.A. Barlaz, et al. 2009. Transport and release of chemicals from plastics to the environment and to wildlife. *Philosophical Transactions of the Royal Society* 364 (1526): 2027–2045.

Tian, Y.A., B. Gao, C. Silvera-Batista and J.K. Ziegler. 2010. Transport of engineered nanoparticles in saturated porous media. *Journal of Nanoparticle Research* 12(7): 2371–2380.

Turner, A. and L.A. Holmes. 2015. Adsorption of trace metals by microplastic pellets in fresh water. *Environmental Chemistry* 12(5): 600–610.

Velez, J.F.M., Y. Shashoua, K. Syberg and F.R. Khan. 2018. Considerations on the use of equilibrium models for the characterization of HOC-microplastic interactions in vector studies. *Chemosphere* 210: 359–365.

Wang, P. and A.A. Keller. 2009. Natural and engineered nano and colloidal transport: Role of zeta potential in prediction of particle deposition. *Langmuir* 25(12): 6856–6862.

Xie, B., Y. Jiang, Z. Zhang, G. Cao, H.M. Sun, N. Wang and S.S. Wang. 2018. Co-transport of Pb (II) and Cd (II) in saturated porous media: Effects of colloids, flow rate and grain size. *Chemical Speciation and Bioavailability* 30: 55–63.

9 Microbial Plastisphere

Microbial Habitation of Particulate Plastics in Terrestrial and Aquatic Environments

Nanthi S. Bolan, M.B. Kirkham, B. Ravindran, Anu Kumar, and Weixin Ding

CONTENTS

9.1 Plastic Pollution 135
9.2 Biofilm and Plastisphere 136
9.3 Biofilm Formation and Microbial Habitat on Particulate Plastics 136
9.4 Distribution of Microbial Species in Particulate Plastics 137
9.5 Factors Affecting Microbial Habitation of Particulate Plastics 140
9.6 Implications of Microbial Habitation of Particulate Plastics 141
9.7 Conclusions 142
References 143

9.1 PLASTIC POLLUTION

By 2018, the annual global production of plastics reached 359 million metric tons (Garside 2019), the greatest percentage of which is in the form of polyethylene. Polyethylene makes up 64% of plastics that are single use or discarded after a first use (Geyer et al. 2017). The final destination for a significant volume of plastics is the water bodies, such as oceans and seas. Current estimates suggest there are 5.25 trillion plastic particles littering the oceans (Eriksen et al. 2014). However, the majority of marine plastics have land-based original sources (Andrady 2011; Bradney et al. 2019; Wijesekara et al. 2018) and enter the marine environment through direct dumping by humans, redistribution from landfills, losses in transport, and inputs from wastewater treatment plants (Bradney et al. 2019).

Microplastics, including microbeads and microfibers from clothes, cosmetics, and sanitary products, are now common constituents of sewage systems, and they frequently bypass the screening mechanisms designed to remove larger waste items (Rios Mendoza et al. 2018). Plastics disposal has become a cause for concern, because the detrimental impacts of plastic pollution in terrestrial (i.e., soil) and aquatic (i.e., marine) ecosystems have been shown to increase exponentially in the past few decades (Eriksen et al. 2014). Plastic degradation can take hundreds of years, and, hence, plastics can be retained in the environment longer than other natural substrates. These long lasting substrates can provide a habitat for the colonization and possible migration of microbial communities, especially in aquatic environments, such as oceans and rivers (Law 2017).

Zooplankton in aquatic ecosystems, such as oceans and rivers was found to outnumber plastic particles five to one; however, plastic outweighed zooplankton six to one (Cole et al. 2013). This discrepancy is of particular importance when considering marine wildlife. Many animals, including marine birds, sea turtles, and cetaceans, often mistake plastic objects for food. Plastic that is ingested remains in the digestive tract and leads to lower feeding stimuli, gastrointestinal blockage, and reproductive problems (Azzarello and Van Vleet 1987; Webb et al. 2013). Other dangers associated with plastic debris include entanglement in large objects, biosorption of toxic chemicals, such as polychlorinated biphenyls, and transportation of non-indigenous harmful algal blooms (Webb et al. 2013).

9.2 BIOFILM AND PLASTISPHERE

Particulate plastic fragments in terrestrial (e.g., soil) and aquatic (e.g., marine) environments are known to accumulate and host various microorganisms. The hydrophobic nature of plastic surfaces facilitates the adsorption of dissolved organic carbon in soil and aquatic ecosystems (Bradney et al. 2019). This stimulates the growth of microorganisms, which then transform into biofilms that help support a wide range of metabolic activities. The amalgam of microbes of the biofilm differs from that of the surrounding environment, for example, seawater and sediment. The assemblage of ecosystems colonizing the particulate plastic environment is referred to as "plastisphere" (Kirstein et al. 2019).

The hard surface of waterborne plastic provides an ideal environment for the formation of biofilms for opportunistic microbial colonizers. Although plastics are extremely resistant to decay, variability in composition determines their specific buoyancy and surface rugosity, which is likely to dictate the extent of microbial colonization and the ability of microbes for long distance dispersal (Hossain et al. 2019). Microorganisms attached to plastics are a distinct biological community with different physical and chemical characteristics from free-living microorganisms (Zettler et al. 2013).

9.3 BIOFILM FORMATION AND MICROBIAL HABITAT ON PARTICULATE PLASTICS

Biofilm formation on most biotic and abiotic surfaces including particulate plastic fragments follows a series of steps (Harrison et al. 2018) (Figure 9.1). Free-floating microorganisms including bacteria come in proximity with an appropriate substrate and "attach" themselves to the surface.

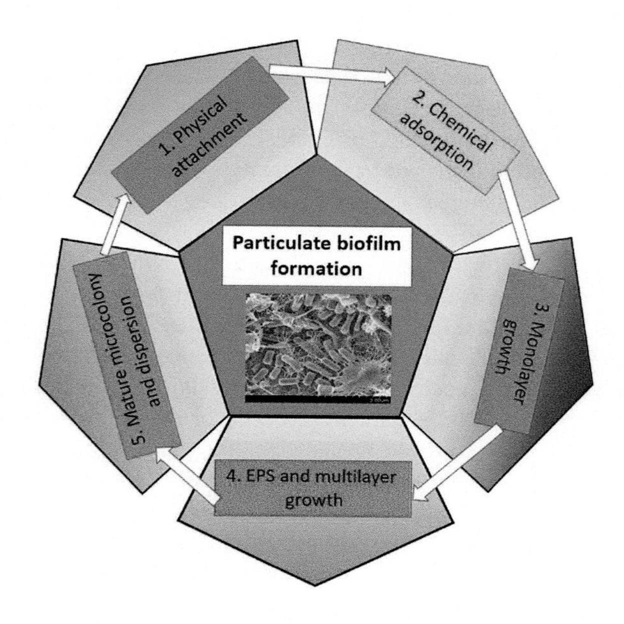

FIGURE 9.1 Various stages of biofilm formation on particulate plastics.

This attachment process includes both physical adhesion and chemical adsorption processes (Cunliffe et al. 1999; Donlan 2002; Rummel et al. 2017). While the physical adhesion process involves the release of an adhesive substance, such as extracellular polymeric substances (EPSs), the chemical adsorption process involves surface charge interactions between the microorganisms and the substrate (Tuson and Weibel 2013; Decho and Gutierrez 2017). In the case of biofilm formation on particulate plastics, both the microorganisms and the particulate plastics carry surface charges, which are involved in the adsorption of microorganisms to the plastic surface. Surface functional groups on microorganisms and particulate plastics contribute to their surface charge. EPSs are composed of sugars, proteins, and nucleic acids, which promote the microorganisms in a biofilm to stay together (Donlan 2002; Vu et al. 2009). The initial attachment of microorganisms to a particulate plastic surface through physical adhesion and chemical adsorption is succeeded by a cycle of rapid growth in which a 2D microbial monolayer is formed. Microorganisms then layer themselves and EPSs upon the first layers, creating a globular or bulbous complex, 3D structure. Nutrients and waste products are exchanged via water channels that form in a crosshatch pattern through the biofilm. Finally, after this process, some of the cells that reside in the biofilm detach and adhere to new surfaces.

Microorganisms that congregate into a biofilm have distinct advantages (Nadell et al. 2009). Communities of microorganisms present in biofilms have higher tolerance of stress factors, which include poor water supply, extreme pH and salinity conditions, and the presence of toxic substances, such as antibiotics, anti-microbials, and heavy metals. The EPS covering layer is a capable, defensive barrier, thereby preventing dehydration and desiccation, or it can act as a guard against ultraviolet light. In addition, anti-microbials and metals are either immobilized or neutralized when they are exposed to the EPS, thereby modifying their toxicity and harmfulness to microorganisms (Ayangbenro and Babalola 2017; Singh et al. 2017). Most microorganisms excrete EPSs that protect metal-sensitive biochemical components by binding to toxic metal ions (Gupta and Diwan 2017). The components of EPSs, including proteins, polysaccharides, and nucleic acids, are involved in the adsorption and complexation of metals (Guibaud et al. 2003; Pal and Paul 2008). For example, complexation of lead (Pb), cadmium (Cd), mercury (Hg), and other metals by EPSs has been observed for bacteria (Chakravarty and Banerjee 2012; Perez et al. 2008).

Microorganisms that have co-habitation in biofilms benefit from other microbial members of the biofilm. Photosynthetic bacteria or algae, i.e., autotrophs, produce their own sustenance in the form of organic carbon (carbon containing) material, while heterotrophs require external sources of carbon provided by the biofilm via the autotrophs present.

9.4 DISTRIBUTION OF MICROBIAL SPECIES IN PARTICULATE PLASTICS

The diversity and activity of the microbial communities on various biotic and abiotic surfaces, including particulate plastics, are often examined using a spectrum of spectroscopic and analytical techniques, including scanning electron and atomic-force microscopy, flow cytometry, enzymatic activity, and metagenomics analysis (Douterelo et al. 2014; Rosenthal et al. 2017). Recently genomic sequencing using 16S and 18S ribosomal RNA was used for identifying specific bacteria (Janda and Abbott 2007; Jacquin et al. 2019; Kettner et al. 2019; Reller et al. 2007). Identifying species associated with plastics using metagenomics analysis will make it possible to determine zones of bacterial colonization and to better understand the dispersion and migration of alien toxic and pathogenic species. For example, Harrison et al. (2014) demonstrated bacterial attachment onto low-density polyethylene (LDPE) within sediments by scanning electron microscopy and catalyzed-reporter-deposition-fluorescence in situ hybridization. They observed that dramatic increases in the abundance of 16S rRNA genes from LDPE-associated bacteria occurred within 7 days and varied between sediment types. Terminal-restriction fragment length polymorphism analysis demonstrated rapid adaptation of LDPE-associated bacterial assemblages, which differed significantly from surrounding sediments. Additionally, terminal-restriction fragment length polymorphism analysis revealed successional convergence of the LDPE-associated communities from the different sediments was dominated by the genera *Arcobacter* and *Colwellia*.

A number of studies have reported a vast array of organisms is associated with particulate plastics in aquatic ecosystem (Jacquin et al. 2019) (Table 9.1). For example, Bryant et al. (2016) surveyed the "great pacific garbage patch" for the metabolic activities of associated microbial communities that reside in the seawater and accumulate on plastic. They noticed that net community oxygen production (net community oxygen production = gross primary production – community respiration) on plastic was positive (i.e., net autotrophic), whereas net community oxygen production in surrounding seawater was close to zero, indicating the microbial functional difference between plastic debris and seawater. Scanning electron microscopy and metagenomic sequencing of particulate plastic-associated communities revealed the dominance of *Bryozoa*, *Cyanobacteria*, *Alphaproteobacteria*, and *Bacteroidetes* (Figure 9.2). Bacteria inhabiting plastics were taxonomically and functionally distinct from the surrounding picoplankton and were well adapted to a solid substrate-associated habitat. Similarly, Reisser et al. (2014) used scanning electron microscopy to characterize the biodiversity of organisms on the surface of floating microplastics in samples from Australia-wide coastal and oceanic zones, as well as tropical to temperate zones. They noticed that the particulate plastics

TABLE 9.1
List of Known Microbial Genera Occurring on Microplastics

Group	Genera
Bacteria	*Acinetobacter, Albidovulum, Alteromonas, Amoebophilus, Bacteriovorax, Bdellovibrio, Blastopirellula, Devosia, Erythrobacter, Filomicrobium, Fulvivirga, Haliscomenobacter, Hellea, Henriciella, Hyphomonas, Idiomarina, Labrenzia, Lewinella, Marinoscillum, Microscilla, Muricauda, Nitrotireductor, Oceaniserpentilla, Parvularcula, Pelagibacter, Phycisphaera, Phormidium, Pleurocapsa, Prochlorococcus, Pseudoalteromonas, Pseudomonas, Psychrobacter, Rhodovulum, Rivularia, Roseovarius, Rubrimonas, Sediminibacterium, Synechococcus, Thalassobius, Thiobios, Tenacibaculum, Thalassobius, Vibrio*
Diatoms	*Amphora, Achananthes, Chaetoceros, Cocconeis, Cyclotella, Cymbella, Grammatophora, Haslea, Licmophora, Mastogloia, c, Microtabella, Minidiscus, Navicula, Nitzschia, Pleurosigma, Sellaphora, Stauroneis, Thalassionema, Thalassiosira*
Coccoliths	*Calcidiscus, Emiliania, Gephyrocapsa, Umbellosphaera, Umbilicosphaera, Coccolithus, Calciosolenia*
Bryozo	*Membranipora, Jellyella, Bowerbankia, Filicrisia*
Hydroids	*Clytia, Gonothyraea, Obelia*
Polychaete	*Spirorbis, Hydroides*
Dinoflagellates	*Alexandrium, Ceratium*
Insect eggs	*Halobates*
Barnacles	*Lepas*
Rhodophyta	*Fosliella*
Foraminifera	*Discorbis*
Radiolaria	*Circorrhegma*
Ciliate	*Ephelota*

Source: Reisser, J. et al., *PLoS One*, 9, e100289, 2014.

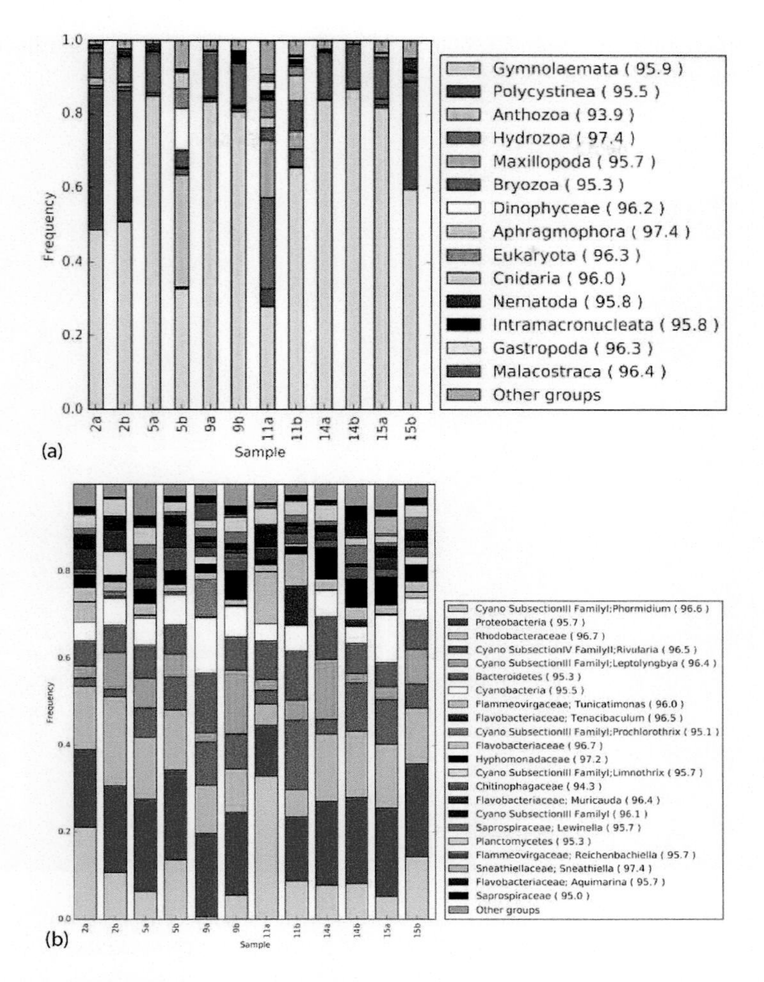

FIGURE 9.2 (a) Bar chart displaying the abundance of eukaryotic classes within reads mapping to small-subunit (SSU) rRNA genes. Reads were assigned to the lowest common ancestor (LCA) of top hits to the SILVA database. Clade abundances in each sample are relative to the total number of reads per sample mapping to a eukaryotic SSU rRNA gene. Clades with abundances of >1% in at least one sample are shown. The average percent identities of sample reads to their top hit within each taxonomic group are displayed in parentheses. (From Bryant, J.A. et al., *mSystems*, 1, e00024–16, 2016.) (b) Bar chart showing the relative abundance of prokaryotic groups based on reads mapping to prokaryotic SSU rRNA genes. Reads were assigned to the LCA of top hits to the SILVA database. Where possible, reads were assigned prokaryotic genera. Broader taxonomic groups are made up of reads that could not be assigned to a genus or had abundances of <3% in all libraries. Clade counts in each sample were normalized to the total number of SSU rRNA gene reads mapped to bacterial taxa in each sample. The average percent identities of sample reads to their top hit within each taxonomic group are in parentheses.

were colonized by a diverse group of plastic colonizers, represented by 14 genera. They also noticed a range of organisms including "epiplastic" coccolithophores (seven genera), bryozoans, barnacles (*Lepas* spp.), a dinoflagellate (*Ceratium*), an isopod (*Asellota*), a marine worm, marine insect eggs (*Halobates* sp.), as well as rounded, elongated, and spiral cells that came from bacteria, cyanobacteria, and fungi (Table 9.1; Figure 9.3).

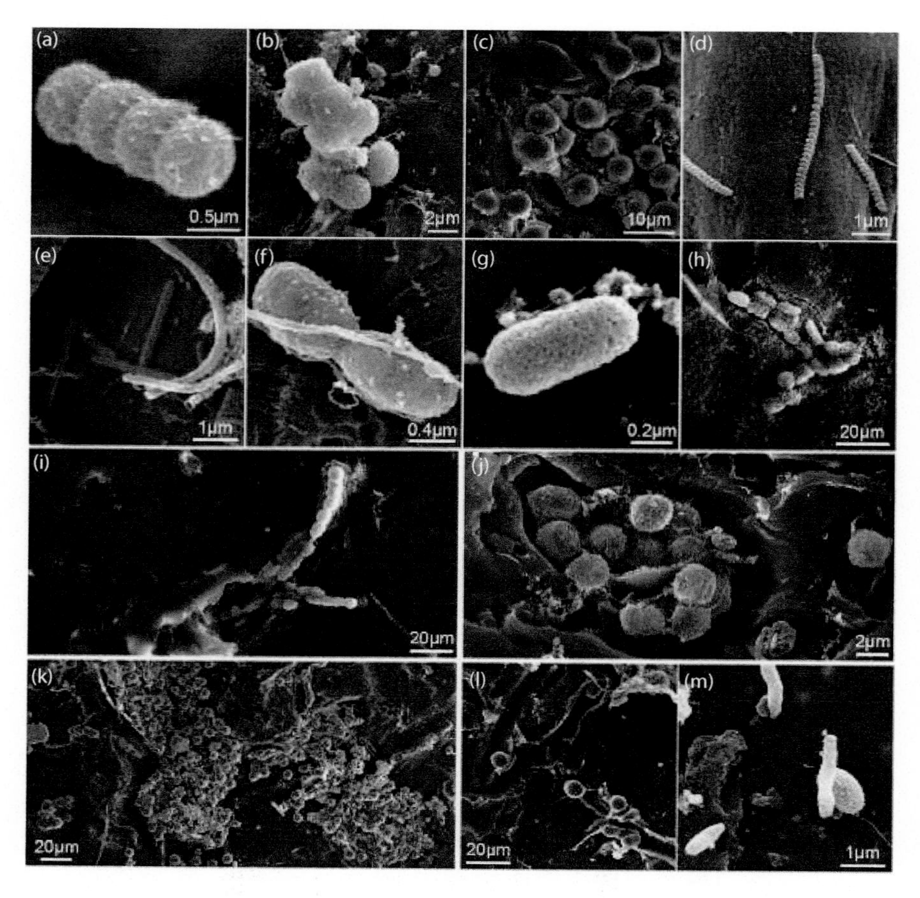

FIGURE 9.3 Microbial colonization of particulate plastics: (a–c) rounded cells, (d) spiral "spirochaete" cell, (e–h) elongated cells, (i–m) pits and grooves on plastics with rounded cells. (From Reisser, J. et al., *PLoS ONE*, 9(6), e100289, 2014.)

9.5 FACTORS AFFECTING MICROBIAL HABITATION OF PARTICULATE PLASTICS

The microbial composition in the biofilms formed on particulate plastics depends on a number of factors including the nature and "weathering age" of particulate plastics and the environmental conditions of the terrestrial and aquatic media (Anderson et al. 2016; Jacquin et al. 2019). Plastispheres are shown to be different depending on the type of plastic. Microorganisms living in microplastics present distinct patterns from surrounding free-living and sediment-associated colonies in seawater, implying that plastic serves as a unique microbial habitat substrate in the ocean. For example, weathered particulate plastics provide greater surface area for the colonization of microbial communities than pristine particulate plastics (Curren and Leong 2019; Scherer et al. 2018). Weathered particulate plastics in terrestrial and aquatic media are generally enriched with dissolved organic carbon, which facilitates the growth of microorganisms (Bradney et al. 2019). Multiple environmental conditions of terrestrial and aquatic media help determine the extent to which a biofilm grows on particulate plastics and the microbial composition of biofilms. For example, microorganisms that produce a copious amount of EPSs can grow rapidly into dense biofilms even in the absence of adequate amounts of essential nutrients required for microbial growth (Donlan 2002; Flemming 2016). However, for microorganisms that require oxygen, the availability of oxygen can limit the level of microbial growth. In a comparatively nutrient-poor medium, the microplastics are generally colonized by bacterial communities that differ significantly from natural communities. Plastic pollution in a nutrient-poor environment has a much higher ecological relevance, and, in these environments, the development of plastic-specific bacterial populations is promoted (Beaumont et al. 2019; Smith et al. 2018).

9.6 IMPLICATIONS OF MICROBIAL HABITATION OF PARTICULATE PLASTICS

The microbial habitation of particulate plastics has several implications that include (Figure 9.4) (Jacquin et al. 2019; Oberbeckmann et al. 2018): (i) degradation of particulate plastics; (ii) horizontal gene transfer between microorganisms; (iii) toxicity of contaminants associated with plastics; and (iv) migration or spreading of microorganisms especially in aquatic ecosystems, such as oceans and rivers. Bacteria and fungi have the ability to degrade efficiently polyethylene, polypropylene, and other polymers. The microbially induced enzymatic degradation of polymers into monomers has been reported (Banerjee et al. 2014). The biodegradation of plastics is thought to be accelerated by certain organisms. The products produced, however, can be potentially hazardous chemicals. But this process also could be potentially useful, as these microorganisms could be utilized to degrade plastics faster, including plastics that are resilient to degradation. However, as plastic degradation occurs, larger plastic particles break down into smaller pieces, eventually into microplastics, and chances are higher that they will enter into the food cycle by plankton consumption. As plankton is consumed by larger organisms, and these are consumed by again larger organisms, the risk remains that the plastic will eventually accumulate in higher food chain fish eaten by humans, and, thus, contaminants will be introduced into the food chain (Toussaint et al. 2019).

There is growing evidence that the plastisphere can promote gene exchange (Jacquin et al. 2019; Zettler et al. 2013); so determining the potential of plastisphere biofilms for providing the surface area for anti-microbial resistance gene transfer is important. Bacteria living on microplastics have high rates of gene exchange leading to mutation, perhaps due to the fact that the high surface area particulate plastics give microorganisms the chance to grow quickly and result in a great diversity of microorganisms. Mutation usually results from a long adaptation process, but it has been shown that gene transfer and mutation can happen rapidly on particulate plastics because of the diversity of microorganisms within a limited space (Steenackers et al. 2016).

Microplastics interact with a large range of organic and inorganic contaminants present in the environment. Adsorption and desorption processes of pollutants in microplastics in terrestrial and aquatic ecosystems are influenced by factors, such as weathering and surface area, interaction with dissolved organic matter, and microbial activity (i.e., biofilm formation). The association of

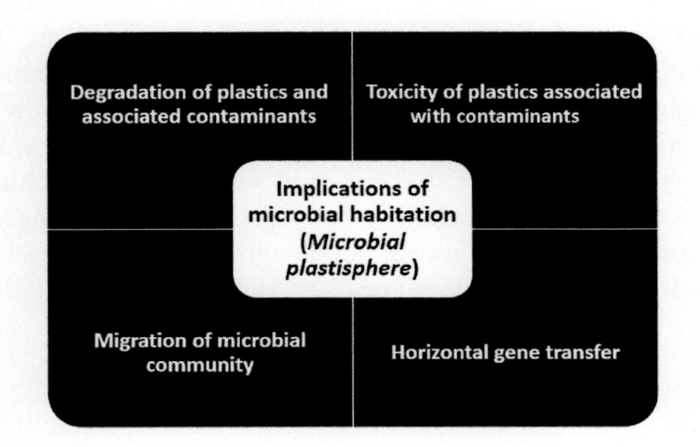

FIGURE 9.4 Implications of biofilm formation on particulate plastics.

dissolved organic matter with microplastics in these ecosystems plays a crucial role in promoting microplastics as a vector for contaminant transport. The surface area of microplastics, which can be altered through weathering, not only aids adsorption of contaminants, but also facilitates the transfer of the pollutants through leaching.

Microorganisms in the biofilm play an active role in the bioremediation of contaminants associated with plastics. Bioremediation, in general, is the use of living organisms, or their products (e.g., enzymes), to degrade contaminants. Biofilms are used in treating wastewater (i.e., the denitrification process of nitrate removal), heavy metal contaminants, such as chromate, and organic contaminants. Microorganisms can either degrade them or change their bioavailability and mobility, and, therefore, make them less harmful to the environment and to humans. For instance, the members of the genus *Erythrobacter* are interesting because they are able to degrade toxic polycyclic aromatic hydrocarbons that are found in the environment as a result of human activities, and they accumulate on microplastics due to their chemical properties (Smith et al. 2018). Roeselers et al. (2008) reviewed the potential applications of microbial mats or biofilms in wastewater treatment, bioremediation, fish-feed production, biohydrogen production, and soil remediation. Microbial mats consisting of biofilms provide several advantages that include low cost, durability, ability to function in extreme conditions including both fresh and salt water, and tolerance to high concentrations of contaminants including metals.

Microplastics in the environment are potential vectors for contaminants, which often desorb when ingested by marine species and can accumulate in the food chain. Microorganisms in the plastisphere may either mitigate this problem through biodegradation or enhance it by increased biofilm binding. The plastisphere system can also harbor toxic microbial species, which are transferred to humans through filter feeders that fish eat, thereby impacting the food chain.

Abiotic particulate plastics provide a more robust vehicle than biodegradable biotic substrates for the transport of organisms over long distances. This long distance transportation can facilitate the movement of microorganisms to different ecosystems and potentially introduce invasive and exotic species. Microorganisms associated with particulate plastics include autotrophs, heterotrophs, and symbionts, leading to both collaborative and competitive existence. The ecosystem created by the plastisphere differs from other naturally occurring floating substrates, such as feathers and algae due to the long-term persistence of plastic substrates.

9.7 CONCLUSIONS

Particulate plastic fragments in terrestrial and aquatic environments host various microorganisms. The hydrophobic nature of plastic surfaces stimulates the rapid growth of microorganisms and the subsequent formation of biofilms. The assemblage of ecosystems colonizing the plastic environment is referred to as "plastisphere." Polymer type, environmental conditions including nutrient status and salinity, and season affect the microbial composition of biofilms. The microbial habitation of particulate plastics facilitates the movement of microorganisms, especially in the aquatic environment, and impacts the toxicity of contaminants associated with these particulate plastic fragments. There is currently a lack of fundamental understanding about the mechanisms by which microorganisms, particularly pathogenic bacteria and viruses, can "hitchhike" on microplastic particles and be transported to beaches, bathing waters, shellfish harvesting waters, and high benthic diversity zones. Consequently, it is not yet possible to determine the risk from these potential pathways or to establish environmental monitoring guidelines for future policy or environmental regulations.

REFERENCES

Anderson, J.C., Park, B.J., Palace, V.P. (2016). Microplastics in aquatic environments: Implications for Canadian ecosystems. *Environmen. Pollut.*, 218, 269–280.

Andrady, A.L. (2011). Microplastics in the marine environment. *Marine Pollution Bulletin*, 62(8), 1596–1605.

Ayangbenro, A.S., Babalola, O.O. (2017). A New strategy for heavy metal polluted environments: A review of microbial biosorbents. *Int. J. Environ. Res. Public Health*, 2017 Jan; 14(1), 94.

Azzarello, M.Y., Van Vleet, E.S. (1987). Marine birds and plastic pollution. *Mar. Ecol. Prog. Ser.*, 6 May 1987; 37(2/3), 295–303.

Banerjee, A., Chatterjee, L., Madras, G. (2014). Enzymatic degradation of polymers: A brief review. *Mat. Sci. Technol.*, 30(5), 567–573.

Beaumont, N.J., Aanesen, M., Austen, M.C., Börger, T., Clark, J.R., Cole, M., Hooper, T., Lindeque, P.K., Pascoe, C., Wyles, K.J. (2019). Global ecological, social and economic impacts of marine plastic. *Mar. Pollut. Bull.*, 142, 189–195.

Bradney, L., Wijesekara, H., Palansooriya, K. N., Obadamudalige, N., Bolan, N. S., Ok, Y. S., Kirkham, M. B. (2019). Particulate plastics as a vector for toxic trace-element uptake by aquatic and terrestrial organisms and human health risk. *Environment International*, 131.

Bryant, J.A., Clemente, T.M., Viviani, D.A., Fong, A.A., Thomas, K.A., Kemp, P., Karl, D.M., White, A.E., DeLong, E.F. (2016). Diversity and activity of communities inhabiting plastic debris in the North Pacific Gyre. *mSystems* 1(3), e00024–16. doi:10.1128/mSystems.00024-16.

Chakravarty, R., Banerjee, P.C. (2012). Mechanism of cadmium binding on the cell wall of an acidophilic bacterium. *Bioresour Technol.*, 108, 176–183.

Cole, M., Lindeque, P., Fileman, E., Halsband, C., Goodhead, R., Moger, J., Galloway, T.S. (2013). Microplastic ingestion by zooplankton. *Environ. Sci. Technol.*, 47(12), 6646–6655.

Cunliffe, D., Smart, C.A., Alexander, C., Vulfson, E.N. (1999). Bacterial adhesion at synthetic surfaces. *Appl. Environ. Microbiol.*, Nov; 65(11), 4995–5002.

Curren, E., Leong, S.C.Y. (2019). Profiles of bacterial assemblages from microplastics of tropical coastal environments. *Sci. Total Environ.*, Mar 10; 655, 313–320.

Decho, A.W., Gutierrez, T. (2017). Microbial extracellular polymeric substances (EPSs) in ocean systems. *Front. Microbiol.*, 26. doi:10.3389/fmicb.2017.00922.

Donlan, R.M. (2002). Biofilms: Microbial life on surfaces. *Emerg Infect Dis.*, Sep; 8(9), 881–890.

Douterelo, I., Boxall, J.B., Deines, P., Sekar, R., Fish, K.E., Biggs, C.A. (2014). Methodological approaches for studying the microbial ecology of drinking water distribution systems. *Water Res.*, 65(15), 134–156.

Eriksen, M., Lebreton, L.C.M., Carson, H.S., Thiel, M., Moore, C.J., Borerro, J.C., et al. (2014). Plastic pollution in the world's oceans: More than 5 trillion plastic pieces weighing over 250,000 tons afloat at sea. *PLoS ONE*, 9(12), e111913. doi:10.1371/journal.pone.0111913.

Flemming, H.C. (2016). EPS—Then and now. *Microorganisms*, Dec; 4(4), 41.

Garside, M. (2019). Global plastic production 1950–2018. (https://www.statista.com/statistics/282732/global-production-of-plastics-since-1950/).

Geyer, R., Rambeck, J.R., Law, K.L. (2017). Production, use, and fate of all plastics ever made. *Sci. Adv.*, Jul; 3(7): e1700782.

Goldstein, M.C., Carson, H.S., Eriksen, M. (2014). Relationship of diversity and habitat area in North Pacific plastic-associated rafting communities. *Mar. Biol.*, 161, 1441–1453.

Goldstein, M.C., Rosenberg, M., Cheng, L. (2012). Increased oceanic microplastic debris enhances oviposition in an endemic pelagic insect. *Biol. Lett.*, 8, 817–820.

Gregory, M.R. (1978). Accumulation and distribution of virgin plastic granules on New Zealand beaches. *New Zeal. J. Mar. Fresh.*, 12, 399–414.

Gregory, M.R. (1983). Virgin plastic granules on some beaches of eastern Canada and Bermuda. *Mar. Environ. Res.*, 10, 73–92.

Guibaud, G., Tixier, N., Bouju, A. and Baudu, M. (2003). Relation between extracellular polymers' composition and its ability to complex Cd, Cu and Pb. *Chemosphere* 52, 1701–1710.

Gupta, P. and Diwan, B. (2017). Bacterial Exopolysaccharide mediated heavy metal removal: a review on biosynthesis, mechanism and remediation strategies. *Biotechnology Reports*, 13, 58–71.

Harrison, J.P., Hoellein, T.J., Sapp, M., Tagg, A.S., Ju-Nam, Y., Ojeda, J.J. (2018). Microplastic-associated biofilms: A comparison of freshwater and marine environments. In: Wagner M., Lambert S. (Eds.), *Freshwater Microplastics. The Handbook of Environmental Chemistry*, Vol. 58. Springer, Cham, Switzerland, pp. 181–201.

Harrison, J.P., Schratzberger, M., Sapp, M., Osborn, A.M. (2014). Rapid bacterial colonization of low-density polyethylene microplastics in coastal sediment microcosms. *BMC Microbiol.* 14, 232. doi: 10.1186/s12866-014-0232-4.

Hossain, M.R., Jiang, M., Wei, Q.H., Leff, L.G. (2019). Microplastic surface properties affect bacterial colonization in freshwater. Biotechnology for environmental and sustainable applications. *J. Basic Microbiol.*, 59(1), 54–61.

Jacquin, J., Cheng, J., Odobel, C., Pandin, C., Conan, P., Pujopay, M., Barbe, V., Meistertzheim, A.L., Ghiglione, J.F. (2019). Microbial ecotoxicology of marine plastic debris: A review on colonization and biodegradation by the "Plastisphere." *Front Microbiol.*, 10, 865.

Janda, M.J., Abbott, S.L. (2007). 16S rRNA gene sequencing for bacterial identification in the diagnostic laboratory: Pluses, perils, and pitfalls. *J. Clin. Microbiol.*, Sep; 45(9), 2761–2764.

Kettner, M.T., Oberbeckmann, S., Labrenz, M. Grossart, H.P. (2019). The eukaryotic life on microplastics in brackish ecosystems. *Front. Microbiol.*, 20. doi:10.3389/fmicb.2019.00538.

Kirstein, I.V., Wichels, A., Gullans, E., Krohne, G., Gerdts, G. (2019). The plastisphere – Uncovering tightly attached plastic "*specific*" microorganisms. *PLoS ONE*, 14(4), e0215859.

Law, K.L. (2017). Plastics in the marine environment. *Annu. Rev. Mar. Sci.*, 9, 205–229.

Nadell, C.D., Xavier, J.B., Foster, K.R. (2009). The sociobiology of biofilms. *FEMS Microbiol Rev.*, 33(2009), 206–224.

Oberbeckmann, S., Kreikemeyer, B., Labrenz, M. (2018). Environmental factors support the formation of specific bacterial assemblages on microplastics. *Front. Microbiol.*, 19 January 2018. doi:10.3389/fmicb.2017.02709.

Pal, A., Paul, A.K. (2008). Microbial extracellular polymeric substances: central elements in heavy metal bioremediation. *Indian J. Microbiol.* 48, 49–64.

Perez, M.P., García-Ribera, R., Quesada, T., Aguilera, M., Ramos-Cormenzana, A., Monteoliva-Sánchez, M. (2008). Biosorption of heavy metals by the exopolysaccharide produced by Paenibacillus jamilae. *World J. Microbiol. Biotechnol.* 24, 2699–2704.

Reisser, J., Shaw, J., Hallegraeff, G., Proietti, M., Barnes, D.K.A., et al. (2014). Millimeter-sized marine plastics: A new pelagic habitat for microorganisms and invertebrates. *PLoS ONE*, 9(6), e100289. doi:10.1371/journal.pone.0100289.

Reller, L.B., Weinstein, M.P., Petti, C.A. (2007). Detection and identification of microorganisms by gene amplification and sequencing. *Clin. Infect. Dis.*, 15 April 2007; 44(8), 1108–1114.

Rios Mendoza, L.M., Karapanagioti, H., Álvarez, N.R. (2018). Micro(nanoplastics) in the marine environment: Current knowledge and gaps. *Curr. Opin. Environ. Sci. Health*, 1, 47–51.

Roeselers, G., Loosdrecht, M.C.M., Muyzer, G. (2008). Phototrophic biofilms and their potential applications. *J Appl Phycol* 20, 227–235.

Rosenthal, K., Oehling, V., Dusny, C., Schmid, A. (2017). Beyond the bulk: Disclosing the life of single microbial cells. *FEMS Microbiol. Rev.*, November 2017, 41(6), 751–780.

Rummel, C.D., Jahnke, A., Grorokhova, E., Kühnel, D., Schmitt-Jansen, M. (2017). Impacts of biofilm formation on the fate and potential effects of microplastic in the aquatic environment. *Environ. Sci. Technol. Lett.*, 47, 258–267.

Scherer, C., Weber, A., Lambert, S., Wagner, M. (2018). Interactions of microplastics with freshwater biota. In: Wagner, M., Lambert, S. (Eds.), *Freshwater Microplastics. The Handbook of Environmental Chemistry*, Vol. 58. Springer, Cham, Switzerland, pp. 153–180.

Singh, S., Singh, S.K., Chowdhury, I., Singh, R. (2017). Understanding the mechanism of bacterial biofilms resistance to antimicrobial agents. *Open Microbiol J.*, 11, 53–62.

Smith, M., Love, D.C., Rochman, C.M., Neff, R.A. (2018). Microplastics in seafood and the implications for human health. *Curr. Environ. Health Rep.*, 5(3), 375–386.

Steenackers, H.P., Parijs, I., Foster, K.R., Vanderleyden, J. (2016). Experimental evolution in biofilm populations. *FEMS Microbiol Rev.*, 1 May 2016; 40(3), 373–397.

Toussaint, B., Raffael, B., Angers-Loustau, A., Gilliland, D., Kestens, V., Petrillo, M., Rio-Echevarria, I.M., Van den Eede, G. (2019). Review of micro- and nanoplastic contamination in the food chain. *Food Additives & Contaminants: Part A*, 36(5), 639–673. doi:10.1080/19440049.2019.1583381.

Tuson, H.H., Weibel, D.B. (2013). Bacteria-surface interactions. *Soft Matter.*, 9(18), 4368–4380.

Vu, B., Chen, M., Crawford, R.J., Ivanova, E.P. (2009). Bacterial extracellular polysaccharides involved in biofilm formation. *Molecules*, 14(7), 2535–2554.

Webb, H.K., Arnott, J., Crawford, R.J., Ivanova, E.P. (2013). Plastic degradation and its environmental implications with special reference to poly(ethylene terephthalate). *Polymers*, 5(1), 1–18.

Wijesekara, H., Bolan, N.S., Bradney, L., Obadamudalige, N., Seshadri, B., Kunhikrishnan, A., Dharmarajan, R., Ok, Y.S., Rinklebe, J., Kirkham, M.B., Vithanage, M. (2018). Trace element dynamics of biosolids-derived microbeads. *Chemosphere,* 199, 331–339.

Zettler, E.R., Mincer, T.J., Amaral-Zettler, L.A. (2013). Life in the "Plastisphere": Microbial communities on plastic marine debris. *Environ. Sci. Technol.*, 47, 7137–7146.

10 Aggregation Behavior of Particulate Plastics and Its Implications

Xinjie Wang, Yang Li, Jiajun Duan, Yuan Liu, Shengdong Liu, Enxiang Shang, Nanthi S. Bolan, and Yining Wang

CONTENTS

10.1 Aggregation Behavior of Particulate Plastics 147
10.2 Factors Affecting Aggregation 151
10.2.1 Physicochemical Properties 151
10.2.2 Environmental Conditions 152
10.2.3 Solid Constituents 153
10.3 Methods for Studying Aggregation Behavior 153
10.3.1 Experimental Techniques for Studying Aggregation 153
10.3.1.1 Visualizing and Imaging Methods 153
10.3.1.2 Light-Scattering Methods 154
10.3.1.3 Other Methods 154
10.3.2 Modeling and Simulation of Aggregation Processes 154
10.3.2.1 Fractal Dimension 154
10.3.2.2 DLVO Theory 155
10.3.2.3 Aggregation Kinetics 155
10.4 Environmental Implication of Aggregation 156
10.4.1 Toxicity to Organisms 157
10.4.2 Transport of Microplastics 157
10.4.3 Transport of Contaminants 157
10.5 Conclusions 158
References 158

10.1 AGGREGATION BEHAVIOR OF PARTICULATE PLASTICS

Particulate plastics include microplastics (MPs) with the size of 0.1 μm–5 mm and nanoplastics (NPs) with the size smaller than 100 nm (Alimi et al. 2018). After particulate plastics enter into the aquatic environment, they may aggregate settling, and degrade (Zhu et al. 2019). Among these behaviors, aggregation has a huge impact on the mobility, stability, and bioavailability of particulate plastics in natural waters (Besseling et al. 2017; Carr et al. 2016; Da Costa et al. 2016). Aggregation refers to the process in which two particles move toward each other, collide, and attach together, which leads to destabilization in colloidal systems (Alimi et al. 2018). There are two types of aggregation, including homoaggregation with the same type of particles and heteroaggregation with various types of particles (Alimi et al. 2018). Heteroaggregation is more prone to occur than homoaggregation for particulate plastics due to the large number of natural colloids, including suspended sediment (SS), and organisms like algae and seaweeds in natural waters (Cai et al. 2018; Lowry et al. 2012; Oriekhova and Stoll 2018; Pelley and Tufenkji 2008).

A comprehensive literature review on the aggregation of particulate plastics in waters is presented in Table 10.1. Many previous works focused on the dispersion and stability of polystyrene (PS), with various particle sizes and surface coatings under different water chemical conditions, because it was ubiquitously detected in both freshwater and seawater (De Haan et al. 2019; Dris et al. 2016; Zhang and Liu 2018). Considering the increasing production, application, and release of particulate plastics with other compositions, it is of great importance to elucidate the aggregation behavior of these particulate plastics. For example, polyethylene (PE) is also the most applied and widely detected plastic particle in packaging (Napper et al. 2015). To the best of our knowledge, its dispersion state as a function of plastic particle physicochemical properties and environmental conditions has not been studied.

Compared with the homoaggregation of MPs/NPs, the heteroaggregation of MPs /NPs with other solid constituents, as well as the effect of water chemical conditions, has not been fully investigated. The hetero aggregation of particulate plastics should not only focus on other solid constituents, but also the constituents that originate from particulate plastics (Alimi et al. 2018). For example, polypropylene and high-density polyethylene (HDPE) can form heteroaggregates with microalgae, while they may release endocrine disrupters into water and influence the aggregation of MPs (Suhrhoff and Scholz-Böttcher 2016; Zhou et al. 2018). In natural waters, particulate plastics could adsorb some environmental contaminants including organic pollutants or heavy metals (Holmes et al. 2012; Hüffer and Hofmann 2016). In addition, particulate plastics may undergo biological degradation, photodegradation, or mechanical erosion (Galloway et al. 2017; Liu et al. 2019c; Song et al. 2017). These transformation processes can disturb the hydrodynamic diameter of particulate plastics under environmentally relevant conditions. Thus, it is urgent to focus on developing reliable detection techniques for investigating the aggregation behavior of particulate plastics.

This chapter provides a critical overview of the existing literature investigating the aggregation of MPs in water. First, we review the main influencing factors of aggregation, including the physicochemical properties of particulate plastics (e.g., size, shape, and surface coating), environmental conditions (e.g., ionic strength, ion valence, and pH), and the solid constituents. Next, the experimental and modeling methods for researching the aggregation process are summarized. The experimental methods including scanning electron microscopy (SEM), dynamic light scattering (DLS), and flow cytometry (FCM) technique are applied to characterize the aggregation of particulate plastics. Modeling methods are described by the fractal dimension, the Derjaguin–Landau–Verwey–Overbeek (DLVO) theory, and aggregation kinetics. Finally, the environmental implications of particulate plastic aggregation, focusing especially on toxicity, their own transport, and their co-transport with contaminants are discussed.

TABLE 10.1
Review of Recent Studies Concerning the Aggregation of Particulate Plastics

Plastic Materials and Surface Properties	Size	Concentration	Environmental Conditions	Main Findings	References
PS	0.1 μm by producer	10, 50 mg·L^{-1}	NaCl, $CaCl_2$, $FeCl_3$ 10 mg·L^{-1} NOM	1. PS NPs remained stable in 1–100 mM NaCl, 0.1–15 mM $CaCl_2$ and 0.01 mM $FeCl_3$ solution, aggregated in 0.1–1 mM $FeCl_3$ solution. 2. The bridging effect between NOM and ferri ion enhanced the aggregation of PSNPs in 1 mM $FeCl_3$ solution.	Cai et al. (2018)
	12.1 μm by SEM	20 mg·L^{-1}	NaCl, $NaNO_3$, KNO_3, $CaCl_2$, $BaCl_2$ pH = 6 0.01–15 mg C·L^{-1} HA	Followed the DLVO theory: HA decreased the aggregation rates of PS and enhanced the stability of PS.	Li et al. (2018)
	2 μm by producer	5.5×10^6 particles·L^{-1}	Cryptophyte at $1–4 \times 10^5$ cells·mL^{-1}	Microorganisms promoted the formation of biofilms on MPs. Extracellular polymer matrix can cause MPs to become sticky and promote the heterogeneous aggregation.	Long et al. (2015)
	90 nm by SEM	10 mg·L^{-1}	NaCl, 10–50 mg·L^{-1} HA, 100 mg·L^{-1} suspended sediment	Followed DLVO theory 1. In 500 mM NaCl solution, the interaction between PS and SS formed heteroaggregates and promoted the settling of PS. 2. HA has a minor effect on the interaction between PS and SS because of the steric force rendered by HA.	Li et al. (2019)
	0.05–0.1 μm by producer	10 mg·L^{-1}	NaCl, KCl, $CaCl_2$, $BaCl_2$, $MgCl_2$ UV (365 nm)	1. In the aging process, aggregation of PS NPs was inhibited by increasing the negative charge on surfaces and the released organic content in NaCl solution. 2. The aggregation of PS NPs was promoted due to the interaction between Ca^{2+} and carboxyl group on the surface of the aged PS NPs in $CaCl_2$ solution.	Liu et al. (2019c)
Sulfonate-PS	115 nm by TEM	1.5×10^{11} particles·L^{-1}	KCl, $MgCl_2$, $LaCl_3$	1. Followed the Schulze-Hardy rule. 2. The aggregation rate increased as the valence of cation ions increased in KCl, $MgCl_2$ and $LaCl_3$ solutions.	Schneider et al. (2011)

(*Continued*)

TABLE 10.1 (*Continued*)
Review of Recent Studies Concerning the Aggregation of Particulate Plastics

Plastic Materials and Surface Properties	Size	Concentration	Environmental Conditions	Main Findings	References
PS, PS-COOH	320, 364 nm by TEM	3×10^{-3}, 1.67×10^{-3} $g \cdot L^{-1}$	NaCl, anionic surfactant NaDBS, cationic surfactant DB	NaDBS and DB formed surfactant coatings on MP surface and improved the stability of MPs due to the stronger electrostatic repulsion between MPs.	Jódar-Reyes et al. (2006b)
PS-COOH	90–244 nm by electron microscopy	0.001%–0.005% (wt)	NaCl, $BaCl_2$, $LaCl_3$	The increase of carboxyl group density on PS surface significantly improves the CCC values.	Sakota and Okaya (1977)
PS-COOH, $PS\text{-}NH_2$	40, 50 nm by TEM	0–50 $mg \cdot L^{-1}$	Natural seawater	PS-COOH rapidly formed aggregates while $PS\text{-}NH_2$ slightly formed aggregates.	Della Torre et al. (2014)
PS, PS-COOH, $n\text{-}PS\text{-}NH_2$, $p\text{-}PS\text{-}NH_2$	100 nm by TEM	100–300 $mg \cdot L^{-1}$	NaCl, 0–50 $mg \cdot L^{-1}$ HA	HA had a greater impact on the aggregation of $p\text{-}PS\text{-}NH_2$ than PS, $n\text{-}PS\text{-}NH_2$ and PS-COOH because the neutralization effect of HA was greater for $p\text{-}PS\text{-}NH_2$ with positive charge.	Wu et al. (2019a)
Amidine latex Sulfate latex	200, 270 nm by DLS	N/A	KCl pH = 4, 5.8	The charged plastic particles could be stabilized by polyelectrolytes with opposite charge adsorbing onto the particle surface due to the repulsive forces rendered by the electrolytes.	Hierrezuelo et al. (2010)
Polypropylene and HDPE	400 μm by SEM	1 $g \cdot L^{-1}$	4×10^6 $cells \cdot mL^{-1}$ freshwater microalgae (*Chlamydomonas reinhardtii*)	TEPs were secreted by microalgae and promoted the formation of heteroaggregates of microalgae with polypropylene and HDPE.	Lagarde et al. (2016)

Abbreviations: NaDBS: sodium dodecylbenzenesulfonate; DB: domiphen bromide or dodecyldimethyl-2-phenoxyethyl ammonium bromide.

10.2 FACTORS AFFECTING AGGREGATION

Many factors affect the aggregation kinetics of MPs/NPs in aquatic environments (Jódar-Reyes et al. 2006b; Liu et al. 2019c; Sakota and Okaya 1977; Zhang et al. 2013). The physical (e.g., size and shape) and chemical (e.g., hydrophobicity and surface functional groups) properties of MPs/NPs will be changed to some extent under the influence of organic/inorganic components, pH, light irradiation, solid constituents, and other factors (Jódar-Reyes et al. 2006b; Liu et al. 2019c; Sakota and Okaya 1977; Zhang et al. 2013). Therefore, the physicochemical properties of MPs/NPs, the environmental conditions, and the coexisting solid constituents could affect the aggregation of MPs/NPs (Jódar-Reyes et al. 2006b; Liu et al. 2019c; Sakota and Okaya 1977; Wu et al. 2019b; Zhang et al. 2013). Recent studies pointing out the factors affecting the aggregation of particulate plastics are summarized in Figure 10.1.

10.2.1 Physicochemical Properties

The size and shape of particulate plastics could affect their aggregation rates by affecting the specific surface areas and the forces between the approaching particles (Derjaguin and Landau 1993). Actually, the shapes of particulate plastics in natural waters are irregular, including fibers, pellets, and fragments of various geometries (Chubarenko et al. 2016), which could influence the interaction forces between the plastic particles and their aggregation rates. Interfacial properties depend substantially on the size scale. A decrease in size scale results in higher relative surface energy and thus destabilizes the system. Smaller MPs aggregate more readily than larger MPs because aggregation will lower the free energy of the system (Waychunas et al. 2005; Wilkinson and Lead 2007). According to the DLVO theory, the interaction energy barrier between the particles decreases as the particle size decreases (Derjaguin and Landau, 1993; Verwey, 1947). Therefore, the stability of particulate plastics is enhanced with increasing particle size.

Surface coatings are commonly used to stabilize particles by increasing repulsive forces and steric hindrance between two approaching particles (Wegner et al. 2012). For instance, linear

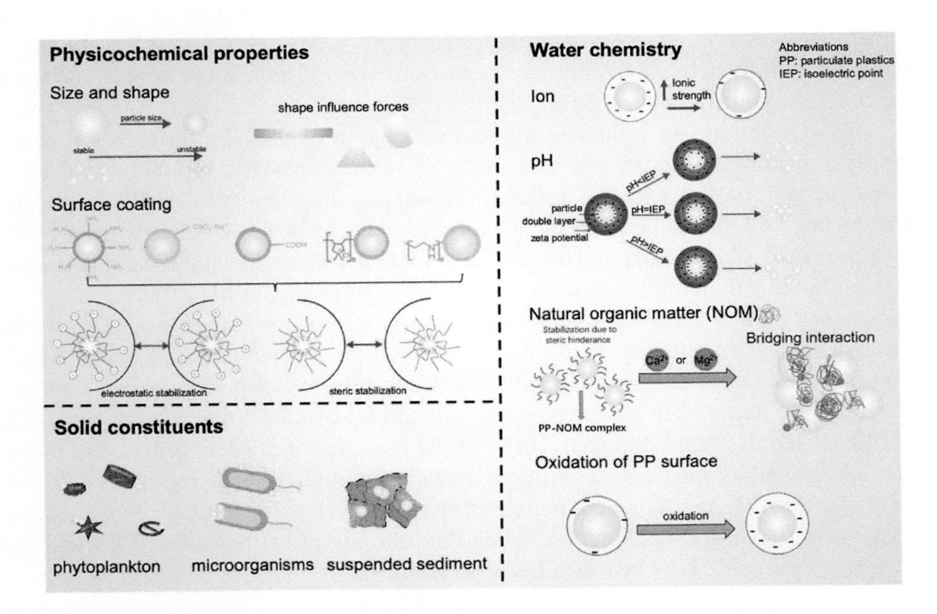

FIGURE 10.1 Effect factors on particulate plastics aggregation in water. IEP represents the isoelectric point and PP represents the plastic particles.

poly(ethylene imine) and poly(diallyldimethyl ammonium chloride) surface coatings improve the stability of sulfate-modified PS MPs in NaCl solution (Sakota and Okaya 1977). Hierrezuelo et al. (2010) showed that the aggregation rate of amidine-modified PS with a diameter of 200 nm was faster than that of the sulfate-modified PS with a diameter of 270 nm. Wu et al. (2019a) found that humic acid (HA) had greater destabilizing effects on p-PS-NH_2 with positive charge than PS, n-PS-NH_2, and PS-COOH with negative charge. This was mainly because negatively charged HA was absorbed at the particle surface, and the neutralization effects were greater for p-PS-NH_2. Della Torre et al. (2014) observed that 40 nm PS-COOH rapidly formed aggregates of 1764 $\pm$ 409 nm, but 50 nm PS-NH_2 only formed aggregates of 89 $\pm$ 2 nm in water at the same time. Thus, chemical properties of particulate plastics, such as surface modification, have an important effect on the stability of particulate plastics.

Some investigators have observed that the roughness, composition, and crystal structure affect the aggregation kinetics of other colloids, such as silver nanoparticles (Liang et al. 2019). Thus, these factors may also exert effects on the stability of particulate plastics in water environments.

10.2.2 Environmental Conditions

Ionic strength and ion valence strongly affect the aggregation behavior of particulate plastics in water (Alimi et al. 2018). The change of surface charge by inorganic ions is the main reason. This phenomenon usually follows the Schulze-Hardy rule, which shows that aggregation ability mainly depends on the ionic valence of the charge opposite to that of colloidal particles (Li et al. 2018; Wu et al. 2019a). The mechanism of its influence is mainly due to the compression of the electric double layer and charge shielding on the surface of particulate plastics, which weaken the repulsive forces between particles (Li et al. 2018; Wu et al. 2019a). In previous studies, the critical coagulation concentrations (CCC) of particulate plastics in divalent electrolytes were 10–15 times higher than those in monovalent electrolytes (Liu et al. 2019c). Cai et al. (2018) demonstrated that PS NPs remained stable in 1–100 mM NaCl, 0.1–15 mM $CaCl_2$, and 0.01 mM $FeCl_3$ solutions, whereas they aggregated in 0.1–1 mM $FeCl_3$ solution. Schneider et al. (2011) showed that the aggregation kinetics of sulfonate-modified PS with a diameter of 115 nm followed the Schulze-Hardy rule, and the aggregation rate increased in the order of $K^+ < Mg^{2+} < La^{3+}$. When the electrolyte concentration exceeds the CCC, the repulsive interactions are insignificant or absent, and the aggregation rate changes slightly (Li et al. 2018).

The pH of the solution can influence the surface charge of PS NPs through ionization and adsorption, which may determine the magnitude of the electrostatic repulsion and affect the aggregation (Liu et al. 2019c). Liu et al. (2019c) showed that the stability of PS NPs increased with increasing pH, because the zeta potentials of PS NPs tended to be more negative with increasing pH. Jódar-Reyes et al. (2006b) demonstrated that PS-COOH was more stable in NaCl solution at pH = 9.0 than that at pH = 7.0, because of a more negative charge at pH = 9.0.

Natural organic matter (NOM) is ubiquitous in natural waters, which can adsorb onto the surfaces of particulate plastics by hydrophobic interaction, ligand exchange, and electrostatic interaction (Baalousha et al. 2008). Adsorbed NOM can impart a negative surface charge and a steric stabilization to particles in aquatic systems, which improve the colloidal stability of PS MPs (Li et al. 2018; Wu et al. 2019a; Zhu et al. 2019). In the presence of divalent or trivalent metal ions, NOM can exert destabilizing effects on particulate plastics through the bridging effect (Wegner et al. 2012). Cai et al. (2018) observed that the increased steric hindrance rendered by NOM inhibited PS NP aggregation in 0.1 mM $FeCl_3$ solution, while the bridging effect between NOM and ferri ion enhanced aggregation of PS NPs in 1 mM $FeCl_3$ solution.

Surfactants are widely used in domestic and industrial products, and they are the most frequently detected organic pollutants in wastewater (Amat et al. 2007). Electrostatic repulsion plays a dominant role in colloidal stability when ionic surfactants cover latex particles (Jódar-Reyes et al. 2006b).

The charge neutralization effect of a cationic surfactant at a low concentration (< 0.48 mM) resulted in aggregation of PS-COOH (Jódar-Reyes et al. 2006a). Jódar-Reyes et al. (2006b) showed that sodium dodecylbenzenesulfonic and dodecyldimethyl-2-phenoxyethyl ammonium bromide formed surfactant coatings on MP surfaces and improved the stability of MPs due to the stronger electrostatic repulsion between MPs.

Many oxygen-containing functional groups (e.g., carbonyl and carboxyl groups) are formed on the surface of particulate plastics after light irradiation, heat processes, and advanced oxidation processes (Liu et al. 2019b,c; Zhu et al. 2019). Oxidation of plastic particle surfaces increases their negative charge, preventing their self-aggregation (Liu et al. 2019c). In the aging process, aggregation of PS NPs was inhibited by increasing the negative charge on their surfaces in NaCl solutions (Liu et al. 2019c). Aggregation of PS NPs was promoted due to the complexation reaction between Ca^{2+} and the carboxyl group on the surface of the aged PS NPs in $CaCl_2$ solution (Liu et al. 2019c; Qu et al. 2010).

10.2.3 Solid Constituents

Microorganisms, phytoplankton, and SS widely exist in natural waters (Long et al. 2015; Zhang and Chen 2019). They can easily adhere to the surfaces of MPs/NPs and form heteroaggregates (Long et al. 2015; Zhang and Chen 2019). Phytoplankton and bacterioplankton can excrete viscous polymer exopolysaccharides, which may coagulate and form large, sticky, discrete organic particles called transparent exopolymer particles (TEPs) (Long et al. 2015). Some researchers found that TEPs attached to polypropylene or HDPE and formed TEPs-MPs heteroaggregates (Long et al. 2015). The interaction between TEPs with polypropylene was stronger than that with HDPE, because TEPs had stronger cohesive and sticking properties toward polypropylene (Lagarde et al. 2016). The attachment of microorganisms to MPs would lead to the formation of a biofilm on the MP surface (Wu et al. 2019b). Microorganisms can cause MPs to become viscous and promote the formation of heterogeneous aggregates (Wu et al. 2019b). In addition, the interactions between SS and MPs have a great impact on the migration and transformation of particulate plastics (Long et al. 2015, 2017). Li et al. (2019) indicated that SS could form heteroaggregates with PS NPs and promoted the settling of plastic particles in NaCl solution. However, the interaction between PE MPs and SS was minor, and PE MPs floated on the water surface for 8 months after the addition of 500 $mg{\cdot}L^{-1}$ SS (Li et al. 2019).

10.3 METHODS FOR STUDYING AGGREGATION BEHAVIOR

10.3.1 Experimental Techniques for Studying Aggregation

In fundamental research, various experimental methods have been used to study the aggregation behavior of NPs or MPs (Schwaferts et al. 2019). The technologies include imaging measurement, light-scattering measurement, and other methods, such as FCM and UV-visible spectrophotometers (Arias-Andres et al. 2018; Cai et al. 2018; Zhu et al. 2019). A brief introduction of these analytical methods will be given in the following section.

10.3.1.1 Visualizing and Imaging Methods

Visualization is the most direct and viable method to analyze the MP or NP aggregates (Schwaferts et al. 2019). This method is frequently achieved with the naked eye or electron microscopy (Schwaferts et al. 2019). The homoaggregation of PE MPs with diameters of 1–5 mm or their heteroaggregates with SS can be discriminated by using the naked eye or microscopes, and the results can be recorded with a digital camera (Zhu et al. 2019). Heteroaggregation behavior of particulate plastics with solid constituents and homoaggregation behavior of particulate plastics are frequently visualized using SEM and transmission electron microscopy (TEM) (Cai et al. 2018;

Oriekhova and Stoll 2018). These two techniques have been applied to provide high-resolution images of NP homoaggregates and heteroaggregates of NPs with Fe_2O_3, alginate biomolecules, and soil (Liu et al. 2019a; Oriekhova and Stoll 2018; Zhu et al. 2019).

Cryogenic SEM (Cryo-SEM) is another useful imaging way to analyze the aggregation process of MPs/NPs (Cai et al. 2018). By quickly freezing samples in liquid ethane or liquid nitrogen at around –200°C, the morphology of MPs/NPs can be maintained, and the integrity is better than samples prepared for SEM/TEM measurements (Cai et al. 2018). Cai et al. (2018) used this method to determine the aggregation state of PS NPs in the mixture of a NOM and salt solution. The results showed that NOM surrounded nearby PS NPs in the presence of 10 $mg \cdot L^{-1}$ NOM and 0.1 mM $FeCl_3$ (Cai et al. 2018). When the concentration of $FeCl_3$ increased to 1 mM, PS NP-NOM clusters were formed through a bridging effect (Cai et al. 2018).

Atomic force microscopy has been identified as one of the most powerful techniques for characterization of plastic surfaces especially at the nanoscale (Last et al. 2010; Stawikowska and Livingston 2013). Atomic force microscopy was used to observe the morphology of surfaces of particulate plastics after formation of particulate plastic-bacteria heteroaggregates (Kumari et al. 2018). Fluorescence spectroscopy is also helpful in detecting plastic materials that have a fluorescent character (Zhu et al. 2019). Fluorescent blue PS NP aggregates can be clearly imaged by fluorescence spectroscopy, and the fluorescent intensity of PS NPs can be used to measure the concentration of plastic particles and investigate their settling kinetics in the presence of SS, HA, or NaCl (Zhu et al. 2019).

10.3.1.2 Light-Scattering Methods

Light-scattering measurements cannot only analyze the hydrodynamic size distribution of aggregates, but also the aggregation kinetics of MPs and NPs (Schwaferts et al. 2019). DLS can measure the hydrodynamic diameter of particulate plastics in the range from 0.6 nm to 6 μm (Baalousha et al. 2012). A laser diffraction particle size analyzer, based on static laser scattering, measures particle sizes at the nano-meter, micrometer, and millimeter scale (Schwaferts et al. 2019). DLS and laser diffraction were used to determine the aggregation profile of MPs and NPs in biological matrixes, such as planktonic crustaceans and Pacific oysters (Gambardella et al. 2017; González-Fernández et al. 2018). In addition, these two techniques can be used to analyze the aggregation rate and kinetics of NPs and MPs under different water-chemical conditions (ionic strength, pH, and NOM) (Summers et al. 2018).

10.3.1.3 Other Methods

FCM analyzes plastic samples with natural or staining fluorescence characteristics in suspension when they pass through a light beam (Phinney and Cucci 1989). In recent studies, FCM has been used to detect and analyze the aggregation behavior of particulate plastics (Arias-Andres et al. 2018; Long et al. 2017). Long et al. (2015, 2016, 2017) used FCM to verify the heteroaggregation between fluorescent MPs and phytoplankton (e.g., the diatom *Chaetoceros neogracile* and the algae in the cryptophyceae class, *Rhodomonas salina*) and measure the concentration of MPs in the aggregates. Arias-Andres et al. (2018) utilized FCM to demonstrate that bacteria could attach to MPs and form biofilms on the particle surfaces. Besides the FCM approach, UV-visible spectrophotometry was used to quantify the amount of PS MPs adsorbed on algal surfaces (Bhattacharya et al. 2010).

10.3.2 Modeling and Simulation of Aggregation Processes

10.3.2.1 Fractal Dimension

It was demonstrated that the colloid aggregates in natural and artificial systems were fractals, and their fractal dimensions were between zero to three (Johnson et al. 1996; Li and Logan 2001). A fractal scaling model has been developed to illustrate the structure of aggregates, aggregate size-density relationships, aggregation rate, and settling velocity of plastic aggregates (Johnson et al. 1996; Li and Logan 2001). The fractal dimension (D_f) is defined by a power-law relationship between fractal aggregate mass and aggregate radius (Kim and Berg 2000).

Fractal dimension is found to be related to the aggregation rate (Kim and Berg 2000). The lower the aggregation rate, the more time particles have to collocate into a denser and more compact structure, and result in a higher fractal dimension value (Kim and Berg 2000). Long et al. (2015) used D_f to check the aggregate structures between 2 μm PS microbeads and phytoplankton or algal species. They found that the aggregate became slightly less fractal after exposure to MPs (Long et al. 2015). The relationship between fractal aggregate size and sinking rates can be established by calculating D_f (Long et al. 2015). D_f also has been used to explain the compactness of MP homoaggregates in salt solution and heteroaggregation of MPs with chitosan or sodium alginate (Ramirez et al. 2016).

10.3.2.2 DLVO Theory

Based on colloidal science principles, the well known DLVO theory can be used to understand the aggregation behavior of MPs and NPs under various conditions (Rajagopalan 1997). DLVO theory has also been used to rationalize forces between interfaces and interpret plastic particle attachment (Rajagopalan 1997). The classic DLVO theory assumes that the interaction forces contain van der Waals and electrostatic forces (Rajagopalan 1997). Many works have employed DLVO theory to interpret the homoaggregation of MPs/NPs in water (Cai et al. 2018; Liu et al. 2019c; Zhu et al. 2019). The equations of DLVO are different based on different interaction geometries, such as particle to particle, particle to plate, and plate to plate (Elimelech et al. 2013).

When particulate plastics are released into the environment, the surface properties of particulate plastics are typically modified by various natural processes including photooxidation and biodegradation, which may influence the attachment of particulate plastics (Alimi et al. 2018). In addition, particulate plastics are more likely to aggregate with other substantial constituents or biotas with various functional groups on their surfaces (Zhu et al. 2019). When particulate plastics interact with these particles, other crucial mechanisms (e.g., steric interaction and elastic repulsion) should be considered to interpret the aggregation processes of particulate plastics (Rajagopalan 1997). In this way, extended DLVO models are more suitable for modeling the aggregation of NPs/MPs in water (Liu et al. 2019a,c). The extended DLVO models include Lewis acid-base interactions, osmotic pressure, depletion attraction, steric forces, or other non-DLVO forces (Elimelech et al. 2013). For example, Liu et al. (2019a) found that DLVO failed to explain the UV-aged PS NP–soil interactions. After taking into account the hydrophobic effect, the observed experimental results for the aggregation kinetics were reasonable (Liu et al. 2019a). Dong et al. (2019) used DLVO and extended DLVO models to understand the interaction forces between MPs and sand in simulated seawater.

10.3.2.3 Aggregation Kinetics

In the aggregation process, a particle collides with other particles or surfaces via diffusion (Rajagopalan 1997). Upon collision, particles may attach to each other successfully, and the probability of this process can be expressed by the attachment efficiency (α). The α values ranging from 0 to 1 can be calculated by normalizing the aggregation rate of a given water-chemistry condition (k) by the aggregation rate under a completely destabilized suspension (non-repulsive) (k_{fast}) (Chen and Elimelech 2006; Saleh et al. 2008). The values of k and α are calculated by Equations 10.1 and 10.2:

$$k \infty \frac{1}{N_0}\left(\frac{dD_h(t)}{dt}\right)_{t\to 0} \tag{10.1}$$

$$\alpha = \frac{k}{k_{fast}} = \frac{\frac{1}{N_0}\left(\frac{dD_h(t)}{dt}\right)_{t\to 0}}{\frac{1}{(N_0)_{fast}}\left(\frac{dD_h(t)}{dt}\right)_{t\to 0,fast}} \tag{10.2}$$

where N_0 refers to particle concentration, D_h refers to the hydrodynamic diameter, and t refers to time. The α values also can be simulated according to the DLVO theory (Elimelech et al. 2013). The equation is defined as the reciprocal of the stability ratio (Chen and Elimelech 2006; Chen et al. 2006; Saleh et al. 2008).

Aggregation kinetics of MPs or NPs have been extensively investigated by α. In addition, aggregation kinetics are controlled by two types of aggregation processes, including reaction-limited cluster aggregation ($\alpha < 1$) or diffusion-limited cluster aggregation ($\alpha = 1$) (Chen and Elimelech 2006). In the reaction-limited cluster aggregation regime, the aggregate growth rates are low and depend on ionic strength (Chen and Elimelech 2007). In the diffusion-limited cluster aggregation regime, because the surface charge of the particle is completely screened, the aggregation kinetics are only related to the diffusion of the plastic particle (Chen and Elimelech 2007).

The aggregation kinetics could predict and elucidate the aggregation behavior of NPs and MPs in complicated, environmental conditions (Li et al. 2019; Liu et al. 2019c). For example, by calculating the aggregation rate from aggregation kinetics, the CCC values of PS NPs can be obtained, which were around 300 to 450 mM in monovalent electrolytes (e.g., NaCl and KCl) and around 30 mM in divalent electrolytes (e.g., $CaCl_2$ and $BaCl_2$) (Liu et al. 2019c; Yu et al. 2019).

10.4 ENVIRONMENTAL IMPLICATION OF AGGREGATION

Recently, the transport of particulate plastics in water has been of concern. The consequences of aggregation on environmental quality can be various and substantial. Aggregation of particulate plastics is important to understand the distribution and potential exposure of particulate plastics in natural waters. One of significant concerns regarding aggregation of particulate plastics is their changed particle sizes, which influences the toxicity to organisms, their own transformations, and their co-transport with other contaminants. Figure 10.2 summarizes the environmental implications of aggregation of particulate plastics in natural waters.

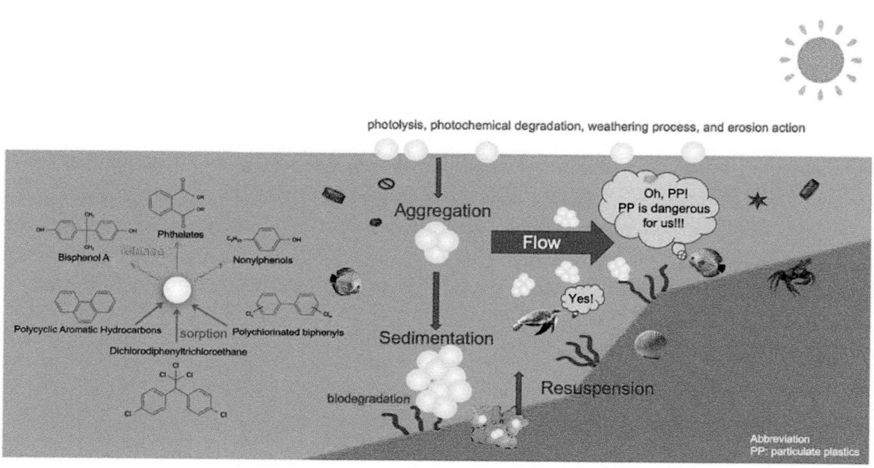

FIGURE 10.2 Environmental implications of aggregation of particulate plastics in water.

10.4.1 Toxicity to Organisms

In natural waters, aggregates of particulate plastics have dimensions from nano-meters to centimeters or larger, and this results in their various toxicity levels towards organisms (Browne et al. 2008; Ward and Kach 2009). If particulate plastics are aggregated slightly, Brownian motion can keep them suspend in the water column for hours or even days, which hinders the normal feeding activity of small creatures or plankton (Bergami et al. 2017). Particulate plastics with a large degree of aggregation were found to settle to the seabed and were trapped in sediments (Bergami et al. 2017). Due to the adsorption of a large number of toxic pollutants on the surface of aggregates of particulate plastics, the detoxification effect for organisms living in the water surface occurred, but toxic effects were shown for the benthic organisms (Zhang et al. 2017). In addition, MPs and NPs were found to adhere to the surface of organisms and form heteroaggregates, which hindered the growth and development of the organisms (Artham et al. 2009). The heteroaggregation of particulate plastics and algae could make the algae precipitate or hinder the photosynthesis of the algae by physically blocking the light and air flow (Bhattacharya et al. 2010; Zhang et al. 2017). In addition, the toxic reactions resulting from aggregation may have a lasting impact through the aquatic food chain (Bhattacharya et al. 2010; Zhang et al. 2017).

10.4.2 Transport of Microplastics

The distribution of particulate plastics in water depends on the aggregation of particulate plastics (Bhattacharya et al. 2010; Da Costa et al. 2016). Nano- and microsized particulate plastics floating on the water surface can form aggregates with microbial communities, which may affect the buoyancy and density of the plastics in the ocean, change their sinking rates, and the depth that they fall in water (Alimi et al. 2018). Aggregates of particulate plastics located at different water depths undergo different degrees of weathering from ultraviolet radiation, biological action, mechanical wear, and pulverization, which results in different mass loss or degradation rates of the aggregates (Alimi et al. 2018). Sunlight plays a dominant role in the migration and transformation of the aggregates of particulate plastics in the upper water body (Chatani et al. 2014). It has been demonstrated that the aggregates of particulate plastics absorb sunlight and generate free radicals through impurities, which leads particulate plastics to break into low molecular weight fragments (Zhao et al. 2018). Aggregates of particulate plastics that have settled to the seabed are not affected by illumination, but the complex hydrodynamic and biological effects affect the physicochemical properties and environmental implication of the bottom plastic particles (Zhao et al. 2018).

10.4.3 Transport of Contaminants

The environmental effects of contaminant transportation by particulate plastics can be divided into two aspects. On the one hand, monomers, residual substances (e.g., solvents), additives, and non-intentionally added substances (e.g., bisphenol A, triclosan, flame retardants, bisphenol ketone, phthalate, and organotin) can be leached by MPs (Suhrhoff and Scholz-Böttcher 2016). These substances can be mutagenic and carcinogenic to organisms, which is of significant environmental concern (Suhrhoff and Scholz-Böttcher 2016). On the other hand, the sorption of inorganic and organic contaminants to MPs is considered as another important process (Holmes et al. 2012; Hüffer and Hofmann 2016). For example, the concentration of organic pollutants on particulate plastics was reported to be six orders of magnitude higher than the background concentration of the surrounding seawater (Lee et al. 2014). Aggregation of particulate plastics affects the transportation of contaminants, including the adsorbed contaminants and the released additives from particulate plastics (Yang et al. 2018). Compared to larger aggregates, small aggregates of particulate plastics may also be subjected to faster rates of degradation (Nguyen et al. 2019). Dimethyltin, monomethyltin, dibutyltin, and monobutyltin were released from polyvinyl chloride MPs under UV and visible light irradiation (Nguyen et al. 2019). And the smaller-sized polyvinyl chloride MPs were found to release more organic compounds, which probably posed a higher toxicity for organisms (Nguyen et al. 2019). MP aggregates can rapidly absorb contaminants, thereby changing the fluidity of organic and inorganic contaminants and their potential for transfer to living organisms (Nguyen et al. 2019).

10.5 CONCLUSIONS

Aggregation of particulate plastics consists of homoaggregation and heteroaggregation in aquatic environments. The aggregation process of particulate plastics can be changed to some extent under the influence of physicochemical properties (e.g., size, shape, surface coating, and composition), water environment conditions (e.g., pH, ionic strength, NOM, and light), and other solid constituents in the waters. Current methods for evaluating the aggregation behavior of particulate plastics mainly include imaging measurements, light-scattering measurements, and other methods, such as FCM and UV-visible spectrophotometry. Modeling methods using fractal dimension, the DLVO theory, and aggregation kinetics can be used effectively to predict or illustrate the aggregation process of particulate plastics. Specifically, the models can calculate the attachment efficiency or elucidate the aggregation mechanisms of particulate plastics in complicated environmental conditions. Aggregation of particulate plastics affects the transport of contaminants, including the additives and absorbed contaminants that are released into the environment. Aggregation behavior of particulate plastics has a great influence on their transformation, and it also affects where they are located in the water column and their modes of action for a range of biota.

REFERENCES

Alimi, O. S., Budarz, J. F., Hernandez, L. M., Tufenkji, N. 2018. Microplastics and nanoplastics in aquatic environments: Aggregation, deposition, and enhanced contaminant transport. *Environmental Science & Technology* 52: 1704–1724.

Amat, A., Arques, A., Miranda, M. A., Vincente, R., Segui, S. 2007. Degradation of two commercial anionic surfactants by means of ozone and/or UV irradiation. *Environmental Engineering Science* 24: 790–794.

Arias-Andres, M., Klumper, U., Rojas-Jimenez, K., Grossart, H. P. 2018. Microplastic pollution increases gene exchange in aquatic ecosystems. *Environmental Pollution* 237: 253–261.

Artham, T., Sudhakar, M., Venkatesan, R., Madhavan Nair, C., Murty, K. V. G. K., Doble, M. 2009. Biofouling and stability of synthetic polymers in sea water. *International Biodeterioration & Biodegradation* 63: 884–890.

Baalousha, M., Ju-Nam, Y., Cole, P., Gaiser, B., Fernandes, T., Hriljac, J., Jepson, M., Stone, V., Tyler, C., Lead, J. 2012. Characterization of cerium oxide nanoparticles – Part 1: Size measurements. *Environmental Toxicology and Chemistry* 31: 983.

Baalousha, M., Manciulea, A., Cumberland, S., Kendall, K., Lead, J. R. 2008. Aggregation and surface properties of iron oxide nanoparticles: Influence of pH and natural organic matter. *Environmental Toxicology and Chemistry* 27: 1875–1882.

Bergami, E., Pugnalini, S., Vannuccini, M. L., Manfra, L., Faleri, C., Savorelli, F., Dawson, K. A., Corsi, I. 2017. Long-term toxicity of surface-charged polystyrene nanoplastics to marine planktonic species Dunaliella tertiolecta and Artemia franciscana. *Aquatic Toxicology* 189: 159–169.

Besseling, E., Quik, J. T. K., Sun, M., Koelmans, A. A. 2017. Fate of nano-and microplastic in freshwater systems: A modeling study. *Environmental Pollution* 220: 540–548.

Bhattacharya, P., Lin, S., Turner, J. P., Ke, P. C. 2010. Physical adsorption of charged plastic nanoparticles affects algal photosynthesis. *The Journal of Physical Chemistry C* 114: 16556–16561.

Browne, M. A., Dissanayake, A., Galloway, T. S., Lowe, D. M., Thompson, R. C. 2008. Ingested microscopic plastic translocates to the circulatory system of the mussel, *Mytilus edulis* (L.). *Environmental Science & Technology* 42: 5026–5031.

Cai, L., Hu, L., Shi, H., Ye, J., Zhang, Y., Kim, H. 2018. Effects of inorganic ions and natural organic matter on the aggregation of nanoplastics. *Chemosphere* 197: 142–151.

Carr, S. A., Liu, J., Tesoro, A. G. 2016. Transport and fate of microplastic particles in wastewater treatment plants. *Water Research* 91: 174–182.

Chatani, S., Kloxin, C. J., Bowman, C. N. 2014. The power of light in polymer science: Photochemical processes to manipulate polymer formation, structure, and properties. *Polymer Chemistry* 5: 2187–2201.

Chen, K. L., Elimelech, M. 2006. Aggregation and deposition kinetics of fullerene (C_{60}) nanoparticles. *Langmuir* 22: 10994–11001.

Chen, K. L., Elimelech, M. 2007. Influence of humic acid on the aggregation kinetics of fullerene (C_{60}) nanoparticles in monovalent and divalent electrolyte solutions. *Journal of Colloid and Interface Science* 309: 126–134.

Chen, K. L., Mylon, S. E., Elimelech, M. 2006. Aggregation kinetics of alginate-coated hematite nanoparticles in monovalent and divalent electrolytes. *Environmental Science & Technology* 40: 1516–1523.

Chubarenko, I., Bagaev, A., Zobkov, M., Esiukova, E. 2016. On some physical and dynamical properties of microplastic particles in marine environment. *Marine Pollution Bulletin* 108: 105–112.

Da Costa, J. P., Santos, P. S. M., Duarte, A. C., Rocha-Santos, T. 2016. (Nano)plastics in the environment – Sources, fates and effects. *Science of the Total Environment* 566–567: 15–26.

De Haan, W. P., Sanchez-Vidal, A., Canals, M. 2019. Floating microplastics and aggregate formation in the Western Mediterranean Sea. *Marine Pollution Bulletin* 140: 523–535.

Della Torre, C., Bergami, E., Salvati, A., Faleri, C., Cirino, P., Dawson, K., Corsi, I. 2014. Accumulation and embryotoxicity of polystyrene nanoparticles at early stage of development of sea urchin embryos Paracentrotus lividus. *Environmental Science & Technology* 48: 12302–12311.

Derjaguin, B. V., Landau, L. D. 1993. Theory of the stability of strongly charged lyphobic sols of the adhesion of strongly charged particles in solutions of electrolytes. *Progress in Surface Science* 43: 30–59.

Dong, Z., Zhu, L., Zhang, W., Huang, R., Lv, X., Jing, X., Yang, Z., Wang, J., Qiu, Y. 2019. Role of surface functionalities of nanoplastics on their transport in seawater-saturated sea sand. *Environmental Pollution* 255: 113–177.

Dris, R., Gasperi, J., Saad, M., Mirande, C., Tassin, B. 2016. Synthetic fibers in atmospheric fallout: A source of microplastics in the environment? *Marine Pollution Bulletin* 104: 290–293.

Elimelech, M., Gregory, J., Jia, X. 2013. *Particle Deposition and Aggregation: Measurement, Modelling and Simulation*, ed. Butterworth-Heinemann, Oxford, New York.

Galloway, T. S., Cole, M., Lewis, C. 2017. Interactions of microplastic debris throughout the marine ecosystem. *Nature Ecology & Evolution* 1: 116.

Gambardella, C., Morgana, S., Ferrando, S., Bramini, M., Piazza, V., Costa, E., Garaventa, F., Faimali, M. 2017. Effects of polystyrene microbeads in marine planktonic crustaceans. *Ecotoxicology and Environmental Safety* 145: 250–257.

González-Fernández, C., Tallec, K., Le Goïc, N., Lambert, C., Soudant, P., Huvet, A., Suquet, M., Berchel, M., Paul-Pont, I. 2018. Cellular responses of Pacific oyster (*Crassostrea gigas*) gametes exposed in vitro to polystyrene nanoparticles. *Chemosphere* 208: 764–772.

Hierrezuelo, J., Sadeghpour, A., Szilagyi, I., Vaccaro, A., Borkovec, M. 2010. Electrostatic stabilization of charged colloidal particles with adsorbed polyelectrolytes of opposite charge. *Langmuir* 26: 15109–15111.

Holmes, L. A., Turner, A., Thompson, R. C. 2012. Adsorption of trace metals to plastic resin pellets in the marine environment. *Environmental Pollution* 160: 42–48.

Hüffer, T., Hofmann, T. 2016. Sorption of non-polar organic compounds by micro-sized plastic particles in aqueous solution. *Environmental Pollution* 214: 194–201.

Jódar-Reyes, A. B., Martín-Rodríguez, A., Ortega-Vinuesa, J. L. 2006a. Effect of the ionic surfactant concentration on the stabilization/destabilization of polystyrene colloidal particles. *Journal of Colloid and Interface Science* 298: 248–257.

Jódar-Reyesń, A. B., Ortega-Vinuesa, J. L., Martín-Rodríguez, A. 2006b. Electrokinetic behavior and colloidal stability of polystyrene latex coated with ionic surfactants. *Journal of Colloid and Interface Science* 297: 170–181.

Johnson, C. P., Li, X., Logan, B. E. 1996. Settling velocities of fractal aggregates. *Environmental Science & Technology* 30: 1911–1918.

Kim, A. Y., Berg, J. C. 2000. Fractal aggregation: Scaling of fractal dimension with stability ratio. *Langmuir* 16: 2101–2104.

Kumari, A., Chaudhary, D. R., Jha, B. 2018. Destabilization of polyethylene and polyvinylchloride structure by marine bacterial strain. *Environmental Science and Pollution Research* 26: 1507–1516.

Lagarde, F., Olivier, O., Zanella, M., Daniel, P., Hiard, S., Caruso, A. 2016. Microplastic interactions with freshwater microalgae: Hetero-aggregation and changes in plastic density appear strongly dependent on polymer type. *Environmental Pollution* 215: 331–339.

Last, J. A., Russell, P., Nealey, P. F., Murphy, C. J. 2010. The applications of atomic force microscopy to vision science. *Investigative Ophthalmology & Visual Science* 51: 6083–6094.

Lee, H., Shim, W. J., Kwon, J.-H. 2014. Sorption capacity of plastic debris for hydrophobic organic chemicals. *Science of The Total Environment* 470–471: 1545–1552.

Li, S., Liu, H., Gao, R., Abdurahman, A., Dai, J., Zeng, F. 2018. Aggregation kinetics of microplastics in aquatic environment: Complex roles of electrolytes, pH, and natural organic matter. *Environmental Pollution* 237: 126–132.

Li, X., Logan, B. E. 2001. Permeability of fractal aggregates. *Water Research* 35: 3373–3380.

Li, Y., Wang, X., Fu, W., Xia, X., Liu, C., Min, J., Zhang, W., Crittenden, J. C. 2019. Interactions between nano/micro plastics and suspended sediment in water: Implications on aggregation and settling. *Water Research* 161: 486–495.

Liang, Y., Zhou, J., Dong, Y., Klumpp, E., Šimůnek, J., Bradford, S. A. 2019. Evidence for the critical role of nanoscale surface roughness on the retention and release of silver nanoparticles in porous media. *Environmental Pollution* 258: 113803.

Liu, J., Zhang, T., Tian, L., Liu, X., Qi, Z., Ma, Y., Ji, R., Chen, W. 2019a. Aging significantly affects mobility and contaminant-mobilizing ability of nanoplastics in saturated loamy sand. *Environmental Science & Technology* 53: 5805–5815.

Liu, P., Qian, L., Wang, H., Zhan, X., Lu, K., Gu, C., Gao, S. 2019b. New insights into the aging behavior of microplastics accelerated by advanced oxidation processes. *Environmental Science & Technology* 53: 3579–3588.

Liu, Y., Hu, Y., Yang, C., Chen, C., Huang, W., Dang, Z. 2019c. Aggregation kinetics of UV irradiated nanoplastics in aquatic environments. *Water Research* 163: 114870.

Long, M., Moriceau, B., Gallinari, M. 2016. On the potential role of phytoplankton aggregates in microplastics sedimentation. *MICRO 2016. Fate and Impact of Microplastics in Marine Ecosystems* 71.

Long, M., Moriceau, B., Gallinari, M., Lambert, C., Huvet, A., Raffray, J., Soudant, P. 2015. Interactions between microplastics and phytoplankton aggregates: Impact on their respective fates. *Marine Chemistry* 175: 39–46.

Long, M., Paul-Pont, I., Hegaret, H., Moriceau, B., Lambert, C., Huvet, A., Soudant, P. 2017. Interactions between polystyrene microplastics and marine phytoplankton lead to species-specific hetero-aggregation. *Environmental Pollution* 228: 454–463.

Lowry, G. V., Gregory, K. B., Apte, S. C., Lead, J. R. 2012. Transformations of nanomaterials in the environment. *Environmental Science & Technology* 46: 6893–6899.

Napper, I. E., Bakir, A., Rowland, S. J., Thompson, R. C. 2015. Characterisation, quantity and sorptive properties of microplastics extracted from cosmetics. *Marine Pollution Bulletin* 99: 178–185.

Nguyen, B., Claveau-Mallet, D., Hernandez, L. M., Xu, E. G., Tufenkji, N. 2019. Separation and analysis of microplastics and nanoplastics in complex environmental samples. *Accounts of Chemical Research* 52: 858–866.

Oriekhova, O., Stoll, S. 2018. Heteroaggregation of nanoplastic particles in the presence of inorganic colloids and natural organic matter. *Environmental Science: Nano* 5: 792–799.

Paraskevopoulou, D., Achilias, D. S., Paraskevopoulou, A. 2012. Migration of styrene from plastic packaging based on polystyrene into food simulants. *Polymer International* 61: 141–148.

Pelley, A. J., Tufenkji, N. 2008. Effect of particle size and natural organic matter on the migration of nano- and microscale latex particles in saturated porous media. *Journal of Colloid And Interface Science* 321: 74–83.

Phinney, D. A., Cucci, T. L. 1989. Flow cytometry and phytoplankton. *Cytometry* 10: 511–521.

Qu, X., Hwang, Y. S., Alvarez, P. J. J., Bouchard, D., Li, Q. 2010. UV irradiation and humic acid mediate aggregation of aqueous fullerene (nC_{60}) nanoparticles. *Environmental Science & Technology* 44: 7821–7826.

Rajagopalan, R. 1997. *Particle Deposition and Aggregation: Measurement, Modelling and Simulation*, ed Elimelech, M.; Gregory, J.; Jia, X.; William, R., 209–210. Butterworth-Heinemann, Oxford, New York.

Ramirez, L., Gentile, S. R., Zimmermann, S. 2016. Comparative study of the effect of aluminum chloride, sodium alginate and chitosan on the coagulation of polystyrene micro-plastic particles. *Journal of Colloid Science and Biotechnology* 5: 190–198.

Sakota, K., Okaya, T. 1977. Electrolyte stability of carboxylated latexes prepared by several polymerization processes. *Journal of Applied Polymer Science* 21: 1025–1034.

Saleh, N. B., Pfefferle, L. D., M., E. 2008. Aggregation kinetics of multiwalled carbon nanotubes in aquatic systems: Measurements and environmental implications. *Environmental Science & Technology* 42: 7963–7969.

Schneider, C., Hanisch, M., Wedel, B., Jusufi, A., Ballauff, M. 2011. Experimental study of electrostatically stabilized colloidal particles: Colloidal stability and charge reversal. *Journal of Colloid and Interface Science* 358: 62–67.

Schwaferts, C., Niessner, R., Elsner, M., Ivleva, N. P. 2019. Methods for the analysis of submicrometer- and nanoplastic particles in the environment. *Trends in Analytical Chemistry* 112: 52–65.

Song, Y. K., Hong, S. H., Jang, M., Han, G. M., Jung, S. W., Shim, W. J. 2017. Combined effects of UV exposure duration and mechanical abrasion on microplastic fragmentation by polymer type. *Environmental Science & Technology* 51: 4368–4376.

Stawikowska, J., Livingston, A. G. 2013. Assessment of atomic force microscopy for characterisation of nanofiltration membranes. *Journal of Membrane Science* 425: 58–70.

Suhrhoff, T. J., Scholz-Böttcher, B. M. 2016. Qualitative impact of salinity, UV radiation and turbulence on leaching of organic plastic additives from four common plastics—A lab experiment. *Marine Pollution Bulletin* 102: 84–94.

Summers, S., Henry, T., Gutierrez, T. 2018. Agglomeration of nano-and microplastic particles in seawater by autochthonous and de novo-produced sources of exopolymeric substances. *Marine Pollution Bulletin* 130: 258–267.

Verwey, E. J. W. 1947. Theory of the stability of lyophobic colloids. *Journal of Physical and Colloid Chemistry* 51: 631–636.

Ward, J. E., Kach, D. J. 2009. Marine aggregates facilitate ingestion of nanoparticles by suspension-feeding bivalves. *Marine Environmental Research* 68: 137–142.

Waychunas, G. A., Kim, C. S., Banfield, J. F. 2005. Nanoparticulate iron oxide minerals in soils and sediments: Unique properties and contaminant scavenging mechanisms. *Journal of Nanoparticle Research* 7: 409–433.

Wegner, A., Besseling, E., Foekema, E. M., Kamermans, P., Koelmans, A. A. 2012. Effects of nanopolystyrene on the feeding behavior of the blue mussel (*Mytilus edulis* L.). *Environmental Toxicology and Chemistry* 31: 2490–2497.

Wilkinson, K. J., Lead, J. R. 2007. *Environmental Colloids and Particles: Behaviour, Separation and Characterisation*, ed John Wiley & Sons, New York.

Wu, J., Jiang, R., Lin, W., Ouyang, G. 2019a. Effect of salinity and humic acid on the aggregation and toxicity of polystyrene nanoplastics with different functional groups and charges. *Environmental Pollution* 245: 836–843.

Wu, P., Huang, J., Zheng, Y., Yang, Y., Zhang, Y., He, F., Chen, H., Quan, G., Yan, J., Li, T., Gao, B. 2019b. Environmental occurrences, fate, and impacts of microplastics. *Ecotoxicology and Environmental Safety* 184: 109612.

Yang, J., Xu, L., Lu, A., Luo, W., Li, J., Chen, W. 2018. Progress in the source and toxicology of micro (nano) plastics in the environment. *Environmental Chemistry* 37: 383–396.

Yu, S., Shen, M., Li, S., Fu, Y., Zhang, D., Liu, H., Liu, J. 2019. Aggregation kinetics of different surface-modified polystyrene nanoparticles in monovalent and divalent electrolytes. *Environmental Pollution* 255: 113302–113302.

Zhang, C., Chen, X., Wang, J., Tan, L. 2017. Toxic effects of microplastic on marine microalgae Skeletonema costatum: Interactions between microplastic and algae. *Environmental Pollution* 220: 1282–1288.

Zhang, G. S., Liu, Y. F. 2018. The distribution of microplastics in soil aggregate fractions in southwestern China. *Science of the Total Environment* 642: 12–20.

Zhang, W., Rattanaudompol, U. S., Li, H., Bouchard, D. 2013. Effects of humic and fulvic acids on aggregation of aqu/nC_{60} nanoparticles. *Water Research* 47: 1793–1802.

Zhang, Z., Chen, Y. 2019. Effects of microplastics on wastewater and sewage sludge treatment and their removal: A review. *Chemical Engineering Journal*: 122955.

Zhao, S., Ward, J. E., Danley, M., Mincer, T. J. 2018. Field-based evidence for microplastic in marine aggregates and mussels: Implications for trophic transfer. *Environmental Science & Technology* 52: 11038–11048.

Zhou, Q., Jin, Z., Li, J., Wang, B., Wei, X., Chen, J. 2018. A novel air-assisted liquid-liquid microextraction based on in-situ phase separation for the HPLC determination of bisphenols migration from disposable lunch boxes to contacting water. *Talanta* 189: 116–121.

Zhu, K., Jia, H., Zhao, S., Xia, T., Guo, X., Wang, T., Zhu, L. 2019. Formation of environmentally persistent free radicals on microplastics under light irradiation. *Environmental Science & Technology* 53: 8177–8186.

Section III

Ecotoxicity of Particulate Plastics

11 Environmentally Toxic Components of Particulate Plastics

Sanchita Mandal, Nanthi S. Bolan, Binoy Sarkar, Hasintha Wijesekara, Lauren Bradney, and M.B. Kirkham

CONTENTS

11.1 Introduction ... 165
11.2 Manufacture of Plastics ... 168
  11.2.1 Thermoplastics ... 168
  11.2.2 Thermosets ... 169
11.3 Presence of Chemicals in Plastics and Their Toxicity ... 170
  11.3.1 Hazardous Metals ... 170
  11.3.2 Bisphenols ... 171
  11.3.3 Phthalates ... 171
  11.3.4 Per- and Polyfluoroalkyl Substances ... 172
  11.3.5 Additives ... 172
11.4 Leaching of Chemicals from Plastics ... 173
11.5 Toxicity of the Chemical Components of Plastics ... 175
11.6 Conclusions ... 175
References ... 176

11.1 INTRODUCTION

The plastic revolution in our daily life enormously increased within the last few decades. It was estimated that around 348 million metric tons of plastics were produced globally in 2017, with around 40% used for packaging purposes (Statista 2018). This increasing amount of plastics ultimately get introduced to the terrestrial (i.e., soil) and aquatic (i.e., marine) ecosystems by generating a considerable amount of litter. About 10 million tons of plastic litter is migrating into the marine ecosystems each year and creating a "plastic footprint" in the environment (Boucher and Billard 2019). The concerns about plastic production and usages are multi-directional, such as: (a) accumulation of a huge amount of non-degradable plastics in the environment, (b) release of toxic chemicals during manufacturing and exposure of plastics to the environment, (c) generation of secondary micro- and nanoplastics during the production and usage of plastics, and (d) improper disposal of plastics into the soil and ocean (Jambeck et al. 2015; Thompson et al. 2009; Galloway 2015; Galloway and Lewis 2016; Biryol et al. 2017; Caporossi and Papaleo 2017; Gallo et al. 2018; Hahladakis et al. 2018). Table 11.1 summarizes information on selected synthetic polymers made of plastics, their origin, and characteristics.

Crude oil is the main source of synthetic plastics, whereas coal and natural gas are also used for plastic production. During the refining process of crude oil, several toxic by-products are produced, such as petrol, paraffin, lubricating oils, and petroleum gases. Plastics are diverse in nature and made

TABLE 11.1
Summarized Information on Selected Synthetic Polymers Made of Plastics, Their Origin, and Characteristics

Chemical Compound	Chemical Formula	Origin or Source to Environment	Characteristics	References
Low-density polyethylene (LDPE)	$(C_2H_4)_n$	Squeeze bottles, toys, carrier bags, chemical tank linings, heavy duty sacks, general packaging, and gas and water pipes	Low density 0.91–0.94 g/cm^3, non-biodegradable, most common plastics	Lambert et al. (2014); Lassen et al. (2015)
High-density polyethylene (HDPE)	$(C_2H_4)_n$	Chemical drums, toys, household and kitchenware, cable insulation, and carrier bags	High density 0.92–0.99 g/cm^3, non-biodegradable	Lambert et al. (2014); Lassen et al. (2015)
Acrylic	Acrylate polymers: based with acrylic acid: CH_2=CHCOOH	Most used fibers in textiles: knitware and plastic flakes	High density 1.16 g/cm^3	Lassen et al. (2015)
Polyethylene terephthalate (PET)	$(C_{10}H_8O_4)_n$	Drinks bottles, oven-ready meal trays cable lining	High density 1.41 g/cm^3	Lambert et al. (2014); Lassen et al. (2015)
Polypropylene (PP)	$(C_3H_6)_n$	Food containers and microwavable meal trays	Low density 0.90–0.91 g/cm^3	Lambert et al. (2014); Lassen et al. (2015)
Polystyrene (PS)	$(C_8H_8)_n$	Food containers, stuffed animals, and protective packaging	High density 1.04–1.13 g/cm^3	Lassen et al. (2015)
Polyvinyl chloride (PVC)	$(C_2H_3Cl)_n$	Water pipes, cable insulation, packaging, and healthcare applications	High density 1.39–1.43 g/cm^3, non-biodegradable	Lassen et al. (2015)

Source: Adapted from Wijesekara et al. (2018).

of a range of different polymers and additives along with other components, such as adhesives and/or coatings. Furthermore, plastics can also contain the residues from the materials used for plastic production, such as solvents and additives. To understand and characterize the potential risks associated with the production, usage, disposal, and recycling of plastics and plastic packaging materials, comprehensive information of all the chemicals involved is needed. A study by Groh et al. (2018) compiled a list of chemicals associated with plastic manufacturing and those present in the final packaging materials. The identified most common hazardous chemicals from plastic productions are: monomers, intermediates, solvents, surfactants, plasticizers, stabilizers, biocides, flame retardants, accelerators, and colorants (Groh et al. 2018). However, achieving a completely comprehensive list of chemicals used in plastic manufacturing is not a straightforward task because plastics are made of a large variety of polymers and additives, and furthermore, the identity of the chemicals in finished plastic packaging is seldom measured. It was estimated that among 906 chemicals likely to be associated with plastic packaging, 63 rank the highest for human hazards, and 68 for environmental hazards, according to the harmonized hazardous classifications assigned by the European Chemicals Agency (Table 11.2) (EU 2018; European Commission 2018). Most of the recent studies have assessed the ultimate fate and impact of plastics and their leachates or adsorbed contaminants without being able to separate the individual effects of hazardous chemicals, or, on the contrary,

TABLE 11.2
List of Hazardous Chemicals Associated with Plastic Production and Packaging

Function	Chemical Group	Chemical Name
Colorants	Dye, azo	1.8-Dihydroxynaphthalene-3,6-disulfonic acid-[2-(4-Azo)]-N-(5-methyl-3-isoxazolyl) benzenesulfonamide
	Pigment	Cobalt (II) diacetate
Fire retardants	Boron	Sodium tetraborate, pentahydrate
		Boric acid
		Sodium tetraborate, anhydrous
		Sodium borate, decahydrate
	Organophosphate	Tris(2-chloroethyl) phosphate
		Triphenyl phosphate
		Trixylyl phosphate
	Others	Tetrachlorophthalic anhydride
		2,2′,6,6′-tetrabromobisphenol A
Foaming agents	Simple hydrocarbon	Isobutane
	Biopolymer	Azodicarbonamide
	Starch	Sodium bicarbonate
		Powdered sodium bicarbonate
Plasticizers	Chlorinated paraffin	Alkanes
		Medium-chain chlorinated paraffins, >17 carbon atoms
	Phthalate	DEHP (Di-ethylhexyl phthalates)
		DINP (Di-isononyl phthalate)
		Diallyl phthalate
		Dimethoxyethyl phthalate
		Diethyl phthalate
		Dioctyl phthalate
		Diundecyl phthalate
Lubricants	Perflourinated hydrocarbons	Polytetrafluoroethylene (PTFE)
	Synthetic hydrocarbon	Polyalpha-olefin (PAO) synthetic esters
Monomers	Acrylic	Isooctyl acrylate
		Acrylonitrile
	Amine	Aziridine
		m-Phenylenediamine
	Bisphenol	Bisphenol A
		Bisphenol F
		Bisphenol S
		Naphthalene

(Continued)

TABLE 11.2 (*Continued*)
List of Hazardous Chemicals Associated with Plastic Production and Packaging

Function	Chemical Group	Chemical Name
Antioxidants	Hydrocarbon polymers	Bis(2,4-di-tert-butylphenyl) pentaerythritol diphosphate and magnesium aluminum hydroxy carbonate hydrate
		2H-benzimidazole-2-thione, 1,3-di-hydro-4(or 5)-methyl
		Tris(mono-nonylphenyl) phosphite with up to 1% triisopropanol amine
		Benzenamine, N-phenyl, reaction products with 2,4,4-trimethylpentene
		1,3-Diethyl-2-thiourea
		Disodium 4,4′-bis(2-sulfonatostyryl) biphenyl
		1,2-Bis(2-methylphenyl) guanidine

they examined only one specific substance/chemical without considering the integrated amount of chemicals/substances present in the plastic wastes (Auta et al. 2017; Anderson et al. 2016; Gallo et al. 2018). Therefore, this chapter aims to understand the manufacturing processes of plastics and to present an overview of environmentally hazardous components and chemicals in various plastics and their final states after manufacturing.

11.2 MANUFACTURE OF PLASTICS

Plastics are commonly derived from various natural and organic materials, such as cellulose, coal, natural gas, and crude oil. Due to the complex nature of crude oil and the presence of thousands of components, it needs to be processed and refined before it can be used. The main production of plastic usually begins with the distillation of crude oil in the oil refineries. During the distillation process, heavy crude oil gets separated into lighter component groups, and they are termed as fractions (Gervet 2007). Each of these fractions is a mixture of hydrocarbon chains, which is made up of carbon and hydrogen atoms. According to the size and the structure of the molecules, the hydrocarbon chains are converted to fuels and other derivatives, such as lubricants, wax, petroleum products, and so on. One of the components of these hydrocarbon fractions is naphtha, which is considered as the most crucial raw material for plastic production. Figure 11.1 represents the production steps of plastics from crude oil in oil refineries.

The two main techniques for plastic production are polymerization and polycondensation, and these require specific catalysts. In the polymerization reactor, monomers, such as propylene and ethylene are combined in order to develop a long polymer chain. Therefore, each polymer has its own structure and size depending on the different varieties of basic monomers used to form the chains. Plastics have a wide range of types based on different base chemistries, additives and derivatives formulated to obtain desired functional properties. Thus, there may be thousands of different plastic types. To simplify their classification, plastics are mainly grouped into two polymer families, namely, thermoplastics and thermosets.

11.2.1 THERMOPLASTICS

Thermoplastics are the most common types of plastics used nowadays. The main feature is that it can undergo several melts and solidification processes without significant degradation. Thermoplastics are normally supplied as small sheets and pellets; therefore, numerous desired shapes can be made

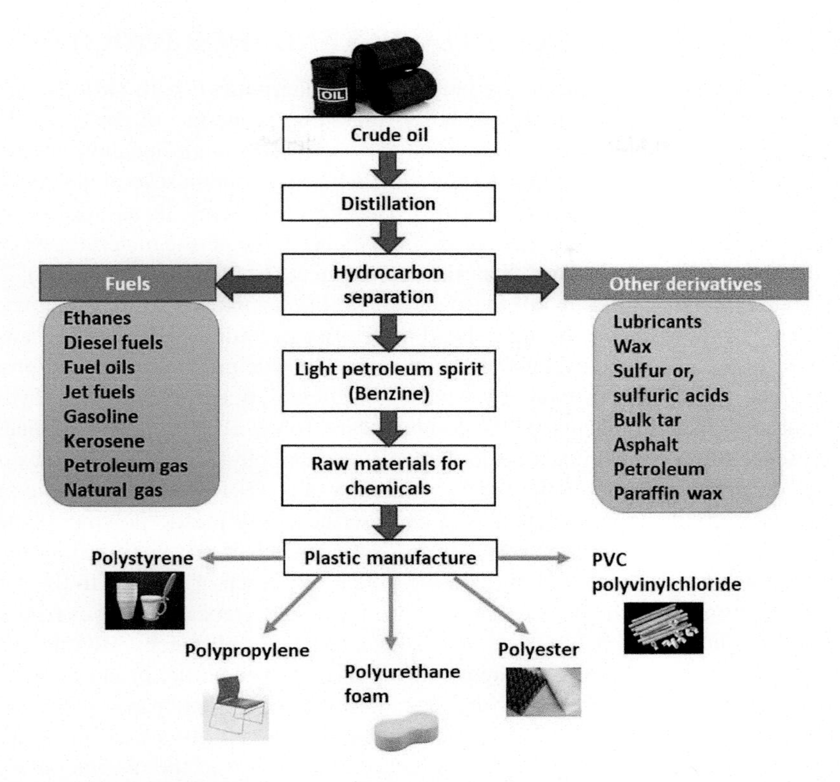

FIGURE 11.1 Crude oil refining process for plastic production.

after processing. Due to the absence of chemical bonding during the processing, the production process is highly reversible, this makes melting and 'recycling or reusing' of thermoplastic highly feasible. The recent development in reactive thermoplastic technology has enabled to produce potential recyclable plastics by taking consideration of the thermoplastic liquid impregnation processes, and these processes offer a strong matrix system with the potential for recycling of plastics (Verrey et al. 2006). Different impregnation routes for thermoplastic production are available. Melt impregnation results in a low penetration of the matrix into the bundle. Powder impregnation leads to the agglomeration and inhomogeneous distribution of powder particles. The solution or liquid impregnation technique in a suspension overcomes the melt or powder impregnation problems (Schulte and Von Lacroix 2000). The common thermoplastic materials are acrylic, acrylonitrile butadiene styrene, polylactic acid, polycarbonate, polyethene, polypropylene, polyvinyl chloride, and so on.

11.2.2 Thermosets

In comparison with thermoplastics, thermosetting plastics remain in a permanent solid form after curing. Materials cross-link during the curing process due to having optimum heat, light, and radiation. When the connection of carbon atoms forms two- and three-dimensional structures instead of one-dimensional polymer chains, the prepared polymer is termed as the thermoset plastic (Dodiuk and Goodman 2013). Thermoset plastics are characterized as non-meltable plastics due to their solidified structure. An irreversible chemical bond formation occurs during the curing process; therefore, recycling and reusing of thermosetting plastics and returning these materials to their original state is not possible. The common thermosetting plastics are cyanate ester, epoxy, polyester, polyurethane, silicon, vulcanized rubber, and so forth.

11.3 PRESENCE OF CHEMICALS IN PLASTICS AND THEIR TOXICITY

In the plastics processing industries, plastic polymers are incorporated with various additives that are used to improve the performance, functionality, and aging properties of the basic plastic polymer. These chemical compounds are introduced to the polymer through injection, molding, extrusion, blow molding, and vacuum molding processes, which help to obtain several desired shapes and functionalities of plastics (Hahladakis et al. 2018). The harmful chemicals associated with plastic production are divided into three broad categories: ingredients of the plastic materials, by-products of manufacturing, and chemicals adsorbed from the environment (Hahladakis et al. 2018). The potential toxicological responses caused by all these categories of chemicals also include a wide variety of additives. Some of these chemicals are defined as priority hazardous pollutants because of their persistent nature in organisms and food webs. For making plastic polymers and end-products, the most commonly used chemicals include heavy metals, pesticides, polycyclic aromatic hydrocarbons (PAHs), and polychlorinated biphenyl (PCB) compounds (Groh et al. 2018, 2019; Hahladakis et al. 2018). The presence of these chemicals could disrupt important physiological processes in animals and humans, and cause diseases. Another study by Lithner et al. (2011), identified and compiled the chemicals present in 55 different thermoplastic and thermosetting plastic polymers, and asserted their possible health hazards according to the United Nations Globally Harmonized System.

The chemicals present in plastic polymers could potentially leach or migrate to the surrounding medium, or slowly migrate to the plastic surfaces. All types and grades of plastics from macro- to nanoplastics are subjected to leaching and/or adsorb hazardous substances. Bhunia et al. (2013) have elaborately reviewed the process of chemicals' migration from plastic products and found that a range of toxic chemicals from plastic packaging can migrate during microwave and conventional heating under different storage conditions. The frequent loss of additives used in plastic production by leaching also influences the polymers' weathering behavior in the marine environment to a large scale, and this weathering further facilitates the leaching or adsorption of many hazardous substances from the environment. Groh et al. (2018) identified different types of hazardous chemicals associated with plastic packaging and plastic products. The authors (Groh et al. 2018) classified those hazardous chemicals as monomers, intermediates, solvents, surfactants, stabilizers, plasticizers, flame retardants, accelerators, biocides, and colorants. Finally, the database identified 906 different substances which are directly associated with plastic packaging in terms of being used during production and in the final products (Groh et al. 2018). The authors also listed around 3377 other chemicals which are partly correlated with plastic production and packaging (Groh et al. 2018). Plastics contain and leach a cocktail of toxic chemicals, and assessing the risks of plastic-derived contaminants or contaminants adsorbed on plastics individually or in a group is extremely challenging (Bittner et al. 2014; Tang et al. 2016; Zhang et al. 2018).

Environmentally unsafe and hazardous chemicals identified in plastics are not only used as the main ingredients (which are monomers), but also used as a range of biocidal agents to prevent molds, give flame retardant properties for increasing the fire resistance, and enable plasticizer properties to enhance flexibility (Groh et al. 2018; Lithner et al. 2011). The hazardous chemicals in plastics can further be grouped into hazardous metals, bispehnols, phthalates, per- and polyfluoroalkyl substances (PFASs), additives, and others. A brief description of these various groups of pollutants associated with plastics are discussed in the following sections.

11.3.1 HAZARDOUS METALS

A group of metals, such as cadmium (Cd), lead (Pb), antimony (Sb), and tin (Sn) (as organotin) are used in plastic products to improve the durability and workability. These metallic elements are highly toxic and harmful for human health and the environment. According to the current restriction of hazardous substances used in the plastic manufacturing sector, plastics containing Cd, Pb, mercury (Hg), and hexavalent chromium (Cr(VI)) may not be suitable for recycling if the concentrations of these elements are higher than 1000 mg kg^{-1} (EU, 2011). The presence of these heavy metals can cause a serious hazard to human health because most of them are known carcinogens, can cause permanent changes to the genetic makeup of cells, and can have significant adverse effects on the fertility function among humans and wild animals. A study by Tang et al. (2015) observed that surface soils and sediments are contaminated with medium to high levels of Cd and Hg, up to 0.355 and 0.408 mg kg^{-1} in soils, and 1.53 and 2.10 mg kg^{-1} in sediments. A follow-up study by the authors also found that plastic waste processing is the major source of heavy metal toxicity and pollution in road dust in China (Tang et al. 2016). Due to the non-biodegradable nature of the metallic contaminants, they tend to persist in the soils, sediments, and water for numerous years, and can biomagnify in the food chain. The adverse effects of toxic heavy metals in the environment have been well documented in the literature over the past few decades (Bolan et al. 2014; Wuana et al. 2011; Wu et al. 2016).

11.3.2 Bisphenols

Three highlighted bisphenols that are commonly used in the manufacture of clear polycarbonate plastics are bisphenol A (BPA), bisphenol S, and bisphenol F. The epoxy resin of BPA, also known as bisphenol A diglyceride ether, has been identified as causing cytotoxic effects in living tissues, which may increase the cell division rate in living organisms (Hahladakis et al. 2018; Lau and Wong 2000). Bisphenols are used in the plastic manufacturing industry as additives in rigid plastics and in the manufacture of other plastic-related materials, including lining inside food and drinks cans. The bisphenols are known as endocrine disrupting chemicals, and a recent CHEM Trust report also highlighted that the use of the bisphenol group of chemicals to be restricted by regulators (Geyer et al. 2017). Bisphenol A diglyceride ether is a major monomer of epoxy resin used in plastic can lining and added to polymers to serve as antioxidants. However, it can potentially be transferred to the food while heating and storage, which could be detrimental for humans because it may contain unreactive BPA, and their low level of exposure is dangerous to human health (Bhunia et al. 2013; Vandenberg et al. 2012). BPA has been frequently detected in the aqueous environments, such as surface water, groundwater, and seawater in various parts of the world, mainly due to the leaching from BPA-based plastics or microplastics (Wei et al. 2019) and incomplete removal by wastewater treatment plants (Im and Loffler 2016). BPA has been widely applied commercially in polycarbonate plastics, electronic equipment, sports healthcare products, and food storage since 1957. The global market of BPA in 2017 was estimated to be 8.15 million tons, and the global gross production was estimated to be 7.10 million tons (Im and Loffler 2016; Li 2019). However, it has been observed that BPA exhibits obvious estrogen disrupting activity, and is vulnerable to leaching out of products. Thus, BPA can have direct exposure to human beings, which has resulted in mounting public concern (Zheng et al. 2018). Therefore, BPA was banned in many countries in the past decade, especially its use in baby food storage materials and beverage containers (Lee et al. 2019).

11.3.3 Phthalates

There can be 14 different hazardous phthalate compounds present in plastics during and after the manufacturing of the products. Phthalates are a group of chemicals that are used as plasticizers (a kind of additive used to enhance the flexibility and durability of a polymer matrix) in plastics during their manufacturing process. Phthalates are associated with several health effects on people including reproductive disruption and metabolic diseases, such as obesity. It was reported that a plastic cooking oil container is a more suitable medium for phthalate migration than a mineral water container (Xu et al. 2010). A higher temperature and longer contact time accelerate the migration of phthalates from plastics. The European Union is currently finalizing a restriction on the use of specific phthalates that are commonly used as plasticizers in plastic packaging because of the health concerns associated with these chemicals.

11.3.4 Per- and Polyfluoroalkyl Substances

The PFASs are a group of chemicals including perfluorooctanoic acid, perflurooctanesulfonate, and so on, which are used in the plastic packaging industry to impart flame retardancy to the final products. Aqueous film forming foam used in firefighting is a point source of PFAS input to soil and groundwater, while plastics in biosolids and landfills act as a non-point source of these harmful chemicals (Ghisi et al. 2018; Hepburn et al. 2019; Washington et al. 2019). The persistence and stable nature of these chemicals make their use attractive in the plastic manufacturing industries. However, due to their persistence, they cannot be broken down, and thus can accumulate in the environment over time. The PFAS chemicals are also used in food packaging products. Therefore, these chemicals can enter the human body through the food chain. Due to the dumping of consumer goods often containing plastics and/or coated with PFAS compounds, leachates of municipal solid waste landfills have shown high concentrations of PFAS (Hepburn et al. 2019). These compounds have the high potential to enter the environment and contaminate the ground and surface water through leachate movement. One of the PFAS chemicals identified in the database as perfluorooctanoic acid is listed as one of the substances of very high concern in the United States, Australia, and European Union due to its reproductive toxicity and extreme environmental persistence (Busch et al. 2010; Gallen et al. 2016, 2018). For example, Gallen et al. (2016, 2017) reported that leachates from operating young landfills in Australia that accept municipal solid wastes have higher mean concentrations of PFAS than old closed landfills. In a study concentrating on 27 landfills across Australia, Gallen et al. (2017) found that perfluorohexanoate was the predominantly detected PFAS in these samples, with a mean concentration of 1700 ng/L, and at least five PFAS compounds were detected ubiquitously. The PFAS chemicals have been shown to accumulate in soils and sediments in alarming quantities in various countries, and due to their highly persistent nature and strong binding with sediments, they can migrate offsite, and have been detected even in the Arctic and Antarctic regions (Butt et al. 2010; Kwok et al. 2013).

11.3.5 Additives

Table 11.2 shows a list of hazardous chemicals in various plastic materials grouped based on different additives used in their production. In the plastic production industries, the basic polymer is incorporated into a plastic compound accompanied by the addition of numerous additives. The additives are chemicals used to increase the performance (e.g., blow molding, vacuum molding, extrusion, shaping of polymer, etc.), functionality, and aging properties of the polymers (Hahladakis et al. 2018). Some polymers incorporate additives during the plastic manufacturing itself, whereas other polymers include them during the processing into their finished products. Additives are also used to protect the polymers from degradation effects, such as light, heat, or bacteria. Some of the common additives used in the plastic manufacturing industries are listed below:

- *Antioxidants*: Antioxidants are embedded in various polymer resins to delay the degradation rate of plastic polymers when exposed to ultraviolet light (Bhunia et al. 2013). These additives are incorporated for plastic processing and outside applications, where weathering resistance of the product is needed. The most common antioxidants used in the plastic food packaging industry are arylamines (Table 11.2).
- *Colorants*: The colorants are used for coloring different parts of the plastic products. Most common colorants are based on two chemical groups: dye azo: 4-methyl-m-phenylenediamine, 4,4-methylenedianiline, etc., and pigments: cobalt (II) diacetate.
- *Foaming agents*: Foaming agents are used in expanded polystyrene cups, building boards, and polyurethane carpet underlayment. These are also defined as functional additives in the plastics manufacturing industry. The common examples of foaming agents are azodicarbonamide, sodium bicarbonate, isobutane, potassium bicarbonate, etc.

- *Plasticizers*: Plasticizers are commonly used for improving the flexibility, durability, and stretchability of polymer films. Some commonly used plasticizers are phthalic esters, such as bis(2-ethylhexyl) phthalate (DEHP) which is used in polyvinyl chloride (PVC) formulations.
- *Lubricants*: Lubricants are generally used to process rigid plastic materials, such as polyvinyl chloride, acrylonitrile-butadiene-styrene, and polypropylene. The use of lubricant materials improves the processability of the polymers. However, using lubricants cannot change or improve the color or heat stability of the materials (Rosen and Hall 1982).
- *Anti-stats*: Anti-stats are generally used to reduce dust collection by static electricity attraction. These additives prevent the buildup of static charges in plastic materials and enhance the practical usability of various plastic products, such as the easy packaging and transport of plastics, reduced risk of electrical discharges, improved processing efficiency, and reduced dust pick up (Wozniak 1990).
- *Anti-microbials*: To control the growth of microorganisms in plastic materials, such as public phones, flooring, and escalator rails, anti-microbials are used in a wide range. Anti-microbials are normally used in materials where infection is a concern, such as hospital-type furniture (D'Arcy 2001). One of the leading anti-microbial products is 10,10′-oxybisphenox arsine.
- *Flame retardants*: Flame retardants are organic and inorganic materials used to make plastic and wood-based materials flameproof. For the reason of safety and safe handling of materials, the use of flame retardants is common and necessary in many plastic products (Wensing et al. 2005). These additives also improve the safety of wire and cable coverings and cultured marbles. Some examples of flame retardants are aluminum hydroxide, magnesium hydroxide, boron compounds, and organochlorines, such as chlorendic acid derivatives and chlorinated paraffin (Table 11.2).

11.4 LEACHING OF CHEMICALS FROM PLASTICS

Due to their large molecular size, polymers are usually considered to be biochemically inert materials, which prohibit their penetration through the cell membrane. Therefore, apparently, they do not pose a threat to the environment. However, plastic materials carry toxic and hazardous chemicals, which are small in molecular size (MW < 1000), can penetrate cells, and disrupt the endocrine systems in humans and animals (Teuten et al. 2009). A cocktail of toxic chemical contaminants can be leached out from the plastic debris (Figure 11.2; Table 11.2). The release of chemicals from plastics to the environment is an undesirable process for both the manufacturer of plastics and the environment because leaching of additives shortens the polymer life, and wildlife and humans get exposed to the toxic chemicals' pollution. However, the release of hazardous substances, monomers, degradation products, by-products, and additives from plastics can occur during all phases of their life cycle. The additives are not usually chemicals bound to the plastic structure; therefore, they are able to leach out from the polymer matrix very easily. This leaching process is further facilitated due to the low molecular weight of the additives. Some additives can form a large proportion of the plastic mass, and the total amount of all leachable chemicals increases significantly. For example, PVC may contain more than 40% by weight of plasticizers which are mostly phthalates. Furthermore, unreacted residual monomers or small oligomers can also be found in plastic products since polymerization reactions are seldom complete. The leaching of chemicals from plastics also depends on several factors, and can be accelerated with induced situations, such as the product is exposed to common-use stresses such as ultraviolet radiation in sunlight, microwave radiation, and/or moist heat via boiling or dishwashing.

Another example of potential leaching of chemicals from plastics is BPA polymers. BPA is commonly known as the monomeric building blocks of polycarbonate plastics. However, the polymerization process of BPA leaves some monomers unbound. Therefore, BPA molecules could be released from bottles and food containers to foods and drinks over time. The chemical's releasing

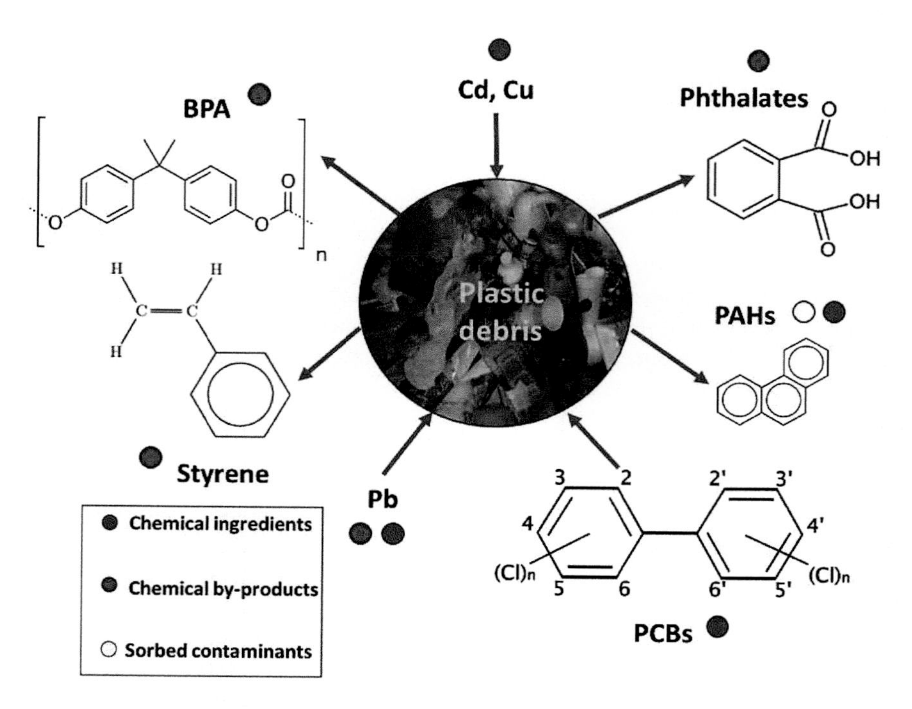

FIGURE 11.2 Cocktail of toxic chemical contaminants present in plastic debris.

process is further enhanced by repeated washing and reusing of the same bottles/containers, storing in them an acidic, and basic items help to break down the polymers too. Therefore, the use of reusable water bottles, baby bottles, and inner linings of food cans could enhance the risk of BPA leaching into food and drinks over time, which is particularly accelerated with elevated temperature (Olea et al. 1996; Kang et al. 2003; Wagner and Oehlmann, 2009; Halden 2010). Unlike BPA, phthalates are not designed to be covalently bound to the polymer matrix, which makes them highly susceptible to leaching.

DEHP widely used in medical devices is another principle phthalate causing serious human health hazards due to their potential leaching losses. It was reported that chemicals leaching from phthalates in medical devices could enter the body fluids and subsequently migrate to human tissues, which is a great concern (Jaeger and Rubin 1970). The leaching of chemicals from medical devices was found to be a function of temperature, storage time, mechanical stress, and chemical composition and geometry of medical devices (Tickner et al. 2001). Björnsdotter (2015) conducted a study where five virgin plastic pellets made of polyethene (PE), polypropylene, polystyrene, PVC, and high-density polyethene were used to test the leaching of additives and residual monomers, while keeping them in artificial seawater for 24 hours. Potential leachable compounds were observed from all the plastic pellets except polypropylene. However, the leachable compounds from PVC were not identified, whereas other monomers like styrene and oligomers were recognized. Another recent study by Hahladakis et al. (2018) investigated the leaching of additives from low-density plastic polymers like polyethylene, polyethylene terephthalate, polystyrene, and PVC, and leaching of these additives was observed in all cases. The rate of additive's leaching from plastics was high at the beginning and reduced over time. However, the leached additives were cumulated, and gave the highest concentration at the end of the experiment after 57 or 78 days (Suhrhoff et al. 2016). This study (Suhrhoff et al. 2016) interestingly found that almost 89%–98% of additives were leached from polyethylene bags, and printed PE and PVC had higher leaching ratios than polyethylene terephthalate. The physical parameters, such as ultraviolet radiation and salinity are important to understand the leaching pattern of additives from plastics.

11.5 TOXICITY OF THE CHEMICAL COMPONENTS OF PLASTICS

Examples of harmful effects from plastic-derived chemicals and monomers include: toxic to the reproduction system (e.g., DEHP and BPA), allergenic (formaldehyde, acrylonitrile, and methyl methacrylate), mutagenic (benzene, phenol, and 1,3-butadiene), carcinogenic (vinyl chloride, acrylonitrile, and 1,3-butadiene), high acute toxicity (phosgene and toluene diisocyanate), and environmentally hazardous when exposed for longer periods of time with long-term effects (pentabromodiphenyl ether and acrylonitrile) (Lithner et al. 2009). Depending on the monomer composition, some chemicals are more toxic than others, for example, polyurethanes, polyacrylonitriles, and PVC were the most toxic and environmentally hazardous polymers compared to polypropylene, ethylene-vinyl acetate, polyvinyl acetate. Due to the potential leaching risk of toxic BPA from baby bottles made from polycarbonate, it was banned in Canada (2008) and in Europe (2011) to reduce the human health hazard (Lithner 2011).

Among the monomers used in plastics, several of them are considered highly harmful for the environment. For example, BPA disrupts the endocrine system, whereas styrene and polyvinyl chloride monomers have both shown to be mutagenic and carcinogenic in nature. Studies have found that the additives and chemicals used in plastics are highly toxic for wildlife and humans. BPA was first created in 1891, has been used commercially from the 1950s, and became most commonly used chemicals for plastics all over the world. It was estimated that around 3.6 million tons of BPA was produced annually in 2006, and 5 million tons in 2010 (Brunelle 2005; Huang et al. 2012). The main problem with BPA materials is they can be adsorbed through skin contact, which means that humans can regularly be exposed through the leaching of these chemicals out of packaging into foods and drinks (Zhang et al. 2018). Over the last 20 years, various studies investigated and observed the link between BPA to several adverse human health effects. If the amount of BPA is high in the human body, it can impact the fetus and young children, who have underdeveloped systems for detoxifying chemicals (Meeker et al. 2009). Studies by Zhang et al. (2018) and Zhao et al. (2018) observed that the effects of high doses of BPA in mice impacted the kidney and liver functions and mammary gland development. Several other studies on human exposure to BPA observed the development of obesity, insulin resistance, metabolic syndrome, diabetes, and atherosclerosis (Lind and Lind 2011; Shankar and Teppala 2011; Wang et al. 2012; Rochester and Bolden 2015; Sui et al. 2018).

For example, the adverse effects of BPA were explored by monitoring the toxicokinetics of BPA during pregnancy using an in vivo approach conducted in pregnant sheep (Gingrich et al. 2019). It was found that there were differences in toxicokinetics of BPA between maternal and fetal circulations, and higher persistence was observed in the fetal circulation than maternal circulation. The effect of BPA on the nervous system of zebrafish embryos was studied, and it was found that high-concentration BPA exposure (3.0 mg/L) could downregulate the expression of six neurodevelopment genes (Gu et al. 2019).

BPA exposure could also affect rodents' gene expression. Rezg et al. (2019) reported that BPA exposure might lead to metabolic syndrome by disrupting the intestinal glucose absorption and glucose metabolism in the liver. Apart from reproductive and hormone disruptive effects, BPA may lead to the migration of cancer cells. For example, nanomolar BPA could induce the in vitro migration of human non-small lung cancer cells through pathways mediated by upregulation of transforming growth factor beta (Song et al. 2019). BPA can also be chlorinated in drinking water supply systems and transformed into more toxic compounds (Zheng et al. 2018). In conclusion, BPA and its metabolites are demonstrated to be toxic to human health and the ecosystem, and thus its release into the environment through plastic wastes require appropriate management practices.

11.6 CONCLUSIONS

Exposure to plastic, plasticizers, and other additives of polymers are unavoidable in today's world. This chapter discusses the increasing usage and potential environmental and human health toxicity of various plastic materials. The elevated usages of plastic products could cause serious environmental pollution because of the use of toxic chemicals during their production, and the possible leaching of chemicals from plastic products at various stages of their manufacturing, transport, storage, usage, and disposal. Hazardous chemicals are commonly used as additives during plastic manufacturing, and the risk of releasing these hazardous substances from plastic wastes is a great concern for human health due to their potential carcinogenic and mutagenic effects. However, the most concerning issue is the presence of toxic chemical contaminants as a cocktail in plastic debris, containing a mixture of BPA, phthalates, polycyclic aromatic hydrocarbons, polychlorinated biphenyls, and so on, which make the removal of these chemicals from water and soils extremely difficult. Similarly, human and environmental risk assessment of a mixture of highly toxic substances is a paramount task. The principal concern from a human health perspective is the endocrine-disruption properties of plastic materials, such as BPA and DEHP. Therefore, appropriate technological, societal, and policy changes are needed to reduce the amount of plastic production and consumption, find alternative material including biodegradable plastics, encourage the use of reusable plastic materials, and implement systematic waste collection and disposal practices.

REFERENCES

Anderson, J.C., Park, B.J., Palace, V.P. 2016. Microplastics in aquatic environments: Implications for Canadian ecosystems. *Environmental Pollution* 218: 269–280.

Auta, H., Emenike, C., Fauziah, S. 2017. Distribution and importance of microplastics in the marine environment: A review of the sources, fate, effects, and potential solutions. *Environment International* 102: 165–176.

Bhunia, K., Sablani, S.S., Tang, J., Rasco, B. 2013. Migration of chemical compounds from packaging polymers during microwave, conventional heat treatment, and storage. *Comprehensive Reviews in Food Science and Food Safety* 12(5): 523–545.

Biryol, D., Nicolas, C.I., Wambaugh, J., Phillips, K., Isaacs, K. 2017. High-throughput dietary exposure predictions for chemical migrants from food contact substances for use in chemical prioritization. *Environment International* 108: 185–194.

Bittner, G.D., Yang, C.Z., Stoner, M.A. 2014. Estrogenic chemicals often leach from BPA-free plastic products that are replacements for BPA-containing polycarbonate products. *Environmental Health* 13(1): 41.

Björnsdotter, M. 2015. Leaching of residual monomers, oligomers and additives from polyethylene, polypropylene, polyvinyl chloride, high-density polyethylene and polystyrene virgin plastics. Bachelor thesis, School of Science and Technology, Örebro University, Sweden, 18 pp.

Bolan, N., Kunhikrishnan, A., Thangarajan, R., Kumpiene, J., Park, J., Makino, T., Kirkham, M.B., Scheckel, K. 2014. Remediation of heavy metal(loid)s contaminated soils – To mobilize or to immobilize? *Journal of Hazardous Materials* 266: 141–166.

Boucher, J., Billard, G. 2019. The challenges of measuring plastic pollution. *Field Actions Science Reports: The Journal of Field Actions* 19: 68–75.

Brunelle, D.J. 2005. *Advances in Polycarbonates: An Overview.* ACS Symposium Series. Oxford University Press, 1–5.

Busch, J., Ahrens, L., Sturm, R., Ebinghaus, R. 2010. Polyfluoroalkyl compounds in landfill leachates. *Environmental Pollution* 158(5): 1467–1471.

Butt, C.M., Berger, U., Bossi, R., Tomy, G.T. 2010. Levels and trends of poly-and perfluorinated compounds in the arctic environment. *Science of the Total Environment* 408(15): 2936–2965.

Caporossi, L., Papaleo, B. 2017. Bisphenol A and metabolic diseases: Challenges for occupational medicine. *International Journal of Environmental Research and Public Health* 14: 959.

D'Arcy, N. 2001. Antimicrobials in plastics: A global review. *Plastics, Additives and Compounding* 3(12): 12–15.

Dodiuk, H., Goodman, S.H. 2013. *Handbook of Thermoset Plastics.* William Andrew, San Diego, CA.

EU. 2011. Directive 2011/65/EU of the European Parliament and of the Council of 8 June 2011, on the restriction of the use of certain hazardous substances in electrical and electronic equipment (recast). *Official Journal of the European Communities* 174: 88–110.

EU. 2018. Directive 2018/852 of the European EU 2018. Parliament and of the Council of 30 May 2018 amending Directive 94/62/EC on packaging and packaging waste (Text with EEA relevance). *Official Journal of the European Union* 64: 141–154.

European Commission. 2018. Communication from the Commission to the European Parliament, the Council, the European Economic and Social Committee and the Committee of the Regions: A European strategy for plastics in a circular economy. *Official Journal of the European Union* 61: 61–68.

Gallen, C., Drage, D., Kaserzon, S., Baduel, C., Gallen, M., Banks, A., Broomhall, S., Mueller, J. 2016. Occurrence and distribution of brominated flame retardants and perfluoroalkyl substances in Australian landfill leachate and biosolids. *Journal of Hazardous Materials* 312: 55–64.

Gallen, C., Drage, D., Eaglesham, G., Grant, S., Bowman, M., Mueller, J. 2017. Australia-wide assessment of perfluoroalkyl substances (PFASs) in landfill leachates. *Journal of Hazardous Materials* 331: 132–141.

Gallen, C., Eaglesham, G., Drage, D., Nguyen, T.H., Mueller, J. 2018. A mass estimate of perfluoroalkyl substance (PFAS) release from Australian wastewater treatment plants. *Chemosphere* 208: 975–983.

Gallo, F., Fossi, C., Weber, R., Santillo, D., Sousa, J., Ingram, I., Nadal, A., Romano, D. 2018. Marine litter plastics and microplastics and their toxic chemicals components: The need for urgent preventive measures. *Environmental Sciences Europe* 30: 1–14.

Galloway, T.S. 2015. Micro- and Nano-plastics and Human Health. In: Bergmann, M., Gutow, L., Klages, M. (Eds.), *Marine Anthropogenic Litter.* Springer International Publishing, Cham, pp. 343–366.

Galloway, T.S., Lewis, C.N. 2016. Marine microplastics spell big problems for future generations. *Proceedings of the National Academy of Sciences* 113: 2331–2333.

Gervet, B. 2007. *The Use of Crude Oil in Plastic Making Contributes to Global Warming.* Lulea University of Technology, Lulea, Sweden.

Geyer, R., Jambeck, J.R., Law, K.L. 2017. Production, use, and fate of all plastics ever made. *Science Advances* 3(7): e1700782.

Ghisi, R., Vamerali, T., Manzetti, S. 2018. Accumulation of perfluorinated alkyl substances (PFAS) in agricultural plants: A review. *Environmental Research* 169: 326–341.

Gingrich, J., Pu, Y., Ehrhardt, R., Karthikraj, R., Kannan, K., Veiga-Lopez, A. 2019. Toxicokinetics of bisphenol A, bisphenol S, and bisphenol F in a pregnancy sheep model. *Chemosphere* 220: 185–194.

Groh, K.J., Backhaus, T., Carney-Almroth, B., Geueke, B., Inostroza, P.A., Lennquist, A., Maffini, M., Leslie, H.A., Slunge, D., Trasande, L. 2018. Chemicals associated with plastic packaging: Inventory and hazards. *PeerJ Preprints* 6: e27036v1.

Groh, K.J., Backhaus, T., Carney-Almroth, B., Geueke, B., Inostroza, P.A., Lennquist, A., Leslie, H.A., Maffini, M., Slunge, D., Trasande, L. 2019. Overview of known plastic packaging-associated chemicals and their hazards. *Science of the Total Environment* 651: 3253–3268.

Gu, J., Zhang, J., Chen, Y., Wang, H., Guo, M., Wang, L., Wang, Z., Wu, S., Shi, L., Gu, A. 2019. Neurobehavioral effects of bisphenol S exposure in early life stages of zebrafish larvae (*Danio rerio*). *Chemosphere* 217: 629–635.

Hahladakis, J.N., Velis, C.A., Weber, R., Iacovidou, E., Purnell, P. 2018. An overview of chemical additives present in plastics: Migration, release, fate and environmental impact during their use, disposal and recycling. *Journal of Hazardous Materials* 344: 179–199.

Halden, R.U. 2010. Plastics and health risks. *Annual Review of Public Health* 31: 179–194.

Hepburn, E., Madden, C., Szabo, D., Coggan, T.L., Clarke, B., Currell, M. 2019. Contamination of groundwater with per-and polyfluoroalkyl substances (PFAS) from legacy landfills in an urban re-development precinct. *Environmental Pollution* 248: 101–113.

Huang, Y.Q., Wong, C.K.C., Zheng, J.S., Bouwman, H., Barra, R., Wahlström, B., Neretin, L., Wong, M.H. 2012. Bisphenol A (BPA) in China: A review of sources, environmental levels, and potential human health impacts. *Environment International* 42: 91–99.

Im, J., Loffler, F.E. 2016. Fate of bisphenol A in terrestrial and aquatic environments. *Environmental Science & Technology* 50(16): 8403–8416.

Jaeger, R., Rubin, R. 1970. Contamination of blood stored in plastic packs. *The Lancet* 296(7664): 151.

Jambeck, J.R., Geyer, R., Wilcox, C., Siegler, T.R., Perryman, M., Andrady, A., Narayan, R., Law, K.L. 2015. Plastic waste inputs from land into the ocean. *Science* 347, 768 –771.

Kang, J.-H., Kito, K., Kondo, F. 2003. Factors influencing the migration of bisphenol A from cans. *Journal of Food Protection* 66(8): 1444–1447.

Kwok, K.Y., Yamazaki, E., Yamashita, N., Taniyasu, S., Murphy, M.B., Horii, Y., Petrick, G., Kallerborn, R., Kannan, K., Murano, K. 2013. Transport of perfluoroalkyl substances (PFAS) from an arctic glacier to downstream locations: Implications for sources. *Science of the Total Environment* 447: 46–55.

Lambert, S., Sinclair, C., Boxall, A. 2014. Occurrence, degradation, and effect of polymer-based materials in the environment. *Reviews of Environmental Contamination and Toxicology* 227: 1–53.

Lassen, C., Hansen, S.F., Magnusson, K., Hartmann, N.B., Jensen, P.R., Nielsen, T.G., Brinch, A. 2015. Microplastics: Occurrence, effects and sources of releases to the environment in Denmark. The Danish Environmental Protection Agency. Environmental Project No. 1793, 2015.

Lau, O.-W., Wong, S.-K. 2000. Contamination in food from packaging material. *Journal of Chromatography A* 882(1–2): 255–270.

Lee, E.-H., Lee, S.K., Kim, M.J., Lee, S.-W. 2019. Simple and rapid detection of bisphenol A using a gold nanoparticle-based colorimetric aptasensor. *Food Chemistry* 287: 205–213.

Lind, P.M., Lind, L. 2011. Circulating levels of bisphenol A and phthalates are related to carotid atherosclerosis in the elderly. *Atherosclerosis* 218(1): 207–213.

Lithner, D., Damberg, J., Dave, G., Larsson, Å. 2009. Leachates from plastic consumer products – Screening for toxicity with Daphnia magna. *Chemosphere* 74(9): 1195–1200.

Lithner, D., Larsson, Å., Dave, G. 2011. Environmental and health hazard ranking and assessment of plastic polymers based on chemical composition. *Science of the Total Environment* 409(18): 3309–3324.

Meeker, J.D., Sathyanarayana, S., Swan, S.H. 2009. Phthalates and other additives in plastics: Human exposure and associated health outcomes. *Philosophical Transactions of the Royal Society B: Biological Sciences* 364(1526): 2097–2113.

Olea, N., Pulgar, R., Pérez, P., Olea-Serrano, F., Rivas, A., Novillo-Fertrell, A., Pedraza, V., Soto, A.M., Sonnenschein, C. 1996. Estrogenicity of resin-based composites and sealants used in dentistry. *Environmental Health Perspectives* 104(3): 298–305.

Rochester, J.R., Bolden, A.L. 2015. Bisphenol S and F: A systematic review and comparison of the hormonal activity of bisphenol A substitutes. *Environmental Health Perspectives* 123(7): 643–650.

Rezg, R., Abot, A., Mornagui, B., Knauf, C. 2019. Bisphenol S exposure affects gene expression related to intestinal glucose absorption and glucose metabolism in mice. *Environmental Science and Pollution Research* 26(4): 3636–3642.

Rosen, M., Hall, L.K. 1982. Glyceryl monostearate plastic lubricants. Google Patents No. US4363891A.

Schulte, K., Von Lacroix, F. 2000. High-density polyethylene fiber. *Comprehensive Composite Materials: Polymer Matrix Composites* 2: 231.

Shankar, A., Teppala, S. 2011. Relationship between urinary bisphenol A levels and diabetes mellitus. *The Journal of Clinical Endocrinology & Metabolism* 96(12): 3822–3826.

Song, P., Fan, K., Tian, X., Wen, J. 2019. Bisphenol S (BPS) triggers the migration of human non-small cell lung cancer cells via upregulation of TGF-β. *Toxicology in Vitro* 54: 224–231.

Statista. 2018. *Plastic waste in the U.S.* https://www.statista.com/study/60094/plastic-waste-in-the-us/

Suhrhoff, T.J., Scholz-Böttcher, B.M. 2016. Qualitative impact of salinity, UV radiation and turbulence on leaching of organic plastic additives from four common plastics—A lab experiment. *Marine Pollution Bulletin* 102(1): 84–94.

Sui, Y., Park, S.-H., Wang, F., Zhou, C., 2018. Perinatal Bisphenol A exposure increases atherosclerosis in adult male PXR-humanized mice. *Endocrinology* 159: 1595–1608.

Tang, Z., Huang, Q., Yang, Y., Nie, Z., Cheng, J., Yang, J., Wang, Y., Chai, M. 2016. Polybrominated diphenyl ethers (PBDEs) and heavy metals in road dusts from a plastic waste recycling area in North China: Implications for human health. *Environmental Science and Pollution Research* 23(1): 625–637.

Tang, Z., Zhang, L., Huang, Q., Yang, Y., Nie, Z., Cheng, J., Yang, J., Wang, Y., Chai, M. 2015. Contamination and risk of heavy metals in soils and sediments from a typical plastic waste recycling area in North China. *Ecotoxicology and Environmental Safety* 122: 343–351.

Teuten, E.L., Saquing, J.M., Knappe, D.R., Barlaz, M.A., Jonsson, S., Björn, A., Rowland, S.J., Thompson, R.C., Galloway, T.S., Yamashita, R. 2009. Transport and release of chemicals from plastics to the environment and to wildlife. *Philosophical Transactions of the Royal Society B: Biological Sciences* 364(1526): 2027–2045.

Thompson, R.C., Swan, S.H., Moore, C.J., Saal, F.S.V. 2009. Our plastic age. *Philosophical Transactions of the Royal Society B: Biological Sciences* 364(1526): 1973–1976.

Tickner, J.A., Schettler, T., Guidotti, T., McCally, M., Rossi, M. 2001. Health risks posed by use of Di-2-ethylhexyl phthalate (DEHP) in PVC medical devices: A critical review. *American Journal of Industrial Medicine* 39(1): 100–111.

Vandenberg, L.N., Colborn, T., Hayes, T.B., Heindel, J.J., Jacobs Jr, D.R., Lee, D.-H., Shioda, T., Soto, A.M., vom Saal, F.S., Welshons, W.V. 2012. Hormones and endocrine-disrupting chemicals: Low-dose effects and nonmonotonic dose responses. *Endocrine Reviews* 33(3): 378–455.

Verrey, J., Wakeman, M.D., Michaud, V., Månson, J.A.E. 2006. Manufacturing cost comparison of thermoplastic and thermoset RTM for an automotive floor pan. *Composites Part A: Applied Science and Manufacturing* 37(1): 9–22.

Wagner, M., Oehlmann, J. 2009. Endocrine disruptors in bottled mineral water: Total estrogenic burden and migration from plastic bottles. *Environmental Science and Pollution Research* 16(3): 278–286.

Wang, T., Li, M., Chen, B., Xu, M., Xu, Y., Huang, Y., Lu, J., Chen, Y., Wang, W., Li, X. 2012. Urinary bisphenol A (BPA) concentration associates with obesity and insulin resistance. *The Journal of Clinical Endocrinology & Metabolism* 97(2): 223–227.

Washington, J.W., Rankin, K., Libelo, E.L., Lynch, D.G., Cyterski, M. 2019. Determining global background soil PFAS loads and the fluorotelomer-based polymer degradation rates that can account for these loads. *Science of the Total Environment* 651: 2444–2449.

Wei, W., Huang, Q.-S., Sun, J., Wang, J.-Y., Wu, S.-L., Ni, B.-J. 2019. Polyvinylchloride microplastics affect methane production from the anaerobic digestion of waste activated sludge through leaching toxic bisphenol-A. *Environmental Science & Technology*.

Wensing, M., Uhde, E., Salthammer, T. 2005. Plastics additives in the indoor environment—flame retardants and plasticizers. *Science of the Total Environment* 339(1–3): 19–40.

Wijesekara, H., Bolan, N.S., Bradney, L., Obadamudalige, N., Seshadri, B., Kunhikrishnan, A., Dharmarajan, R., Ok, Y.S., Rinklebe, J., Kirkham, M.B., Vithanage, M. 2018. Trace element dynamics of biosolids-derived microbeads. *Chemosphere* 199: 331–339.

Wozniak, D.S. 1990. *Anti-Stat for Polyvinyl Chloride Polymers*. Google Patents.

Wu, X., Cobbina, S.J., Mao, G., Xu, H., Zhang, Z., Yang, L. 2016. A review of toxicity and mechanisms of individual and mixtures of heavy metals in the environment. *Environmental Science and Pollution Research* 23(9): 8244–8259.

Wuana, R.A., Okieimen, F.E. 2011. Heavy metals in contaminated soils: A review of sources, chemistry, risks and best available strategies for remediation. *Isrn Ecology* 2011.

Xu, Q., Yin, X., Wang, M., Wang, H., Zhang, N., Shen, Y., Xu, S., Zhang, L., Gu, Z. 2010. Analysis of phthalate migration from plastic containers to packaged cooking oil and mineral water. *Journal of Agricultural and Food Chemistry* 58(21): 11311–11317.

Zhang, Z., Lin, L., Gai, Y., Hong, Y., Li, L., Weng, L. 2018. Subchronic bisphenol S exposure affects liver function in mice involving oxidative damage. *Regulatory Toxicology and Pharmacology* 92: 138–144.

Zheng, S., Shi, J., Zhang, J., Yang, Y., Hu, J., Shao, B. 2018. Identification of the disinfection byproducts of bisphenol S and the disrupting effect on peroxisome proliferator-activated receptor gamma (PPARγ) induced by chlorination. *Water Research* 132: 167–176.

12 Particulate Plastics as Vectors of Heavy Metal(loid)s

Hasintha Wijesekara, Lauren Bradney, Sanchita Mandal, Binoy Sarkar, Hocheol Song, Nanthi S. Bolan, and M.B. Kirkham

CONTENTS

12.1 Introduction 181
12.2 Sources of Trace Metals and Heavy Metal(loid)s in the Environment 183
12.3 Interactions between Particulate Plastics and Heavy Metal(loid)s 185
  12.3.1 Particulate Plastics as a Source of Heavy Metal(loid)s 185
  12.3.2 Particulate Plastics as a Sink for Heavy Metal(loid)s 185
12.4 Bioavailability of Particulate Plastic-Derived Heavy Metal(loid)s 188
12.5 Summary and Conclusions 189
References 190

12.1 INTRODUCTION

Particulate plastics (i.e., micro- or nanoplastics) in the terrestrial and aquatic environments are a group of synthetic polymer fragments or beads ranging in diameter from roughly 5 mm down to the nano-meter scale (Frias and Nash 2019). Based on the origin, particulate plastics can be categorized into primary particulate plastics and secondary particulate plastics (Duis and Coors 2016). Primary particulate plastics are manufactured and used in a wide variety of industrial products (e.g., cosmetics and clothing) or as raw materials for industrial processes (Fendall and Sewell 2009; Sundt et al. 2014). Secondary particulate plastics are plastic fragments derived from the physical- and chemical-assisted breakdown of large plastic products, such as plastic packaging, plastic bags, or bottles (Duis and Coors 2016). Research has identified that both primary and secondary particulate plastics can persist in the environment for decades (Henry et al. 2019; Rillig 2012). Land-based sources, such as landfills, compost and biosolid (i.e., treated sewage sludge) application, disposal of waste water treatment plants' effluents, and atmospheric deposition are considered as main pathways to transport particulate plastics into the environment (Kilponen 2016; Rochman 2018; Sundt et al. 2014; UNEP 2016; Zubris and Richards 2005). There has been renewed interest in large-scale application of biowastes (i.e., composts and biosolids) on soil systems that are mainly intended to enhance soil fertility. However, biowastes can contain particulate plastics that have originated from sources, such as plastic bags and plastic-coated paper products (Figure 12.1). Although medium- and large-sized plastic materials are

This chapter is a condensed version of the following review paper: L. Bradney, H. Wijesekara, K.N. Palansooriya, N. Obadamudalige, N.S. Bolan, Y.S. Ok, J. Rinklebe, Ki-Hyun Kim, M.B. Kirkham. 2019. Particulate plastics as a vector for toxic trace-element uptake by aquatic and terrestrial organisms and human health risk. *Environmental International*, 131, 104937. doi:10.1016/j.envint.2019.104937.

FIGURE 12.1 (a) A few types of particulate plastics found under stereomicroscopy in a biosolid sample collected from a wastewater treatment plant in Sydney, New South Wales, Australia; (b) appearance of various particulate plastics: purchased pristine polyethylene particulate plastics (PPP), extracted particulate plastics from biosolids (BSPP), sediments (SEPP), and soils (SPP).

generally segregated during the composting process through sieving and handpicking, a significant portion of small- and micro-sized plastics remain in the final biowaste products. Because of the subsequent milling of biowastes (i.e., specifically for composts derived from municipal solid wastes and co-composting), most plastics end up as particulate plastics, thereby entering into the soil with their land application. The weathering of plastic film mulch used over agricultural fields is also a major source of particulate plastics in soils (Rochman 2018). The accumulated particulate plastics in soil can be transported to the aquatic environment through soil erosion, which is one of the main processes that allows the transport of particulate plastics from the terrestrial to the aquatic ecosystem (Wijesekara et al. 2018; Zubris and Richards 2005).

Particulate plastics can contain several compounds or components, such as polymers, petroleum hydrocarbons, additives, adhesives, coatings, residues (i.e., solvents) and impurities, oligomers or degradation products, and heavy metal(loid)s. Thus, particulate plastics give rise to a number of environmental impacts. Trophic transfer of particulate plastics can cause bioaccumulation and biomagnification (Bradney et al. 2019; Ng et al. 2018). In addition, a wide range of organic (e.g., polychlorinated biphenyls, polycyclic aromatic hydrocarbons, and organochlorine pesticides) and inorganic [e.g., heavy metal(loid)s] contaminants can cause toxicity within aquatic and terrestrial organisms through trophic transfer, as well as to humans through food chain contamination (Bradney et al. 2019; Teuten et al. 2009). The transport of these contaminants occurs as a result of adsorption-desorption processes, as affected by the pollutant concentrations in a given area. For example, research related to particulate plastics interactions with organic contaminants has reported that polychlorinated biphenyls, polycyclic aromatic hydrocarbons, petroleum hydrocarbons, organochlorine pesticides, and bisphenol A have all been transported by particulate plastics into the marine ecosystem (Teuten et al. 2009). However, a limited number of studies have focused on identifying particulate plastics as vectors of heavy metal(loid)s (Brennecke et al. 2016; Holmes et al. 2014; Turner and Holmes 2015; Wang et al. 2018; Wijesekara et al. 2018). Therefore, this chapter provides an overview of particulate plastics as a source and sink for heavy metal(loid)s in the environment. In addition, a major focus was given to identifying mechanisms involved in forming heavy metal(loid)s-particulate plastic complexes and their interactions and transportation in the environment.

12.2 SOURCES OF TRACE METALS AND HEAVY METAL(LOID)S IN THE ENVIRONMENT

Heavy metal(loid)s are elements that have properties in between metals and non-metals. The most common heavy metal(loid)s are arsenic (As), cadmium (Cd), chromium (Cr), copper (Cu), mercury (Hg), and selenium (Se). They reach the environment through various pedogenic and anthropogenic processes and, consequently, occur in nature (Figure 12.2). For example, soil parent materials (including igneous and sedimentary rocks) act as a major source for the heavy metal(loid)s. The majority of As is derived from a geogenic origin, for example, coal has been estimated to release about 45,000 tons of As annually, while human activities release approximately 50,000 tons (Bolan et al. 2014). Although the anthropogenic As source is increasingly becoming important, the recent episode of extensive As-contamination of groundwaters and food crops in Bangladesh and India (especially in West Bengal) is of geological origin (Bolan et al. 2014; Hossain 2006). Industrial processes, manufacturing, and applications of products, such as alloying agents, wood preservatives (i.e., chromated copper arsenate), pesticides, pharmaceuticals, transistors, and laser lightning are examples of anthropogenic As input to the environment. Use of As-containing feed additives in the poultry industry has been estimated to contribute between 250 and 350 tons of As annually in poultry wastes in the United States (Nachman et al. 2005).

Cadmium is a major toxicant with widespread exposure and can cause several adverse health effects in humans (e.g., Itai–Itai disease in Japan) (James and Meliker 2013). Disposal of domestic and industrial waste materials can be identified as one of the major sources of Cd enrichment in soils.

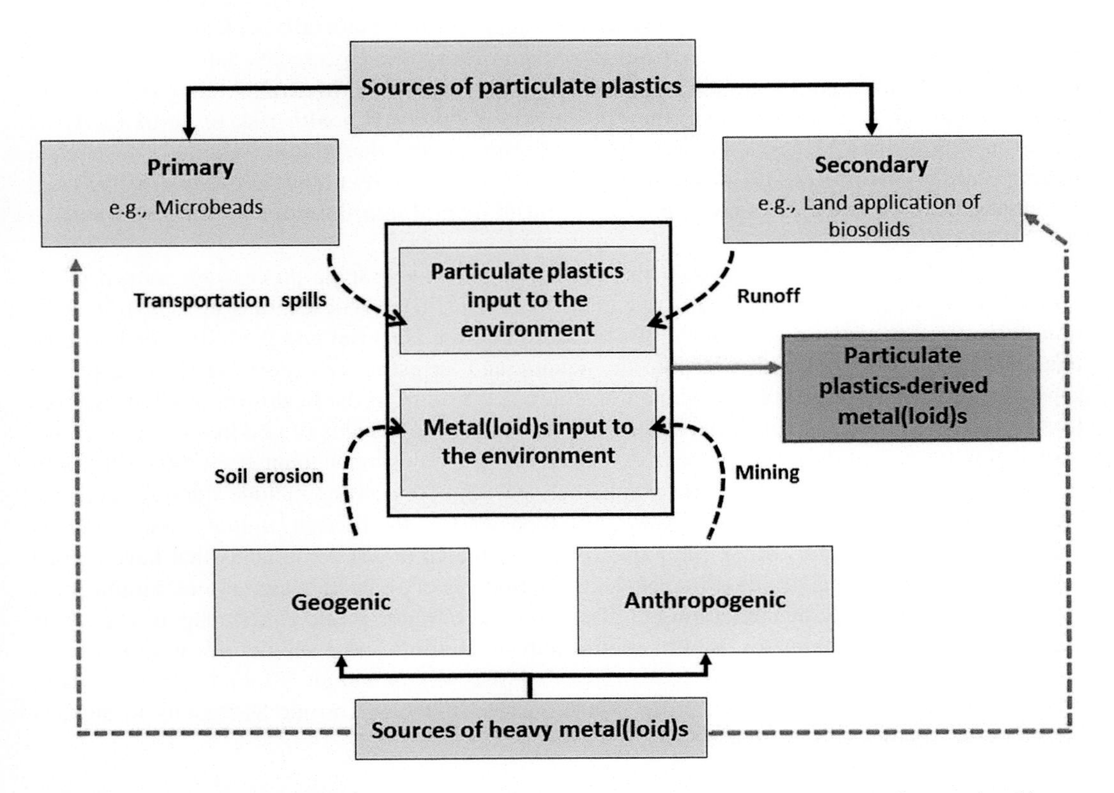

FIGURE 12.2 Conceptual diagram illustrating the sources and sinks of particulate plastic-associated heavy metal(loid)s.

For example, land application of biosolids (i.e., treated sewage sludge) can act as a major source of Cd input to agricultural soils and rehabilitated mine sites (Wijesekara et al. 2016). Use of cadmium-containing phosphorus (P) fertilizers is also a major source of Cd in agricultural soils. The Cd in most P fertilizers originates mainly from the phosphate rocks with varying Cd contents (Bolan et al. 2014).

Large quantities of Cr are used in tannery (i.e., mainly Cr(III)) and timber (i.e., mainly Cr(VI)) treatment, and, therefore, high concentrations of Cr are found in the effluents derived in these facilities. In addition, landfill leachate contains considerable concentrations of Cr, thereby becoming a major source of Cr contamination in aquatic and terrestrial environments (Wijesekara et al. 2014). Copper is widely used in agriculture, horticulture, and animal industries. For example, Cu-containing fungicides (e.g., copper oxychloride and "Bordeaux" mixture) and growth promoters are used in piggery and poultry industries (Bolan et al. 2014). Continuous application of biosolids can lead to the accumulation of Cu in soils, too (Wijesekara et al. 2016). Further, chromated copper-arsenate-treated timber used as fence posts and in vineyards can also release a significant amount Cu, Cr, and As into the environment (Bolan et al. 2014).

Mercury exists in four forms in the environment: elemental/metallic; inorganic [i.e., mercurous (Hg^+), mercuric (Hg^{2+}) salts]; organic (e.g., methylmercury/CH_3Hg, ethylmercury/$C_2H_5Hg^+$, dimethylmercury/CH_3HgCH_3); and mercury vapor (Hg^0). The major environmental compartments of Hg are atmospheric, terrestrial, and aquatic environments. Globally, it has been estimated that the atmosphere, soil, and oceans and lakes hold between 4400–5300, 250–1000, and 270–450 tons of Hg, respectively (Obrist et al. 2018). Mercury can be found in batteries, measuring devices (e.g., thermometers and barometers), electric switches and relays, lamps and some types of light bulbs, dental amalgam, and some types of cosmetics and pharmaceuticals (e.g., skin-lightening products). Mercury is also released via some industrial processes (e.g., coal burning and mining) and during waste incineration. Naturally, volcanic activity and weathering of rocks release Hg into the environment. Inorganic Hg species can be transformed by bacteria into methylmercury through the methylation process. Methylmercury is known to bioaccumulate in fish and shellfish, thereby causing a potential health risk to humans. Apart from the well known Minamata disease, it has been reported that among selected subsistence fishing populations in Brazil, Canada, China, Colombia, and Greenland, between 1.5/1000 and 17/1000 children showed cognitive impairment (mild mental retardation) due to Hg contamination (WHO 2017).

Lead (Pb)-based batteries, paints, fuels, chemicals for photographic processing, stained glass, jewelry, toys, and pipes contain Pb in large contents (WHO 2018; Wijesekara et al. 2014). In addition, mine spoil is also a major source of Pb contamination. Disposal and landfilling of Pb-based materials result in the release of Pb into the aquatic and terrestrial environments. Lead exposure can have serious consequences for human health, in particular, to the health of children. At high levels of Pb, exposure can attack the brain and central nervous systems of children to cause coma, convulsions, and even death. It has been estimated that Pb exposure accounted for 540,000 deaths worldwide in 2016 (WHO 2018). Environmental sources of zinc (Zn) include fluorescent tubes, batteries, galvanized products, and a variety of food wastes. In addition, military smoke bombs contain significant amounts of Zn (i.e., zinc oxide or zinc chloride) compounds that have caused several military personnel deaths (Plum et al. 2010). Sources of Se include agriculture, mining, coal combustion, fly-ash files, manufacturing of glass, paint, petroleum, pesticides, shampoos (i.e., Se is a component in some shampoos used to control human dandruff and a supplement used in canine diet), and electrical components (Wright and Welbourn 2002). Selenium is found in some soils in high concentrations, and it is taken up by plants, thereby inducing chronic Se toxicity in humans (MacFarquhar et al. 2010; Nuttall 2006).

12.3 INTERACTIONS BETWEEN PARTICULATE PLASTICS AND HEAVY METAL(LOID)S

Particulate plastics impact the dynamics of metal(loid)s in terrestrial and aquatic environments through both acting as a source and sink for metal(loid)s. As proposed by Bradney et al. (2019), the interaction of heavy metal(loid)s and particulate plastics can occur through three main processes. Primarily, heavy metal(loid)s can directly adsorb to particulate plastics through physisorption. Heavy metal(loid)s can also adsorb onto charged sites or neutral regions of the surface of the particulate plastics. Adsorption of heavy metal(loid)s onto particulate plastics through hydrous oxides of iron (Fe) and manganese (Mn) is also possible (Ashton et al. 2010). Agricultural practices (i.e., usage of plastic film mulching and utilization of biowastes and composts) and municipal solid waste management practices (i.e., landfill leachate) are potential occasions for particulate plastic and heavy metal(loid) interactions. Such interactions of heavy metal(loid)s and particulate plastics result in their transport in the aquatic and terrestrial environments (Holmes et al. 2012, 2014; Munier and Bendell 2018; Turner and Holmes 2015). Further, particulate plastic's hydrophobic nature and large surface area can be identified as influencing factors for the adsorption of heavy metal(loid)s.

12.3.1 Particulate Plastics as a Source of Heavy Metal(loid)s

In environmental toxicology, the "source" can be defined where pollution is emitted to the environment, or where the pollution originates. Particulate plastics act as a "source" when they release a pollutant into the environment. As particulate plastics contain heavy metal(loid)s in their composition at the first phases of their life cycles (i.e., raw materials, during manufacturing), particulate plastics become a source of heavy metal(loid)s into the environment. During some manufacturing processes, heavy metal(loid)s are mixed with particulate plastics as catalysts and also to develop special qualities, such as heat stabilizing properties. Heavy metal(loid)-containing heat stabilizers are often added to particulate plastics. For example, barium-zinc additives are known to be very effective heat stabilizers for polyvinyl chloride (PVC) (Sastri 2010). Lead and Cd are also found in different types of plastic additives (Hahladakis et al. 2018; Munier and Bendell 2018). However, the input of different additives (i.e., pigments and lubricants) during plastic production can also result in metal(loid)s within particulate plastics (Hahladakis et al. 2018). Table 12.1 summarizes some selected studies reporting the occurrence of heavy metal(loid)s in particulate plastics in the environment.

12.3.2 Particulate Plastics as a Sink for Heavy Metal(loid)s

The "sink" can be identified where pollution is removed from or released to (or retained in) the environment. Particulate plastics act as a "sink" when they remove or retain a pollutant from the environment. Particulate plastics are capable of accumulating heavy metal(loid)s from their surroundings. Therefore, the particulate plastics become a sink for heavy metal(loid)s in the environment (Holmes et al. 2012, 2014; Turner and Holmes 2015). Particulate plastic-related environmental reactions are mainly influenced by the above processes (see Section 14.4). A number of studies have provided explanations for environmental reactions between heavy metal(loid)s and particulate plastics, thereby identifying particulate plastics as a sink for heavy metal(loid)s. For example, Holmes et al. (2012) measured the concentration of a number of heavy metal(loids) in plastic production pellets at four beaches in Southwest England. The highest mean concentrations of Cu, Zn, and Pb were 1.32, 23.3, and 1.64 ng/g, respectively. In addition, Cr and Cd concentrations found were 751 and 76.7 ng/g, respectively (Holmes et al. 2012).

TABLE 12.1
Selected Trace Metal and Heavy Metal(loid) Concentrations Associated with Particulate Plastics in the Environment

Source	Type of Particulate Particle	Ag	Cd	Cr	Cu	Hg	Pb	Zn	References
Freshwater (Watergate Bay)	Beached pellets	<3	5	42.5	47	<3	109	196	Turner and Holmes (2015)
Seawater (Southwest England)	Virgin and beached PE pellets		1.09–76.7[a]	44–751[a]	0.064–1.32[b]		0.149–1.64[b]	0.299–23.3[b]	Holmes et al. (2012)
River water	Virgin pellets[c]		0.0894	Nd	1.58		0.922		Holmes et al. (2014)
	Beached pellets[c]		2.21	1.79	nd		13.2		
Seawater	Virgin pellets[c]		0.00383	5.72	4.11		nd		
	Beached pellets[c]		0.0904	8.48	nd		3.3		
Seawater (Portugal)	Virgin PS beads[b]				11.70			29.33	Brennecke et al. (2016)
	Aged PVC[b] fragments				3.11			6.12	
Coastlines (England)	Plastic production pellets (3–5 mm)								
Soar Mill		2.4[a]	1.7[a]	19[a]	0.06[b]		0.15[b]	0.55[b]	Ashton et al. (2010)
Thurlestone		4.9	10	63	0.14		0.41	0.42	
Bovisand		6	3.6	69	0.29		0.73	0.94	
Saltram		30	5	151	0.61		1.08	2.34	
San Diego Bay									
Coronado Cays	PET (12 months)[a]		2	103.5			732.5	5802	
	HDPE		nd	134.5			534.5	6525.5	Rochman et al. (2014)
Shelter Island	PET		nd	81.5			765.5	2743.5	
	HDPE		nd	96			925	3899	
Nimitz Marine Facility	PET		3.5	424			1419	6046	
	HDPE		nd	79			594	2573	

[a] $ng\ g^{-1}$.
[b] $\mu g\ g^{-1}$.
[c] $mol\ g^{-1}$.

Abbreviations: nd = not detected; PET = polyethylene terephthalate; HDPE = high-density polyethylene; PS = polystyrene; PE = polyethylene; PVC = polyvinyl chloride.

Turner and Holmes (2015) studied the adsorption of Cd, Cr, Cu, Hg, Pb, and Zn in beached pellets in river water. The adsorption capacities of the above metals were as low as <3 ng/g for heavy metal(loid)s. Wijesekara et al. (2018) reported the concentrations of Cu, Zn, and Cd in biosolid-derived particulate plastics as 180.64, 178.03, and 2.34 ng/g, respectively. Also, the concentrations of As, Pb, and Se adsorbed to the same particulate plastics varied within the range of <1 to 1.72 ng/g.

Natural organic matter (NOM), including dissolved organic matter (DOM), have been identified for particulate plastic-heavy metal(loid) interactions, thereby facilitating particulate plastics to become a sink for heavy metal(loid)s. In fact, when DOM is adsorbed to the particulate plastics, it increases the plastic's heavy metal(loid) adsorption and complexation ability, thereby becoming an efficient carrier for heavy metal(loid)s (Holmes et al. 2014; Wijesekara et al. 2018). Pristine particulate plastics are not effective in the retention of heavy metal(loid)s (Bradney et al. 2019). In contrast, the DOM-enriched particulate plastics are found to be effective in the retention of heavy metal(loid)s in soil environments. For example, Wijesekara et al. (2018) examined Cu toxicity to soil microorganisms as impacted by different particulate plastic inputs to the soil. Pristine polyethylene (PPP) and surface-modified (BSPP) particulate plastics were used in a laboratory incubation study. The BSPP samples were prepared by mixing PPP samples with dissolved organic carbon extracted from biosolid samples. Both pristine and Cu-contaminated soil samples were treated with the two types of particulate plastics. These soil samples were subsequently analyzed for bioavailable Cu concentration, soil basal respiration, microbial biomass carbon, and dehydrogenase activity. Results showed that bioavailability of Cu in soil decreased with the addition of both pristine and surface-modified plastics. The particulate plastic addition resulted in an increased soil respiration, dehydrogenase activity, and microbial biomass carbon in both uncontaminated and contaminated soils. Improved soil aeration, together with reduced Cu bioavailability through particulate plastic-assisted Cu immobilization, would have increased the microbial activity in the experimental soils. The same study documented that Cu-adsorbed particulate plastics reduced the soil respiration and dehydrogenase activity by ~26% and ~39%, respectively, compared to the control (Wijesekara et al. 2018). In the terrestrial environment, Hodson et al. (2017) observed that earthworms could ingest Zn-bearing particulate plastics, thereby increasing the Zn bioavailability. Table 12.2 summarizes some selected studies on the impacts of particulate plastic-derived heavy metal(loid)s in the environment.

Beyond the adverse effects of particulate plastic-derived heavy metal(loid)s in terrestrial environments, a few studies have shown the adverse effects among bird species in marine environments. For example, Fife et al. (2015) studied the uptake of heavy metal(loid)s through particulate plastics in dovekies, an Arctic seabird in White Bay, Newfoundland, Canada. They observed particulate plastic debris in nine gizzard samples among 65 samples analyzed and observed a high level of hepatic (i.e., liver) Hg in the dovekies which feed at high trophic levels. Lavers and Bond (2016) evaluated the vector effect of ingested particulate plastics for heavy metal(loid) contents within the stomach and feathers of seabirds. They found that a greater amount of ingested particulate plastics caused an increase in levels of heavy metal(loid)s, such as Pb in the feathers of juvenile Laysan albatrosses and Bonin petrels. Lavers et al. (2014) also observed a positive relationship between the ingested particulate plastics and the Cr and Ag contents in the stomach and feathers of flesh-footed shearwater (*Puffinus carneipes*) fledglings in Lord Howe Island, New South Wales, Australia. High levels of ingested particulate plastics and metal(loid)s in birds caused a reduction in their body mass, head bill length, and wing chord, within seabird populations in Australian marine environments (Lavers et al. 2014). Table 12.2 summarizes some selected studies reporting the impacts of particulate plastic-derived heavy metal(loid)s in the aquatic and terrestrial organisms.

TABLE 12.2
Selected References on the Occurrence and Impacts of Plastic-Derived Trace Metal and Heavy Metal(loid)s in Aquatic and Terrestrial Organisms

Animals (Environments)	Species (Common Name)	Species (Scientific Name)	Association between Particulate Plastics and Heavy Metal(loid)s in the Organisms	References
Fish (Marine)	Shrimp scad Bartail flathead	*Alepes djedaba* *Platycephalus indicus*	Positive relationship between particulate plastics and metal(loid)s (e.g., Hg and Se) in fish muscles	Akhbarizadeh et al. (2018)
Fish (Freshwater)	Zebrafish	*Danio rerio*	Exposure to the Ag-incubated particulate plastic beads (~75% of the Ag bound to particulate plastic beads) significantly reduced Ag uptake and significantly increased the proportion of intestinal Ag	Khan et al. (2015)
	Rainbow trout	*Oncorhynchus mykiss*	Co-exposure of particulate plastics and Ag and, Ag-incubated particulate plastics had no effect on the uptake of Ag in the anterior/mid intestine of rainbow trout	Khan et al. (2017)
Seabird (Terrestrial, Marine, and Atmospheric)	Dovekies	*Alle*	Plastic debris in 9 of 65 (14%) of gizzards samples Greater hepatic Hg levels in dovekies feeding at higher trophic levels	Fife et al. (2015)
	Laysan albatross Bonin petrel	*Phoebastria immutabilis* *Pterodroma hypoleuca*	Increased levels of ingested particulate plastics caused to increase the concentrations of Pb in feathers	Lavers and Bond (2016)
	Flesh-footed shearwaters	*Puffinus carneipes*	High concentrations of Cr and Ag were positively related with the mass of ingested particulate plastic	Lavers et al. (2014)
Earthworm (Terrestrial)	Common earthworms	*Lumbricus terrestris*	Particulate plastics could increase Zn bioavailability *Lumbricus terrestris* ingested the Zn-bearing particulate plastics	Hodson et al. (2017)

12.4 BIOAVAILABILITY OF PARTICULATE PLASTIC-DERIVED HEAVY METAL(LOID)S

In order to understand the potential risks associated with heavy metal(loid)-particulate plastic concentrations, the bioavailable fraction must be identified and quantified. The bioavailable fraction is the proportion of total metal(loid)s that are available for incorporation into organisms. The total metal(loid) fraction does not necessarily correspond with metal(loid) bioavailability. As observed by Lavers et al. (2014), there were a number of impacts and consequences (e.g., reduction in body mass, head bill length, and wing chord) in seabirds due to the exposure of particulate plastic-containing metal(loid)s. To better understand such a contamination issue, it is important to examine the adverse impacts in relation to the bioavailability of heavy metal(loid)s (Sleight et al. 2017). A range of interacting factors, such as environmental temperature, pH, salinity, DOM level, type of heavy metal(loid)s or plastic involved, size of particulate plastics, aging, and variables more specific to the organisms (e.g., digestive system functions), influence the bioavailability of heavy metal(loid)s within organisms

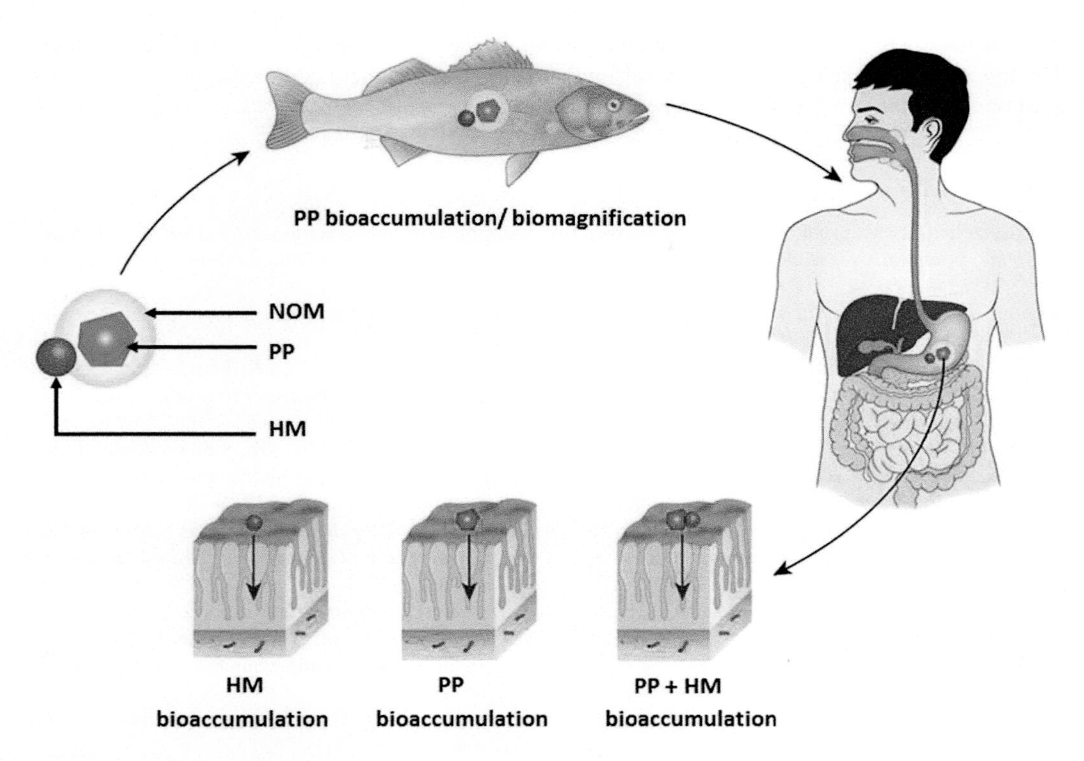

FIGURE 12.3 Conceptual diagram illustrating the impacts of particulate plastic (PP)-derived heavy metal(loid)s (HM) to aquatic organisms and humans. NOM stands for natural organic matter.

(Mota 2017; Sleight et al. 2017; Triebskorn et al. 2019). Studies have shown that the age of particulate plastics can influence their metal(loid) adsorption capacities. For example, Kedzierski et al. (2018) found that aged PVC adsorbed Cu and Ag, thereby leading to higher metal(loid)s bioaccumulation than new PVC beads. Brennecke et al. (2016) also found that aged PVC adsorbed Cu and Zn. Increased surface area and increased retention time between metal(loid)s and particulate plastics have been suggested as reasons for the above observations. Figure 12.3 illustrates the bioaccumulation and biomagnification of particulate plastic derived heavy metal(loid)s in the environment.

12.5 SUMMARY AND CONCLUSIONS

This chapter aimed to provide an overview of particulate plastics as a source and sink for heavy metal(loid)s in the environment. Additionally, an attempt was also made to understand the processes involved in the bioaccumulation of particulate plastic-derived heavy metal(loid)s in terrestrial and aquatic organisms. Particulate plastics affect heavy metal(loid) dynamics both in terrestrial and aquatic environments by acting either as a source of input or as a sink for the retention of heavy metal(loid)s. Heavy metal(loid)s, such as Hg and Cr are added as catalysts during the manufacture of plastic-based consumer products, and the levels of heavy metal(loid) s in plastic-based consumer products depend on the end-use purposes of these products. These heavy metal(loid)s are leached to the soil and water environments when particulate plastics undergo weathering processes. Particulate plastics are also effective in the retention of the heavy metal(loid)s, especially in the presence of NOM. The NOM provides functional groups for the retention of metal(loid)s. Retention of heavy metal(loid)s facilitates their transport in the aquatic environment, while reducing the bioavailability of heavy metal(loid)s in terrestrial environments. The long-term release of the heavy metal(loid)s and their subsequent ecotoxicity need to be examined under various soil and climatic conditions.

REFERENCES

Akhbarizadeh, R., Moore, F., Keshavarzi, B. 2018. Investigating a probable relationship between microplastics and potentially toxic elements in fish muscles from northeast of Persian Gulf. *Environmental Pollution*, **232**, 154–163.

Ashton, K., Holmes, L., Turner, A. 2010. Association of metals with plastic production pellets in the marine environment. *Marine Pollution Bulletin*, **60**(11), 2050–2055.

Bolan, N., Kunhikrishnan, A., Thangarajan, R., Kumpiene, J., Park, J., Makino, T., Kirkham, M.B., Scheckel, K. 2014. Remediation of heavy metal(loid)s contaminated soils – To mobilize or to immobilize? *Journal of Hazardous Materials*, **266**, 141–166.

Bradney, L., Wijesekara, H., Palansooriya, K.N., Obadamudalige, N., Bolan, N.S., Ok, Y.S., Rinklebe, J., Kim K., Kirkham, M.B. 2019. Particulate plastics as a vector for toxic trace-element uptake by aquatic and terrestrial organisms and human health risk. *Environmental International*, **131**, 1–18.

Brennecke, D., Duarte, B., Paiva, F., Caçador, I., Canning-Clode, J. 2016. Microplastics as vector for heavy metal contamination from the marine environment. *Estuarine, Coastal and Shelf Science*, **178**, 189–195.

Duis, K., Coors, A. 2016. Microplastics in the aquatic and terrestrial environment: Sources (with a specific focus on personal care products), fate and effects. *Environmental Sciences Europe*, **28**(1), 2.

Fendall, L.S., Sewell, M.A. 2009. Contributing to marine pollution by washing your face: Microplastics in facial cleansers. *Marine Pollution Bulletin*, **58**(8), 1225–1228.

Fife, D.T., Robertson, G.J., Shutler, D., Braune, B.M., Mallory, M.L. 2015. Trace elements and ingested plastic debris in wintering dovekies (Alle alle). *Marine Pollution Bulletin*, **91**(1), 368–371.

Frias, J.P.G.L., Nash, R. 2019. Microplastics: Finding a consensus on the definition. *Marine Pollution Bulletin*, **138**, 145–147.

Hahladakis, J.N., Velis, C.A., Weber, R., Iacovidou, E., Purnell, P. 2018. An overview of chemical additives present in plastics: Migration, release, fate and environmental impact during their use, disposal and recycling. *Journal of Hazardous Materials*, **344**, 179–199.

Henry, B., Laitala, K., Klepp, I.G. 2019. Microfibres from apparel and home textiles: Prospects for including microplastics in environmental sustainability assessment. *Science of the Total Environment*, **652**, 483–494.

Hodson, M.E., Duffus-Hodson, C.A., Clark, A., Prendergast-Miller, M.T., Thorpe, K.L. 2017. Plastic bag derived-microplastics as a vector for metal exposure in terrestrial invertebrates. *Environmental Science & Technology*, **51**(8), 4714–4721.

Holmes, L.A., Turner, A., Thompson, R.C. 2012. Adsorption of trace metals to plastic resin pellets in the marine environment. *Environmental Pollution*, **160**, 42–48.

Holmes, L.A., Turner, A., Thompson, R.C. 2014. Interactions between trace metals and plastic production pellets under estuarine conditions. *Marine Chemistry*, **167**, 25–32.

Hossain, M.F. 2006. Arsenic contamination in Bangladesh—An overview. *Agriculture, Ecosystems & Environment*, **113**(1), 1–16.

James, K.A., Meliker, J.R. 2013. Environmental cadmium exposure and osteoporosis: A review. *International Journal of Public Health*, **58**(5), 737–745.

Kedzierski, M., D'Almeida, M., Magueresse, A., Le Grand, A., Duval, H., César, G., Sire, O., Bruzaud, S., Le Tilly, V. 2018. Threat of plastic ageing in marine environment. Adsorption/desorption of micropollutants. *Marine Pollution Bulletin*, **127**, 684–694.

Khan, F.R., Boyle, D., Chang, E., Bury, N.R. 2017. Do polyethylene microplastic beads alter the intestinal uptake of Ag in rainbow trout (*Oncorhynchus mykiss*)? Analysis of the MP vector effect using in vitro gut sacs. *Environmental Pollution*, **231**, 200–206.

Khan, F.R., Syberg, K., Shashoua, Y., Bury, N.R. 2015. Influence of polyethylene microplastic beads on the uptake and localization of silver in zebrafish (*Danio rerio*). *Environmental Pollution*, **206**, 73–79.

Kilponen, J. 2016. Microplastics and harmful substances in urban runoffs and landfill leachates – Possible emission sources to marine environment. In: *Faculty of Technology*, Vol. Environmental Technology, Lahti University of Applied Sciences.

Lavers, J.L., Bond, A.L. 2016. Ingested plastic as a route for trace metals in Laysan albatross (*Phoebastria immutabilis*) and Bonin petrel (*Pterodroma hypoleuca*) from midway atoll. *Marine Pollution Bulletin*, **110**(1), 493–500.

Lavers, J.L., Bond, A.L., Hutton, I. 2014. Plastic ingestion by flesh-footed shearwaters (*Puffinus carneipes*): Implications for fledgling body condition and the accumulation of plastic-derived chemicals. *Environmental Pollution*, **187**, 124–129.

MacFarquhar, J.K., Broussard, D.L., Melstrom, P., Hutchinson, R., Wolkin, A., Martin, C., Burk, R.F. et al., 2010. Acute selenium toxicity associated with a dietary supplement. *Archives of Internal Medicine*, **170**(3), 256–261.

Mota, A.H.S.N.D. 2017. The potential of microplastic pellets as a vector to metal contamination in two sympatric marine species. https://run.unl.pt/handle/10362/30788.

Munier, B., Bendell, L.I. 2018. Macro and micro plastics sorb and desorb metals and act as a point source of trace metals to coastal ecosystems. PloS *One*, **13**(2), e0191759.

Nachman, K.E., Graham, J.P., Price, L.B., Silbergeld, E.K. 2005. Arsenic: A roadblock to potential animal waste management solutions. *Environmental Health Perspective*, **113**, 1123–1124.

Ng, E.-L., Huerta Lwanga, E., Eldridge, S.M., Johnston, P., Hu, H.-W., Geissen, V., Chen, D. 2018. An overview of microplastic and nanoplastic pollution in agroecosystems. *Science of The Total Environment*, **627**, 1377–1388.

Nuttall, K.L. 2006. Review: Evaluating selenium poisoning. *Annals of Clinical & Laboratory Science*, **36**(4), 409–420.

Obrist, D., Kirk, J.L., Zhang, L., Sunderland, E.M., Jiskra, M., Selin, N.E. 2018. A review of global environmental mercury processes in response to human and natural perturbations: Changes of emissions, climate, and land use. *Ambio*, **47**(2), 116–140.

Plum, L.M., Rink, L., Haase, H. 2010. The essential toxin: Impact of zinc on human health. *International Journal of Environmental Research and Public Health*, **7**, 1342–1365.

Rillig, M.C. 2012. Microplastic in terrestrial ecosystems and the soil? *Environmental Science & Technology*, **46**(12), 6453–6454.

Rochman, C.M. 2018. Microplastics research—from sink to source. *Science*, **360**(6384), 28.

Rochman, C.M., Hentschel, B.T., Teh, S.J. 2014. Long-term sorption of metals is similar among plastic types: Implications for plastic debris in aquatic environments. *Plos One*, **9**(1), e85433.

Sastri, V.R. 2010. Materials used in medical devices. In: *Plastics in Medical Devices: Properties, Requirements, and Applications*, ed. Saatri, V.R. Boston, MA, William Andrew Publishing, pp. 21–32.

Sleight, V.A., Bakir, A., Thompson, R.C., Henry, T.B. 2017. Assessment of microplastic-sorbed contaminant bioavailability through analysis of biomarker gene expression in larval zebrafish. *Marine Pollution Bulletin*, **116**(1), 291–297.

Sundt, P., Schulze, P.-E., Syversen, F. 2014. Sources of microplastics pollution to the marine environment. *Mepex for the Norwegian Environment Agency*, Report no: M-321|2015, 1–86.

Teuten, E.L., Saquing, J.M., Knappe, D.R.U., Barlaz, M.A., Jonsson, S., Björn, A., Rowland, S.J. et al. 2009. Transport and release of chemicals from plastics to the environment and to wildlife. *Philosophical Transactions of the Royal Society B: Biological Sciences*, **364**(1526), 2027–2045.

Triebskorn, R., Braunbeck, T., Grummt, T., Hanslik, L., Huppertsberg, S., Jekel, M., Knepper, T.P. et al. 2019. Relevance of nano- and microplastics for freshwater ecosystems: A critical review. *TrAC Trends in Analytical Chemistry*, **110**, 375–392.

Turner, A., Holmes, L.A. 2015. Adsorption of trace metals by microplastic pellets in fresh water. *Environmental Chemistry*, **12**(5), 600–610.

UNEP. 2016. Marine plastic debris and microplastics – Global lessons and research to inspire action and guide policy change. United Nations Environment Programme, Nairobi http://hdl.handle.net/20.500.11822/7720.

Wang, F., Wong, C.S., Chen, D., Lu, X., Wang, F., Zeng, E.Y. 2018. Interaction of toxic chemicals with microplastics: A critical review. *Water Research*, **139**, 208–219.

WHO. 2017. Mercury and health. https://www.who.int/news-room/fact-sheets/detail/mercury-and-health

WHO. 2018. Lead poisoning and health. https://www.who.int/en/news-room/fact-sheets/detail/lead-poisoning-and-health

Wijesekara, S.S.R.M.D.H.R., Mayakaduwa, S.S., Siriwardana, A.R., Silva, N.D., Basnayake, B.F.A., Kawamoto, K., Vithanage, M. 2014. Fate and transport of pollutants through a municipal solid waste landfill leachate in Sri Lanka. *Journal of Environmental Earth Sciences*, **77**(5), 1707–1719.

Wijesekara, H., Bolan, N.S., Vithanage, M., Xu, Y., Mandal, S., Brown, S.L., Hettiarachchi, G.M. et al. 2016. Utilization of biowaste for mine spoil rehabilitation. *Advances in Agronomy*, **138**, 97–173.

Wijesekara, H., Bolan, N.S., Bradney, L., Obadamudalige, N., Seshadri, B., Kunhikrishnan, A., Dharmarajan, R. et al. 2018. Trace element dynamics of biosolids-derived microbeads. *Chemosphere*, **199**, 331–339.

Wright, D.A., Welbourn, P. 2002. Metals and other inorganic chemicals. In: *Environmental Toxicology*, ed. Wright, D.A., and Welbourn, P. Cambridge University Press, UK, pp. 306–307.

Zubris, K.A.V., Richards, B.K. 2005. Synthetic fibers as an indicator of land application of sludge. *Environmental Pollution*, **138**(2), 201–211.

13 Water Relations and Cadmium Uptake of Wheat Grown in Soil with Particulate Plastics

M.B. Kirkham

CONTENTS

13.1 Introduction 193
13.2 Materials and Methods 194
13.3 Results and Discussion 197
13.3.1 Germination 197
13.3.2 Height 199
13.3.3 Evapotranspiration Rate 199
13.3.4 Stomatal Resistance 200
13.3.5 Fresh and Dry Weights and Cd in Plants 201
13.3.6 Soil Analyses 202
13.4 Conclusions 204
Acknowledgments 204
References 204

13.1 INTRODUCTION

Plastic debris litters marine and terrestrial habitats worldwide (Teuten et al. 2007). The plastics degrade over time, producing microplastics, defined as synthetic organic polymer particles with a size between 100 nm and 5 mm (Duis and Coors 2016). The more general term, particulate plastics, also is used to define synthetic polymer particles, and they include both microplastics and nanoplastics (less than 100 nm) (Bradney et al. 2020). Here, the term particulate plastics will be used. Particulate plastics enter the environment either directly (e.g., use of microbeads in cosmetics) or after plastic products have broken down due to weathering or other degradation processes. Plastic products that are washed down the drain (e.g., cosmetics; defoliants used in hand cleaners, facial cleaners, and toothpaste) (Duis and Coors 2016) or fibers washed off clothes during laundering (Browne et al. 2011) end up in sewage treatment plants, settle during wastewater treatment, and concentrate in the biosolids (Carr et al. 2016; Murphy et al. 2016; New York State Office of the Attorney General, 2015; Van Wezel et al. 2016; Wijesekara et al. 2018). When the biosolids are applied to soil, the soil will be contaminated with the plastic fragments.

Agriculture uses plastics in many different ways. Plastic films are laid down to prevent evaporation from the soil surface and to suppress weeds (Duis and Coors 2016). Plastic films also are used as anti-transpirants (Das and Raghavendra 1979). Synthetic polymer particles have been used for at least 70 years as soil conditioners. Most of the early soil conditioners were some form of a polyacrilonitrile polymer (Kirkham and Runkles 1952). The Monsanto Chemical Company was one of the first companies to make them and marketed the product under the name of Krilium, which was a vinyl acetate polymer (Kirkham and Runkles 1952). Today, polyacrylamide is widely used to facilitate tillage, promote soil aeration, control soil erosion, and increase infiltration (Chen 2020). New plastic products are being developed all the time to be used in agriculture. For example, a subsurface water retention technology, which uses plastic films placed at various depths below a plant's root zone to retain soil water, has been developed at Michigan State University (Smucker et al. 2018). The film is made of an impermeable, low-density polyethylene.

Particulate plastics in the aquatic environment have been widely studied and reviewed (Andrady 2011; Browne et al. 2011; Duis and Coors 2016; GESAMP 2015; Hidalgo-Ruz 2012; Lusher 2015; Thompson 2015; van Cauwenberghe et al. 2015; Wright et al. 2013). Particulate plastics in the terrestrial environment have been less studied, and essentially no information exists concerning the water relations of plants grown in soil with particulate plastics. It is important to know if particulate plastics control the hydration status of plants. Therefore, the first objective of this experiment was to determine how a particulate plastic, polyethylene glycol (molecular weight 8000), affects the growth, evapotranspiration rate, and stomatal resistance of wheat. Wheat was chosen because more land is devoted worldwide to the production of wheat than to any other commercial crop, and, on a global basis, wheat provides more nourishment for people than any other food source (Briggle and Curtis 1987).

Polyethylene glycol was chosen as a model particulate plastic. Ethylene glycol is the simplest polyhydric alcohol, $OH{\cdot}CH_2{\cdot}CH_2{\cdot}OH$ or $C_2H_4(OH)_2$. It is sometimes shortened to just "glycol." Glycol and ethylene glycol are the same. It is a colorless, syrupy liquid, prepared by heating any of certain ethylene compounds with an alkali carbonate and used as antifreeze (Webster's New World Dictionary of the American Language 1959). Polyethylene glycols with high molecular weights have been used in hydroponics for decades to control the osmotic potential of the medium (Comeau et al. 2010; Janes 1961; Jarvis and Jarvis 1965; Lagerwerff et al. 1961; Michel and Kaufmann 1973; Money 1989; Ranjbarfordoei et al. 2000). The early literature also called polyethylene glycol "Carbowax" (Lagerwerff et al. 1961; Michel 1970, 1971). Polyethylene glycols are available in a series of average molecular weights from approximately 300–20,000 (Jackson 1962). The higher molecular weight polyethylene glycols, like polyethylene glycol (PEG) 8000, are used because they are less likely than the lower molecular weight forms to be absorbed by plants (Kaufman and Eckard 1971; Lawlor 1970; Michel 1970). However, studies do report the uptake of higher molecular weight polyethylene glycols (Lawlor 1970; Macklon and Weatherley 1965; Resnik 1970).

Polyethylene glycols have limited use as osmotic agents in hydroponics. But, because polyethylene glycols have a low toxicity, they are used in a variety of commercial products (https://en.wikipedia.org/wiki/Polyethylene_glycol). They are used in medicines (e.g., as laxatives), in industry (e.g., as lubricating coatings and to reduce foaming), and in cosmetic and toiletry products (e.g., skin creams, lipsticks, and toothpastes) (https://en.wikipedia.org/wiki/Polyethylene_glycol). They are used in cancer therapy (Cruje and Chithrani 2014). Polyethylene glycol 8000 and polyethylene glycols used as laxatives (e.g., Clearlax Polyethylene Glycol 3350, made by McKesson, San Francisco, CA) are flaky powders.

Particulate plastics can be a vector for toxic trace element uptake (Bradney et al. 2019). They have a high surface area and polarity, and these factors aid the sorption of inorganic contaminants (Bradney et al. 2020). Therefore, a second objective of this experiment was to determine the uptake of Cd to see if the polyethylene glycol enhanced its uptake. Cadmium was chosen because it is a toxic element that is almost ubiquitously present at low levels in the environment due to anthropogenic influences (Clemens and Ma 2016).

13.2 MATERIALS AND METHODS

The experiment was carried out between May 31 and June 28, 2019 in a greenhouse in Manhattan, KS (39° 12′N, 96°35′W, 325 m above sea level), with natural light supplemented by cool, white fluorescent lamps (Sylvania Gro-Lux, F15T8/GRO/WS, OSRAM Sylvania, Inc., Wilmington, MA). Lights were on from 07:00 hours to 22:00 hours and turned off between 22:00 hours and

07:00 hours. Thirty-six pots (72 mm tall; 60 mm diameter) with no drainage holes were used in the experiment. The growth medium was a commercial potting soil (Moisture Control Potting Mix, Miracle-Gro Lawn Products, Inc., Marysville, OH).

The polyethylene glycol 8000 used in this experiment came from Sigma Chemical Company (now Sigma-Aldrich), St. Louis, Missouri. Sigma Chemical Company originally labeled this product (Product No. P-2139) as polyethylene glycol 6000. But in the 1980s, the company changed the molecular weight specification for this product from approximately 6000 to approximately 8000. While there was no change in this product, improved methods of analysis showed the higher value to be more correct. The smallest flakes of the polyethylene 8000 glycol used in this experiment went through a United States Standard Testing Sieve with a 500 micrometer opening (Sieve No. 35; ASTM E-11 specification, or specification of Committee E11 on Quality and Statistics of American Society for Testing Materials International). The largest particles were irregular in shape, and the largest ones were 11 mm in length, as measured by a ruler.

On May 31, 2019, the following procedures were done. Untreated soil (soil with no PEG) was placed in 12 of the pots. Into another 12 pots was placed soil with 2% PEG 8000. Thus, 20 g of dry PEG 8000 was mixed thoroughly into 1000 g of soil to get the 2% PEG 8000 treatment, which will henceforth be called the dry-PEG treatment. These 24 pots then were irrigated with 50 mL tap water. Into the final 12 pots was placed the untreated soil, but these pots then got 50 mL of a 2% PEG 8000 solution. This treatment was called the wet-PEG treatment. To make the 2% solution, 20 g of dry PEG 8000 was added to 1000 mL tap water, and the mixture was heated over low heat to get the PEG 8000 into solution. After cooling, the 2% PEG solution was irrigated onto the soil. After the water or 2% PEG 8000 solution had been added to the 36 pots, five seeds of winter wheat (*Triticum aestivum* L. 'Everest') were planted in each pot. 'Everest' was the top wheat cultivar planted in the eastern third of Kansas in the 2019 crop year (United States Department of Agriculture 2019). The 36 pots then were divided in half. Half of the pots were irrigated with 25 mL of water and the other half of the pots were irrigated with 25 mL of a 100 μg/mL solution of Cd using cadmium sulfate (anhydrous $CdSO_4$) (J.T. Baker Chemical Co., Phillipsburg, NJ). Therefore, there were six treatments in the experiment with six pots for each treatment:

- No PEG: Water irrigation
- No PEG: Cadmium irrigation
- Dry PEG: Water irrigation
- Dry PEG: Cadmium irrigation
- Wet PEG: PEG Solution irrigation
- Wet PEG: PEG Solution with cadmium irrigation

Day 0 of the experiment was the day of planting (May 31, 2019). Germination was measured until seedlings stopped emerging. Throughout the experiment, pan evaporation rate, evapotranspiration rate, height, and stomatal resistance were measured. To determine evapotranspiration rate, pots were weighed daily using a scale (Pelouze Model SP5, Rubbermaid Incorporated, Sanford Brands, Oak Brook, IL). After seedlings emerged, height was measured daily, using a ruler, from the soil surface to the tip of the most recently matured leaf. When tips of leaves began to senescence and turn yellow, only the green part of the leaf was measured. Stomatal resistance was measured when the leaves were wide enough to cover the sensor, which was on June 16, using a steady-state diffusion porometer (Model SC-1, METER Group, Pullman, WA). Measurements were taken on the adaxial surface of the leaf, because there are more stomata on the adaxial surface of wheat leaves than on the abaxial surface (Kirkham 2014, p. 434). Each day, it took about two hours to measure the stomatal resistance of the leaves (36 measurements; one leaf per pot). Measurements usually started about 11:00 hours and ended about 13:00 hours. To make sure there was no bias due to time of measurement of stomatal resistances, measurements were taken starting with Pot 1 on one day and the next day measurements were taken starting with Pot 36. No measurements were taken between June 11 and June 14, because I was out of town at a meeting.

After the first irrigations of the pots on May 31, pots were irrigated with 25 mL of the appropriate solution (i.e., tap water; 2% PEG solution; 100 μg/mL Cd solution; or a solution with 2% PEG and 100 μg/mL Cd) on the following days: June 8, June 15, June 18, June 22, and June 25. The water content in the pots was monitored using a soil moisture meter (HoldAll Moisture Meter, Panacea Products Corp., Columbus, OH; Product code 0-70686 26002 9). The soil moisture meter uses a bimetallic tip to generate electricity in the presence of moisture. The electricity causes a pointer to move across a scale going from 1 to 10, which indicated the degree of wetness (Faber et al. 1994). Between readings of 8 and 10, the soil was "wet;" between readings of 4 and 7, the soil was "moist;" and between readings of 1 and 3, the soil was "dry." The soil was watered when most of the pots recorded a reading of 8.

After measurements were taken on June 28, the plants were cut just above the soil surface and the fresh weight of leaves and culms (together called the "shoots") was measured using an analytical balance (SCIENTECH Model No. SA 210, Boulder, CO). Roots were not extracted from the soil due to the small number of roots in the treatments that received wet PEG and/or Cd. Shoots were dried for 64 hours in a forced air drying oven set at 57°C, and then dry weight was measured using the same analytical balance.

The shoots were submitted to the Soil Testing Laboratory at Kansas State University for analysis of Cd. They were digested using a nitric-perchloric acid digest (Kirkham 2000) and analyzed for Cd using inductively couple plasma-atomic emission spectroscopy, also referred to as inductively couple plasma-optical emission spectrometry. The detection limit for Cd measured in the inductively couple plasma-atomic emission spectroscopy was 0.005 mg/kg. Quality assurance (quality control) was done by duplicating 10% of the samples, and the standard reference material came from the National Institute of Standards and Technology (1515, apple leaves).

At harvest, roots were removed from the soil and the soil was submitted to the soil testing laboratory for analysis. The soil in three pots of the same treatment was combined to have enough soil for the analyses. Total concentrations of Cd in the soil were determined using a nitric acid digest (Wahla and Kirkham 2008). Extractable concentrations of Cd in the soil were determined using diethylenetriaminepentaacetic acid (Lindsay and Norvell 1978). Total and extractable concentrations of Cd were measured using the inductively couple plasma-atomic emission spectroscopy. The pH of the soil was determined using the methods described by Watson and Brown (1998). Exchangeable concentrations of K, Ca, Mg, and Na in the soil were determined using the method of Warncke and Brown (1998), and the cation exchange capacity (CEC) was calculated by summing these values, as described by Warncke and Brown (1998). Total nitrogen and total carbon in the soil were determined using a LECO TruSpec CN Carbon/Nitrogen combustion analyzer, which reports total levels (inorganic and organic) of C and N on a weight percent basis, according to the TruSpec CN instrument method "Carbon and Nitrogen in Soil and Sediment," published by the LECO Corporation, St. Joseph, MI, in 2005.

The polyethylene glycol was analyzed for total Cd using the same method to analyze the soil for total Cd. The total Cd in the polyethylene glycol was 0.00 mg/kg. It was less than the detectable concentration (0.005 mg/kg).

The bulk potting soil also was analyzed for pH; CEC; total N; total C; exchangeable K, Ca, Mg, Na; total Cd; and extractable Cd using the same methods that were used for the soil in the experiment. The potting soil had large pieces of bark in it. These large pieces were sieved out using a 2-mm opening sieve before analysis, following the procedure of Gelderman and Mallarino (1998) for soil sample preparation. In addition to the analyses that were done on the experimental soil, the potting soil also was analyzed for electrical conductivity (EC) using the method of Whitney (1998); particle-size analysis using the hydrometer method (Gee and Bauder 1986); extractable phosphorus using the Mehlich-3 test (Frank et al. 1998), NO_3-N and NH_4-N following the methods of Alpkem Corporation (1986a,b); and organic matter using the loss-on-ignition method (Combs and Nathan 1998). The experimental soils were not analyzed for the additional methods used to characterize the potting soil, because there was not enough soil for all the analyses. The physical and chemical

TABLE 13.1
Physical and Chemical Characteristics of the Potting Soil

Sand (%)	64
Silt (%)	28
Clay (%)	8
pH	5.8
Electrical conductivity (dS/m)	4.15
Organic matter (%)	51.1
CEC (meq/100 g)	60.6
Total N (%)	0.72
Total C (%)	29.5
NH_4-N (mg/kg)	52.7
NO_3-N (mg/kg)	637
Mehlich-3 P (mg/kg)	491
K (mg/kg)	3073
Ca (mg/kg)	5875
Mg (mg/kg)	1720
Na (mg/kg)	453
Total Cd (mg/kg)	0.39
Extractable Cd (mg/kg)	0.30

characteristics of the potting soil are given in Table 13.1. The sieved potting soil was a sandy loam, based on the soil triangle (Soil Science Society of America 2008). The soil was "moderately saline," based on the rankings of Whitney (1998).

The average maximum and minimum air temperatures during the 28-day study were about 29°C and 22°C, respectively. Throughout the experiment, wet and dry bulb temperatures were measured at the time of measurements with an hygrometer (Taylor Humidiguide, Model No. 5534, Taylor Instrument Companies, Rochester, NY; now Taylor Precision Products, Oak Brook, IL), and the relative humidity was calculated from them. Dry bulb temperatures varied from 23°C to 28°C, and the relative humidity varied from 62% to 87%. The pan evaporation rate was determined by weighing daily a round metal pan (23.0 cm in diameter and 3.7 cm in height) containing tap water using the same scale used to determine the evapotranspiration rate. Average pan evaporation rate during the experiment was 1 mm/day.

The experimental design was a strip-plot design. The whole area where the pots sat (approximately 45 × 60 cm) was divided into horizontal strips for the PEG treatments and vertical strips for the Cd treatments. There were two horizontal strips for each of the three PEG treatments (no PEG, dry PEG, and wet PEG), for a total of six horizontal strips. There were two vertical strips for the two Cd treatments (no Cd and Cd irrigated onto the pots). Therefore, there were 12 cells in the experiment with three pots receiving the same treatment in each cell. Due to the duplication of the horizontal rows, there were six pots in the experiment receiving the same treatment. The horizontal strips were rotated daily throughout the experiment to expose plants to uniform environmental conditions. Means and standard errors were calculated.

13.3 RESULTS AND DISCUSSION

13.3.1 GERMINATION

Germination of the wheat seeds in the three different PEG treatments is shown in Figure 13.1. Due to the variability in germination (which took from 3 to 15 days after planting), the germination of seeds in three pots (a cell) were averaged together, and then the two sets of three pots under the

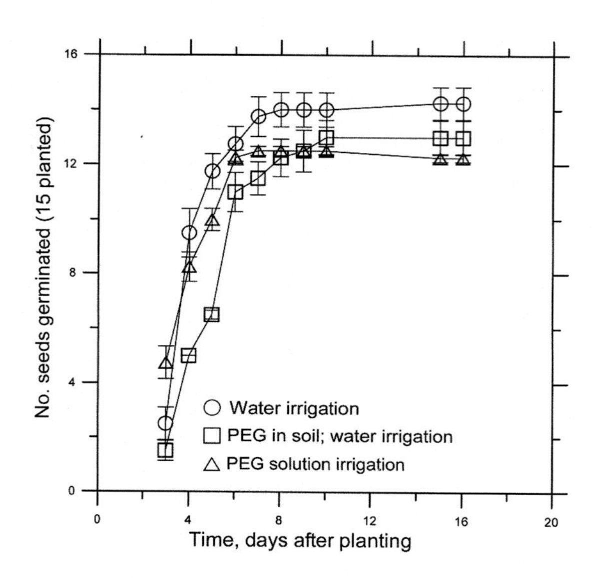

FIGURE 13.1 Germination of wheat seeds in pots of soil with no polyethylene glycol 8000 (PEG 8000) and irrigated with tap water (circles); with dry PEG 8000 (2% w:w) mixed into the soil before planting and irrigated with tap water (squares); and with a 2% solution of PEG 8000 irrigated onto soil without PEG in it (triangles). Five seeds were planted in each pot, and three pots under the same treatment were combined to get a data point. See text for details. Mean and standard error are shown (n = 12).

same treatments were averaged together to get a mean and standard error. Because each pot had five seeds, the total possible germination number in a cell was 15 seeds. There was no difference in germination due to Cd, so the pots with and without Cd were averaged together. Thus, each value in Figure 13.1 is the average of 12 pots.

Seeds germinated fastest in the pots with PEG irrigated onto the soil (Figure 13.1, triangles). It is known that PEG promotes germination (Zhang et al. 2015). The PEG solution probably covered the seeds and attracted water to the seeds, which allowed a few seeds to germinate fast. But the PEG solution did not allow the water to evaporate (as will be seen in the section after next), and the soil probably became anaerobic, which hindered further germination. By the end of the experiment, pots with the PEG solution irrigated onto them had the lowest number of seeds germinate.

Four, 5, 6, and 7 days after planting seeds, pots with the dry PEG in the soil had the lowest germination (Figure 13.1, squares). This could be due to an increase in EC caused by the dry PEG added to the soil. The potting soil was moderately saline, even before the powdered PEG was added to it (Table 13.1). But, because EC was not measured in the treated soil, it is unknown what the EC was, under the different treatments. By 10 days after planting, the germination in the pots with the dry PEG added to the soil and in the soil with no PEG was the same. Pots were watered 8 days after planting, which might have reduced the EC, resulting in a final, equal germination in the pots without PEG and in the pots with dry PEG in the soil.

Germination percentages 16 days after planting for the no-PEG, dry-PEG, and wet-PEG treatments were 95%, 87%, and 82%, respectively (Figure 13.1). In the wet-PEG treatment, one plant that had germinated by day 10 died by day 15, so the germination fell for the wet-PEG treatment.

After germination, results for the plants in the no-PEG treatment were similar to those for plants in the dry-PEG treatment. Therefore, for measurements taken daily (height, evapotranspiration, and stomatal resistance), only results for the plants without PEG and the plants irrigated with PEG will be shown.

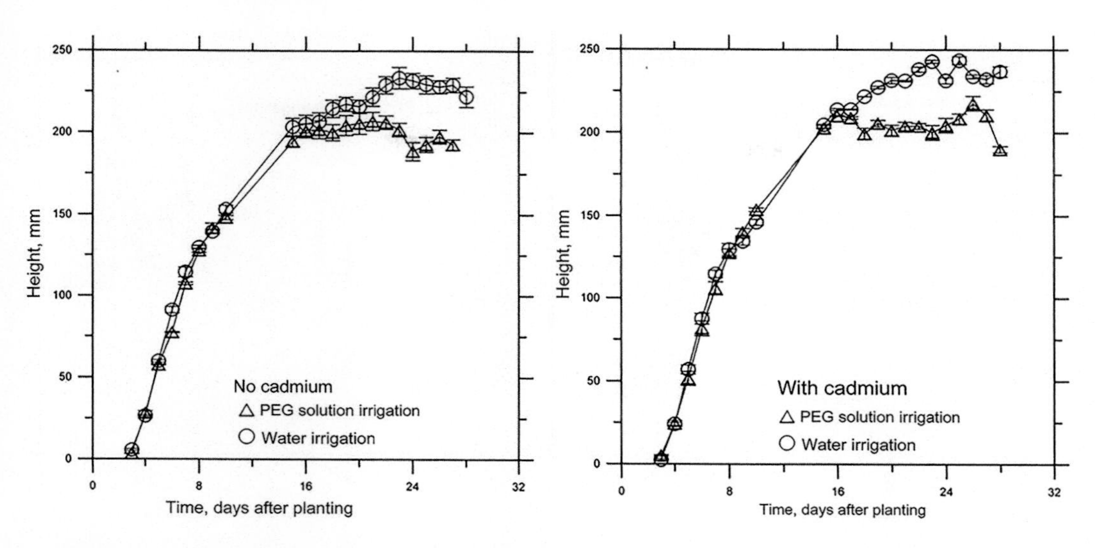

FIGURE 13.2 Height of wheat grown with water or PEG irrigations and without Cd (left) or with (right) Cd. See text for details. Mean and standard error are shown ($n = 6$).

13.3.2 Height

Both without Cd (Figure 13.2, left) and with Cd (Figure 13.2, right), 18 days after planting plants in the wet-PEG treatment were shorter than the plants in the no-PEG treatment. Plants grown with the PEG began to turn yellow at the end of the experiment. Cadmium had little effect on the height of the plants. However, plants grown with PEG irrigations and with Cd rapidly died off at the end of the experiment (Figure 13.2, right, 28 days after planting), and Cd increased the chlorosis of the leaves. The leaves never showed any white efflorescence, which indicated that the PEG was not being taken up by the roots and being transported to the leaves. Lagerwerff et al. (1961) found an efflorescence of white material on the upper surface of leaves of kidney beans (*Phaseolus vulgaris* L.) that they grew in culture solutions with polyethylene glycol with molecular weight 20,000, which they called Carbowax 20,000. They used infrared analysis to show that the efflorescence was identical to the Carbowax 20,000. This suggested that the Carbowax 20,000 passed through the plant system without undergoing any breakdown of its basic structural unit. They concluded, based on these results and others (Lagerwerff and Ogata 1960), that it was improbable that Carbowax 20,000 by itself exerted any physiological effects on plant growth, and its main effects were to increase the osmotic pressure of the culture medium. However, Leshem (1966) found no efflorescence of white material on the leaves of Aleppo pine (*Pinus halepensis* Mill.), when he grew seedlings in solutions containing Carbowax with molecular weights of 400 and 1500, even though the Carbowaxes resulted in injury to the leaves.

13.3.3 Evapotranspiration Rate

Both with and without Cd, the evapotranspiration rate of plants in the wet-PEG treatment was less than that of plants in the no-PEG treatment (Figure 13.3). When all measurement days were averaged together, plants grown without Cd in the wet-PEG and no-PEG treatments had a mean evapotranspiration rate with a standard error of 2.2 ± 0.1 mm/day and 2.5 ± 0.1 mm/day, respectively (Figure 13.3, left); for plants grown with Cd, these values were 2.3 ± 0.1 mm/day and 2.6 ± 0.1 mm/day, respectively (Figure 13.3, right). Therefore, for both plants grown without and with Cd, the evapotranspiration rate was reduced 0.3 mm/day due to the wet-PEG treatment. The ups and downs in the lines in Figure 13.3 represent different environmental conditions. Evapotranspiration rate was higher when the air temperature was higher than when it was lower, and it was higher on sunny days compared to cloudy days.

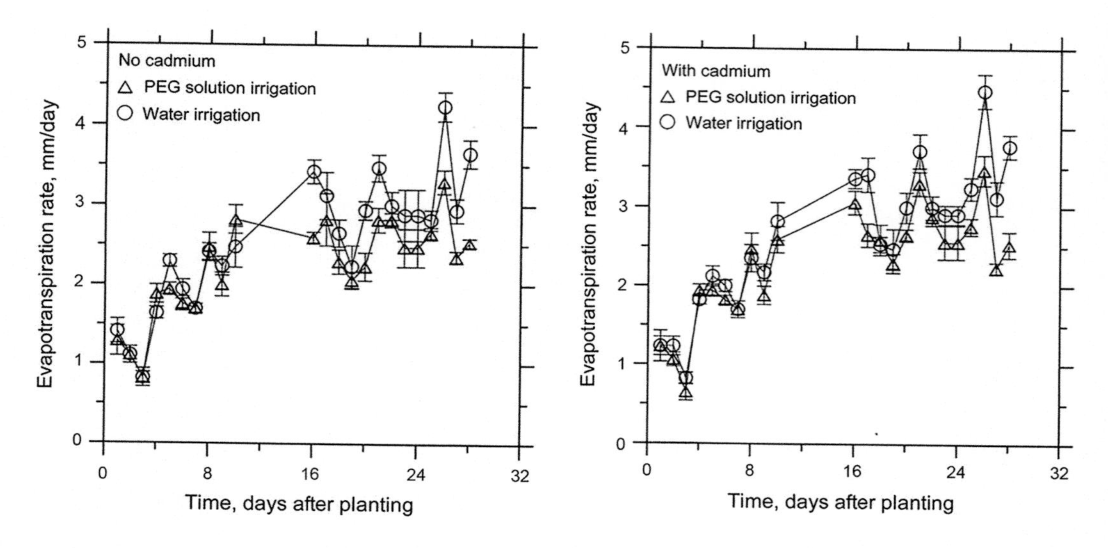

FIGURE 13.3 Evapotranspiration rate of wheat grown with water or PEG irrigations and without Cd (left) or with Cd (right). See text for details. Mean and standard error are shown ($n = 6$).

The reduction in evapotranspiration rate by the presence of the wet PEG may have been due to the clogging of the pores in the soil with the large molecular-weight PEG with associated water molecules. Infiltration was inhibited in the pots receiving the wet PEG. The soil with the wet PEG was saturated for a longer period of time than the soil without the wet PEG. In some pots receiving wet PEG, ponded-water remained on the surface of the pots for 1–2 days after an irrigation. Thus, the soil was anaerobic and the roots lacked oxygen. The decrease in evapotranspiration rate in the treatments with wet PEG reflected the reduced growth in these pots (Figure 13.2).

13.3.4 Stomatal Resistance

The stomatal resistance of plants in the wet-PEG treatment was higher than that of plants in the no-PEG treatment (Figure 13.4). When all measurements days were averaged together, plants

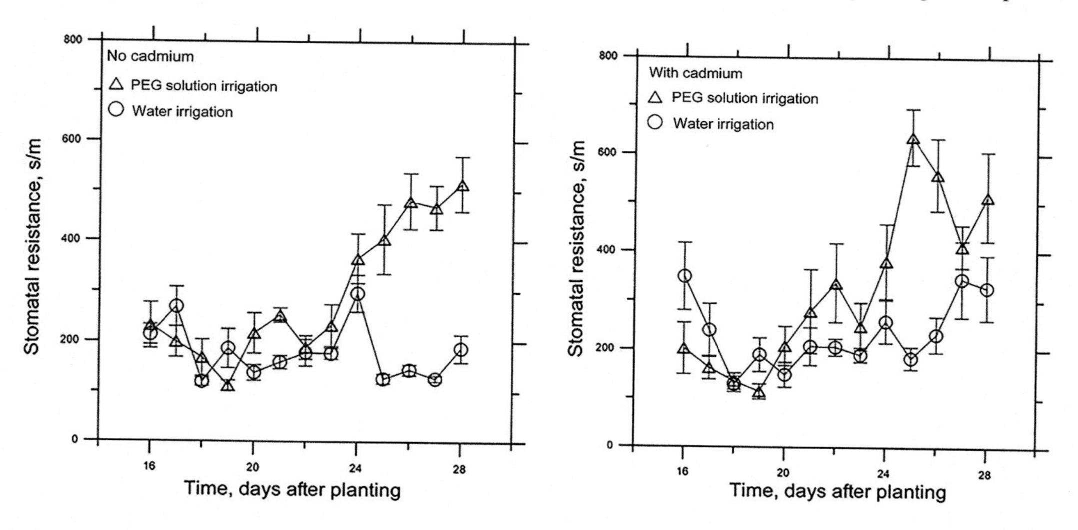

FIGURE 13.4 Stomatal resistance of wheat grown with water or PEG irrigations and without Cd (left) or with Cd (right). See text for details. Mean and standard error are shown ($n = 6$).

grown without Cd in the wet-PEG treatment and the no-PEG treatment had a mean stomatal resistance with standard error of 295 ± 37 s/m and 178 ± 15 s/m, respectively (Figure 13.4, left); for plants grown with Cd, these values were 322 ± 47 s/m and 231 ± 71 s/m, respectively (Figure 13.4, right). Plants grown with Cd and irrigated with PEG had high stomatal resistances at the end of the experiment (>500 s/m). This reflected the fact that they were rapidly senescing.

13.3.5 Fresh and Dry Weights and Cd in Plants

In general, plants in the different treatments had similar fresh and dry weights (Table 13.2). However, Cd concentration in the plants varied with treatment. Plants not irrigated with Cd had similar concentrations of Cd in them. The average concentration of the three treatments without Cd (10.6 mg/kg, no-PEG treatment; 5.7 mg/kg, dry-PEG treatment; and 7.3 mg/kg, wet-PEG treatment) was 7.9 mg/kg. Plants with no PEG in the soil and irrigated with Cd, and plants grown in soil with dry PEG added before planting and irrigated with Cd, had similar concentrations of Cd (129.6 and 130.5 mg/kg, respectively, for an average of 130.0 mg/kg). Thus, the Cd irrigations increased the Cd in the plants by 16 times (130.0 vs. 7.9 mg/kg). However, the Cd concentration in the plants irrigated with wet PEG and Cd was 204.8 mg/kg. The irrigation with wet PEG containing Cd increased the Cd in the plants by 1.5 times compared to the irrigation with water containing Cd (204.8 vs. 130.0 mg/kg).

The early literature concerning the use of polyethylene glycols as osmotic agents in hydroponic experiments pointed out the need to purify them to remove toxic trace elements, such as aluminum (Lagerwerff and Eagle 1961; Lagerwerff et al. 1961). Toxic effects of polyethylene glycols on the physiology of plants in these early studies, aside from their effects as osmotic agents, were observed – e.g., damage to roots, stems, and leaves of pine (*Pinus halepensis* Mill.) seedlings (Leshem 1966); inhibition of the rate of elongation of root hairs of redtop grass (*Agrostis alba* L.) seedlings (Jackson 1962); and death of maize (*Zea mays* L.) leaves (Lawlor 1970), but the cause was not known. Large molecular-size polyethylene glycols, like PEG 20,000, required long dialysis to remove toxic aluminum (Ruf et al. 1963). The polyethylene glycol used in this experiment (PEG 8000) was highly purified. Its cadmium concentration was below detection limits (<0.005 mg/kg). Because the polyethylene glycols have high affinities for trace elements, as these early studies showed, it is not surprising that in this experiment, when PEG 8000 was used in soil with a toxic trace element (Cd), the plants would show physiological toxicities (reduced height and evapotranspiration rate and increased stomatal resistance). While there was no white efflorescence on the leaves, indicating that the PEG was not being taken up by the plants, it still might have gotten in the transpiration stream. Lawlor (1970) suggested that PEGs of higher molecular weights could

TABLE 13.2
Fresh Weight, Dry Weight, and Cd Concentration of Wheat Plants Grown with Water or PEG Irrigations and with and without Cd

Treatment	Fresh Weight, g	Dry Weight, g	Cd in Leaves, mg/kg
No PEG, Water irrigation	0.983 ± 0.105	0.133 ± 0.014	10.6 ± 4.3
No PEG, Cd in water irrigation	1.143 ± 0.048	0.156 ± 0.009	129.6 ± 9.0
Dry PEG, Water irrigation	0.919 ± 0.119	0.127 ± 0.017	5.7 ± 3.9
Dry PEG, Cd in water irrigation	1.031 ± 0.084	0.140 ± 0.010	130.5 ± 24.4
Wet PEG irrigation, No Cd	0.924 ± 0.115	0.119 ± 0.012	7.3 ± 2.2
Wet PEG irrigation, With Cd	0.811 ± 0.067	0.122 ± 0.010	204.8 ± 33.7

Note: The dry PEG treatment had dry PEG mixed into the soil before planting. See text for details. Mean and standard error are shown ($n = 6$).

enter the transpiration pathway and block it. Cadmium attached to PEG could be taken up as a plant transpires, thereby facilitating the movement of Cd to the leaves. Even without the presence of particulate plastics in irrigation water, Cd increases stomatal resistance at concentrations that are permitted in irrigation water (0.05 μg/mL) (Kirkham 1978).

The results suggest that if particulate plastics are in water used to irrigate plants on soil contaminated with Cd, the Cd will be more readily taken up than if the particulate plastics are not present. The experiment was terminated before the wheat grew to maturity and produced grain used for bread. Future research is needed to see if particulate plastics in irrigation water increase the potential for bread to be contaminated with Cd. Already, dietary intake of plant-derived food represents a major fraction of health-threatening human exposure to cadmium (Clemens and Ma 2016).

13.3.6 Soil Analyses

Table 13.3 shows the chemical characteristics of the soil under the different PEG and Cd treatments after the roots were removed from the soil at harvest. Because the soil in three pots (one cell) was combined for the soil measurements, each value in Table 13.3 is the average of two measurements. The pH was similar among the different treatments. The total N tended to be highest in the soil with no PEG. Total C was highest in the soil that received the PEG irrigations, because carbon in the PEG was being added to the soil. The CEC for all treatments was high (>50 m.e./100 g), which reflected the large amount of organic matter in the soil (51.1%, Table 13.1). Buckman and Brady (1969, p. 99) gave the CEC of sandy loams, which ranged from 2.3 to 17.1 m.e./100 g for the soils that they surveyed. In this experiment, the CEC varied with treatment, and it was lowest in the soil that received the wet-PEG irrigations. The exchangeable concentrations of K, Ca, Mg, and Na also were lowest in the soil that received the wet-PEG irrigations, which reflects the fact that the CEC is calculated by summing the exchangeable concentrations of K, Ca, Mg, and Na (Warncke and Brown 1998).

Lagerwerff et al. (1961) postulated that the large Carbowax 20,000 molecules in their study participate in a process of solvation, such that aqueous solutions of Carbowax do not observe van't Hoff's law (See Kirkham 2014, p. 290, for van't Hoff's law.). Solvation is an interaction of a solute with the solvent (Sienko and Plane 1957, p. 196). A large molecule (for example, a PEG molecule) with its associated cluster of water molecules is a solvated molecule and is believed to be in solution as a unit (Sienko and Plane 1957, p. 197). The large solvated molecules may cover the soil surface and reduce the exposure of the exchangeable cations, thereby perhaps reducing their exchangeable concentrations.

The total and extractable concentrations of Cd in the soil that received Cd irrigations were high (>760 mg/kg) (Table 13.3). For three treatments (no PEG, water irrigation; no PEG, Cd irrigation; and dry PEG added to the soil before planting with Cd irrigation), the extractable concentrations of Cd were higher than the total concentrations.

The soil that received no Cd irrigations had high concentrations of total and extractable concentrations of Cd. Based on the literature survey by Kirkham (2008), under non-contaminated conditions, the average total concentration of Cd in soil is 0.5 mg/kg with a range of 0.01–2.53 mg/kg, and the average extractable concentration of Cd in soil is 0.06 mg/kg with a range of 0.01 to 0.5 mg/kg. The potting soil had normal concentrations of total and extractable Cd (0.39 and 0.30 mg/kg, respectively) (Table 13.1). Under non-contaminated conditions, the maximum concentration of Cd in plants is considered to be 0.20 mg/kg (Liphadzi and Kirkham 2006), but a normal range of 0.1–5 mg/kg Cd has been recorded for leaves of maize (Kirkham 1975). Both the plants (Table 13.2) and the soil (Table 13.3) grown with no Cd irrigations had higher than normal amounts of Cd. Because the sieved potting soil had normal amounts of Cd, the Cd in the experimental plants and soil not treated with Cd must have come from the bark that was sieved out of the potting soil before analysis.

TABLE 13.3
Chemical Characteristics of Soil after Harvest of Wheat Grown with Water or Polyethylene Glycol (PEG) Irrigations and with and without Cd

Treatment	pH	CEC, me/100 g	Total N, %	Total C, %	K, mg/kg	Ca, mg/kg	Mg, mg/kg	Na, mg/kg	Total Cd, mg/kg	Ext. Cd, mg/kg
No PEG, Water irrigation	6.1 ± 0.1	62.9 ± 0.7	0.70 ± 0.01	28.6 ± 0.4	2567 ± 11	6612 ± 77	2018 ± 37	816 ± 2	12.2 ± 0.0	15.1 ± 0.1
No PEG, Cd in water irrigation	5.9 ± 0.1	60.7 ± 0.01	0.73 ± 0.03	28.8 ± 0.4	2412 ± 28	6492 ± 85	1988 ± 7	788 ± 8	893.8 ± 36.0	938.9 ± 1.4
Dry PEG, Water irrigation	6.3 ± 0	62.6 ± 2.6	0.63 ± 0.05	28.7 ± 2.4	2574 ± 235	6707 ± 292	2031 ± 105	833 ± 54	7.0 ± 4.8	2.7 ± 1.9
Dry PEG, Cd in water irrigation	6.0 ± 0.1	58.6 ± 0.9	0.66 ± 0.02	28.7 ± 1.4	2432 ± 19	6451 ± 69	1919 ± 25	777 ± 4	763.2 ± 61.1	802.3 ± 73.1
Wet PEG irrigation, No Cd	6.0 ± 0.1	55.5 ± 1.5	0.67 ± 0.04	31.3 ± 0.1	2253 ± 101	6022 ± 36	1809 ± 21	652 ± 20	8.2 ± 3.2	6.1 ± 2.8
Wet PEG irrigation, With Cd	6.0 ± 0.1	51.3 ± 0.2	0.61 ± 0.02	31.3 ± 1.3	2086 ± 29	5613 ± 86	1667 ± 30	654 ± 20	898.1 ± 9.4	888.8 ± 25.0

Note: The dry PEG treatment had dry PEG mixed into the soil before planting. See text for details. Mean and standard error are shown ($n = 2$).

13.4 CONCLUSIONS

Polyethylene glycol irrigated onto soil reduced the plant growth and evapotranspiration rate of wheat and increased its stomatal resistance. The observations of Bradney et al. (2019) were confirmed. That is, polyethylene glycol was a potent vector for the transport of the toxic trace element, Cd, to leaves. When soil was irrigated with a PEG solution containing Cd, wheat plants had 1.5 times more Cd in the shoots (204.8 mg/kg) than plants irrigated with Cd without PEG (130.0 mg/kg). The results suggest that if particulate plastics are in water used to irrigate plants on soil contaminated with Cd, the Cd will be more readily taken up than if the particulate plastics are not present.

ACKNOWLEDGMENTS

I thank Professor V.E. Kirkham of the University of Pennsylvania for suggesting the experiment. I thank Jane E. Lingenfelser, head of Crop Performance Testing at Kansas State University, for providing the wheat seeds. I thank Kathy Lowe and Jacob Thomas of the Soil Testing Laboratory at Kansas State University for plant and soil analyses, respectively. The research was funded by the State of Kansas Organized Research Grant No. 381041, OR-19. This is contribution no. 20-075-B from the Kansas Agricultural Experiment Station, Manhattan, Kansas 66506.

REFERENCES

Alpkem Corporation. 1986a. RFA™ Methodology No. A303-S021. Ammonia nitrogen. Clackamas, OR 97015.

Alpkem Corporation. 1986b. RFA™ Methodology No. A303-S170. Nitrate + nitrite nitrogen. Clackamas, OR 97015.

Andrady, A.L. 2011. Microplastics in the marine environment. *Marine Pollution Bull.* 62:1596–1605. doi:10.1016/j.marpolbul.2011.05.030.

Bradney, L., H. Wijesekara, K.N. Palansooriya, N. Obadamudalige, N.S. Bolan, Y.S. Ok, J. Rinklebe, K.-H. Kim, and M.B. Kirkham. 2019. Particulate plastics as a vector for toxic trace-element uptake by aquatic and terrestrial organisms and human health risk. *Environ. Int.* 131:104937 (Open Access) (no page numbers) https://doi.org/10.1016/j.envint.2019.104937.

Bradney, L., H. Wijesekara, N.S. Bolan, and M.B. Kirkham. 2020. Sources of particulate plastics in terrestrial ecosystems. In: N. Bolan, M.B. Kirkham, C. Halsband, and Y.S. Ok (Editors). *Particulate Plastics in the Terrestrial and Aquatic Environments.* CRC Press, Taylor & Francis Group, Boca Raton, FL.

Briggle, L.W., and B.C. Curtis. 1987. Wheat worldwide, pp. 1–32. In: E.G. Heyne (Editor). *Wheat and Wheat Improvement.* 2nd ed. Amer. Soc. Agronomy, Crop Sci. Soc. Amer., and Soil Sci. Soc. Amer., Madison, WI.

Browne, M.A., P. Crump, S.J. Niven, E.L. Teuten, A. Tonkin, T. Galloway, and R.D. Thompson. 2011. Accumulations of microplastic on shorelines worldwide: Sources and sinks. *Environ. Sci. Technol.* 45:9175–9179. https://doi.org/10.1021/es201811s.

Buckman, H.O., and N.C. Brady. 1969. *The Nature and Properties of Soils.* 7th ed. Macmillan, New York. 653 p.

Carr, S.A., J. Liu, and A.G. Tesoro. 2016. Transport and fate of microplastic particles in wastewater treatments plants. *Water Res.* 91:174–182. doi:10.1016/j.watres.2016.01.002.

Chen, F.M. 2020. Polyacrylamide (PAM) as a source of particulate plastics in the terrestrial environment. In: N. Bolan, M.B. Kirkham, C. Halsband, and Y.S. Ok (Editors). *Particulate Plastics in the Terrestrial and Aquatic Environments.* CRC Press, Taylor & Francis Group, Boca Raton, FL.

Clemens, S., and J.F. Ma. 2016. Toxic heavy metal and metalloid accumulation in crop plants and foods. *Annu. Rev. Plant Biol.* 67:489–512.

Combs, S.M., and M.V. Nathan. 1998. Soil organic matter, pp. 53–58. In: Brown, J.R. (Editor). *Recommended Chemical Soil Test Procedures for the North Central Region.* SB 1001. Missouri Agricultural Experiment Station, Columbia, MO.

Comeau, A., L. Nodichao, J. Collin, M. Baum, J. Samsatly, D. Hamidou, F. Langevin, A. Laroche, and E. Picard. 2010. New approaches for the study of osmotic stress induced by polyethylene glycol (PEG) in cereal species. *Cereal Res. Commun.* 38:471–481.

Cruje, C., and D.B. Chithrani. 2014. Polyethylene glycol density and length affects [sic] nanoparticle uptake by cancer cells. *J. Nanomedicine Res.* 1(1):00006. doi:10.15406/jnmr:2014.01.00006.

Das, V.S.R., and A.S. Raghavendra. 1979. Antitranspirants for improvement of water use efficiency of crops. *Outlook Agr.* 10(2):92–98.

Duis, K., and A. Coors. 2016. Microplastics in the aquatic and terrestrial environment: Sources (with a specific focus on personal care products), fate and effects. *Environ. Sci. Europe* 28(2): Unpaged (25 pages). doi: 10.1186/s12302-015-0069-y.

Faber, B., J. Downer, and L. Yates. 1994. Portable soil meters. *Amer. Nurseryman* 179(2):93–94.

Frank, K., D. Beegle, and J. Denning. 1998. Phosphorus, pp. 21–26. In: Brown, J.R. (Editor). *Recommended Chemical Soil Test Procedures for the North Central Region. SB 1001.* Missouri Agricultural Experiment Station, Columbia, MO.

Gee, G.W., and J.W. Bauder. 1986. Particle-size analysis, pp. 383–411. In: Klute, A. (Editor). *Methods of Soil Analysis. Part 1. Physical and Mineralogical Methods.* 2nd ed. American Society of Agronomy and Soil Science Society of America, Madison, WI.

Gelderman, B.H., and A.P. Mallarino. 1998. Soil sample preparation, pp. 5–6. In: Brown, J.R. (Editor). *Recommended Chemical Soil Test Procedures for the North Central Region.* SB 1001. Missouri Agricultural Experiment Station, Columbia, MO.

GESAMP. 2015. Sources, fate and effects of microplastics in the marine environment: A global assessment. Peter Kershaw (Editor). IMO/FAO/UNESCO-IOC/UNIDO/WMO/IAEA/UN/UNEP/UNDP Joint Group of Experts on the Scientific Aspects of Marine Environmental Protection. *Rep. Stud.* GESAMP No. 90, 96 p.

Hidalgo-Ruz, V., L. Gutow, R.C. Thompson, and M. Thiel. 2012. Microplastics in the marine environment: A review of the methods used for identification and quantification. *Environ. Sci. Technol.* 46:3060–3075. doi:10.1021/es2031505.

Jackson, W.T. 1962. Use of Carbowaxes (polyethylene glycols) as osmotic agents. *Plant Physiol.* 37:513–519.

Janes, B.E. 1961. Use of polyethylene glycol as a solvent to increase the osmotic pressure of nutrient solutions in studies on the physiology of water in plants. *Plant Physiol.* 36 (supple.): xxiv–xxv.

Jarvis, P.G., and M.S. Jarvis. 1965. The water relations of tree seedlings. V. Growth and root respiration in relation to osmotic potential of the root medium, pp. 167–182. In: Slavík, B. (Editor). *Water Stress in Plants.* Dr. W. Junk Publishers, The Hague.

Kaufman, M.R., and A.N. Eckard. 1971. Evaluation of water stress control with polyethylene glycols by analysis of guttation. *Plant Physiol.* 47:453–456.

Kirkham, D., and J. Runkles. 1952. Evaluation of new soil conditioners. (Iowa) *State Hort. Soc.* 87:41–46.

Kirkham, M.B. 1975. Trace elements in corn grown on long-term sludge disposal site. *Environ. Sci. Technol.* 9:765–768.

Kirkham, M.B. 1978. Water relations of cadmium-treated plants. *J. Environ. Quality* 7:334–336.

Kirkham, M.B. 2000. EDTA-facilitated phytoremediation of soil with heavy metals from sewage sludge. *Int. J. Phytoremediation* 2:159–172.

Kirkham, M.B. 2008. Trace elements, pp. 786–790. In: Chesworth, W. (Editor). *Encyclopedia of Soil Science. Encyclopedia of Earth Sciences Series.* Springer, Dordrecht, the Netherlands.

Kirkham, M.B. 2014. *Principles of Soil and Plant Water Relations.* 2nd ed. Elsevier Academic Press, Amsterdam, the Netherlands. 579 p.

Lagerwerff, J.V., and G. Ogata. 1960. Plant growth as a function of interacting activities of water and ions under saline conditions. *Trans. Seventh Int. Congr. Soil Sci.* 3:475–480.

Lagerwerff, J.V., and H.E. Eagle. 1961. Osmotic & specific effects of excess salts on beans. *Plant Physiol.* 36:472–477.

Lagerwerff, J.V., G. Ogata, and H.E. Eagle. 1961. Control of osmotic pressure of culture solutions with polyethylene glycol. *Science* 133:1486–1987.

Lawlor, D.W. 1970. Absorption of polyethylene glycols by plants and their effects on plant growth. *New Phytol.* 69:501–513.

Leshem, B. 1966. Toxic effects of Carbowaxes (polyethylene glycols) on *Pinus halepensis* Mill. seedlings. *Plant Soil* 24:322–324.

Lindsay, W.L., and W.A. Norvell. 1978. Development of a DTPA soil test for zinc, iron, manganese, and copper. *Soil Sci. Soc. Amer. J.* 42:421–428.

Liphadzi, M.S., and M.B. Kirkham. 2006. Physiological effects of heavy metals on plant growth and function, pp. 243–269. In: Huang, B. (Ed.), *Plant-Environment Interactions.* 3rd ed. CRC, Taylor & Francis Group, Boca Raton, FL.

Lusher, A. 2015. Microplastics in the marine environment: Distribution, interactions and effects, pp. 245–307. In: Bergmann, M., L. Gatow, and M. Klages (Editors). *Marine Anthropogenic Litter*, Springer, Cham, Switzerland.

Macklon, A.E.S., and P.E. Weatherley. 1965. Controlled environment studies of the nature and origins of water deficits in plants. *New Phytol.* 64:414–427.

Michel, B.E. 1970. Carbowax 6000 compared with mannitol as a suppressant of cucumber hypocotyl elongation. *Plant Physiol.* 45:507–509.

Michel, B.E. 1971. Further comparisons between Carbowax 6000 and mannitol as suppressants of cucumber hypocotyl elongation. *Plant Physiol.* 48:513–516.

Michel, B.E., and M.R. Kaufmann. 1973. The osmotic potential of polyethylene glycol 6000. *Plant Physiol.* 51:914–916.

Money, N.P. 1989. Osmotic pressure of aqueous polyethylene glycols: Relationship between molecular weight and vapor pressure deficit. *Plant Physiol.* 91:766–769.

Murphy, F., C. Ewins, F. Carbonnier, and B. Quinn. 2016. Wastewater treatment works (WwTW) as a source of microplastics in the aquatic environment. *Environ. Sci. Technol.* 50:5800–5808.

New York State Office of the Attorney General. 2015. *Discharging Microbeads to Our Waters: An Examination of Wastewater Treatment Plants in New York.* New York State Office of the Attorney General, Albany, NY. 11 pages.

Ranjbarfordoei, A., R. Samson, P. van Damme, and R. Lemeur. 2000. Effects of drought stress induced by polyethylene glycol on pigment content and photosynthetic gas exchange of *Pistacia khinjuk* and *P. mutica*. *Photosynthetica* 38:443–447.

Resnik, M.E. 1970. Effect of mannitol and polyethylene glycol on phosphorus uptake by maize plants. *Ann Bot.* 34:497–504.

Ruf, R.H., Jr., R.E. Eckert, Jr., and R.O. Gifford. 1963. Osmotic adjustment of cell sap to increases in root medium osmotic stress. *Soil Sci.* 96:326–330.

Sienko, J.J., and R.A. Plane. 1957. *Chemistry.* McGraw-Hill, New York. 621 p.

Smucker, A.J.M., B.C. Levene, and M. Ngouajio. 2018. Increasing vegetable production on transformed sand to retain twice the soil water holding capacity in plant root zone. *J. Hort.* 5(4): Unpaged (7 pages). doi: 10.4172/2376-0354.1000246.

Soil Science Society of America. 2008. *Glossary of Soil Science Terms* 2008. Soil Science Society of America, Madison, WI. 88 p.

Teuten, E.L., S.J. Rowland, T.S. Galloway, and R.C. Thompson. 2007. Potential for plastics to transport hydrophobic contaminants. *Environ. Sci. Technol.* 41:7759–7764.

Thompson, R.C. 2015. Microplastics in the marine environment: Sources, consequences and solutions, pp. 185–200. In: Bergmann, M., L. Gatow, and M. Klages (Editors). *Marine Anthropogenic Litter*, Springer, Cham, Switzerland.

United States Department of Agriculture. 2019. Kansas Wheat Varieties. Released March 2019. *United States Department of Agriculture, National Agricultural Statistics Service.* National Operations Center, St. Louis, MO. 4 pages.

Van Cauwenberghe, L., L. Devriese, F. Galgani, J. Robbens, and C.R. Janssen. 2015. Microplastics in sediments: A review of techniques, occurrence and effects. *Marine Environ. Res.* 111:5–17.

Van Wezel, A., I. Carusm, and S.A.E. Kools. 2016. Release of primary microplastics from consumer products to wastewater in the Netherlands. *Environ. Toxicol. Chem.* 35:1627–1631.

Wahla, I.H., and M.B. Kirkham. 2008. Heavy metal displacement in salt-water-irrigated soil during phytoremediation. *Environ. Pollution* 155:271–283.

Warncke, D., and J.R. Brown 1998. Potassium and other basic cations, pp. 31–33. In: Brown, J.R. (Editor). *Recommended Chemical Soil Test Procedures for the North Central Region.* SB 1001. Missouri Agricultural Experiment Station, Columbia, MO.

Watson, M.E., and J.R. Brown. 1998. pH and lime requirement, pp. 13–16. In: Brown, J.R. (Editor). *Recommended Chemical Soil Test Procedures for the North Central Region.* SB 1001. Missouri Agricultural Experiment Station, Columbia, MO.

Webster's New World Dictionary of the American Language, College Edition. 1959. World Publishing Co., Cleveland and New York. 1724 p.

Whitney, D.A. 1998. Soil salinity, pp. 59–60. In: Brown, J.R. (Editor). *Recommended Chemical Soil Test Procedures for the North Central Region.* SB 1001. Missouri Agricultural Experiment Station, Columbia, MO.

Wijesekara, H., N.S. Bolan, L. Bradney, N. Obadamudalige, B. Seshadri, A. Kunhikrishnan, R. Dharmarajan, Y.S. Ok, J. Rinklebe, M.B. Kirkham, and M. Vithanage. 2018. Trace element dynamics of biosolids-derived microbeads. *Chemosphere* 199:331–339.

Wright, S.L., R.C. Thompson, and T.S. Galloway. 2013. The physical impacts of microplastics on marine organisms: A review. *Environ. Pollution* 178:483–492.

Zhang, F., J. Yu, C.R. Johnston, Y. Wang, K. Zhu, F. Lu, Z. Zhang, and J. Zoul. 2015. Seed priming with polyethylene glycol induces physiological changes in sorghum (*Sorghum bicolour* L. Moench) seedlings under suboptimal soil moisture environments. *PLoS One* 10(10):e0140620. (no page numbers) https://doi.org/10.1371/journal.pone.0140620.

14 Microplastics as Vectors of Chemicals and Microorganisms in the Environment

Yini Ma, Lin Wang, Ting Wang, Qianqian Chen, and Rong Ji

CONTENTS

14.1 Introduction 209
14.2 Determination of the Organic Pollutants and Microbial Communities Associated with MPs 211
14.2.1 Organic Pollutants on MPs in the Environment 211
14.2.2 Microbial Communities on Biofilms of MPs 213
14.3 MPs as Sites of Biofilm Formation 214
14.3.1 Biofilm Formation on MPs 214
14.3.2 Factors Affecting the Microbial Community of the "Plastisphere" 214
14.3.3 MPs as Sites for Harmful Microorganisms 216
14.4 MPs as Vectors in Biota: Effects on Bioaccumulation 216
14.4.1 Adsorption of Organic Pollutants on MPs 216
14.4.2 MPs as Vectors Affecting the Bioaccumulation of Adsorbed Organic Pollutants 217
14.4.3 Vector Effects of MPs on Adsorbed Organic Pollutants in the Food Web 218
14.4.4 MPs as Vectors for Microorganisms and Gut Pathogens 219
14.5 MPs as Vectors in the Environment: Effects on Transport 219
14.5.1 Transport in Waters 219
14.5.2 Transport in Sediments and Soils 220
14.5.3 Environmental Processes Affecting MPs as Vectors 221
14.6 Conclusions and Outlook 222
Acknowledgments 223
References 223

14.1 INTRODUCTION

Microplastics (MPs), defined as plastic particles <5 mm, are widely present in the open ocean, marine sediments, lakes, and soils, as well as different types of animals, even in salts (Andrady 2011; Horton et al. 2017; Peng et al. 2018; Seth and Shriwastav 2018). A growing body of studies has reported the concentrations of MPs, especially in marine environments (Figure 14.1). According to Eriksen et al. (2014) roughly 5.25 trillion pieces of plastic debris weighing over 250,000 tons in total are floating on the sea surface, with ~80% having originated from land-based sources (Andrady 2011). Other studies have reported concentrations of MPs ranging from 4.8×10^{-6} to 8.6×10^{3} n m^{-3} in the open ocean (Noren 2007; Spear et al. 1995) and from 0.047 to 62,100 n kg^{-1} along the coast and in marine sediments (Liebezeit and Dubaish 2012; Zhao et al. 2015). The highest concentrations of MPs are usually detected in water and sediments adjacent to areas with a large human population or with a high density of plastic-related industries.

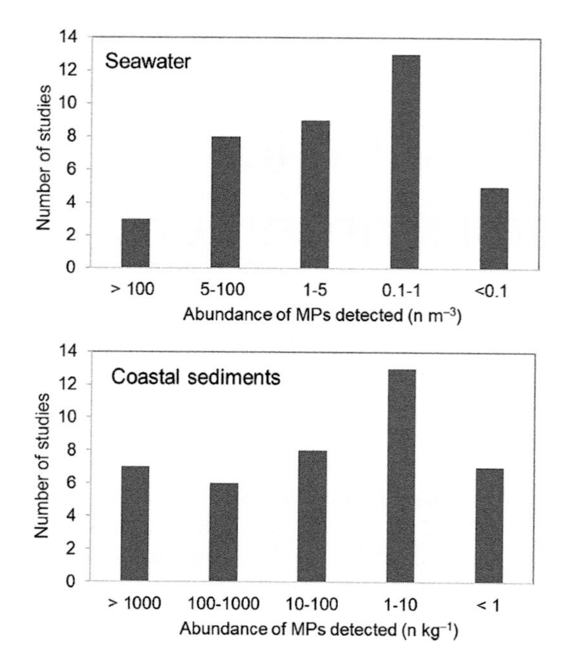

FIGURE 14.1 Summary of literature reports estimating MPs concentrations in seawater and in coastal sediments (*n* stands for number).

MPs in the environment may be primary or secondary, depending on their origin. Primary plastics are produced as micro/nano-sized particles that are added to products, such as those developed for personal consumer care and household use (Gregory 1996). These plastic particles, together with microfibers that are washed from clothes during their laundering, are released into the environment through sewage effluents. Secondary MPs are generated from the fragmentation and weathering of larger plastic debris in the environment, as a result of phototransformation or physical or biological disturbances (Cooper and Corcoran 2010). The physical and chemical properties of secondary MPs are usually distinct from those of primary MPs and include a much higher surface area and more hydrophilic and highly oxidized surfaces.

One of the most important transport strategies of aquatic organisms, especially microbes and algae, is attachment onto floating particulates (Thiel and Gutow 2005). Compared to natural particulate organic matter, MPs have more durable and persistent surfaces, which promote microbial colonization and biofilm formation (Zettler et al. 2013). Consequently, plastic debris represent a new floating substrate for microbial colonization and serves as a vector for phytoplankton, zooplankton, and bacteria, including pathogens (Carson et al. 2013). Studies have shown that by rafting on plastic debris, non-native species can reach new habitats; moreover, the substrates are also conducive to the reproduction of the transported organisms (Oberbeckmann et al. 2015). Because plastic debris trapped in gyres cannot easily escape, buoyant plastic particles tend to accumulate and persist in these areas of the world's oceans.

In addition to the basic polymers used in plastic production, chemical compounds may be added, either by physical mixing or chemical reaction, to improve the performance, functionality, and resistance of the plastic. The most commonly used additives are plasticizers, flame retardants, antioxidants, acid scavengers, light and heat stabilizers, lubricants, pigments, anti-static agents, and slip compounds (Hahladakis et al. 2018). During processes, such as weathering, fragmentation, and pulverization, these additives may be released from the plastic products into the surrounding environment at significant rates following the physical and/or chemical disintegration of the underlying matrices (Chua et al. 2014; Gaylor et al. 2013). Some of the released additives, e.g., phthalates, bisphenol A (BPA), and polybrominated diphenyl ethers (PBDEs), have adverse effects on animals, plants, and human health (Lithner et al. 2009, 2011; Sun et al. 2019; Teuten et al. 2009; Thompson et al. 2009). Thus, MPs in the environment also act as carriers for organic additives, facilitating both their spread and accumulation in the environment as well as in different organisms.

In addition to plastic additives, MPs may adsorb various organic (Frias et al. 2010) and inorganic contaminants (Hodson et al. 2017) from the surrounding environment, owing to the large specific surface area and relatively high hydrophobicity of the particles. Due to the "passive sampler" function of MPs, the concentrations of certain organic pollutants, especially persistent organic pollutants, on the surface of MPs in the ocean may be several orders of magnitude higher than in the surrounding environment (Cole et al. 2011; Thompson et al. 2009). These adsorbed pollutants may be transported around the ocean, even to remote regions, such as the Arctic (Zarfl and Matthies 2010). Following the ingestion of the MPs by animals, the attached chemicals can be desorbed and accumulate in various tissues, potentially causing toxic effects (Besseling et al. 2013; Duis and Coors 2016; Tanaka et al. 2013). The adsorption and desorption equilibrium of organic pollutants on MPs may influence their environmental fate (Ma et al. 2016).

A recently introduced concept is that of the "plastisphere", defined as the ecological community of microbial organisms living on ocean plastic. In this chapter, the term "plastisphere" is used to refer to the more complex system of plastic-attached biofilms and adsorbed pollutants, nutrients, and minerals. Our review summarizes the recent recognition of MPs as: (1) islands of microbial colonization and pollutant accumulation and (2) vectors of microorganisms living on surfaces of the particles and of pollutants within the MPs or adsorbed on their surfaces. The possible interactions between the microbial community and pollutants in the plastisphere are considered herein.

14.2 DETERMINATION OF THE ORGANIC POLLUTANTS AND MICROBIAL COMMUNITIES ASSOCIATED WITH MPs

14.2.1 Organic Pollutants on MPs in the Environment

Many different types of organic pollutants may be associated with MPs collected from seawater, freshwater, sediments, soils, and biota. The first step in their identification is the separation and purification of the MPs from the surrounding matrixes, achieved mainly via floatation and filtration. Polypropylene (PP) and polyethylene (PE) MPs have relatively low densities, whereas MPs made of other compounds, such as polyvinyl chloride (PVC) are much heavier. Thus, MPs of different densities can be floated out using high-density solutions, in which the densities range from 1.05 to 1.85 g cm^{-3} and the different solutes include NaCl (Fries et al. 2013), potassium formate (Zhang et al. 2016), sodium polytungstate (Corcoran et al. 2009) and NaI (Claessens et al. 2013). An important consideration if the goal of the study is the detection of MP-associated organic pollutants is that the flotation solution does not cause the oxidation of the targeted compounds or their dissolution from the MPs. After the MPs have been separated, they can be identified microscopically by Fourier-transform infrared (FTIR) or Raman spectroscopy. However, these techniques are best for MPs >10 μm in size. Methods specifically suited for the separation and detection of MPs of submicroscale or nanoscale size are still needed.

The purified MPs can then be extracted with organic solvents for analysis by gas chromatography–mass spectrometry (GC-MS) or LC-MS/MS (Antunes et al. 2013; Ghosal et al. 2018). However, during the extraction process, plastics may dissolve in the organic solvent such that it can be difficult to determine whether the organic pollutants were adsorbed on the surface or were an additive of the plastic during its production. For example, polystyrene (PS) dissolves easily in several solvents, including dichloromethane and toluene, while PE is relatively resistant to most common organic solvents at room temperature. Thus, when analyzing organic pollutants on MPs, dissolution of the plastics should be avoided by choosing the correct organic extraction solvent.

The organic pollutants detected on MPs obtained from various environmental media (water, coasts, and sediments) and biota (fish and birds) have been the focus of several studies (Table 14.1). In most, the MPs studied consisted of PP, PE, and PS and the associated organic pollutants included polycyclic aromatic hydrocarbons (PAHs), polychlorinated biphenyls (PCBs), PBDEs, 2,2-bis (p-chlorophenyl)-1,1,1-trichloroethane (DDT), hexachlorobenzene (HCB), BPA, and nonylphenols (NPs). The concentration of the detected pollutant ranged from 0.19 to as high as 44,800 ng (g MPs)$^{-1}$ (Table 14.1). The concentration of PAHs was the highest, possibly due to oil spills from the maritime oil drilling industry. High concentrations of pollutants adsorbed on MPs in the oceans reflect the proximity of the sampled waters to cities, port facilities (Antunes et al. 2013), agricultural activities, and coal burning (Zhang et al. 2015). Due to the limitation of separation techniques, determination of MPs and their co-existence with organic pollutants in soil and terrestrial environment are still rare. Moreover, the currently used flotation/filtration separation plus microscopic identification by Fourier-transform infrared or Raman spectroscopy fit best for MPs with sizes larger than 10 μm. It is urgent to have breakthroughs in the separation and detection methods of MPs with

TABLE 14.1
Detection of Microplastics (MPs) and Organic Pollutants in the Environment

Location	Medium		MPs	Pollutant Type and Concentration ($ng \cdot g^{-1}$)		Reference
North Pacific Ocean	Seawater		PP, PE	PAHs	1–846	Rios and Jones (2015)
				PCBs	1–223	
Pacific Ocean	Seawater		PE, PP, PS	Deca-BDE	NA[a]	Ghosal et al. (2018)
China	Reservoir		PS, PP, PE, PC, PVC	Pharmaceutical intermediates	NA	Di and Wang (2018)
Portugal	Sea coast		Plastic resin	PAHs	53–44,800	Antunes et al. (2013)
				PCBs	2–223	
				DDT	0.42–41	
China	Beach		Plastic resin	PAHs	136.3–2384.2	Zhang et al. (2015)
				PCBs	21.5–323.2	
				DDT	1.2–127.0	
Brazil	Beach		PE, PP	PAHs	737–39,763	Fisner et al. (2017)
Malta	Sediment		NA	PAHs	7×10^3– 5.6×10^6	Romeo et al. (2015)
				PCBs	<5	
Switzerland	Lake, beach, sediment	Birds, fish	PE, PP, PS	PCBs, OCPs, PAHs, PBDEs, phthalates, NP, BPA	NA	Faure et al. (2015)
Norway	Sea	Birds	NA	PCBs, DDT, PBDEs		Herzke et al. (2016)
Mexico	Gulf	Whale shark	NA	PCBs	8.42	Fossi et al. (2017)
				DDT	1.31	
				PBDEs	0.29	
				HCBs	0.19	

[a] Not available from the reference cited.

sub-microscale, even nanoscale sizes. Recent studies revealed that pyrolysis-gas chromatography/mass spectrometry (Pyr-GC/MS) may be a possible approach for the simultaneous detection of nanoplastics and coexisting organic pollutants (Claessens et al. 2013).

14.2.2 Microbial Communities on Biofilms of MPs

The hydrophobic surface of plastic debris suspended in the water column stimulates the rapid formation of biofilms, giving rise to an artificial "microbial reef". A global survey showed that up to 50% of floating debris, including plastics and natural organic particles, are colonized by marine organisms (Barnes 2002). Scanning electron microscopy is the technique most commonly used to identify microbial colonizers, but more recent studies using next-generation DNA sequencing have provided a more detailed picture of the microbial life on plastic debris, identifying bacteria, archaea, fungi, and microbial eukaryotes (Table 14.2). Those studies also introduced the

TABLE 14.2
Detection of Microbial Communities on MPs in the Environment

Location	MP Type	Microbial Type	Reference
Mediterranean Sea	PS	Bacteria, archaea	Briand et al. (2012)
Australia—wide coast and ocean	PE, PP, expanded PS	Bacteria, cyanobacteria, fungi, microalgae	Reisser et al. (2014)
Baltic Sea	PE, PS	Eukaryotes, prokaryotes	Kettner et al. (2019)
East China Sea, Estuary	PE, PP, PS	Bacteria	Jiang et al. (2018)
East Lothian, United Kingdom, Forth Estuary	Not available	Bacteria (*Escherichia coli* and *Vibrio* spp.)	Rodrigues et al. (2019)
Guanabara Bay	PE, PP, PET	Bacteria	Silva et al. (2019)
Mediterranean Sea	PE, polyester, PHBV	Bacteria, archaea, microbial eukaryotes	Dussud, Hudec et al. (2018)
Mediterranean Sea	PP, PS	Bacteria, archaea	Dussud, Meistertzheim et al. (2018)
North Sea, United Kingdom	PET	Bacteria, archaea	Oberbeckmann et al. (2014)
North Sea	PET	Bacteria, archaea, eukaryotes, fungi	Oberbeckmann et al. (2016)
North Sea	PE, PS	Bacteria, archaea	Oberbeckmann et al. (2017)
North Sea	PE	Bacteria, archaea	De Tender, Devriese et al. (2017)
Northeast Atlantic	Expanded PS	Bacteria	Summers et al. (2018)
North Atlantic	PE, PP	Bacteria, eukaryotes	Zettler et al. (2013)
North Atlantic sub-tropical gyre	PE	Bacteria, archaea, eukaryotes	Didier et al. (2017)
North Pacific sub-tropical gyre	PE, PP	Bacteria, eukaryotes	Bryant et al. (2016)
Eastern North Pacific gyre	PE, PP, PS	Bacteria, microalgae, radiolarians	Carson et al. (2013)
Spurn Point, Humber Estuary, United Kingdom	LDPE	Bacteria (*Arcobacter* and *Colwellia*)	Harrison et al. (2014)
Southern Baltic Sea, Germany	PS	Bacteria	Kesy et al. (2016)
North and Baltic Sea	PE, PP, PS	Bacteria (pathogenic *Vibrio* spp.)	Kirstein et al. (2016)
Sea off the Swedish west coast	LDPE	Filamentous algae, bryozoans	Karlsson et al. (2018)
Vitória Bay, Brazil	Expanded PS	Bacteria, fungi	Baptista Neto et al. (2019)

term "plastisphere" (De Tender, Schlundt et al. 2017; Zettler et al. 2013). The types of plastics detected were mostly PE, PS, PP, and polyethylene terephthalate (PET), consistent with the field sampling results of marine plastic debris (Browne et al. 2011; Eriksen et al. 2014; Zettler et al. 2013). The first colonizers on the analyzed plastics were mostly degraders of natural organic matter (Ogonowski et al. 2018), with subsequent colonization by a wide range of species, including plastic degraders (Kettner et al. 2019). Although the microorganisms on MPs originated from the surrounding environment, the structure of the microbial community on the studied plastic-associated biofilms largely differed from it. Moreover, differences in microbial community structure depending on the type of MP have also been reported and were attributed to the different surface properties, chemical structure, and degradability of the particles. For example, clear differences in bacterial abundance, diversity, and activity were found between non-biodegradable and biodegradable plastics (Dussud, Hudec et al. 2018).

To date, most studies have focused on plastics in surface waters, especially in the oceans, whereas the "plastisphere" in deep waters and on the surface of sediments has received much less attention. Nonetheless, studies of both are needed to understand microbial colonization on MPs with higher densities. Unlike surface water, the bioturbation in sediments may influence microbial colonization in benthic environments. In addition, benthic organisms feed on MPs, such that the particles are mixed with the indigenous gut microbial community, including during subsequent excretion as part of the digestive process. In terrestrial environments, the distinct chemical structure of the natural organic particles in soils may determine the types of MP colonization. Although less likely to transport microbes compared to waters, sub-micron or nano-sized MPs may also function as vectors, shuttling microbes between the soil system and groundwater, which further alters the ecological aspect of these MP-polluted soils.

14.3 MPs AS SITES OF BIOFILM FORMATION

14.3.1 Biofilm Formation on MPs

The initial attachment of microorganisms to surfaces exposed to an environment containing dissolved organic solutes is often mediated by a "conditioning film" made up of organic and inorganic substances adsorbed to the substrate surface (Bakker et al. 2003; Bradshaw et al. 1997). Artificial surfaces (e.g., acryl, glass, and steel) that are covered by nutrients are rapidly (within hours) colonized by bacterial groups, such as members of the γ-Proteobacteria (initial phase 0–9 hours, e.g., *Pseudomonas, Acinetobacter*), followed by those of the α-Proteobacteria (24–36 hours, e.g., *Loktanella, Methylobacterium*) (Lee et al. 2008). Properties of the substratum, such as hydrophobicity and roughness, play a more important role than biological processes in the early stage of colonization (Lee et al. 2008; Mercier et al. 2017), while the opposite is true during later stages, when species richness and diversity increase with the increasing biological complexity of the biofilm (Dussud, Meistertzheim et al. 2018). Biofilm development on PE-based plastic bags, PET bottles, PVC fragments, or PS fragments in seawater may last several weeks (Briand et al. 2012; Dang and Lovell 2000; Lobelle and Cunliffe 2011; Oberbeckmann et al. 2014). Analogous studies in freshwater and terrestrial systems are scarce (Arias-Andres et al. 2018), except for composting systems, where plastic degrading organisms and enzymes were isolated (Bonhomme et al. 2003; Mercier et al. 2017; Muller et al. 2001; Yoshida et al. 2016).

14.3.2 Factors Affecting the Microbial Community of the "Plastisphere"

As noted above, the microbial community associated with MPs is distinct from that of the surrounding environment (Oberbeckmann et al. 2015; Zettler et al. 2013). The differences have been attributed to environmental conditions and the activities of the background communities, which may lead to substantial differences in biomass build up and in the heterotrophic activities of the MP-associated biofilms. Arias-Andre et al. (2018) found that the total biofilm biomass on MPs was higher in oligo-mesotrophic and dystrophic lakes than in a eutrophic lake. However, only the biofilms in the oligo-mesotrophic lake had a higher functional richness than the ambient water. In the study of Oberbeckmann et al. (2014) the microbial communities on PET fragments sampled from Northern European waters varied according to both season and location, possibly related to the community shifts in the surrounding waters and to environmental factors, such as water temperature and dissolved oxygen concentration. Among the microbial groups identified on the different biofilms were bacteria belonging to Bacteroidetes, Proteobacteria, and Cyanobacteria and eukaryotes belonging to Bacillariophyceae and Phaeophyceae (Oberbeckmann et al. 2014). In a culture-based study, Foulon et al. (2016) found a much higher percentage of colonized particles in Zobell culture medium than in other growth media. The difference may have been related to the availability of nutrients supporting the formation of pili and exopolysaccharide for surface adhesion.

Besides the background microbial community and environmental factors, the chemical composition and physical properties of the plastics may affect the biofilm's structure. Carson et al. (2013) compared the microbial communities that had colonized PE, PS, and PP particles collected from the North Pacific. The highest microbial abundances occurred on PS foams, possibly due to their porous structure and rough surfaces. Another finding of the above-cited study of Oberbeckmann et al. (2014) was that polymer composition not only influenced the abundance of attached microorganisms, but it also shaped biofilm community structure. The authors identified clear differences at the genus level between the community on PET bottles and fragments of other polymers. Their further investigation revealed that this plastic-specific pattern developed only under conditions of low nutrient concentrations and high salinity (Oberbeckmann et al. 2017). Ogonowski et al. (2018) similarly reported that plastic-associated communities were distinctly different from the communities growing on non-plastic substrates. For example, the abundance of *Burkholderiales* was two-fold higher on plastic than on non-plastic substrates, while the latter had a significantly higher proportion of *Actinobacteria* and *Cytophagia*. The variation in community structure was strongly linked to the hydrophobicity of the substrate.

Most of the abovementioned studies on biofilm formation were conducted in samples from the natural environment, i.e., with unknown and uncontrolled conditions. Therefore, it was not possible to determine whether the dynamics of microbial community structure were due solely to the composition of the plastic polymer matrix or influenced by the pollutants and nutrients adsorbed from the marine environment. The bacterial genus *Erythrobacter*, which is able to utilize PAHs as metabolic substrates, was found specifically on MPs, consistent with the widespread presence of PAHs on MPs isolated from different marine environments (Oberbeckmann et al. 2017) (Table 14.2). A better understanding of the factors controlling biofilm formation on MPs will require long-term experiments performed in controlled environments.

In addition to offering microorganisms a habitat for growth and survival, including protection from prey, the ability to utilize the plastic substrate itself may promote the colonization of MPs. This would explain the abovementioned difference in the communities on non-biodegradable vs. biodegradable plastics (Dussud, Hudec et al. 2018). In that study, a more abundant and active microbial community was detected on biodegradable plastics, such as poly(3-hydroxybutyrate-co-3-hydroxyvalerate) (PHBV) than on non-biodegradable PE plastics. Nonetheless, there are bacteria able to degrade otherwise non-degradable plastic polymers, such as mixtures of high- and low-density PE (LDPE and HDPE) (e.g., *Bacillus sphaericus* GC subgroup IV and *Bacillus cereus* subgroup A) and nylon (*Bacillus cereus*, *Bacillus sphaericus*, *Vibrio furnisii*, and *Brevundimonas vesicularis*) to obtain carbon (Sudhakar et al. 2007, 2008). Fungi have also been shown to degrade plastics, for example, PS (Tian et al. 2017) and polyester polyurethane (Russell et al. 2011).

14.3.3 MPs as Sites for Harmful Microorganisms

The dispersal of harmful microorganisms attached to plastic was first suggested by Maso and colleagues (2003). Those authors identified potentially harmful microalgae, such as *Ostreopsis* sp., *Coolia* sp., and *Alexandrium*, on plastic debris collected from the Catalan coast (northwestern Mediterranean), where obvious green-yellow patches on the water surface were seen. *Cyanobacteria* have been widely found on MPs collected in other marine areas (Dussud, Meistertzheim et al. 2018; Oberbeckmann et al. 2014). Zettler et al. (2013) were among the groups of researchers suggesting that plastic particles can act as vectors for the dispersal of human pathogens (*Vibrio* spp.). Using a culture-independent approach, they detected sequences affiliated with *Vibrio* spp. on marine plastic debris from the North Atlantic. Similarly, Kirstein et al. (2016) discovered potentially pathogenic *Vibrio parahaemolyticus* on PE, PP, and PS MPs from the North Sea and Baltic Sea. The genus *Vibrio* harbors several pathogenic species that are an integral part of the marine community and contribute to biofilm formation on various biotic surfaces (Pruzzo et al. 2008; Romalde et al. 2014). However, while many studies have detected pathogenic species on plastic debris, others have not. In a study conducted at intertidal locations around the Yangtze Estuary in China, pathogenic *Vibrio* species were detected on only a few of the MP samples (Jiang et al. 2018). Oberbeckmann et al. (2017) also found no evidence of an enrichment of potential pathogens on PE and PS incubated for 2 weeks in waters representing an environmental gradient ranging from marine (Baltic Sea coastal) to wastewater treatment plant conditions. These inconsistent patterns suggest that pathogenic species are microbial "hitchhikers" and opportunists, but the factors determining their colonization on MPs need further investigation (Keswani et al. 2016).

14.4 MPs AS VECTORS IN BIOTA: EFFECTS ON BIOACCUMULATION

14.4.1 Adsorption of Organic Pollutants on MPs

MPs, especially nanoplastics, adsorb significant amounts of hydrophobic organic contaminants (HOCs) in the environment (Lee et al. 2014; Sudhakar et al. 2008; Velzeboer et al. 2014), further affecting the bioavailability and food-web bioaccumulation of these compounds (Besseling et al. 2013; Rios et al. 2007; Teuten et al. 2009). The adsorption behavior of various types of organic pollutants on MPs has been the focus of several studies. Among the pollutants identified in early work were hydrophobic compounds, such as PAHs, PCBs, and PBDEs, whereas recent studies have reported bisphenols and personal care products (Bakir et al. 2012, 2014a,b; Guo et al. 2012; Lee et al. 2014; Li, Zhang et al. 2018; Ma et al. 2016; Seidensticker et al. 2017, 2018; Velzeboer et al. 2014; Wang and Wang 2018; Wu et al. 2016; Zhang et al. 2018; Zuo et al. 2019). For most pollutants, the adsorption constant is related to their hydrophobicity (log K_{ow}) (Bradford et al. 2002; Dussud, Meistertzheim et al. 2018; Lee et al. 2014; Muller et al. 2018; Razanajatovo et al. 2018; Wu et al. 2019). For example, Lee et al. (2014) analyzed eight PAHs, four hexachlorocyclohexanes, and two chlorinated benzenes using a third-phase partitioning method. The results showed good linear correlations between the measured partition coefficients of MPs and seawater (K_{MPSW}) vs. the K_{ow} values of these compounds. Similarly, for less hydrophobic compounds, such as personal care products, Wu et al. (2016) showed that the adsorption coefficient K_d of carbamazepine, 17α-ethinyl estradiol (EE2), triclosan (TCS), and 4-methylbenzylidene camphor (4MBC) also followed their K_{ow} values, with carbamazepine (log K_{ow} = 2.45) < EE2 (log K_{ow} = 3.67) < TCS (log K_{ow} = 4.76) < 4 MBC (log K_{ow} = 5.1). Thus, when combined with the previously obtained results on adsorption, log K_{ow} may be a good parameter to predict the vector effect of MPs on different types of environmental pollutants. However, an important aspect ignored in most adsorption studies is the different surfaces of MPs present in the environment, especially secondary MPs, compared to the homogeneous MPs purchased for laboratory experiments (Liu, Qian et al. 2019). The aging and degradation of plastic polymers involve polymer radical formation, oxygen addition, hydrogen abstraction, and chain scission or cross-linking (Cai et al. 2018; Gewert et al. 2015). The oxidized functional groups formed on the surface layer of MPs during aging facilitate the adsorption of polar pollutants (Llorca et al. 2018; Muller et al. 2018; Wang et al. 2018; Wang and Wang 2018). Zhang et al. (2018) found that, compared to virgin PS foams, the PS foams collected from beaches had a higher degree of oxidation, as well as larger surface and microporous areas. These PS foams adsorbed twice as much oxytetracycline as did virgin PS foams. Another study on hydrophobic fuel aromatics showed that aging had no effect on the sorption behavior of PPs, while aged PSs sorbed less benzene, toluene, ethyl benzene, and xylene, due to the formation of an oxidized surface layer (Muller et al. 2018). In addition to the physical and chemical aging of MPs, the biological colonization alters their surface properties, either directly, via biofilm formation, or indirectly, via the effects of the extracellular polymeric substances secreted by biofilm-forming microorganisms. Thus, the adsorption of pollutants on MPs in the real environment remains a challenging field of study.

14.4.2 MPs as Vectors Affecting the Bioaccumulation of Adsorbed Organic Pollutants

The vector effects of MPs leading to bioaccumulation depend largely on the interactions between MPs and available pollutants. The size of MPs directly influences their pollutant adsorption capacity and therefore the bioaccumulation of the pollutants. For example, Ma et al. (2016) showed significantly greater bioaccumulation of phenanthrene-derived residues in daphnia exposed to 50-nm than to 10-μm PS particles. The difference was attributed to the stronger adsorption of phenanthrene on smaller PS particles. The size of the MPs also affects the ability of the ingested particles to become translocated into tissues and organs together with the adsorbed pollutants (Devriese et al. 2017). Several studies using fluorescent MPs have claimed that MPs may penetrate cells of the digestive tract and become translocated to organs and enter the circulation. For example, 3.0- and 9.6-μm MPs were translocated within 3 days to the circulatory system of mussels, where they persisted for >48 days (Browne et al. 2008). The translocation of PS microspheres (8–10 μm) to the gills of the shore crab (*Carcinus maenas*) was also reported (Watts et al. 2014). Although the ability of MPs to penetrate the gut wall and become distributed in animal tissues and organs still need further investigation, it is clear that the bioaccumulation and fate of the MPs themselves determine the fate of the carried pollutants.

Conversely, the strong adsorption of organic compounds by MPs may reduce the freely available concentration of pollutants as well as their bioaccumulation. For example, MPs decreased the uptake of PBDEs by the amphipod *Allorchestes compressa* (Chua et al. 2014), and the accumulation of PCBs in sediment-ingesting worms (*Lytechinus variegatus*) was lowered by 76% in the presence vs. the absence of MPs, due to the reduced bioavailability of PCBs mediated by the plastic particles (Beckingham and Ghosh 2017). A study on the biological accumulation of PCBs in *A. marina* showed less bioaccumulation in the presence of 0.74% and 7.4% MPs than in the presence of 0.074% MPs. This difference was attributed to the decrease in the free PCB concentration in the sediment due to the adsorption on MPs, which in turn was related to the MP concentration (Besseling et al. 2013). In addition to the "carrying" and "diluting" effects of MPs, Koelmans et al. (2013) proposed a third route by which MPs reduce the bioaccumulation of pollutants: via the ingestion of clean MPs that bind to pollutants that have already accumulated in the gut of the feeding organism. Mixing of the pollutants with the MPs then facilitates their excretion. However, this "cleansing" effect requires further investigation, especially regarding its occurrence in open environmental systems.

The desorption kinetics of pollutants on MPs are another important factor in the uptake and release of adsorbed HOCs and the subsequent transfer of these chemicals to other organs. The role of gut surfactants in the release of adsorbed HOCs from MPs during gut passage has been investigated in several studies. Bakir et al. (2014a) reported that the amount of desorption of DDT, phenanthrene, perfluorooctanoic acid, and di-2-ethylhexyl phthalate from PVC and PE was up to 30-fold greater under gut conditions than in seawater, especially under gut conditions simulating warm blooded organisms. Hodson et al. (2017) found that the desorption of the heavy metal Zn from MPs in synthetic earthworm gut fluids was higher than in soil, suggesting a greater Zn bioavailability in the gut. In a model-based study assuming an extremely high intake of MPs, Lee et al. (2019) concluded that MP ingestion can increase the total uptake of pentachlorobenzene and hexachlorobenzene, due to the accelerated desorption of these compounds from the particles into the gut fluid. Other than environmental conditions (gut surfactants vs. seawater), the desorption tendency of pollutants depends on the properties of the MPs to which they are bound, including the size of the rubbery sub-fraction (Liu et al. 2018; Zuo et al. 2019), extent of surface oxidation by aging (Liu, Zhang, et al. 2019), and properties of the adsorbed pollutants, such as their K_{ow} and chemical structure (Liu et al. 2018; Razanajatovo et al. 2018; Wardrop et al. 2016).

Despite convincing studies showing that MPs can adsorb organic pollutants, and that the particles can be ingested by organisms, direct evidence for the vector effects of MPs is still lacking. In fact, most studies conducted in natural systems have concluded that food and water are the main sources of pollutants in many organisms, because the abundance of MPs in the environment is much lower than that of other types of pollutant carriers, such as natural organic matter and sediment particles (Beckingham and Ghosh 2017; Koelmans et al. 2016). In addition, laboratory simulations using synthetic gut fluid may not accurately mimic the conditions in organisms. In the study of Hodson et al. (2017), although simulated gut fluid resulted in the increased desorption of Zn, in the earthworm *Lumbricus terrestris*, there was no evidence of a change in Zn accumulation, mortality, or biomass. Bakir et al. (2016) used a modeling approach to examine the transfer of organic pollutants adsorbed on MPs to a benthic invertebrate, a fish, and a seabird. In all three organisms, the intakes from food and water were the main route of exposure to phenanthrene, di-2-ethylhexyl phthalate, and DDT, whereas the contribution of plastics to bioaccumulation was negligible. These results suggested that MP ingestion does not provide a quantitatively important pathway for the transfer of adsorbed chemicals from seawater to biota via the gut. In addition, according to the modeling results of Lee et al. (2019), the predicted steady-state bioaccumulation factor of pollutants decreases with the increasing ingestion of MPs. Therefore, further studies on the vector effects of MPs must be conducted using environmentally relevant concentrations of pollutants and environmentally relevant conditions. Furthermore, it must be taken into account that hydrophobic plastic additives in MPs may be present at a concentration higher than the adsorbed pollutants on the MP surface.

14.4.3 Vector Effects of MPs on Adsorbed Organic Pollutants in the Food Web

Pollutants carried by MPs may also be transferred along the food chain during feeding activities. Laboratory studies have demonstrated the transfer of MPs from mesozooplankton (the copepod *Eurytemora*) to macrozooplankton (mysid shrimp) (Setala et al. 2014) and from mussel (*Mytilus edulis*) to crab (*Carcinus maenas*) (Farrell and Nelson 2013). Nano-sized MPs adhering to phytoplankton were shown to be transferred along an artificial food chain to more than two trophic levels: from algae through zooplankton to fish (Cedervall et al. 2012). Since MPs <10 μm in size have been shown to enter the digestive tract and adhere to animal tissues and organs, the carried pollutants could be transferred to and accumulate in predators. Batel et al. (2016) used fluorescent MPs loaded or not with benzo[*a*]pyrene to study the transfer of MPs and PAHs from *Artemia nauplii* (brine shrimp) to zebrafish. They observed that most of the MP particles passed through the fish gut together with the digestive fluid, with only a small fraction of the MP particles passing the mucus layer and thus attaching to the villi of enterocytes. When the MPs were loaded with benzo[*a*]pyrene, a significant bioaccumulation of benzo[*a*]pyrene was observed in *Artemia nauplii*, and then in the intestinal tract of the zebrafish. However, it is to be noticed that a very significant fluorescence signal was seen throughout the nauplius bodies. Thus, benzo[*a*]pyrene adhering to MP particles is desorbed, and then taken up by *A. nauplii*, such that its transfer to zebrafish is particle-independent. The vector effects of MPs, therefore, occurred largely during the ingestion of the particles by *A. nauplii*, not the later transport to zebrafish. Further studies of the fate of MPs that accumulate in predators and their prey are needed.

Other than laboratory incubation experiments using artificial food chains, the trophic level transfer of MPs and associated pollutants can be evaluated in modeling studies. Diepens and Koelmans (2018) used the model "MICROWEB" to predict the effects of MPs on the trophic transfer and biomagnification of HOCs. The results showed that in the absence of MPs (0% plastic), PCB concentration in the lipids of biota increased along the food chain, whereas in the presence of a high concentration of MPs in the diet, biomagnification was attenuated, consistent with a "cleansing" effect of the particles. However, opposite results were obtained with PAHs, indicating that the vector effects of MPs through different trophic levels are compound specific. The mechanisms underlying this behavior merit further investigation.

14.4.4 MPs as Vectors for Microorganisms and Gut Pathogens

The bioaccumulation of adsorbed toxic substances is not the sole issue of concern, as potentially pathogenic microorganisms attached to MPs may pose a threat to marine and terrestrial food webs and to human health. As mentioned in Section 14.3.3, microbial pathogens were detected on the surface of MPs found in different areas. Frere et al. (2018) also detected *Vibrio-splendidus*-related species harboring putative oyster pathogens on most of the MPs (77%) in samples collected from the Bay of Brest. These findings demonstrated the risks of pathogen transport and therefore of disease emergence. Biofilms that colonize the "plastisphere" may also act as a reservoir for fecal indicator organisms, such as *Escherichia coli*. Rodrigues et al. (2019) sampled beachcast plastic resin pellets at five public bathing beaches, and at each one, found pellets colonized by *E. coli*. When marine animals ingest these MPs, whether incidentally or intentionally, during feeding, the pathogens may enter their intestinal tracts. Although the specific gut conditions, including high concentration of gut surfactants and digestive enzymes, are able to kill most harmful microorganisms, some microbial species, including members of the genus *Vibrio*, are very resistant to the actions of gut fluid (Plante et al. 2008; Plante and Shriver 1998). Since fecal material in marine water represents a significant source of organic matter for benthic species (Sauchyn and Scheibling 2009), the fecal microbiome of the marine benthic organisms likely includes potential pathogens. Thus, knowledge of the colonization and persistence of gut pathogens on MPs in the environment and along the intestinal tract is needed to better understand the environmental risks posed by these organisms, including: (1) the dynamics of fecally contaminated plastic-associated biofilm that form on excreted MP particles after their passage through the guts of pelagic and benthic organisms and (2) the potential consequences of these contaminants for marine pelagic and benthic food webs.

14.5 MPs AS VECTORS IN THE ENVIRONMENT: EFFECTS ON TRANSPORT

14.5.1 Transport in Waters

Most of the plastic debris in the ocean originate on land and is transported by rivers (Horton et al. 2017). Schmidt et al. (2017) analyzed a global compilation of data on plastic debris in the water columns of a wide range of rivers and calculated that global annual input of plastic debris from rivers into the sea are in the range of $0.41–4 \times 10^6$ t. Siegfried et al. (2017) analyzed the composition and quantity of point-source MP fluxes from European rivers to the sea and determined that about two-thirds of the MPs flow into the Mediterranean and Black Seas. Particles with low densities (e.g., LDPE and PS) make up the majority of plastic debris transported by rivers, while particles with densities greater than water (e.g., PVC and PET) gradually settle to the sediments, except in very turbulent waters. These high-density plastics are less likely to be carried over long distances. Cozar et al. (2014) estimated that the current plastic load in surface waters of the open ocean is 10,000–40,000 t. Oceanographic models have shown that particles along the United States Eastern Seaboard can be transported to the North Atlantic sub-tropical gyre in <60 days (Law et al. 2010).

As discussed above, MPs in the water column act as passive samplers, adsorbing organic pollutants, such as PAHs, DDT, and PCBs (Table 14.1). Although the concentrations of these pollutants in marine systems are lower than in the terrestrial environment, given the long lifetime and high adsorption capacity of plastics, the pollutants in seawater may be concentrated on the MP particle surfaces (Rios and Jones 2015). On land, plastic debris often contain pollutants, such as phthalates, flame retardants, and heavy metals. These pollutants are transported along with their parent plastics into the ocean. Based on the large volume of plastics that enters the ocean, the amounts of these pollutants carried by such "plastic currents" may be considerable. Zarfl and Matthies (2010) estimated the flux of PCBs to the Arctic (250 g to 130 kg y^{-1}) and that of the plastic additive PBDE to the Arctic Ocean (25 g to 5.9 kg y^{-1}). Modeling studies have indicated a size selection for the transport of plastic debris from the coast to offshore ocean waters, showing that mesoplastics (approximately >5 mm) are more likely to be trapped in nearshore waters, whereas MPs account for the majority of the plastic debris in offshore ocean waters (Isobe et al. 2014). Mesoplastics that are washed ashore may break down into MPs that may then be washed back out to sea and reach the ocean.

In addition to pollutants, MPs may transport pathogenic microorganisms, which, as discussed above, may be part of the biofilm community that forms on plastic debris (Kirstein et al. 2016; Silva et al. 2019). Oberbeckmann et al. (2017) noted the abundant colonization of bacteria associated with antibiotic resistance genes on MPs from a wastewater treatment plant. Antibiotic resistance in bacteria can be acquired from other bacteria through horizontal gene transfer (Koonin et al. 2001). Moreover, because biofilms protect bacteria from antibiotics, they are a hotspot for horizontal gene transfer among resident bacteria. Consequently, the transport of MPs in aquatic environments may not only accelerate the diffusion of microorganisms, but also promote gene exchange among different species, thereby accelerating the migration of resistant bacteria and the transmission of resistance genes in the environment. However, few studies have investigated the possible shifts in the microbial community during transport and the gene transfer activity on these plastic "islands."

14.5.2 Transport in Sediments and Soils

Due to the size limitation of the pore spaces in sediment and soils, the transport of MPs is much less than in aquatic environments. Large particles (>10 μm) are hardly transported in soil except via the tunnels created by dead plant roots or animal burrowing. However, smaller particles, especially nano-sized spheres, can be transported through saturated and unsaturated sediments, sands, or soils (Bradford et al. 2002; Torkzaban et al. 2008; Zhuang et al. 2005). The heterogeneity, surface coating, and biofilm formation of granular media, the flow rate of water, the degree of soil saturation, and the feeding and burrowing activities of soil animals, such as earthworms are among the many factors that significantly influence the retention vs. the transport of MPs in soil (Alimi et al. 2018; Huerta-Lwanga et al. 2016, 2017; Mitzel et al. 2016; Shani et al. 2008). Further investigations using environmentally relevant media are needed to obtain a comprehensive understanding of the effects of geochemistry and environmental conditions on the transport of MPs in soils and sediments.

The adsorption capacity of MPs for HOCs may be equal to or even higher than that of the surrounding sediments and soils (Liu et al. 2018; Velzeboer et al. 2014). Liu et al. (2018) reported that 100-nm PS MPs significantly enhanced the transport of organic pollutants (PBDEs and NP) in a saturated loamy soil column. In the study of Johari et al. (2010) poly(ethylene glycol)-modified urethane acrylate facilitated the mobility of phenanthrene in porous media and increased its bioavailability. Factors such as the surface properties and composition of MPs may influence the adsorption and desorption equilibrium of organic pollutants on MPs and thus the transport of the associated pollutants. A laboratory study using columns packed with leaf compost showed that phenanthrene transport in the column was significantly higher in the presence of more hydrophobic sulfided PS MPs than in the presence of carboxylated PS MPs (Jaradat et al. 2009). Aging can significantly alter the physicochemical properties of plastics, in turn influencing their fate, transport, and vector activity. Liu et al. (2019) demonstrated that plastic aging induced by UV or O_3 exposure greatly enhanced the mobility and thus the contaminant-mobilizing ability of spherical PS nanoplastics in water-saturated loamy sand.

The ionic strength of the medium affects the stability and aggregation of MPs and thus influences their potential to transport organic pollutants in soils (Sojitra et al. 1995). At low ionic strength, both pyrene and phenanthrene showed an earlier breakthrough in the presence of PS particles; at high ionic strength, the increased retention of PS particles in the granular medium resulted in an increased retardation of the pollutants. In addition, the salinity and ionic strength of the surrounding aqueous environment can affect the water solubility and sorption behavior of organics, thereby modifying their co-transport with MPs.

The vector effects of MPs on the transport of organic pollutants differ depending on the sorption-desorption equilibrium. Liu et al. (2018) found that low concentrations of PS nanoplastics significantly enhanced the transport of non-polar (pyrene) and weakly polar (2,2′,4,4′-tetrabromodiphenyl ether) compounds, but had no effect on the transport of the polar compounds BPA, BPF, and NP. Desorption hysteresis was proposed as the mechanism accounting for the behavior of non-polar or weakly polar pollutants, because these compounds tend to adsorb in the inner matrices of the glassy polymeric structure of PS, where they become physically entrapped. By contrast, polar compounds exhibit surface adsorption and thus do not undergo desorption hysteresis. Aging of the MPs by oxidation changed their surface properties and resulted in the irreversible adsorption of a polar pollutant (NP), in turn, enhancing its transport (Liu, Zhang et al. 2019).

14.5.3 Environmental Processes Affecting MPs as Vectors

During the transport of MPs from their sources to sinks in sub-tropical gyres (Cozar et al., 2014), virgin plastic particles adsorb pollutants (①) (numbers in circles refer to Figure 14.2), form biofilms (②), and facilitate the capture of clays (Johansen et al. 2019). The density of floating MP particles increases with its residence time in the environment. By biofouling, plastic debris become more hydrophilic and extracellular polymeric substances generated by the attached biofilm further increase the stickiness of the debris. Within a few days, MP particles rapidly form aggregates with biogenic particles (③). Aggregate formation may be significantly accelerated by the microbial biofilms on the surfaces of the plastics (Michels et al. 2018). Collisions between MPs in the water column are also affected by turbulence, the differential settlement of MPs, and animal feeding activities (Thornton 2002). In addition, because the density of seawater increases with depth, drifting or slowly sinking plastic debris may remain suspended at various depths. However, over time, most MPs, together with their adsorbed pollutants, slowly sink and reach the sediments (④) (Chen et al. 2019).

Furthermore, biofouled plastic debris undergo rapid defouling (⑦) when submerged, mediated by the oxygen concentration, light, or pH, and thus may return to the water surface (⑧) (Song and Andrady 1991). The microbial composition in the biofilm may change during defouling (⑥). This vertical movement and redistribution of the "plastisphere" and its carried pollutants may have distinct impacts on organisms at different depths of the water column. However, the effects of aggregation, mineral association, sedimentation, and biofouling or defouling on the concentration of pollutants carried by MPs, the microbial community composition of the MPs, and the consequences for aquatic organisms are complex and still poorly understood.

Transport of the MPs from freshwater into the sea also impacts the behavior of MPs (Figure 14.2). The increase of salinity from freshwater to seawater, as well as changes in the dissolved organic matter composition, influence the aggregation of MPs and the transport of associated pollutants

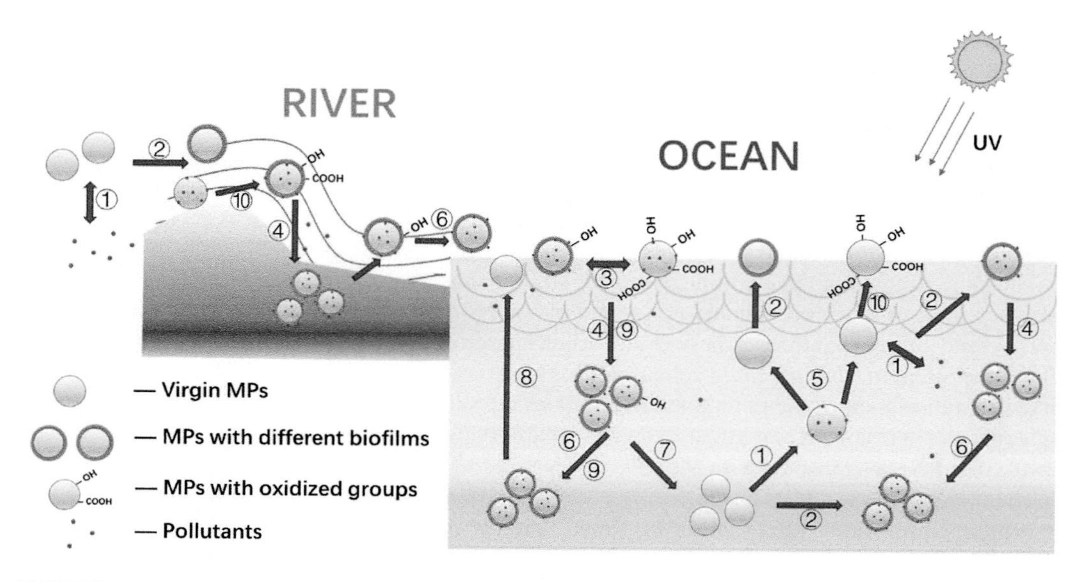

FIGURE 14.2 The environmental processes of MPs as vectors for microorganisms and pollutants in river-ocean systems. Indication of different processes showing in numbers: (1) adsorption of pollutants, (2) biofilm formation, (3) heterogeneous agglomeration, (4) sedimentation, (5) desorption of pollutants, (6) alteration of biofilm composition, (7) defouling, (8) reflotation, (9) consumption of oxidized groups on the surface, and (10) surface oxidation.

(Alimi et al. 2018; Grolimund and Borkovec 2005; Li, Liu, et al. 2018). The microbial community on the MPs also responds to the shift from river to estuarine, and then to marine waters, and specifically to the changes in temperature, salinity, pH, plankton abundance, and surrounding microorganisms (Jiang et al. 2018). The increased salinity and ionic strength may affect the adsorption-desorption equilibrium of pollutants on MPs. For example, Hu et al. (2017) reported a significant increase in the adsorption of lubrication oil on nano-PE and micro-PS with increasing salinity, although in another study, the salinity gradient of riverine to brackish and marine waters did not significantly alter the adsorption and desorption of phenanthrene and DDT on MPs (Bakir et al. 2014b).

In the presence of O_2, MPs exposed to sunlight in the environment undergo photo-oxidation (Figure 14.2) (⑩), which increases the roughness and cracking of the MP surface and generates reactive functional groups. Fourier-transform infrared spectroscopy revealed the presence of ester carbonyl, carboxyl, and ester bonds in beached PE films (Fotopoulou and Karapanagioti 2015). These oxidized groups are preferentially utilized by microbes (⑨) as carbon sources compared to the virgin plastic matrix (Eyheraguibel et al. 2017). Microorganisms on MPs may also affect the degradation of pollutants. In analyses of LDPE slice surfaces prepared from particles obtained from marine and river sediments, microorganisms were found to have enhanced the biotransformation of DDT and some PAH congeners (Wu et al. 2017). An understanding of the mechanisms underlying the interactions between microorganisms and contaminant degraders in the "plastisphere" and the impact of different environmental conditions awaits further investigation.

14.6 CONCLUSIONS AND OUTLOOK

The growing number of studies of the "plastisphere" has revealed the complexity of this habitat, whose adsorbed pollutants, nutrients, and minerals support a broad diversity of biofilm-forming microorganisms. In this chapter, we discussed the methods used in the detection of the microbial community and pollutants associated with MPs in the environment, the interactions of MPs with microorganisms and pollutants, and MPs as vectors for pollutants and microbes in the environment. MPs facilitate the transport, redistribution, bioaccumulation, and spreading not only of pollutants, but also of microbes and their genes, thereby impacting aquatic and terrestrial food webs. The role of MP properties, environmental factors, and chemical structure of pollutants in the vector effects of MPs have been discussed.

The diverse communities of active microbes and the high concentrations of organic pollutants adsorbed on MPs contribute substantially to the potential impact of plastics on global biogeochemical cycles and ecosystem health. Thus, future directions of research need to focus on the following aspects: (1) improved methods for the detection and quantification of sub-micron/nano-sized MPs and organic pollutants in complex matrixes; (2) determination of the ecosystem effects of MPs using particles with environmentally relevant characteristics, i.e., with associated biofilms, adsorbed pollutants, and degraded surfaces instead of virgin spherical MPs; and (3) long-term studies on the vector effects of MPs in open, dynamic systems, taking into account the properties of the plastic matrix, the microbial community composition, pollutant concentrations, and their potential interactions in the "plastisphere."

ACKNOWLEDGMENTS

This work was supported by the National Key Research and Development Program of China (2016YFC1402203) and National Natural Science Foundation of China (21876079).

REFERENCES

Alimi, O. S., J. Farner Budarz, L. M. Hernandez, and N. Tufenkji. 2018. Microplastics and nanoplastics in aquatic environments: Aggregation, deposition, and enhanced contaminant transport. *Environmental Science & Technology* 52 (4):1704–1724.

Andrady, A. L. 2011. Microplastics in the marine environment. *Marine Pollution Bulletin* 62 (8):1596–1605.

Antunes, J. C., J. G. L. Frias, A. C. Micaelo, and P. Sobral. 2013. Resin pellets from beaches of the Portuguese coast and adsorbed persistent organic pollutants. *Estuarine Coastal and Shelf Science* 130 (4):62–69.

Arias-Andres, M., M. T. Kettner, T. Miki, and H. P. Grossart. 2018. Microplastics: New substrates for heterotrophic activity contribute to altering organic matter cycles in aquatic ecosystems. *Science of the Total Environment* 635:1152–1159.

Bakir, A., I. A. O'Connor, S. J. Rowland, A. J. Hendriks, and R. C. Thompson. 2016. Relative importance of microplastics as a pathway for the transfer of hydrophobic organic chemicals to marine life. *Environmental Pollution* 219:56–65.

Bakir, A., S. J. Rowland, and R. C. Thompson. 2012. Competitive sorption of persistent organic pollutants onto microplastics in the marine environment. *Marine Pollution Bulletin* 64 (12):2782–2789.

Bakir, A., S. J. Rowland, and R. C. Thompson. 2014a. Enhanced desorption of persistent organic pollutants from microplastics under simulated physiological conditions. *Environmental Pollution* 185 (4):16–23.

Bakir, A., S. J. Rowland, and R. C. Thompson. 2014b. Transport of persistent organic pollutants by microplastics in estuarine conditions. *Estuarine Coastal and Shelf Science* 140 (3):14–21.

Bakker, D. P., J. W. Klijnstra, H. J. Busscher, and H. C. van der Mei. 2003. The effect of dissolved organic carbon on bacterial adhesion to conditioning films adsorbed on glass from natural seawater collected during different seasons. *Biofouling* 19 (6):391–397.

Baptista Neto, J. A., C. Gaylarde, I. Beech, A. C. Bastos, V. da Silva Quaresma, and D. G. de Carvalho. 2019. Microplastics and attached microorganisms in sediments of the Vitória bay estuarine system in SE Brazil. *Ocean & Coastal Management* 169:247–253.

Barnes, D. K. 2002. Biodiversity: Invasions by marine life on plastic debris. *Nature* 416 (6883):808–809.

Batel, A., F. Linti, M. Scherer, L. Erdinger, and T. Braunbeck. 2016. The transfer of benzo[*a*]pyrene from microplastics to *Artemia nauplii* and further to zebrafish via a trophic food web experiment—CYP1A induction and visual tracking of persistent organic pollutants. *Environmental Toxicology & Chemistry* 35 (7):1656–1666.

Beckingham, B., and U. Ghosh. 2017. Differential bioavailability of polychlorinated biphenyls associated with environmental particles: Microplastic in comparison to wood, coal and biochar. *Environmental Pollution* 220 (Pt A):150–158.

Besseling, E., A. Wegner, E. M. Foekema, M. J. van den Heuvel-Greve, and A. A. Koelmans. 2013. Effects of microplastic on fitness and PCB bioaccumulation by the lugworm *Arenicola marina* (L.). *Environmental Science & Technology* 47 (1):593–600.

Bonhomme, S., A. Cuer, A. M. Delort, J. Lemaire, M. Sancelme, and G. Scott. 2003. Environmental biodegradation of polyethylene. *Polymer Degradation and Stability* 81 (3):441–452.

Bradford, S. A., S. R. Yates, M. Bettahar, and J. Simunek. 2002. Physical factors affecting the transport and fate of colloids in saturated porous media. *Water Resources Research* 38 (12):1327–1338.

Bradshaw, D. J., P. D. Marsh, G. K. Watson, and C. Allison. 1997. Effect of conditioning films on oral microbial biofilm development. *Biofouling* 11 (3):217–226.

Briand, J. F., I. Djeridi, D. Jamet, S. Coupe, C. Bressy, M. Molmeret, B. Le Berre, F. Rimet, A. Bouchez, and Y. Blache. 2012. Pioneer marine biofilms on artificial surfaces including antifouling coatings immersed in two contrasting French Mediterranean coast sites. *Biofouling* 28 (5):453–463.

Browne, M. A., P. Crump, S. J. Niven, E. Teuten, A. Tonkin, T. Galloway, and R. Thompson. 2011. Accumulation of microplastic on shorelines worldwide: Sources and sinks. *Environmental Science & Technology* 45 (21):9175–9179.

Browne, M. A., A. Dissanayake, T. S. Galloway, D. M. Lowe, and R. C. Thompson. 2008. Ingested microscopic plastic translocates to the circulatory system of the mussel, *Mytilus edulis* (L). *Environmental Science & Technology* 42 (13):5026–5031.

Bryant, J. A., T. M. Clemente, D. A. Viviani, A. A. Fong, K. A. Thomas, P. Kemp, D. M. Karl, A. E. White, and E. F. DeLong. 2016. Diversity and activity of communities inhabiting plastic debris in the North Pacific Gyre. *mSystems* 1 (3):e00024-16.

Cai, L., J. Wang, J. Peng, Z. Wu, and X. Tan. 2018. Observation of the degradation of three types of plastic pellets exposed to UV irradiation in three different environments. *Science of the Total Environment* 628–629:740–747.

Carson, H. S., M. S. Nerheim, K. A. Carroll, and M. Eriksen. 2013. The plastic-associated microorganisms of the North Pacific Gyre. *Marine Pollution Bulletin* 75 (1–2):126–132.

Cedervall, T., L. A. Hansson, M. Lard, B. Frohm, and S. Linse. 2012. Food chain transport of nanoparticles affects behaviour and fat metabolism in fish. *PLoS One* 7 (2):e32254.

Chen, X., X. Xiong, X. Jiang, H. Shi, and C. Wu. 2019. Sinking of floating plastic debris caused by biofilm development in a freshwater lake. *Chemosphere* 222:856–864.

Chua, E. M., J. Shimeta, D. Nugegoda, P. D. Morrison, and B. O. Clarke. 2014. Assimilation of polybrominated diphenyl ethers from microplastics by the marine amphipod, *Allorchestes compressa. Environmental Science & Technology* 48 (14):8127–8134.

Claessens, M., L. Van Cauwenberghe, M. B. Vandegehuchte, and C. R. Janssen. 2013. New techniques for the detection of microplastics in sediments and field collected organisms. *Marine Pollution Bulletin* 70 (1–2):227–233.

Cole, M., P. Lindeque, C. Halsband, and T. S. Galloway. 2011. Microplastics as contaminants in the marine environment: A review. *Marine Pollution Bulletin* 62 (12):2588–2597.

Cooper, D. A., and P. L. Corcoran. 2010. Effects of mechanical and chemical processes on the degradation of plastic beach debris on the island of Kauai, Hawaii. *Marine Pollution Bulletin* 60 (5):650–654.

Corcoran, P. L., M. C. Biesinger, and M. Grifi. 2009. Plastics and beaches: A degrading relationship. *Marine Pollution Bulletin* 58 (1):80–84.

Cozar, A., F. Echevarria, J. I. Gonzalez-Gordillo, X. Irigoien, B. Ubeda, S. Hernandez-Leon, A. T. Palma, et al. 2014. Plastic debris in the open ocean. *Proceedings of the National Academy of Sciences of the United States of America* 111 (28):10239–10244.

Dang, H., and C. R. Lovell. 2000. Bacterial primary colonization and early succession on surfaces in marine waters as determined by amplified rRNA gene restriction analysis and sequence analysis of 16S rRNA genes. *Applied and Environmental Microbiology* 66 (2):467–475.

De Tender, C., L. I. Devriese, A. Haegeman, S. Maes, J. Vangeyte, A. Cattrijsse, P. Dawyndt, and T. Ruttink. 2017. Temporal dynamics of bacterial and fungal colonization on plastic debris in the North Sea. *Environmental Science & Technology* 51 (13):7350–7360.

De Tender, C., C. Schlundt, L. I. Devriese, T. J. Mincer, E. R. Zettler, and L. A. Amaral-Zettler. 2017. A review of microscopy and comparative molecular-based methods to characterize "Plastisphere" communities. *Analytical Methods* 9 (14):2132–2143.

Devriese, L. I., B. De Witte, A. D. Vethaak, K. Hostens, and H. A. Leslie. 2017. Bioaccumulation of PCBs from microplastics in Norway lobster (*Nephrops norvegicus*): An experimental study. *Chemosphere* 186:10–16.

Di, M., and J. Wang. 2018. Microplastics in surface waters and sediments of the Three Gorges Reservoir, China. *Science of the Total Environment* 616–617:1620–1627.

Didier, D., M. Anne, and T. H. Alexandra. 2017. Plastics in the North Atlantic garbage patch: A boat-microbe for hitchhikers and plastic degraders. *Science of the Total Environment* 599–600:1222.

Diepens, N. J., and A. A. Koelmans. 2018. Accumulation of plastic debris and associated contaminants in aquatic food webs. *Environmental Science & Technology* 52 (15):8510–8520.

Duis, K., and A. Coors. 2016. Microplastics in the aquatic and terrestrial environment: Sources (with a specific focus on personal care products), fate and effects. *Environmental Sciences Europe* 28 (1):2.

Dussud, C., C. Hudec, M. George, P. Fabre, P. Higgs, S. Bruzaud, A. M. Delort et al. 2018. Colonization of non-biodegradable and biodegradable plastics by marine microorganisms. *Frontiers in Microbiology* 9:1571.

Dussud, C., A. L. Meistertzheim, P. Conan, M. Pujo-Pay, M. George, P. Fabre, J. Coudane, et al. 2018. Evidence of niche partitioning among bacteria living on plastics, organic particles and surrounding seawaters. *Environmental Pollution* 236:807–816.

Eriksen, M., L. C. Lebreton, H. S. Carson, M. Thiel, C. J. Moore, J. C. Borerro, F. Galgani, P. G. Ryan, and J. Reisser. 2014. Plastic pollution in the world's oceans: More than 5 trillion plastic pieces weighing over 250,000 tons afloat at sea. *PLoS One* 9 (12):e111913.

Eyheraguibel, B., M. Traikia, S. Fontanella, M. Sancelme, S. Bonhomme, D. Fromageot, J. Lemaire, G. Lauranson, J. Lacoste, and A. M. Delort. 2017. Characterization of oxidized oligomers from polyethylene films by mass spectrometry and NMR spectroscopy before and after biodegradation by a *Rhodococcus rhodochrous* strain. *Chemosphere* 184:366–374.

Farrell, P., and K. Nelson. 2013. Trophic level transfer of microplastic: *Mytilus edulis* (L.) to *Carcinus maenas* (L.). *Environmental Pollution* 177:1–3.

Faure, F., C. Demars, O. Wieser, M. Kunz, and L. F. de Alencastro. 2015. Plastic pollution in Swiss surface waters: Nature and concentrations, interaction with pollutants. *Environmental Chemistry* 12 (5):582–591.

Fisner, M., A. Majer, S. Taniguchi, M. Bicego, A. Turra, and D. Gorman. 2017. Colour spectrum and resin-type determine the concentration and composition of polycyclic aromatic hydrocarbons (PAHs) in plastic pellets. *Marine Pollution Bulletin* 122 (1–2):323–330.

Fossi, M. C., M. Baini, C. Panti, M. Galli, B. Jimenez, J. Munoz-Arnanz, L. Marsili, M. G. Finoia, and D. Ramirez-Macias. 2017. Are whale sharks exposed to persistent organic pollutants and plastic pollution in the Gulf of California (Mexico)? First ecotoxicological investigation using skin biopsies. *Comparative Biochemistry and Physiology Part C: Toxicology & Pharmacology* 199:48–58.

Fotopoulou, K. N., and H. K. Karapanagioti. 2015. Surface properties of beached plastics. *Environmental Science and Pollution Research International* 22 (14):11022–11032.

Foulon, V., F. Le Roux, C. Lambert, A. Huvet, P. Soudant, and I. Paul-Pont. 2016. Colonization of polystyrene microparticles by *vibrio crassostreae*: Light and electron microscopic investigation. *Environmental Science & Technology* 50 (20):10988–10996.

Frere, L., L. Maignien, M. Chalopin, A. Huvet, E. Rinnert, H. Morrison, S. Kerninon, A.-L. Cassone, C. Lambert, J. Reveillaud, and I. Paul-Pont. 2018. Microplastic bacterial communities in the Bay of Brest: Influence of polymer type and size. *Environmental Pollution* 242:614–625.

Frias, J. P., P. Sobral, and A. M. Ferreira. 2010. Organic pollutants in microplastics from two beaches of the Portuguese coast. *Marine Pollution Bulletin* 60 (11):1988–1992.

Fries, E., J. H. Dekiff, J. Willmeyer, M. T. Nuelle, M. Ebert, and D. Remy. 2013. Identification of polymer types and additives in marine microplastic particles using pyrolysis-GC/MS and scanning electron microscopy. *Environmental Science-Processes & Impacts* 15 (10):1949–1956.

Gaylor, M. O., E. Harvey, and R. C. Hale. 2013. Polybrominated diphenyl ether (PBDE) accumulation by earthworms (*Eisenia fetida*) exposed to biosolids-, polyurethane foam microparticle-, and Penta-BDE-amended soils. *Environmental Science & Technology* 47 (23):13831–13839.

Gewert, B., M. M. Plassmann, and M. MacLeod. 2015. Pathways for degradation of plastic polymers floating in the marine environment. *Environmental Science-Processes & Impacts* 17 (9):1513–1521.

Ghosal, S., M. Chen, J. Wagner, Z. M. Wang, and S. Wall. 2018. Molecular identification of polymers and anthropogenic particles extracted from oceanic water and fish stomach—A Raman micro-spectroscopy study. *Environmental Pollution* 233:1113–1124.

Gregory, M. R. 1996. Plastic "scrubbers" in hand cleansers: A further (and minor) source for marine pollution identified. *Marine Pollution Bulletin* 32 (12):867–871.

Grolimund, D., and M. Borkovec. 2005. Colloid facilitated transport of strongly sorbing contaminants in natural porous media: Mathematical modeling and laboratory column experiments. *Environmental Science & Technology* 39 (17):6378–6386.

Guo, X., X. Wang, X. Zhou, X. Kong, S. Tao, and B. Xing. 2012. Sorption of four hydrophobic organic compounds by three chemically distinct polymers: Role of chemical and physical composition. *Environmental Science & Technology* 46 (13):7252–7259.

Hahladakis, J. N., C. A. Velis, R. Weber, E. Iacovidou, and P. Purnell. 2018. An overview of chemical additives present in plastics: Migration, release, fate and environmental impact during their use, disposal and recycling. *Journal of Hazardous Materials* 344:179–199.

Harrison, J. P., M. Schratzberger, M. Sapp, and A. M. Osborn. 2014. Rapid bacterial colonization of low-density polyethylene microplastics in coastal sediment microcosms. *BMC Microbiology* 14 (1):232.

Herzke, D., T. Anker-Nilssen, T. H. Nost, A. Gotsch, S. Christensen-Dalsgaard, M. Langset, K. Fangel, and A. A. Koelmans. 2016. Negligible impact of ingested microplastics on tissue concentrations of persistent organic pollutants in northern fulmars off coastal Norway. *Environmental Science & Technology* 50 (4):1924–1933.

Hodson, M. E., C. A. Duffus-Hodson, A. Clark, M. T. Prendergast-Miller, and K. L. Thorpe. 2017. Plastic bag derived-microplastics as a vector for metal exposure in terrestrial invertebrates. *Environmental Science & Technology* 51 (8):4714–4721.

Horton, A. A., A. Walton, D. J. Spurgeon, E. Lahive, and C. Svendsen. 2017. Microplastics in freshwater and terrestrial environments: Evaluating the current understanding to identify the knowledge gaps and future research priorities. *Science of the Total Environment* 586:127–141.

Hu, J. Q., S. Z. Yang, L. Guo, X. Xu, T. Yao, and F. Xie. 2017. Microscopic investigation on the adsorption of lubrication oil on microplastics. *Journal of Molecular Liquids* 227:351–355.

Huerta-Lwanga, E., H. Gertsen, H. Gooren, P. Peters, T. Salanki, M. van der Ploeg, E. Besseling, A. A. Koelmans, and V. Geissen. 2016. Microplastics in the terrestrial ecosystem: Implications for *Lumbricus terrestris* (Oligochaeta, Lumbricidae). *Environmental Science & Technology* 50 (5):2685–2691.

Huerta-Lwanga, E., H. Gertsen, H. Gooren, P. Peters, T. Salánki, M. van der Ploeg, E. Besseling, A. A. Koelmans, and V. Geissen. 2017. Incorporation of microplastics from litter into burrows of *Lumbricus terrestris*. *Environmental Pollution* 220:523–531.

Isobe, A., K. Kubo, Y. Tamura, S. Kako, E. Nakashima, and N. Fujii. 2014. Selective transport of microplastics and mesoplastics by drifting in coastal waters. *Marine Pollution Bulletin* 89 (1–2):324–330.

Jaradat, A. Q., K. Fowler, S. J. Grimberg, and T. M. Holsen. 2009. Transport of colloids and associated hydrophobic organic chemicals through a natural media filter. *Journal of Environmental Engineering-ASCE* 135 (1):36–45.

Jiang, P., S. Zhao, L. Zhu, and D. Li. 2018. Microplastic-associated bacterial assemblages in the intertidal zone of the Yangtze Estuary. *Science of the Total Environment* 624:48–54.

Johansen, M. P., T. Cresswell, J. Davis, D. L. Howard, N. R. Howell, and E. Prentice. 2019. Biofilm-enhanced adsorption of strong and weak cations onto different microplastic sample types: Use of spectroscopy, microscopy and radiotracer methods. *Water Research* 158:392–400.

Johari, W. L., P. J. Diamessis, and L. W. Lion. 2010. Mass transfer model of nanoparticle-facilitated contaminant transport in saturated porous media. *Water Research* 44 (4):1028–1037.

Karlsson, T. M., M. Hassellov, and I. Jakubowicz. 2018. Influence of thermooxidative degradation on the in situ fate of polyethylene in temperate coastal waters. *Marine Pollution Bulletin* 135:187–194.

Keswani, A., D. M. Oliver, T. Gutierrez, and R. S. Quilliam. 2016. Microbial hitchhikers on marine plastic debris: Human exposure risks at bathing waters and beach environments. *Marine Environmental Research* 118:10–19.

Kesy, K., S. Oberbeckmann, F. Müller, and M. Labrenz. 2016. Polystyrene influences bacterial assemblages in Arenicola marina-populated aquatic environments *in vitro*. *Environmental Pollution* 219:219–227.

Kettner, M. T., S. Oberbeckmann, M. Labrenz, and H. P. Grossart. 2019. The eukaryotic life on microplastics in brackish ecosystems. *Frontiers in Microbiology* 10:538.

Kirstein, I. V., S. Kirmizi, A. Wichels, A. Garin-Fernandez, R. Erler, M. Loder, and G. Gerdts. 2016. Dangerous hitchhikers? Evidence for potentially pathogenic *Vibrio* spp. on microplastic particles. *Marine Environmental Research* 120:1–8.

Koelmans, A. A., A. Bakir, G. A. Burton, and C. R. Janssen. 2016. Microplastic as a vector for chemicals in the aquatic environment: Critical review and model-supported reinterpretation of empirical studies. *Environmental Science & Technology* 50 (7):3315–3326.

Koelmans, A. A., E. Besseling, A. Wegner, and E. M. Foekema. 2013. Plastic as a carrier of POPs to aquatic organisms: A model analysis. *Environmental Science & Technology* 47 (14):7812–7820.

Koonin, E. V., K. S. Makarova, and L. Aravind. 2001. Horizontal gene transfer in prokaryotes: Quantification and classification. *Annual Review of Microbiology* 55 (1):709–942.

Law, K. L., S. Moret-Ferguson, N. A. Maximenko, G. Proskurowski, E. E. Peacock, J. Hafner, and C. M. Reddy. 2010. Plastic accumulation in the North Atlantic subtropical gyre. *Science* 329 (5996):1185–1188.

Lee, H., H. J. Lee, and J. H. Kwon. 2019. Estimating microplastic-bound intake of hydrophobic organic chemicals by fish using measured desorption rates to artificial gut fluid. *Science of the Total Environment* 651 (Pt 1):162–170.

Lee, H., W. J. Shim, and J. H. Kwon. 2014. Sorption capacity of plastic debris for hydrophobic organic chemicals. *Science of the Total Environment* 470–471 (2):1545–1552.

Lee, J. W., J. H. Nam, Y. H. Kim, K. H. Lee, and D. H. Lee. 2008. Bacterial communities in the initial stage of marine biofilm formation on artificial surfaces. *Journal of Microbiology* 46 (2):174–182.

Li, J., K. Zhang, and H. Zhang. 2018. Adsorption of antibiotics on microplastics. *Environmental Pollution* 237:460–467.

Li, S., H. Liu, R. Gao, A. Abdurahman, J. Dai, and F. Zeng. 2018. Aggregation kinetics of microplastics in aquatic environment: Complex roles of electrolytes, pH, and natural organic matter. *Environmental Pollution* 237:126–132.

Liebezeit, G., and F. Dubaish. 2012. Microplastics in beaches of the East Frisian islands Spiekeroog and Kachelotplate. *Bulletin of Environmental Contamination and Toxicology* 89 (1):213–217.

Lithner, D., J. Damberg, G. Dave, and K. Larsson. 2009. Leachates from plastic consumer products—Screening for toxicity with *Daphnia magna*. *Chemosphere* 74 (9):1195–1200.

Lithner, D., A. Larsson, and G. Dave. 2011. Environmental and health hazard ranking and assessment of plastic polymers based on chemical composition. *Science of the Total Environment* 409 (18):3309–3324.

Liu, J., Y. Ma, D. Zhu, T. Xia, Y. Qi, Y. Yao, X. Guo, R. Ji, and W. Chen. 2018. Polystyrene nanoplastics-enhanced contaminant transport: Role of irreversible adsorption in glassy polymeric domain. *Environmental Science & Technology* 52 (5):2677–2685.

Liu, J., T. Zhang, L. Tian, X. Liu, Z. Qi, Y. Ma, R. Ji, and W. Chen. 2019. Aging significantly affects mobility and contaminant-mobilizing ability of nanoplastics in saturated loamy sand. *Environmental Science & Technology* 53 (10):5805–5815.

Liu, P., L. Qian, H. Wang, X. Zhan, K. Lu, C. Gu, and S. Gao. 2019. New insights into the aging behavior of microplastics accelerated by advanced oxidation processes. *Environmental Science & Technology* 53 (7):3579–3588.

Llorca, M., G. Schirinzi, M. Martinez, D. Barcelo, and M. Farre. 2018. Adsorption of perfluoroalkyl substances on microplastics under environmental conditions. *Environmental Pollution* 235:680–691.

Lobelle, D., and M. Cunliffe. 2011. Early microbial biofilm formation on marine plastic debris. *Marine Pollution Bulletin* 62 (1):197–200.

Ma, Y., A. Huang, S. Cao, F. Sun, L. Wang, H. Guo, and R. Ji. 2016. Effects of nanoplastics and microplastics on toxicity, bioaccumulation, and environmental fate of phenanthrene in fresh water. *Environmental Pollution* 219:166–173.

Maso, M., E. Garces, F. Pages, and J. Camp. 2003. Drifting plastic debris as a potential vector for dispersing harmful algal bloom (HAB) species. *Scientia Marina* 67 (1):107–111.

Mercier, A., K. Gravouil, W. Aucher, S. Brosset-Vincent, L. Kadri, J. Colas, D. Bouchon, and T. Ferreira. 2017. Fate of eight different polymers under uncontrolled composting conditions: Relationships between deterioration, biofilm formation, and the material surface properties. *Environmental Science & Technology* 51 (4):1988–1997.

Michels, J., A. Stippkugel, M. Lenz, K. Wirtz, and A. Engel. 2018. Rapid aggregation of biofilm-covered microplastics with marine biogenic particles. *Proceedings of the Royal Society B: Biological Sciences* 285 (1885):9.

Mitzel, M. R., S. Sand, J. K. Whalen, and N. Tufenkji. 2016. Hydrophobicity of biofilm coatings influences the transport dynamics of polystyrene nanoparticles in biofilm-coated sand. *Water Research* 92:113–120.

Muller, A., R. Becker, U. Dorgerloh, F. G. Simon, and U. Braun. 2018. The effect of polymer aging on the uptake of fuel aromatics and ethers by microplastics. *Environmental Pollution* 240:639–646.

Muller, R. J., I. Kleeberg, and W. D. Deckwer. 2001. Biodegradation of polyesters containing aromatic constituents. *Journal of Biotechnology* 86 (2):87–95.

Noren, F. 2007. *Small Plastic Particles in Coastal Swedish Waters*. Lysekil, Sweden: KIMO Sweden.

Oberbeckmann, S., B. Kreikemeyer, and M. Labrenz. 2017. Environmental factors support the formation of specific bacterial assemblages on microplastics. *Frontiers in Microbiology* 8:2709.

Oberbeckmann, S., M. G. J. Loder, and M. Labrenz. 2015. Marine microplastic- associated biofilms: A review. *Environmental Chemistry* 12 (5):551–562.

Oberbeckmann, S., M. G. Loeder, G. Gerdts, and A. M. Osborn. 2014. Spatial and seasonal variation in diversity and structure of microbial biofilms on marine plastics in Northern European waters. *FEMS Microbiology Ecology* 90 (2):478–492.

Oberbeckmann, S., A. M. Osborn, and M. B. Duhaime. 2016. Microbes on a bottle: Substrate, season and geography influence community composition of microbes colonizing marine plastic debris. *PLoS One* 11 (8):e0159289.

Ogonowski, M., A. Motiei, K. Ininbergs, E. Hell, Z. Gerdes, K. I. Udekwu, Z. Bacsik, and E. Gorokhova. 2018. Evidence for selective bacterial community structuring on microplastics. *Environmental Microbiology* 20 (8):2796–2808.

Peng, G., P. Xu, B. Zhu, M. Bai, and D. Li. 2018. Microplastics in freshwater river sediments in Shanghai, China: A case study of risk assessment in mega-cities. *Environmental Pollution* 234:448–456.

Plante, C. J., K. M. Coe, and R. G. Plante. 2008. Isolation of surfactant-resistant bacteria from natural, surfactant-rich marine habitats. *Applied and Environmental Microbiology* 74 (16):5093–5099.

Plante, C. J., and A. G. Shriver. 1998. Patterns of differential digestion of bacteria in deposit feeders: A test of resource partitioning. *Marine Ecology Progress Series* 163:253–258.

Pruzzo, C., L. Vezzulli, and R. R. Colwell. 2008. Global impact of *Vibrio cholerae* interactions with chitin. *Environmental Microbiology* 10 (6):1400–1410.

Razanajatovo, R. M., J. Ding, S. Zhang, H. Jiang, and H. Zou. 2018. Sorption and desorption of selected pharmaceuticals by polyethylene microplastics. *Marine Pollution Bulletin* 136:516–523.

Reisser, J., J. Shaw, G. Hallegraeff, M. Proietti, D. K. Barnes, M. Thums, C. Wilcox, B. D. Hardesty, and C. Pattiaratchi. 2014. Millimeter-sized marine plastics: A new pelagic habitat for microorganisms and invertebrates. *PLoS One* 9 (6):e100289.

Rios, L. M., and P. R. Jones. 2015. Characterisation of microplastics and toxic chemicals extracted from microplastic samples from the North Pacific Gyre. *Environmental Chemistry* 12 (5):611–617.

Rios, L. M., C. Moore, and P. R. Jones. 2007. Persistent organic pollutants carried by synthetic polymers in the ocean environment. *Marine Pollution Bulletin* 54 (8):1230–1237.

Rodrigues, A., D. M. Oliver, A. McCarron, and R. S. Quilliam. 2019. Colonisation of plastic pellets (nurdles) by *E. coli* at public bathing beaches. *Marine Pollution Bulletin* 139:376–380.

Romalde, J. L., A. L. Dieguez, A. Lasa, and S. Balboa. 2014. New *Vibrio* species associated to molluscan microbiota: A review. *Frontiers in Microbiology* 4:413.

Romeo, T., M. D'Alessandro, V. Esposito, G. Scotti, D. Berto, M. Formalewicz, S. Noventa, et al. 2015. Environmental quality assessment of Grand Harbour (Valletta, Maltese Islands): A case study of a busy harbour in the Central Mediterranean Sea. *Environmental Monitoring and Assessment* 187 (12):747.

Russell, J. R., J. Huang, P. Anand, K. Kucera, A. G. Sandoval, K. W. Dantzler, D. Hickman, et al. 2011. Biodegradation of polyester polyurethane by endophytic fungi. *Applied and Environmental Microbiology* 77 (17):6076–6084.

Sauchyn, L. K., and R. E. Scheibling. 2009. Degradation of sea urchin feces in a rocky subtidal ecosystem: Implications for nutrient cycling and energy flow. *Aquatic Biology* 6 (1–3):99–108.

Schmidt, C., T. Krauth, and S. Wagner. 2017. Export of plastic debris by rivers into the sea. *Environmental Science & Technology* 51 (21):12246–12253.

Seidensticker, S., P. Grathwohl, J. Lamprecht, and C. Zarfl. 2018. A combined experimental and modeling study to evaluate pH-dependent sorption of polar and non-polar compounds to polyethylene and polystyrene microplastics. *Environmental Sciences Europe* 30 (1):30.

Seidensticker, S., C. Zarfl, O. A. Cirpka, G. Fellenberg, and P. Grathwohl. 2017. Shift in mass transfer of wastewater contaminants from microplastics in the presence of dissolved substances. *Environmental Science & Technology* 51 (21):12254–12263.

Setala, O., V. Fleming-Lehtinen, and M. Lehtiniemi. 2014. Ingestion and transfer of microplastics in the planktonic food web. *Environmental Pollution* 185:77–83.

Seth, C. K., and A. Shriwastav. 2018. Contamination of Indian sea salts with microplastics and a potential prevention strategy. *Environmental Science and Pollution Research International* 25 (30):30122–30131.

Shani, C., N. Weisbrod, and A. Yakirevich. 2008. Colloid transport through saturated sand columns: Influence of physical and chemical surface properties on deposition. *Colloids and Surfaces A: Physicochemical and Engineering Aspects* 316 (1–3):142–150.

Siegfried, M., A. A. Koelmans, E. Besseling, and C. Kroeze. 2017. Export of microplastics from land to sea. A modelling approach. *Water Research* 127:249–257.

Silva, M. M., G. C. Maldonado, R. O. Castro, J. de Sa Felizardo, R. P. Cardoso, R. M. D. Anjos, and F. V. Araujo. 2019. Dispersal of potentially pathogenic bacteria by plastic debris in Guanabara Bay, RJ, Brazil. *Marine Pollution Bulletin* 141:561–568.

Sojitra, I., K. T. Valsaraj, D. D. Reible, and L. J. Thibodeaux. 1995. Transport of hydrophobic organics by colloids through porous-media.1. Experimental results. *Colloids and Surfaces A: Physicochemical and Engineering Aspects* 94 (2–3):197–211.

Song, Y., and A. L. Andrady. 1991. Fouling of floating plastic debris under Biscayne Bay exposure conditions. *Marine Pollution Bulletin* 22 (12):608–613.

Spear, L. B., D. G. Ainley, and C. A. Ribic. 1995. Incidence of plastic in seabirds from the tropical Pacific, 1984–1991: Relation with distribution of species, sex, age, season, year and body-weight. *Marine Environmental Research* 40 (2):123–146.

Sudhakar, M., M. Doble, P. S. Murthy, and R. Venkatesan. 2008. Marine microbe-mediated biodegradation of low- and high-density polyethylenes. *International Biodeterioration & Biodegradation* 61 (3):203–213.

Sudhakar, M., C. Priyadarshini, M. Doble, P. S. Murthy, and R. Venkatesan. 2007. Marine bacteria mediated degradation of nylon 66 and 6. *International Biodeterioration & Biodegradation* 60 (3):144–151.

Summers, S., T. Henry, and T. Gutierrez. 2018. Agglomeration of nano- and microplastic particles in seawater by autochthonous and de novo-produced sources of exopolymeric substances. *Marine Pollution Bulletin* 130:258–267.

Sun, B., Y. Hu, H. Cheng, and S. Tao. 2019. Releases of brominated flame retardants (BFRs) from microplastics in aqueous medium: Kinetics and molecular-size dependence of diffusion. *Water Research* 151:215–225.

Tanaka, K., H. Takada, R. Yamashita, K. Mizukawa, M. A. Fukuwaka, and Y. Watanuki. 2013. Accumulation of plastic-derived chemicals in tissues of seabirds ingesting marine plastics. *Marine Pollution Bulletin* 69 (1–2):219–222.

Teuten, E. L., J. M. Saquing, D. R. Knappe, M. A. Barlaz, S. Jonsson, A. Bjorn, S. J. Rowland, et al. 2009. Transport and release of chemicals from plastics to the environment and to wildlife. *Philosophical Transactions: Royal Society. Biological Sciences* 364 (1526):2027–2045.

Thiel, M., and L. Gutow. 2005. The ecology of rafting in the marine environment. II: The rafting organisms and community. In *Oceanography and Marine Biology: An Annual Review, Vol. 43*, edited by R. N. Gibson, R. J. A. Atkinson and J. D. M. Gordon, pp. 279–418, Taylor & Francis Group, Boca Raton, FL.

Thompson, R. C., C. J. Moore, F. S. vom Saal, and S. H. Swan. 2009. Plastics, the environment and human health: Current consensus and future trends. *Philosophical Transactions: Royal Society: Biological Sciences* 364 (1526):2153–2166.

Thornton, D. C. O. 2002. Diatom aggregation in the sea: Mechanisms and ecological implications. *European Journal of Phycology* 37 (2):149–161.

Tian, L., B. Kolvenbach, N. Corvini, S. Wang, N. Tavanaie, L. Wang, Y. Ma, S. Scheu, P. F. Corvini, and R. Ji. 2017. Mineralisation of ^{14}C-labelled polystyrene plastics by *Penicillium variabile* after ozonation pre-treatment. *New Biotechnology* 38 (Pt B):101–105.

Torkzaban, S., S. A. Bradford, M. T. van Genuchten, and S. L. Walker. 2008. Colloid transport in unsaturated porous media: The role of water content and ionic strength on particle straining. *Journal of Contaminant Hydrology* 96 (1–4):113–127.

Velzeboer, I., C. J. Kwadijk, and A. A. Koelmans. 2014. Strong sorption of PCBs to nanoplastics, microplastics, carbon nanotubes, and fullerenes. *Environmental Science & Technology* 48 (9):4869–4876.

Wang, F., C. S. Wong, D. Chen, X. Lu, F. Wang, and E. Y. Zeng. 2018. Interaction of toxic chemicals with microplastics: A critical review. *Water Research* 139:208–219.

Wang, W., and J. Wang. 2018. Different partition of polycyclic aromatic hydrocarbon on environmental particulates in freshwater: Microplastics in comparison to natural sediment. *Ecotoxicology and Environmental Safety* 147:648–655.

Wardrop, P., J. Shimeta, D. Nugegoda, P. D. Morrison, A. Miranda, M. Tang, and B. O. Clarke. 2016. Chemical pollutants sorbed to ingested microbeads from personal care products accumulate in fish. *Environmental Science & Technology* 50 (7):4037–4044.

Watts, A. J., C. Lewis, R. M. Goodhead, S. J. Beckett, J. Moger, C. R. Tyler, and T. S. Galloway. 2014. Uptake and retention of microplastics by the shore crab *Carcinus maenas*. *Environmental Science & Technology* 48 (15):8823–8830.

Wu, C., K. Zhang, X. Huang, and J. Liu. 2016. Sorption of pharmaceuticals and personal care products to polyethylene debris. *Environmental Science and Pollution Research International* 23 (9):8819–8826.

Wu, C. C., L. J. Bao, L. Y. Liu, L. Shi, S. Tao, and E. Y. Zeng. 2017. Impact of polymer colonization on the fate of organic contaminants in sediment. *Environmental Science & Technology* 51 (18):10555–10561.

Wu, P., Z. Cai, H. Jin, and Y. Tang. 2019. Adsorption mechanisms of five bisphenol analogues on PVC microplastics. *Science of the Total Environment* 650 (Pt 1):671–678.

Yoshida, S., K. Hiraga, T. Takehana, I. Taniguchi, H. Yamaji, Y. Maeda, K. Toyohara, K. Miyamoto, Y. Kimura, and K. Oda. 2016. A bacterium that degrades and assimilates poly(ethylene terephthalate). *Science* 351 (6278):1196–1199.

Zarfl, C., and M. Matthies. 2010. Are marine plastic particles transport vectors for organic pollutants to the Arctic? *Marine Pollution Bulletin* 60 (10):1810–1814.

Zettler, E. R., T. J. Mincer, and L. A. Amaral-Zettler. 2013. Life in the "plastisphere": Microbial communities on plastic marine debris. *Environmental Science & Technology* 47 (13):7137–7146.

Zhang, H., J. Wang, B. Zhou, Y. Zhou, Z. Dai, Q. Zhou, P. Chriestie, and Y. Luo. 2018. Enhanced adsorption of oxytetracycline to weathered microplastic polystyrene: Kinetics, isotherms and influencing factors. *Environmental Pollution* 243 (Pt B):1550–1557.

Zhang, K., J. Su, X. Xiong, X. Wu, C. Wu, and J. Liu. 2016. Microplastic pollution of lakeshore sediments from remote lakes in Tibet Plateau, China. *Environmental Pollution* 219:450–455.

Zhang, W., X. Ma, Z. Zhang, Y. Wang, J. Wang, J. Wang, and D. Ma. 2015. Persistent organic pollutants carried on plastic resin pellets from two beaches in China. *Marine Pollution Bulletin* 99 (1–2):28–34.

Zhao, S. Y., L. X. Zhu, and D. J. Li. 2015. Characterization of small plastic debris on tourism beaches around the South China Sea. *Regional Studies in Marine Science* 1:55–62.

Zhuang, J., J. Qi, and Y. Jin. 2005. Retention and transport of amphiphilic colloids under unsaturated flow conditions: Effect of particle size and surface property. *Environmental Science & Technology* 39 (20):7853–7859.

Zuo, L. Z., H. X. Li, L. Lin, Y. X. Sun, Z. H. Diao, S. Liu, Z. Y. Zhang, and X. R. Xu. 2019. Sorption and desorption of phenanthrene on biodegradable poly(butylene adipate co-terephtalate) microplastics. *Chemosphere* 215:25–32.

15 Ecological Impacts of Particulate Plastics in Marine Ecosystems

Claudia Halsband and Andy M. Booth

CONTENTS

15.1 Introduction 231
15.2 Microalgae and Marine Bacteria 232
15.3 Zooplankton 233
15.4 Benthic Organisms 235
15.5 Fish 236
15.6 Seabirds 237
15.7 Other Higher Marine Species 238
15.8 Food Web Effects 238
15.9 Conclusions and Recommendations 240
Acknowledgments 241
References 241

15.1 INTRODUCTION

Plastic debris is a global threat to marine wildlife, as it is present in all marine habitats, from the sea surface through the water column and down to the sediments (Kanhai et al. 2018; Law et al. 2014; Maes et al. 2017), from beaches and the littoral zones to the deep sea (de Carvalho and Baptista Neto 2016; Naji et al. 2017; Van Cauwenberghe et al. 2013), as well as from the tropics to the poles. Adverse effects of macroplastic (>5 mm in size) have been shown in fish, turtles, seabirds, and marine mammals (Gall and Thompson 2015). Entanglement in lost or discarded fishing gear and other debris has been observed for many species (Laist 1997), but also lethal and sub-lethal effects resulting directly from the ingestion of plastic debris (Baulch and Perry 2014; Kühn et al. 2015; Schuyler et al. 2014). Microplastic particles (MPs; <1 mm) are less obvious and more difficult to study, but readily bioavailable through direct and passive ingestion to a wide range of aquatic organisms owing to their small size. Small organisms, such as marine zooplankton and early life stages of large organisms are especially likely to interact with MPs (Desforges et al. 2015), but also marine filter feeders ranging in size from minute plankton to whales. While the size of the ingested plastic depends on the type, body size, and life stage of the organism (Cole et al. 2013; Ramos et al. 2012; Ryan et al. 2009), existing knowledge concerning the impacts and mechanisms of toxicity of larger plastic litter items on marine animals may be transferable to those organisms that ingest small MPs, but this remains to be demonstrated (Ross and Morales-Caselles 2015). For example, obstruction or blockage of the intestinal tract or digestive organs can prevent further ingestion of food or diminish the triggering of feeding stimuli (pseudo-satiation) in both large and small organisms (Kühn et al. 2015).

Despite an increasing number of studies showing the uptake of plastic particles (PPs) by a broad range of marine species inhabiting most environmental compartments and representing many trophic levels, a definitive understanding of the impacts of PPs is still lacking. The potential impacts of adhered, adsorbed, and ingested PPs are driven by their mechanical and chemical effects, the latter being influenced by the presence of additive chemicals and adsorbed organic pollutants. Mechanical effects include attachment of the particles to external surfaces, thereby hindering mobility and clogging of external appendages, such as swimming legs or feeding apparatus, or internal clogging of the digestive tract, while chemical effects can include inflammation, hepatic stress, and/or decreased growth (Setälä et al. 2016). Indirect effects such as the deviation of resources and energy from growth and reproduction to mitigate against plastic ingestion have also been reported. Effects at the immunological and molecular level, such as changes in gene expression, transcriptomics, and proteomics have been studied in invertebrates and fish (Avio et al. 2015; Choi et al. 2018; Espinosa et al. 2017; Sussarellu et al. 2016). The impacts of plastic debris on marine organisms have recently been comprehensively reviewed (Auta et al. 2017; Law 2017; Lusher 2015; Rochman et al. 2016; Solomon and Palanisami 2016; Van Cauwenberghe et al. 2015; Wright et al. 2013b), including toxicological effects at the molecular level in vertebrates and invertebrates (Alimba and Faggio 2019) and the effects of polystyrene MPs across trophic levels (Gambardella et al. 2018). Here, we present an overview of the current knowledge on the impacts and toxicity of particulate plastics to a range of marine functional groups, including both acute and sub-lethal endpoints.

15.2 MICROALGAE AND MARINE BACTERIA

Phytoplankton represent the base of the marine food chain, channeling energy from sunlight and CO_2 to all higher trophic levels. Marine bacteria also play an essential role in food webs and biogeochemical cycles as users of dissolved organic matter and contributors to the recycling of detrital-based energy back into the food web. Although these groups provide fundamental ecosystem functions, very few MP toxicity studies focus on microalgae or bacteria. This scarcity of toxicity studies may point to a perceived lack of relevance, where no toxicological responses resulting from interactions between MPs and similar sized microalgae are expected. Marine bacteria, typically less than 1 μm in size, are often smaller than MPs, and thus interaction is limited potentially to colonization of MP surfaces. The impacts of the smallest fraction of plastic litter, nanoplastics (NPs), have, therefore, been more frequently investigated with marine microorganisms. The roles of particle size (50, 500 nm, and 6 μm) and physicochemical properties (negatively charged and uncharged) were investigated for the toxicity of polystyrene MPs to the marine diatom *Thalassiosira pseudonana* and the marine flagellate *Dunaliella tertiolecta* (Sjollema et al. 2016). None of the treatments tested had an effect on the photosynthetic activity of the algae after 72 hours, while growth was reduced by up to 45% by uncharged polystyrene particles, but only at high concentrations of 250 mg/L. Significant growth inhibition of approximately 40% was also caused by polyvinyl chloride (PVC) MPs (1 μm; exposure up to 96 hours) at lower concentrations of 50 mg/L, together with decreased chlorophyll content and photosynthetic efficiency in the marine diatom *Skeletonema costatum* (Zhang et al. 2017). In contrast, larger PVC debris of 1 mm size had no effect on the growth of microalgae. It is suggested that interactions between MPs and microalgae, such as adsorption and aggregation, accounted for the observed toxic effects, rather than indirect negative effects through shading (Zhang et al. 2017). Similar results were obtained in a small number of experiments that studied interactions of algae with plastic NPs. A 72 hours exposure of *Scenedesmus obliquus* to polystyrene NPs resulted in a reduction in both growth and cellular chlorophyll content (Besseling et al. 2014). NP polystyrene was also observed to adsorb to *Scenedesmus* cells, causing reduced photosynthesis and increased production of reactive oxygen species (Bhattacharya et al. 2010). Dose-dependent adverse effects of micro PVC were found for *Karenia mikimotoi* growth, chlorophyll content, and photosynthetic efficiency (Zhao et al. 2019). In contrast, neither single MP effects nor combined effects with copper on population growth were found for *Tetraselmis chuii* (Davarpanah and Guilhermino 2015). The marine bacterium *Vibrio fischeri* did not exhibit an acute toxic response following exposure to two poly(methylmethacrylate)-based NPs with different surface chemistries (medium and hydrophobic) at concentrations ranging from 0.01 to 1000 mg/L (Booth et al. 2016). Recent studies with carbon-based nano-materials have highlighted potential problems in the accurate quantification of chlorophyll in algae tests conducted with particulate materials (Farkas and Booth 2017; Hund-Rinke et al. 2016). In addition to shading of algal cells during the exposure period, shading of the algal cells during UV fluorescence measurements can underestimate algal growth. If chlorophyll is extracted prior to analysis, this may adsorb strongly to any hydrophobic particles (e.g., MPs and NPs) present in the samples. This issue should be carefully considered in future studies looking at MP and NP interactions and effects on microalgae species.

15.3 ZOOPLANKTON

Marine zooplankton, especially microzooplankton species, are the main grazers of microalgae, and consequently their natural prey is in the same µm-size range as MP (Cole et al. 2013). Zooplankton represent an important trophic link as the primary consumers of phytoplankton, while they serve as food for many planktivorous predators from carnivorous zooplankton to baleen whales. They are very abundant and distributed across the world's oceans. Estimates of zooplankton abundance versus MP concentrations show that encounter rates are potentially high. In the North Pacific gyre, where plastic debris concentrations are high, plastic abundance was only 20% that of plankton abundance, but the mass of plastic was around six times higher than plankton biomass (Moore et al. 2001). In addition, refined MP measurements of the small size fraction (<250 µm) show that the smallest MPs are the most numerous (Bergmann et al. 2017b).

Copepods, small planktonic crustaceans of 0.5–5 mm size, are often the lowest trophic level to exhibit direct ingestion of MPs, and some species are easy to use in laboratory experiments. Hence, an increasing number of studies have been conducted with marine zooplankton in general, and copepods in particular. Gastropod veliger larvae and copepods responded with changes in swimming behavior and a reduction in clearance rates to the presence of non-organic particles, while rotifers and ciliates did not show such sensitivity (Hansen et al. 1991). Ingestion of plastic has been shown for many different plankton taxa under controlled laboratory conditions (Cole et al. 2013; Lee et al. 2013; Setälä et al. 2014; Vroom et al. 2017). Particle size and feeding strategy are important factors in determining the extent of plastic ingestion (Cole et al. 2013; Lee et al. 2013; Setälä et al. 2014). The mesh size of the feeding apparatus of filter feeders only lets certain particle sizes pass through. Avoidance strategies against MP uptake during feeding were shown for pristine plastic, where the number of algal cells grazed decreased in the presence of similar-sized MPs (Coppock et al. 2019). Prey selectivity was also shown to be significantly altered in copepods exposed to nylon fibers (Cole et al. 2019).

While some authors concluded that ingested MPs do not induce acute toxicity in marine zooplankton (Beiras et al. 2018) or act as vectors for hydrophobic organic chemical exposure (Beiras and Tato 2019), other investigations reported negative impacts on organism function and health. Exposure of the copepod *Centropages typicus* to natural assemblages of algae with and without MPs showed that high concentrations of 7 µm MPs (>4000 mL^{-1}) significantly decreased algal feeding (Cole et al. 2013). The survival and fecundity of the copepod *Tigriopus japonicus* was also negatively impacted at chronic exposure (96 hours) to high concentrations of polystyrene MPs (1.25–25 g mL^{-1}) (Lee et al. 2013) (Figure 15.1). However, no acute toxicity was observed in either nauplii or adult copepods (*Tigriopus japonicas*) exposed to high concentrations of polystyrene MPs at sizes of 0.05, 0.5, and 6 µm (Lee et al. 2013). At lower exposure concentrations (75 particles mL^{-1}) of 20 µm polystyrene MPs, energetic depletion and reduced reproduction were observed in the copepod *Calanus helgolandicus* (Cole et al. 2015). They concluded that MPs competed with food items for ingestion, and that the copepods did not actively decline ingestion of non-nutritious particles. A follow-up study showed that exposure to nylon fibers decreased ingestion of similar-shaped

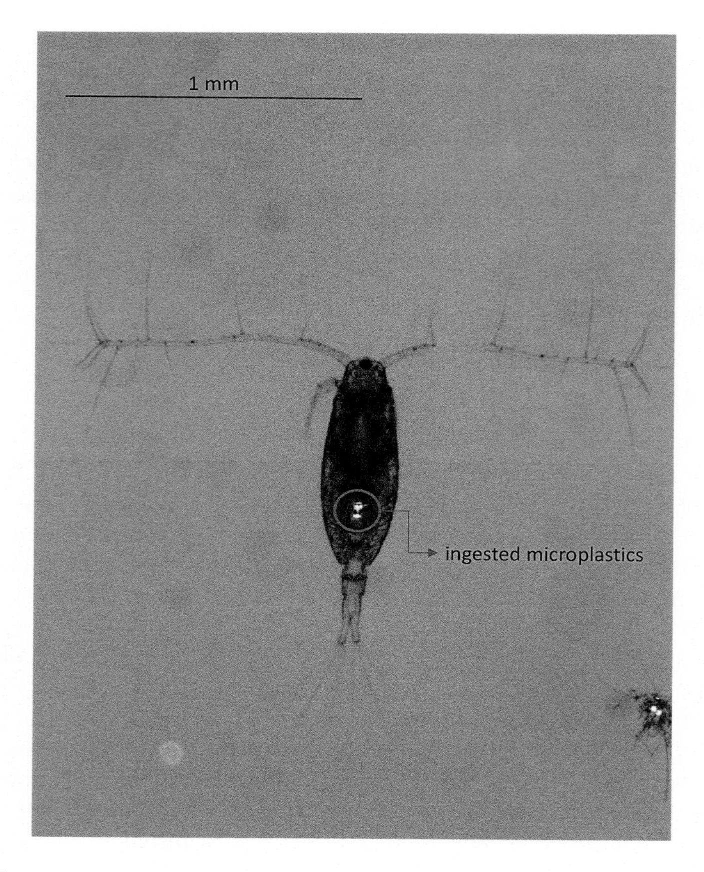

FIGURE 15.1 Copepod (*Acartia longiremis*) with ingested 15 μm polyethylene particles in the gut. (© Renske Vroom.)

chain-forming algae, while the presence of nylon fragments induced a reduction in grazing on algae similar in shape and size (Coppock et al. 2019). Copepods therefore seem to be able to sense the quality of the particles available for ingestion and try to avoid inert particles. Furthermore, exposure of *C. finmarchicus* to nylon fibers and fragments resulted in stymied lipid accumulation and premature molting in juvenile organisms (Cole et al. 2019). Despite the potential for selection against unwanted particles, copepods ingested MPs in most laboratory exposures reported (Cole et al. 2013, 2015; Lee et al. 2013; Vroom et al. 2017) (Figure 15.1). An explanation for the lack of avoidance of plastic ingestion may lie in the formation of nutritious biofilms, containing bacteria, microalgae, and invertebrates, around the plastic surface (Vroom et al. 2017). Colonization of plastic particles by microbes happens rapidly upon entering the marine environment (Reisser et al. 2014) and may disguise them as food items. That plastic ingestion takes place under natural environmental conditions was confirmed in situ for copepods and euphausiids (Desforges et al. 2015). A laboratory study on plastic leachates showed differential toxicity to the harpacticoid copepod *Nitocra spinipes* depending on the plastic product and the simulated weathering conditions applied (Bejgarn et al. 2015).

A small number of other marine zooplankton species has also been studied. A particle size-dependent effect was observed for polystyrene microbeads (0.05, 0.5, and 6 μm) exposed to the monogonot rotifer (*Brachionus koreanus*), with smaller particles eliciting increased responses in a range of endpoints including in vitro studies, growth rate, fecundity lifespan, and reproduction time (Jeong et al. 2016). The early life stages of many benthic organisms also pass through a planktonic stage, and a number of these have been subjected to MP and NP uptake and effects studies. The larvae of the sea urchin *Tripneustes gratilla* exposed to polyethylene MPs (5 days) exhibited reduced body width at the highest exposure tested abundance (300 particles/mL), but no effects at environmentally relevant concentrations (Kaposi et al. 2014). Larvae of the sea urchin *Paracentrotus lividus* were also exposed to polystyrene NPs with different surface functionalization (NH_2 and COOH). No embryotoxicity was observed for PS-COOH up to 50 μg mL^{-1}, whereas PS-NH_2 caused severe developmental defects (EC_{50} 2.61 μg mL^{-1} 48 post fertilization) (Della Torre et al. 2014). The results indicate that differences in surface chemistry can significantly influence MP and NP toxicity. Embryotoxic effects on the same species were also observed for PVCs (Oliviero et al. 2019), where some colors induced more toxicity than others. The authors hypothesize that the amount and combination of heavy metal content in the coloring agents caused these differences. Larvae of the blue mussel (*Mytilus galloprovincalis*) showed changes in the transcription of genes related to shell biogenesis and immunomodulation in response to polystyrene MPs (Capolupo et al. 2018). Bioavailability and effect studies on zooplankton were summarized and evaluated with the result that 45% of investigations find negative effects on physiological and fitness endpoints compared to 14% with neutral results (Botterell et al. 2019).

15.4 BENTHIC ORGANISMS

Many of the parent polymer materials in MPs have densities higher than that of seawater, while the processes of aging and biofouling are considered to increase the sedimentation of MPs that are low density (Andrady 2011). Repackaging of ingested MPs within the fecal material of marine organisms is also thought to promote their transport to the seafloor (Cole et al. 2016; Coppock et al. 2019). As a result, the sediment surface is expected to act as a sink for MPs, indicating that benthic species are at the highest risk to the highest exposure levels and potential impacts.

A positive relationship was observed between the polystyrene MP concentration in sediment and both uptake of MP particles and weight loss by the lugworm *Arenicola marina* (Besseling et al. 2013). Furthermore, a reduction in feeding activity was observed at a PS (polystyrene) dose of 7.4% dry weight (Besseling et al. 2013). Plastic ingestion may also result in additive leaching from the plastic to the intestines. However, it is important to note that a modeling study on nonylphenol and bisphenol A concluded that leached chemical concentrations are likely to be much lower compared to environmental background concentrations, suggesting this may represent a negligible pathway for exposure (Koelmans et al. 2014). *A. marina* exposed to PVC MPs for 4 weeks fed less, had reduced lipid reserves, and exhibited increased phagocytic activity and inflammatory response (Wright et al. 2013a). PVC MPs were also found to increase susceptibility of *A. marina* to oxidative stress (Browne et al. 2013).

Although a reduction in food uptake was observed for the marine isopod *Idotea emarginata* exposed to polyethylene MPs over a 7-week period, there were no effects on survival, intermolt duration, or growth (Hämer et al. 2014). Similarly, no adverse effects were reported for the gooseneck barnacle (*Lepas* spp.), despite a third of stomachs examined containing MPs (Goldstein and Goodwin 2013). Beach hoppers (*Platorchestia smithi*) were observed to readily ingest marine-contaminated MPs, causing an impact on their survival (Tosetto et al. 2016). Individuals also displayed reduced jump height and an increase in weight, although there was no significant difference in time taken to relocate shelter post disturbance. In contrast, the benthic marine amphipod *Corophium volutator* did not exhibit acute toxicity to two poly(methylmethacrylate)-based NPs with different surface chemistries (medium and hydrophobic) at concentrations ranging from 0.01 to 500 mg/L, and showed no effects in a reburial test conducted after 10 days of exposure (Booth et al. 2016). Furthermore, no significant effects were observed in adult sandhoppers (*Talitrus saltator*) exposed to polyethylene MPs for 24 hours, followed by a 7 day depuration period (Ugolini et al. 2013), while polystyrene MPs did not elicit physical or behavioral effects in the common littoral crab (*Carcinus maenas* (L.) over a 21-day exposure (Farrell and Nelson 2013). Langoustines (*Nephrops norvegicus*) exposed to 3–5 mm long polypropylene fibers exhibited reduced body mass, blood proteins, and storage lipids compared to controls (Welden and Cowie 2016).

Mussels are sedentary filter feeders commonly used as laboratory test species and have been utilized in a number of studies investigating MP and NP ingestion and toxicity. In the presence of polystyrene NPs, the blue mussel *Mytilus edulis* exhibited reduced filtering activity and production of pseudo-feces, which is suggested to indicate a purging response to the low nutritional value of the MPs (Wegner et al. 2012). *M. edulis* also accumulated high-density polyethylene, which led to an inflammatory response within 6 hours of ingestion and destabilization of the lysosomal membrane after 96-hours exposure (von Moos et al. 2012). Suspended PVC MPs (1–50 µm) exposed to the Asian green mussel *Perna viridis* for two 2-hour-time-periods per day caused a decrease in mussel filtration and respiration rates after 44 days, and a decline in survival after 91 days (Rist et al. 2016). The authors suggest these negative effects resulted from prolonged periods of valve closure as a reaction to MP presence. In contrast to the above studies, ingestion of polystyrene MPs (3.0 or 9.6 µm; 3 or 12 hours exposure and 48 days depuration) caused no significant effects on the oxidative status of hemolymph, viability or phagocytic activity of hemocytes, or filter-feeding activity of *M. edulis* (Browne et al. 2008). Polystyrene MPs (2 and 6 µm in diameter; 0.023 mg L^{-1}) exposed to adult Pacific oysters (*Crassostrea gigas*) for 2 months during their reproductive cycle elicited significant decreases in oocyte number (–38%), diameter (–5%), and sperm velocity (–23%) (Sussarellu et al. 2016). Furthermore, D-larval yield and larval development of offspring derived from the exposed parents decreased by 41% and 18%, respectively. Leachates of both virgin plastic pellets and beached pellets induced toxicity in brown mussels (*Perna perna*) (Gandara e Silva et al. 2016).

Corals ingested and retained MPs for 24 hours or more in a study with a variety of different PP types (Allen et al. 2017). In contrast to the results with copepods described above, biofouled particles were less often ingested than unfouled ones. The authors concluded that leaching plastic additives triggered chemoreceptors, which in turn stimulated a feeding response.

A review of the effects of MPs and NPs on benthic invertebrates summarizes both the physical and chemical impacts on this group, including the effects of leachates and plastic-contaminant interactions (Haegerbaeumer et al. 2019).

15.5 FISH

Many studies have reported the presence of MPs in a broad range of marine fish. However, there are a very limited number of studies specifically investigating the effects of MP exposure on species at higher trophic levels. This is likely to reflect, at least partially, the additional costs, resources, and ethical considerations necessary for conducting toxicity studies with larger vertebrate species. Polyethylene MPs (1–5 µm) exposed to the juvenile common goby (*Pomatoschistus microps*), an inhabitant of brackish coastal waters, significantly reduced acetylcholinesterase activity, but no significant effects were found for glutathione S-transferase activity or lipid peroxidation (Oliveira et al. 2013). In a follow-up study, wild-caught *P. microps* under simultaneous exposure to MPs and *Artemia* showed a significant reduction of the predatory performance (65%) and efficiency (up to 50%) (de Sá et al. 2015). This reduction in performance depended upon the catch location of the fish, indicating that developmental conditions may influence the prey selection capability of fish. The results indicate that MP-induced reduction of food intake may decrease individual and population fitness. Experimental in vivo and in vitro studies with gilthead seabream (*Sparus aurata*) and European sea bass (*Dicentrarchus labrax*) revealed subtle increases in immune activities upon MP exposure, caused by oxidative stress in their leucocytes (Espinosa et al. 2017, 2018). The authors emphasize the possible implications of their short-term exposures for fish exposed to MPs and contaminants more chronically, which may be the case in highly polluted seas as, e.g., the Mediterranean, where two seabream species with MP-contaminated intestines have been recorded (Savoca et al. 2019). Neurotoxicity and oxidative stress and damage from MPs alone and in interactions with mercury were found in the European seabass (*Dicentrarchus labrax*) (Barboza et al. 2018). Measurements of MPs and hydrophobic organic contaminants (HOCs) in wild-caught Baltic herring did, however, not show direct links between the levels of HOCs and MP ingestion

(Ogonowski et al. 2019). HOC desorption rates from MPs to artificial gut fluids indicated that HOCs that are near phase equilibrium between microplastics and environmental media do not accumulate further via ingestion of MPs, even when high proportions of ingested HOCs come from the ingested MPs (Lee et al. 2019).

15.6 SEABIRDS

Encounters between seabirds and marine plastic debris have been widely documented over the past decades (Amélineau et al. 2016; Furness 1983; van Franeker et al. 2011; Wilcox et al. 2015). Interactions include entanglement (Kühn et al. 2015), ingestion (Amélineau et al. 2016; Poon et al. 2017; Trevail et al. 2015), egestion in guano (Provencher et al. 2018), usage of plastic items in nest material (Votier et al. 2011), and transfer of plastic-contaminated food to chicks (Carey 2011) (Figure 15.2). Differential responses by seabird species with different foraging strategies imply that some birds confuse certain plastic debris with food based on shape and/or color (Santos et al. 2016). In addition, biofilms forming on the plastic exude olfactory cues that are detected by the birds and classified as indicators of food (Savoca et al. 2016). Contamination of either their habitat or the birds themselves does, however, not automatically prove concrete impacts (Law 2017; Rochman et al. 2016). Ingestion of plastic may nonetheless impact seabirds through pseudo-satiation and subsequent lack of food uptake, physical damage of intestines from sharp objects, and exposure to toxic additives. Sub-lethal effects include lower body condition, reproductive output, or longevity, but are difficult to measure (Law 2017). Transfer of plastic-derived chemicals into seabird tissue was hypothesized for short-tailed shearwaters (*Puffinus tenuirostris*), inferred from polybrominated diphenyl ethers that were measured in both plastic particles found in the stomachs and in stomach oil and tissues of the abdominal adipose of the birds (Tanaka et al. 2013, 2015). The toxic effects of these compounds were, however, not quantified. Herzke et al. (2016) only found negligible amounts of persistent organic pollutants from MPs in tissues of northern fulmars, where the level of contamination was higher in the natural prey than in the ingested plastic. The potential toxicity of plastic additives, therefore, depends on the concentration of the additive, the residence time in the ingesting organism, and external factors, such as prey and background concentrations of the target contaminant.

FIGURE 15.2 Gannet (*Morus bassanus*) feeding chick with marine debris. (© Bo Eide.)

15.7 OTHER HIGHER MARINE SPECIES

Mechanical effects of marine debris lost or discarded at sea on marine wildlife have been projected already in the mid-1980s (Laist 1987), but the extent of the problem and the risks to marine populations are not quantified (Simmonds 2012). Marine reptiles (e.g., turtles) and marine mammals (seals, walruses, polar bears, and whales) all encounter plastic debris in their habitats and may suffer from entanglement and subsequent drowning, impaired ability to move and catch food, or injuries from cuts or abrasion (Bergmann et al. 2017a; Laist 1987; Law 2017). Entanglement impacts on pinniped species have been recently reviewed (Butterworth and Sayer 2017). Ingestion effects are similar to those in seabirds and others, including blocked digestive tracks and potential uptake of plastic additives into tissues. Ingestion of plastic debris, including plastic bags, was determined as the cause of death for green turtles in Florida, as well as southern Brazil (Bjorndal et al. 1994; Bugoni et al. 2001). A case of mortality due to the ingestion of large amounts of plastic debris from greenhouses has been documented for a sperm whale in the Mediterranean (de Stephanis et al. 2013). A modeling and mapping exercise in British Columbia (Canada) showed that overlap of debris hotspots and marine mammal distributions were often remote, decreasing the chances for opportunistic reporting of interaction incidents (Williams et al. 2011). Baleen whales may also be at risk to ingest large quantities of MPs during filter-feeding. Although no direct negative effects from such exposure have been reported so far, phthalates have been suggested as biomarkers for this type of impact (Fossi et al. 2012). Similar ecotoxicological sampling of endangered whale sharks (*Rhincodon typus*) attempted to correlate plastic additives in skin biopsies to a possible exposure to macro- and MPs (Fossi et al. 2017).

15.8 FOOD WEB EFFECTS

Despite an increasing number of studies reporting the impacts of plastics at the species level, effects on assemblages and communities within real habitats remain effectively unknown (Browne et al. 2015; Carbery et al. 2018). Plastic particles undergo a number of transformations after they enter the marine environment, which influence their pathways and behavior within food webs. Plastic exposed to marine conditions starts degrading, and, although this process is generally slow (Gewert et al. 2015), the surface structure of the material changes, becomes more brittle and cracked, and provides a habitat for microbes and other rafting organisms (Barnes and Milner 2005). These species colonize the MPs in a distinct succession of taxa, forming a biofilm on and within the surface structure (Reisser et al. 2014). The biofilm alters the properties of the particles (e.g., their density and sinking velocity) and also disguises the MP as nutritionally valuable particles that get eaten by planktonic grazers (Cole et al. 2016; Vroom et al. 2017), increasing its bioavailability. Although ingested MPs are temporarily removed from the water column, their incorporation into zooplankton intestines provides potential for trophic transfer and biomagnification once these organisms get eaten by higher trophic levels. In the East China Sea, retention of MPs by zooplankton varied from 0.13 ind.$^{-1}$ (copepods) to 0.35 ind.$^{-1}$ (pteropods), equivalent to approximately 20 pieces m^{-3} (Sun et al. 2018). This plastic incorporated into plankton biomass is expected to be much more bioavailable than free-floating plastic particles in the water column. Alternatively, MPs are repackaged in fecal material and excreted, again in the form of nutritious organic material that may be consumed by detritivores in the water column and the benthos. The repackaging in biomass or fecal pellets further alters the MP properties. The sinking velocities of copepod fecal pellets loaded with MPs were significantly slower than without MPs (Cole et al. 2016; Coppock et al. 2019).

Trophic transfer from prey to predator has been shown for several predator-prey pairs (Farrell and Nelson 2013; Nelms et al. 2018; Welden et al. 2018), but a systematic food web analysis is still lacking. A recent systematic meta-analysis of microplastic effects across a range of marine and freshwater taxa confirms the expectation that MP exposure generally induces no or negative effects in organism performance, health, and fitness (Foley et al. 2018) (Figure 15.3). Cascading effects in predator-prey interactions and biomagnification along marine food chains remain to

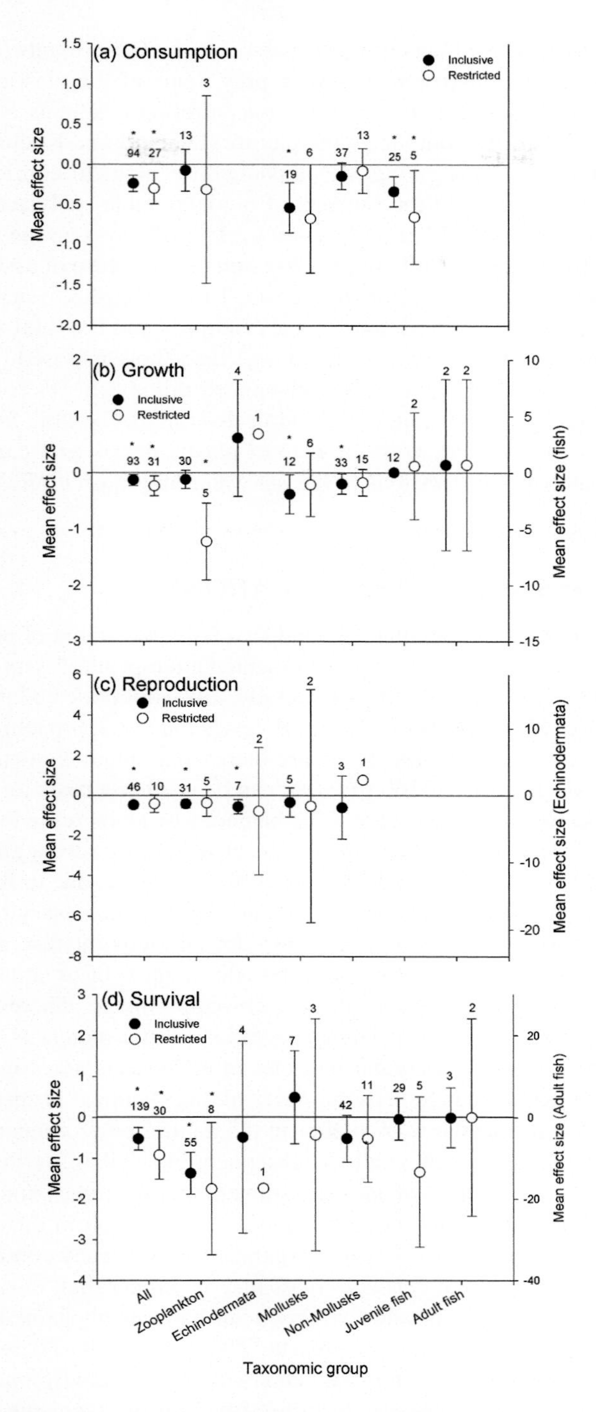

FIGURE 15.3 Mean effect sizes (±95% confidence interval) for four different response categories (a–d) and taxonomic groups. Closed circles indicate the "inclusive" dataset of all 43 studies analyzed; open circles indicate a "restricted" subset of data (the most extreme treatments; outliers removed). Effect sizes calculated as Hedges' g (Hedges and Olkin 1985) for (a–c), and ln(odds ratio) (Gurevitch et al. 2001) for (d). Sample sizes noted above each bar. "Juvenile" fish may include juvenile or larval fish. Significant effects are marked with *. Note different scale for fish in (b) and (d) and Echinodermata in (c). (From Foley, C.J. et al., *Sci. Total Environ.*, 631–632, 550–559, 2018.)

be demonstrated. Toxicological effects of plastic-associated contaminants depend on their background concentrations in the environment and in prey items of the plastic-exposed consumers (Herzke et al. 2016; Koelmans 2015), and their concentration gradients and synergistic versus antagonistic effect combinations complicate the picture (Diepens and Koelmans 2018; Koelmans et al. 2013). The first study investigating MP impacts on ecological communities was recently published (Green 2016). It reported the impacts of biodegradable (polylactic acid) and conventional (high-density polyethylene) MPs (0.8 μg–80 μg L^{-1}; 60 days) on the health and biological functioning of European flat oysters (*Ostrea edulis*), and the structure of associated macro-faunal assemblages were assessed in an outdoor mesocosm. Effects on the oysters were minimal, but benthic assemblage structures differed, and species richness and the total number of organisms were ~1.2 and 1.5 times greater in control mesocosms than those exposed to high doses of MPs. The study indicates that repeated exposure to high concentrations of MPs could alter assemblages in marine habitats by reducing the abundance and/or diversity of benthic fauna. Finally, the dispersal of alien and invasive species with the help of plastic debris as a carrier may represent a similarly important pathway for invasions in ballast water and mariculture (Gall and Thompson 2015; Rech et al. 2018).

15.9 CONCLUSIONS AND RECOMMENDATIONS

There remains insufficient data to adequately conduct a risk assessment of plastic particles of different sizes, reflected by the fact that the first documented attempt at assessing the risk of MPs was only conducted in 2018 (Everaert et al. 2018). Crucially, this work indicated that in areas where the highest concentrations of MPs have been measured, levels may be approaching those necessary to elicit negative effects on certain species. There are increasing amounts of environmental concentration data for larger MP particles (>100 μm) that can be used as a basis for estimating the exposure component of risk. However, this needs to be balanced by an increase in good-quality-effects data for species representing different trophic levels. Determining robust environmental concentration data for small MPs (<100 μm) and NPs remains a challenge, due to limitations for analytical approaches for these particles. Generating this knowledge is necessary for estimating current and future risk levels. Effects studies focusing on relevant endpoints are also urgently needed, and data generated must be interpreted within a broader context of organisms being exposed to multiple stressors, including other chemicals and climatic change. There is extremely limited knowledge concerning the toxicity of partially degraded plastic materials. Virtually all studies reported having utilized "pristine" polymer particles in MP effects assessment. However, degradation processes are known to change both the physical and chemical properties of plastic, and it is this type of material that organisms encounter in the natural environment. While the issues of pollutant adsorption to plastic particles and its subsequent bioavailability to organisms following particle ingestion have been the focus of an increasing number of studies, findings remain unclear. Further work is needed to understand the importance of this process in the context of more environmentally relevant exposure scenarios, where organisms may already contain levels of chemical pollutants and where environmental concentrations may be higher than those associated with the particles. The bioavailability of plastic additive chemicals has recently gained significant attention, but knowledge remains relatively limited at present. There is a need to determine how the many different types of additives behave when plastic enters the marine environment and whether their release from plastic into natural waters or in the digestive tracts of organisms is an important exposure route. Perhaps one of the greatest challenges in conducting effects assessment of MPs and NPs on marine organisms is linking exposure concentrations used in laboratory studies to environmentally relevant exposure concentrations, with the latter being poorly understood, especially in the case of NPs. Furthermore, ensuring consistent exposure levels in effects studies can be challenging owing to the propensity for most plastic particles either to float or sink rather than remain dispersed homogeneously in the water column.

Finally, distribution, toxicity, bioaccumulation, and biomagnification patterns of PPs need to be combined in ecosystem-wide risk assessments and ecological models in order to be able to establish the current state of PP impacts on marine ecosystems and provide impact predictions for future pollution levels and mitigation scenarios.

ACKNOWLEDGMENTS

This work was supported by the Research Council of Norway project "MICROFIBRE" (Grant 268404) to AB, and JPI Oceans/Research Council of Norway project "PLASTOX" (Grant 257479) to AB. The Fram Centre (Hazardous Substances, Plastic in the Arctic) has supported this work through the projects MARS and ArcticFibre to CH and AB.

REFERENCES

Alimba, C.G. and Faggio, C. (2019) Microplastics in the marine environment: Current trends in environmental pollution and mechanisms of toxicological profile. *Environmental Toxicology and Pharmacology* 68, 61–74.

Allen, A.S., Seymour, A.C. and Rittschof, D. (2017) Chemoreception drives plastic consumption in a hard coral. *Marine Pollution Bulletin* 124(1), 198–205.

Amélineau, F., Bonnet, D., Heitz, O., Mortreux, V., Harding, A.M.A., Karnovsky, N., Walkusz, W., Fort, J. and Grémillet, D. (2016) Microplastic pollution in the Greenland Sea: Background levels and selective contamination of planktivorous diving seabirds. *Environmental Pollution* 219, 1131–1139.

Andrady, A.L. (2011) Microplastics in the marine environment. *Marine Pollution Bulletin* 62(8), 1596–1605.

Auta, H.S., Emenike, C.U. and Fauziah, S.H. (2017) Distribution and importance of microplastics in the marine environment: A review of the sources, fate, effects, and potential solutions. *Environment International* 102, 165–176.

Avio, C.G., Gorbi, S., Milan, M., Benedetti, M., Fattorini, D., d'Errico, G., Pauletto, M., Bargelloni, L. and Regoli, F. (2015) Pollutants bioavailability and toxicological risk from microplastics to marine mussels. *Environmental Pollution* 198, 211–222.

Barboza, L.G.A., Vieira, L.R., Branco, V., Figueiredo, N., Carvalho, F., Carvalho, C. and Guilhermino, L. (2018) Microplastics cause neurotoxicity, oxidative damage and energy-related changes and interact with the bioaccumulation of mercury in the European seabass, *Dicentrarchus labrax* (Linnaeus, 1758). *Aquatic Toxicology* 195, 49–57.

Barnes, D.K.A. and Milner, P. (2005) Drifting plastic and its consequences for sessile organism dispersal in the Atlantic Ocean. *Marine Biology* 146(4), 815–825.

Baulch, S. and Perry, C. (2014) Evaluating the impacts of marine debris on cetaceans. *Marine Pollution Bulletin* 80(1–2), 210–221.

Beiras, R., Bellas, J., Cachot, J., Cormier, B., Cousin, X., Engwall, M., Gambardella, C., et al. (2018) Ingestion and contact with polyethylene microplastics does not cause acute toxicity on marine zooplankton. *Journal of Hazardous Materials* 360, 452–460.

Beiras, R. and Tato, T. (2019) Microplastics do not increase toxicity of a hydrophobic organic chemical to marine plankton. *Marine Pollution Bulletin* 138, 58–62.

Bejgarn, S., MacLeod, M., Bogdal, C. and Breitholtz, M. (2015) Toxicity of leachate from weathering plastics: An exploratory screening study with Nitocra spinipes. *Chemosphere* 132, 114–119.

Bergmann, M., Lutz, B., Tekman, M.B. and Gutow, L. (2017a) Citizen scientists reveal: Marine litter pollutes Arctic beaches and affects wild life. *Marine Pollution Bulletin* 125(1), 535–540.

Bergmann, M., Wirzberger, V., Krumpen, T., Lorenz, C., Primpke, S., Tekman, M.B. and Gerdts, G. (2017b) High quantities of microplastic in arctic deep-sea sediments from the HAUSGARTEN Observatory. *Environmental Science & Technology* 51(19), 11000–11010.

Besseling, E., Wang, B., Lürling, M. and Koelmans, A.A. (2014) Nanoplastic affects growth of S. obliquus and reproduction of D. magna. *Environmental Science & Technology* 48(20), 12336–12343.

Besseling, E., Wegner, A., Foekema, E.M., van den Heuvel-Greve, M.J. and Koelmans, A.A. (2013) Effects of microplastic on fitness and PCB bioaccumulation by the lugworm *Arenicola marina* (L.). *Environmental Science & Technology* 47(1), 593–600.

Bhattacharya, P., Lin, S.J., Turner, J.P. and Ke, P.C. (2010) Physical adsorption of charged plastic nanoparticles affects algal photosynthesis. *Journal of Physical Chemistry C* 114(39), 16556–16561.

Bjorndal, K.A., Bolten, A.B. and Lagueux, C.J. (1994) Ingestion of marine debris by juvenile sea turtles in coastal Florida habitats. *Marine Pollution Bulletin* 28(3), 154–158.

Booth, A.M., Hansen, B.H., Frenzel, M., Johnsen, H. and Altin, D. (2016) Uptake and toxicity of methylmethacrylate-based nanoplastic particles in aquatic organisms. *Environmental Toxicology and Chemistry* 35(7), 1641–1649.

Botterell, Z.L.R., Beaumont, N., Dorrington, T., Steinke, M., Thompson, R.C. and Lindeque, P.K. (2019) Bioavailability and effects of microplastics on marine zooplankton: A review. *Environmental Pollution* 245, 98–110.

Browne, M.A., Dissanayake, A., Galloway, T.S., Lowe, D.M. and Thompson, R.C. (2008) Ingested microscopic plastic translocates to the circulatory system of the mussel, *Mytilus edulis* (L.). *Environmental Science & Technology* 42(13), 5026–5031.

Browne, M.A., Niven, S.J., Galloway, T.S., Rowland, S.J. and Thompson, R.C. (2013) Microplastic moves pollutants and additives to worms, reducing functions linked to health and biodiversity. *Current Biology* 23(23), 2388–2392.

Browne, M.A., Underwood, A.J., Chapman, M.G., Williams, R., Thompson, R.C. and van Franeker, J.A. (2015) Linking effects of anthropogenic debris to ecological impacts. *Proceedings of the Royal Society B: Biological Sciences* 282(1807), 20142929.

Bugoni, L., Krause, L.G. and Virgínia Petry, M. (2001) Marine debris and human impacts on sea turtles in Southern Brazil. *Marine Pollution Bulletin* 42(12), 1330–1334.

Butterworth, A. and Sayer, S. (2017) The welfare impact on pinnipeds of marine debris and fisheries. *Marine Mammal Welfare: Human Induced Change in the Marine Environment and its Impacts on Marine Mammal Welfare*. Butterworth, A. (ed.), pp. 215–239, Springer International Publishing, Cham, Switzerland.

Capolupo, M., Franzellitti, S., Valbonesi, P., Lanzas, C.S. and Fabbri, E. (2018) Uptake and transcriptional effects of polystyrene microplastics in larval stages of the Mediterranean mussel *Mytilus galloprovincialis*. *Environmental Pollution* 241, 1038–1047.

Carbery, M., O'Connor, W. and Palanisami, T. (2018) Trophic transfer of microplastics and mixed contaminants in the marine food web and implications for human health. *Environment International* 115, 400–409.

Carey, M.J. (2011) Intergenerational transfer of plastic debris by short-tailed shearwaters (*Ardenna tenuirostris*). *Emu: Austral Ornithology* 111(3), 229–234.

Choi, J.S., Jung, Y.-J., Hong, N.-H., Hong, S.H. and Park, J.-W. (2018) Toxicological effects of irregularly shaped and spherical microplastics in a marine teleost, the sheepshead minnow (*Cyprinodon variegatus*). *Marine Pollution Bulletin* 129(1), 231–240.

Cole, M., Coppock, R., Lindeque, P.K., Altin, D., Reed, S., Pond, D.W., Sørensen, L., Galloway, T.S. and Booth, A.M. (2019) Effects of nylon microplastic on feeding, lipid accumulation, and moulting in a coldwater copepod. *Environmental Science & Technology* 53(12), 7075–7082.

Cole, M., Lindeque, P., Fileman, E., Halsband, C. and Galloway, T.S. (2015) The impact of polystyrene microplastics on feeding, function and fecundity in the marine copepod *Calanus helgolandicus*. *Environmental Science & Technology* 49(2), 1130–1137.

Cole, M., Lindeque, P., Fileman, E., Halsband, C., Goodhead, R., Moger, J. and Galloway, T.S. (2013) Microplastic ingestion by zooplankton. *Environmental Science & Technology* 47(12), 6646–6655.

Cole, M., Lindeque, P.K., Fileman, E., Clark, J., Lewis, C., Halsband, C. and Galloway, T.S. (2016) Microplastics alter the properties and sinking rates of zooplankton faecal pellets. *Environmental Science & Technology* 50(6), 3239–3246.

Coppock, R.L., Galloway, T.S., Cole, M., Fileman, E.S., Queirós, A.M. and Lindeque, P.K. (2019) Microplastics alter feeding selectivity and faecal density in the copepod, Calanus helgolandicus. *Science of the Total Environment* 687, 780–789.

Davarpanah, E. and Guilhermino, L. (2015) Single and combined effects of microplastics and copper on the population growth of the marine microalgae *Tetraselmis chuii*. *Estuarine, Coastal and Shelf Science* 167, 269–275.

de Carvalho, D.G. and Baptista Neto, J.A. (2016) Microplastic pollution of the beaches of Guanabara Bay, Southeast Brazil. *Ocean & Coastal Management* 128, 10–17.

de Sá, L.C., Luís, L.G. and Guilhermino, L. (2015) Effects of microplastics on juveniles of the common goby (*Pomatoschistus microps*): Confusion with prey, reduction of the predatory performance and efficiency, and possible influence of developmental conditions. *Environmental Pollution* 196, 359–362.

de Stephanis, R., Giménez, J., Carpinelli, E., Gutierrez-Exposito, C. and Cañadas, A. (2013) As main meal for sperm whales: Plastics debris. *Marine Pollution Bulletin* 69(1), 206–214.

Della Torre, C., Bergami, E., Salvati, A., Faleri, C., Cirino, P., Dawson, K.A. and Corsi, I. (2014) Accumulation and embryotoxicity of polystyrene nanoparticles at early stage of development of sea urchin embryos *Paracentrotus lividus*. *Environmental Science and Technology* 48(20), 12302–12311.

Desforges, J.-P.W., Galbraith, M. and Ross, P.S. (2015) Ingestion of microplastics by zooplankton in the northeast Pacific Ocean. *Archives of Environmental Contamination and Toxicology 69*, 320–330.

Diepens, N.J. and Koelmans, A.A. (2018) Accumulation of plastic debris and associated contaminants in aquatic food webs. *Environmental Science & Technology* 52, 8510–8520.

Espinosa, C., Cuesta, A. and Esteban, M.Á. (2017) Effects of dietary polyvinylchloride microparticles on general health, immune status and expression of several genes related to stress in gilthead seabream (*Sparus aurata* L.). *Fish & Shellfish Immunology* 68, 251–259.

Espinosa, C., García Beltrán, J.M., Esteban, M.A. and Cuesta, A. (2018) *In vitro* effects of virgin microplastics on fish head-kidney leucocyte activities. *Environmental Pollution* 235, 30–38.

Everaert, G., Van Cauwenberghe, L., De Rijcke, M., Koelmans, A.A., Mees, J., Vandegehuchte, M. and Janssen, C.R. (2018) Risk assessment of microplastics in the ocean: Modelling approach and first conclusions. *Environmental Pollution* 242, 1930–1938.

Farkas, J. and Booth, A.M. (2017) Are fluorescence-based chlorophyll quantification methods suitable for algae toxicity assessment of carbon nanomaterials? *Nanotoxicology* 11(4), 569–577.

Farrell, P. and Nelson, K. (2013) Trophic level transfer of microplastic: *Mytilus edulis* (L.) to *Carcinus maenas* (L.). *Environmental Pollution* 177, 1–3.

Foley, C.J., Feiner, Z.S., Malinich, T.D. and Höök, T.O. (2018) A meta-analysis of the effects of exposure to microplastics on fish and aquatic invertebrates. *Science of the Total Environment* 631–632, 550–559.

Fossi, M.C., Baini, M., Panti, C., Galli, M., Jiménez, B., Muñoz-Arnanz, J., Marsili, L., Finoia, M.G. and Ramírez-Macías, D. (2017) Are whale sharks exposed to persistent organic pollutants and plastic pollution in the Gulf of California (Mexico)? First ecotoxicological investigation using skin biopsies. *Comparative Biochemistry and Physiology Part C: Toxicology & Pharmacology* 199, 48–58.

Fossi, M.C., Panti, C., Guerranti, C., Coppola, D., Giannetti, M., Marsili, L. and Minutoli, R. (2012) Are baleen whales exposed to the threat of microplastics? A case study of the Mediterranean fin whale (*Balaenoptera physalus*). *Marine Pollution Bulletin* 64(11), 2374–2379.

Furness, B.L. (1983) Plastic particles in three procellariiform seabirds from the Benguela Current, South Africa. *Marine Pollution Bulletin* 14(8), 307–308.

Gall, S.C. and Thompson, R.C. (2015) The impact of debris on marine life. *Marine Pollution Bulletin* 92(1), 170–179.

Gambardella, C., Morgana, S., Bramini, M., Rotini, A., Manfra, L., Migliore, L., Piazza, V., Garaventa, F. and Faimali, M. (2018) Ecotoxicological effects of polystyrene microbeads in a battery of marine organisms belonging to different trophic levels. *Marine Environmental Research* 141, 313–321.

Gandara e Silva, P.P., Nobre, C.R., Resaffe, P., Pereira, C.D.S. and Gusmão, F. (2016) Leachate from microplastics impairs larval development in brown mussels. *Water Research* 106, 364–370.

Gewert, B., Plassmann, M.M. and MacLeod, M. (2015) Pathways for degradation of plastic polymers floating in the marine environment. *Environmental Science: Processes & Impacts* 17(9), 1513–1521.

Goldstein, M.C. and Goodwin, D.S. (2013) Gooseneck barnacles (*Lepas* spp.) ingest microplastic debris in the North Pacific subtropical gyre. *PeerJ* 1, e184.

Green, D.S. (2016) Effects of microplastics on European flat oysters, *Ostrea edulis* and their associated benthic communities. *Environmental Pollution* 216, 95–103.

Gurevitch, J., Curtis, P.S. and Jones, M.H. 2001. Meta-analysis in ecology. *Advances in Ecological Research* 32, 199–247.

Haegerbaeumer, A., Mueller, M.-T., Fueser, H. and Traunspurger, W. (2019) Impacts of Micro- and nano-sized plastic particles on benthic invertebrates: A literature review and gap analysis. *Frontiers in Environmental Science* 7, 17.

Hämer, J., Gutow, L., Köhler, A. and Saborowski, R. (2014) Fate of microplastics in the marine isopod idotea emarginata. *Environmental Science & Technology* 48(22), 13451–13458.

Hansen, B., Hansen, P.J. and Nielsen, T.G. (1991) Effects of large nongrazable particles on clearance and swimming behaviour of zooplankton. *Journal of Experimental Marine Biology and Ecology* 152(2), 257–269.

Hedges, L.V. and Olkin, I. (1985) *Statistical Methods for Meta-Analysis*. Academic Press, San Diego, CA.

Herzke, D., Anker-Nilssen, T., Nøst, T.H., Götsch, A., Christensen-Dalsgaard, S., Langset, M., Fangel, K. and Koelmans, A.A. (2016) Negligible impact of ingested microplastics on tissue concentrations of persistent organic pollutants in northern fulmars off coastal Norway. *Environmental Science & Technology* 50(4), 1924–1933.

Hund-Rinke, K., Baun, A., Cupi, D., Fernandes, T.F., Handy, R., Kinross, J.H., Navas, J.M., et al. (2016) Regulatory ecotoxicity testing of nanomaterials: Proposed modifications of OECD test guidelines based on laboratory experience with silver and titanium dioxide nanoparticles. *Nanotoxicology* 10(10), 1442–1447.

Jeong, C.-B., Won, E.-J., Kang, H.-M., Lee, M.-C., Hwang, D.-S., Hwang, U.-K., Zhou, B., Souissi, S., Lee, S.-J. and Lee, J.-S. (2016) Microplastic size-dependent toxicity, oxidative stress induction, and p-JNK and p-p38 activation in the monogonont rotifer (*Brachionus koreanus*). *Environmental Science & Technology* 50(16), 8849–8857.

Kanhai, L.D.K., Gårdfeldt, K., Lyashevska, O., Hassellöv, M., Thompson, R.C. and O'Connor, I. (2018) Microplastics in sub-surface waters of the Arctic Central Basin. *Marine Pollution Bulletin* 130, 8–18.

Kaposi, K.L., Mos, B., Kelaher, B.P. and Dworjanyn, S.A. (2014) Ingestion of microplastic has limited impact on a marine larva. *Environmental Science & Technology* 48(3), 1638–1645.

Koelmans, A.A. (2015) Modeling the role of microplastics in bioaccumulation of organic chemicals to marine aquatic organisms: A critical review. *Marine Anthropogenic Litter.* Bergmann, M., Gutow, L. and Klages, M. (eds.), pp. 309–324, Springer, Cham, Switzerland.

Koelmans, A.A., Besseling, E. and Foekema, E.M. (2014) Leaching of plastic additives to marine organisms. *Environmental Pollution* 187, 49–54.

Koelmans, A.A., Besseling, E., Wegner, A. and Foekema, E.M. (2013) Plastic as a carrier of POPs to aquatic organisms: A model analysis. *Environmental Science & Technology* 47(14), 7812–7820.

Kühn, S., Bravo Rebolledo, E.L. and van Franeker, J.A. (2015) Deleterious effects of litter on marine life. *Marine Anthropogenic Litter.* Bergmann, M., Gutow, L. and Klages, M. (eds.), pp. 75–116, Springer International Publishing, Cham, Switzerland.

Laist, D.W. (1987) Overview of the biological effects of lost and discarded plastic debris in the marine environment. *Marine Pollution Bulletin* 18(6, Supplement 2), 319–326.

Laist, D.W. (1997) Impacts of marine debris: Entanglement of marine life in marine debris including a comprehensive list of species with entanglement and ingestion records. *Marine Debris: Sources, Impacts, and Solutions.* Coe, J.M. and Rogers, D.B. (eds.), pp. 99–139, Springer New York, New York.

Law, K.L. (2017) Plastics in the marine environment. *Annual Review of Marine Science* 9(1), 205–229.

Law, K.L., Moret-Ferguson, S.E., Goodwin, D.S., Zettler, E.R., De Force, E., Kukulka, T. and Proskurowski, G. (2014) Distribution of surface plastic debris in the Eastern Pacific Ocean from an 11-year data set. *Environmental Science & Technology* 48(9), 4732–4738.

Lee, H., Lee, H.-J. and Kwon, J.-H. (2019) Estimating microplastic-bound intake of hydrophobic organic chemicals by fish using measured desorption rates to artificial gut fluid. *Science of the Total Environment* 651, 162–170.

Lee, K.-W., Shim, W.J., Kwon, O.Y. and Kang, J.-H. (2013) Size-dependent effects of micro polystyrene particles in the marine copepod *Tigriopus japonicus. Environmental Science & Technology* 47(19), 11278–11283.

Lusher, A. (2015) Microplastics in the marine environment: Distribution, interactions and effects. *Marine Anthropogenic Litter.* Bergmann, M., Gutow, L. and Klages, M. (eds.), pp. 245–308, Springer International Publishing, Cham, Switzerland.

Maes, T., Van der Meulen, M.D., Devriese, L.I., Leslie, H.A., Huvet, A., Frère, L., Robbens, J. and Vethaak, A.D. (2017) Microplastics baseline surveys at the water surface and in sediments of the North-East Atlantic. *Frontiers in Marine Science* 4(135), 1–13.

Moore, C.J., Moore, S.L., Leecaster, M.K. and Weisberg, S.B. (2001) A comparison of plastic and plankton in the North Pacific central gyre. *Marine Pollution Bulletin* 42(12), 1297–1300.

Naji, A., Esmaili, Z., Mason, S.A. and Dick Vethaak, A. (2017) The occurrence of microplastic contamination in littoral sediments of the Persian Gulf, Iran. *Environmental Science and Pollution Research* 24(25), 20459–20468.

Nelms, S.E., Galloway, T.S., Godley, B.J., Jarvis, D.S. and Lindeque, P.K. (2018) Investigating microplastic trophic transfer in marine top predators. *Environmental Pollution* 238, 999–1007.

Ogonowski, M., Wenman, V., Barth, A., Hamacher-Barth, E., Danielsson, S. and Gorokhova, E. (2019) Microplastic intake, its biotic drivers, and hydrophobic organic contaminant levels in the Baltic herring. *Frontiers in Environmental Science* 7, 134.

Oliveira, M., Ribeiro, A., Hylland, K. and Guilhermino, L. (2013) Single and combined effects of microplastics and pyrene on juveniles (0+ group) of the common goby *Pomatoschistus microps* (Teleostei, Gobiidae). *Ecological Indicators* 34, 641–647.

Oliviero, M., Tato, T., Schiavo, S., Fernández, V., Manzo, S. and Beiras, R. (2019) Leachates of micronized plastic toys provoke embryotoxic effects upon sea urchin *Paracentrotus lividus. Environmental Pollution* 247, 706–715.

Poon, F.E., Provencher, J.F., Mallory, M.L., Braune, B.M. and Smith, P.A. (2017) Levels of ingested debris vary across species in Canadian Arctic seabirds. *Marine Pollution Bulletin* 116(1–2), 517–520.

Provencher, J.F., Vermaire, J.C., Avery-Gomm, S., Braune, B.M. and Mallory, M.L. (2018) Garbage in guano? Microplastic debris found in faecal precursors of seabirds known to ingest plastics. *Science of the Total Environment* 644, 1477–1484.

Ramos, J.A.A., Barletta, M. and Costa, M.F. (2012) Ingestion of nylon threads by Gerreidae while using a tropical estuary as foraging grounds. *Aquatic Biology* 17(1), 29–34.

Rech, S., Salmina, S., Borrell Pichs, Y.J. and García-Vazquez, E. (2018) Dispersal of alien invasive species on anthropogenic litter from European mariculture areas. *Marine Pollution Bulletin* 131, 10–16.

Reisser, J., Shaw, J., Hallegraeff, G., Proietti, M., Barnes, D.K.A., Thums, M., Wilcox, C., Hardesty, B.D. and Pattiaratchi, C. (2014) Millimeter-sized marine plastics: A new pelagic habitat for microorganisms and invertebrates. *Plos One* 9(6), e100289.

Rist, S.E., Assidqi, K., Zamani, N.P., Appel, D., Perschke, M., Huhn, M. and Lenz, M. (2016) Suspended micro-sized PVC particles impair the performance and decrease survival in the Asian green mussel *Perna viridis*. *Marine Pollution Bulletin* 111(1–2), 213–220.

Rochman, C.M., Browne, M.A., Underwood, A.J., van Franeker, J.A., Thompson, Richard C. and Amaral-Zettler, L.A. (2016) The ecological impacts of marine debris: Unraveling the demonstrated evidence from what is perceived. *Ecology* 97(2), 302–312.

Ross, P.S. and Morales-Caselles, C. (2015) Out of sight, but no longer out of mind: Microplastics as a global pollutant. *Integrated Environmental Assessment and Management* 11(4), 721–722.

Ryan, P.G., Moore, C.J., van Franeker, J.A. and Moloney, C.L. (2009) Monitoring the abundance of plastic debris in the marine environment. *Philosophical Transactions of the Royal Society B: Biological Sciences* 364(1526), 1999–2012.

Santos, R.G., Andrades, R., Fardim, L.M. and Martins, A.S. (2016) Marine debris ingestion and Thayer's law: The importance of plastic color. *Environmental Pollution* 214, 585–588.

Savoca, M.S., Wohlfeil, M.E., Ebeler, S.E. and Nevitt, G.A. (2016) Marine plastic debris emits a keystone infochemical for olfactory foraging seabirds. *Science Advances* 2(11), e1600395.

Savoca, S., Capillo, G., Mancuso, M., Bottari, T., Crupi, R., Branca, C., Romano, V., Faggio, C., D'Angelo, G. and Spanò, N. (2019) Microplastics occurrence in the Tyrrhenian waters and in the gastrointestinal tract of two congener species of seabreams. *Environmental Toxicology and Pharmacology* 67, 35–41.

Schuyler, Q., Hardesty, B.D., Wilcox, C. and Townsend, K. (2014) Global analysis of anthropogenic debris ingestion by sea turtles. *Conservation Biology* 28(1), 129–139.

Setälä, O., Fleming-Lehtinen, V. and Lehtiniemi, M. (2014) Ingestion and transfer of microplastics in the planktonic food web. *Environmental Pollution* 185, 77–83.

Setälä, O., Norkko, J. and Lehtiniemi, M. (2016) Feeding type affects microplastic ingestion in a coastal invertebrate community. *Marine Pollution Bulletin* 102(1), 95–101.

Simmonds, M.P. (2012) Cetaceans and marine debris: The great unknown. *Journal of Marine Biology* 2012, 8.

Sjollema, S.B., Redondo-Hasselerharm, P., Leslie, H.A., Kraak, M.H.S. and Vethaak, A.D. (2016) Do plastic particles affect microalgal photosynthesis and growth? *Aquatic Toxicology* 170, 259–261.

Solomon, O.O. and Palanisami, T. (2016) Microplastics in the marine environment: Current status, assessment methodologies, impacts and solutions. *Journal of Pollution Effects & Control* 4(3), 161.

Sun, X., Liu, T., Zhu, M., Liang, J., Zhao, Y. and Zhang, B. (2018) Retention and characteristics of microplastics in natural zooplankton taxa from the East China Sea. *Science of the Total Environment* 640–641, 232–242.

Sussarellu, R., Suquet, M., Thomas, Y., Lambert, C., Fabioux, C., Pernet, M.E.J., Le Goïc, N., et al. (2016) Oyster reproduction is affected by exposure to polystyrene microplastics. *Proceedings of the National Academy of Sciences* 113(9), 2430–2435.

Tanaka, K., Takada, H., Yamashita, R., Mizukawa, K., Fukuwaka, M.-A. and Watanuki, Y. (2015) Facilitated leaching of additive-derived PBDEs from plastic by seabirds' stomach oil and accumulation in tissues. *Environmental Science & Technology* 49(19), 11799–11807.

Tanaka, K., Takada, H., Yamashita, R., Mizukawa, K., Fukuwaka, M.A. and Watanuki, Y. (2013) Accumulation of plastic-derived chemicals in tissues of seabirds ingesting marine plastics. *Marine Pollution Bulletin* 69(1–2), 219–222.

Tosetto, L., Brown, C. and Williamson, J.E. (2016) Microplastics on beaches: Ingestion and behavioural consequences for beach hoppers. *Marine Biology* 163(10), 199.

Trevail, A.M., Gabrielsen, G.W., Kühn, S. and Van Franeker, J.A. (2015) Elevated levels of ingested plastic in a high Arctic seabird, the northern fulmar (*Fulmarus glacialis*). *Polar Biology* 38(7), 975–981.

Ugolini, A., Ungherese, G., Ciofini, M., Lapucci, A. and Camaiti, M. (2013) Microplastic debris in sandhoppers. *Estuarine, Coastal and Shelf Science* 129, 19–22.

Van Cauwenberghe, L., Devriese, L., Galgani, F., Robbens, J. and Janssen, C.R. (2015) Microplastics in sediments: A review of techniques, occurrence and effects. *Marine Environmental Research* 111, 5–17.

Van Cauwenberghe, L., Vanreusel, A., Mees, J. and Janssen, C.R. (2013) Microplastic pollution in deep-sea sediments. *Environmental Pollution* 182, 495–499.

van Franeker, J.A., Blaize, C., Danielsen, J., Fairclough, K., Gollan, J., Guse, N., Hansen, P.-L. et al. (2011) Monitoring plastic ingestion by the northern fulmar *Fulmarus glacialis* in the North Sea. *Environmental Pollution* 159(10), 2609–2615.

von Moos, N., Burkhardt-Holm, P. and Köhler, A. (2012) Uptake and effects of microplastics on cells and tissue of the blue mussel *Mytilus edulis* L. after an experimental exposure. *Environmental Science & Technology* 46(20), 11327–11335.

Votier, S.C., Archibald, K., Morgan, G. and Morgan, L. (2011) The use of plastic debris as nesting material by a colonial seabird and associated entanglement mortality. *Marine Pollution Bulletin* 62(1), 168–172.

Vroom, R.J.E., Koelmans, A.A., Besseling, E. and Halsband, C. (2017) Aging of microplastics promotes their ingestion by marine zooplankton. *Environmental Pollution* 231, Part 1, 987–996.

Wegner, A., Besseling, E., Foekema, E.M., Kamermans, P. and Koelmans, A.A. (2012) Effects of nanopolystyrene on the feeding behavior of the blue mussel (*Mytilus edulis* L.). *Environmental Toxicology and Chemistry* 31(11), 2490–2497.

Welden, N.A., Abylkhani, B. and Howarth, L.M. (2018) The effects of trophic transfer and environmental factors on microplastic uptake by plaice, Pleuronectes plastessa, and spider crab, *Maja squinado. Environmental Pollution* 239, 351–358.

Welden, N.A.C. and Cowie, P.R. (2016) Long-term microplastic retention causes reduced body condition in the langoustine, Nephrops norvegicus. *Environmental Pollution* 218, 895–900.

Wilcox, C., Van Sebille, E. and Hardesty, B.D. (2015) Threat of plastic pollution to seabirds is global, pervasive, and increasing. *Proceedings of the National Academy of Sciences* 112(38), 11899–11904.

Williams, R., Ashe, E. and O'Hara, P.D. (2011) Marine mammals and debris in coastal waters of British Columbia, Canada. *Marine Pollution Bulletin* 62(6), 1303–1316.

Wright, S.L., Rowe, D., Thompson, R.C. and Galloway, T.S. (2013a) Microplastic ingestion decreases energy reserves in marine worms. *Current Biology* 23(23), R1031–R1033.

Wright, S.L., Thompson, R.C. and Galloway, T.S. (2013b) The physical impacts of microplastics on marine organisms: A review. *Environmental Pollution* 178, 483–492.

Zhang, C., Chen, X., Wang, J. and Tan, L. (2017) Toxic effects of microplastic on marine microalgae *Skeletonema costatum*: Interactions between microplastic and algae. *Environmental Pollution* 220, 1282–1288.

Zhao, T., Tan, L., Huang, W. and Wang, J. (2019) The interactions between micro polyvinyl chloride (mPVC) and marine dinoflagellate *Karenia mikimotoi*: The inhibition of growth, chlorophyll and photosynthetic efficiency. *Environmental Pollution* 247, 883–889.

16 Sub-Lethal Responses to Microplastic Ingestion in Invertebrates

Toward a Mechanistic Understanding Using Energy Flux

Charlene Trestrail, Jeff Shimeta, and Dayanthi Nugegoda

CONTENTS

16.1 Introduction 247
16.1.1 Microplastic Exposure Routes in Invertebrates 248
16.1.2 Biological Responses to Microplastics 249
16.1.3 Toward a Mechanistic Understanding of Sub-Lethal Responses: Tracking Energy Flux 261
16.1.4 The Tenets of DEB Theory 262
16.1.5 Applying DEB Theory to Explain Microplastic-Induced Sub-Lethal Responses 264
16.2 The Effects of Microplastic Ingestion on Energy Acquisition 265
16.2.1 Dietary Dilution 265
16.2.2 Feeding Behavior 266
16.2.2.1 Mechanisms behind Altered Feeding Behavior 267
16.2.3 Digestion 267
16.2.4 Assimilation 269
16.3 The Effects of Microplastic Ingestion on Energy Allocation 269
16.3.1 Increasing Maintenance Costs 269
16.3.1.1 Increasing Maintenance Costs of the Antioxidant System 270
16.3.1.2 Increasing Maintenance Costs of the Immune System 271
16.4 Conclusions and Suggestions for Future Work 272
References 273

16.1 INTRODUCTION

Microplastics contaminate terrestrial and aquatic environments across the globe (Browne et al. 2010; Horton et al. 2017; Lebreton et al. 2017; Wagner and Lambert 2018). Microplastic pollution is so ubiquitous, that it has been reported in locations as geographically remote as Antarctic islands (Cincinelli et al. 2017; Lavers et al. 2017) and the deep sea (Van Cauwenberghe et al. 2013). Concentrations of microplastics vary between ecosystems. They are particularly prevalent in ocean ecosystems, where they have become the dominant form of pollution (Thompson 2004), with an estimated 51 trillion microplastics present at the sea surface (van Sebille et al. 2015). The widespread distribution and high abundance of microplastics mean that many organisms will encounter microplastics in their environment.

While there is no consensus on what constitutes a microplastic (Frias and Nash 2019), they are commonly defined as plastic particles measuring 1 μm–5 mm at their widest point (Arthur et al. 2009; Koelmans et al. 2015; Rios Mendoza et al. 2018). This small size makes microplastics available for consumption by invertebrates. Laboratory studies have observed microplastic consumption in several invertebrate taxa and across various feeding modes (Allen et al. 2017; Avio et al. 2015; Cole et al. 2013; Hämer et al. 2014; Sussarellu et al. 2016; Van Cauwenberghe et al. 2015; Vroom et al. 2017; Welden and Cowie 2016a). The discovery of microplastics in the digestive tracts of invertebrates collected from the field confirms that microplastic consumption is occurring in natural settings (McNeish et al. 2018; Van Cauwenberghe et al. 2015; Waite et al. 2018).

It is important to understand how invertebrates respond to microplastics because invertebrates account for the majority of organisms on earth (Mora et al. 2011). Many invertebrate species are important sources of protein for humans and agricultural animals (Ramos-Elorduy 1997; Sánchez-Muros et al. 2014), and the sale of invertebrates can contribute substantially to national economies (Chávez 2007; Purcell 2014). More importantly, invertebrates play key roles in a plethora of ecosystem services, including nutrient cycling, pollination and seed dispersal, and maintaining water quality (Losey and Vaughan 2006; Prather et al. 2013). Ecosystem functions are directly affected when invertebrate populations are reduced (Mulder et al. 1999). Since invertebrates are sensitive to a range of anthropogenic contaminants (Carsten von der Ohe and Liess 2004; Hickey and Martin 1995), their response to microplastics warrants serious consideration.

16.1.1 Microplastic Exposure Routes in Invertebrates

Ingestion is the most widely studied route of exposure to microplastics for invertebrates and for animals in general. Why invertebrates deliberately ingest microplastics is unclear, and it is particularly perplexing when it occurs in invertebrates that display strong particle selection (Cruz-Rivera and Hay 2000; Ward and Shumway 2004). It is thought that the size and surface chemistry of microplastics are similar to those of legitimate food particles, causing some invertebrates to misinterpret microplastics as food (Allen et al. 2017; Cole et al. 2013). Bivalves may have no choice, but to consume captured microplastics since, in these invertebrates, particle selection is a passive process driven by size and surface charge (Rosa et al. 2018; Ward and Shumway 2004); captured microplastics that meet the right criteria will pass pre-ingestion sorting and be consumed.

Microplastic ingestion can also occur incidentally, such as when gastropods ingest microplastics attached to the surface of the seaweed they graze on (Gutow et al. 2016), or when deposit-feeding annelids ingest mouthfuls of sediment that contain microplastics (Wright et al. 2013). The tendency for microplastics to adhere to the appendages of prey organisms (Bergami et al. 2016; Cole et al. 2013) will facilitate incidental microplastic ingestion by predatory invertebrates. Even invertebrates whose feeding modes would not normally facilitate microplastic capture, such as detritivores and scavengers, may incidentally ingest microplastics when they consume the flesh of organisms that have themselves consumed microplastics (Watts et al. 2014). Incidental microplastic ingestion can facilitate the transfer of microplastics to higher trophic levels (Farrell and Nelson 2013; Murray and Cowie 2011; Setälä et al. 2014), increasing the likelihood that invertebrates at all trophic levels will consume microplastics over the course of their lives.

For aquatic invertebrates, exterior-facing membranes, such as gill surfaces present another exposure route to microplastics. Inspiration draws suspended microplastics into the gills, and very small microplastics can pass across the gill surfaces (von Moos et al. 2012; Watts et al. 2014). Transport across epithelial boundaries may be a significant exposure pathway for nano-sized microplastics, which can permeate lipid membranes (Rossi et al. 2014; Salvati et al. 2011). However, it is unclear whether this exposure route is significant for microplastics larger than 1 μm and requires further investigation. Therefore, in this chapter, we focus only on biological responses to ingested microplastics.

16.1.2 Biological Responses to Microplastics

The standard way to assess the biological effect of a contaminant is by assessing toxicity. This is done by quantifying the dose-dependent relationship between contaminant concentration and the percentage of the test population it kills over a given exposure period (Trevan 1927). Lethality tests are usually the first tests conducted on new toxicants, but this technique appears to yield limited information when applied to microplastic ingestion.

Some attempts to conduct lethality tests for microplastics in invertebrate species were unsuccessful because microplastic ingestion did not cause lethality (Jemec et al. 2016; Lee et al. 2013). Other studies did determine lethal concentrations of microplastic ingestion (Table 16.1); in all cases, invertebrate species were exposed to strikingly high concentrations of microplastics before mortality occurred. For example, 10 days' continuous exposure to 4.64×10^4 microplastics mL^{-1} were required before 50% of the test amphipods, *Hyalella azteca*, were killed (Au et al. 2015). Similarly, only 30% of the test shrimp, *Neomysis japonica*, were killed by 3 days' exposure to 1×10^6 µg mL^{-1} microplastic spheres (Wang et al. 2017). These lethal concentrations are several orders of magnitude higher than estimates of environmental microplastic concentrations (see other chapters within this book). Therefore, under current environmental conditions, microplastic ingestion is unlikely to result in an acute lethal response.

Despite these results, significant increases in mortality outside of a dose-dependent relationship have been reported for several invertebrate species (Table 16.2). Clearly, microplastics can cause death under the right conditions. The fact that this is not detected in a dose-dependent manner during lethality tests suggests that microplastics cause death through an indirect mechanism, and that lethality itself may be an inappropriate endpoint for assessing these responses. Perhaps it is unsurprising that traditional, lethal measures of toxicity have limited applicability to plastics; after all, the carbon polymer chain of microplastics is chemically inert (Lithner et al. 2011; Teuten et al. 2009), so microplastics are unlikely to exert the same effect as chemical toxicants.

In contrast to lethality, life history processes, such as growth and reproduction are more sensitive to toxicant exposure (Dhawan et al. 1999; Manfra et al. 2016; McLoughlin et al. 2000); thus, endpoints associated with these life history processes will detect biological responses to a toxicant at a concentration below that which causes lethality. Although not as extreme as death, sub-lethal

TABLE 16.1
Lethal Concentrations (LCs) of Microplastics Reported for Invertebrates

Species	Plastic Type	Shape	Diameter (µm)	Duration (d)	LC_x	Source
Hyella azteca	PE	Sphere	10–27	10	LC_{50}: 4.64×10^4 microplastics mL^{-1}	Au et al. (2015)
	PP	Fibers	20–75[a]	10	LC_{50}: 71.4 microplastics mL^{-1}	
Neomysis japonica	PS	Sphere	5	3	LC_{30}: 1×10^6 µg mL^{-1}	Wang et al. (2017)
Tigriopus japonicus	PS	Sphere	0.05	14	LC_{50}: 2.15 µg mL^{-1}[b]	Lee et al. (2013)
			0.05	14	LC_{50}: 0.16 µg mL^{-1}[c]	
			0.5	14	LC_{50}: 23.5 µg mL^{-1}[c]	

Note: X indicates the percentage of the test population killed. LC values have been converted to number/weight of microplastics mL^{-1} to aid comparison.

Abbreviations: PE, polyethylene; PP, polypropylene; PS, polystyrene.

[a] Length.

[b] First generation exposed (F0).

[c] Second generation exposed (F1).

TABLE 16.2
Microplastic-Induced Changes in Invertebrate Energy Acquisition and Energy Expenditure

Organism		Energy Acquisition Markers				Energy Reserve Markers	Somatic Sink Markers (*k* Fraction)					Reproductive Sink Markers (1 – *k* Fraction)			
							Maintenance					Maintenance	Reproductive Growth		
Taxonomy	Species	Particle Selection	Feeding Rate	Gut Residence Time	Energy Assimilation	Energy Content	Survival	Metabolism	AOS	Behavior	Somatic Growth	Immune System	Maturity	Reproductive Buffer	Reference
Annelida															
Clitellata	*Eisenia fetida*							↑ Metabolic enzyme activity	↑AOS enzyme activity ↑Lipid damage						Rodríguez-Seijo et al. (2018)
	Lumbricus terrestris			NC g cast worm^{-1} 60 d^{-1}			↓				↓ mg worm^{-1} d^{-1}			NC # cocoons worm^{-1} mo^{-1} NC Cocoon mass	Huerta Lwanga et al. (2016)
Polychaeta	*Perinereis aibuhitensis*						↓				↓ Segment regeneration rate				Leung and Chan (2018)
	Arenicola marina		↓ No. ccasts worm^{-1} wk^{-1} ↓* % feeding individuals	↓ No. egestion events worm^{-1} h^{-1}		NC [Carbohydrate] ↓ [Lipid] ↓* [Protein] ↓ Total energy reserves					NC Mass	↑ Phagocytic cell count			Wright et al. (2013)
	Arenicola marina		↓ No. heaps worm^{-1} d^{-1}				NC				↓ Mass				Besseling et al. (2013)
	Arenicola marina					NC [Carbohydrate] NC [Lipid] ↑ [Protein] NC Cellular energy allocation		NC Energy consumption							Van Cauwenberghe et al. (2015)

(Continued)

TABLE 16.2 (*Continued*)
Microplastic-Induced Changes in Invertebrate Energy Acquisition and Energy Expenditure

Organism		Energy Acquisition Markers				Energy Reserve Markers	Somatic Sink Markers (k Fraction)					Reproductive Sink Markers ($1 - k$ Fraction)			
							Maintenance					Maintenance	Reproductive Growth		
Taxonomy	Species	Particle Selection	Feeding Rate	Gut Residence Time	Energy Assimilation	Energy Content	Survival	Metabolism	AOS	Behavior	Somatic Growth	Immune System	Maturity	Reproductive Buffer	Reference
	Arenicola marina		↑* No. casts worm^{-1} d^{-1}				NC		↓ Reducing capacity			↑* Phagocytic activity			Browne et al. (2013)
	Arenicola marina		↓ No. total casts ↓ Cast size				NC	↑ Respiration			NC Mass				Green et al. (2016)
Arthropoda															
Branchiopoda Order: Cladocera	*Daphnia magna* (adults & offspring)		↓ mg carbon				↓ Adult survival NC Offspring survival				↓ Growth rate		NC Age at maturity	NC Time between broods ↓ No. broods individual^{-1} ↓ No. offspring individual^{-1} NC offspring size	Ogonowski et al. (2016)
	Daphnia magna						↓				NC Body size				Jemec et al. (2016)
Insecta	*Chironomus tepperi* (larvae)						↓				↓ Body size ↓ Development rate				Ziajahromi et al. (2018)
Malacostraca Order: Amphipoda	*Hyalella azteca*			↑ Egestion time (hr)			↓				↓ mg amphipod^{-1}			↓ No. neonates female^{-1}	Au et al. (2015)

(*Continued*)

TABLE 16.2 (*Continued*)
Microplastic-Induced Changes in Invertebrate Energy Acquisition and Energy Expenditure

Organism		Energy Acquisition Markers				Energy Reserve Markers	Somatic Sink Markers (k Fraction)					Reproductive Sink Markers ($1-k$ Fraction)			
							Maintenance					Maintenance	Reproductive Growth		
Taxonomy	Species	Particle Selection	Feeding Rate	Gut Residence Time	Energy Assimilation	Energy Content	Survival	Metabolism	AOS	Behavior	Somatic Growth	Immune System	Maturity	Reproductive Buffer	Reference
	Gammarus fossarum		NC mg consumed mg wet weight^{-1} d^{-1}		↓ % energy assimilated						NC mg wet weight				Blarer and Burkhardt-Holm (2016)
	Gammarus fossarum		↓ mg consumed mg wet weight^{-1} d^{-1}		↓ % energy assimilated						↓ mg wet weight				Straub et al. (2017)
	Gammarus pulex		NC mg consumed mg wet weight^{-1} d^{-1}			NC [Lipid] NC [Glycogen]					NC Molt period (d)				Weber et al. (2018)
	Platorchestia smithi						↓			↓ Jump height ↑* Jump frequency ↑* Shelter searching time	↑ mg wet weight				Tosetto et al. (2016)
Malacostraca Order: Decapoda	*Carcinus maenas*		↓ g consumed d^{-1}			↓ Scope for growth									Watts et al. (2015)
	Nephrops norvegicus		↓* g consumed g wet weight^{-1} mo^{-1}			↓ [Lipid]		↓* [Hemolymph protein]			↓ Mass				Welden and Cowrie (2016b)

(*Continued*)

TABLE 16.2 (*Continued*)
Microplastic-Induced Changes in Invertebrate Energy Acquisition and Energy Expenditure

Organism		Energy Acquisition Markers				Energy Reserve Markers	Somatic Sink Markers (k Fraction)					Reproductive Sink Markers ($1 - k$ Fraction)			
							Maintenance					Maintenance	Reproductive Growth		
Taxonomy	Species	Particle Selection	Feeding Rate	Gut Residence Time	Energy Assimilation	Energy Content	Survival	Metabolism	AOS	Behavior	Somatic Growth	Immune System	Maturity	Reproductive Buffer	Reference
Malacostraca Order: Isopoda	*Idotea emarginata* (juveniles & adults)		↑*↓* mg consumed wet weight^{-1} d^{-1}				NC				NC Body size (mm) NC Intermolt duration				Hämer et al. (2014)
	Porcellio scaber		NC mg consumed mg wet weight^{-1} d^{-1}	NC mg feces wet weight^{-1} d^{-1}	NC Energy assimilation rate NC Energy assimilation efficiency	NC [Carbohydrate] NC [Lipid] NC [Protein]	NC				NC Mass				Jemec Kokalj et al. (2018)
Maxillopoda Order: Calanoida	*Calanus helgolandicus*	↓ Size of cells consumed	↓* Cells consumed copepod^{-1} d^{-1} ↓ µg C consumed copepod^{-1} d^{-1}				↓*	NC O_2 consumption						NC Egg production rate ↓ Egg diameter ↓ Egg hatching success	Cole et al. (2015)
Maxillopoda Order: Harpacticoida	*Tigriopus japonicus* (larvae & adults)		↑ No. particles copepod^{-1} h^{-1}				↓							↓ No. nauplius brood^{-1} female^{-1} ↓ Larval development rate	Lee et al. (2013)

(*Continued*)

TABLE 16.2 (*Continued*)
Microplastic-Induced Changes in Invertebrate Energy Acquisition and Energy Expenditure

Organism		Energy Acquisition Markers				Energy Reserve Markers	Somatic Sink Markers (*k* Fraction)					Reproductive Sink Markers (1 − *k* Fraction)			
							Maintenance					Maintenance	Reproductive Growth		
Taxonomy	Species	Particle Selection	Feeding Rate	Gut Residence Time	Energy Assimilation	Energy Content	Survival	Metabolism	AOS	Behavior	Somatic Growth	Immune System	Maturity	Reproductive Buffer	Reference
Chordata															
Ascidiacea	*Ciona robusta* (larvae & juveniles)						NC				↓ Rate of larval development NC No. phenotypic abnormalities				Messinetti et al. (2018)
Echinodermata															
Echinoidea	*Paracentrotus lividus* (larvae)						NC			↑↓ swimming speed			NC No. non-developed larvae		Gambardella et al. (2018)
	Paracentrotus lividus (larvae)						NC				↓↑ Body length ↓↑ Arm length ↓ Body width				Messinetti et al. (2018)
	Tripneustes gratilla (larvae)						↓*				↓ Body width NC Post oral arm length				Kaposi et al. (2014)
Mollusca															
Bivalvia	*Abra nitida*					NC [Carbohydrate] NC [Lipid] ↓ [Protein] NC Total energy NC Condition Index	NC			NC Burrowing activity					Bour et al. (2018)

(*Continued*)

TABLE 16.2 (*Continued*)
Microplastic-Induced Changes in Invertebrate Energy Acquisition and Energy Expenditure

Organism		Energy Acquisition Markers				Energy Reserve Markers	Somatic Sink Markers (k Fraction)					Reproductive Sink Markers ($1 - k$ Fraction)			
							Maintenance					Maintenance	Reproductive Growth		
Taxonomy	Species	Particle Selection	Feeding Rate	Gut Residence Time	Energy Assimilation	Energy Content	Survival	Metabolism	AOS	Behavior	Somatic Growth	Immune System	Maturity	Reproductive Buffer	Reference
	Atactodea striata		↓ mL filtered bivalve^{-1} min^{-1}		NC % energy assimilated			NC O_2 consumption							Xu et al. (2017)
	Corbicula fluminea		↓ mL filtered bivalve^{-1} min^{-1}					NC IDH & ODH activity	NC AOS enzymes ↑ Lipid damage						Oliveira et al. (2018)
	Crassostrea gigas (larvae)		↓* ng carbon consumed larva^{-1} d^{-1}								NC Body size				Cole and Galloway (2015)
	Crassostrea gigas (larvae & adults)		↑ µm^3 filtered bivalve^{-1} h^{-1}		↑ % energy assimilated									↓ No. eggs female^{-1} ↓ Egg size ↓ Sperm velocity ↓ Hatching success ↓ Larval development rate	Sussarellu et al. (2016)
	Crassostrea virginica								NC Lysosome stability						Gaspar et al. (2018)

(*Continued*)

TABLE 16.2 (*Continued*)
Microplastic-Induced Changes in Invertebrate Energy Acquisition and Energy Expenditure

Organism		Energy Acquisition Markers				Energy Reserve Markers	Somatic Sink Markers (k Fraction)					Reproductive Sink Markers ($1 - k$ Fraction)			
							Maintenance					Maintenance	Reproductive Growth		
Taxonomy	Species	Particle Selection	Feeding Rate	Gut Residence Time	Energy Assimilation	Energy Content	Survival	Metabolism	AOS	Behavior	Somatic Growth	Immune System	Maturity	Reproductive Buffer	Reference
	Ennucula tenius					↓* [Carbohydrate] ↓ [Lipid] ↓* [Protein] ↓ Total energy NC Condition Index	NC			NC Burrowing activity					Bour et al. (2018)
	Mytilus edulis		↓ Cells mg dry weight^{-1} h^{-1}												Green et al. (2017)
	Mytilus edulis					NC [Carbohydrate] NC [Lipid] NC [Protein] NC Cellular energy allocation		↑ Energy consumption							Van Cauwenberghe et al. (2015)

(*Continued*)

TABLE 16.2 (*Continued*)
Microplastic-Induced Changes in Invertebrate Energy Acquisition and Energy Expenditure

Organism		Energy Acquisition Markers				Energy Reserve Markers	Somatic Sink Markers (*k* Fraction)					Reproductive Sink Markers (1 − *k* Fraction)			
							Maintenance					Maintenance	Reproductive Growth		
Taxonomy	Species	Particle Selection	Feeding Rate	Gut Residence Time	Energy Assimilation	Energy Content	Survival	Metabolism	AOS	Behavior	Somatic Growth	Immune System	Maturity	Reproductive Buffer	Reference
	Mytilus galloprovincialis								NC AOS enzyme activity NC DNA damage ↑* Lipid damage NC Oxyradical scavenging capacity ↓ Lysosome stability			↑↓Phagocytic activity			Pittura et al. (2018)
	Mytilus galloprovincialis								↑ Oxidative stress response[a]		↓	↑ Immune response[a]			Détrée et al. (2018)
	Mytilus galloprovincialis							↑ & ↓ Metabolism[a]	↑ Oxidative stress response[a]			↑ Immune response[a]			Détrée et al. (2017)
	Mytilus galloprovincialis								NC AOS enzyme activity NC Lipid damage						Gonçalves et al. (2019)

(*Continued*)

TABLE 16.2 (*Continued*)

Microplastic-Induced Changes in Invertebrate Energy Acquisition and Energy Expenditure

Organism		Energy Acquisition Markers				Energy Reserve Markers	Somatic Sink Markers (*k* Fraction)					Reproductive Sink Markers (1 − *k* Fraction)			
							Maintenance					Maintenance	Reproductive Growth		
Taxonomy	Species	Particle Selection	Feeding Rate	Gut Residence Time	Energy Assimilation	Energy Content	Survival	Metabolism	AOS	Behavior	Somatic Growth	Immune System	Maturity	Reproductive Buffer	Reference
	Mytilus galloprovincialis								↑* DNA damage NC Lipid damage ↓* AOS enzyme activity ↓* TOSC			↓ Granulocytes: hyalinocytes ↓* Phagocytic activity ↓ Hemocyte membrane stability			Avio et al. (2015)
	Ostrea edulis		↑ Cells mg dry weight^{-1} h^{-1}												Green et al. (2017)
	Ostrea edulis		NC % algae filtered bivalve^{-1} h^{-1}					↑* O_2 consumption			NC Shell growth rate				Green (2016)
	Perna viridis		↓ L filtered bivlave^{-1} h^{-1}				↓	↓ O_2 consumption			↓ No. byssus threads produced				Rist et al. (2016)
	Pinctada margaritifera		NC Cells g dry weight^{-1} h^{-1}		↓ % energy assimilated	↓ Scope for growth		NC O_2 consumption			NC Shell growth rate		↓ Gametogenesis NC Gonadal development[b]		Gardon et al. (2018)

(*Continued*)

TABLE 16.2 (*Continued*)
Microplastic-Induced Changes in Invertebrate Energy Acquisition and Energy Expenditure

Organism		Energy Acquisition Markers				Energy Reserve Markers	Somatic Sink Markers (*k* Fraction)					Reproductive Sink Markers (1 − *k* Fraction)			
							Maintenance					Maintenance	Reproductive Growth		
Taxonomy	Species	Particle Selection	Feeding Rate	Gut Residence Time	Energy Assimilation	Energy Content	Survival	Metabolism	AOS	Behavior	Somatic Growth	Immune System	Maturity	Reproductive Buffer	Reference
	Scrobicularia plana					NC Condition Index			↑↓AOS enzyme activity NC DNA damage ↑*↓ Lipid damage						O'Donovan et al. (2018)
Gastropoda	*Crepidula onyx* (larvae & juveniles)						NC				↓ Larval growth rate ↑* Settlement rate ↓ Juvenile growth rate				Lo and Chan (2018)
	Potampoyrgus antipodarum					NC Condition Index					↓* Shell growth rate NC Growth rate			NC No. offspring individual^{-1} ↓ Offspring growth rate ↑ % deformed offspring	Imhof and Laforsch (2016)

(*Continued*)

TABLE 16.2 (*Continued*)
Microplastic-Induced Changes in Invertebrate Energy Acquisition and Energy Expenditure

Organism		Energy Acquisition Markers				Energy Reserve Markers	Somatic Sink Markers (k Fraction)					Reproductive Sink Markers ($1 - k$ Fraction)			
							Maintenance					Maintenance	Reproductive Growth		
Taxonomy	Species	Particle Selection	Feeding Rate	Gut Residence Time	Energy Assimilation	Energy Content	Survival	Metabolism	AOS	Behavior	Somatic Growth	Immune System	Maturity	Reproductive Buffer	Reference
Rotifera															
Monogononta	*Brachionus koreanus*						↓ Lifespan		↑ ROS production ↑↓ AOS enzyme activity				↑ Time to produce first brood	↓ No. offspring	Jeong et al. (2016)
	Brachionus plicatilis						NC			↑ Swimming speed					Gambardella et al. (2018)

Note: Results are categorized according to markers of energy acquisition, energy reserve density, and energy expenditure in the somatic and reproductive sinks. Cells have been left blank when markers where not measured by the study. Arrows indicate the direction of change in the response variable; NC indicates no change occurred. Where both change markers appear, i.e., (↑↓), it indicates that both results were observed in the study, at different sampling times. Concentration is indicated with [], and * indicates that the observed change was not statistically significant ($p > 0.05$).

Abbreviations: AOS, antioxidant system; IDH, isocitrate dehydrogenase; ODH, octopine dehydrogenase; ROS, reactive oxygen species; TOSC, total oxyradical scavenging capacity.

[a] Expression of related genes.

[b] Several individuals showed gonadal regression, suggesting that maturity structure was being dismantled and redeployed.

responses are ecologically significant because such responses reduce an organism's fitness and thereby influence population dynamics over several generations (Johnston et al. 2014; Klok and de Roos 1996).

Studies have reported that the growth rate of somatic structures is inhibited by microplastic ingestion. The cnidarian, *Hydra*, developed fewer hydranths after exposure to irregularly shaped polyethylene microplastics (Murphy and Quinn 2018), and the regeneration of amputated segments of the polychaete, *Perinereis aibuhitensis*, was slowed by exposure to polystyrene microspheres (Leung and Chan 2018). After consuming microplastics, mussels produced fewer byssus threads after being detached from the substrate (Rist et al. 2016), limiting their ability to recover from physical disturbances.

Reproductive growth, such as the maternal provisioning of gametes, is also sensitive to microplastic ingestion. Consumption of microplastics reduced the number and size of eggs produced by female invertebrates, and these eggs can have reduced hatching success (Cole et al. 2015; Heindler et al. 2017; Lee et al. 2013; Sussarellu et al. 2016). In the freshwater crustacean, *Daphnia*, ingestion of microplastics lengthened the time required by females to produce broods (Ogonowski et al. 2016). Male reproductive output is also affected by microplastics, with Sussarellu et al. (2016) demonstrating that male oysters that had consumed microplastics produced slower swimming sperm. Such poorly provisioned gametes can produce smaller offspring (Sussarellu et al. 2016) with abnormal morphology (Imhof and Laforsch 2016).

Most invertebrate taxa have complex life cycles, with individuals passing through several larval stages before metamorphosing into juveniles. These larval stages are also sensitive to microplastics, with studies showing microplastic consumption reduces the larval maturation rates of ascidians, sea urchins, gastropods, bivalves, and arthropods (Kaposi et al. 2014; Lo and Chan 2018; Messinetti et al. 2018; Sussarellu et al. 2016; Ziajahromi et al. 2018). Suppressed growth during larval stages can affect an organism for the rest of its life. For example, Lo and Chan (2018) reported that microplastic-exposed veliger larvae of slipper snails continued to experience suppressed growth after metamorphosis, even though microplastic exposure ceased after this transition. Furthermore, the effects of microplastic ingestion can transcend generations: microplastic ingestion reduces copepod nauplii growth rates, but these reductions are more severe if the mother of the nauplii has also consumed microplastics (Lee et al. 2013).

16.1.3 Toward a Mechanistic Understanding of Sub-Lethal Responses: Tracking Energy Flux

Sub-lethal responses occur in such a diverse range of biological processes, that they can appear at the outset to be unrelated. This is far from true; these processes are linked by energy. Growth, metamorphosis, and reproduction are energetically demanding processes, and cannot all be provisioned optimally when an organism has limited energy available. Similarly, sub-lethal effects that carry across life stages and down generations are also linked by energy. Death, too, is connected to energy flux and can occur when an organism has insufficient energy to fuel its needs. There is a strong possibility that disrupting energy flows in an organism is the mechanism through which ingested microplastics generate sub-lethal responses. This possibility can be explored using bioenergetic models, which estimate an organism's energy acquisition and energy allocation to biological processes (Chipps and Wahl 2008).

The dynamic energy budget (DEB) theory proposed by Kooijman (2010) has become one of the most widely accepted approaches to modeling species' bioenergetic responses (Sarà et al. 2014). Unlike earlier, static energy budgets that only modeled energy fluxes for an individual in a single state (Kooijman 2012; van der Meer 2006), however, the DEB theory models how organisms acquire and allocate energy over time and with varying environmental conditions. The DEB theory is, therefore, an ideal bioenergetics framework for interpreting microplastic-induced responses because it encompasses different life stages and dynamic environments (Nisbet et al. 2012). The DEB theory has been applied to over a thousand species, and the model generally fits the collected data very well (Augustine and Kooijman 2019; Marques et al. 2018). The DEB theory has proven robust in its explanation of universally observed patterns in metabolism and the stoichiometry of organisms (Kooijman 2012). It has the added advantage of linking different levels of biological organization, from cells to populations.

The DEB theory has been adopted in several scientific fields that focus on the whole life history of an animal, such as conservation (Arnall et al. 2019) and aquaculture (Stavrakidis-Zachou et al. 2019). The theory has also been widely adopted by ecotoxicologists to quantify the link between anthropogenic chemicals and the biological responses of an organism (Ashauer et al. 2011; Baas et al. 2018; Kooijman et al. 2009). Interestingly, the DEB theory has been applied to the toxicology of nanoparticles, which are the only other toxicant to exert a significant physical presence like microplastics (Holden et al. 2013; Muller et al. 2014). Yet, despite its widespread use, the potential for the DEB model to explain responses to microplastic ingestion remains largely ignored (Sussarellu et al. 2016 provide an exception). In this chapter, we use the DEB framework to identify mechanisms by which microplastic ingestion can generate sub-lethal responses, and we collate the experimental evidence for these potential mechanisms.

16.1.4 The Tenets of DEB Theory

Before we apply the DEB model to identify processes that can be influenced by microplastic ingestion, we will first briefly explain the principles behind the model. What follows is a biologically focused description of the DEB theory that qualitatively describes an organism's acquisition and allocation of energy to various biological processes throughout its life cycle. For simplicity, we have deliberately omitted mathematical descriptions of the DEB model; we recommend that readers who are interested in this facet of the model peruse the references herein.

Organisms can be described as thermodynamically open systems that require external energy sources and constant energy flow (Sokolova et al. 2012). All animals acquire energy from the environment in the form of food. The DEB theory posits that, once food is consumed and digested, a portion of the energy contained within the food is assimilated into the organism and stored in an energy reserve in the form of proteins, lipids, and carbohydrates (Kooijman 2010; Sarà et al. 2014; Sousa et al. 2010). The unassimilated digestate is expelled from the body as feces. The concept of the reserve as the first location of assimilated energy distinguishes the DEB theory from static budget models, which assume that ingested energy is available immediately for use and only excess energy is stored in the reserve (Sarà et al. 2014).

The energy reserve is mobilized to fuel the various biological processes within the organism. However, the mobilized energy is unequally distributed amongst these processes, giving rise to competition at three points downstream of reserve mobilization (Figure 16.1). Understanding this competition enables us to mechanistically understand and predict biological responses to environmental stress. The first point of competition occurs when the mobilized reserves are partitioned between two energy sinks (Kooijman 2000), which we refer to as the "somatic sink" and the "reproductive sink." A fixed portion of the mobilized energy, κ, is allocated to the high-priority somatic sink, and will be used to fuel somatic maintenance and somatic growth. The lower priority reproductive sink is allocated the remaining portion of the mobilized energy, $1 - \kappa$, to fuel maturity maintenance, and either maturation growth or the reproductive buffer (defined below). Since κ is a fixed fraction, the competition for energy between the somatic and reproductive sinks is indirect: the reproductive sink will be allocated smaller portions of energy not because κ increases, but because there is less energy in the reserve and thus less energy available after the κ fraction has been subtracted.

The second and third instances of competition for energy arise from the way the κ and $1 - \kappa$ energy portions are unevenly divided between maintenance processes and growth *within* the sinks. Maintenance always has priority over growth; thus, within each sink, energy is first used to pay maintenance costs. Any remaining energy is allocated to growth processes. According to this order

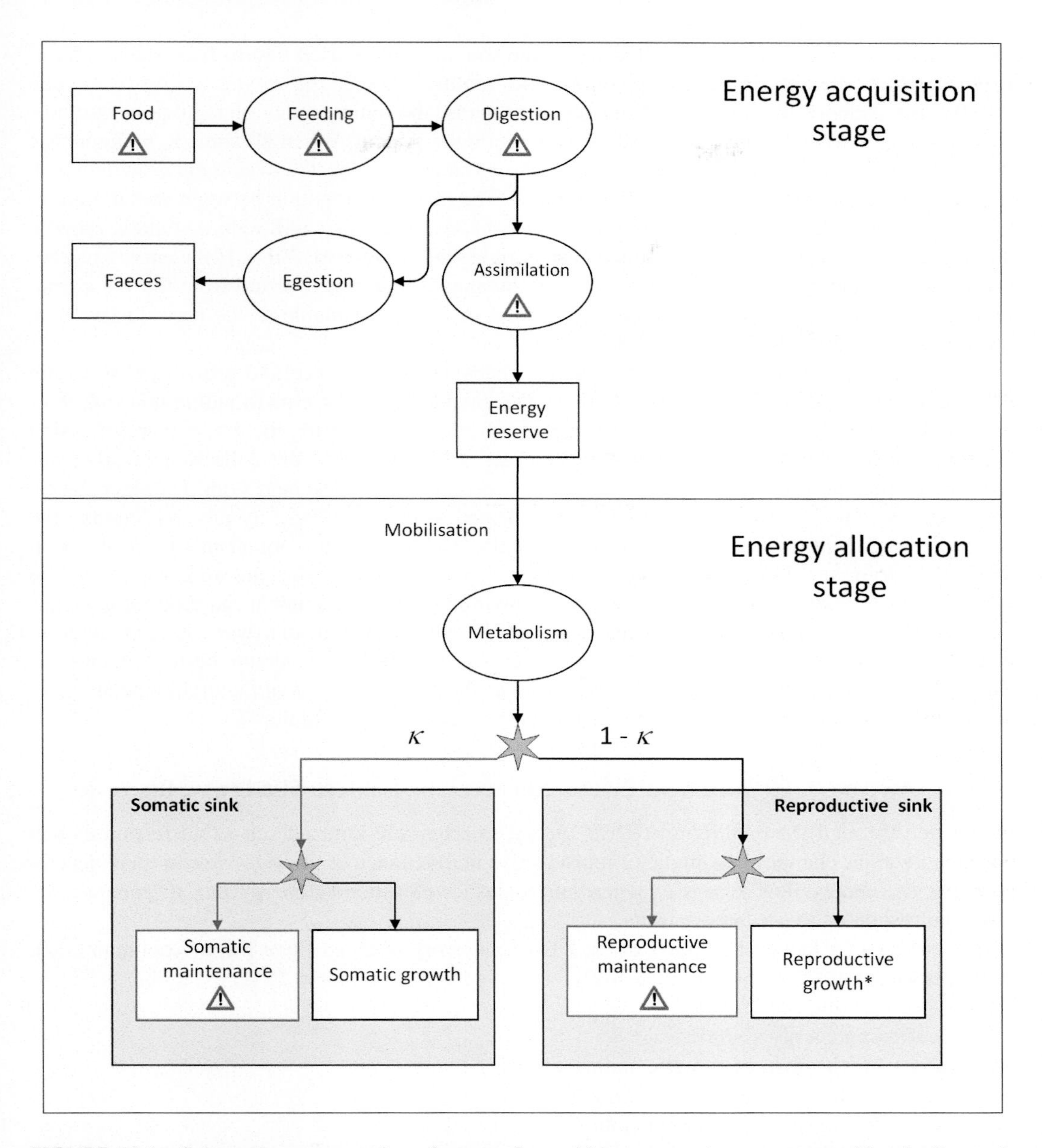

FIGURE 16.1 Schematic representation of energy flow within an organism, as proposed by the Dynamic Energy Budget model (Kooijman 2000). All animals acquire energy from the environment in the form of food. Some of this energy is assimilated into the organism's energy reserve, from where it is mobilized and unequally distributed, in κ and $1 - \kappa$ fractions, between the somatic and reproductive sinks, respectively. Within each sink, the energy fraction is further divided between maintenance and growth processes. Reproductive growth (*) in juveniles refers to maturation, while reproductive growth in adults refers to the reproductive buffer used to produce gametes. Stars indicate points of energy competition between the two sinks and between the processes inside the sinks; red arrows and boxes indicate high-priority processes, which receive energy first, and black arrows and boxes indicate lower priority processes, which receive whatever energy remains of the respective portion after high-priority processes (red boxes) have been fueled. Exclamation symbols indicate parts of the energy flow process which may be influenced by microplastic ingestion. (Modified from McKenzie, D. J. et al., *Conser. Physiol.*, 4, cow046, 2016.)

of priorities, the κ energy fraction allocated to the somatic sink is used first to fuel somatic maintenance, which refers to all the biological processes required to keep the organism alive (Kooijman 2000). This includes the structural turnover of proteins, the maintenance of regulatory systems, such as osmoregulation, and the functions of the immune system. When all somatic maintenance costs are paid, the unspent energy remaining of the κ fraction is allocated to somatic growth, which increases the organism's structure and body mass. The direct competition between somatic maintenance and growth means that increasing somatic maintenance costs will reduce somatic growth, and in extreme situations, somatic growth ceases altogether (Sarà et al. 2014). If the energy reserve contains insufficient energy to cover somatic maintenance, energy is diverted from the $1 - \kappa$ fraction to cover the loss, leaving a smaller proportion of energy available to the reproductive sink (Desforges et al. 2017; Sousa et al. 2010).

As in the somatic sink, the same competition between maintenance and growth exists within the reproductive sink, with the slight difference that maintenance and growth within this sink refer to reproductive rather than somatic structures. In the reproductive sink, the $1 - \kappa$ energy fraction is spent first to pay reproductive maintenance costs, which refers to the collection of processes required to maintain the organism's current level of gonadal development (van der Meer 2006). What energy remains of the $1 - \kappa$ fraction is allocated to the lower priority process, maturation, which is the growth of sexual structures in a juvenile that increase the organism's level of sexual maturity (Kooijman 2000). Maturation continues until an organism reaches the adult stage of its life cycle, at which point maturation has reached its maximum, and the organism can produce gametes. Maturation ceases after adulthood is reached, and the energy portion that was formerly given to maturation is instead shunted into a different type of growth: the reproductive buffer, which fuels the production of gametes (Sousa et al. 2010). Hence, the growth process of maturation occurs only in juveniles, and the growth of the reproductive buffer occurs only in adults.

16.1.5 Applying DEB Theory to Explain Microplastic-Induced Sub-Lethal Responses

Interpreted through the paradigm of DEB theory, microplastic-induced sub-lethal responses can be categorized as changes in somatic or reproductive maintenance or growth. We can then identify points in the energy flow processes where microplastics can modify energy flux to generate the observed responses in life history traits.

Toxicants can affect energy flows in the DEB framework in at least five ways (Kooijman 2010; Van Leeuwen et al. 2010) by non-exclusively:

1. Decreasing energy assimilation
2. Increasing the costs of somatic maintenance
3. Increasing the costs of somatic growth
4. Increasing the costs of reproduction
5. Causing deleterious effects on embryos

In the following sections of this chapter, we review the evidence for how the microplastics induce sub-lethal responses in invertebrates according to the categories listed above. To assist our review, we have conceptually divided the DEB model into two stages (see Figure 16.1). We refer to the upper half of the DEB energy flow diagram as the "energy acquisition stage," which encompasses the characteristics of food and biological responses that influence the amount of energy assimilated and ultimately stored in the energy reserve. We refer to the lower half of the DEB model as the "energy allocation stage," which encompasses the distribution of energy from the reserve to the competing somatic and reproductive energy sinks, and to the competing maintenance and growth processes within these sinks.

We start our review by considering the evidence that microplastics can decrease energy assimilation (Section 16.2). We expand our attention to include not only the process of energy assimilation itself, but also the feeding and digestion behaviors that precede it in the energy acquisition stage. In Section 16.3, we assess the potential for microplastics to increase the maintenance costs associated with two systems crucial for maintaining life: the antioxidant system and the immune system. Finally, we conclude this chapter by identifying ways in which future microplastics research can contribute to a mechanistic understanding of biological responses to microplastic ingestion.

16.2 THE EFFECTS OF MICROPLASTIC INGESTION ON ENERGY ACQUISITION

In this section, we explore how microplastics can influence energy flow in an organism by affecting processes upstream of the energy reserve, in the energy acquisition phase (Figure 16.1). This includes the potential of microplastics to inhibit the process of energy assimilation, as identified by Kooijman (2010) and Van Leeuwen et al. (2010), as well as the feeding behaviors and digestion that contribute to effective energy assimilation. Together, these processes influence the density of an organism's energy reserve.

As animals, invertebrates must acquire energy from the environment in the form of food, which first requires them to expend energy searching for and capturing food. Further energy is expended to mechanically and chemically digest food into units that are small enough to be assimilated into the energy reserve. Ideally, the amount of energy obtained from food is equal to, or larger than, the energy invested in feeding and digestion. However, this is not always the case. The quality and amount of food an invertebrate can acquire is influenced by environmental conditions, such as season: the energy reserve acts as a buffer to dampen environmental food fluctuations, enabling organisms to survive periods where energy assimilation is insufficient to meet metabolic needs.

Microplastics can influence energy acquisition by altering the following: (1) the energy content of ingested food, (2) feeding behaviors, (3) digestion efficiency, and/or (4) energy assimilation efficiency. As specified by the DEB framework, a reduction in one or more of these processes ultimately diminishes the reserve density of an organism and limits the amount of energy available for downstream processes. This exacerbates competition between the somatic and reproductive sinks, and between maintenance and growth processes within the sinks. Here, we review the literature relating to changes in parameters in the energy acquisition stage.

16.2.1 Dietary Dilution

Microplastics can reduce the density of the energy reserve in invertebrates through dietary dilution, which occurs when nutritionally inert microplastics displace real food and thereby dilute the energy content of the ingested meal (McCauley and Bjorndal 1999; Vegter et al. 2014). Experiments with indigestible materials have shown that dietary dilution can displace enough energy to reduce significantly energy reserves (Slansky and Wheeler 1991) and affect the downstream processes of growth, maturation, and reproduction (Lyn et al. 2011; Rollo and Hawryluk 1988; Tooci et al. 2009).

Although some researchers have highlighted the potential for plastic-related dietary dilution (Machovsky-Capuska et al. 2019; Nelms et al. 2016), this particular interaction between plastic and animals remains unexplored. Few researchers consider dietary dilution when discussing their experimental results, and dietary dilution has not been an explicit focus of any plastic studies of which we are aware. Determining the effects of dietary dilution on the density of the energy reserve may be a particularly difficult area to explore because dietary dilution can elicit compensatory feeding in some organisms. Therefore, these two issues need to be considered in conjunction. The use of positive controls, such as silica, would also be useful, as it would allow researchers to determine whether dietary dilution by microplastics has similar effects to dietary dilution by other indigestible particles.

16.2.2 Feeding Behavior

Microplastics can affect an organism's energy reserve by influencing feeding behavior, and hence altering the amount of energy ingested from the environment. One way that this can occur is by influencing the type of particle selected for consumption. To date, only one study has investigated how microplastics alter food selectivity (Cole et al. 2015). It found that the presence of microplastics caused marine copepods to significantly shift their particle choice to smaller size classes, resulting in a reduced carbon intake per day and fewer maternal resources allocated to egg production.

The second way that microplastics can alter feeding behavior is through influencing ingestion rates. The effect of microplastics on ingestion rates has been reported for several species, although the results are inconsistent (Table 16.2). Exposing the oyster, *Crassostrea gigas*, to polystyrene microplastics for 60 days resulted in increased filtration rates (Sussarellu et al. 2016). A similar increase was observed when the oyster, *Ostrea edulis*, was exposed to polylactic acid (PLA) and high-density polyethylene (HDPE) microplastics in a 50-day mesocosm experiment (Green et al. 2017). However, a 60-day mesocosm experiment detected no change in oyster filtration rates, even though it was conducted simultaneously at the same research facility and used the same type and concentration of microplastics (Green 2016). Increased filtration rates could occur to compensate for the low nutritional value of the ingested meal, and such compensatory feeding has been observed in several invertebrate taxa subjected to experimental nutritional dilution (Berner et al. 2005; Duarte et al. 2011; Flores et al. 2013; Lares and McClintock 1991). By increasing ingestion rates, organisms increase the likelihood that their food intake will meet their energetic requirements, despite any nutritional dilution caused by microplastics.

However, increased feeding rates have not been detected for many other invertebrates tested. A much larger body of research has found that microplastic exposure, when it influences feeding behavior at all, suppresses feeding activity, a phenomenon that was also observed in vertebrates (Ryan 1988). Work with Mytilidae mussels showed that, after 50 days' exposure, both HDPE and PLA microplastics reduced the filtration rate of *Mytilus* (Green et al. 2017), and the same response was observed in the mussel *Perna viridis* after 44 days' exposure to polyvinyl chloride (PVC) microplastics (Rist et al. 2016). Microplastics can suppress bivalve feeding rates rapidly: only one day of microplastic exposure was required to halve the filtration rates of the marine clam, *Atactodea striata* (Xu et al. 2017). This response is particularly striking since microplastics accounted for less than 0.001% of the algae cells offered to the clams. Startling responses have also been reported for the freshwater clam, *Corbicula fluminea*, which suffered a 95% reduction in clearance rates after ingesting microplastics (Oliveira et al. 2018).

Little is known about how microplastic ingestion affects the consumption of genuine food particles in larvae. Our current understanding is based on a single study, which found that microplastics reduced the daily carbon intake of oyster larvae (Cole and Galloway 2015). This response is an interesting contrast to the increased feeding rates reported for adult oysters (Green et al. 2017; Sussarellu et al. 2016). These different responses could be driven by the differences in feeding modes and particle acquisition of the two life stages. Interestingly, only one size and concentration of microplastics tested by Cole and Galloway (2015) reduced the feeding rate of the oyster larvae, suggesting that responses to microplastic ingestion could be size-dependent and observable only after a threshold concentration has been breached.

Regardless of the direction of the response, feeding rates appear to be particularly sensitive to microplastics, and under the right combination of experimental conditions are significantly affected by relatively small amounts of plastic. Microfiber concentrations in food as low as 0.3% w/w reduced crab food consumption over 4 weeks' exposure (Watts et al. 2015), and it was correlated with reduced energy reserves. A reduction in clam feeding rates was triggered by microplastics that accounted for less than 0.001% of the food particles available (Xu et al. 2017), while compensatory feeding in clams occurred when microplastics constituted only 0.2% of the volume of algal food supplied (Sussarellu et al. 2016). It seems that small amounts of ingested microplastics can have large effects on energy acquisition by altering feeding behavior.

16.2.2.1 Mechanisms behind Altered Feeding Behavior

Changes in feeding behavior are likely driven by post ingestion mechanisms. This hypothesis is favored by several researchers who suggest that suppressed feeding activity is the result of ingested microplastics generating feelings of satiation (Farrell and Nelson 2013; Guilhermino et al. 2018; Oliveira et al. 2018; von Moos et al. 2012). Indications that post ingestion mechanisms are at work come from the unusual methodology used by Rist et al. (2016). Whereas most studies expose organisms to microplastics and food particles simultaneously, Rist et al. fed mussels with algae cells *prior* to microplastic exposure each day. This indicates that the reduced filtration rates they observed were not driven by pre-ingestion particle selection, but rather from some internal signal generated from ingested microplastics in the digestive system. This theory also explains why clam filtration rates remained depressed 6 days after depuration (Oliveira et al. 2018), a response only possible if post ingestion mechanisms are at work.

For invertebrates, gastric distension is one factor that contributes to satiety (Breed and Moore 2015; Susswein and Kupfermann 1975). Microplastics consumed as part of a meal may contribute disproportionally to gastric distension because, unlike food particles, their volume is not easily condensed by chewing, and is unaffected by enzymatic digestion. Additionally, microplastic-induced gastric distension may last for longer periods compared to the distension caused by a microplastic-free meal, because microplastics can have a longer gut retention time than food particles (discussed in Section 16.2.3).

Chemical factors, such as the nutrient status of ingested food and the presence of oxidizable fuels, also contribute to satiety (You and Avery 2012) and could be influenced by the chemical signature of microplastics. The prominent physical presence of microplastics makes it easy to forget that, at the molecular level, plastics are essentially carbon chains, a chemical signature shared by carbohydrates. The chemical signature of plastics, or of the myriad of additives added to them, may be interpreted by digestive systems as an indicator of nutritional content. The research by Straub et al. (2017) is useful in untangling the physical and chemical effects of microplastics, because it is one of the few microplastic studies to use a positive physical control. They found that microplastics and silica particles both reduced the feeding rate of the amphipod, *Gammarus fossarum*; however, microplastics had the more significant effect. Similarly, ingesting PVC microplastics reduced the feeding rate of a marine polychaete, but the ingestion of silica did not generate the same response (Wright et al. 2013). Clearly, microplastics research would benefit from the more widespread use of positive controls in laboratory experiments, as this would allow researchers to untangle physical or chemical effects of microplastic ingestion.

16.2.3 Digestion

Although the structure of the digestive system varies among invertebrate taxa (Wright and Ahearn 2011), the digestion process is generally similar, and very much like the digestion process in vertebrates. After an invertebrate ingests food, the bolus traverses the digestive tract, where it undergoes mechanical and chemical extracellular digestion to break it into compounds that are small enough to be absorbed and assimilated into the energy reserve. The presence of plastic in the digestive system can alter the efficiency of the digestion process, and in turn reduce the amount of energy from ingested food that is in a physical state which can be absorbed across the gut lumen and assimilated into the energy reserve.

Plastic-induced reductions in digestion efficiency are most salient in vertebrates, such as birds and marine mammals, because they occur at a macroscopic scale that is easy to perceive. Large plastic items that are ingested by these animals obstruct the digestive tract, restricting the passage of food through the gut lumen, which reduces digestion efficiency to such an extent that it can result in starvation (Baird and Hooker 2000; Jacobsen et al. 2010; Pierce et al. 2004). Gut blockage is correlated with reduced energy stores (Auman et al. 1997; Danner et al. 2009; Lavers et al. 2014; Spear et al. 1995) and growth rates (Sievert and Sileo 1993). Given the evidence implicating macroplastic interference in digestive efficiency, it is reasonable to conjecture that the same mechanism may be operating, at a smaller spatial scale, in invertebrates that consume microplastics.

To date, there have been no reports of microplastics completely blocking the gut passages of invertebrate digestive systems. However, microplastics tend to aggregate (Karami 2017; Ribeiro et al. 2017), and the formation of polystyrene and polyethylene microplastic aggregations around the delicate digestive ducts and tubules of bivalves is well documented (Avio et al. 2015; Browne et al. 2008; Pittura et al. 2018; von Moos et al. 2012). Since tubules are the principle location for nutrient absorption in bivalves, microplastic aggregations in tubules will reduce the surface area of tubules touching the digestate, and thus reduce the surface area available for absorption. Microplastic aggregates larger than 80 μm^2 formed in *Daphnia* and were correlated with increased interbrood period and reduced reproduction (Ogonowski et al. 2016), potentially because the energy reserve was less dense, and so less energy was available to go into the reproductive buffer. Microplastics in the size ranges 1–4 μm and 10–27 μm formed aggregates in the gut of larval *Chironomus tepperi* and were correlated with significant reductions in growth and survival (Ziajahromi et al. 2018). Interestingly, larger-sized microplastics also reduced growth and survival, but did not aggregate in the gut. This suggests that reductions in digestion efficiency may be one of several mechanisms at work.

Elongated, thread-like microfibers may be more likely than spherical or irregularly shaped microplastics to obstruct the gastrointestinal tract because they tend to form intertwined aggregations within the digestive system (Murray and Cowie 2011; Welden and Cowie 2016a,b). A laboratory experiment in which microfibers were fed to langoustine found that, after 8 months' exposure, langoustine had lost a significant amount of body mass and lipid stores, but this was not as severe as the loss which occurred in starved langoustine (Welden and Cowie 2016b). This suggests that the accumulated microfibers in the digestive system interfered with digestion and absorption.

There is concern that ingesting plastic can reduce digestion efficiency by blocking the production of digestive enzymes (Azzarello and Van Vleet 1987; Galgani et al. 2010; Kolandhasamy et al. 2018), although this scientific question has been scarcely investigated. The block could occur because ingesting microplastics causes abdominal distention (Choi et al. 2018), and this influences the production of digestive enzymes. Studies exposing fish to microplastics have generated conflicting results, with increases and decreases in digestive enzyme activities being reported, depending on the enzyme family (Romano et al. 2018; Wen et al. 2018). Consuming polystyrene microplastics significantly reduced the amount of trypsin produced by herring larvae, compared with larvae that consumed similarly sized food particles (Hjelmeland et al. 1988). Regarding invertebrates, only one study to date has explored the effects of microplastic ingestion on digestive enzyme production. Korez et al. (2016) exposed the marine isopod, *Idotea emarginata*, to microplastic-contaminated food, and found that the digestive enzyme production varied with type of enzyme, tissue, and food source. This topic clearly needs further investigation, using a wider range of taxa, to determine whether microplastics can influence digestive efficiency by affecting gastric enzyme production. The microplastics themselves cannot be digested because invertebrates lack the enzymes required to break down plastic; and for this reason, plastic is generally considered to be bioinert (Andrady 2011).

Microplastics, and the food that is ingested simultaneously with them, can be retained in the digestive system for longer than usual periods. Polychaetes retained ingested sediment containing microplastics for 1.5 times longer than they retained clean sediment (Wright et al. 2013). The research team theorizes that the microplastic-contaminated bolus was being retained for longer so that it could be subjected to more rigorous digestion. This was also correlated with a reduced feeding rate and a reduction in body energy reserves. Microplastic fibers, compared with food particles, are also retained for longer periods by amphipods and appear to do so in a dose-dependent manner (Au et al. 2015). In *Daphnia*, the number of microplastics in the gut did not decrease at all over the observation time (Aljaibachi and Callaghan 2018). *Mytilus* larvae retained microplastics in the gut for up to 192 hours (Capolupo et al. 2018), although this did not lead to any changes in consumption of algae or development rate. Our knowledge of gut retention for microplastics in terrestrial invertebrates is limited to two studies. The egestion rates of a terrestrial isopod (Jemec Kokalj et al. 2018) and an oligochaete (Huerta Lwanga et al. 2016) showed no responses to consuming microplastics. This suggests that in these organisms, at least, gut retention was not influenced by the presence of microplastics.

Potentially, changes in digestive efficiency can have inputs back into the feeding stage. For example, changes in digestive efficiency caused by abiotic environmental conditions can trigger compensatory feeding (Stumpp et al. 2013). This link makes it difficult to determine at exactly which stage the microplastic is directly affecting feeding.

16.2.4 Assimilation

The small compounds resulting from digestion, such as proteins and lipids, are absorbed by epithelial cells lining the gut lumen. This process assimilates the compounds into the organism's energy reserve, from which point they can be used to fuel biological processes. This process of absorption and assimilation can be affected by microplastics. Alone, and particularly in aggregations, microplastics in the lumen that happen to touch the gut epithelium reduce the amount of surface area in contact with digestate, thereby limiting the absorption rate. Ingested microfibers reduced the assimilation efficiency of amphipods (Blarer and Burkhardt-Holm 2016). Exposing amphipods to microplastic spheres did not elicit the same effect, which implies that the shape of the microplastic may be what influences assimilation efficiency. The authors theorize that the shape of the plastics, particularly the sharp edges of the microfibers, could be damaging the digestive tract or cause inflammation, which in turn reduces assimilation efficiency.

It has been suggested that microplastics may damage the gut by causing cuts and abrasions, which may be particularly true when microplastics form aggregates in the digestive tract (Ogonowski et al. 2016; von Moos et al. 2012). This type of damage can reduce energy assimilation because damaged cells are unable to absorb nutrients. PVC pellets consumed by the European sea bass caused damage to the intestinal epithelium, and the severity of the damage increased with prolonged microplastic exposure (Pedà et al. 2016). Similarly, in *Danio rerio*, microplastics severely damaged the microvilli and enterocytes (Lei et al. 2018), and both structures are crucial to nutrient absorption. In invertebrates, changes in intestinal histology has been reported in *Mytilus*, where HDPE microplastics elicited a strong inflammatory response in the digestive system and caused the formation of granulocytomas 6 hours after microplastic ingestion (von Moos et al. 2012).

Some of the most interesting evidence of the potential of microplastics to reduce absorption efficiency comes from an experiment using the marine isopod, *Idotea emarginata* (Hämer et al. 2014). After 6 weeks of consuming microplastics with their food, the isopods showed no effects in their mortality or growth rates. The authors speculate that these results could be due to the isopods' unique digestive physiology: a fine-filtering system in the digestive tract allows only fluids and particles smaller than 1 μm to pass into the midgut, the main organ for nutrient absorption in this species. This structure would prohibit microplastics from physically damaging the structure of the midgut, and from reducing epithelial contact with digestate.

16.3 THE EFFECTS OF MICROPLASTIC INGESTION ON ENERGY ALLOCATION

The second stage of the DEB model, which we refer to as the "energy allocation phase," occurs downstream of the energy reserve (Figure 16.1). In this phase, the energy contained in biomolecules that make up the energy reserve (Kooijman 2010; Sarà et al. 2014; Sousa et al. 2010) is mobilized via the metabolism. The energy is then allocated unevenly to the somatic and reproductive sinks, where it is used to fuel biological processes according to a priority ranking: the maintenance processes within each sink are fueled first, and any energy remaining within each sink is used for growth processes.

16.3.1 Increasing Maintenance Costs

Microplastics can exacerbate the competition for energy between maintenance and growth processes, resulting in sub-lethal reductions in growth and reproduction. This can occur if microplastics increase the energy requirements for maintenance processes. According to the DEB theory, maintenance costs include all the processes that keep the organism alive and maintain existing structures. Maintenance processes occur at all organizational levels of an organism, and they include the movement and activity of the organism, structural repair and turnover of biomolecules, operation of homeostatic systems, and maintenance of transmembrane gradients (Sokolova et al. 2012; Sousa et al. 2010). Since both somatic and reproductive structures need to be maintained, maintenance costs are present in both the somatic and reproductive sinks and will have priority access to the energy allocated to those sinks (see Section 16.1.4 for an in-depth description of energy allocation). Increasing maintenance costs in either sink will, therefore, reduce the energy available for somatic or reproductive growth. Increased maintenance costs can arise when environmental stresses disrupt an organisms' internal environment, placing pressure on homeostatic systems (Sokolova et al. 2012).

Here, we focus on two key maintenance systems in animals: the antioxidant system, which defends against oxidative damage, and the immune system, which fights invasion by pathogens. Since both systems are energetically expensive (Calow 1991; Flye-Sainte-Marie et al. 2009; Gillespie et al. 2012; Lochmiller and Deerenberg 2000; Pytharopoulou et al. 2006; Wang et al. 2012), changes in the functioning of these systems could result in severe implications for energy flux within an organism. Changes in these systems are also relatively easy to measure, and so have been investigated as part of several microplastic studies. In the following sections, we focus on the potential for microplastics to increase the maintenance costs associated with running these systems.

16.3.1.1 Increasing Maintenance Costs of the Antioxidant System

All aerobic organisms possess an antioxidant system to eliminate the reactive oxygen species (ROS) that are generated as by-products of aerobic respiration (Halliwell and Gutteridge 2007). Since ROS oxidatively damage proteins, lipids, and nucleic acids (Cadet and Wagner 2013; Halliwell and Gutteridge 2007; Niki et al., 2005), the intracellular ROS concentration, and the corresponding levels of oxidative damage, are maintained within a narrow, steady-state concentration (Lushchak 2011). This state of redox homeostasis is maintained by the enzymes and molecules of the antioxidant system, which minimize oxidative damage to biomolecules by neutralizing or sacrificially reacting with ROS (Pamplona and Costantini 2011).

Being a homeostatic mechanism, the energy required to fuel the antioxidant system comes from the portion of energy allocated to maintenance. Therefore, allocating additional energy to the antioxidant system reduces the amount of energy remaining for growth or reproduction (Petes et al. 2008). The antioxidant system and redox homeostasis, in general, are sensitive to toxicants and thus may also be affected by microplastics (Cacciatore et al. 2015; Pires et al. 2016; Valko et al. 2005).

Our understanding of how microplastics increase the energy required to fuel the antioxidant system can be guided by how other toxicants affect the antioxidant system. First, there is strong evidence suggesting that microplastics can increase intracellular ROS concentrations. Jeong et al. (2016) directly measured ROS formation in rotifers and demonstrated that ROS formation was positively correlated with the concentration of microplastics ingested. Further, higher ROS concentration reduced the rates of somatic growth and maturation in the rotifers, indicating that ROS diverted energy toward maintenance in both the somatic and reproductive sinks. Ultimately, higher ROS concentrations increase maintenance costs because oxidatively damaged biomolecules need to be repaired or replaced (Figure 16.2), which diverts energy to maintenance and away from growth (Augustine and Kooijman 2019; Michalek-Wagner and Willis 2001). Microplastic-induced oxidative damage has been detected in bivalves and annelids, suggesting that this response could be widespread among taxa (Détrée and Gallardo-Escárate 2017; Oliveira et al. 2018; Rodríguez-Seijo et al. 2018). The mechanism by which microplastics increase ROS concentrations is undetermined. It may occur because consuming microplastics can increase an organism's metabolism (Green 2016; Green et al. 2016; Rodríguez-Seijo et al. 2018). Van Cauwenberghe et al. (2015) observed that the digestive gland in *Mytilus* mussels, which is where microplastics are present after a meal, experienced a 25%

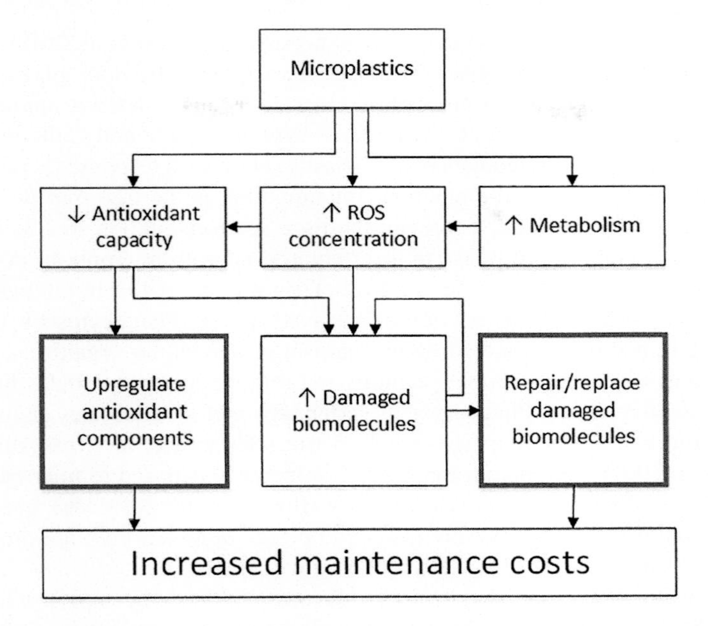

FIGURE 16.2 The pathways by which microplastics can increase maintenance costs related to the antioxidant system. Microplastics could increase ROS concentrations or decrease antioxidant capacity, leading to increased energetic costs (indicated by bold boxes), and hence increased maintenance costs.

increase in metabolism when mussels were exposed to microplastics. Alternatively, microplastics may increase ROS concentrations because they contain additives and plasticizers that can leach out of the polymer matrix and trigger a chemical reaction which generates ROS.

Second, laboratory studies have reported that microplastic ingestion reduces an organism's antioxidant capacity (Browne et al. 2013), requiring organisms to expend energy to synthesize more antioxidant molecules and enzymes (Lushchak 2011). The upregulation of antioxidant components after microplastic ingestion has been reported for a range of invertebrates, including annelids, arthropods, and mollusks (Jeong et al. 2017; Magara et al. 2018; Rodríguez-Seijo et al. 2018). Whether antioxidant capacity is reduced by microplastics directly or occurs simply because antioxidant components are diminished by reacting sacrificially with higher concentrations of ROS (Figure 16.2), remains unclear.

16.3.1.2 Increasing Maintenance Costs of the Immune System

The diversity of life forms across invertebrate taxa means that immune systems across these taxa are also very diverse (Loker et al. 2004). Broadly speaking, however, the immune system functions similarly in all invertebrates to identify and eliminate pathogens that enter the organism's body. This role is carried out by several types of free-circulating cells in the blood (Canesi and Procházková 2014). To mount an effective response to pathogens, the cells of the immune system drastically increase their metabolism to fuel a range of energetically expensive processes (Wolowczuk et al. 2008). Since the immune system is crucial to keeping the organism alive, the energy required to fuel immune responses is a type of maintenance cost. Under the DEB theory, the immune system is classified as a maintenance cost of the reproductive sink. Therefore, increased energy required for the immune system takes priority over allocating energy to reproductive growth in the form of maturation or gamete production.

The notion that environmental stresses can influence the immune system is gaining traction, and it is a core concept of the burgeoning field of ecological immunity, which seeks to quantify the link between environmental stresses and immune responses (Canesi and Procházková 2014).

The immune system responds to the presence of nanoparticles (Chen et al. 2018), and researchers have highlighted that immune responses could also be triggered by microplastics (Lusher et al. 2017). Initial evidence suggested that microplastic ingestion caused *Mytilus* mussels to upregulate the expression of genes associated with the immune system (Détrée and Gallardo-Escárate 2017, 2018). However, the pathway through which microplastics elicit such responses is poorly understood.

One potential pathway is via inflammation. Inflammation is the first line of immune defense in invertebrates (Rowley 1996), and, as previously discussed (see Section 16.2.3), inflammation of the gut epithelium is a common response in invertebrates fed with microplastics (Avio et al. 2015; Rodríguez-Seijo et al. 2017; von Moos et al. 2012; Wright et al. 2013). Inflammation is typically characterized by the phagocytic activity of immune system cells. Phagocytosis is universal among the animal kingdom; it started as a feeding mechanism in unicellular organisms and has evolved into a key mechanism by which animals capture and remove foreign particles that have entered the body cavity (Rowley 1996). Phagocytes in vertebrates and invertebrates engulf microplastics measuring from the nanoscale to approximately 5 μm (Bergami et al. 2019; Blank et al. 2013; Tomazic-Jezic et al. 2001). Polychaetes that were continuously exposed to microplastics displayed higher phagocytic cell counts and increased activity (Browne et al. 2013; Wright et al. 2013), both responses that increase the energy requirements and place additional pressure on energy reserves that are affected by suppressed feeding activity (Wright et al. 2013).

Other evidence suggests that microplastics decrease phagocytic activity in invertebrates. Bergami et al. (2019) reported that, although nanoplastics were engulfed by sea urchin immune cells, the phagocytic activity of these cells was then reduced significantly. Similarly, microplastics decreased phagocytic activity in *Mytilus* mussels (Avio et al. 2015) and reduced the viability of immune cells in the estuarine ragworm, *Hediste diversicolor* (Revel et al. 2018). These inhibited immune responses could allow invading pathogens to cause higher levels of damage, which require energy to repair, before they can be neutralized by the immune system. Further, a drop in immune capacity may trigger the creation of more immune cells to make up the difference. This process, too, would increase the energy requirements of the immune system.

Another way that microplastics can increase the activity and hence energy requirements of the immune system are through the damage they cause to the epithelial cells of the gut lumen (discussed in Section 16.2.3). The gut is potentially filled with a multitude of pathogens, and breaches in the gut barrier allow pathogens to enter the main body cavity. In response, they will be phagocytosed by immune cells (Rowley and Powell 2007). The tissue damage itself will also activate components of the immune system, namely, hemocytes, which are involved in wound repair (Rowley and Powell 2007).

Larger foreign bodies cannot be phagocytosed by a single immune cell, and are instead encapsulated into nodules, called granulocytomas, because the nodules are mainly composed of granulocytes. von Moos et al. (2012) observed that microplastics in the gut lumen became encapsulated in nodules formed by granulocytes. What happens to this relatively large, encapsulated material is largely unknown, but it could be that energy is expended to move the encapsulated mass out of the body (as observed in Olsen et al. 2015).

16.4 CONCLUSIONS AND SUGGESTIONS FOR FUTURE WORK

The number of studies reporting biological responses to microplastic ingestion is booming, and this research field will continue to grow as concern over microplastic pollution builds social and scientific momentum. The adoption of an overarching paradigm would enable a mechanistic understanding of the effects of microplastics, which can lead to predictive modeling and would guide the direction of future studies. We proposed the DEB bioenergetic model, widely adopted to explain toxic and sub-lethal effects, as an appropriate framework through which to interpret the effects of microplastics on biota.

In this chapter, we synthesized a range of sub-lethal microplastic responses reported for invertebrate species and applied the DEB bioenergetic framework to highlight the energetic links that exist between these often-disparate responses. Using the DEB framework, we identified stages where microplastics can influence energy flux in an invertebrate and thereby generate the sub-lethal responses reported in the literature. We have been limited in our ability to draw robust conclusions about the extent to which microplastics can alter energy flux in the energy acquisition stage and the energy allocation stages of an invertebrate's daily existence. This is because many studies do not include biomarkers at several stages of the DEB model, thus restricting our ability to determine if microplastics affect more than one stage.

The understanding of biological responses to microplastic pollution would be enhanced greatly if researchers:

- Adopted the DEB framework to interpret the biological responses to microplastics
- Incorporate a range of biomarkers to monitor energy flow through several stages of the DEB model
- Experiment on invertebrates for which the DEB data already exist to enable comparisons

REFERENCES

Aljaibachi, R., & Callaghan, A. (2018). Impact of polystyrene microplastics on *Daphnia magna* mortality and reproduction in relation to food availability. *PeerJ*, *6*, e4601. http://doi.org/10.7717/peerj.4601

Allen, A. S., Seymour, A. C., & Rittschof, D. (2017). Chemoreception drives plastic consumption in a hard coral. *Marine Pollution Bulletin*, *124* (1): 198–205. http://doi.org/10.1016/j.marpolbul.2017.07.030

Andrady, A. L. (2011). Microplastics in the marine environment. *Marine Pollution Bulletin*, *62* (8): 1596–1605. http://doi.org/10.1016/j.marpolbul.2011.05.030

Arnall, S. G., Mitchell, N. J., Kuchling, G., Durell, B., Kooijman, S. A. L. M., & Kearney, M. R. (2019). Life in the slow lane? A dynamic energy budget model for the western swamp turtle, *Pseudemydura umbrina*. *Journal of Sea Research*, *143*, 89–99. http://doi.org/10.1016/j.seares.2018.04.006

Arthur, C., Baker, J., Bamford, H., Barnea, N., Lohmann, R., McElwee, K., Morishige, C., & Thompson, R. (2009). Executive summary. In C. Arthur, J. Baker, & H. Bamford (Eds.), *Proceedings of the international research workshop on the occurrence, effects and fate of microplastic marine debris* (pp. 7–17). Silver Spring, MD: NOAA Technical Memorandum NOS-OR&R-30. Retrieved from https://marinedebris.noaa.gov/proceedings-international-research-workshop-microplastic-marine-debris

Ashauer, R., Agatz, A., Albert, C., Ducrot, V., Galic, N., Hendriks, J., Jager, T. et al. (2011). Toxicokinetic-toxicodynamic modeling of quantal and graded sublethal endpoints: A brief discussion of concepts. *Environmental Toxicology and Chemistry*, *30* (11): 2519–2524. http://doi.org/10.1002/etc.639

Au, S. Y., Bruce, T. F., Bridges, W. C., & Klaine, S. J. (2015). Responses of *Hyalella azteca* to acute and chronic microplastic exposures. *Environmental Toxicology and Chemistry*, *34* (11): 2564–2572. http://doi.org/10.1002/etc.3093

Augustine, S., & Kooijman, S. (2019). A new phase in DEB research. *Journal of Sea Research*, *143*, 1–7. http://doi.org/10.1016/j.seares.2018.06.003

Auman, H. J., Ludwig, J. P., Giesy, J. P., & Colborn, T. (1997). Plastic ingestion by Laysan albatross chicks on Sand Island, Midway Atoll, in 1994 and 1995. In G. Robinson & R. Gales (Eds.), *Albatross Biology and Conservation* (pp. 239–244). Chipping Norton: Surrey Beatty & Sons.

Avio, C. G., Gorbi, S., Milan, M., Benedetti, M., Fattorini, D., D'Errico, G., Pauletto, M., Bargelloni, L., & Regoli, F. (2015). Pollutants bioavailability and toxicological risk from microplastics to marine mussels. *Environmental Pollution*, *198*, 211–222. http://doi.org/10.1016/j.envpol.2014.12.021

Azzarello, M., & Van Vleet, E. (1987). Marine birds and plastic pollution. *Marine Ecology Progress Series*, *37*, 295–303. http://doi.org/10.3354/meps037295

Baas, J., Augustine, S., Marques, G. M., & Dorne, J.-L. (2018). Dynamic energy budget models in ecological risk assessment: From principles to applications. *Science of the Total Environment*, *628–629*, 249–260. http://doi.org/10.1016/j.scitotenv.2018.02.058

Baird, R. W., & Hooker, S. K. (2000). Ingestion of plastic and unusual prey by a juvenile harbour porpoise. *Marine Pollution Bulletin*, *40* (8): 719–720.

Bergami, E., Bocci, E., Vannuccini, M. L., Monopoli, M., Salvati, A., Dawson, K. A., & Corsi, I. (2016). Nano-sized polystyrene affects feeding, behavior and physiology of brine shrimp *Artemia franciscana* larvae. *Ecotoxicology and Environmental Safety, 123*, 18–25. http://doi.org/10.1016/j.ecoenv.2015.09.021

Bergami, E., Krupinski Emerenciano, A., González-Aravena, M., Cárdenas, C. A., Hernández, P., Silva, J. R. M. C., & Corsi, I. (2019). Polystyrene nanoparticles affect the innate immune system of the Antarctic sea urchin *Sterechinus neumayeri. Polar Biology, 42* (4): 743–757. http://doi.org/10.1007/s00300-019-02468-6

Berner, D., Blanckenhorn, W. U., & Körner, C. (2005). Grasshoppers cope with low host plant quality by compensatory feeding and food selection: N limitation challenged. *Oikos, 111* (3): 525–533. http://doi.org/10.1111/j.1600-0706.2005.14144.x

Besseling, E., Wegner, A., Foekema, E. M., van den Heuvel-Greve, M. J., & Koelmans, A. A. (2013). Effects of microplastic on fitness and PCB bioaccumulation by the lugworm *Arenicola marina* (L.). *Environmental Science & Technology, 47* (1): 593–600. http://doi.org/10.1021/es302763x

Blank, F., Stumbles, P. A., Seydoux, E., Holt, P. G., Fink, A., Rothen-Rutishauser, B., Strickland, D. H., & von Garnier, C. (2013). Size-dependent uptake of particles by pulmonary antigen-presenting cell populations and trafficking to regional lymph nodes. *American Journal of Respiratory Cell and Molecular Biology, 49* (1): 67–77. http://doi.org/10.1165/rcmb.2012-0387OC

Blarer, P., & Burkhardt-Holm, P. (2016). Microplastics affect assimilation efficiency in the freshwater amphipod *Gammarus fossarum. Environmental Science and Pollution Research, 23*, 23522–23532. http://doi.org/10.1007/s11356-016-7584-2

Bour, A., Haarr, A., Keiter, S., & Hylland, K. (2018). Environmentally relevant microplastic exposure affects sediment-dwelling bivalves. *Environmental Pollution, 236*, 652–660. http://doi.org/10.1016/j.envpol.2018.02.006

Breed, M. D., & Moore, J. (2015). Homeostasis and time budgets. In M. D. Breed & J. Moore (Eds.), *Animal Behavior* (2nd ed., pp. 109–144). London: Academic Press. http://doi.org/10.1016/b978-0-12-801532-2.00004-0

Browne, M. A., Dissanayake, A., Galloway, T. S., Lowe, D. M., & Thompson, R. C. (2008). Ingested microscopic plastic translocates to the circulatory system of the mussel, *Mytilus edulis* (L.). *Environmental Science and Technology, 42* (13): 5026–5031. http://doi.org/10.1021/es800249a

Browne, M. A., Galloway, T. S., & Thompson, R. C. (2010). Spatial patterns of plastic debris along estuarine shorelines. *Environmental Science and Technology, 44* (9): 3404–3409. http://doi.org/10.1021/es903784e

Browne, M. A., Niven, S. J., Galloway, T. S., Rowland, S. J., & Thompson, R. C. (2013). Microplastic moves pollutants and additives to worms, reducing functions linked to health and biodiversity. *Current Biology, 23* (23): 2388–2392. http://doi.org/10.1016/j.cub.2013.10.012

Cacciatore, L. C., Nemirovsky, S. I., Verrengia Guerrero, N. R., & Cochón, A. C. (2015). Azinphos-methyl and chlorpyrifos, alone or in a binary mixture, produce oxidative stress and lipid peroxidation in the freshwater gastropod *Planorbarius corneus. Aquatic Toxicology, 167*, 12–19. http://doi.org/10.1016/j.aquatox.2015.07.009

Cadet, J., & Wagner, J. R. (2013). DNA base damage by reactive oxygen species, oxidizing agents, and UV radiation. *Cold Spring Harbor Perspectives in Biology, 5* (2): a012559. http://doi.org/10.1101/cshperspect.a012559

Calow, P. (1991). Physiological costs of combating chemical toxicants: Ecological implications. *Comparative Biochemistry and Physiology Part C: Comparative Pharmacology, 100* (1–2): 3–6. http://doi.org/10.1016/0742-8413(91)90110-F

Canesi, L., & Procházková, P. (2014). The invertebrate immune system as a model for investigating the environmental impact of nanoparticles. In D. Boraschi & A. Duschl (Eds.), *Nanoparticles and the Immune System: Safety and Effects* (pp. 91–112). Oxford: Academic Press.

Capolupo, M., Franzellitti, S., Valbonesi, P., Lanzas, C. S., & Fabbri, E. (2018). Uptake and transcriptional effects of polystyrene microplastics in larval stages of the Mediterranean mussel *Mytilus galloprovincialis. Environmental Pollution, 241*, 1038–1047. http://doi.org/10.1016/j.envpol.2018.06.035

Carsten von der Ohe, P., & Liess, M. (2004). Relative sensitivity distribution of aquatic invertebrates to organic and metal compounds. *Environmental Toxicology and Chemistry, 23* (1): 150. http://doi.org/10.1897/02-577

Chávez, E. A. (2007). Socio-economic assessment for the management of the Caribbean spiny lobster. In *Proceedings of the 60th Gulf and Caribbean Fisheries Institute* (pp. 193–196). Punta Cana, Dominican Republic: Gulf and Caribbean Fisheries Institute. Retrieved from http://aquaticcommons.org/15422/

Chen, L., Liu, J., Zhang, Y., Zhang, G., Kang, Y., Chen, A., Feng, X., & Shao, L. (2018). The toxicity of silica nanoparticles to the immune system. *Nanomedicine*, *13* (15): 1939–1962. http://doi.org/10.2217/nnm-2018-0076

Chipps, S. R., & Wahl, D. H. (2008). Bioenergetics modeling in the 21st century: Reviewing new insights and revisiting old constraints. *Transactions of the American Fisheries Society*, *137* (1): 298–313. http://doi.org/10.1577/t05-236.1

Choi, J. S., Jung, Y. J., Hong, N. H., Hong, S. H., & Park, J. W. (2018). Toxicological effects of irregularly shaped and spherical microplastics in a marine teleost, the sheepshead minnow (*Cyprinodon variegatus*). *Marine Pollution Bulletin*, *129* (1): 231–240. http://doi.org/10.1016/j.marpolbul.2018.02.039

Cincinelli, A., Scopetani, C., Chelazzi, D., Lombardini, E., Martellini, T., Katsoyiannis, A., Fossi, M. C., & Corsolini, S. (2017). Microplastic in the surface waters of the Ross Sea (Antarctica): Occurrence, distribution and characterization by FTIR. *Chemosphere*, *175*, 391–400. http://doi.org/10.1016/j.chemosphere.2017.02.024

Cole, M., & Galloway, T. S. (2015). Ingestion of nanoplastics and microplastics by Pacific oyster larvae. *Environmental Science and Technology*, *49* (24): 14625–14632. http://doi.org/10.1021/acs.est.5b04099

Cole, M., Lindeque, P., Fileman, E., Halsband, C., & Galloway, T. S. (2015). The impact of polystyrene microplastics on feeding, function and fecundity in the marine copepod *Calanus helgolandicus*. *Environmental Science and Technology*, *49* (2): 1130–1137. http://doi.org/10.1021/es504525u

Cole, M., Lindeque, P., Fileman, E., Halsband, C., Goodhead, R., Moger, J., & Galloway, T. S. (2013). Microplastic ingestion by zooplankton. *Environmental Science and Technology*, *47* (12): 6646–6655. http://doi.org/10.1021/es400663f

Cruz-Rivera, E., & Hay, M. E. (2000). Can quantity replace quality? Food choice, compensatory feeding, and fitness of marine mesograzers. *Ecology*, *81* (1): 201–219.

Danner, G. R., Chacko, J., & Brautigam, F. (2009). Voluntary ingestion of soft plastic fishing lures affects brook trout growth in the laboratory. *North American Journal of Fisheries Management*, *29* (2): 352–360. http://doi.org/10.1577/m08-085.1

Desforges, J.-P. W., Sonne, C., & Dietz, R. (2017). Using energy budgets to combine ecology and toxicology in a mammalian sentinel species. *Scientific Reports*, *7* (1): 46267. http://doi.org/10.1038/srep46267

Détrée, C., & Gallardo-Escárate, C. (2017). Polyethylene microbeads induce transcriptional responses with tissue-dependent patterns in the mussel *Mytilus galloprovincialis*. *Journal of Molluscan Studies*, *83* (2): 220–225. http://doi.org/10.1093/mollus/eyx005

Détrée, C., & Gallardo-Escárate, C. (2018). Single and repetitive microplastics exposures induce immune system modulation and homeostasis alteration in the edible mussel *Mytilus galloprovincialis*. *Fish & Shellfish Immunology*, *83*, 52–60. http://doi.org/10.1016/j.fsi.2018.09.018

Dhawan, R., Dusenbery, D. B., & Williams, P. L. (1999). Comparison of lethality, reproduction, and behavior as toxicological endpoints in the nematode *Caenorhabditis elegans*. *Journal of Toxicology and Environmental Health – Part A*, *58*, 451–462. http://doi.org/10.1080/009841099157179

Duarte, C., Acuña, K., Navarro, J. M., & Gómez, I. (2011). Intra-plant differences in seaweed nutritional quality and chemical defenses: Importance for the feeding behavior of the intertidal amphipod *Orchestoidea tuberculata*. *Journal of Sea Research*, *66* (3): 215–221. http://doi.org/10.1016/j.seares.2011.07.007

Farrell, P., & Nelson, K. (2013). Trophic level transfer of microplastic: *Mytilus edulis* (L.) to *Carcinus maenas* (L.). *Environmental Pollution*, *177*, 1–3. http://doi.org/10.1016/j.envpol.2013.01.046

Flores, L., Larrañaga, A., & Elosegi, A. (2013). Compensatory feeding of a stream detritivore alleviates the effects of poor food quality when enough food is supplied. *Freshwater Science*, *33* (1): 134–141. http://doi.org/https://doi.org/10.1086/674578

Flye-Sainte-Marie, J., Jean, F., Paillard, C., & Kooijman, S. A. L. M. (2009). A quantitative estimation of the energetic cost of brown ring disease in the Manila clam using Dynamic Energy Budget theory. *Journal of Sea Research*, *62* (2–3): 114–123. http://doi.org/10.1016/j.seares.2009.01.007

Frias, J. P. G. L., & Nash, R. (2019). Microplastics: Finding a consensus on the definition. *Marine Pollution Bulletin*, *138*, 145–147. http://doi.org/10.1016/j.marpolbul.2018.11.022

Galgani, F., Fleet, D., Van Franeker, J., Katsanevakis, S., Maes, T., Mouat, J., Oosterbaan, L. et al. (2010). *Marine Strategy Framework Directive: Task Group 10 Report Marine Litter. JRC Scientific and Technical Reports*. Italy. Retrieved from http://ec.europa.eu/environment/marine/pdf/9-Task-Group-10.pdf

Gambardella, C., Morgana, S., Bramini, M., Rotini, A., Manfra, L., Migliore, L., Piazza, V., Garaventa, F., & Faimali, M. (2018). Ecotoxicological effects of polystyrene microbeads in a battery of marine organisms belonging to different trophic levels. *Marine Environmental Research*, *141*, 313–321. http://doi.org/10.1016/j.marenvres.2018.09.023

Gardon, T., Reisser, C., Soyez, C., Quillien, V., & Le Moullac, G. (2018). Microplastics affect energy balance and gametogenesis in the pearl oyster *Pinctada margaritifera*. *Environmental Science & Technology*, *52* (9): 5277–5286. http://doi.org/10.1021/acs.est.8b00168

Gaspar, T. R., Chi, R. J., Parrow, M. W., & Ringwood, A. H. (2018). Cellular bioreactivity of micro- and nano-plastic particles in oysters. *Frontiers in Marine Science*, *5*, Article 345. http://doi.org/10.3389/fmars.2018.00345

Gillespie, K. M., Xu, F., Richter, K. T., Mcgrath, J. M., Markelz, R. J. C., Ort, D. R., Leakey, A. D. B., & Ainsworth, E. A. (2012). Greater antioxidant and respiratory metabolism in field-grown soybean exposed to elevated O_3 under both ambient and elevated CO_2. *Plant, Cell and Environment*, *35* (1): 169–184. http://doi.org/10.1111/j.1365-3040.2011.02427.x

Gonçalves, C., Martins, M., Sobral, P., Costa, P. M., & Costa, M. H. (2019). An assessment of the ability to ingest and excrete microplastics by filter-feeders: A case study with the Mediterranean mussel. *Environmental Pollution*, *245*, 600–606. http://doi.org/10.1016/j.envpol.2018.11.038

Green, D. S. (2016). Effects of microplastics on European flat oysters, *Ostrea edulis* and their associated benthic communities. *Environmental Pollution*, *216*, 95–103. http://doi.org/10.1016/j.envpol.2016.05.043

Green, D. S., Boots, B., O'Connor, N. E., & Thompson, R. (2017). Microplastics affect the ecological functioning of an important biogenic habitat. *Environmental Science & Technology*, *51* (1): 68–77. http://doi.org/10.1021/acs.est.6b04496

Green, D. S., Boots, B., Sigwart, J., Jiang, S., & Rocha, C. (2016). Effects of conventional and biodegradable microplastics on a marine ecosystem engineer (*Arenicola marina*) and sediment nutrient cycling. *Environmental Pollution*, *208*, 426–434. http://doi.org/10.1016/j.envpol.2015.10.010

Guilhermino, L., Vieira, L. R., Ribeiro, D., Tavares, A. S., Cardoso, V., Alves, A., & Almeida, J. M. (2018). Uptake and effects of the antimicrobial florfenicol, microplastics and their mixtures on freshwater exotic invasive bivalve *Corbicula fluminea*. *Science of the Total Environment*, *622–623*, 1131–1142. http://doi.org/10.1016/j.scitotenv.2017.12.020

Gutow, L., Eckerlebe, A., Giménez, L., & Saborowski, R. (2016). Experimental evaluation of seaweeds as a vector for microplastics into marine food webs. *Environmental Science and Technology*, *50* (2): 915–923. http://doi.org/10.1021/acs.est.5b02431

Halliwell, B., & Gutteridge, J. M. C. (2007). *Free Radicals in Biology and Medicine* (4th ed.). New York: Oxford University Press.

Hämer, J., Gutow, L., Köhler, A., & Saborowski, R. (2014). Fate of microplastics in the marine isopod *Idotea emarginata*. *Environmental Science & Technology*, *48* (22): 13451–13458. http://doi.org/10.1021/es501385y

Heindler, F. M., Alajmi, F., Huerlimann, R., Zeng, C., Newman, S. J., Vamvounis, G., & van Herwerden, L. (2017). Toxic effects of polyethylene terephthalate microparticles and di(2-ethylhexyl)phthalate on the calanoid copepod, *Parvocalanus crassirostris*. *Ecotoxicology and Environmental Safety*, *141*, 298–305. http://doi.org/10.1016/j.ecoenv.2017.03.029

Hickey, C. W., & Martin, M. L. (1995). Relative sensitivity of five benthic invertebrate species to reference toxicants and resin-acid contaminated sediments. *Environmental Toxicology and Chemistry*, *14* (8): 1401–1409. http://doi.org/10.1002/etc.5620140817

Hjelmeland, K., Pedersen, B. H., & Nilssen, E. M. (1988). Trypsin content in intestines of herring larvae, *Clupea harengus*, ingesting inert polystyrene spheres or live crustacea prey. *Marine Biology*, *98* (3): 331–335. http://doi.org/10.1007/BF00391108

Holden, P. A., Nisbet, R. M., Lenihan, H. S., Miller, R. J., Cherr, G. N., Schimel, J. P., & Gardea-Torresdey, J. L. (2013). Ecological nanotoxicology: Integrating nanomaterial hazard considerations across the subcellular, population, community, and ecosystems levels. *Accounts of Chemical Research*, *46* (3): 813–822. http://doi.org/10.1021/ar300069t

Horton, A. A., Walton, A., Spurgeon, D. J., Lahive, E., & Svendsen, C. (2017). Microplastics in freshwater and terrestrial environments: Evaluating the current understanding to identify the knowledge gaps and future research priorities. *Science of the Total Environment*, *586*, 127–141. http://doi.org/10.1016/j.scitotenv.2017.01.190

Huerta Lwanga, E., Gertsen, H., Gooren, H., Peters, P., Salánki, T., Van Der Ploeg, M., Besseling, E., Koelmans, A. A., & Geissen, V. (2016). Microplastics in the terrestrial ecosystem: Implications for *Lumbricus terrestris* (Oligochaeta, Lumbricidae). *Environmental Science and Technology*, *50* (5): 2685–2691. http://doi.org/10.1021/acs.est.5b05478

Imhof, H. K., & Laforsch, C. (2016). Hazardous or not – Are adult and juvenile individuals of *Potamopyrgus antipodarum* affected by non-buoyant microplastic particles? *Environmental Pollution*, *218*, 383–391. http://doi.org/10.1016/j.envpol.2016.07.017

Jacobsen, J. K., Massey, L., & Gulland, F. (2010). Fatal ingestion of floating net debris by two sperm whales (*Physeter macrocephalus*). *Marine Pollution Bulletin*, *60* (5): 765–767. http://doi.org/10.1016/j.marpolbul.2010.03.008

Jemec, A., Horvat, P., Kunej, U., Bele, M., & Kržan, A. (2016). Uptake and effects of microplastic textile fibers on freshwater crustacean *Daphnia magna*. *Environmental Pollution*, *219*, 201–209. http://doi.org/10.1016/j.envpol.2016.10.037

Jemec Kokalj, A., Horvat, P., Skalar, T., & Kržan, A. (2018). Plastic bag and facial cleanser derived microplastic do not affect feeding behaviour and energy reserves of terrestrial isopods. *Science of the Total Environment*, *615*, 761–766. http://doi.org/10.1016/j.scitotenv.2017.10.020

Jeong, C.-B., Kang, H.-M., Lee, M.-C., Kim, D.-H., Han, J., Hwang, D.-S., Souissi, S. et al. (2017). Adverse effects of microplastics and oxidative stress-induced MAPK/Nrf2 pathway-mediated defense mechanisms in the marine copepod *Paracyclopina nana*. *Scientific Reports*, *7* (1): 41323. http://doi.org/10.1038/srep41323

Jeong, C.-B., Won, E.-J., Kang, H.-M., Lee, M.-C., Hwang, D.-S., Hwang, U.-K., Zhou, B., Souissi, S., Lee, S.-J., & Lee, J.-S. (2016). Microplastic size-dependent toxicity, oxidative stress induction, and p-JNK and p-p38 activation in the monogonont rotifer (*Brachionus koreanus*). *Environmental Science & Technology*, *50* (16): 8849–8857. http://doi.org/10.1021/acs.est.6b01441

Johnston, A. S. A., Holmstrup, M., Hodson, M. E., Thorbek, P., Alvarez, T., & Sibly, R. M. (2014). Earthworm distribution and abundance predicted by a process-based model. *Applied Soil Ecology*, *84*, 112–123. http://doi.org/10.1016/j.apsoil.2014.06.001

Kaposi, K. L., Mos, B., Kelaher, B. P., & Dworjanyn, S. A. (2014). Ingestion of microplastic has limited impact on a marine larva. *Environmental Science & Technology*, *48* (3): 1638–1645. http://doi.org/10.1021/es404295e

Karami, A. (2017). Gaps in aquatic toxicological studies of microplastics. *Chemosphere*, *184*, 841–848. http://doi.org/10.1016/j.chemosphere.2017.06.048

Klok, C., & de Roos, A. M. (1996). Population level consequences of toxicological influences on individual growth and reproduction in *Lumbricus rubellus* (Lumbricidae, Oligochaeta). *Ecotoxicology and Environmental Safety*, *33* (2): 118–127. http://doi.org/10.1006/eesa.1996.0015

Koelmans, A. A., Besseling, E., & Shim, W. J. (2015). Nanoplastics in the aquatic environment. Critical review. In M. Bergmann, L. Gutow L, & M. Klages (Eds.), *Marine Anthropogenic Litter* (pp. 325–340). Cham, Switzerland: Springer. http://doi.org/10.1007/978-3-319-16510-3_12

Kolandhasamy, P., Su, L., Li, J., Qu, X., Jabeen, K., & Shi, H. (2018). Adherence of microplastics to soft tissue of mussels: A novel way to uptake microplastics beyond ingestion. *Science of the Total Environment*, *610–611*, 635–640. http://doi.org/10.1016/j.scitotenv.2017.08.053

Kooijman, S. (2000). *Dynamic Energy and Mass Budgets in Biological Systems* (2nd ed.). Cambridge: Cambridge University Press. http://doi.org/10.1017/CBO9780511565403

Kooijman, S. (2010). *Dynamic Energy Budget Theory for Metabolic Organization* (3rd ed.). Cambridge: Cambridge University Press.

Kooijman, S. (2012). Energy budgets. In A. Hastings & L. Gross (Eds.), *Encyclopedia of Theoretical Ecology* (1st ed., pp. 249–258). Los Angeles, CA: University of California Press.

Kooijman, S., Baas, J., Bontje, D., Broerse, M., van Gestel, C. A. M., & Jager, T. (2009). Ecotoxicological applications of Dynamic Energy Budget theory. In *Ecotoxicology Modeling. Emerging Topics in Ecotoxicology: Principles, Approaches and Perspectives* (2nd ed., pp. 237–259). Boston, MA: Springer. http://doi.org/10.1007/978-1-4419-0197-2_9

Korez, Š., Gutow, L., & Saborowski, R. (2016). Effects of microplastics on digestive enzymes in the marine isopod *Idotea emarginata*. In J. Baztan, B. Jorgensen, S. Pahl, R. C. Thompson, & J.-P. Vanderlinden (Eds.), *Fate and Impact of Microplastics in Marine Ecosystems. From the Coastline to the Open Sea* (pp. 131–132). Lanzarote, Canary Islands: Elsevier. http://doi.org/10.1016/B978-0-12-812271-6.00128-9

Lares, M. T., & McClintock, J. B. (1991). The effects of temperature on the survival, organismal activity, nutrition, growth, and reproduction of the carnivorous, tropical sea urchin *Eucidaris tribuloides*. *Marine Behaviour and Physiology*, *19* (2): 75–96. http://doi.org/10.1080/10236249109378798

Lavers, J. L., Bond, A. L., & Hutton, I. (2014). Plastic ingestion by flesh-footed Shearwaters (*Puffinus carneipes*): Implications for fledgling body condition and the accumulation of plastic-derived chemicals. *Environmental Pollution*, *187*, 124–129. http://doi.org/10.1016/j.envpol.2013.12.020

Lavers, J. L., Bond, A. L., & Karl, D. M. (2017). Exceptional and rapid accumulation of anthropogenic debris on one of the world's most remote and pristine islands. *Proceedings of the National Academy of Sciences of the United States of America*, *114* (23): 6052–6055. http://doi.org/10.1073/pnas.1619818114

Lebreton, L. C. M., van der Zwet, J., Damsteeg, J.-W., Slat, B., Andrady, A., & Reisser, J. (2017). River plastic emissions to the world's oceans. *Nature Communications*, *8*, 15611. http://doi.org/10.1038/ncomms15611

Lee, K.-W., Shim, W. J., Kwon, O. Y., & Kang, J.-H. (2013). Size-dependent effects of micro polystyrene particles in the marine copepod *Tigriopus japonicus*. *Environmental Science & Technology*, *47* (19): 11278–11283. http://doi.org/10.1021/es401932b

Lei, L., Wu, S., Lu, S., Liu, M., Song, Y., Fu, Z., Shi, H., Raley-Susman, K. M., & He, D. (2018). Microplastic particles cause intestinal damage and other adverse effects in zebrafish *Danio rerio* and nematode *Caenorhabditis elegans*. *Science of The Total Environment*, *619–620*, 1–8. http://doi.org/10.1016/j.scitotenv.2017.11.103

Leung, J., & Chan, K. Y. K. (2018). Microplastics reduced posterior segment regeneration rate of the polychaete *Perinereis aibuhitensis*. *Marine Pollution Bulletin*, *129* (2): 782–786. http://doi.org/10.1016/j.marpolbul.2017.10.072

Lithner, D., Larsson, Å., & Dave, G. (2011). Environmental and health hazard ranking and assessment of plastic polymers based on chemical composition. *Science of The Total Environment*, *409* (18): 3309–3324. http://doi.org/10.1016/j.scitotenv.2011.04.038

Lo, H. K. A., & Chan, K. Y. K. (2018). Negative effects of microplastic exposure on growth and development of *Crepidula onyx*. *Environmental Pollution*, *233*, 588–595. http://doi.org/10.1016/j.envpol.2017.10.095

Lochmiller, R. L., & Deerenberg, C. (2000). Trade-offs in evolutionary immunology: Just what is the cost of immunity? *Oikos*, *88* (1): 87–98. http://doi.org/10.1034/j.1600-0706.2000.880110.x

Loker, E. S., Adema, C. M., Zhang, S.-M., & Kepler, T. B. (2004). Invertebrate immune systems – not homogeneous, not simple, not well understood. *Immunological Reviews*, *198* (1): 10–24. http://doi.org/10.1111/j.0105-2896.2004.0117.x

Losey, J. E., & Vaughan, M. (2006). The economic value of ecological services provided by insects. *BioScience*, *56* (4): 311–323. http://doi.org/10.1641/0006-3568(2006)56[311:tevoes]2.0.co;2

Lushchak, V. I. (2011). Environmentally induced oxidative stress in aquatic animals. *Aquatic Toxicology*, *101* (1): 13–30. http://doi.org/10.1016/j.aquatox.2010.10.006

Lusher, A. L., Hollman, P., & Mendoza-Hill, J. (2017). *Microplastics in fisheries and aquaculture. Status of knowledge on their occurrence and implications for aquatic organisms and food safety. FAO Fisheries and Aquaculture Technical Paper. No. 615*. Rome, Italy. Retrieved from http://www.fao.org/3/a-i7677e.pdf

Lyn, J. C., Naikkhwah, W., Aksenov, V., & Rollo, C. D. (2011). Influence of two methods of dietary restriction on life history features and aging of the cricket *Acheta domesticus*. *Age (Dordrecht, Netherlands)*, *33* (4): 509–522. http://doi.org/10.1007/s11357-010-9195-z

Machovsky-Capuska, G. E., Amiot, C., Denuncio, P., Grainger, R., & Raubenheimer, D. (2019). A nutritional perspective on plastic ingestion in wildlife. *Science of the Total Environment*, *656*, 789–796. http://doi.org/10.1016/j.scitotenv.2018.11.418

Magara, G., Elia, A. C., Syberg, K., & Khan, F. R. (2018). Single contaminant and combined exposures of polyethylene microplastics and fluoranthene: Accumulation and oxidative stress response in the blue mussel, *Mytilus edulis*. *Journal of Toxicology and Environmental Health, Part A*, *81* (16): 761–773. http://doi.org/10.1080/15287394.2018.1488639

Manfra, L., Canepa, S., Piazza, V., & Faimali, M. (2016). Lethal and sublethal endpoints observed for *Artemia* exposed to two reference toxicants and an ecotoxicological concern organic compound. *Ecotoxicology and Environmental Safety*, *123*, 60–64. http://doi.org/10.1016/j.ecoenv.2015.08.017

Marques, G. M., Augustine, S., Lika, K., Pecquerie, L., Domingos, T., & Kooijman, S. A. L. M. (2018). The AmP project: Comparing species on the basis of dynamic energy budget parameters. *PLOS Computational Biology*, *14* (5): e1006100. http://doi.org/10.1371/journal.pcbi.1006100

McCauley, S. J., & Bjorndal, K. A. (1999). Conservation implications of dietary dilution from debris ingestion: Sublethal effects in post-hatchling loggerhead sea turtles. *Conservation Biology*, *13* (4): 925–929. http://doi.org/10.1046/j.1523-1739.1999.98264.x

McKenzie, D. J., Axelsson, M., Chabot, D., Claireaux, G., Cooke, S. J., Corner, R. A., De Boeck, G. et al. (2016). Conservation physiology of marine fishes: State of the art and prospects for policy. *Conservation Physiology*, *4* (1): cow046. http://doi.org/10.1093/conphys/cow046

McLoughlin, N., Yin, D., Maltby, L., Wood, R. M., & Yu, H. (2000). Evaluation of sensitivity and specificity of two crustacean biochemical biomarkers. *Environmental Toxicology and Chemistry*, *19* (8): 2085–2092. http://doi.org/10.1002/etc.5620190818

McNeish, R. E., Kim, L. H., Barrett, H. A., Mason, S. A., Kelly, J. J., & Hoellein, T. J. (2018). Microplastic in riverine fish is connected to species traits. *Scientific Reports*, *8* (1): 1–12. http://doi.org/10.1038/s41598-018-29980-9

Messinetti, S., Mercurio, S., Parolini, M., Sugni, M., & Pennati, R. (2018). Effects of polystyrene microplastics on early stages of two marine invertebrates with different feeding strategies. *Environmental Pollution*, *237*, 1080–1087. http://doi.org/10.1016/j.envpol.2017.11.030

Michalek-Wagner, K., & Willis, B. L. (2001). Impacts of bleaching on the soft coral *Lobophytum compactum*. I. Fecundity, fertilization and offspring viability. *Coral Reefs*, *19* (3): 231–239. http://doi.org/10.1007/s003380170003

Mora, C., Tittensor, D. P., Adl, S., Simpson, A. G. B., & Worm, B. (2011). How many species are there on Earth and in the ocean? *PLoS Biology*, *9* (8): e1001127. http://doi.org/10.1371/journal.pbio.1001127

Mulder, C., Koricheva, J., Huss-Danell, K., Högberg, P., & Joshi, J. (1999). Insects affect relationships between plant species richness and ecosystem processes. *Ecology Letters*, *2* (4): 237–246. http://doi.org/10.1046/j.1461-0248.1999.00070.x

Muller, E. B., Hanna, S. K., Lenihan, H. S., Miller, R. J., & Nisbet, R. M. (2014). Impact of engineered zinc oxide nanoparticles on the energy budgets of *Mytilus galloprovincialis*. *Journal of Sea Research*, *94*, 29–36. http://doi.org/10.1016/j.seares.2013.12.013

Murphy, F., & Quinn, B. (2018). The effects of microplastic on freshwater *Hydra attenuata* feeding, morphology & reproduction. *Environmental Pollution*, *234*, 487–494. http://doi.org/10.1016/j.envpol.2017.11.029

Murray, F., & Cowie, P. R. (2011). Plastic contamination in the decapod crustacean *Nephrops norvegicus* (Linnaeus, 1758). *Marine Pollution Bulletin*, *62* (6): 1207–1217. http://doi.org/10.1016/j.marpolbul.2011.03.032

Nelms, S. E., Duncan, E. M., Broderick, A. C., Galloway, T. S., Godfrey, M. H., Hamann, M., Lindeque, P. K., & Godley, B. J. (2016). Plastic and marine turtles: A review and call for research. *ICES Journal of Marine Science: Journal Du Conseil*, *73* (2): 165–181. http://doi.org/10.1093/icesjms/fsv165

Niki, E., Yoshida, Y., Saito, Y., & Noguchi, N. (2005). Lipid peroxidation: Mechanisms, inhibition, and biological effects. *Biochemical and Biophysical Research Communications*, *338* (1): 668–676. http://doi.org/10.1016/j.bbrc.2005.08.072

Nisbet, R. M., Jusup, M., Klanjscek, T., & Pecquerie, L. (2012). Integrating dynamic energy budget (DEB) theory with traditional bioenergetic models. *Journal of Experimental Biology*, *215* (6): 892–902. http://doi.org/10.1242/jeb.059675

O'Donovan, S., Mestre, N. C., Abel, S., Fonseca, T. G., Carteny, C. C., Cormier, B., Keiter, S. H., & Bebianno, M. J. (2018). Ecotoxicological effects of chemical contaminants adsorbed to microplastics in the clam *Scrobicularia plana*. *Frontiers in Marine Science*, *5*, 1–15. http://doi.org/10.3389/fmars.2018.00143

Ogonowski, M., Schür, C., Jarsén, Å., & Gorokhova, E. (2016). The effects of natural and anthropogenic microparticles on individual fitness in *Daphnia magna*. *PLOS ONE*, *11* (5): e0155063. http://doi.org/10.1371/journal.pone.0155063

Oliveira, P., Barboza, L. G. A., Branco, V., Figueiredo, N., Carvalho, C., & Guilhermino, L. (2018). Effects of microplastics and mercury in the freshwater bivalve *Corbicula fluminea* (Müller, 1774): Filtration rate, biochemical biomarkers and mercury bioconcentration. *Ecotoxicology and Environmental Safety*, *164*, 155–163. http://doi.org/10.1016/j.ecoenv.2018.07.062

Olsen, T. B., Christensen, F. E. G., Lundgreen, K., Dunn, P. H., & Levitis, D. A. (2015). Coelomic transport and clearance of durable foreign bodies by starfish (*Asterias rubens*). *The Biological Bulletin*, *228* (2): 156–162. http://doi.org/10.1086/BBLv228n2p156

Pamplona, R., & Costantini, D. (2011). Molecular and structural antioxidant defenses against oxidative stress in animals. *American Journal of Physiology-Regulatory, Integrative and Comparative Physiology*, *301* (4): R843–R863. http://doi.org/10.1152/ajpregu.00034.2011

Pedà, C., Caccamo, L., Fossi, M. C., Gai, F., Andaloro, F., Genovese, L., Perdichizzi, A., Romeo, T., & Maricchiolo, G. (2016). Intestinal alterations in European sea bass *Dicentrarchus labrax* (Linnaeus, 1758) exposed to microplastics: Preliminary results. *Environmental Pollution*, *212*, 251–256. http://doi.org/10.1016/j.envpol.2016.01.083

Petes, L. E., Menge, B. A., & Harris, A. L. (2008). Intertidal mussels exhibit energetic trade-offs between reproduction and stress resistance. *Ecological Monographs*, *78* (3): 387–402.

Pierce, K. E., Harris, R. J., Larned, L. S., & Pokras, M. A. (2004). Obstruction and starvation associated with plastic ingestion in a northern gannet *Morus bassanus* and a greater shearwater *Puffinus gravis*. *Marine Ornithology*, *32* (2): 187–189.

Pires, A., Almeida, Â., Calisto, V., Schneider, R. J., Esteves, V. I., Wrona, F. J., Soares, A. M. V. M., Figueira, E., & Freitas, R. (2016). *Hediste diversicolor* as bioindicator of pharmaceutical pollution: Results from single and combined exposure to carbamazepine and caffeine. *Comparative Biochemistry and Physiology, Part C*, *188*, 30–38. http://doi.org/10.1016/j.cbpc.2016.06.003

Pittura, L., Avio, C. G., Giuliani, M. E., D'Errico, G., Keiter, S. H., Cormier, B., Gorbi, S., & Regoli, F. (2018). Microplastics as vehicles of environmental PAHs to marine organisms: Combined chemical and physical hazards to the Mediterranean mussels, *Mytilus galloprovincialis*. *Frontiers in Marine Science, 5*, 103. http://doi.org/10.3389/fmars.2018.00103

Prather, C. M., Pelini, S. L., Laws, A., Rivest, E., Woltz, M., Bloch, C. P., Del Toro, I. et al. (2013). Invertebrates, ecosystem services and climate change. *Biological Reviews, 88* (2): 327–348. http://doi.org/10.1111/brv.12002

Purcell, S. W. (2014). Value, market preferences and trade of beche-de-mer from Pacific island sea cucumbers. *PLoS ONE, 9* (4): e95075. http://doi.org/10.1371/journal.pone.0095075

Pytharopoulou, S., Kouvela, E. C., Sazakli, E., Leotsinidis, M., & Kalpaxis, D. L. (2006). Evaluation of the global protein synthesis in *Mytilus galloprovincialis* in marine pollution monitoring: Seasonal variability and correlations with other biomarkers. *Aquatic Toxicology, 80* (1): 33–41. http://doi.org/10.1016/j.aquatox.2006.07.010

Ramos-Elorduy, J. (1997). Insects: A sustainable source of food? *Ecology of Food and Nutrition, 36* (2–4): 247–276. http://doi.org/10.1080/03670244.1997.9991519

Revel, M., Yakovenko, N., Caley, T., Guillet, C., Châtel, A., & Mouneyrac, C. (2018). Accumulation and immunotoxicity of microplastics in the estuarine worm *Hediste diversicolor* in environmentally relevant conditions of exposure. *Environmental Science and Pollution Research, 27*, 3574–3583. http://doi.org/10.1007/s11356-018-3497-6

Ribeiro, F., Garcia, A. R., Pereira, B. P., Fonseca, M., Mestre, N. C., Fonseca, T. G., Ilharco, L. M., & Bebianno, M. J. (2017). Microplastics effects in *Scrobicularia plana*. *Marine Pollution Bulletin, 122*, 379–391. http://doi.org/10.1016/j.marpolbul.2017.06.078

Rios Mendoza, L. M., Karapanagioti, H., & Álvarez, N. R. (2018). Micro(nanoplastics) in the marine environment: Current knowledge and gaps. *Current Opinion in Environmental Science & Health, 1*, 47–51. http://doi.org/10.1016/j.coesh.2017.11.004

Rist, S. E., Assidqi, K., Zamani, N. P., Appel, D., Perschke, M., Huhn, M., & Lenz, M. (2016). Suspended micro-sized PVC particles impair the performance and decrease survival in the Asian green mussel *Perna viridis*. *Marine Pollution Bulletin, 111* (1–2): 213–220. http://doi.org/10.1016/j.marpolbul.2016.07.006

Rodríguez-Seijo, A., da Costa, J. P., Rocha-Santos, T., Duarte, A. C., & Pereira, R. (2018). Oxidative stress, energy metabolism and molecular responses of earthworms (*Eisenia fetida*) exposed to low-density polyethylene microplastics. *Environmental Science and Pollution Research, 25*, 33599–33610. http://doi.org/10.1007/s11356-018-3317-z

Rodríguez-Seijo, A., Lourenço, J., Rocha-Santos, T. A. P., da Costa, J., Duarte, A. C., Vala, H., & Pereira, R. (2017). Histopathological and molecular effects of microplastics in *Eisenia andrei* Bouché. *Environmental Pollution, 220*, 495–503. http://doi.org/10.1016/j.envpol.2016.09.092

Rollo, C. D., & Hawryluk, M. D. (1988). Compensatory scope and resource allocation in two species of aquatic snails. *Ecology, 69* (1): 146–156. http://doi.org/10.2307/1943169

Romano, N., Ashikin, M., Teh, J. C., Syukri, F., & Karami, A. (2018). Effects of pristine polyvinyl chloride fragments on whole body histology and protease activity in silver barb *Barbodes gonionotus* fry. *Environmental Pollution, 237*, 1106–1111. http://doi.org/10.1016/j.envpol.2017.11.040

Rosa, M., Ward, J. E., & Shumway, S. E. (2018). Selective capture and ingestion of particles by suspension-feeding bivalve molluscs: A review. *Journal of Shellfish Research, 37* (4): 727–746. http://doi.org/10.2983/035.037.0405

Rossi, G., Barnoud, J., & Monticelli, L. (2014). Polystyrene nanoparticles perturb lipid membranes. *The Journal of Physical Chemistry Letters, 5* (1): 241–246. http://doi.org/10.1021/jz402234c

Rowley, A. F. (1996). The evolution of inflammatory mediators. *Mediators of Inflammation, 5* (1): 3–13. http://doi.org/10.1155/S0962935196000014

Rowley, A. F., & Powell, A. (2007). Invertebrate immune systems – Specific, quasi-specific, or nonspecific? *The Journal of Immunology, 179* (11): 7209–7214. http://doi.org/10.4049/jimmunol.179.11.7209

Ryan, P. G. (1988). Effects of ingested plastic on seabird feeding: Evidence from chickens. *Marine Pollution Bulletin, 19* (3): 125–128. http://doi.org/10.1016/0025-326X(88)90708-4

Salvati, A., Åberg, C., dos Santos, T., Varela, J., Pinto, P., Lynch, I., & Dawson, K. A. (2011). Experimental and theoretical comparison of intracellular import of polymeric nanoparticles and small molecules: Toward models of uptake kinetics. *Nanomedicine: Nanotechnology, Biology and Medicine, 7* (6): 818–826. http://doi.org/10.1016/j.nano.2011.03.005

Sánchez-Muros, M.-J., Barroso, F. G., & Manzano-Agugliaro, F. (2014). Insect meal as renewable source of food for animal feeding: A review. *Journal of Cleaner Production, 65*, 16–27. http://doi.org/10.1016/j.jclepro.2013.11.068

Sarà, G., Rinaldi, A., & Montalto, V. (2014). Thinking beyond organism energy use: A trait-based bioenergetic mechanistic approach for predictions of life history traits in marine organisms. *Marine Ecology*, *35* (4): 506–515. http://doi.org/10.1111/maec.12106

Setälä, O., Fleming-Lehtinen, V., & Lehtiniemi, M. (2014). Ingestion and transfer of microplastics in the planktonic food web. *Environmental Pollution*, *185*, 77–83. http://doi.org/10.1016/j.envpol.2013.10.013

Sievert, P. R., & Sileo, L. (1993). The effects of ingested plastic on growth and survival of albatross chicks. In R. Vermee, K. T. Briggs, K. H. Morgan, & D. Siegel-Causey (Eds.), *The Status, Ecology, and Conservation of Marine Birds of the North Pacific* (pp. 212–217). Ottawa, Canada: Canadian Wildlife Service Special Publication.

Slansky, F., & Wheeler, G. S. (1991). Food consumption and utilization responses to dietary dilution with cellulose and water by velvetbean caterpillars, *Anticarsia gemmatalis*. *Physiological Entomology*, *16* (1): 99–116. http://doi.org/10.1111/j.1365-3032.1991.tb00547.x

Sokolova, I. M., Frederich, M., Bagwe, R., Lannig, G., & Sukhotin, A. A. (2012). Energy homeostasis as an integrative tool for assessing limits of environmental stress tolerance in aquatic invertebrates. *Marine Environmental Research*, *79*, 1–15. http://doi.org/10.1016/j.marenvres.2012.04.003

Sousa, T., Domingos, T., Poggiale, J.-C., & Kooijman, S. A. L. M. (2010). Dynamic energy budget theory restores coherence in biology. *Philosophical Transactions of the Royal Society B: Biological Sciences*, *365* (1557): 3413–3428. http://doi.org/10.1098/rstb.2010.0166

Spear, L. B., Ainley, D. G., & Ribic, C. A. (1995). Incidence of plastic in seabirds from the tropical pacific, 1984–1991: Relation with distribution of species, sex, age, season, year and body weight. *Marine Environmental Research*, *40* (2): 123–146. http://doi.org/10.1016/0141-1136(94)00140-K

Stavrakidis-Zachou, O., Papandroulakis, N., & Lika, K. (2019). A DEB model for European sea bass (*Dicentrarchus labrax*): Parameterisation and application in aquaculture. *Journal of Sea Research*, *143*, 262–271. http://doi.org/10.1016/j.seares.2018.05.008

Straub, S., Hirsch, P. E., & Burkhardt-Holm, P. (2017). Biodegradable and petroleum-based microplastics do not differ in their ingestion and excretion but in their biological effects in a freshwater invertebrate *Gammarus fossarum*. *International Journal of Environmental Research and Public Health*, *14* (7): 774. http://doi.org/10.3390/ijerph14070774

Stumpp, M., Hu, M., Casties, I., Saborowski, R., Bleich, M., Melzner, F., & Dupont, S. (2013). Digestion in sea urchin larvae impaired under ocean acidification. *Nature Climate Change*, *3* (12): 1044–1049. http://doi.org/10.1038/nclimate2028

Sussarellu, R., Suquet, M., Thomas, Y., Lambert, C., Fabioux, C., Pernet, M. E. J., Le Goïc, N. et al. (2016). Oyster reproduction is affected by exposure to polystyrene microplastics. *Proceedings of the National Academy of Sciences*, *113* (9): 2430–2435. http://doi.org/10.1073/pnas.1519019113

Susswein, A. J., & Kupfermann, I. (1975). Localization of bulk stimuli underlying satiation in *Aplysia*. *Journal of Comparative Physiology A*, *101* (4): 309–328. http://doi.org/10.1007/BF00657048

Teuten, E. L., Saquing, J. M., Knappe, D. R. U., Barlaz, M. A., Jonsson, S., Björn, A., Rowland, S. J. et al. (2009). Transport and release of chemicals from plastics to the environment and to wildlife. *Philosophical Transactions of the Royal Society B: Biological Sciences*, *364* (1526): 2027–2045. http://doi.org/10.1098/rstb.2008.0284

Thompson, R. C. (2004). Lost at sea: Where is all the plastic? *Science*, *304* (5672): 838–838. http://doi.org/10.1126/science.1094559

Tomazic-Jezic, V. J., Merritt, K., & Umbreit, T. H. (2001). Significance of the type and the size of biomaterial particles on phagocytosis and tissue distribution. *Journal of Biomedical Materials Research*, *55* (4): 523–529. http://doi.org/10.1002/1097-4636(20010615)55:4<523::AID-JBM1045>3.0.CO;2-G

Tooci, S., Shivazad, M., Eila, N., & Zarei, A. (2009). Effect of dietary dilution of energy and nutrients during different growing periods on compensatory growth of Ross broilers. *African Journal of Biotechnology*, *8* (22): 6470–6475. http://doi.org/10.5897/AJB09.774

Tosetto, L., Brown, C., & Williamson, J. E. (2016). Microplastics on beaches: Ingestion and behavioural consequences for beach hoppers. *Marine Biology*, *163* (10): 199. http://doi.org/10.1007/s00227-016-2973-0

Trevan, J. W. (1927). The error of determination of toxicity. *Proceedings of the Royal Society B: Biological Sciences*, *101* (712): 483–514. http://doi.org/10.1098/rspb.1927.0030

Valko, M., Morris, H., & Cronin, M. (2005). Metals, toxicity and oxidative stress. *Current Medicinal Chemistry*, *12* (10): 1161–1208. http://doi.org/10.2174/0929867053764635

Van Cauwenberghe, L., Claessens, M., Vandegehuchte, M. B., & Janssen, C. R. (2015). Microplastics are taken up by mussels (*Mytilus edulis*) and lugworms (*Arenicola marina*) living in natural habitats. *Environmental Pollution*, *199*, 10–17. http://doi.org/10.1016/j.envpol.2015.01.008

Van Cauwenberghe, L., Vanreusel, A., Mees, J., & Janssen, C. R. (2013). Microplastic pollution in deep-sea sediments. *Environmental Pollution, 182*, 495–499. http://doi.org/10.1016/j.envpol.2013.08.013

van der Meer, J. (2006). An introduction to Dynamic Energy Budget (DEB) models with special emphasis on parameter estimation. *Journal of Sea Research, 56* (2): 85–102. http://doi.org/10.1016/j.seares.2006.03.001

Van Leeuwen, I. M. M., Vera, J., & Wolkenhauer, O. (2010). Dynamic energy budget approaches for modelling organismal ageing. *Philosophical Transactions of the Royal Society B: Biological Sciences, 365* (1557): 3443–3454. http://doi.org/10.1098/rstb.2010.0071

van Sebille, E., Wilcox, C., Lebreton, L., Maximenko, N., Hardesty, B. D., van Franeker, J. A., Eriksen, M., Siegel, D., Galgani, F., & Law, K. L. (2015). A global inventory of small floating plastic debris. *Environmental Research Letters, 10* (12): 124006. Retrieved from http://stacks.iop.org/1748-9326/10/i=12/a=124006

Vegter, A. C., Barletta, M., Beck, C., Borrero, J., Burton, H., Campbell, M. L., Costa, M. F. et al. (2014). Global research priorities to mitigate plastic pollution impacts on marine wildlife. *Endangered Species Research, 25* (3): 225–247. http://doi.org/10.3354/esr00623

von Moos, N., Burkhardt-Holm, P., & Köhler, A. (2012). Uptake and effects of microplastics on cells and tissue of the blue mussel *Mytilus edulis* L. after an experimental exposure. *Environmental Science & Technology, 46* (20): 11327–11335. http://doi.org/10.1021/es302332w

Vroom, R. J. E., Koelmans, A. A., Besseling, E., & Halsband, C. (2017). Aging of microplastics promotes their ingestion by marine zooplankton. *Environmental Pollution, 231*, 987–996. http://doi.org/10.1016/j.envpol.2017.08.088

Wagner, M., & Lambert, S. (Eds.). (2018). *Freshwater Microplastics: Emerging Environmental Contaminants?* Cham, Switzerland: Springer International Publishing. http://doi.org/10.1007/978-3-319-61615-5

Waite, H. R., Donnelly, M. J., & Walters, L. J. (2018). Quantity and types of microplastics in the organic tissues of the eastern oyster *Crassostrea virginica* and Atlantic mud crab *Panopeus herbstii* from a Florida estuary. *Marine Pollution Bulletin, 129* (1): 179–185. http://doi.org/10.1016/j.marpolbul.2018.02.026

Wang, M., Wang, X., Luo, X., Zheng, H., (2017). Short-term toxicity of polystryrene microplastics on mysid shrimps Neomysis japonica. In *3rd International Conference on Energy Materials and Environment Engineering. IOP Conference Series: Earth and Environmental Science.* IOP Publishing Ltd, Bangkok, Thailand, p. 012136. https://doi.org/10.1088/1755-1315/61/1/012136

Wang, X., Wang, L., Zhang, H., Ji, Q., Song, L., Qiu, L., Zhou, Z., Wang, M., & Wang, L. (2012). Immune response and energy metabolism of *Chlamys farreri* under *Vibrio anguillarum* challenge and high temperature exposure. *Fish & Shellfish Immunology, 33* (4): 1016–1026. http://doi.org/10.1016/j.fsi.2012.08.026

Ward, E. J., & Shumway, S. E. (2004). Separating the grain from the chaff: Particle selection in suspension- and deposit-feeding bivalves. *Journal of Experimental Marine Biology and Ecology, 300* (1–2): 83–130. http://doi.org/10.1016/j.jembe.2004.03.002

Watts, A. J. R., Lewis, C., Goodhead, R. M., Beckett, S. J., Moger, J., Tyler, C. R., & Galloway, T. S. (2014). Uptake and retention of microplastics by the shore crab *Carcinus maenas. Environmental Science & Technology, 48* (15): 8823–8830. http://doi.org/10.1021/es501090e

Watts, A. J. R., Urbina, M. A., Corr, S., Lewis, C., & Galloway, T. S. (2015). Ingestion of plastic microfibers by the crab *Carcinus maenas* and its effect on food consumption and energy balance. *Environmental Science & Technology, 49* (24): 14597–14604. http://doi.org/10.1021/acs.est.5b04026

Weber, A., Scherer, C., Brennholt, N., Reifferscheid, G., & Wagner, M. (2018). PET microplastics do not negatively affect the survival, development, metabolism and feeding activity of the freshwater invertebrate *Gammarus pulex. Environmental Pollution, 234*, 181–189. http://doi.org/10.1016/j.envpol.2017.11.014

Welden, N. A. C., & Cowie, P. R. (2016a). Environment and gut morphology influence microplastic retention in langoustine, *Nephrops norvegicus. Environmental Pollution, 214*, 859–865. http://doi.org/10.1016/j.envpol.2016.03.067

Welden, N. A. C., & Cowie, P. R. (2016b). Long-term microplastic retention causes reduced body condition in the langoustine, *Nephrops norvegicus. Environmental Pollution, 218*, 895–900. http://doi.org/10.1016/j.envpol.2016.08.020

Wen, B., Zhang, N., Jin, S.-R., Chen, Z.-Z., Gao, J.-Z., Liu, Y., Liu, H.-P., & Xu, Z. (2018). Microplastics have a more profound impact than elevated temperatures on the predatory performance, digestion and energy metabolism of an Amazonian cichlid. *Aquatic Toxicology, 195*, 67–76. http://doi.org/10.1016/j.aquatox.2017.12.010

Wolowczuk, I., Verwaerde, C., Viltart, O., Delanoye, A., Delacre, M., Pot, B., & Grangette, C. (2008). Feeding our immune system: Impact on metabolism. *Clinical and Developmental Immunology, 2008*, 639803. http://doi.org/10.1155/2008/639803

Wright, S. H., & Ahearn, G. A. (2011). Nutrient absorption in invertebrates. In R. Terjung (Ed.), *Comprehensive Physiology* (pp. 1137–1205). Hoboken, NJ: John Wiley & Sons. http://doi.org/10.1002/cphy.cp130216

Wright, S. L., Rowe, D., Thompson, R. C., & Galloway, T. S. (2013). Microplastic ingestion decreases energy reserves in marine worms. *Current Biology, 23* (23): R1031-1033. http://doi.org/10.1016/j.cub.2013.10.068

Xu, X.-Y., Lee, W. T., Chan, A. K. Y., Lo, H. S., Shin, P. K. S., & Cheung, S. G. (2017). Microplastic ingestion reduces energy intake in the clam *Atactodea striata*. *Marine Pollution Bulletin, 124* (2): 798–802. http://doi.org/10.1016/j.marpolbul.2016.12.027

You, Y. J., & Avery, L. (2012). Appetite control: Worm's-eye-view. *Animal Cells and Systems, 16* (5): 351–356. http://doi.org/10.1080/19768354.2012.716791

Ziajahromi, S., Kumar, A., Neale, P. A., & Leusch, F. D. L. (2018). Environmentally relevant concentrations of polyethylene microplastics negatively impact the survival, growth and emergence of sediment-dwelling invertebrates. *Environmental Pollution, 236*, 425–431. http://doi.org/10.1016/j.envpol.2018.01.094

17 Particulate Plastics and Human Health

Nanthi S. Bolan, M.B. Kirkham, Shiv Shankar Bolan, Daniel C.W. Tsang, Yiu Fai Tsang, and Hailong Wang

CONTENTS

17.1 Pathways of Human Exposure to Particulate Plastics ... 285
17.2 Impacts of Life Cycle Stages of Plastics on Human Health ... 286
17.3 Cellular Uptake of Particulate Plastics and Associated Contaminants ... 288
17.4 Human Health Impacts of Plastics ... 289
17.5 Conclusions ... 292
References ... 292

17.1 PATHWAYS OF HUMAN EXPOSURE TO PARTICULATE PLASTICS

Once particulate plastics reach the environment in the form of micro- and nanoplastics, they contaminate and accumulate in agricultural soils and the water sources and supply. They are transported through the terrestrial and aquatic food chain. Particulate plastics in the environment can leach toxic additives, or chemo-concentrate toxins already present in the environment, making them bioavailable again for direct or indirect human exposure. Particulate plastics have been measured in organisms at various trophic levels, leading to the accumulation of associated contaminants and eventual biomagnification. This is likely to occur in organisms at higher trophic levels, and the contaminants could ultimately reach the food chain, thereby affecting human health.

Particulate plastics, such as micro and nanofibers and beads can reach the body through dermal absorption, ingestion, or inhalation. They may lead to a wide range of detrimental health impacts as a result of their size, and capacity to penetrate tissues and cells, especially in the case of micro and nanoplastics, and also as a result of the vast range of toxic chemicals that are associated with plastics.

Routes of human exposure of particulate plastics include oral uptake through drinking water and diet containing particulate plastics, inhalation of micro and nanoplastics from the atmosphere, and dermal uptake of nanoplastics and the associated chemicals derived from cosmetics. The presence of particulate plastics in terrestrial (i.e., soil) and aquatic (i.e., freshwater) ecosystems has been documented, including in catchments used as sources of drinking water. They represent a pathway of human exposure to particulate plastics, especially in the case of micro and nanoplastics particles, which can pass through filtration systems in water and wastewater treatment plants (Kay et al. 2018). Wastewater treatment plants, which were not originally designed for treating micro and nanoplastics, have been identified as one of the largest sources of particulate plastics in the water environments (Boucher and Friot 2017; Ziajahromi et al. 2016).

Aquatic organisms may be exposed to contamination by particulate plastics, either through water enriched with particulate plastics or by feeding on other organisms containing particulate plastics, and they may function as a source of human exposure. Plastic particles have been detected in a range of organisms starting from the lowest levels of the food chain, such as zooplanktonic organisms (Cole et al. 2011, 2013), to the highest levels in both invertebrates (e.g., crustacea and mollusks) and vertebrates (e.g., fish) (de Sá et al. 2018; Smith et al. 2018). Seafood consumption represents one of the major pathways for particulate plastic exposure to humans (Zhang et al. 2019). Humans consume the whole soft tissues of bivalves, which may contain microscopic plastic debris. For benthopelagic fish, particulate plastics are in higher concentration in sea fish than in freshwater organisms.

Recently, particulate particles larger than 149 μm were measured in 17 salt brands from eight countries (Karami et al. 2017). Particulate plastics can also be ingested accidentally through personal care products, including toothpaste and lipstick containing microbeads (Cheung and Fok 2016; Cole et al. 2013; Duis and Coors 2016; Smith et al. 2018). Similarly, recently, there have been concerns about the adulteration of rice with particulate plastics. For example, synthetic plastic rice and sold as real rice has been reported in many countries, including India, Nigeria, Gambia, Ghana, and Philippines, but the reports were not confirmed (https://en.wikipedia.org/wiki/Artificial_rice).

Human interactions with water contaminated with particulate plastics during washing or through facial scrubs containing particulate plastics may lead to direct dermal uptake of particulate plastics (Thompson et al. 2009). However, due to the size of particulate plastics, and because uptake of particles across the skin requires entry through the striatum corneum that is limited to particles below 100 nm, dermal absorption is likely to occur only with nanoplastics. For example, nanoparticles present in skin care products, including sunscreen lotions, could penetrate through human skin (Filon et al. 2015). Human exposure to particulate plastics through inhalation could occur after micro and nanoplastics become airborne. This could also occur potentially from aerosols and dust released during wave action in aquatic environments, and particulate-enriched recycled water and biosolid application to the land. While few studies can be found on the presence of particulate plastics in the air compared to terrestrial and aquatic environments, several reports have demonstrated the presence of nanoplastics in the atmosphere. If some particulate plastics pass through during wastewater treatment processes (Kay et al. 2018), then a significant component of the plastic particles are entrapped in the biosolids. Application of biosolids to improve soil health has been shown to introduce particulate plastics to the terrestrial environment (Wang et al. 2018; Wijesekara et al. 2018).

17.2 IMPACTS OF LIFE CYCLE STAGES OF PLASTICS ON HUMAN HEALTH

Particulate plastics impact human health at various stages of the life cycle of the plastics. These life cycle stages of particulate plastics include extraction of fossil fuel feedstocks to synthesize plastics, manufacture of plastic products, utilization of plastic products by consumers, and management of plastic wastes (Figure 17.1).

Extraction and transport of fossil fuel feedstocks, which include crude oil and natural gas used for plastic manufacture, lead to the release of toxic substances to the air, water, and soil. For example, increasingly, natural gas is used as a major feedstock material for plastic manufacture. The first stop in the processing of plastics from natural gas is the cracker plant. Cracker plants turn either

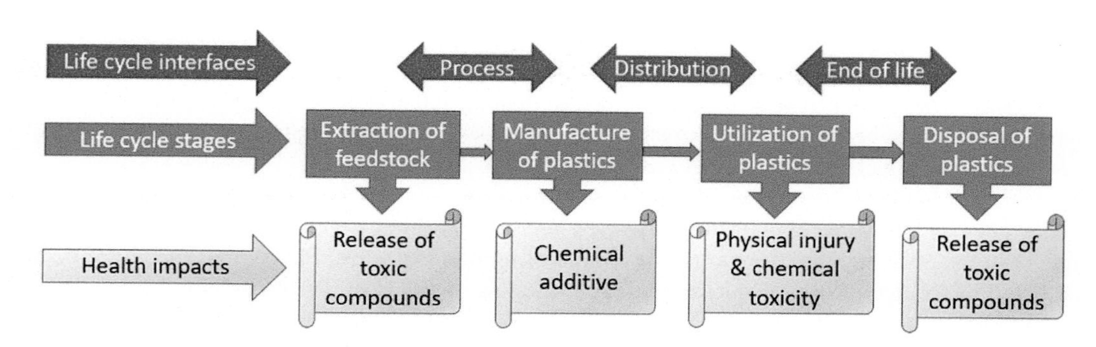

FIGURE 17.1 Life cycle impacts of particulate plastics on human health.

the crude oil-based product (naphtha) or the natural gas-based product (ethane) into ethylene, a starting point for a variety of chemical products. Ethane is significantly cheaper than naphtha, and crackers using the natural gas have a significant advantage. However, a large number of chemicals are used for hydraulic fracturing for natural gas, which are released into the atmosphere, impacting human health.

During the refining and synthesis of plastic resins from fossil fuel, a number of additives are added to improve the performance, functionality, and anti-aging properties of the basic plastic polymers (Hahladakis et al. 2018). Some of the common additives used in the plastic manufacturing industries include antioxidants, colorants, foaming agents, plasticizers, lubricants, and flame retardants (Table 5.3 in Chapter 5). For making plastic polymers and end products, the most commonly used chemicals include heavy metals (e.g., cadmium and lead), pesticides (e.g., triclosan), polycyclic aromatic hydrocarbons (e.g., benzene disulfonyl hydrazide), and polychlorinated biphenyls, and endocrine disrupting chemicals (e.g., bisphenol A) (Hahladakis et al. 2018). For example, 4-methyl-m-phenylenediamine, 4,4-methylenedianiline, and cobalt (II) diacetate are used as colorants, and bis(2-ethylhexyl) phthalate and poly- and perfluoroalkyl substances (PFAS) are used as plasticizers and flame retardants.

Consumer utilization of plastic products leads to ingestion and inhalation of large amounts of both micro and nanoplastic particles and the associated toxic compounds (Figure 17.2). Some of the chemicals added during plastic manufacture are released during the weathering of plastic products, thereby reaching the food chain. For example, food packaging using plastic products may lead to the release of some of the additive chemicals directly to the food. Plastic wrapping contains phthalates, one of the most widespread endocrine disruptors.

Particulate plastic waste management practices include incineration, gasification, and pyrolysis. These practices can result in the release of toxic metals, such as cadmium and mercury, organic compounds (e.g., dioxins and furans), gaseous acidic compounds, and other toxic contaminants to atmospheric, terrestrial, and aquatic environments. These practices can lead to direct and indirect exposure to toxic compounds for industrial personnel and neighboring communities. Exposure includes through inhalation of contaminated air, aerosols or dust, direct dermal contact with contaminated soil or water, and ingestion of foods derived from terrestrial and aquatic environments polluted with these toxic compounds. Most plastic additives are not strongly bound to the polymer matrix and can easily leach into the surrounding environment. As plastic particles continue to weather and degrade due to photochemical and microbial actions, fresh surface areas are exposed, allowing continued release of toxic additives from the inner core to the surface of the particle and subsequently to the surrounding environment.

FIGURE 17.2 Peeling of plastics due to continuous use of plastic products including plastic containers and plastic cutting board. (Courtesy of Bolan.)

17.3 CELLULAR UPTAKE OF PARTICULATE PLASTICS AND ASSOCIATED CONTAMINANTS

The toxicity of ingested contaminants, including particulate plastics and the associated chemicals, is determined ultimately by the extent to which they are solubilized in the gut (bioaccessibility), their permeability through intestinal epithelial cells and subsequent circulation in the blood (bioavailability), and their assimilation and metabolic action in any tissues that subsequently absorb them (bioactivity). The bioaccessibility–bioavailability–bioactivity continuum (Figure 17.3) plays a critical role in the toxicity of heavy metal(loid)s to biota (Bolan et al. 2017; Jaishankar et al. 2014). Bioaccessibility is usually evaluated in vitro by physiologically based extraction tests and gastrointestinal digestion procedures. Thus, for example, bioaccessibility can be used as a conservative estimate for bioavailability, because bioaccessibility is a theoretical maximum possible for bioavailability. Bioavailability, which expresses the fraction of the bioaccessible compounds that enter the blood circulation, refers to the rate and extent to which the compound permeates through the intestinal epithelial cells (Jaishankar et al. 2014). Bioactivity refers to the physiological and metabolic interactions between the compound and the human tissue or organ, which disturb homeostasis (Rehman et al. 2018). Toxicity of these contaminants can be mitigated by reducing their permeability in the intestine, thereby reducing the amount of contaminant entering the systemic circulation (Egorova and Ananikov 2017; Jaishankar et al. 2014).

In the case of particulate plastics, the cellular uptake of both particulate plastics and the associated chemicals leads to toxicity and human health issues. A special concern of particulate plastics, especially in the case of nanoplastics, is their ability to penetrate skin cells and the gut epithelium. Although oral uptake of nanoparticles has been investigated only recently, the gastrointestinal translocation of engineered nanoparticles, including carbon nanotubes, titanium dioxide, zinc oxide, and elemental silver, has been demonstrated extensively (Bergin and Witzmann 2013; Li et al. 2017). After translocation across the gut barrier, nanoparticles can reach the systemic circulation, depending on their size, shape, and surface properties including charge and functional groups. Translocation of particulate plastics across the gastrointestinal tract has been demonstrated for aquatic organisms, such as crabs and mussels (Anderson et al. 2016; Smith et al. 2018). The presence of particulate plastics in tissues beyond the gastrointestinal tract, such as livers, has been reported for fish fed with particulate plastics (Critchell and Hoogenboom 2018). The active and passive uptake of contaminants associated with particulate

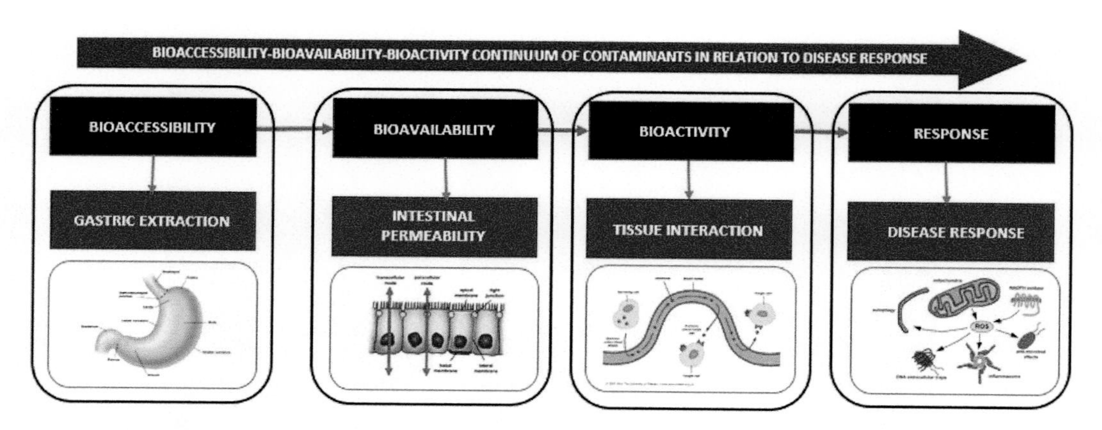

FIGURE 17.3 The bioaccessibility–bioavailability–bioactivity continuum of contaminants derived from particulate plastics in relation to toxicity response.

plastics, which include heavy metals and organic chemicals, through intestinal epithelial cells has been reported (Kole et al. 2017).

17.4 HUMAN HEALTH IMPACTS OF PLASTICS

Particulate plastics can lead to a range of detrimental health impacts, including inflammation, genotoxicity, oxidative stress, apoptosis, and necrosis, which are linked to a range of harmful health outcomes including cancer, cardiovascular diseases, inflammatory bowel disease, diabetes, rheumatoid arthritis, autoimmune conditions, neurodegenerative diseases, and strokes (Table 17.1). Adverse effects from particulate plastics may result from the combined processes of the plastic's intrinsic toxicity (e.g., physical damage), chemical composition (i.e., leaching of chemical additives), and

TABLE 17.1
Health Impacts for Particulate Plastic Sources and Associated Chemicals

Plastic Type/Additives	Common Uses	Adverse Health Effects
Polyvinyl chloride (PVC)	Food packaging, plastic wrap, containers for toiletries, cosmetics, crib bumpers, floor tiles, pacifiers, shower curtains, toys, water pipes, garden hoses, auto upholstery, and inflatable swimming pools	Can cause cancer, birth defects, genetic changes, chronic bronchitis, ulcers, skin diseases, deafness, vision failure, indigestion, and liver dysfunction
Phthalates (diethylhexyl phthalate [DEHP], diisononyl phthalate [DINP] and others)	Softened vinyl products manufactured with phthalates include vinyl clothing, emulsion paint, footwear, printing inks, non-mouthing toys and children's products, product packaging and food wrap, vinyl flooring, blood bags and tubing, intravenous (IV) containers and components, surgical gloves, breathing tubes, general purpose labware, inhalation masks, and many other medical devices	Endocrine disruption, linked to asthma, developmental, and reproductive effects. Medical waste with PVC and phthalates is regularly incinerated causing public health effects from the release of dioxins and mercury, including cancer, birth defects, hormonal changes, declining sperm counts, infertility, endometriosis, and immune system impairment
Polycarbonate, with Bisphenol A	Water bottles	Linked to very low doses of bisphenol A exposure to cancers, impaired immune function, early onset of puberty, obesity, diabetes, and hyperactivity, among other problems
Polystyrene	Many food containers for meats, fish, cheeses, yogurt, foam and clear clamshell containers, foam and rigid plates, clear bakery containers, packaging "peanuts," foam packaging, audio cassette housings, compact disc (CD) cases, disposable cutlery, building insulation, flotation devices, ice buckets, wall tile, paints, serving trays, throw away hot drink cups, and toys	Can irritate eyes, nose, and throat and can cause dizziness and unconsciousness. Migrates into food and stores in body fat. Elevated rates of lymphatic and hematopoietic cancers for workers
Polyethylene terephthalate (PET)	Water and soda bottles, carpet fiber, chewing gum, coffee stirrers, drinking glasses, food containers and wrappers, heat-sealed plastic packaging, kitchenware, plastic bags, squeeze bottles, and toys	Suspected human carcinogen

(*Continued*)

TABLE 17.1 (*Continued*)
Health Impacts for Particulate Plastics Sources and Associated Chemicals

Plastic Type/Additives	Common Uses	Adverse Health Effects
Polyester	Bedding, clothing, disposable diapers, food packaging, tampons, and upholstery	Can cause eye and respiratory tract irritation and acute skin rashes
Urea-formaldehyde	Particle board, plywood, building insulation, and fabric finishes	Formaldehyde is a suspected carcinogen and has been shown to cause birth defects and genetic changes. Inhaling formaldehyde can cause cough, swelling of the throat, watery eyes, breathing problems, headaches, rashes, and tiredness
Polyurethane foam	Cushions, mattresses, and pillows	Bronchitis, coughing, skin and eye problems. Can release toluene diisocyanate, which can produce severe lung problems
Acrylic	Clothing, blankets, carpets made from acrylic fibers, adhesives, contact lenses, dentures, floor waxes, food preparation equipment, disposable diapers, sanitary napkins, and paints	Can cause breathing difficulties, vomiting, diarrhea, nausea, weakness, headache, and fatigue
Tetrafluoro-ethylene	Non-stick coating on cookware, clothes irons, ironing board covers, plumbing, and tools	Can irritate eyes, nose, and throat, and can cause breathing difficulties
Fire retardants – PFAS compounds	Plastic packaging industry to impart flame retardancy to the final products	Potential developmental or reproductive effects, and the International Agency for Research on Cancer has classified PFOA as a group 2B carcinogen, i.e., possibly carcinogenic to humans
Heavy metals – mercury (Hg), cadmium (Cd), lead (Pb), antimony (Sb), and tin (Sn)	Improve the durability and workability	Most of these heavy metals are known carcinogens, can cause permanent changes to the genetic makeup of cells, and can have significant adverse effects on the fertility function among humans and wild animals

Source: Thompson, R.C. et al., *Philos. Trans. Royal Soc. B*, 364, 2153–2166, 2009; Azoulay, D. et al., *Plastic & Health: The Hidden Costs of a Plastic Planet*, 2019; https://ecologycenter.org/factsheets/adverse-health-effects-of-plastics/.

ability to retain and release toxic contaminants for the uptake by the organisms (Koch and Calafat 2009; Wright et al. 2013) (Figure 17.4). Particulate plastics could also serve as a vector for pathogenic microorganisms through "biofilm" formation, thereby leading to potential release and dispersion of a species into a new ecosystem (Oberbeckmann et al. 2016; Rummel et al. 2017).

Thus, particulate plastics could potentially induce physical damage through the particles themselves and biochemical stress through the presence of the particulate plastics alone or through the leaching of additives including inorganic and organic chemicals. The physical presence of particulate plastics may result in toxicity due to their inherent ability to induce intestinal blockage or tissue abrasion (Revel et al. 2018; Wright et al. 2013). For example, Revel et al. (2018) demonstrated oxidative stress through generation of reactive free oxygen species in cerebral and epithelial human cells, resulting from exposure to particulate plastics. Results from a recent study using mice indicated that ingestion of particulate plastics leads to the accumulation of particles in the liver, kidney, and gut (Deng et al. 2017). The kinetics and distribution of particulate plastics were related to their

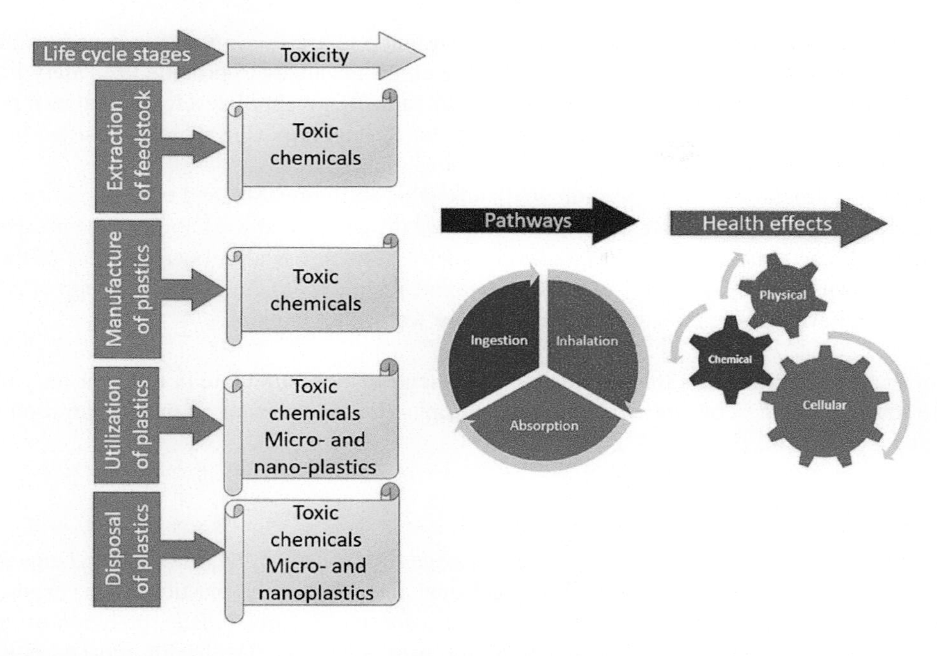

FIGURE 17.4 Toxicity sources, pathways, and health effects of particulate plastics.

particle size and surface characteristics including surface charge. The biochemical biomarkers and metabolomic profiles in mice livers suggested that exposure to particulate plastics induced alterations in oxidative stress and energy and lipid metabolism, leading to neurotoxicity. These results raise concerns about the potential cellular toxicity of particulate plastics on human liver cells.

Generally, human health risks associated with the use of particulate plastics are ascribed to the presence of the wide range of toxic chemicals added during plastic manufacture and the subsequent scavenging of contaminants from the environment (Koelmans et al. 2017; Thompson et al. 2009). Due to their high specific surface area (i.e., surface area per unit mass or volume) and hydrophobicity, particulate plastics may serve as a vector for exposure and dispersion of toxic environmental contaminants including organic compounds and inorganic heavy metals. This transfer of chemicals from particulate plastics into organisms through ingestion and contact at low trophic levels raises the possibility that these toxic chemicals may biomagnify in higher level predators, including humans (Andrady 2011; Browne et al. 2011, 2013; Fossi et al. 2014; Halden 2010). Measured cellular effects can include alterations of immunological responses, lysosomal compartmentalization, peroxisomal proliferation, antioxidant system activation, neurotoxic effects, genotoxicity, and changes in gene expression profiles. The four most common groups of plastic additives that have been associated with human health impacts include:

- Bisphenol A, which is often used in food and beverages plastic containers, such as water bottles. With continued use of these containers, the polymer chains of bisphenol A break down and can enter the human body through drinking contaminated water or ingesting food sources, such as fish and crabs that are exposed to the toxins released from particulate plastics. Specifically, bisphenol A is a known endocrine disturbing substance that interferes with human hormonal function.
- Phthalates, primarily used in polyvinylchloride and many other plastic products to make them flexible (He et al. 2015). These chemicals are also used in children's toys, flooring, clothes, and a range of other items. For example, diethylhexyl phthalate (DEHP), present in some plastic products, is a toxic carcinogen.

- Flame retardants including PFAS are heat resistant and are used in plastic-based electric and electronic products, upholstery, and other consumer items to provide fire safety benefits. Toxicological studies of PFAS compounds indicate potential developmental or reproductive effects, and the International Agency for Research on Cancer has classified them as a group 2B carcinogen, i.e., possibly carcinogenic to humans.
- A group of metals, such as cadmium (Cd), lead (Pb), antimony (Sb), and tin (Sn) (as organotin). They are used in plastic products to improve durability and workability. The presence of these heavy metals can cause a serious hazard to human health, because most of them are known carcinogens. They can cause permanent changes to the genetic makeup of cells and can have significant, adverse effects on the fertility of humans.

Most of these toxic chemicals in particulate plastics tend to bioaccumulate in lipids or fatty tissues and are found in higher concentrations at higher trophic-level organisms in the human food chain (Duis and Coors 2016; Gallo et al. 2018).

17.5 CONCLUSIONS

Particulate plastics impact human health at various stages of their life cycle. First, toxic chemicals are released during feedstock material extraction and manufacturing of the plastics. Then, exposure to chemical additives occurs during regular use of plastic consumer items. Finally, pollution of the terrestrial and aquatic environments results from plastic waste disposal. The health impacts arise from the direct effects of intake of particulate plastics (mostly physical effects) and the indirect effects of chemical contaminants associated with plastics during their manufacture and subsequent release of these chemicals into soil and aquatic environments. Despite their pervasive presence and potentially significant impacts across an array of pathways including ingestion, inhalation, and dermal absorption, research into the impacts and movement of particulate plastics at various trophic levels is limited. The potential transfer of plastics and associated toxic chemicals to crops and animals demands urgent and sustained investigation. Health impact assessments should focus on the plastic components of products, chemical additives, and their behavior at every stage of the plastic life cycle. Examining the relationship between toxic chemicals added during plastic manufacture, and subsequent accumulation in the environment, and adverse human effects introduces a number of challenges. These include the changing patterns of manufacture of plastics including the addition of various chemicals, the subsequent use of plastic-based consumer items, and the confidential nature of industrial specifications. These challenges make exposure assessments of particulate plastics particularly difficult. Risk assessment processes need to evaluate the health impacts of cumulative exposure to the mixtures of a vast array of chemicals used in consumer goods like food packaging and those found in the environment.

REFERENCES

Adverse Health Effects of Plastic. https://ecologycenter.org/factsheets/adverse-health-effects-of-plastics/ (accessed 5 January, 2020).

Anderson, J.C., Park, B.J. and Palace, V.P. (2016). Microplastics in aquatic environments: Implications for Canadian ecosystems. *Environmental Pollution*, 218, 269–280.

Andrady, A.L. (2011). Microplastics in the marine environment. *Marine Pollution Bulletin*, 62(8), 1596–1605.

Azoulay, D., Villa, P., Arellano, Y., Gordon, M., Moon, D., Miller, K. and Thompson, K. (2019). *Plastic & Health: The Hidden Costs of a Plastic Planet.*www.ciel.org/plasticandhealth

Bergin, I.L. and Witzmann, F.A. (2013). Nanoparticle toxicity by the gastrointestinal route: Evidence and knowledge gaps. *International Journal of Biomedical Nanoscience and Nanotechnology*, 3(1–2). doi:10.1504/IJBNN.2013.054515

Bolan, S., Kunhikrishnan, A., Seshadri, B., Choppala, G., Naidu, R., Bolan, N.S. and Kirkham, M.B. (2017). Sources, distribution, bioavailability, toxicity, and risk assessment of heavy metal(loid)s in complementary medicines. *Environment International*, 108, 103–118.

Boucher, J. and Friot, D. (2017). *Primary Microplastics in the Oceans: A Global Evaluation of Sources.* Gland, Switzerland: IUCN.

Browne, M.A., Crump, P., Niven, S.J., Teuten, E.L., Tonkin, A., Galloway, T., et al. (2011). Accumulations of microplastic on shorelines worldwide: Sources and sinks. *Environmental Science and Technology*, 45, 9175–9179.

Browne, M.A., Niven, S.J., Galloway, T.S., Rowland, S.J. and Thompson, R.C. (2013). Microplastic moves pollutants and additives to worms, reducing functions linked to health and biodiversity. *Current Biology*, 23(23), 2388–2392.

Cheung, P.K. and Fok, L. (2016). Evidence of microbeads from personal care product contaminating the sea. *Marine Pollution Bulletin*, 109(1), 582–585.

Cole, M., Lindeque, P., Goodhead, R., Moger, J., Halsband-Lenk, C. and Galloway, T.S. (2013). Microplastic ingestion by zooplankton. *Environmental Science and Technology*, 47, 6646–6655.

Cole, M., Lindeque, P., Halsband-Lenk, C. and Galloway, T.S. (2011). Microplastic as a contaminant in the marine environment: A review. *Marine Pollution Bulletin*, 62, 2588–2597.

Critchell, K. and Hoogenboom, M.O. (2018). Effects of microplastic exposure on the body condition and behaviour of planktivorous reef fish (*Acanthochromis polyacanthus*). *PLoS One*, 13(3), e0193308. doi:10.1371/journal.pone.0193308

de Sá, L.C., Oliveira, M., Ribeiro, F., Rocha, L.T. and Futter, M.N. (2018). Studies of the effects of microplastics on aquatic organisms: What do we know and where should we focus our efforts in the future? *Science of the Total Environment*, 645, 1029–1039.

Deng, Y.F., Zhang, Y., Lemos, B. and Ren, H. (2017). Tissue accumulation of microplastics in mice and biomarker responses suggest widespread health risks of exposure. *Scientific Reports*, 7, 46687. doi: 10.1038/srep46687

Duis, K. and Coors, A. (2016). Microplastics in the aquatic and terrestrial environment: Sources (with a specific focus on personal care products), fate and effects. *Environmental Sciences Europe*, 28(1), 2. doi:10.1186/s12302-015-0069-y

Egorova, K.S. and Ananikov, V.P. (2017). Toxicity of metal compounds: Knowledge and myths. *Organometallics*, 36(21), 4071–4090.

Filon, L.F., Mauro, M., Adami, G., Bovenzi, M. and Crosera, M. (2015). Nanoparticles skin absorption: New aspects for a safety profile evaluation. *Regulatory Toxicology and Pharmacology*, 72(2), 310–322.

Fossi, M.C., Coppola, D., Baini, M., Giannetti, M., Guerranti, C., Marsili, L., et al. (2014). Large filter feeding marine organisms as indicators of microplastic in the pelagic environment: The case studies of the Mediterranean basking shark (*Cetorhinus maximus*) and fin whale (*Balaenoptera physalus*). *Marine Environmental Research*, 100, 17–24.

Gallo, F., Fossi, C., Weber, R., Santillo, D., Sousa, J., Ingram, I., et al. (2018). Marine litter plastics and microplastics and their toxic chemicals components: The need for urgent preventive measures. *Environmental Sciences Europe*, 30, 1–14.

Hahladakis, J.N., Velis, C.A., Weber, R., Iacovidou, E. and Purnell, P. (2018). An overview of chemical additives present in plastics: Migration, release, fate and environmental impact during their use, disposal and recycling. *Journal of Hazardous Materials*, 344, 179–199.

Halden, R.U. (2010). Plastics and health risks. *Annual Review of Public Health*, 31, 179–194.

He, L., Gielen, G., Bolan, N.S., Zhang, X., Qin, H., Huang, H. and Wang, H. (2015). Contamination and remediation of phthalic acid esters in agricultural soils in China: A review. *Agronomy for Sustainable Development*, 35, 519–534.

Jaishankar, M., Tseten, T., Anbalagan, N., Mathew, B.B. and Beeregowda, K.N. (2014). Toxicity, mechanism and health effects of some heavy metals. *Interdisciplinary Toxicology*, 7(2), 60–72.

Karami, A., Golieskardi, A., Choo, C.K., Larat, V., Galloway, T.S. and Salamatinia, B. (2017). The presence of microplastics in commercial salts from different countries. *Scientific Reports*, 7, Article number: 46173.

Kay, P., Hiscoe, R., Moberley, I., Bajic, L. and McKenna, N. (2018). Wastewater treatment plants as a source of microplastics in river catchments. *Environmental Science and Pollution Research*, 25(20), 20264–20267.

Koch, H.M. and Calafat, A.M. (2009). Human body burdens of chemicals used in plastics manufacture. *Philosophical Transactions of the Royal Society B*, 364, 2063–2078.

Koelmans, A.A., Besseling, E., Foekema, E., Kooi, M., Mintenig, S., Ossendorp, B.C., et al. (2017). Risks of plastic debris: Unravelling fact, opinion, perception, and belief. *Environmental Science and Technology*, 51(20), 11513–11519.

Kole, P.J., Löhr, A.J., Van Belleghem, F.G.A.J. and Ragas, Ad. M.J. (2017). Wear and tear of tyres: A stealthy source of microplastics in the environment. *International Journal of Environmental Research and Public Health*, 14(10), 1265. doi:10.3390/ijerph14101265

Li, D., Zhuang, J., He, H., Jiang, S., Banerjee, A., Lu, Y., Wu, W., Mitragotri, S., Gan, L. and Qi, J. (2017). Influence of particle geometry on gastrointestinal transit and absorption following oral administration. *ACS Applied Materials & Interfaces*, 9(49), 42492–42502.

Oberbeckmann, S., Osborn, A.M. and Duhaime, M.B. (2016). Microbes on a bottle: Substrate, season and geography influence community composition of microbes colonizing marine plastic debris. *PLoS One*, 11(8), e0159289. doi:10.1371/journal.pone.0159289

Rehman, K., Fatima, F., Waheed, I. and Mohammad Sajid, H.A. (2018). Prevalence of exposure of heavy metals and their impact on health consequences. *Journal of Cellular Biochemistry*, 119, 157–184.

Revel, M., Châtel, A. and Mouneyrac, K. (2018). Micro(nano)plastics: A threat to human health? *Current Opinion in Environmental Science & Health*, 1, 17–23.

Rummel, C.D., Jahnke, A., Gorokhova, E., Kuhnel, D. and Schmitt-Jansen, M. (2017). Impacts of biofilm formation on the fate and potential effects of microplastic in the aquatic environment. *Environmental Science and Technology Letters*, 4, 258–267.

Smith, M., Love, D.C., Rochman, C.M. and Neff, R.A. (2018). Microplastics in seafood and the implications for human health. *Current Environmental Health Reports*, 5, 375–386.

Thompson, R.C., Moore, C.J., vom Saal, F.S. and Swan, S.H. (2009). Plastics, the environment and human health: Current consensus and future trends. *Philosophical Transactions of the Royal Society B*, 364, 2153–2166.

Wang, Z., Taylor, S.E., Sharma, P. and Flury, M. (2018). Poor extraction efficiencies of polystyrene nano- and microplastics from biosolids and soil. *PLoS ONE*, 13(11), e0208009. https://doi.org/10.1371/journal.pone.0208009

Wijesekara, H., Bolan, N., Bradney, L., Obadamudalige, N., Seshadri, B., Kunhikrishnan, A. and Vithanage, M. (2018). Trace element dynamics of biosolids-derived microbeads. *Chemosphere*, 199, 331–339.

Wright, S.L., Thompson, R.C. and Galloway, T.S. (2013). The physical impacts of microplastics on marine organisms: A review. *Environmental Pollution*, 178, 483–492.

Ziajahromi, S., Neale, P.A. and Leusch, F.D. (2016). Wastewater treatment plant effluent as a source of microplastics: Review of the fate, chemical interactions and potential risks to aquatic organisms. *Water Science and Technology*, 74(10), 2253–2269.

Zhang, F., Man, Y.B., Mo, W.Y., Man, K.Y. and Wong, M.H. (2019). Direct and indirect effects of microplastics on bivalves, with a focus on edible species: A mini-review. *Critical Reviews in Environmental Science and Technology*. doi:10.1080/10643389.2019.1700752.

Section IV

Case Studies of Particulate Plastics in the Environment

18 Status of Particulate Marine Plastics in Sri Lanka

Research Gaps and Policy Needs

T.W.G.F. Mafaziya Nijamdeen, Thilakshani Atugoda, P.B. Terney Pradeep Kumara, A.J.M. Gunasekara, and Meththika Vithanage

CONTENTS

18.1 Introduction 298
18.1.1 Maritime Zones of Sri Lanka 298
18.1.2 Marine Litter Status at the National Level 300
18.2 Origin, Pathways, and Trends 300
18.2.1 Origin and Sources of Marine Debris in Sri Lanka 300
18.2.2 Typology of Marine Litter in Sri Lanka 302
18.2.3 Amount of Marine Debris in Sri Lanka Beaches 303
18.2.4 Microplastic Debris 303
18.3 Impacts of Marine Litter on Sri Lanka 305
18.3.1 Social-Human Health and Food Safety 305
18.3.2 Ecosystem Impacts with Economic Consequences 306
18.3.3 Impact on Marine Ecosystem and Biodiversity 306
18.4 Management Agencies, Policies, Strategies, and Activities Taken to Minimize Marine Litter 307
18.4.1 Management Agencies and Their Responsibilities 307
18.4.2 Management Policies and Strategies and Their Effectiveness 308
18.4.3 Management Activities Done for Land-Based, Beach-Based, and Marine-Based Litter 309
18.4.4 Target under the Agenda 2030 and UN Sustainable Development Goals for Marine Litter Management 311
18.4.5 National Marine Litter Monitoring Program 312
18.5 Knowledge Needs and a Proposal for Setting Priorities 312
18.5.1 Inadequate Institutional Frameworks and Stakeholder Involvement 312
18.5.2 Policy and Legislation Gaps 312
18.5.3 Lack of Infrastructures for Waste Collection, Transportation, and Recycling 312
18.5.4 Education and Awareness 313
18.5.5 Strengthening Management and Control Efforts and Financing Mechanisms 313
18.6 The Way Forward 313
18.6.1 The Dispersion of Management Authorities of Marine Debris and the Absence of Interagency Cooperation between Authorities 313
18.6.2 The Absence of Scientific Survey and Statistics for Managing Marine Debris ... 315
18.6.2.1 Beach Surveys 315
18.6.2.2 Floating Debris Survey 317

18.6.2.3 Seabed Survey 317
18.6.2.4 Survey on Debris Ingested in Marine Organisms 317
18.6.3 Poor Management of Wastes Flown from Land to Ocean 317
18.6.4 International and Regional Cooperation 318
18.7 Recommendations 320
18.7.1 Introduce Integrated National Marine Debris Management Policy, Strategy, and Management Plan 320
18.7.2 Establishment of a New Legal Framework 320
18.7.3 Establishment and Revamping of the Institutional Structure 320
18.7.4 Intensive Management of Marine Debris Sources 321
18.7.5 Marine Debris Collection, Disposal, and Recycling 321
18.7.6 Research, Surveys, and Proper Modeling Systems on Marine Debris 321
18.7.7 Development of Education Awareness Program to Manage Marine Litter 321
18.7.8 Regional and International Cooperation 322
References 322

18.1 INTRODUCTION

Marine plastic pollution has become a significant problem worldwide. It has been estimated that about 8 million Mt of plastic debris enter into the sea annually (Ocean Conservancy 2019; Eriksen et al. 2014). Effects of debris accumulation in marine environments are acute on islands because of their proximity to debris concentrating zones and gyres. Hence, island nations and their marine ecosystems are under severe threat (Lavers and Bond 2017; Monteiro et al. 2018). However, transboundary marine pollution and land-based debris dispersion are inevitable for coastal states and islands (Ourmieres et al. 2018).

In spite of being a small island, Sri Lanka ranks fifth among the top 20 countries contributing to the marine debris into the world's oceans; this debris mostly consists of plastics (Jambeck et al. 2015). Nevertheless, no proper data are available to reveal the current status of marine debris or plastic debris in Sri Lanka (Jambeck et al. 2015). Marine litter has become a major concern of the island in the past few decades because five out of nine provinces of Sri Lanka are bordered by a coastal belt. Population densities of these coastal provinces are comparatively high; 35% of the total population has coastline inhabitants and 65% of the industries are in coastal areas. Moreover, 80% of the tourism industry is based on coastal habitation. Marine litter can be derived from either land-based or sea-based sources. Studies show that more than 90% of the litter on Sri Lankan coasts originates from the land (CZMP 2018).

A one-week survey along Sri Lankan beaches showed that plastic debris accumulation is high on the eastern coast, and it mainly occurs in river mouths and urban beach areas (Jang et al. 2018). Unavailability of long-term monitoring of marine plastic debris makes it difficult to develop prediction models to study the trends of plastic pollution and initiate proper mitigation plans to solve this issue. There is a vast need to evaluate the severity of plastic pollution in marine environments. This chapter focuses on the current status of marine debris, mainly particulate marine plastics, in Sri Lanka and probable research gaps that can be filled in the future in order to enhance policy needs.

18.1.1 Maritime Zones of Sri Lanka

During the last few decades, many initiatives have been taken by the United Nations (UN) to establish zones of peace in different parts of the world to address various legal issues. There are land-based zones and marine zones, depending on the necessities (Hancock and Mitchell 2007). Being an island, Sri Lanka has jurisdiction over certain areas of the sea where it can control marine pollution.

Sri Lanka is situated in the Indian Ocean. The land area of the island is 65,610 sq. km, and its coastal stretch is approximately 1620 km. Sri Lanka and the Indian sub-continent share the same continental shelf and are separated by the 30 m deep Palk Strait (Bandara 1989). Defining the maritime seascape of islands has become troublesome in recent years, mainly due to complicated and overlapping sovereignty and jurisdictional claims (Schofield 2009). There are certain proclaimed areas of national maritime jurisdiction under the Maritime Zones Law No. 22 of 1976, Sri Lanka is in accordance with the provisions of the United Nations Convention on the Law of the Sea (Figure 18.1).

Maritime jurisdiction of Sri Lanka covers the following major areas:

1. Internal waters are waters in the landward side of the main coast or the baseline from limits of the territorial sea. This includes the embayment and areas of coastal sea and all inland waters
2. Historic waters include the Gulf of Mannar, Palk Bay, Palk Strait, and areas on the basis of traditional use by Sri Lankans
3. The territorial sea extends to 12 nautical miles (21,500 sq. km). Sri Lanka proclaims sovereign rights over these areas; this includes the passage of foreign ships and aircraft through the water and air within this area

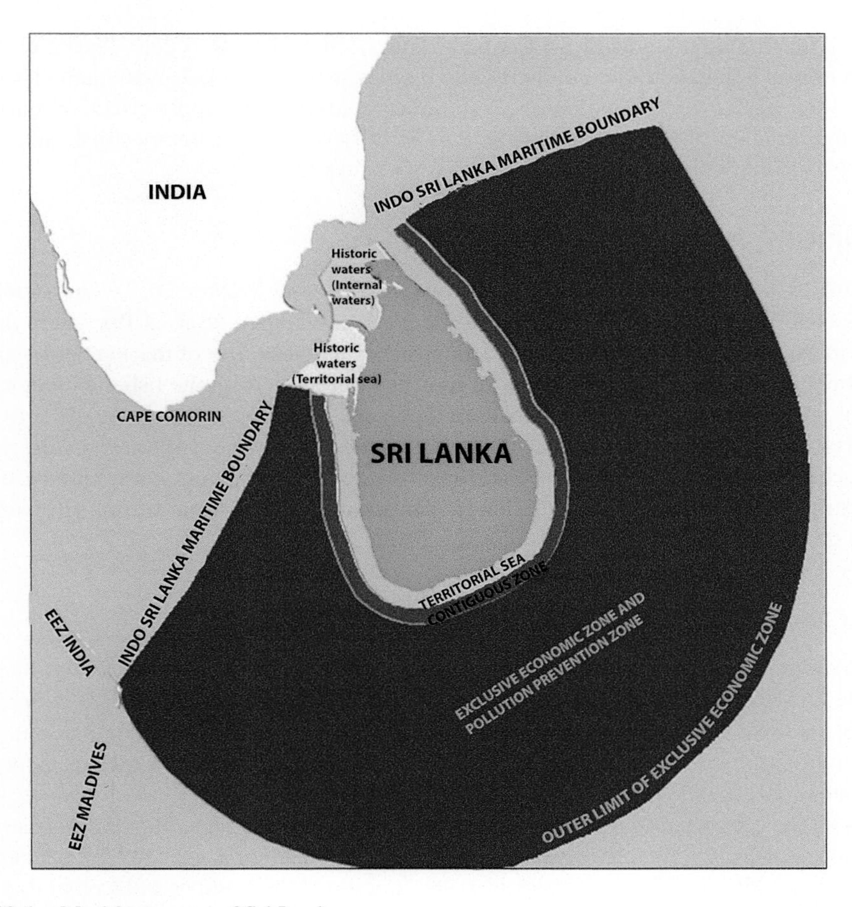

FIGURE 18.1 Maritime zones of Sri Lanka.

4. The contiguous zone extends to 24 nautical miles. Sri Lanka asserts its rights to take measures necessary to secure the enforcement of, and avert the violation of, its laws regarding security, immigration, health, sanitation, or customs
5. The exclusive economic zone extends to 200 nautical miles from the baseline (517,000 sq. km). Within this zone, Sri Lanka proclaims, among other factors, sovereign management of natural resources (living and non-living) and the right to govern and supervise scientific research (Dela 2002)

18.1.2 Marine Litter Status at the National Level

In Sri Lanka, major debris input to the beaches comes from the mouths of rivers (MEPA 2017). Local governmental authorities are responsible for the regulation of land-based wastes and their management. Present municipal waste generation in Sri Lanka is around 6500 to 7000 Mt/day, whereas waste generation may vary from 0.4 to 1 kg per day per capita based on the living standards. Municipal solid waste collection of Sri Lanka is 3500 Mt/day. The collection capacity is nearly 50% of the total waste generated (Menikpura et al. 2012).

Apart from the inland-based sources, there are sea-based sources that also include transboundary pollution. Therefore, marine litter accumulation has to be studied in association with global marine litter. As an island in the Indian Ocean, the Sri Lankan climate is governed by two active monsoonal winds (southwest and northeast monsoons). Thus, it is inclined to debris accumulation including foreign debris floating on the sea. This debris reaches the coast around Sri Lanka with water currents propelled by monsoons (Jang et al. 2018).

The total marine litter status of Sri Lanka has not been evaluated to date. Plastic waste and its typologies are not being monitored properly, and there is an acute need to understand different types of marine litter to reduce its accumulation. In order to develop and apply effective marine debris pollution management plans, it is mandatory to identify the origin, composition, and density of marine debris along with its potential pathways to coastal sites.

18.2 ORIGIN, PATHWAYS, AND TRENDS

Marine litter is a global environmental concern (Xanthos and Walker 2017). On a global scale, about 61% to 87% of this litter is predicted to be plastics (Galgani et al. 2019). There are several ways that marine litter can be formed. The composition and pathways of marine debris are diverse (Kandziora et al. 2019). Sea-based sources of marine litter arise from the fisheries sector and merchant ships, cruise liners, and recreational activities, such as diving, boating, leisure activities, and water sports; coastal-based activities include enhancement of tourism, fisheries, hotels, restaurants on the beaches, houses, fishery harbors, and illegal waste dumping sites; and inland-based sources include untreated municipal sewage discharge, stormwater, and riverine transport of waste from dumpsites on marshy land and wetlands (CZMP 2018).

18.2.1 Origin and Sources of Marine Debris in Sri Lanka

In Sri Lanka, improper municipal solid waste management is speculated to be a significant factor responsible for debris accumulation on both land and sea. There are 23 municipal councils and 41 urban councils in Sri Lanka responsible for municipal solid waste collection and disposal. However, the most common practice is open dumping with no proper environmental management practices (Hasintha et al. 2014).

Studies on the composition of Sri Lankan municipal solid waste show that it consists of kitchen wastes (up to 74.6%), and the other wastes are composed of garden wastes (4.8%), paper and cardboard (7.8%), soft plastic (4.2%), hard plastic (0.9%), textiles (1.0%), rubber and leather (0.4%), metal (0.9%), glass (1.7%), ceramics (0.5%), hazardous wastes (0.4%), E-wastes (0.2%), and miscellaneous

(2.7%) (SATREPS 2014). From the above results, it is evident that more than 87% of the waste materials are biodegradable. Unmanaged wastes are 27% of the total, and they contribute significantly to marine debris. Even though nationwide data are unavailable, results from two beaches in Negombo show that most of the debris (99%) was domestic, leaving only 1% foreign. Packaging materials (55%) were the primary source, together with consumer products (26%). Marine debris from fisheries also comprised 18% of the debris. Plastics composed 79% of the total material, which is similar to the worldwide situation.

Rivers are major carriers of inland debris to beaches (León et al. 2018). Sri Lanka has 149 rivers and streams, most of which start from the central highlands (Silva 2004). These rivers flow through the urban areas and reach the sea. Studies show that the Kelani River alone carries more than 6000 plastic pieces per day (SACEP 2007). Beach typology also plays an essential role in the amount of marine debris accumulation. The distance of urban cities or a river mouth from a beach, the persistence of a physical barrier, and the time of the monsoons are contributory factors for plastic litter accumulation at Sri Lankan beaches (Figure 18.2) (Jang et al. 2018).

FIGURE 18.2 Beaches polluted with plastic debris at Mirissa (top), Gurunagar (middle), and Kudawella (bottom), Sri Lanka.

Tourism is another major cause of accumulation of marine plastic debris. Small restaurants and hotels do not have well designed waste management systems, and they dump solid waste directly into the sea and lagoons. Ferry- and boat-repairing workplaces produce wastes of polystyrene and fiberglass, and they are not properly disposed of. Beaches can be further polluted when they are littered with plastics by beachgoers (Derraik 2002).

Studies show that the east coast beaches in Sri Lanka have higher debris accumulation than other beaches (Figure 18.2). It is thought that most of this debris originated from countries associated with the Bay of Bengal. Furthermore, the Bengal gyre might be responsible for transboundary debris transportation through Bengal currents (Jang et al. 2018). Figure 18.3 shows ocean water currents and winds close to Sri Lanka.

18.2.2 Typology of Marine Litter in Sri Lanka

Marine litter in Sri Lanka has a wide range of substances created from land-based, beach-, or ocean-based production. The major litter categories are polythene, plastics, polystyrene, rubber, wood, metals, discarded medical and sanitary equipment, paper and cardboard, glass, tins, fishery sector litter, carton packages for food and beverages, waste from households, building materials, pottery and ceramics, used batteries, cigarette butts, and filters (SACEP 2007).

Because Sri Lanka is located in a strategic position that is ideal for shipping, more than 4500 ships per year arrive at four major ports and other smaller ports. International shipping routes and lanes are close to Sri Lanka, including a major international shipping route that connects the East and West. Three hundred to 350 ships sail every day on this route, and it is located near the southern coast of Sri Lanka (SLPA 2015). Surveys by the Marine Environment Protection Authority (MEPA) show that marine litter and illegal ship discharges, including oil spills, litter these shipping routes. According to the MARPOL convention, harbors should provide sufficient waste collection and disposal facilities. Sri Lanka is a participant country of the MARPOL Convention 1973/78 (Marine Pollution Prevention Act No. 35 2008).

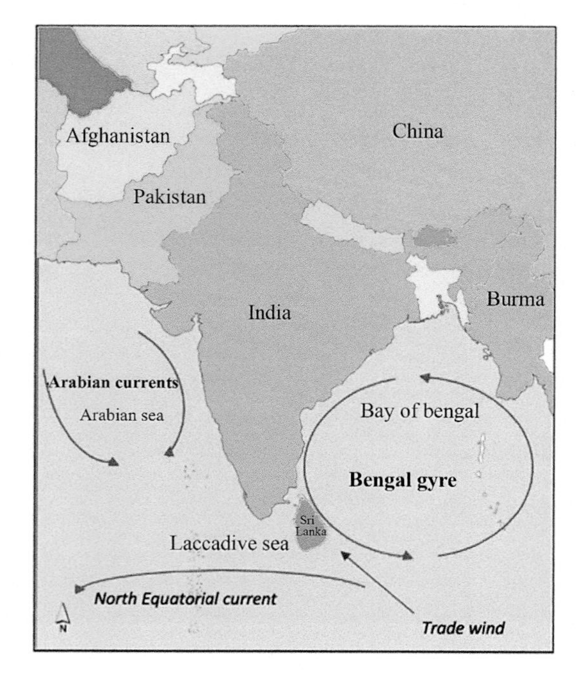

FIGURE 18.3 Ocean currents and trade wind passing Sri Lankan coasts.

Coastal cleanup data for five years (2008–2013) for Sri Lankan beaches reveal that around 75% of the marine litter is from recreational activities, and it includes bags (plastic and paper), balloons, beverage bottles (plastic, 2 L or less, and glass), beverage cans, caps, lids, clothing, shoes, eating utensils, cups, plates, food wrappers, containers, pull tabs, six-pack holders, shotgun shells, wadding, straws, stirrers, and toys (Gunasekara et al. 2014). The cleanup data were only for beaches, but there can be large amounts of marine litter underneath the water column. However, marine litter in deeper ocean parts of Sri Lankan maritime zones has not yet been appropriately assessed. Marine litter can further be classified into two groups according to its size: large or small. Large marine debris is bigger than 25 mm, whereas small marine debris is 5–25 mm. Surveys reveal that both of these categories have been found on Sri Lankan beaches (Jang et al. 2018).

18.2.3 Amount of Marine Debris in Sri Lanka Beaches

It is essential to assess marine debris, microplastics, and pollutants on beaches and in oceans around Sri Lanka to investigate the current pollution status. Although nationwide data are scarce, studies of the Indian Ocean reveal that microfibers from marine water are highest in Bangladesh, India, Pakistan, and Sri Lanka (Balasubramaniam and Phillott 2016). According to the first report from Sri Lanka (Jang et al. 2018), microplastic debris was found on all beaches investigated, which gives an idea of the composition of the marine debris.

The primary source of marine debris in Sri Lanka is from plastic packaging materials, which are the most abundant and mismanaged pollutants on Sri Lankan beaches (Tables 18.1 and 18.2). Small plastic fragments are the most abundant, and particular attention must be given to regulate them.

18.2.4 Microplastic Debris

Researchers have given various definitions to "microplastics" and "microlitter." Andrady (2011) defines both micro litter and microplastics as minute particles that can pass through sieves ranging in size from 500 to 67 μm. He defines larger plastic as "mesoplastic litter." However, in common usage, the smaller (typically <5 mm) fragments of plastic debris in the ocean are loosely referred to as microplastics. Microplastics now appear to be a ubiquitous pollutant of beaches, surface waters, and marine sediments worldwide (Claessens et al. 2011; Law et al. 2010; Moore et al. 2001). When

TABLE 18.1
Marine Debris Type and Their Percentages on Sri Lankan Beaches

Marine Debris Type	Percentage (%)
Foreign debris	0.9
Packaging material	55.1
End-consumer products	25.1
Fishery-related debris	10.5
Wastes from other industries	0.03

Marine Debris Type	Percentage (%)
Foreign debris	0.9
Packaging material	55.1
End-consumer products	25.1
Fishery-related debris	10.5
Wastes from other industries	0.03

TABLE 18.2
Marine Debris Material Types and Their Sizes along with Percentages on Sri Lankan Beaches

Marine Debris Material Type	Large >25 mm (%)	5–25 mm (%)
Plastics	93.3	98.7
Metals	2.0	0.0
Glass and ceramics	2.9	0.8
Wood	0.7	0.0
Paper	1.0	0.5
Other materials	0.0	0.0

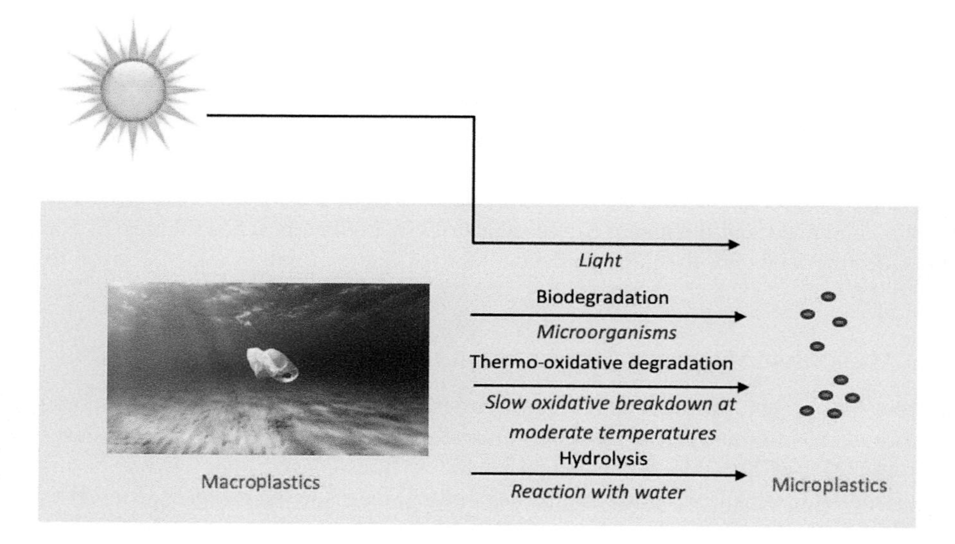

FIGURE 18.4 Possible pathways of plastic degradation in the marine environment.

small plastic fragments that are visible to the naked eye are mixed with sand particles, they are difficult to distinguish. That is, it is impossible to differentiate microplastic fragments from sand grains (Andrady 2011). After degradation, plastics become brittle and fragmented. Microbes, temperature, and light cause them to degrade, as shown in Figure 18.4.

There are two types of microplastics: primary and secondary microplastics, as listed in Table 18.3.

TABLE 18.3
Types of Microplastics

Type of Microplastics	Origin	Examples
Primary microplastics	Industrial, domestic products	Air blasting media (facial cleansers, cosmetics), drug vectors, microplastic scrubbers (paint, rust removal)
Secondary microplastics	Fragmentation of large plastics into smaller debris	Packing material, consumables

TABLE 18.4
Harmful Effects of Microplastics to Humans

	Chemical Effects of Microplastics	Physical Effects of Microplastics
Morphological effects	Inflammation (Gasperi et al. 2018) Complications in digestion (Auta et al. 2017) Alteration of enzyme secretions (Sharma and Chatterjee 2017) Difficulty in breathing (Kaya et al. 2018) Obesity (Galloway 2015) Cancer (Browne et al. 2007)	Heartbeat alterations (Waring et al. 2018) Neurotoxicity (Barboza et al. 2018)
Behavioral effects	Reduced vigor, mobility effects (Sharma and Chatterjee 2017)	Reduced vigor, mobility effects (Sharma and Chatterjee 2017)
Reproductive effects	Infertility (Rusthoven 2018)	Delay in ovulation (Sharma and Chatterjee 2017)

Exposure of humans to microplastics is inevitable, because day-to-day activities involve a variety of them – from cosmetic sprays to handwashes. Microplastics can have toxic effects, and hazardous substances, such as phthalates, adsorbed onto them also are toxic (Zhang et al. 2018). It is important to note that a wide variety of chemicals are used in the production of microplastics, which are harmful to human health (Galloway and Lewis 2016), as documented in Table 18.4.

18.3 IMPACTS OF MARINE LITTER ON SRI LANKA

Marine litter has increased significantly on Sri Lankan beaches in recent years. The most prominent reason speculated for this increase is the use of synthetic packing and container materials. Plastics are often preferred instead of paper due to cost effectiveness. Moreover, plastic bottles are widely used in the mineral water and soft drink industries. The abundance of plastic litter varies from province to province in Sri Lanka (MEPA 2018).

Source reduction is one of the best ways to mitigate marine litter. Well planned litter monitoring systems and international partnerships are crucial for effective management of marine litter in Sri Lanka. As noted, Sri Lanka is fifth on the list of plastic polluting countries in the world (Jang et al. 2018). China produces the highest amount of plastics, followed by Indonesia, the Philippines, Vietnam, and Sri Lanka. In Sri Lanka, 84% of the total waste is mismanaged, of which, the mismanaged plastic waste per year is around 1.59 MMT. This contributes to 0.24–0.64 MMT/year plastic marine debris (Jambeck et al. 2015).

18.3.1 Social-Human Health and Food Safety

Marine litter can cause potential harm to human health and safety in numerous ways. Human health impacts related to wastes have not been widely studied in Sri Lanka. Beachgoers can get directly injured by waste materials, including glass and plastics. Furthermore, hazardous and hospital wastes that are harmful to human health are found on beaches. Even though they are handled with care in hospitals, the fate of the hazardous wastes is either a dumpsite or beach (Figure 18.5). This hazardous waste contains unused antibiotics, pain killers, and other narcotic drugs that could cause severe repercussions to the environment and especially to marine organisms (Ashiq et al. 2019).

Marine litter affects recreational activities and damages boat engines. Moreover, the mismanaged plastic wastes are the major source of marine pollution (Jambeck et al. 2015). Plastic debris in the sea comes in all shapes and sizes, and it is directly connected to human health through physical damage, bioaccumulation of microplastics, and leaching of poisonous chemicals (Zalasiewicz et al. 2016).

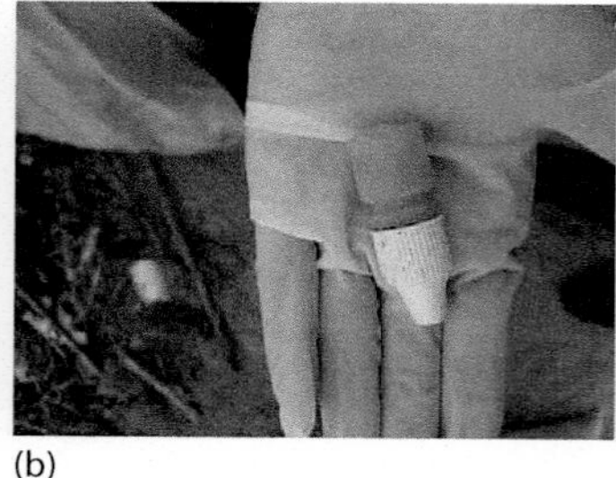

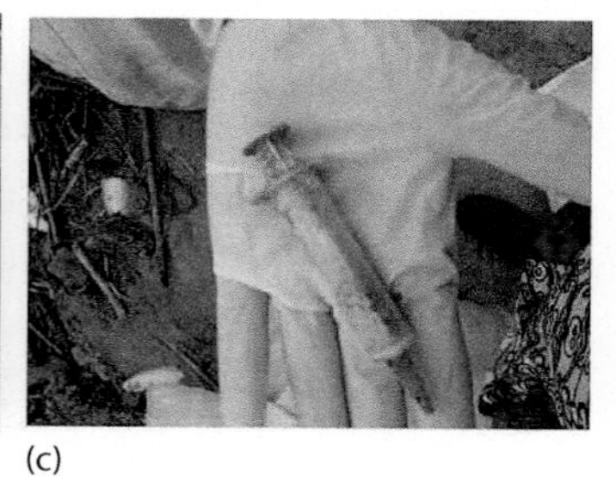

(a) (b) (c)

FIGURE 18.5 Medical wastes, reagent bottles (a, b) and syringe (c) on the Puttalam Beach.

18.3.2 Ecosystem Impacts with Economic Consequences

It is crucial to understand the potential economic consequences of marine debris, because it affects ecosystem functioning. Economic impacts are mainly on the marine environment, biodiversity, human health, and safety. Until now, only a few studies have focused on economic consequences. In this regard, both direct and indirect impacts have to be evaluated (Lee 2015). However, economic consequences of marine litter on the marine industrial sector and tourism have to be carried out to estimate the cost of beach cleanups by municipalities and to prevent marine litter accumulation (Mouat et al. 2010). Studies show that governmental initiatives are the primary motivating factor for removing marine litter. This litter prevention is forced onto municipalities mostly to enhance tourism and related trades (Ofiara 2001). In 2008, the damage caused to marine industries by marine litter in 21 countries across the Asia/Pacific Rim was estimated to be US$1.26 billion per year. Marine litter causes damage to aquaculture, harbors, ports (Lee 2015), fishing gear (Takehama 1990), fishing motors, and ferries, and these losses are direct and can be deduced from market measures (McIlgorm et al. 2011).

The indirect economic impacts are difficult to assess. The connection of litter management and other sectors of the economy are apparent, because marine debris reduces the aesthetic value of an environment, which, in turn, reduces new tourism opportunities and discourages public and private sector investment in tourism (Ofiara 2001). The overall value of marine ecosystems is reduced when they are not managed by the government (McIlgorm et al. 2011).

18.3.3 Impact on Marine Ecosystem and Biodiversity

Marine debris poses many problems to the ecosystem and biodiversity. One such common problem is entanglement. Entanglement can cause smothering, amputation, and drowning of marine animals. They can also confine movement and promote infections and starvation. Most animals die after encountering plastic nets and strings (Laist 1997). In this way, whales, turtles, and birds are mostly affected. Nesting materials are lost for birds, and entangled nets cause asphyxiation or amputation of flippers, tails, and legs, resulting in death (Cassoff et al. 2011; Hoopes et al. 2000). Ingestion of marine debris is another issue, in which marine species confuse marine litter and debris, mainly plastics, with food. Plastic usually blocks internal organ systems. Many studies show that organisms accidentally eat plastics. Common examples are plastic bag ingestion by turtles that mistake the bags for jellyfish; whales that eat marine litter, and birds that consume plastic pieces, including microplastics (Ryan 2008; Mrosovsky et al. 2009).

Owing to their small size, microplastics have become widely spread around coastal areas, and their impact on biodiversity has become a global concern (Barnes et al. 2009). Microplastic ingestion is a great threat to marine organisms due to the small size of the plastic (Thompson et al. 2004). Apart from direct ingestion, extraneous pollutants and contaminants attached to the microplastics can have adverse effects (Cole et al. 2011). Due to their low density and small size, microplastics can be found in benthic and pelagic zones of aquatic environments (Anderson et al. 2016). This microplastic debris tends to accumulate in food chains. Organisms at low trophic levels ingest microplastics, because they are unable to distinguish plastic particles from food particles. Larval stages of fish and crustaceans not only feed on plankton, but also on microplastics (Fendall and Sewell 2009).

18.4 MANAGEMENT AGENCIES, POLICIES, STRATEGIES, AND ACTIVITIES TAKEN TO MINIMIZE MARINE LITTER

There are numerous agencies responsible for waste management in Sri Lanka. These agencies are broadly categorized as central government agencies and local government agencies. In order to manage marine litter, its primary causes, and sources, as well as human actions causing this litter, management agencies need to be identified and policies have to be updated. According to the MEPA, land-based waste is the result of Sri Lanka's growing population, accelerating urbanization, and increasing industrial activities. Marine litter is the result of poor management of wastes by coastal residents and fishermen, a lack of public awareness, insufficient waste management in rivers and canals connected to oceans, and a shortage of collection and disposal facilities, and it has to be adequately managed (MEPA 2018). Therefore, management agencies have a significant role in the proper management of land-based wastes to control the addition of plastics and other litter into the ocean.

18.4.1 Management Agencies and Their Responsibilities

Management agencies have distinct responsibilities, and, in some instances, their responsibilities overlap. The following agencies have responsibilities over marine litter management.

1. *Ministry of Local Government and Provincial Councils (MoLGPC)*
   The major task of the MoLGPC is the implementation of policies and plans. These plans are given to the nine provincial councils (PC) and from there to local authorities (LA). The MoLPGC further coordinates the responsibilities between the central government and the PCs, which aid in the establishment and implementation of national policies. Human resource development, financing, technical assistance, and implementation of research are some other contributions of the MoLPGC. The ministry has internal branches, such as the local loan and development fund to aid financially LAs and the Sri Lankan Institute of Local Governance for administrative capacity building and research for LAs (JICA 2016).
2. *Ministry of Mahaweli Development and Environment (MoMDE)*
   The MoMDE is involved in the formulation of a national policy for waste management. Since 1998, a municipal waste database has been prepared using survey data from LAs (JICA 2016).
3. *Ministry of Megapolis and Western Development (MoMWD)*
   The MoMWD is a new ministry established in 2015. It mainly focuses on waste management, regulation of problems related to slums and shanties, and reducing traffic congestion. The ministry gives suggestions to LAs on solid waste management without direct involvement.
   The projects of this ministry are approved by a decision-making organization, called the committee of secretaries, and it includes representatives from the MoMWD, MoLGPC, Ministry of Water Supply and City Planning, the Sri Lanka Land Reclamation and Development Corporation, and the Provincial Council of Western Province (chief secretary) (JICA 2016).
4. *Ministry of Health, Nutrition, and Indigenous Medicine (MoH)*
   The MoH is involved with the management of medical waste. The ministry has established the healthcare waste management national policy for proper disposal of medical waste.

Public health inspectors, medical officers of health, and divisional secretaries are involved with the handling and management of medical wastes (JICA 2016).

5. *National Solid Waste Management Support Center*
   The National Solid Waste Management Support Center was established in 2007 by the MoLGPC. It aids in the improvement of solid waste management by LAs by providing guidelines. The guidelines include ways to implement proper solid waste management, to provide technical assistance on solid waste management, and to analyze current practices and compare them with other countries. Further, it assists LAs in getting grants and financial assistance from Non-Governmental Organizations (NGOs) and donors (JICA 2016).
6. *Central Environmental Authority (CEA)*
   The CEA is the major authorized body under the National Environmental Act. It regulates solid waste management. Its environmental assessment unit deals with the environmental impact assessment (EIA), required by the National Environmental Act. Furthermore, regulatory activities associated with environmental pollution, laboratory services, and monitoring services are some other services provided. The waste management unit of the CEA deals with litter, the National Post Consumer Plastic Waste Management Project, and sanitary landfill sites (JICA 2016).
7. *Coast Conservation and Coastal Resources Management Department (CC & CRMD)*
   The Coast Conservation and Coastal Resources Management Department regulates developmental activities in coastal zones. Coastal pollution is mitigated by the Coastal Zone Management Aspects Plan (JICA 2016).
8. *MEPA*
   The MEPA regulates marine pollution. It prevents, mitigates, and controls marine pollution from sea-based and land-based sources. In accordance with Section 21 of the Marine Pollution Prevention Act, MEPA is involved with management of wastes from ships (JICA 2016).

Apart from the abovementioned governmental agencies, there are local agencies that are responsible for litter management. In 1987, provincial councils were given the authority to supervise waste management through the 13th Amendment to the Constitution. The Western Provincial Council has a separate Waste Management Authority.

Nine provincial councils guide local authorities of respective regions. Administrative guidance and financial aid for waste management are provided to local authorities. According to the local government legislative system, local authorities can recommend laws through Parliament; however, this is not practiced often (JICA 2016).

18.4.2 Management Policies and Strategies and Their Effectiveness

National Policies are developed to implement better waste management practices and mitigate pollution. The following policies deal with waste management.

1. *National Strategy for Solid Waste Management (2000)*: Three-year action plan is to reduce waste and practice the "3R's" (reduce, reuse, recycle).
2. *Caring for the Environment Phase I (2003–2007) and Phase II (2008–2012) (2003)*: This is a UN-funded program for development and part of the United Nations Development Programme (UNDP) National Environmental Action Plans.
3. *Vision for a New Sri Lanka (2005)*: It was implemented by the Ministry of Finance and has the Ten Year Planning Developmental Framework 2006–2016, which includes solid waste and pollution management initiatives.
4. *Coastal Zone Management Plan (2006)*: The plan is for solid waste management in coastal zones to reduce coastal pollution.

5. *National Policy on Solid Waste Management (2007)*: The policy is for waste reduction, 3R implementation, capacity building, research and development, and for regulation of sanitary landfills using best available technologies and best environmental practices.
6. *Pilisaru Program Phase I (January 20–December 2013) and Phase II (January 2014–December 2018) (2008)*: This is a national program for solid waste management under the Ministry of Environment, CEA. Its aim was to introduce medium and small waste treatment systems in all LAs by 2018.
7. *National Action Plan for Haritha Lanka Program (2009)*: The plan is to establish and implement the Haritha (Green) Lanka Program of the National Council for Sustainable Development (Laws of Sri Lanka 2019).

Local authorities are responsible for handling waste from their residents. This is in accordance with the Municipal Council Ordinance No. 16, Urban Council Ordinance No. 61, and Pradeshiya Sabha Act No. 15. (Pradeshiya Sabhas were introduced in 1987 through the 13th Amendment to the Constitution of Sri Lanka. They are under the provincial councils in the local government system of Sri Lanka.) The Act gives the authority to formulate rules and regulations and to implement them. According to the 13th Amendment of the Constitution of 1987, the PCs supervise LAs. The National Environment Act No. 47 of 1980 was implemented to preserve the environment and reduce pollution (Laws of Sri Lanka 2019).

Moreover, there are many legal frameworks related to solid waste management as follows:

1. Urban Council Ordinance No. 61 of 1939: Section 118, 119, and 120 specify waste management responsibilities of Urban Councils (UCs)
2. Nuisance Ordinance No. 62 of 1939 and No. 57 of 1946: Sections 1–12
3. Municipal Councils Procedure Act No. 15 of 1979-Public Nuisances: Section 98
4. Provincial Councils Act No. 42 of 1987: Amended by Act No. 56 of 1988. It contains provisions for waste management by LAs
5. Pradeshiya Sabha Act No. 15 of 1987: Sections 93 and 94. The Act specifies waste management responsibilities of Pradeshiya Sabhas, which are part of the local government system of Sri Lanka
6. National Environment Act No. 47 of 1980: Sections 12 and 26
7. Prevention of Mosquitoes Breeding Act No. 11 of 2007: The Act prohibits creating conditions favorable to the breeding of mosquitoes
8. National Thoroughfares Act No. 40 of 2008: Section 64 (a) (b) and Section 65
9. Gazette No. 1627/19 National Environmental (Municipal Solid Waste) Regulations, No. 1 of 2009. These regulations give the general rules for solid waste management
10. Marine Pollution Prevention Act No. 35 (2008): Section 21, Section 26, and Section 27
11. Marine Environment Protection Sea Dumping Regulation 2103 (2012): This regulation requires wastewater discharge standards and sea dumping permits (Laws of Sri Lanka, 2019)

Litter management and solid waste management policies are legal requirements. The policies are structured to prevent pollution and to provide awareness of pollution through education.

18.4.3 Management Activities Done for Land-Based, Beach-Based, and Marine-Based Litter

In recent years, addressing the problem of marine litter has become a focus of the Sri Lankan government. In the management process, marine litter has to be reduced in maritime zones and beaches to a point where quantities do not harm the marine environment; applicable programs and measures to minimize amounts of marine litter need to be established (Figure 18.6).

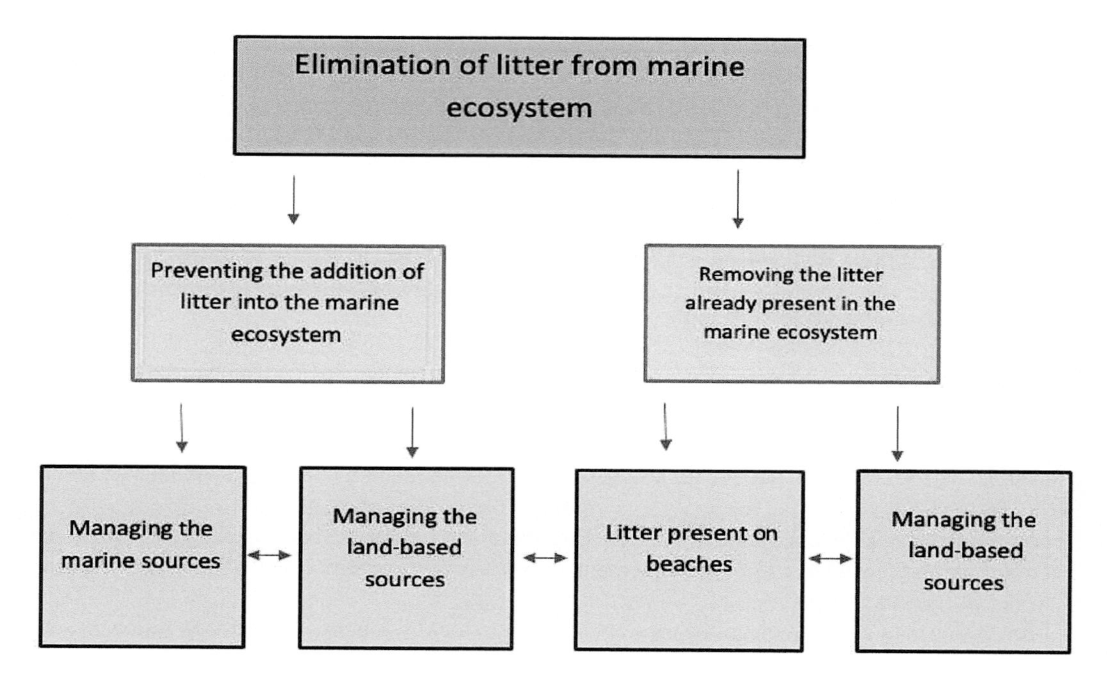

FIGURE 18.6 Sections that need to be considered in managing plastics.

It is evident that land-based sources are the most substantial part of marine plastics. They end up in the oceans directly or indirectly by anthropogenic activities. Improperly disposed of waste becomes marine litter. Natural causes like erosion from running water, wind, and rain can make mismanaged wastes into marine litter (Seabin Group 2016).

Under the Pilisaru Project, established by the Central Environmental Authority, many awareness and education programs are conducted. Mass scale awareness programs are done through media. The MEPA has introduced collection facilities for marine debris, such as plastic waste and fishing nets. This was initiated under the "Marine debris management plan in Negombo" (MEPA 2017).

MEPA has introduced a styrofoam recycling facility. It helps to recycle styrofoam, which is commonly used in the fishing industry. The compressed styrofoam is given to frame manufactures by the Negombo Municipal Council. Parallel to this, MEPA introduced a can compressor and a plastic compressor to recycling centers of the country. These machines make the recycling process more efficient (MEPA 2018).

The handling and management of non-degradable wastes, such as plastic waste, have become a major concern. Because of their low cost and high durability, their usage has significantly increased in recent years. Plastics are imported into Sri Lanka in large quantities (Gunaratna 2012), however, there are several plastic collecting and recycling centers. The CEA recently launched a program called the "National Post Consumer Plastic Management Program." It is funded by a tax on plastic importers.

Since September 1, 2017, the Sri Lankan government has banned the production, usage, and import of plastic bags, such as grocery bags, shopping bags, and disposable cups and spoons. This policy is mainly for reducing plastic wastes and enhancing public safety. Offenders could be fined and jailed for up to 2 years (Laws of Sri Lanka 2019).

18.4.4 Target under the Agenda 2030 and UN Sustainable Development Goals for Marine Litter Management

The UN Sustainable Development Agenda provides a framework to address the issues related to marine litter and is an overarching framework for international, regional, national, and local initiatives (UNEP 2016). The Agenda 2030 UN Sustainable Development Goals consist of 17 goals, 169 targets, and some of them deal with marine plastics (Table 18.5) (UNEP 2016; Alcamo 2019; Barboza et al. 2019).

The ocean-related goals are related to each other in the sustainable development goals (SDGs). For instance, environmentally friendly, land-based waste management systems would reduce simultaneously land-based marine pollution.

When it comes to plastic pollution, a minute percentage of the total plastic waste is recycled around the world. The untreated plastic debris finally ends up in oceans. The United Nations estimates that the ocean is polluted with approximately 51 trillion pieces of microplastic debris (UNSDG 2018). In this way, marine litter hinders achievement of the SDGs. Life cycle assessments of plastics reveal that probable plastic debris leakage into the ocean is found in all phases

TABLE 18.5
Targets under the Agenda 2030 UN Sustainable Development Goals for Marine Litter Management

6.3 By 2030, the proportion of untreated wastewater should be halved

11.6 By 2030, reduce the adverse per capita environmental impact of cities, including by paying particular attention to air quality and municipal and other waste management

12.1 Implement the 10-year framework of programs on sustainable consumption and production, all countries taking action, with developed countries taking the lead, taking into account the development and capabilities of developing countries

12.2 By 2030, achieve the sustainable management and efficient use of natural resources

12.4 By 2020, achieve the environmentally sound management of chemicals and all wastes throughout their life cycle, in accordance with agreed international frameworks, and significantly reduce their release to air, water, and soil in order to minimize their adverse impacts on human health and the environment

12.5 By 2030, substantially reduce waste generation through prevention, reduction, recycling, and reuse

12.b Develop and implement tools to monitor sustainable development impacts for sustainable tourism that creates jobs and promotes local culture and products

14.1 By 2025, prevent and significantly reduce marine pollution of all kinds, in particular, from land-based activities, including marine debris and nutrient pollution

14.2 By 2020, sustainably manage and protect marine and coastal ecosystems to avoid significant adverse impacts, including by strengthening their resilience, and take action for their restoration in order to achieve healthy and productive oceans

14.7 By 2030, increase the economic benefits to small island developing states and least developed countries from the sustainable use of marine resources, including through sustainable management of fisheries, aquaculture, and tourism

14.a Increase scientific knowledge, develop research capacity and transfer marine technology, taking into account the Intergovernmental Oceanographic Commission Criteria and Guidelines on the Transfer of Marine Technology, in order to improve ocean health and to enhance the contribution of marine biodiversity to the development of developing countries, in particular, small island developing states and least developed countries

14.c Enhance the conservation and sustainable use of oceans and their resources by implementing international law as reflected in United Nations Convention on the Law of the Sea (UNCLOS), which provides the legal framework for the conservation and sustainable use of oceans and their resources, as recalled in paragraph 158 of The Future We Want

15.5 Take urgent and significant action to reduce the degradation of natural habitats, halt the loss of biodiversity and, by 2020, protect and prevent the extinction of threatened species

of production, consumption, and disposal. Achieving SDG 12 by eco-friendly plastic management throughout a product's life cycle may help to implement the 2030 Agenda (UNSDG 2018).

18.4.5 National Marine Litter Monitoring Program

Several governmental agencies deal with the issue of monitoring of marine litter. In Sri Lanka, there is no comprehensive marine litter monitoring program. Coastal cleanups have been carried out island wide. Underwater surveys are not carried out near beaches to check the composition of marine debris. It is crucial that an integrated marine litter program be adopted to mitigate this problem. There is a huge knowledge gap in identifying macroplastic and microplastic composition on beaches, and further multidisciplinary research should be encouraged.

18.5 KNOWLEDGE NEEDS AND A PROPOSAL FOR SETTING PRIORITIES

There are numerous knowledge needs related to marine litter management in Sri Lanka. It is mandatory for stake holders, government agencies, expert groups, and researchers to formulate ways to eliminate gaps in knowledge and regulate properly marine debris.

18.5.1 Inadequate Institutional Frameworks and Stakeholder Involvement

The management of marine litter is distributed among multiple sectors and agencies, and most of the agencies do not have full authority over all the components related to the management of marine resources and pollution. This inequality in responsibility results in ineffective management of marine debris (Christodoulou-Varotsi 2018). Previous studies show that marine debris originates from multiple sources (Jang et al. 2018). Accordingly, management should address the prevention of coast-based pollution, as well as regulating inland-based pollution. Lack of serious political will and poor coordination among government agencies have led to an unclear distribution of responsibilities. Moreover, government agencies are expected to share available and up-to-date information and be transparent with each other, while serving a cohort, when dealing with marine pollution.

18.5.2 Policy and Legislation Gaps

As noted, several agencies regulate waste management, both land-based and ocean-based. There are two major acts and a policy that are directly related to the prevention of marine debris.

1. National Policy on Solid Waste Management (2007)
2. Coast Conservation Act
3. Marine Pollution Prevention Act

Even though they are formulated to mitigate pollution, there are no clear mechanisms to manage marine debris along seashores. A national policy integrating the socio-economic and environmental issues related to marine pollution is crucial. Persisting legislation may suffice if it is adequately enforced. However, the absence of an integrated waste management strategy makes it difficult to identify the duties of agencies, stakeholders, and the general public.

18.5.3 Lack of Infrastructures for Waste Collection, Transportation, and Recycling

The adverse effects of marine litter are well understood. However, there are no adequate waste collection methods in coastal environments. The current waste collection facilities focus on daily waste collection, transportation, and dumping. This is mainly done for municipal solid waste (Gunarathne et al. 2018). It is essential to facilitate a proper way to collect and recycle plastic wastes. Beach cleanups are done occasionally, but a continuous cleanup coupled with source reduction can significantly reduce marine debris pollution. There is no specific, annual marine debris monitoring program that surveys the composition and amount of marine debris produced in different districts. Debris is not appropriately monitored in protected areas in marine ecosystems, even though there are numerous marine protected areas in the country. They are not properly managed to eliminate litter accumulation (Perera and de Vos 2007). Managers and governmental agencies are focused more on creating novel legislation than enforcing the existing laws. A well integrated, national level marine debris monitoring program is needed to ensure implementation of existing regulations. New legislation needs to be synthesized, if necessary.

18.5.4 Education and Awareness

Waste disposal mainly depends on personal choices. Most people do not focus on the prolonged, adverse effects of waste generation and accumulation. Sanitation often has a higher priority than the management of wastes. The general public lacks awareness related to marine litter management. Well planned awareness programs that focus on attitudinal changes should be implemented. Various segments of society have to be targeted separately, like schoolchildren, residents, traders, farmers, and civil and public servants. Proper programs should be implemented with coordination with NGO's. Mass media should be utilized to spread the message. Details of the current status of solid waste in Sri Lanka, and most importantly its marine waste accumulation, must be readily available online to anyone at any time. There is a massive need for an emergency service to rectify the issue when illegal dumping is done and with less paperwork. The 3R strategy has to be implemented to cope with the growing problem of plastic pollution and reducing it.

18.5.5 Strengthening Management and Control Efforts and Financing Mechanisms

It is inevitable that the consumption of plastics will increase in the future. Therefore, plastic production and consumption have to be controlled in their life cycle. Proper mechanisms and investments allocated for effective recycling will reduce debris accumulation. Risk analysis of marine debris has to be carried out regularly.

18.6 THE WAY FORWARD

18.6.1 The Dispersion of Management Authorities of Marine Debris and the Absence of Interagency Cooperation between Authorities

The Coast Conservation and Coastal Resources Management Department and the Marine Environment Protection Authority are the main central government regulatory bodies responsible for conserving the maritime zone and the coastal zone of Sri Lanka (Figure 18.7). These agencies are directly involved in the management and pollution abatement of the country's marine and the coastal zone, functioning under the MoMDE

The first act taken toward coast conservation of the island was enacted by the parliament of Sri Lanka in 1981, and it was the Coast Conservation Act No. 57. An amendment to this act was passed in 1988, and it was the Coast Conservation (Amendments) No. 64 of 1988. The CC & CRMD was established under these acts, and the department functions according to the provisions of the act in order to improve the status of the coastal environment and the living standards of coastal communities. Also, the CC & CRMD is responsible for setting mitigation measures for indirect impacts, such as pollution due to incorrect siting of sewage disposal systems related to tourist facilities and inadequate infrastructure (CC & CRMD 2019). The EIA and initial environment examination were introduced into the Coast Conservation Act for any development project that takes place within the

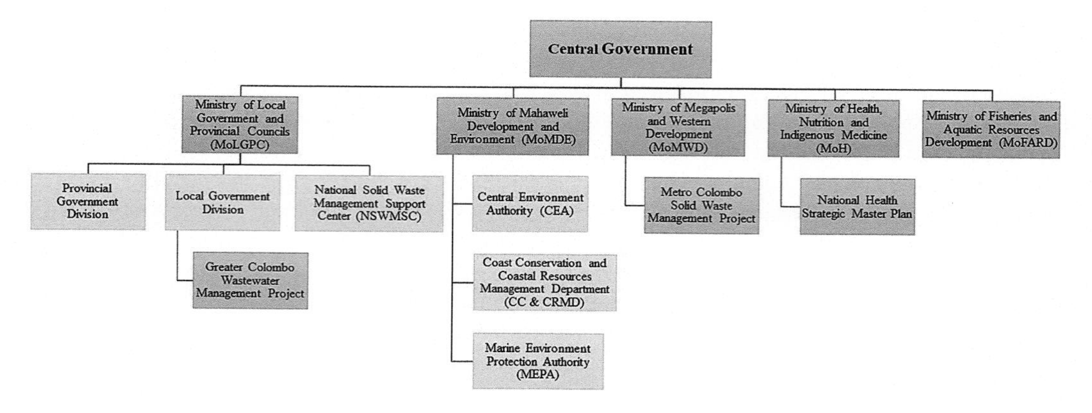

FIGURE 18.7 Structure of marine debris management institutions.

coastal zone of Sri Lanka, and the CC & CMD decides whether the particular project shall undergo an EIA or initial environmental examination or not (Coast Conservation (Amendment) Act No. 49 2011).

The Marine Environment Protection Authority was established under the Marine Pollution Prevention Act No. 35 of 2008. The sole responsibility of MEPA is to function according to the provisions of the Act, which are to protect the marine environment through prevention, control, and management of pollution arising out of ship-based activity and shore-based maritime related activity in the territorial waters of Sri Lanka, its foreshores, or the coastal zone of Sri Lanka (MEPA 2019).

Land-based solid waste generation influences coastal pollution and marine debris. Hence, the inland solid waste managing authorities are aware of this problem. Inland waste management is carried out by several central government agencies, such as the MoLGPC, the MoMDE, the MoMWD, and the MoH. Under the MoLGPC, the national solid waste management policy is implemented through the LAs and the PCs. All LAs, such as municipal councils, Pradeshiya Sabha, and the urban councils, are directly responsible for managing the collection and disposal of solid waste of their particular areas. All the LAs are under the supervision of the PCs and, under the Constitution, all waste management powers are delegated to the nine PCs. Also, the MoLGPC functions as the National Solid Waste Management Support Center to manage the solid waste in LA-administrative areas by providing consultative, financial, and technical assistance and adequate infrastructure for proper solid waste management.

The CEA stands as another central government body that is operated by the MoMDE, and it was established under the provisions of the National Environmental Act. The authority issues Environment Protection Licenses, which are approval by the EIA and initial environmental examination and recommends environmental solutions (CEA 2019).

The MoMWD ensures the development of western urban areas to enhance the environmental quality of its inhabitants and resolving distinctive issues related to urbanization, such as solid waste management and pollution reduction. As a solution for the vast amount of solid waste generated in the Colombo area, the Metro Colombo Solid Waste Management Project was proposed by the ministry (MoMWD 2019).

The Ministry of Fisheries and Aquatic Resources Development prevents marine pollution by taking necessary precautions to address its adverse impacts on fisheries and marine aquatic resources (SLPA 2015).

18.6.2 The Absence of Scientific Survey and Statistics for Managing Marine Debris

In addressing the marine litter pollution, relevant scientific data and information need to be available to achieve good management practices. Generally, policies and strategies are formulated based on the outcomes of the data, which have been comprehensively analyzed. Therefore, the effectiveness of the policies, strategies, and management plans will entirely depend on the accuracy and quantity of the baseline data. Currently, none of the South Asian Seas (SAS) member countries (Bangladesh, India, Maldives, Pakistan, and Sri Lanka) has a consolidated database. Little information from surveys carried for research and study purposes is available concerning the amount and type of marine litter. However, Sri Lanka lacks a national-level surveying system or a database for monitoring the magnitude of litter along its coastal areas or territorial waters. The actual scale of marine littering, and its social, economic, and ecological impacts are still vague. Degradation of natural coastal habitats could become perilous, causing huge economic losses and social unrest if immediate action is not taken. A proper scientific survey and monitoring system of debris could benefit Sri Lanka in numerous ways, as follows (SACEP 2018).

- Extract information on the type and quantities of marine debris and assess its spatial and temporal variation
- Provide insight into problems and threats associated with a specific area in the ocean or coastal area and come up with mitigation measures
- Assess the effectiveness of appropriate legislation and coastal management policies
- Identify potential sources of marine debris whether from fisheries, tourism, shipping, or coastal communities
- Explore public health issues related to marine debris and coastal conditions and increase public awareness
- Identify threats to the natural flora and fauna in marine ecosystems

In the past few decades, marine debris surveys have been carried in the Mediterranean Sea, Caribbean Sea, Baltic Sea, Black Sea, the Pacific Ocean, the Atlantic Ocean, and along the coasts of many countries. These surveys have employed various methods for assessing marine debris in terms of identification, quantification, and sampling the debris, but it has been difficult to compare different surveys. A proper scientific survey on marine litter should be comprehensive with the ability to compare its data with others. It should cover a wide range of the marine environment and include the following factors (SACEP 2007):

- Debris stranded on shorelines and beaches
- Floating debris in the open ocean
- Debris on the seafloor and sediments
- Debris ingested in marine organisms

18.6.2.1 Beach Surveys

Much of the information regarding abundance, distribution, and the origin of debris in the marine zone of Sri Lanka comes from beach surveys of litter stranded along the shoreline in certain areas of the island (Table 18.6). Beach surveys are often focused on cleaning up beaches, and they are conducted by NGOs and governmental bodies with the participation of volunteers and stakeholders. Involving volunteers is effective for improving social awareness of marine pollution and littering. The public learns of its responsibility as citizens to protect the country's natural resources. Not many scientific research studies on shoreline surveys have been reported in Sri Lanka (Angelini et al. 2019).

TABLE 18.6
Marine Debris Surveys and Research Conducted in Sri Lanka

Year and Duration	Institution	Coastal Area	Objective	References
2010–2017	Ocean University of Sri Lanka/NARA/ University of Sri Jayewardenepura	Negombo Lagoon	Identify marine debris and its potential impacts on mangrove ecosystem	Jayapala and Jayasiri (2018)
2017 Four months	Uva Wellassa University of Sri Lanka/NARA	Uswetakeiyawa Kerawalapitiya Dikowita Modera Kollupitiya Bambalapitiya Wellawatta	Identify morphological characterization and quantification of plastic debris	Athawuda et al. (2018)
2017 Monthly	NARA	Colombo Beruwala Hikkaduwa	Quantify microplastic litter in coastal sand in western and the southwestern coasts	Weerakoon et al. (2018)
2016 One week	University of Sri Jayewardenepura/ MEPA	22 beaches along the 1340 km coastline of Sri Lanka	Identify composition and abundance of marine debris stranded on the beaches	Jang et al. (2018)
2016 2 days	University of Sri Jayewardenepura/ MEPA/Korea Maritime Institute (KMI)/Ocean Research Institute (ORI)/Marine Tech Engineering	Negombo Fishery Harbor	To identify distribution of deposited marine debris in the harbor via scan sonar survey and diving survey	MEPA (2017)
–	University of Ruhuna/ MEPA	Thalalla Dondra Harbor Weligama Unawatuna Galle Rathgama Koggala Hikkaduwa Kahawa Ambalangoda Harbor	Spatial distribution of microplastics in the southern coast of Sri Lanka in terms of the abundance, mass, type, and particle size	Bimali Koongolla et al. (2018)
– 12 hours	Ocean University of Sri Lanka	Pareiwella Nilwellla Dickwella Rekawa	Small plastic debris in the ranges between 0.5 to 5 mm	Sathsara et al. (2017)
2008–2013	MEPA/University of Sri Jayewardenepura	–	Analyze collected and categorized marine debris during international coastal cleanup programs in Sri Lanka	Gunasekara et al. (2014)

Beach surveys usually reflect the long-term balance between the inflow and the outflow of the debris of the shoreline. Inflow is influenced by the river outfalls, land-based solid waste, storm-water, floods, and litter, while outflow is influenced by cleanups, export, burial, and tides. In the long term, most of the debris that persists along the shorelines is non-biodegradable, such as plastic, metals, and glass (Ryan et al. 2009). As plastics have been a significant nuisance to the maritime zone, beach surveys are focused on all forms of plastics, as follows: macroplastics, mesoplastics, and microplastics. In a beach survey, data on the abundance, the mass of debris, type of material, source of debris, and the usage are recorded; but, in terms of cleaning, less attention is paid to the classification and sources of debris. Factors influencing the inflow and outflow may also influence beach litter density, both spatially and temporally. By performing regular surveys on a weekly, monthly, or annual basis, the rate of change of density and composition of debris can be traced, thereby revealing the trends of accumulation on beaches. However, past studies conducted from 2008 to 2017 in Sri Lanka have lacked consistency in terms of the research objectives, sampling, and analytical methods, and these differences cause difficulties in comparing data (Bergmann et al. 2015).

18.6.2.2 Floating Debris Survey

Debris taken away from the shore by tides or wind to the sea could submerge and transfer long distances in the open ocean or sink and deposit in the sea floor. Floating macro-litter is usually monitored by simple visual observations from the shore or with the use of net trawls on ships or boats. Aerial surveys are employed to cover a huge area of the sea and are done usually to identify large floating debris resistant to the wind (Ryan et al. 2012).

18.6.2.3 Seabed Survey

Evaluations of seabed litter are fewer than those for the sea surface due to sampling difficulties, inaccessibility to deep ocean areas, and the high cost of operations. Thus, most research has been limited to continental shelf areas. Seabed surveys are carried out using divers, submersible remotely operated vehicles, or trawlers. Benthic trawling is known to be a consistent method, even though it is rather destructive to benthic marine biota (Bergmann et al. 2015).

18.6.2.4 Survey on Debris Ingested in Marine Organisms

Ingestion of small plastic debris by marine organisms is frequent and more threatening than entanglement. Mostly fish and invertebrates feed on small plastic pellets, such as microplastics, which they mistakenly identify as plankton or organic matter. Also seabirds and other organisms along coasts feed on the plastic matter while foraging. Studies on the contents of the gut in dead organisms are a good indicator for the abundance of debris in the marine environment (Ryan et al. 2012).

Even though globally, data are available on marine debris surveys, it is not sufficient to predict the extent of marine pollution in the world. Adoption of standardized methods and protocols, streamlining of data flows, proper interpretation of data, data storage, and access to data would facilitate an effective monitoring system. Sri Lanka needs to prioritize the need for a proper scientific survey and database to enhance the effectiveness of its policies in addressing marine debris pollution.

18.6.3 Poor Management of Wastes Flown from Land to Ocean

Around 80% of marine plastic litter originates from land-based sources (Napper and Thompson 2019). Sri Lanka generates a total of 7000 MT of solid waste per day. Of it, 60% is solid waste generated in the Western Province. About 3500 MT of solid waste is collected by the local authority and disposed of properly. The remaining waste is thrown away along the roadside (Fernando and Lalitha 2019). Of the solid wastes collected in the entire country, 75% is from the five Maritime Provinces. Rapid urbanization and industrialization have led to the generation of more waste. In the long term, open dumps and roadside garbage disposals retain items resistant to degradation, especially plastics.

Open dumping and open burning create numerous environmental problems. The most common solid waste management technique is open dumping in a large area authorized by a relevant government body. In Colombo, unsanitary open dumpsites can be found at Karadiyana, Bluemendhal, Meethotamulla, and Kolonnawa. Altogether, there are nearly 800 dumping sites on the island, and more than 200 of them are located within coastal regions, and some of them are on low lying marshy lands (Leslie Joseph 2003). Even inside the country, most of the dumpsites are situated adjacent to a stream or river, and untreated leachate is often released to a water body. During rainfall or due to wind, debris from open dumps or on river banks may escape with stormwater, or via the air, into streams or rivers, and pollutants are deposited at their outfalls and along the coastline, or they become submerged in the open sea. Small plastic debris, such as microplastics incorporated into personal care products, are present in domestic sewage. In Sri Lanka, municipal sewage is left untreated and discharged into surface waters, which eventually contaminate the marine environment with all forms of plastic debris (Gunasekera 2016).

Even if dumped, there is not enough land to accommodate the enormous amount of solid waste being generated in the cities of Sri Lanka. Local authorities try to locate barren, vacant sites for dumping, but the general public near the sites is not willing to allow it. Also, local authorities have to deal with public protests, because of bad odors and health issues caused by unhygienic conditions that arise due to the open dumping. Many municipalities do not own dumping grounds; therefore, they have to outsource waste collection and handling to private companies. When dumping grounds are insufficient, wastes are disposed on paddy fields and marshy lands, which is the worst case scenario (Fernando and Lalitha 2019). A good example of this was the dumping site at Nawinna along the High Level Road. This area was once a flourishing paddy field, but cultivation came to a standstill when waste from nearby factories started seeping into the field. The field then was used by the Maharagama Municipal Council for open dumping (LBO 2006).

Local authorities lack proper infrastructure and waste management policies to support effective waste management in the country. Municipal councils do not have adequate staff members and compactors to collect and transport solid waste to disposal sites. At the sites, expertise for waste segregating, composting, and recycling are not available, and this leads to poor waste processing. Recycling of plastic waste is the best option for the handling of plastics. However, recycling is not successfully done, because of the time and effort taken to separate plastics from waste. Sometimes plastic waste is handled at recycling centers run by private owners, but there is no reliable process to upkeep such efforts. Promoting more collection centers and profitable ventures could make it more accessible for the public, reducing the gap between the waste generators and recyclers. Most of the domestic non-degradable garbage is supposed to be segregated into plastics, glass, and metal before handing over to the municipal council (MC) waste transport. The major issue faced by the MC is the hesitancy of people to segregate their waste accordingly. The hesitancy of the public to take responsibility for their waste tends to make them get rid of waste by haphazard ways, thus ending at streets, roads, or nearby lands. Even though many awareness and education programs have been conducted, they have not been successful enough to eliminate the negative attitude of the public.

Open dumping occurred in Sri Lanka until the first sanitary landfill was established in 2014 in Dompe, Gampaha. This landfill is the only properly functioning sanitary landfill in the country (Fernandopulle 2018). Therefore, the local authorities in Sri Lankan face a severe problem in solid waste management, because the waste is growing beyond management capacity. It is the responsibility of governmental bodies to act immediately in addressing the issue of proper waste disposal.

18.6.4 International and Regional Cooperation

Because marine debris in oceans is linked to population density, amount of solid waste, and economic status of countries, it has a transboundary nature and global impact. South Asia contributes a significant amount of marine debris (SACEP 2007). Hence, international and regional debris mitigation plans are needed to be carried out to mitigate further environment degradation. The conventions and international programs related to abatement are given in Tables 18.7 and 18.8.

TABLE 18.7
Marine Debris Abatement Related to International Conventions Ratified by Sri Lanka

Convention	Implementing Agency	Obligations
United Nations Convention on the Law of the Sea (UNCLOS) 1982 (Ratified July 1995)	Ministry of Fisheries and Aquatic Resources	Strengthen the sovereign of the Exclusive Economic Zone (EEZ) Protect and preserve the marine environment and adhere to international standards Formulate regulations to control controlling land-based, seabed-based, and open sea-based pollution
International Convention for the Prevention of Pollution from Ships (MARPOL) 1973 (Ratified June 1997)	Marine Environmental Protection Agency (MEPA)	Contains regulations for prevention and minimizing pollution from ships in the sea under six technical annexes. Annex v refers to "garbage" including plastics and other litter, which is prohibited from disposing of Sri Lankan seas
London Convention 1972, Convention on the Prevention of Maritime Pollution by Dumping of Wastes and Other Matters	Marine Environmental Protection Agency (MEPA)	Control all sources of marine pollution and prevent pollution of the sea through regulation of dumping into the sea of waste materials
Global Programme of Action for the Protection of the Marine Environment from Land-based Activities (GPA) (Ratified November 1995)	Ministry of Environment moreover, Renewable Resources	Protect and preserve the marine environment from adverse environmental impacts of land-based activities Take mitigatory measures to prevent seas from destruction of habitat, nutrients, sediment mobilization, persistent organic pollutants, oils, litter, heavy metals, and radioactive substances
Convention on the Conservation of Migratory Species of Wild Animals (also known as CMS or Bonn Convention) 1979 (Ratified June 1990)	Department of Wildlife Conservation	Conservation of wildlife and habitats on a global scale and prevent habitat destruction from pollution (focused mainly on turtle conservation)
Basel Convention	Marine Environmental Protection Agency (MEPA)	Control of transboundary movements of hazardous waste and their disposal

Source: BOBLME, Country report on pollution — Sri Lanka, Bay of Bengal Large Marine Ecosystem, BOBLME-2011-Ecology-14, 2013.

TABLE 18.8
Regional Programs for Marine Debris Abatement

Convention	Implementing Agency	Obligations
South Asia Cooperative Environment Program (SACEP) 1982	SACEP Secretariat	Regional agreement of the South Asia countries: Afghanistan, Bangladesh, Bhutan, India, Maldives, Nepal, Pakistan, and Sri Lanka To protect and manage the marine environment and related coastal ecosystems of the region in an environmentally sound and sustainable manner
South Asian Seas Program (SASP) 1995	SACEP Secretariat	Development of an action plan outlining the strategy and substance of a regionally coordinated program, aimed at the protection of a typical body of water

(*Continued*)

TABLE 18.8 (*Continued*)
Regional Programs for Marine Debris Abatement

Convention	Implementing Agency	Obligations
South Asian Association for Regional Cooperation (SAARC) 1985	Ministry of Foreign Affairs	Governments of Bangladesh, Bhutan, India, Maldives, Nepal, Pakistan, and Sri Lanka formally adopted the SAARC for the promotion of economic and social progress, cultural development within the South Asia region

Source: BOBLME, Country report on pollution — Sri Lanka, Bay of Bengal Large Marine Ecosystem, BOBLME-2011-Ecology-14, 2013.

18.7 RECOMMENDATIONS

18.7.1 Introduce Integrated National Marine Debris Management Policy, Strategy, and Management Plan

Even though Sri Lanka has a series of environmental acts and laws for the maritime zone conservation and pollution abatement, the country needs an integrated national marine debris management policy and strategy. Many approaches taken so far have failed to strengthen the existing frameworks and to mitigate marine pollution. Marine debris arises from various sources and becomes dispersed through numerous pathways. Therefore, it is necessary to engage all governmental and non-governmental stakeholders, such as civil society, local communities, academia, and businesses, in decision-making. Existing laws and legislations for marine pollution prevention need to be aligned with local, regional, and international standards. A policy that is based on the "polluter pays principle" is one mechanism that might be introduced. The policy should support short-term and long-term strategies to achieve management goals.

18.7.2 Establishment of a New Legal Framework

Sri Lanka has a strong legal framework for marine protection, although law enforcement is weak to backup prevention of marine pollution. Apart from the Marine Protection Act and Coast Conservation Act, there are other Acts enacted by the government that contribute to the protection of the marine environment. A main drawback is caused by the instability of the legal framework for the handling of municipal solid wastes. Hence, there is a strong need to review the existing laws to minimize legal disputes. Some laws are applied inconsistently among various sectors, and the laws become problematic whenever legal actions are taken against them. Therefore, dedicated laws should be formulated that adhere to regional and international legal frameworks.

18.7.3 Establishment and Revamping of the Institutional Structure

The MEPA and CC & CMD are the apex government regulatory institutions in Sri Lanka, vested with the laws and regulations to prevent, control, and manage the pollution of Sri Lanka's marine environment empowered by the Marine Pollution Prevention Act and the Coast Conservation and Prevention Act. A proper marine debris management system is an integrated approach of all the institutions responsible for contributing to marine and costal pollution either directly or indirectly. The existing institutional structure should be reviewed for its strengths and weaknesses in order to improve the capacity in marine debris management. The current institutional structure in Sri Lanka is not effective, as their approaches are isolated from one another and not oriented toward a common goal for marine debris management. It is a must that the vision, scope, and goals of regulatory institutions should be reviewed and aligned with the national policy on marine debris management. Also, collaborations with other organizations having environmental, scientific, and legal expertise could provide linkages to improve the management efforts further.

18.7.4 Intensive Management of Marine Debris Sources

Major sources of marine debris are mostly land based. Even though this is a global issue, a large percentage of pollution can be mitigated through local source reduction, recycling, adapting systematic international protocols of waste management and global partnerships. A thorough investigation of marine debris sources and distributing pathways could give a clear understanding of stakeholders and responsible sectors. Good coordination between the regulatory agencies and rigid enforcement of the existing laws shall be effective options to mitigate debris sources. Preventive strategies should be introduced to minimize wastes entering the marine environment. As 80% of marine debris is from land-based sources, a robust solid waste management system should be introduced to the MC. Local authorities and provincial councils should strengthen policies and legislations for proper inland solid waste collection and disposal mechanism. In the meantime, as preventive measures, penalties should be charged for coastal communities and industries (port, fisheries, and tourism) for waste disposal and polluting the maritime zone.

18.7.5 Marine Debris Collection, Disposal, and Recycling

Most marine litter is collected and removed through beach cleaning campaigns organized by governmental institutions and NGO's that involve public volunteers. Cleaning campaigns are not conducted regularly and do not apply to the entire maritime environment of Sri Lanka. Therefore, a formal marine debris collection system that operates regularly and that covers the entire country needs to be introduced through the government intervention. Also, adequate waste reception facilities need to be established along the coasts. Proper infrastructure facilities should be available for disposal and recycling of collected wastes. Unless cleanup efforts are taken, waste could be disseminated into broader and more sensitive areas causing marine degradation.

18.7.6 Research, Surveys, and Proper Modeling Systems on Marine Debris

No extensive studies concerning marine debris have been done on a global scale. And no studies have been done for Sri Lanka on a local scale. Therefore, limited, reliable data are available for Sri Lanka. Fund allocation shall promote further studies on sources of debris and its environmental, social, and economic impact. The origin and transportation of debris can be measured through long-term surveying and monitoring techniques. Data should be compiled and deposited in a national marine debris data system, thereby giving access to authorized parties to make policy decisions and to mode the fate of marine debris. These models need to include geographical and seasonal distribution of marine debris, and they will help to understand what further research is needed.

18.7.7 Development of Education Awareness Program to Manage Marine Litter

The aim of the awareness programs should be focused on the type of participants, such as schoolchildren or adults. It is important to provide a foundation for effective public participation in these awareness programs. The public should be encouraged to comply with the debris management regulations and become involved in implementation programs.

18.7.8 Regional and International Cooperation

Marine debris can move freely from one area to another area through wind and ocean currents and become distributed in much larger areas that may be thousands of miles wide. Asia is home to one fifth of the world's population and has rapid economy growth. The amount of plastics produced and discarded will be significantly higher in the future. Hence, this issue cannot be solved unless countries of the region work collaboratively to mitigate plastic pollution. Regional organizations, such as the SAS program have been introduced to solve problems, and they are led on a global scale by organizations, such as the United Nations Environment Protection Agency.

REFERENCES

Alcamo, Joseph. 2019. "Water Quality and Its Interlinkages with the Sustainable Development Goals." *Current Opinion in Environmental Sustainability* 36: 126–140.

Anderson, Julie C., Bradley J. Park, and Vince P. Palace. 2016. "Microplastics in Aquatic Environments: Implications for Canadian Ecosystems." *Environmental Pollution* 218: 269–280.

Andrady, Anthony L. 2011. "Microplastics in the Marine Environment." *Marine Pollution Bulletin* 62 (8): 1596–1605. doi:10.1016/j.marpolbul.2011.05.030.

Angelini, Zachary, Nancy Kinner, Justin Thibault, Phil Ramsey, and Kenneth Fuld. 2019. "Marine Debris Visual Identification Assessment" *Marine Pollution Bulletin* 142: 69–75. doi:10.1016/j.marpolbul.2019.02.044.

Ashiq, Ahmed, Nadeesh M. Adassooriya, Binoy Sarkar, Anushka Upamali Rajapaksha, Yong Sik Ok, and Meththika Vithanage. 2019. "Municipal Solid Waste Biochar-Bentonite Composite for the Removal of Antibiotic Ciprofloxacin from Aqueous Media." *Journal of Environmental Management*. doi:10.1016/j.jenvman.2019.02.006.

Athawuda, A. M. G. A. D., H. B. Jayasiri, S. C. Jayamanne, WRWMAP Weerakoon, G. G. N. Thushari, and K. P. G. Guruge. 2018. "Plastic Litter Enumeration and Characterization in Coastal Water, off Colombo, Sri Lanka." In International Scientific Sessions of NARA, National Aquatic Resources Research and Development Agency (NARA), Sri Lanka.

Auta, Helen Shnada, C. U. Emenike, and S. H. Fauziah. 2017. "Distribution and Importance of Microplastics in the Marine Environment: A Review of the Sources, Fate, Effects, and Potential Solutions." *Environment International* 102: 165–176.

BOBLME, 2013. Country report on pollution - Sri Lanka, Bay of Bengal Large Marine Ecosystem, BOBLME-2011-Ecology-14.

Balasubramanium, Mathura, and Andrea D. Phillott. 2016. "Preliminary Observations of Microplastics From Beaches in the Indian Ocean." *Indian Ocean Turtle Newsletter* 23: 13.

Bandara, C. M. Madduma. 1989. *A Survey of the Coastal Zone of Sri Lanka*. Colombo, Sri Lanka, Coast Conservation Department.

Barboza, Luís Gabriel A., Andrés Cózar, Barbara C. G. Gimenez, Thayanne Lima Barros, Peter J. Kershaw, and Lúcia Guilhermino. 2019. "Macroplastics Pollution in the Marine Environment." In *World Seas: An Environmental Evaluation* 305–328. Elsevier Academic Press, Cambridge, MA.

Barboza, Luís Gabriel Antão, Luís Russo Vieira, Vasco Branco, Neusa Figueiredo, Felix Carvalho, Cristina Carvalho, and Lucia Guilhermino. 2018. "Microplastics Cause Neurotoxicity, Oxidative Damage and Energy-Related Changes and Interact with the Bioaccumulation of Mercury in the European Seabass, Dicentrarchus Labrax (Linnaeus, 1758)." *Aquatic Toxicology* 195: 49–57.

Barnes, David K. A., Francois Galgani, Richard C. Thompson, and Morton Barlaz. 2009. "Accumulation and Fragmentation of Plastic Debris in Global Environments." *Philosophical Transactions of the Royal Society B: Biological Sciences*. doi:10.1098/rstb.2008.0205.

Bergmann, Melanie, Lars Gutow, and Michael Klages. 2015. *Marine Anthropogenic Litter*. Cham, Switzerland, Springer.

Bimali Koongolla, J., A. L. Andrady, P. B. Terney Pradeep Kumara, and C. S. Gangabadage. 2018. "Evidence of Microplastics Pollution in Coastal Beaches and Waters in Southern Sri Lanka." *Marine Pollution Bulletin*. doi:10.1016/j.marpolbul.2018.10.031.

Browne, Mark, Tamara Galloway, and Richard Thompson. 2007. "Microplastic—An Emerging Contaminant of Potential Concern?" *Integrated Environmental Assessment and Management* 3: 559–561. doi:10.1897/ieam_2007-048.

Cassoff, Rachel M., Kathleen M. Moore, William A. McLellan, Susan G. Barco, David S. Rotstein, and Michael J. Moore. 2011. "Lethal Entanglement in Baleen Whales." *Diseases of Aquatic Organisms* 96 (3): 175–185.

CC & CRMD. 2019. "Coast Conservation and Coastal Resources Management Department Official Web Page, Sri Lanka." www.coastal.gov.lk/.

CEA. 2019. "Central Environmental Authority Official Webpage, Sri Lanka." www.cea.lk/web/en/about-us.

Christodoulou-Varotsi, Iliana. 2018. *Marine Pollution Control: Legal and Managerial Frameworks*. New York, Taylor & Francis Group.

Claessens, Michiel, Steven De Meester, Lieve Van Landuyt, Karen De Clerck, and Colin R. Janssen. 2011. "Occurrence and Distribution of Microplastics in Marine Sediments Along." *Marine Pollution Bulletin* 62 (10): 2199–2204.

Coast Conservation (Amendment) Act No. 49. 2011. *The Gazette of the Democratic Socialist Republic of Sri Lanka*. Colombo, Sri Lanka, Democratic Socialist Repulic of Sri Lanka.

Cole, Matthew, Pennie Lindeque, Claudia Halsband, and Tamara S. Galloway. 2011. "Microplastics as Contaminants in the Marine Environment: A Review." *Marine Pollution Bulletin* 62 (12): 2588–2597. doi:10.1016/j.marpolbul.2011.09.025.

CZMP, 2018. Coastal Zone Management Plan, The Gazette of the Democratic Socialist Republic of Sri Lanka.

Dela, Jinie D. S. 2002. "State of the Environment in Sri Lanka: A Report for SAARC." Colombo, Ministry of Environment and Natural Resources (MoENR).

Derraik, José G. B. 2002. "The Pollution of the Marine Environment by Plastic Debris: A Review." *Marine Pollution Bulletin*. doi:10.1016/S0025-326X(02)00220-5.

Eriksen, Marcus, Laurent C. M. Lebreton, Henry S. Carson, Martin Thiel, Charles J. Moore, Jose C. Borerro, Francois Galgani, Peter G. Ryan, and Julia Reisser. 2014. "Plastic Pollution in the World's Oceans: More than 5 Trillion Plastic Pieces Weighing over 250,000 Tons Afloat at Sea." *PLoS One*. doi:10.1371/journal.pone.0111913.

Fendall, Lisa S., and Mary A. Sewell. 2009. "Contributing to Marine Pollution by Washing Your Face: Microplastics in Facial Cleansers." *Marine Pollution Bulletin* 58 (8): 1225–1228.

Fernando, S., and R. Lalitha. 2019. "Solid Waste Management of Local Governments in the Western Province of Sri Lanka: An Implementation Analysis." *Waste Management*. doi:10.1016/j.wasman.2018.11.030.

Fernandopulle, Sheain. 2018. "Sri Lanka's First Sanitary Landfill in Dompe." *Daily Mirror*. www.dailymirror.lk/article/Sri-Lanka-s-first-sanitary-landfill-in-Dompe-154755.html.

Galgani, Luisa, Ricardo Beiras, François Galgani, Cristina Panti, and Angel Borja. 2019. "Impacts of Marine Litter." *Frontiers in Marine Science* 6. Frontiers: 208.

Galloway, Tamara S., and Ceri N. Lewis. 2016. "Marine Microplastics Spell Big Problems for Future Generations." In *Proceedings of the National Academy of Sciences*. doi:10.1073/pnas.1600715113.

Galloway, Tamara S. 2015. "Micro-and Nano-Plastics and Human Health." In *Marine Anthropogenic Litter*, 343–366. Cham, Switzerland, Springer.

Gasperi, Johnny, Stephanie L. Wright, Rachid Dris, France Collard, Corinne Mandin, Mohamed Guerrouache, Valérie Langlois, Frank J. Kelly, and Bruno Tassin. 2018. "Microplastics in Air: Are We Breathing It In?" *Current Opinion in Environmental Science & Health* 1: 1–5.

Gunarathne, Viraj, Ahamed Ashiq, Maneesha Prasaad Ginige, Shashikala Dilrukshi Premarathna, Ajith de Alwis, Bandunee Athapattu, Anushka Upamali Rajapaksha, and Meththika Vithanage. 2018. "Municipal Waste Biochar for Energy and Pollution Remediation." In *Springer*, edited by Grégorio Crini and Eric Lichtfouse, 227–252. Cham, Switzerland, Springer International Publishing. doi:10.1007/978-3-319-92162-4_7.

Gunaratna, D. N. J. C. J. 2012. "Analysis on Future Trends of Plastic Recycling in Sri Lanka." University of Sri Jayewardenepura, Nugegoda, Sri Lanka.

Gunasekara, Agampodi Jagath Mendis, R. N. Priyadarshana, T. S. Ranasinghe, R. P. Ranaweera, Eranthi Fernando, J. M. Amali Shanika, H. R. D. Subashine, and R. R. M. K. P. Ranatunga. 2014. "Status of Marine Debris Accumulated in Coastal Areas of Sri Lanka." In *Proceedings of the International Forestry and Environment Symposium*, 19: 758.

Gunasekera, Rohan. 2016. "Marine Litter Threatens Sri Lanka Tourism Prospects." *Economynext*. https://economynext.com/Marine_litter_threatens_Sri_Lanka_tourism_prospects-3-5227-.html.

Hancock, Landon E., and Christopher Mitchell. 2007. *Zones of Peace*. Bloomfield, CT, Kumarian Press.

Hasintha, S. S. R. M. D., R. Wijesekara, Sonia S. Mayakaduwa, A. R. Siriwardana, Nalin de Silva, B. F. A. Basnayake, Ken Kawamoto, and Meththika Vithanage. 2014. "Fate and Transport of Pollutants through a Municipal Solid Waste Landfill Leachate in Sri Lanka." *Environmental Earth Sciences*. doi:10.1007/s12665-014-3075-2.

Hoopes, Lisa A., André M. Landry Jr., and Erich K. Stabenau. 2000. "Physiological Effects of Capturing Kemp's Ridley Sea Turtles, Lepidochelys Kempii, in Entanglement Nets." *Canadian Journal of Zoology* 78 (11): 1941–1947.

Jambeck, Jenna R., Roland Geyer, Chris Wilcox, Theodore R. Siegler, Miriam Perryman, Anthony Andrady, Ramani Narayan, and Kara Lavender Law. 2015. "Plastic Waste Inputs from Land into the Ocean." *Science*. doi:10.1126/science.1260352.

Jang, Yong Chang, R. R. M. K. P. Ranatunga, Jin Yong Mok, Kyung Shin Kim, Su Yeon Hong, Young Rae Choi, and A. J. M. Gunasekara. 2018. "Composition and Abundance of Marine Debris Stranded on the Beaches of Sri Lanka: Results from the First Island-Wide Survey." *Marine Pollution Bulletin*. doi:10.1016/j.marpolbul.2018.01.018.

Jayapala, Sajeewanie Jayapala, and Bentotage Jayasiri. 2018. *Marine Debris and Its Potential Impacts on Mangrove Ecosystem in Negombo Lagoon. National Aquatic Resources Research and Development Agency (NARA) International Scientific Sessions 2018.*

JICA. 2016. "Data Collection Survey on Solid Waste Management in Democratic Socialist Republic of Sri Lanka." Colombo, Sri Lanka.

Kandziora, J. H., N. van Toulon, P. Sobral, H. L. Taylor, A. J. Ribbink, J. R. Jambeck, and S. Werner. 2019. "The Important Role of Marine Debris Networks to Prevent and Reduce Ocean Plastic Pollution." *Marine Pollution Bulletin* 141: 657–662.

Kaya, Ahmet Tunahan, Meral Yurtsever, and Senem Çiftçi Bayraktar. 2018. "Ubiquitous Exposure to Microfiber Pollution in the Air." *The European Physical Journal Plus* 133 (11): 488.

Laist, David W. 1997. "Impacts of Marine Debris: Entanglement of Marine Life in Marine Debris Including a Comprehensive List of Species with Entanglement and Ingestion Records." In *Marine Debris*, 99–139. New York, Springer.

Lavers, Jennifer L., and Alexander L. Bond. 2017. "Exceptional and Rapid Accumulation of Anthropogenic Debris on One of the World's Most Remote and Pristine Islands." *Proceedings of the National Academy of Sciences* 114 (23): 6052–6055.

Law, Kara Lavender, Skye Morét-Ferguson, Nikolai A Maximenko, Giora Proskurowski, Emily E. Peacock, Jan Hafner, and Christopher M. Reddy. 2010. "Plastic Accumulation in the North Atlantic Subtropical Gyre." *Science* 329 (5996): 1185–1188.

Laws of Sri Lanka. 2019. "Bulletin 18." Blackhall Publishing. https://srilankalaw.lk/.

LBO. 2006. "Garbage In." *Lanka Business Online*, January 29. www.lankabusinessonline.com/garbage-in/.

Lee, Jeo. 2015. "Economic Valuation of Marine Litter and Microplastic Pollution in the Marine Enviroment: An Initial Assessment of the Case of the United Kingdom." *Centre for Financial Management Studies, DP126*. University of London, United Kingdom.

León, Víctor M., Inés García, Emilia González, Raquel Samper, Verónica Fernández-González, and Soledad Muniategui-Lorenzo. 2018. "Potential Transfer of Organic Pollutants from Littoral Plastics Debris to the Marine Environment." *Environmental Pollution* 236: 442–453.

Leslie Joseph – National Consultant. 2003. "National Report of Sri Lanka on the Formulation of a Transboundary Diagnostic Analysis and Strategic Action Plan for the Bay of Bengal Large Marine Ecosystem Programme." Bay of Bengal Large Marine Ecosystem Programme (BoBLME).

Marine Pollution Prevention Act No. 35. 2008. *The Gazette of the Democratic Socialist Republic of Sri Lanka*. Colombo, Sri Lanka, Democratic Socialist Republic of Sri Lanka.

McIlgorm, Alistair, Harry F. Campbell, and Michael J. Rule. 2011. "The Economic Cost and Control of Marine Debris Damage in the Asia-Pacific Region." *Ocean & Coastal Management* 54 (9): 643–651.

Menikpura, S. N. M., S. H. Gheewala, and S. Bonnet. 2012. Sustainability assessment of municipal solid waste management in Sri Lanka: Problems and prospects. *The Journal of Material Cycles and Waste Management* 14: 181–192.

MEPA. 2017. "Marine Debris Management Plan in Negombo, Marine Environment Protection Authority Sri Lanka." Sri Lanka, Marine Environment Protection Authority (MEPA).

MEPA. 2018. "Status of Marine Debris Management in Sri Lanka, Marine Environment Protection Authority Sri Lanka." Sri Lanka, Marine Environment Protection Authority (MEPA).

MEPA. 2019. "Marine Environment Protection Authority Official Webpage, Sri Lanka." www.mepa.gov.lk/web/. Sri Lanka, Marine Environment Protection Authority (MEPA).

MoMWD. 2019. "Western Region Master Plan 2030, Western Region Megapolis Project, Ministry of Megapolis and Western Development, Sri Lanka." Sri Lanka, Ministry of Megapolis and Western Development (MoMWD).

Monteiro, Raqueline C. P., Juliana A. Ivar do Sul, and Monica F. Costa. 2018. "Plastic Pollution in Islands of the Atlantic Ocean." *Environmental Pollution*. doi:10.1016/j.envpol.2018.01.096.

Moore, Charles James, S. L. Moore, M. K. Leecaster, and S. B. Weisberg. 2001. "A Comparison of Plastic and Plankton in the North Pacific Central Gyre." *Marine Pollution Bulletin* 42 (12): 1297–1300. doi:10.1016/S0025-326X(01)00114-X.

Mouat, John, R. L. Lozano, and H. Bateson. 2010. "Economic Impacts of Marine Litter. KIMO." *Kommunenes Internasjonale Miljøorganisasjon (KIMO)* 367–394.

Mrosovsky, Nicholas, Geraldine D. Ryan, and Michael C. James. 2009. "Leatherback Turtles: The Menace of Plastic." *Marine Pollution Bulletin* 58 (2): 287–289.

Napper, Imogen E., and Richard C. Thompson. 2019. *Marine Plastic Pollution: Other than Microplastic. Waste*. 2nd ed. Elsevier. doi:10.1016/b978-0-12-815060-3.00022-0.

Ocean Conservancy. 2019. "Trash Free Seas." *Ocean Conservancy*. https://oceanconservancy.org/trash-free-seas/plastics-in-the-ocean/.

Ofiara, Douglas D. 2001. "Assessment of Economic Losses from Marine Pollution: An Introduction to Economic Principles and Methods." *Marine Pollution Bulletin* 42 (9): 709–725.

Ourmieres, Yann, Jérémy Mansui, Anne Molcard, François Galgani, and Isabelle Poitou. 2018. "The Boundary Current Role on the Transport and Stranding of Floating Marine Litter: The French Riviera Case." *Continental Shelf Research* 155: 11–20.

Perera, Nishan, and Asha de Vos. 2007. "Marine Protected Areas in Sri Lanka: A Review." *Environmental Management* 40 (5): 727.

Rusthoven, Ian. 2018. "Temporal/Spatial Trends and Concentrations of Microplastics in Streams Throughout the Central Illinois Watersheds." *University Research Symposium*, Illinois State University.

Ryan, Peter G., Charles J. Moore, Jan A. Van Franeker, and Coleen L. Moloney. 2009. "Monitoring the Abundance of Plastic Debris in the Marine Environment." *Philosophical Transactions of the Royal Society B: Biological Sciences*. doi:10.1098/rstb.2008.0207.

Ryan, Peter G. 2008. "Seabirds Indicate Changes in the Composition of Plastic Litter in the Atlantic and South-Western Indian Oceans." *Marine Pollution Bulletin* 56 (8): 1406–1409.

Ryan, Peter G., Charles J. Moore, Jan A. Van Franeker, and Coleen L. Moloney. 2012. "Monitoring the Abundance of Plastic Debris in the Marine Environment," 364 (2009): 1999–2012. doi:10.1098/rstb.2008.0207.

SACEP. 2007. "Marine Litter in the South Asians Seas Region." South Asian Cooperation for Environment Protection (SACEP).

SACEP. 2018. "Towards Litter Free Indian Ocean, Summary of the Regional Marine Litter Action Plan for SAS Region." South Asian Cooperation for Environment Protection (SACEP).

Sathsara, B. M. Y., M. I. U. Manikarachchi, K. P. G. L. Sandaruwan, D. W. L. U. De Silva, G. D. T. M. Jayasinghe, B. K. K. K. Jinadasa, H. D. Wimalasena, M. M. A. S. Maheepala, D. W. L. U. De Silva, and K. H. M. L. Amarala. 2017. "Small Plastic Debris in Beach Sand: A Quantitative Analysis with Regards to Beach Usage." In. NARA.

SATREPS. 2014. "Waste Amount and Composition Surveys (WACS) Implemented in the Central and Southern Provinces of Sri Lanka."

Schofield, Clive. 2009. "The Trouble with Islands: The Definition and Role of Islands and Rocks in Maritime Boundary Delimitation." In *Maritime Boundary Disputes, Settlement Processes, and the Law of the Sea*, 19–38. Martinus Nijhoff Publishers, the Netherlands.

Seabin Group. 2016. "Seabin Project." https://seabinproject.com/.

Sharma, Shivika, and Subhankar Chatterjee. 2017. "Microplastic Pollution, a Threat to Marine Ecosystem and Human Health: A Short Review." *Environmental Science and Pollution Research* 24 (27): 21530–12547.

Silva, E. I. L. 2004. "Quality of Irrigation Water in Sri Lanka–Status and Trends." *Asian Journal of Water, Environment and Pollution* 1 (1, 2): 5–12.

SLPA. 2015. "Sri Lanka Ports Authority Annual Report 2015." Sri Lanka, Sri Lanka Ports Authority.

Takehama, Suichi. 1990. "Estimation of Damage to Fishing Vessels Caused by Marine Debris, Based on Insurance Statistics." In *Proceedings of the Second International Conference on Marine Debris*, 792–809. National Oceanic Atmospheric Administration.

Thompson, Richard C. , Ylva Olsen, Richard P. Mitchell, Anthony Davis, Steven J. Rowland, Anthony W. G. John, Daniel McGonigle, and Andrea E. Russell. 2004. "Lost at Sea: Where Is All the Plastic?" *Science* 304 (5672): 838.

UNEP. 2016. "Marine Plastic Debris and Microplastics, Global Lessons and Research to Inspire Action and Guide Policy Change. United Nations Environment Programme, Nairobi." United Nations Environment Programme.

UNSDG. 2018. "Transformation towards Sustainable and Resilient Societies, United Nations Division for Sustainable Development Goals Department of Economic and Social Affairs."

Waring, Rosemary H., R. M. Harris, and S. C. Mitchell. 2018. "Plastic Contamination of the Food Chain: A Threat to Human Health?" *Maturitas*. 154: 64–68.

Weerakoon, W. R. W. M. A.P., T. B. D. T. Samaranayake, H. B. Jayasiri, and K. Arulananthan. 2018. "Quantitative Analysis of Micro-Plastic Contamination in Beach Sand at the Western and Southwestern Coastal Stretches in Sri Lanka." In *International Scientific Sessions*, National Aquatic Resources Research and Development Agency (NARA).

Xanthos, Dirk, and Tony R. Walker. 2017. "International Policies to Reduce Plastic Marine Pollution from Single-Use Plastics (Plastic Bags and Microbeads): A Review." *Marine Pollution Bulletin* 118 (1–2): 17–26.

Zalasiewicz, Jan, Colin N. Waters, Juliana A. Ivar do Sul, Patricia L. Corcoran, Anthony D. Barnosky, Alejandro Cearreta, Matt Edgeworth, Agnieszka Gałuszka, Catherine Jeandel, and Reinhold Leinfelder. 2016. "The Geological Cycle of Plastics and Their Use as a Stratigraphic Indicator of the Anthropocene." *Anthropocene* 13: 4–17.

Zhang, Haibo, Qian Zhou, Zhiyong Xie, Yang Zhou, Chen Tu, Chuancheng Fu, Wenying Mi, Ralf Ebinghaus, Peter Christie, and Yongming Luo. 2018. "Occurrences of Organophosphorus Esters and Phthalates in the Microplastics from the Coastal Beaches in North China." *Science of the Total Environment* 616: 1505–1512.

19 Case Studies of Particulate Plastic Distribution and Ecotoxicity in Japan

Shosaku Kashiwada

CONTENTS

19.1 Production, Consumption, Recycling, and Waste Management of Plastics in Japan 327
19.1.1 Production of Plastics and Environmental Health Issues in Japan 327
19.1.2 Waste Management and Recycling of Plastics in Japan 328
19.1.3 Distribution of Asian Plastic Production and Sources of Particulate Plastics in the Seas around Japan 329
19.2 National Survey Reports of Drifting Plastic Garbage in the Seas around Japan 330
19.2.1 Fisheries Agency 330
19.2.2 Japan Meteorological Agency 330
19.2.3 Japan Coast Guard 330
19.2.4 Ministry of the Environment 330
19.2.5 Development of Drifting Garbage Prediction Models 335
19.2.5.1 Drift of Plastic Debris with Different Submergence Ratios in the Sea of Japan 339
19.2.5.2 Drift of Disposable Lighters Released from Different Countries Bordering the Sea of Japan 341
19.2.5.3 Origins of Plastic Debris Drifting to Coastlines on the East China Sea 342
19.2.5.4 Drift of Plastic Debris from Japan to the North Pacific Ocean 344
19.2.6 Microplastics 344
19.3 Environmental Fate and Toxicological Risks of Particulate Plastics in the Ocean 348
19.3.1 Chemical Contamination of Plastic Debris Drifting in the Oceans 348
19.3.2 Fate of Chemical-Contaminated Plastics in Fish 349
19.4 Particulate Plastics and Human Health 351
References 352

19.1 PRODUCTION, CONSUMPTION, RECYCLING, AND WASTE MANAGEMENT OF PLASTICS IN JAPAN

19.1.1 Production of Plastics and Environmental Health Issues in Japan

Industrial production of plastics (polyvinyl chloride; PVC) in Japan began in 1941 (Vinyl Environmental Council, Japan). In 1959, production reached 179 × 10^3 t and ranked second in the world's plastic production. Along with the increase in plastic production and consumption, the issue of plastic waste management was discussed in a Japanese diet deliberation in 1970 (House of Councilors proceedings record information 1970). Furthermore, the carcinogen vinyl chloride monomer was reported by the International Agency for Research on Cancer in 1987 (IARC 1987). In 1990, plastic production exceeded 2000 × 10^3 t in Japan (Vinyl Environmental Council, Japan).

In response to the United Nations Conference on Environment and Development (the Earth Summit) in 1992, the Act on the Promotion of Sorted Collection and Recycling of Containers and Packaging (in 1995) and the Basic Act for the Promotion of a Recycling-Oriented Society (in 2000) were promulgated in Japan. During the 1990s in Japan, there were issues regarding dioxin production from vinyl chloride combustion and endocrine disruption by the plasticizer phthalic ester and by bisphenol-A (Japan Environmental Agency 1998). Dioxin production can be suppressed by controlling the combustion conditions (Kasai 2008). The Ministry of the Environment (former Japan Environmental Agency) reported that phthalic ester has demonstrated little endocrine disruption activity (Ministry of the Environment 2003). But bisphenol-A is well known as an endocrine disruptor (Bonefeld-Jørgensen Eva et al. 2007). Because of the overseas relocation of PVC processing companies and developments in China and Korea, in 2016, Japan became the fifth largest plastics producer (Japan Petrochemical Industry Association 2018). In 2017, Japan produced 11,020 × 10^3 t of plastics (Plastics Waste Management Institute Japan 2017) of a global total of 348,000 × 10^3 t (PlasticsEurope 2018); Japan's production accounted for only 3.16% of the world production. Plastics produced in Japan 2017 were classified as polyethylene 24.1%, polypropylene 22.7%, PVC 15.5%, polystyrene 11.3%, thermoplastic resin 17.9%, and thermosetting resin 8.5%. These plastics were produced for packaging and containers 40.7%; electronic equipment and wire cables 18.8%; transportation 12.1%; building materials 11.8%; household goods, clothing, footwear, furniture, and toys 9.5%; agriculture, forestry, and fisheries 1.4%; and other 5.6% (Plastics Waste Management Institute Japan 2017). Japan and the European Union (EU), plus Norway (NO) and Switzerland (CH), are producing plastics by using petroleum naphtha, an intermediate hydrocarbon liquid stream derived from the refining of crude oil. In 2017, to produce 11,075 × 10^3 t of plastic, 45,848 × 10^6 L of petroleum naphtha was used. To produce the same amount of naphtha, 387,377 × 10^3 t of crude oil was needed. This means that 2.86% of the total amount of crude oil was consumed for plastic production in the world (calculated using data from a report by the Japan Petrochemical Industry Association 2018). In Japan, to reduce the use of naphtha for plastic production, polyethylene terephthalate (PET) bottles were thinned, with a weight saving (−23.9%) per PET bottle. Because of this thinning and weight saving, although shipments of PET increased 1.54 times from 14.8 billion in 2004 to 22.7 billion in 2017, carbon dioxide emissions from PET bottle manufacture remained stable at only a 1.04 times increase from 2089 × 10^3 t in 2004 to 2167 × 10^3 t in 2017 (Council for PET Bottle Recycling 2018).

19.1.2 Waste Management and Recycling of Plastics in Japan

Annual household garbage per person in Japan was estimated at 270 kg, of which, plastics account for 35.1 kg (Ministry of the Environment 2016a). The amounts of household garbage and household plastic waste in Japan are generally smaller than those in other countries: for example, there are annual totals of 460 kg garbage and 50.6 kg plastics in the United States; 460 kg garbage and 46.0 kg plastics in Germany; 370 kg garbage and 11.1 kg plastics in Canada (Ministry of the Environment 2016a). Of the total plastic production in Japan (11,020 × 10^3 t, 2017), the amount consumed domestically was 9780 × 10^3 t, and the amount of domestic waste plastics collected was 8280 × 10^3 t (recovery rate 84.7%) (Plastics Waste Management Institute Japan 2017). The amount of plastics consumed per person annually in 2016 was 74 kg/year in Japan – lower than the 132 kg/year in the United States and 96 kg/year in the EU (plus NO and CH) (Japan Petrochemical Industry Association 2018). Collected waste plastics and unused plastics as manufacturing loss (750 × 10^3 t, 2017) were subjected to recycle systems. A total of 9030 × 10^3 t of plastics were subjected to mainly material (23%), chemical (4%), and thermal (58%) recycling in 2017. Some of them were sent to incineration disposal (8%) and landfill (6%) (Plastics Waste Management Institute Japan 2017). In the EU (plus NO and CH) in 2016, of the total plastic production of 60,000 × 10^3 t, the amount collected was 27,100 × 10^3 t (recovery rate, 45.2%). Treatments for waste plastics included recycling (as material and chemical recycling) 31.1%, energy recovery (thermal recycling) 41.6%, and landfill 27.3% (PlasticsEurope 2018). In 2017, a total of 587 × 10^3 t of PET bottles were sold in Japan, accounting for 5.33% of all plastics produced in Japan, the amount of PET bottles collected was 624 × 10^3 t (domestic plus overseas), and the amount recycled was 498 × 10^3 t. The recovery rate calculated using data of sold and recycled PET bottles was 84.8% in Japan (Council for PET Bottle Recycling 2018). This compares with the last decade's ratios of between 82.2% and 89.9% in Japan, which were higher than 41.8% (EU + NO and CH in 2017) and 20.1% (U.S. in 2016) (Council for PET Bottle Recycling 2018). The amount of styrene foams sold was 133 × 10^3 t in Japan, for use as containers (54.2%), buffer materials (30.2%), and construction and civil engineering materials (15.6%) (Japan Expanded Polystyrene Association 2017). The amount of styrene foams collected was 123 × 10^3 t (recovery rate 92.5%) in Japan. Of these collected foams, 54.4% was subjected to material recycling and 36.0% to thermal recycling. Thus, a total of 90.4% of collected styrene foams was recycled, 5.3% was sent to incineration disposal, and 4.3% was landfilled (Japan Expanded Polystyrene Association 2017). By comparison, the EU's plastic packaging recycling rate is about 41% (PlasticsEurope 2018).

19.1.3 Distribution of Asian Plastic Production and Sources of Particulate Plastics in the Seas around Japan

China is the largest plastic-producing country in the world, with 82,267 × 10^3 t of the total world production of 335,000 × 10^3 t in 2016; this was equal to 24.6% of global plastic production. Other producers were the EU (plus NO and CH) (17.9%), the United States (15.2%), South Korea (4.6%), and Japan (3.2%) (Japan Petrochemical Industry Association 2018). Although China has the largest plastic production, there are no reliable statistical data on consumption and recycling in China. There is the same issue in South Korea. However, because generally chemical production and consumption yield the same amounts of waste, we can surmise that China is the greatest contributor of mismanaged plastic waste entering the ocean (Jambeck et al. 2015). Plastic waste escapes from cities to the oceans via rivers, especially with seasonal rains, such as monsoonal rain. More than 74.5% of all plastics entering the oceans via rivers do so between May and October in the Northern Hemisphere. Lebreton et al. reported in 2017 that the highest global input occurs in August, at 229 (range 193 to 375) × 10^3 t, and the lowest occurs in January, at 46 (range 34 to 87) × 10^3 t. These seasonal inputs were driven mainly by large inputs from China, which are regulated by the East Asian monsoon (Lebreton et al. 2017).

It has been calculated that rivers contribute between 410,000 and 4,000,000 t a year to oceanic plastic debris, with 88% to 95% coming from only ten rivers, namely, the Yangtze, Yellow, Hai He, Pearl, Amur, and Mekong in East Asia, the Indus and Ganges Delta in South Asia, and the Niger and Nile in Africa (Schmidt et al. 2017). The first four rivers run through China. Most of the rivers' plastic input comes from Asia, and this emphasizes the need to focus monitoring and mitigation efforts on Asian countries that are undergoing rapid economic development and have poor waste management (Lebreton et al. 2017). As mentioned above, Japan has exerted strong control of waste management and plastics recycling in recent decades. One report (Jambeck et al. 2015) calculated the mass of mismanaged plastic waste by using a hypothesized ratio of plastic debris (macro- and microplastics) to collected plastics in each river basin; their hypothesized ratio was 0.3% to 0.8%. A report recently emerged about the actual ratios in a major and minor river in Japan. Image analysis was used directly to monitor and measure plastic debris running through the Edo River (a major river) entering to the Tokyo Bay and the Ohori River (a minor river). The ratio in each river was 0.025% and 0.002%, respectively (Tokyo University of Science press release). Thus, in Japan, with its well managed waste, these ratios were 10 to 100 times lower than those in the above research. Therefore, the contribution of Japan to mismanaged plastic waste and to sources of particulate plastics in and around Japanese seas is likely lower than those of China and other Asian countries.

19.2 NATIONAL SURVEY REPORTS OF DRIFTING PLASTIC GARBAGE IN THE SEAS AROUND JAPAN

19.2.1 Fisheries Agency

The Fisheries Agency of Japan conducted visual surveys of floating substances in the North Atlantic Ocean and its adjacent waters from 1986 to 1991 (Matsumura and Nasu 1997). Suspended matter consisted of 60% petrochemical products, such as polystyrene foam and plastics, 10% fishing nets, and the remaining 30% natural items, such as logs. Polystyrene foam was found in 77% of the survey blocks; it was suggested that much of it was of terrestrial origin, especially in the case of foam floating off southern Japan, in the East China Sea, off Central America, and in the Gulf of Mexico. Because plastic products were found with high frequency on the Pacific side of Central America and the Caribbean coast, as with the polystyrene foam, it was surmised that they were predominantly of land origin. There were many fishing nets drifting in the Pacific Ocean between 20°N to 30°N and 150°W to 130°W and also between 30°N to 40°N and 140°E to 150°E.

19.2.2 Japan Meteorological Agency

The Japan Meteorological Agency conducts visual observations of floating matter and sea surface oil films and reports the distribution of these floating substances around Japan (Japan Meteorological Agency 2009). The average distribution of floating pollutants over the 20 years from 1981 to 2000 was high in the waters surrounding Japan, and lower in the waters south of 20°N and in the subarctic regions. The southern areas of the Kuroshio Current, which is around 30°N and east of 150°E, have more floating pollutants than areas at the same longitude, but at higher or lower latitudes. However, even at low latitudes, floating pollutants are relatively abundant to the north of New Guinea. This suggests that ocean current systems have a strong influence on the distribution of floating pollutants in the open ocean.

19.2.3 Japan Coast Guard

The Japan Coast Guard has been conducting visual surveys of maritime drift in sea areas around Japan since 1991 (Japan Coast Guard 2006). In the 2005 survey, the survey distance was 500 nautical miles, and the total count of materials drifting on the ocean surface was 254. The number per 10 nautical miles was 5.08 (=2.7/10 km) – lower than the previous year's 12.86 per 10 nautical miles (=7.0/10 km). There were many petrochemical products, including polystyrene foam, plastic bags, and solid plastics, which accounted for more than 60% of the total.

19.2.4 Ministry of the Environment

Japan's Ministry of the Environment has continued its national survey of drifting garbage, including plastics, since Fiscal Year (FY) 2006. In FY 2006, drifting plastic garbage was reported as part of all drifting debris (mainly driftwood) by the Ministry of the Environment (2006a). A survey was performed on the west coast of Nagasaki Prefecture, which is located near where the East China Sea and the Sea of Japan meet. By August 11, 2006, a total of 68,159 items of driftwood were counted along the western shoreline of the prefecture in this survey. Some of the driftwood was identified as pine, euodia (a tree or shrub in the Rutaceae family), Lauraceae, and Sapotaceae that drifted from China, Taiwan, and Japan. Sandals and empty cans (made in China or Taiwan), medical syringes (label-disappeared), and other waste labeled with Chinese or Korean characters were found. In addition, the Ministry of the Environment reported on the status of garbage drifting on Japanese coastlines along the Sea of Japan in 2006 (2006b). The aim was to prepare clear explanatory materials for distribution to neighboring countries to promote the reduction of garbage beached on the Japanese coast of the

Sea of Japan. For this report, disposable lighters, PET bottles, fishing floats, plastic bottles, medical waste, and others were collected. Since then, disposable lighters and PET bottles were employed as model plastic garbage for a survey in the Sea of Japan and the East China Sea, because these items are usually labeled and it is therefore possible to determine the country in which they were produced. The survey of drifting garbage was performed along two Japanese coastlines (Figure 19.1), namely, the island of Fukueshima in Nagasaki Prefecture, located near the strait where the Sea of Japan and the East China Sea meet, and the island of Tobishima in Yamagata Prefecture in the Sea of Japan. Collected garbage was individually categorized and the country of origin was identified from the label (Table 19.1). Disposable lighters and PET bottles were common items of drifting garbage. Much larger numbers of foreign and Japanese disposable lighters were found on Fukueshima than on Tobishima, and larger numbers of PET bottles mainly of Japanese origin were found on Tobishima than on Fukueshima. This was likely because disposable lighters are more likely to sink below the sea surface and thus not be influenced by wind; they are thus more likely to reach islands closer to their sources. In contrast, PET bottles can drift long distances with the wind. The category "Other" consisted mainly of cylinders from Korea used for sea eel fishing. This item of disposable fishing gear is an important source of plastic pollution in the Sea of Japan.

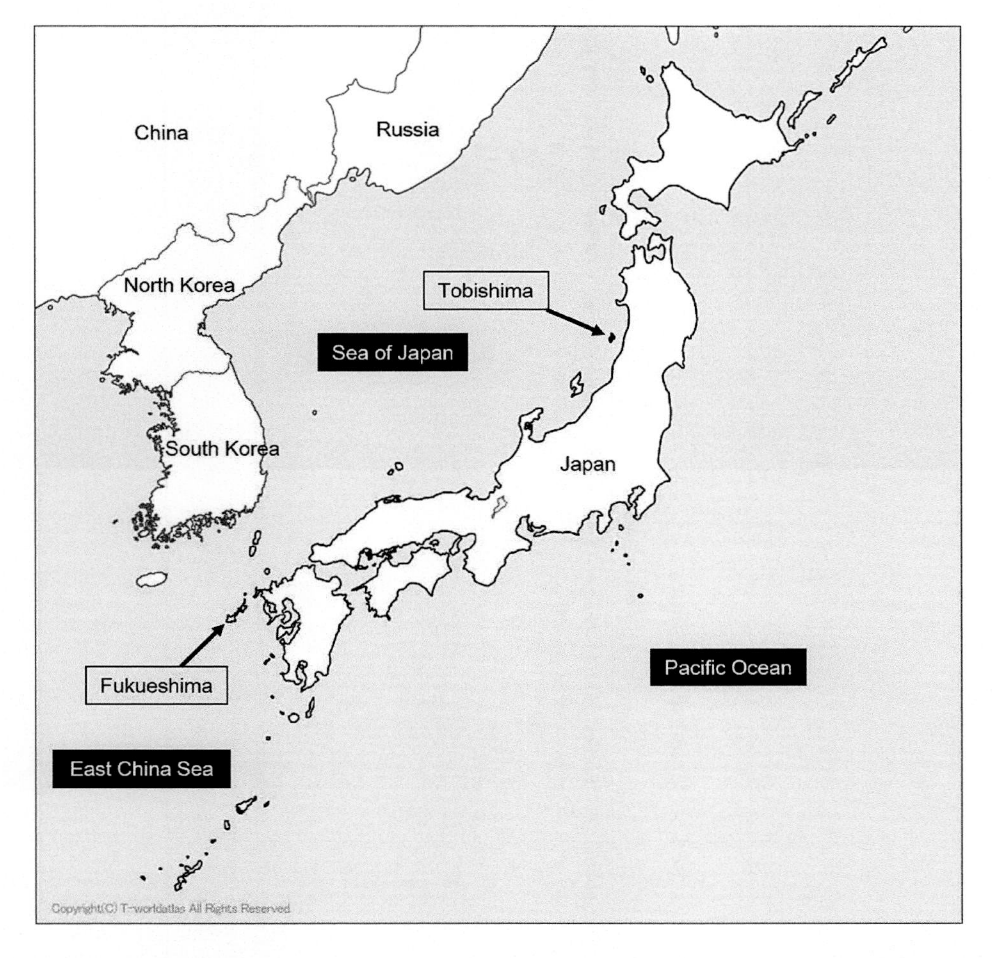

FIGURE 19.1 Location of the survey of drifting garbage in Japan. (Data adapted from Ministry of the Environment, FY 2006 International reduction measures survey report on floating marine litter and beach litter [in Japanese], 2006b.)

TABLE 19.1
Garbage Beached on Japanese Coastlines

Country of	Fukueshima						Tobishima					
Origin	Lighters	PET Bottles	Fishing Floats	Plastic Bottle	Medical Waste	Other	Lighters	PET Bottles	Fishing Floats	Plastic Bottles	Medical Waste	Other
Japan	37.2	12.8	10.3			26.9	6.3	44.8	1.4	0.5	0.9	18.1
Korea	19.2	7.7	24.4	3.8		52.6[a]	2.3	12.7	1.8	5		52.9[a]
China	30.8	2.6	14.1		9	20.5	3.6	6.8	1.8		0.5	0.9
Taiwan	14.1	1.3	62.8			1.3		3.2	6.3			0.9
Hong Kong	2.6							0.5				
Russia		1.3						12.2				1.4
Philippines	2.6											
Unknown	35.9	19.2	46.2		19.2		5.4	29.9		0.9	1.8	

Source: Ministry of the Environment, FY 2006 International reduction measures survey report on floating marine litter and beach litter (in Japanese), 2006b.
Note: Individual items per 100 m are listed.
[a] Mainly cylinders for fishing for sea eels.

After several surveys through 2006–2009, the Ministry of the Environment set up seven stations (Ishigaki Island, Kamisu, Tsushima, Shimonoseki, Hakui, Minami Satsuma, and Awaji) for monitoring beached garbage. They were chosen on the basis of four points: (i) the characteristics of the ocean current (i.e., sea area classification); (ii) the need to ensure estimation accuracy (i.e., coastlines with abundant drifting items); (iii) cost, participation by locals, easy access, and the possible seasonal nature of the survey; and (iv) the need to make comparisons with other survey results (Table 19.2 and Figure 19.2). From September 2010 to February 2013, beached garbage was collected on each

TABLE 19.2
Stations Selected for Monitoring of Beached Garbage

Station	Name	Ocean Current (Sea Area Classification)
A	Ishigaki Island	Upstream of Japan Current
B	Kamisu	Downstream of Chishima Current
C	Tsushima	Tsushima Strait
D	Shimonoseki	Upstream of Tsushima Current
E	Hakui	Midstream of Tsushima Current
F	Minami Satsuma	East China Sea
G	Awaji	Seto Island Sea

Source: Ministry of the Environment, FY 2012 Research report on measures to tackle beach debris (in Japanese), 2012.

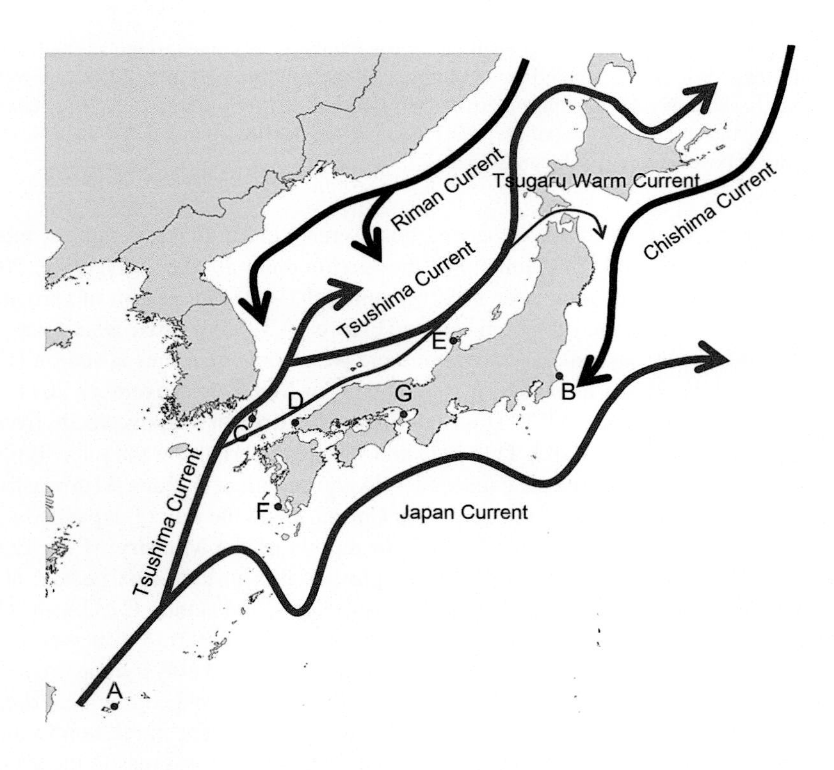

FIGURE 19.2 Map of ocean currents and monitoring stations shown in Table 19.2.

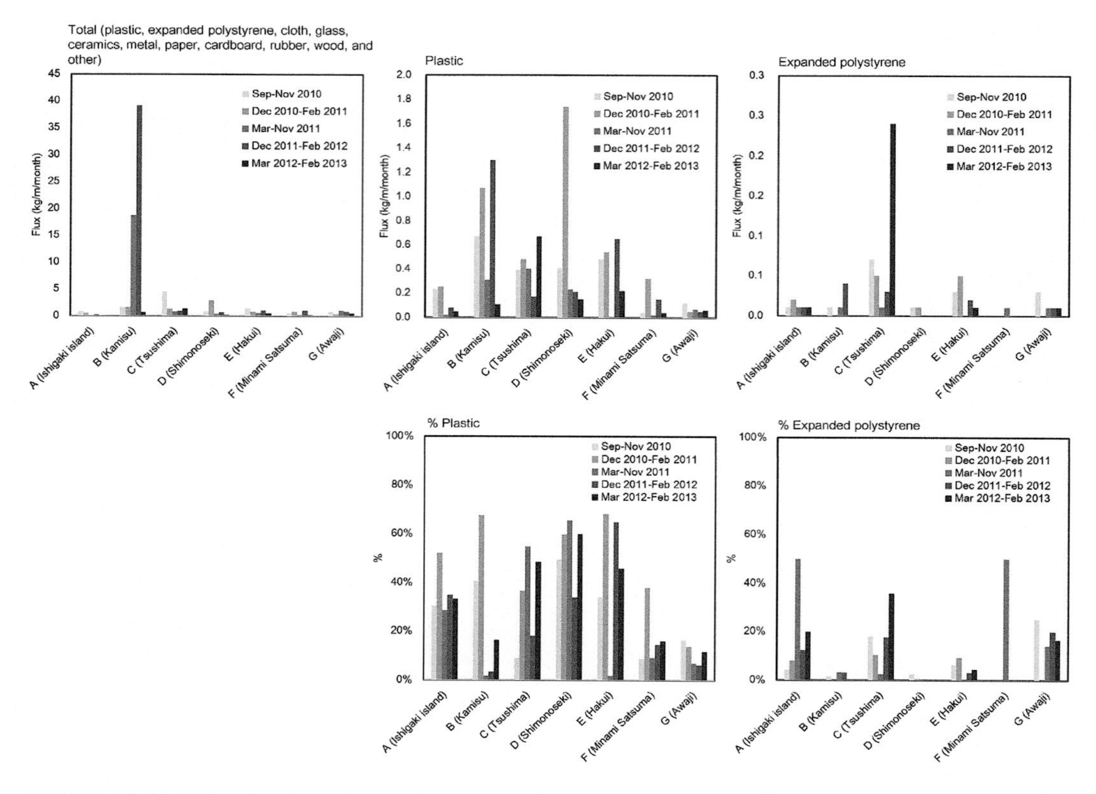

FIGURE 19.3 Flux of garbage beached at each monitoring station, and plastics and expanded polystyrene as percentages of total flux. (Data adapted from Ministry of the Environment, FY 2010 Research report on measures to tackle beach debris [in Japanese], 2010a; Ministry of the Environment, FY 2011 Research report on measures to tackle beach debris [in Japanese], 2011; Ministry of the Environment, FY 2012 Research report on measures to tackle beach debris [in Japanese], 2012.)

coastline and categorized into plastics, expanded polystyrene, cloth, glass, ceramics, metal, paper, cardboard, rubber, wood, and others (Ministry of the Environment 2010a, 2011, 2012, 2013). After the measurement of volume and weight, the flux (kg/m/month) – the rate of drift of garbage from the ocean to the coast – was calculated for each station (Figure 19.3). Total flux, which was calculated over 3 years, showed minor seasonal variation, but there were some extremes at station B (Kamisu). There was a marked increase in total flux at station B from March to November 2011, and again from December 2011 to February 2012. This was due to the drift of massive reeds from the East Japan Great Earthquake disaster (March 11, 2011) and wooden debris from the huge Typhoon Roke (September 13, 2011). The flux of plastic exceeded 1.0 kg/m/month at stations B (downstream of the Chishima Current) and D (upstream of the Tsushima Current), and the flux of expanded polystyrene was highest at station C (in the Tsushima Straits). In reports of the Ministry of the Environment (2010a, 2011, 2012), it was shown that the ratio of plastics flux to total flux peaked at station D (average 53.8%), followed by station E (43%), station A (35.8%), and station C (33.4%). The ratio of expanded polystyrene flux to total flux peaked at station A (average 19.0%), followed by station C (16.9%), station G (15.2%), and station F (10.0%). Plastic and expanded polystyrene fluxes accounted for a high proportion of the total flux. Using these flux data and other geographical data, the flux of drifting garbage in each sea area was calculated (Figure 19.4). Except in relation to the irregular events of the Great East Japan Earthquake and the huge typhoon (this is a reason there was a much larger flux (20.36) downstream of the Japan Current and Chishima Current), the Tsushima Current was a major contributor to garbage flux – mainly of plastics – onto Japanese coasts.

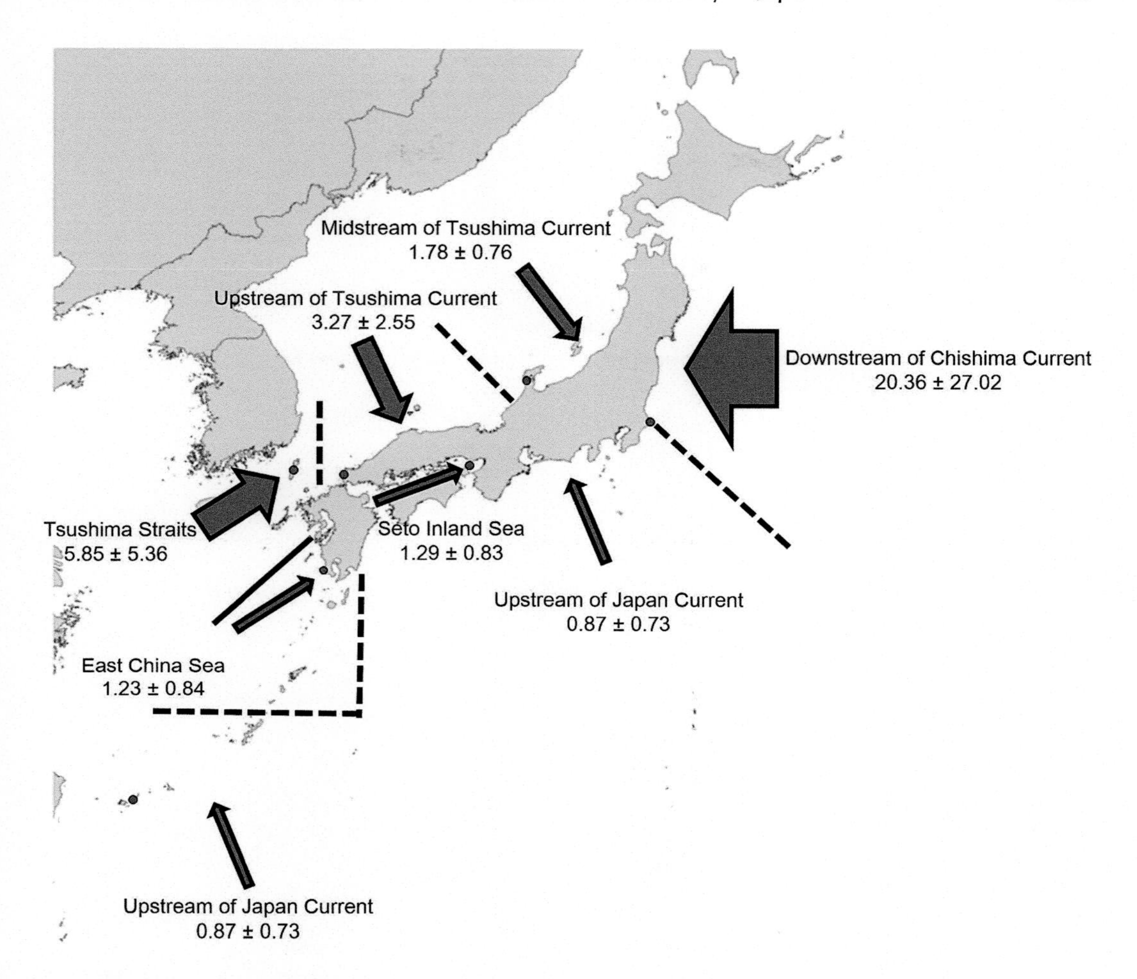

FIGURE 19.4 Calculated flux (kg/m/month) of drifting garbage in areas of sea, as classified by ocean currents. (Data adapted from Ministry of the Environment, FY 2010 Research report on measures to tackle beach debris [in Japanese], 2010a; Ministry of the Environment, FY 2011 Research report on measures to tackle beach debris [in Japanese], 2011; Ministry of the Environment, FY 2012 Research report on measures to tackle beach debris [in Japanese], 2012.)

19.2.5 Development of Drifting Garbage Prediction Models

The Ministry of the Environment reported the development of a model for drifting-garbage prediction in the Sea of Japan in FY 2006 (Ministry of the Environment 2006b), the East China Sea in FY 2007 (Ministry of the Environment 2007), and the North Pacific Ocean in FY 2010 (Ministry of the Environment 2010b). In these reports, disposable lighters were used as model items to develop a drifting-garbage-prediction model for the Sea of Japan and the East China Sea.

Because they were the materials drifting most frequently in the oceans around Japan, disposable lighters and PET bottles were employed as references for the drifting-garbage-prediction model. The floating statuses of disposable lighters and PET bottles were estimated by using collected samples. For disposable lighters, the presence of a metal windshield bracket, the presence of a spark wheel, and the volume level of liquefied gas were factors considered to affect submergence ratios (volume above water: volume submerged). On disposable lighters collected at Fukueshima and Tobishima, there was no metal windshield bracket in 99% and 87%, respectively; no liquefied gas in 87% and 69%; and a missing spark wheel in 62% and 41% (Figure 19.5a).

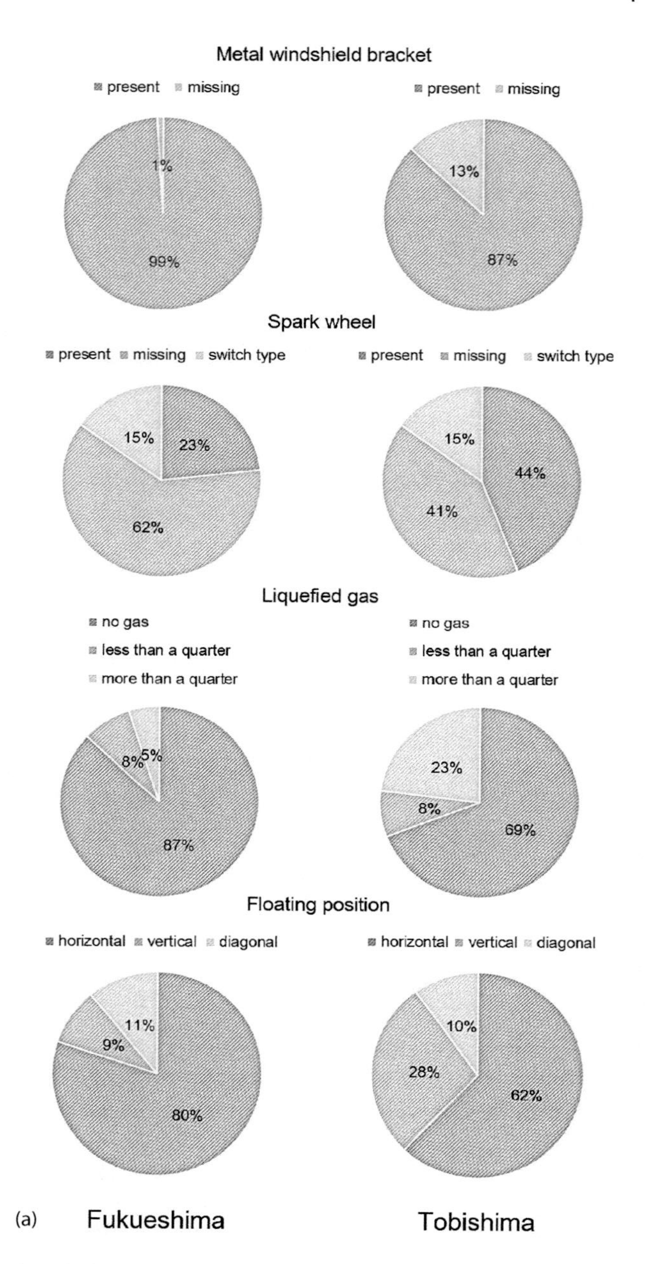

FIGURE 19.5 Properties of disposable lighters. (a) PET bottles. *(Continued)*

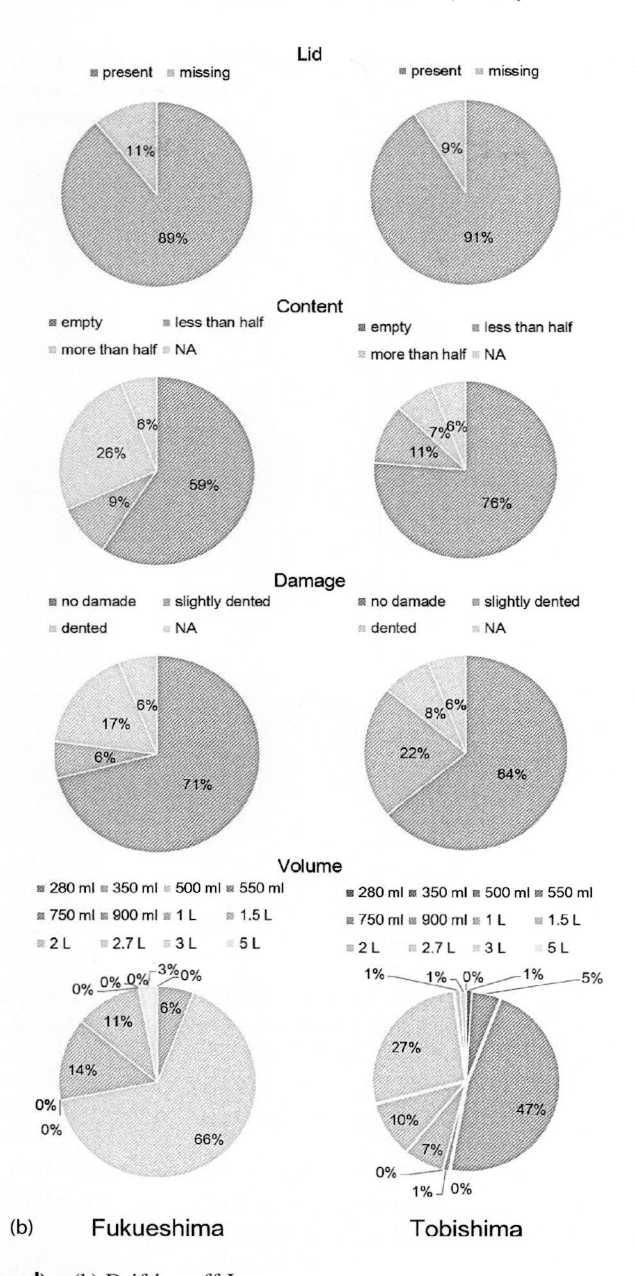

FIGURE 19.5 (Continued) (b) Drifting off Japan. *(Continued)*

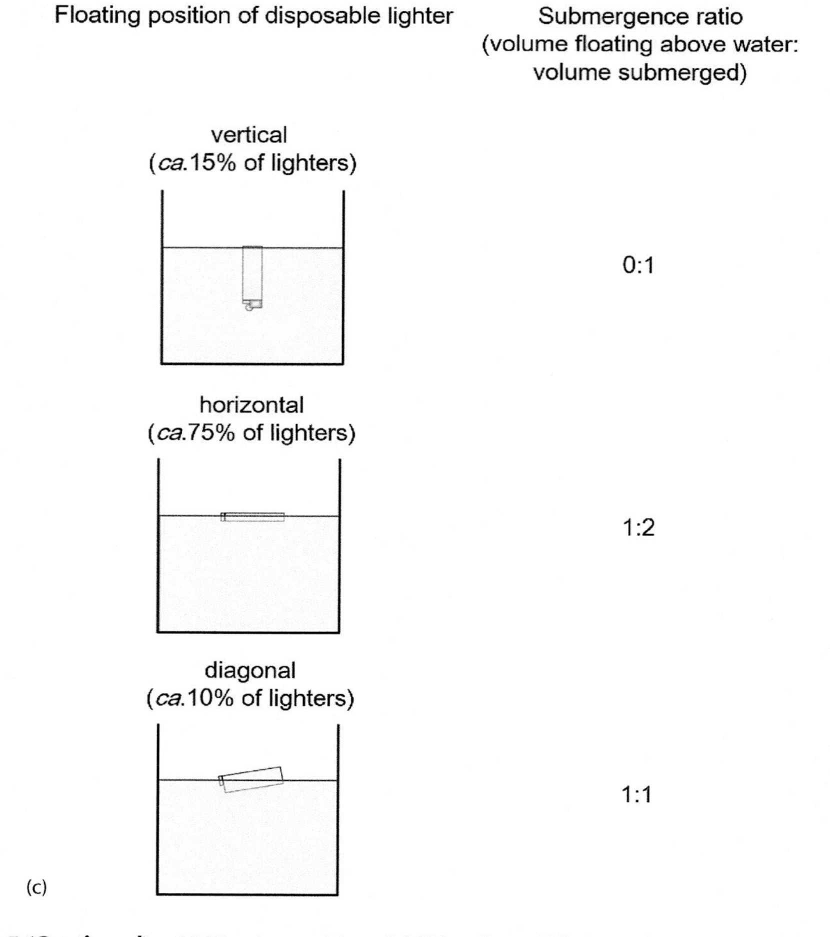

FIGURE 19.5 (Continued) (c) Floating position of drifting disposal lighters. (Data adapted from Ministry of the Environment, FY 2006 International reduction measures survey report on floating marine litter and beach litter [in Japanese], 2006b.)

To assess floating positions and submergence ratios, collected disposable lighters were added to seawater in the laboratory. Floating position was categorized as vertical, diagonal, or horizontal (Figure 19.5c). The most frequently observed floating position of disposable lighters collected at Fukueshima and Tobishima was horizontal (80% and 62%, respectively), followed by vertical (9% and 28%, respectively), and diagonal (11% and 10%, respectively) (Figure 19.5a). For PET bottles, presence of a lid, volume level of contents, and degree of damage of bottle were factors considered to affect submergence ratios. Of PET bottles collected at Fukueshima and Tobishima, 89% and 91%, respectively, were bottles with lids; 59% and 76% were empty; and 71% and 64% were undamaged (Figure 19.5b). The most common drifting PET bottle had a volume of 500 mL (66% and 47%, respectively, Figure 19.5b). PET bottles without lids would be easy to be flooded and thus were more likely to have sunk to the sea floor. This was likely why there was a high percentage of drifting PET bottles with lids. The drifting-garbage-prediction model was created by using the flow prediction model RIAM Ocean Model (RIAMOM) (Lee 2003), developed by Dr. Jong-Hwan Yoon, professor at the Research Institute for Applied Mechanics (RIAM), Center for East Asian Ocean-Atmosphere Research at Kyushu University, Japan.

19.2.5.1 Drift of Plastic Debris with Different Submergence Ratios in the Sea of Japan

The Ministry of the Environment prepared a drift scenario of plastic debris in the Sea of Japan (Ministry of the Environment 2006b). The simulation was done as a 3-year prediction; materials were all introduced over the same area (every 1/6th of a degree) of ocean on the first day of the month (blue dots) for 1 year (12 times in total) (Figure 19.6). To simulate material flows in the ocean, the submergence ratio of disposable lighters was set at 0:1 (volume above water: volume submerged; about 15% of drifting lighters) in the vertical position; 1:2 in the horizontal position (about 75% of drifting lighters); and 1:1 in the diagonal position (about 10% of drifting lighters) (Figure 19.6); the submergence ratios of empty polyethylene tank (polytanks) with lids and PET bottles with lids were set at 10:1 and 100:1, respectively. Simulation prediction maps are given in Figure 19.6. Blue dots indicate drifting materials, and red dots indicate materials that have been beached (Figure 19.6).

- Most of the disposable lighters with a submergence ratio of 0:1 were still drifting in the Sea of Japan 1 year after their introduction; most lighters were drifting toward the Japanese coast, and some were still drifting in bands after 2 years; almost all lighters had completed their drift to the Japanese coast after 3 years.
- Most disposable lighters with a submergence ratio of 1:2 or 1:1 were drifting in bands in the Sea of Japan 1 year after introduction; after 2 years, almost all lighters had completed their drift to the Japanese coast.

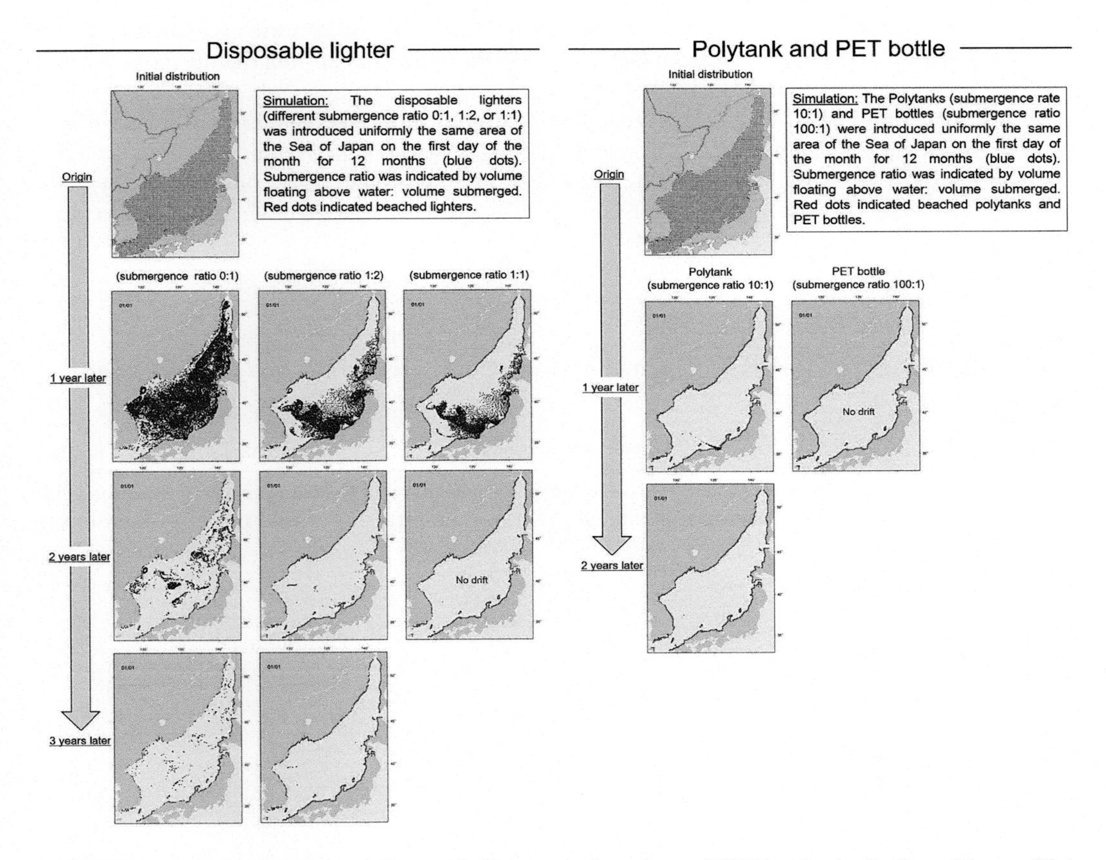

FIGURE 19.6 Drift prediction of disposable lighters, polytanks, and PET bottles in the Sea of Japan. (Data adapted from Ministry of the Environment, FY 2006 International reduction measures survey report on floating marine litter and beach litter [in Japanese], 2006b.)

- Most of the polytanks and PET bottles with submergence ratios of 10:1 and 100:1, respectively, had completed their drift to the Japanese coast 1 year after introduction.
- Materials with more of their volume above water than submerged were affected by wind, and materials with more of their volume submerged were affected by ocean currents. Hence, materials with higher submergence ratios would be faster to be drifted and beached on the seashore than with lower ratios. Drift depended on submergence ratio.

No retention area was detected during the 3-year simulation. However, there was evidence of turning or drifting (or both) of materials in the central area of the Sea of Japan 1 or 2 years after introduction. This is because RIAMOM reproduced medium-sized vortices well, and this successfully indicated the fate of small materials, such as disposable lighters floating nearly entirely submerged.

Using simulation data developed by the drifting-garbage-prediction model, the numbers of beached materials were counted for 3 years. Almost all drifting materials, with different submergence ratios from 0:1 to 100:1, drifted to Hokkaido, Yamagata, Fukui, and Shimane (Table 19.3 and Figure 19.7). Very small numbers of materials drifted to Korea. The distributions of beached materials were visualized by month. A typical distribution (simulation data for December) is shown in Figure 19.8.

- Almost all disposable lighters with a submergence ratio of 0:1 were calculated to have drifted to the whole of the Japanese coast on the Sea of Japan – but mainly to the coast of the Tsugaru Strait – no matter the month.
- Almost all disposable lighters with submergence ratios of 1:2 or 1:1 were also calculated to have drifted to the whole of the Japanese coast on the Sea of Japan – but mainly to the coast of the Tsugaru Strait and Hokkaido – except July, August, and September.
- Almost all polytanks with submergence ratios of 10:1 were calculated to have drifted to the coasts of Tohoku and western Japan between November and March; they then turned northward and drifted to the North Korean coast and Russian coast from June to July of the next year. Thereafter, the polytanks turned eastward and drifted down to Hokkaido via

TABLE 19.3
Predicted Numbers of Materials Beached on Each Coast around the Sea of Japan over 3 Years

Station	Area Name	Disposable Lighters			Polytanks	PET Bottles
		(submergence ratio 0:1)	(submergence ratio 1:2)	(submergence ratio 1:1)	(submergence ratio 10:1)	(submergence ratio 100:1)
A	Hokkaido (Japan)	317	217	212	117	122
B	Yamagata (Japan)	115	131	122	173	137
C	Fukui (Japan)	29	80	188	194	145
D	Shimane (Japan)	18	27	32	61	86
E	Fukuoka (Japan)	0	10	7	26	88
F	Korea (South)	7	2	4	6	10
G	Korea (North)	1	2	2	4	15
H	North Korea (East)	3	0	2	11	16
I	Russia (South)	0	23	33	61	95
J	Russia (North)	0	5	10	71	93

Source: Ministry of the Environment, FY 2006 International reduction measures survey report on floating marine litter and beach litter (in Japanese), 2006b.

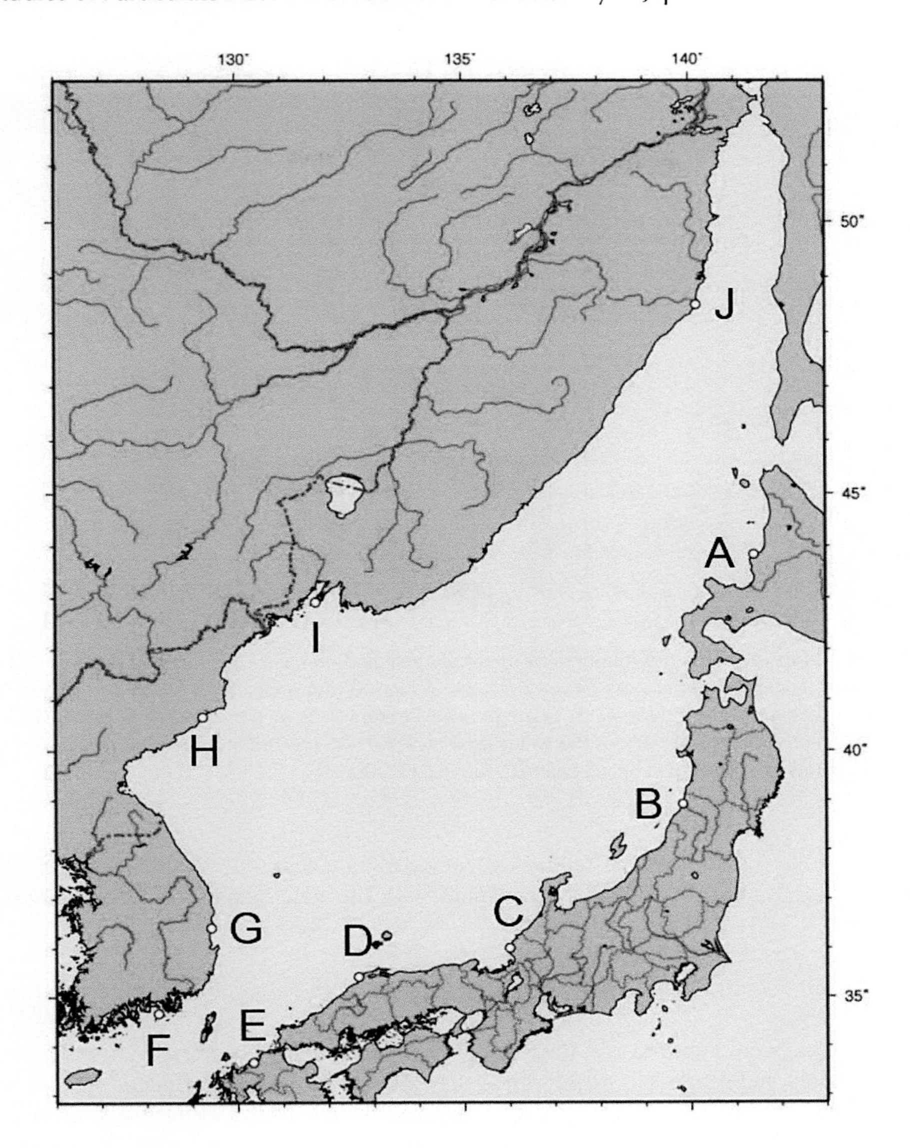

FIGURE 19.7 Map of monitoring sites for prediction numbers of materials beached on each coast around the Sea of Japan over 3 years shown in Table 19.3. (Data adapted from Ministry of the Environment, FY 2006 International reduction measures survey report on floating marine litter and beach litter [in Japanese], 2006b.)

Sakhalin between October and November. PET bottles with a submergence ratio of 100:1 followed two paths – one to the east coast of South Korea and the other to Sakhalin – in September. They then drifted back to the coasts of Tohoku and western Japan.

19.2.5.2 Drift of Disposable Lighters Released from Different Countries Bordering the Sea of Japan

The Ministry of the Environment prepared a simulated drift scenario of disposable lighters, poly-tanks with lids, and PET bottles with lids (Ministry of the Environment 2006b). After several simulations and comparisons with field survey results in the Sea of Japan, a simulation using

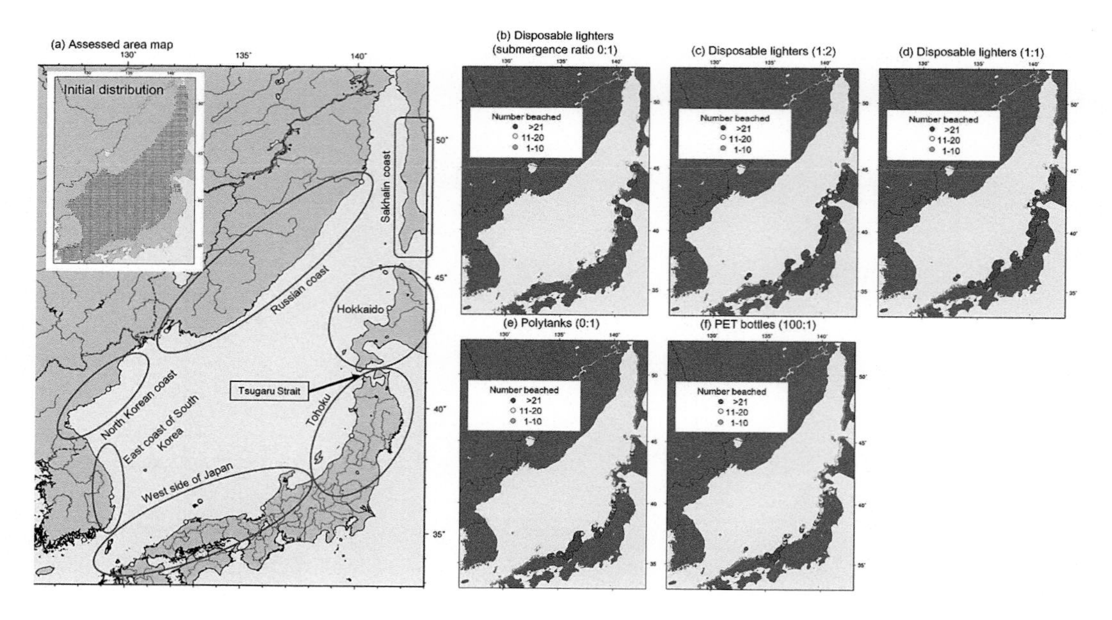

FIGURE 19.8 Maps of predicted numbers of materials beached on each coast around the Sea of Japan for 3 years (simulation data for winter (as December)). (a) Assessed area map, (b) disposable lighters (submergence ratio 0:1), (c) disposable lighters (1:2), (d) disposable lighters (1:1), (e) polytanks (0:1), and (f) PET bottles (100:1). (Data adapted from Ministry of the Environment, FY 2006 International reduction measures survey report on floating marine litter and beach litter [in Japanese], 2006b.)

drifting disposable lighters with different percentage abundances – 15% (submergence ratio 0:1), 75% (1:2), and 10% (1:1) – gave good agreement with the field survey results. The simulation revealed that:

- If the source were Japan, most disposable lighters were assumed to have drifted to the coast of Japan
- If the source were South Korea, the disposable lighters would have drifted to the coasts of South Korea and Japan. No matter when they were released, in particular, most of them would have drifted to the entire coast of Japan in winter; to the western side of Japan in spring; and to the coasts of Tohoku and Hokkaido in autumn
- If the source were North Korea, disposable lighters would have drifted to the coast of North Korea in summer and to the coast of Japan in winter. The range of drift was not the whole coast of Japan, but mainly Tohoku
- If the source were Russia, almost all disposable lighters would have drifted to the coasts of Russia and Japan. They would have drifted to near Vladivostok in spring, summer, and autumn and to Tohoku and Hokkaido in winter
- If the source were China, it was assumed that waste disposable lighters would have passed through the Tsushima Strait and flowed into the Sea of Japan. Most would have drifted to the coast of Japan, and some to the coast of South Korea

19.2.5.3 Origins of Plastic Debris Drifting to Coastlines on the East China Sea

In FY 2006, in another simulation (see Section 19.2.5.1), trash was arranged over in the Sea of Japan (every 1/6th of a degree), and the drift characteristics of disposable lighters were modeled (Figure 19.6. Ministry of the Environment [2006b]). In another report by the Ministry of the

Environment (2007), a route was generated for the drift of disposable lighters from each city located on the coasts of Japan, South Korea, North Korea, China, and Taiwan over 6 years by using RIAMOM. The input frequency of disposable lighters in the ocean was assumed to be one per day for every 200,000 people. The same submergence ratios as used previously were used.

After several calculations using different input conditions (e.g., submergence ratio of plastic items, release place of plastic items, seasonal condition of ocean), the simulation scenario with the best fit with actual survey data was chosen. To confirm the calculations, a field survey of disposable lighter drift was performed from August 2003 to January 2006 in eight areas bordering the East China Sea (Figure 19.9a). A larger number of disposable lighters obviously coming from China and Taiwan drifted onto the southern islands (A to E) than to Kagoshima (F) or areas farther north (Figure 19.9b). In an analysis of drift paths to the west coast of Kagoshima (F), disposable lighters drifting from overseas were numerous in summer. Most of them went northward from the south with the Tsushima warm current, and then branched off south of Fukueshima (see the location in Figure 19.1); the majority drifted onto the west coast of Kagoshima (Figure 19.9c). In an analysis of drift paths to Okinawa (D), disposable lighters drifting there from overseas were numerous in autumn (Figure 19.9d). The origins and drift paths therefore varied. Some lighters drifted far away into the Pacific Ocean. There was little drift onto the north side of Okinawa; however, the majority of lighters drifted onto the south side of Okinawa.

In other calculations, the origins of empty polytanks and PET bottles were simulated by using submergence ratios of 10:1 and 100:1, respectively. Island sites, with the exception of Yonagunishima (A) and Haterumashima (B), received polytanks mainly from Japan. Yonagunishima (A) and Haterumashima (B) were estimated to receive polytanks mainly from China and Taiwan (more than 90% and 75%, respectively) (Data not shown). All islands except for Ishigakishima (C) received PET bottles mainly from Japan. More than 50% of the PET bottles received by Ishigakishima (C) came from China and Taiwan. Although Haterumashima and Ishigakishima are located close to each other (Figure 19.9a), the drifted items were not always same, that is likely depending on the submergence ratio.

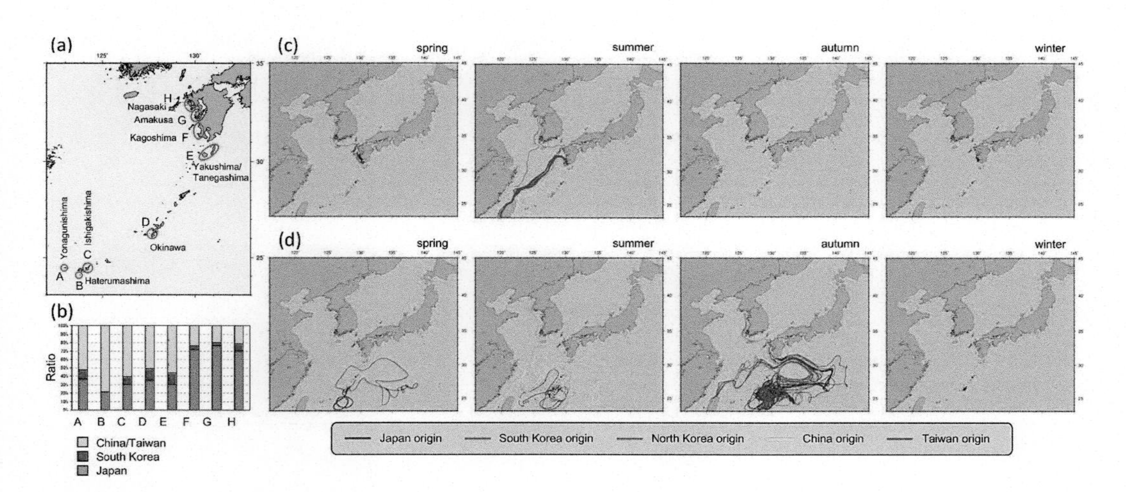

FIGURE 19.9 Research area, disposable lighters found, and best-fit simulated drift scenario in the East China Sea. (a) Area of research. (b) Ratios (%) of disposable lighters found in each research area (A to H, August 2003 to January 2006). (c) Seasonal drift paths of disposable lighters toward the west coast of Kagoshima (F). (d) Seasonal drift paths of disposable lighters toward Okinawa (D). (Data adapted from Ministry of the Environment, FY 2007 International reduction measures survey report on floating marine litter and beach litter [in Japanese], 2007.)

19.2.5.4 Drift of Plastic Debris from Japan to the North Pacific Ocean

A report by the Ministry of the Environment (2010b) used RIAMOM to model the route of drift of plastic debris from all over Japan to the North Pacific Ocean (submergence ratios 0:1, 10:1, and 100:1) over 6 years under wind pressure flow. In this calculation, debris was released over 6 years continuously, and three submergence ratios were used, namely, 0:1 (considered items: fishing nets, and not damaged disposable lighters with vertical position), 10:1 (considered items: PET bottles without lids, fishing buoys, and polytanks), and 100:1 (considered item: expanded polystyrene).

As shown in Section 19.2.5.1 (Figure 19.6), drift depended on the submergence ratio. First, plastic debris with a submergence ratio 0:1 released from Japan drifted onto the Japanese coast (Figure 19.10a). Some drifted to the North Pacific Ocean (1 year later) and rode the North Equatorial Current and the Equatorial Counter Current (3 and 6 years later). Two years after release, the pieces of debris spread out in the North Pacific Ocean, and then they started to drift along the west coast of North America (3 years later). This diffusion did not change, even 6 years after the release (Figure 19.10a). Second, fewer items of plastic debris with a submergence ratio of 10:1 were released from Japan, but they took the same route. However, almost none of these items drifted onto the west coast of North America, even after 6 years (Figure 19.10b). Last, much less plastic debris with a submergence ratio of 100:1 was released from the Japanese archipelago. Even 6 years later, almost nothing had been released, and almost all of it remained on the Japanese coast, although some debris drifted onto the west coast of North America (Figure 19.10c).

19.2.6 Microplastics

Japan's Ministry of the Environment first mentioned the issue of microplastics from plastic debris in the ocean, and their environmental impacts, in FY 2013 (Ministry of the Environment 2013). A national survey of microplastics was performed at the same seven stations as used in the previous survey of beached garbage (Table 19.2). Waste plastics, which account for 70% of artificial marine waste, were classified by size (δ, which is either the length or diameter), namely, as macroscopic plastics (those remaining in their original shape), mesoplastics (δ > 5 mm; fine pieces), microplastics (5 mm > δ > several microns),

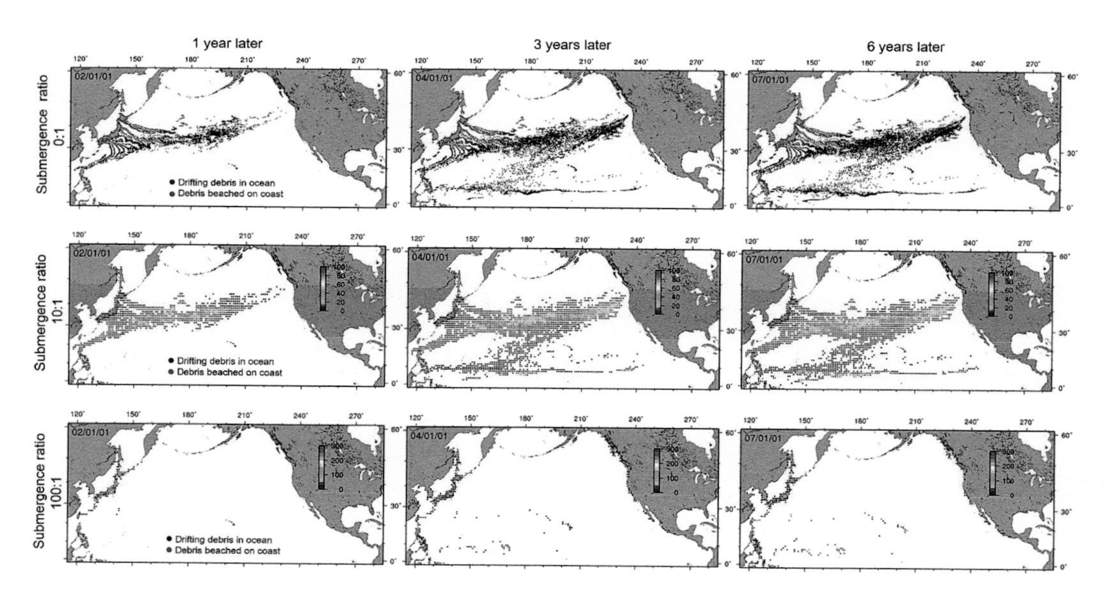

FIGURE 19.10 Scenario of drift of plastic debris from Japan to the North Pacific Ocean. (Data adapted from Ministry of the Environment, FY 2010 Investigation of causes, foreign outflow investigation report on floating marine litter and beach litter [in Japanese], 2010b.)

and nanoplastics (δ < several microns) (Andrady 2011; Cole et al. 2011). Included among micro and nanoplastics are primary products (primary microplastics) that are mixed into products, such as face-washing agents and toothpaste during the manufacturing process, and then leak out after use. In addition, a serious problem occurs with the production of secondary products (secondary microplastics) by the breakdown and weathering of plastic waste along the coast. From a physical perspective, to examine the process of drift of these fine plastic pieces, the Ministry of the Environment conducted a full-scale survey of microplastics floating on the ocean surface (Ministry of the Environment 2014a, 2015a, 2016b). In this survey, from FY 2014 to FY 2016, two training vessels (*Umitaka-maru* and *Shinyou-maru*) from Tokyo University of Marine Science and Technology were employed. Research staff gathered fine plastic pieces, polystyrene pieces, and lint (weathered chemical fibers) from the seas around Japan's four main islands and examined their drift status (Figure 19.11). Measurement points were defined randomly along a circular route of the Japanese archipelago. In addition, in FY 2016, the annual change in buoyant density in Sagami Bay was examined by research staff on the *Seiou-maru*.

Microplastics were collected with a Neuston net with a drainage meter [Japan Meteorological Association Neuston net no. 5552: opening 75 cm square (0.56 m^2); net length 300 cm; mesh size 350 μm]. A 20-minute trawl with the net at 2–3 knots was performed to collect microplastics. Large impurities were removed from the net-collected samples and sieved through a 2.0-mm sieve, followed by a 300-μm sieve. After being sieved, the plastics were classified visually and manually put into petri dishes. Still smaller pieces were sorted by suction filtration, and then separated. The composition of fine pieces smaller than 1 mm was identified by Fourier-transform infrared spectroscopy because it was difficult to identify. The lower size limit for identification was 0.3 to 0.4 mm. Maximum length of the microplastics was measured with an optical microscope and image processing software. At each measurement point, the sizes of all fine pieces were measured. The size range of the measured fine pieces was 0.3 mm < δ < 30 mm. The number per unit volume of seawater was

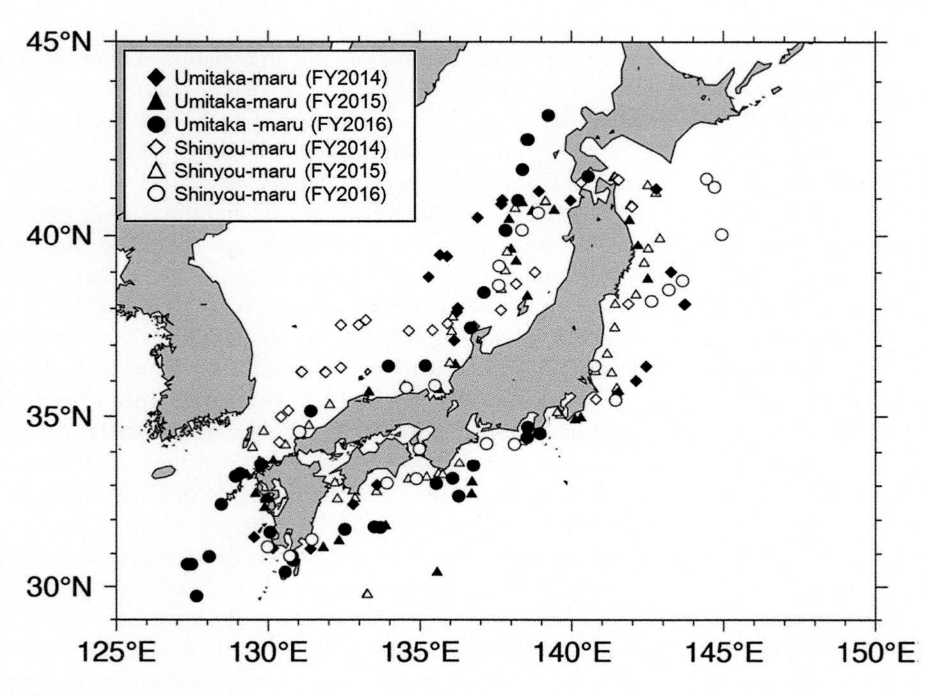

FIGURE 19.11 Map of microplastic surveys from FY 2014 to FY 2016. (Data adapted from Ministry of the Environment, FY 2016 Research report on marine debris floating on the ocean surface and settled on the seabed in offshore areas around Japan [in Japanese], 2016b.)

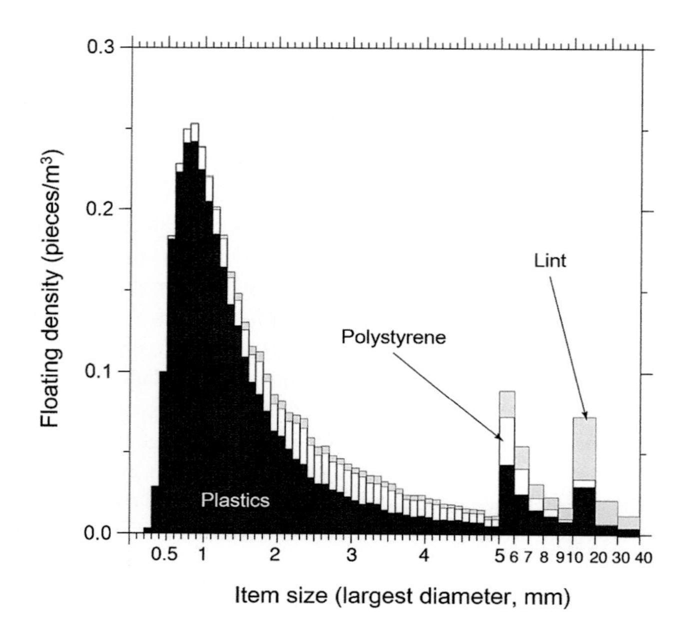

FIGURE 19.12 Size-dependent distributions of plastic (mesoplastics and microplastics), expanded polystyrene, and lint (weathered chemical fiber) per unit volume of seawater (floating density, pieces/m^3). (Data adapted from Ministry of the Environment, FY 2016 Research report on marine debris floating on the ocean surface and settled on the seabed in offshore areas around Japan [in Japanese], 2016b.)

determined from the volume of drained water and the number of fine plastic pieces in each sea area. Not only plastics, but also polystyrene and lint, were detected.

The size (largest diameter or length)-dependent distributions per unit volume of seawater (floating density, pieces/m^3) of plastics (mesoplastics and microplastics), expanded polystyrene, and lint were plotted (Figure 19.12). The floating density of plastic pieces increased as the size decreased, likely because many minute pieces were formed by repeated fragmentation of larger pieces as they deteriorated. However, when the size fell below 0.8 mm, the floating density decreased. In addition, there were two peaks 5–6 and 10–20 mm of plastics, expanded polystyrene, and lint. Unlike plastics, the floating density of expanded polystyrene pieces did not increase as the size decreased. However, the polystyrene floating density ceased at less than 0.5 mm of greatest diameter. The floating density of lint decreased as the size decreased, and no lint was detected with less than 1.0 mm of length.

The floating densities of microplastics, expanded polystyrene, and lint less than 5 mm in size (largest diameter or length) detected from 2014 to 2016 were determined (Table 19.4). The mean floating density (2.7 pieces/m^3) of microplastics was slightly less compared with 3.7 piece/m^3 in FY 2014 (Isobe et al. 2017). In addition, compared with 1.2 pieces/m^3 in FY 2014, the floating density of expanded polystyrene decreased considerably. The floating densities of all three items changed little between 2015 and 2016. To examine the distribution of microplastics around Japan, the spatial distribution of microplastics in the 3 years from 2014 to 2016 were integrated. The floating density of microplastics was higher in the northern part of the Sea of Japan, off Hokuriku and Tohoku (Figure 19.13a), than in southwestern Japan. In addition, relatively high floating densities were observed off the western coast of Japan (the Sea of Japan side) and the Pacific coast of southern Japan. High floating densities were observed from the Tsugaru Strait to the west coast. Considering that most of the surveys were conducted during summer, it was inferred that the microplastics were

TABLE 19.4
Floating Densities (pieces/m³) of Microplastics, Expanded Polystyrene, and Lint (Weathered Chemical Fiber) Less Than 5 mm in Size (Greatest Diameter or Length)

Fiscal Year	2014	2015	2016	Mean in 3 Years
Microplastics	3.7	2.4	2.1	2.7
Expanded polystyrene	1.2	0.2	0.32	0.57
Lint	0.13	0.06	0.09	0.09

Source: Ministry of the Environment, FY 2014 Research report on marine debris floating on the ocean surface and settled on the seabed in offshore areas around Japan (in Japanese), 2014a; Ministry of the Environment, FY 2015 Research report on floating marine litter and seabed marine litter in coastal areas around Japan (in Japanese), 2015a; Ministry of the Environment, FY 2016 Research report on marine debris floating on the ocean surface and settled on the seabed in offshore areas around Japan (in Japanese), 2016b.

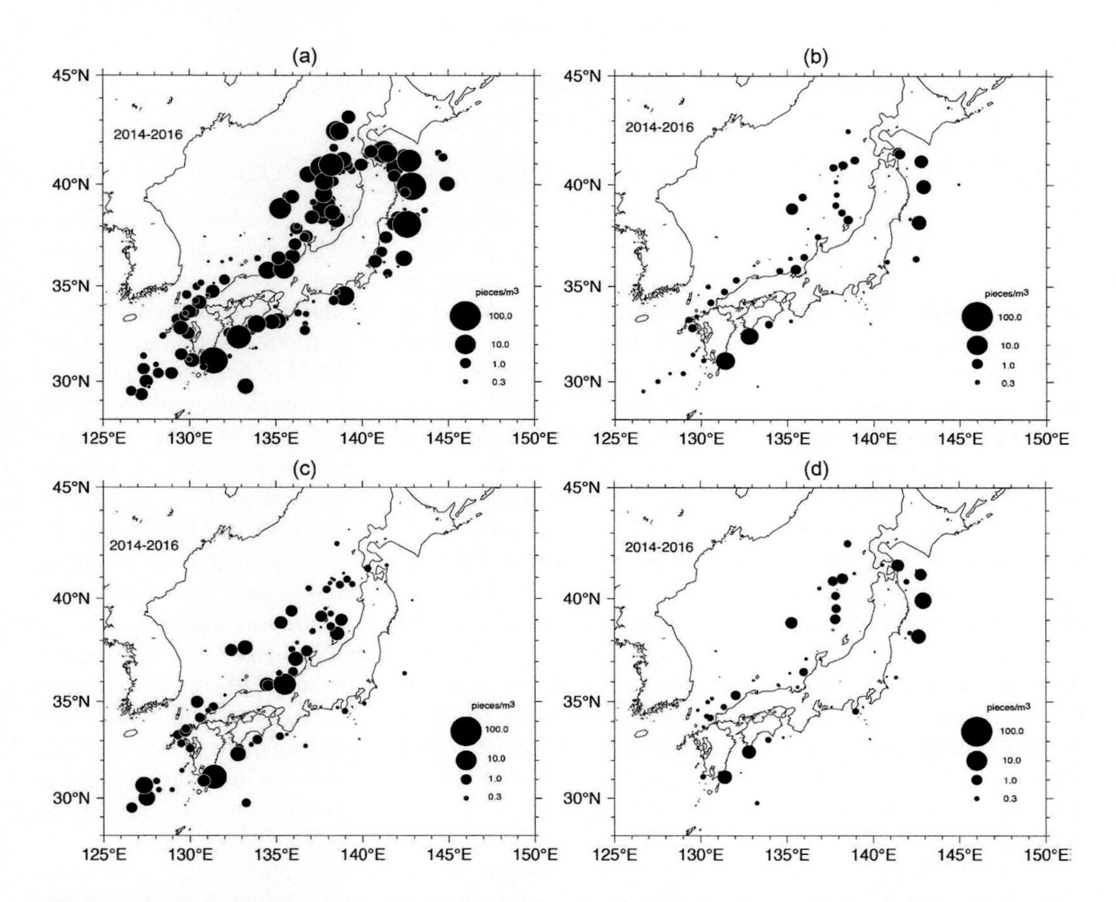

FIGURE 19.13 Spatial distributions of microplastics (a), mesoplastics (b), expanded polystyrene less than 5 mm in largest diameter (c), and lint (weathered chemical fiber) less than 5 mm in length (d). (Data adapted from Ministry of the Environment, FY 2016 Research report on marine debris floating on the ocean surface and settled on the seabed in offshore areas around Japan [in Japanese], 2016b.)

transported to ocean currents at that time of year (the Tsugaru Warm Current passing northward from the Tsugaru Strait to the Pacific Ocean, and the Japan Current traveling east along the south coast of Japan). Although it is impossible to conclude the cause of the very large floating densities of microplastics in the northern part of the Japan Sea from these survey results alone, there is likely to have been selective and intensive dumping of waste plastics somewhere in the northern part of the Sea of Japan. This may represent a transitional distribution of microplastics from winter to summer. It will be important to continue these investigations in the future through changes in the seasons. The distribution of mesoplastic floating densities (Figure 19.13b), although these particles were much less numerous, was roughly similar to that of microplastics.

Interestingly, unlike plastic pieces, which were detected in most areas around the Japanese archipelago, expanded polystyrene pieces were scarcely detected outside the Sea of Japan, except at one point in Kagoshima (Figure 19.13c). It is obvious that polystyrene foam products used intensively in the catchment area of the Tsushima Current flowing into the Sea of Japan or the East China Sea of Japan were the source of the Kagoshima foam. In the future, we will need to specify the source in accordance with the results of visual observations of floating drifting objects. The distribution of lint was similar to the distribution of microplastics, although much more sparse (Figure 19.13d).

19.3 ENVIRONMENTAL FATE AND TOXICOLOGICAL RISKS OF PARTICULATE PLASTICS IN THE OCEAN

19.3.1 Chemical Contamination of Plastic Debris Drifting in the Oceans

Because of the chemical hydrophobic properties of plastics, there is concern about their adsorption of inorganic or organic pollutant chemicals and ecotoxicological risks, as well as about obstruction of the digestive tracts of aquatic organisms by plastic debris through accidental consumption. In addition, there is environmental concern about additives, such as plasticizers and flame retardants contained in plastics. Nakashima et al. (2012) detected chromium (Cr), cadmium (Cd), tin (Sn), antimony (Sb), and lead (Pb) in PVC buoys collected from the Ogushi coast of the Goto Islands of Nagasaki Prefecture. They also measured the elution flux of Pb. The amount of Pb contained in plastic garbage on the Ogushi coast was estimated to be 23 ± 11 g in polyethylene and 284 ± 247 g in PVC. The total found in this and other collected garbage was estimated to be 313 ± 247 g. Pb was present at the highest concentration among the metals analyzed. The detected concentrations of Cr, Cd, and Sb were below the lower limits of quantification. Also, in a metal elution test using collected PVC buoys, the elution flux of Pb was estimated to be $0.45 \pm 0.45 \times 10^{-3}$ g/h and the annual elution was estimated to be 0.6 ± 0.6 g/year. Besides concern about heavy metal contamination, there is concern about contamination by persistent organic pollutants (POPs, such as polychlorinated biphenyls [PCBs], polybrominated diphenyl ethers [PBDEs], hexachlorocyclohexanes [HCHs], and dichlorodiphenyltrichloroethane [DDT]), polycyclic aromatic hydrocarbons, and perfluorooctanoic acid. In a preliminary survey conducted by the Ministry of the Environment in FY 2014, concentrations of PBDEs and PCBs, which are flame retardants and are adsorbed as POPs onto drifting resin pellets, were detected to be much higher in Shimonoseki (10 ng/g-resin pellet and 30 ng/g-resin pellet, respectively) than in other locations (0.17 to 0.40 ng/g-resin pellet and 0.66 to 7.6 ng/g-resin pellet, respectively) (Ministry of Environment 2014b).

In FY 2015 and FY 2016, the Ministry of the Environment conducted a national survey of coastal drift resin pellets for POP analysis (Ministry of the Environment 2015b, 2016c). A total of 28 stations (st.) were set up around Japan, and pellets on the beach were sampled from each coastal location for chemical analysis (Figure 19.14). The concentration of total PCBs (ΣPCBs) ranged from 0.9 ng/g-plastic pellet (Miyakojima, st. 27) to 942.5 ng/g (Hakotsukuri, st. 15) (Table 19.5). Stations 7–11, 14, 15, and 20, where there are representative industrial areas, had more than 100 ng/g. Stations with concentrations exceeding 500 ng/g were located in inner bays, but these high concentrations were not obtained in the FY 2016 survey. Compared with the FY 2015 data, the FY 2016 results were

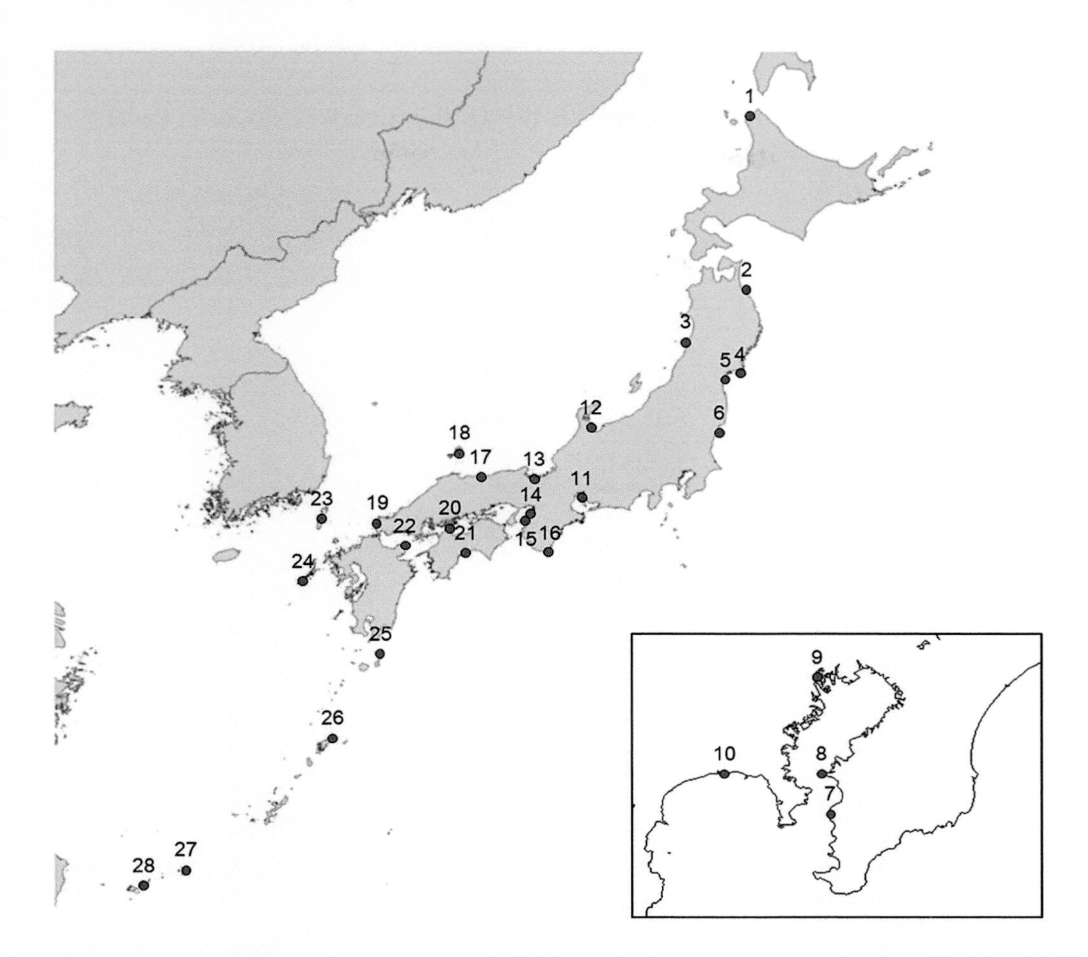

FIGURE 19.14 Map of stations used for sampling of coastal drift resin pellets for persistent organic pollutant analysis shown in Table 19.5 (FY 2015–2016). (Data adapted from Ministry of the Environment, FY 2015 Comprehensive investigation on measures to tackle beach debris [in Japanese], 2015b; Ministry of the Environment, FY 2016 Comprehensive investigation on measures to tackle beach debris [in Japanese], 2016c.)

about the same, except in the inner bays adjacent to urban areas, such as Tokyo Bay and Osaka Bay. The results are roughly comparable to those observed in other developed countries and are consistent with global trends so far (International Pellet Watch). Also, at Wakkanai (st. 1) and Okinoshima (st. 18), where industrial activity was less than in metropolitan areas even before PCB manufacture and use were banned, the concentrations of PCBs in the collected pellets were higher than those at other stations located in similar rural areas (e.g., st. 21). It was possible that in such cases the PCBs were transported from remote sources of contamination. This trend occurred in the case of ΣDDT. The detected range of ΣDDT was from 2.3 ng/g (Sendai, st. 5) to 147.1 ng/g (Imabari, st. 20). These data were similar to the global contamination levels of 0.1 to 1061 ng/g (International Pellet Watch). The detected HCH concentrations ranged from 0.1 ng/g (Odaiba, st. 9 and Kushimoto, st. 16) to 16.6 ng/g (Kanaya, st. 7), although none were detected at Goto (st. 24) (Table 19.5).

19.3.2 Fate of Chemical-Contaminated Plastics in Fish

There has been concern about the distribution and biological effects of particulate plastics in fish, because plastics can absorb chemical contaminants. Although the fate of xenobiotics in fish has

TABLE 19.5
Concentrations of Persistent Organic Pollutants (POPs) in Coastal Drift Resin Pellets (FY 2015–2016)

			POPs (ng/g-plastic pellet)			
Station #	**Location (Prefecture)**	**Year**	**ΣPCBs**	**ΣDDTs**	**ΣHCHs**	**ΣPBDEs**
1	Wakkanai (Hokkaido)	2016	89.5	2.6	1.4	NA
2	Hachionohe (Aomori)	2016	14.4	7.3	0.7	NA
3	Yuza (Yamagata)	2016	14.1	7.6	0.9	NA
4	Kesenuma (Sendai)	2016	7.0	9.5	1.7	NA
5	Sendai (Miyagi)	2015	31.6	4.8	1.3	NA
5	Sendai (Miyagi)	2016	9.0	2.3	0.5	NA
6	Onahama (Fukushima)	2015	9.5	9.6	0.8	NA
7	Kanaya (Chiba)	2015	213.6	24.5	16.6	NA
8	Nunobiki (Chiba)	2015	670.9	NA	NA	0.0
9	Odaiba (Tokyo)	2015	357.6	23.4	0.1	NA
10	Syonan (Kanagawa)	2015	509.1	36.6	2.0	NA
11	Fujimaehigata (Aichi)	2016	187.6	16.0	0.9	NA
12	Toyama (Toyama)	2016	3.1	3.4	0.6	NA
13	Kyoto (Kyoto)	2016	23.9	5.7	1.3	NA
14	Nishikinomana (Osaka)	2015	502.7	21.9	0.8	NA
15	Hakotsukuri (Osaka)	2015	942.5	23.8	0.4	NA
16	Kushimoto (Wakayama)	2015	70.3	28.6	0.1	NA
17	Tottori (Tottori)	2016	25.2	13.7	3.0	NA
18	Okinoshima (Shimane)	2016	52.8	21.0	5.1	NA
18	Okinoshima (Shimane)	2015	52.8	9.6	5.1	NA
19	Shimonoseki (Yamaguchi)	2015	9.4	12.2	6.1	NA
20	Imabari (Ehime)	2015	201.1	147.1	2.2	NA
21	Kouchi (Kouchi) (polyethylene)	2015	6.4	NA	NA	1.2
21	Kouchi (Kouchi) (polypropylene)	2015	4.5	NA	NA	0.9
22	Kunisaki (Oita)	2015	63.4	3.0	0.4	NA
23	Tsushima (Nagasaki)	2016	1.2	NA	NA	1.1
24	Goto (Nagasaki)	2016	5.2	5.8	ND	NA
25	Tanegashima (Kagoshima)	2015	9.4	17.4	0.2	NA
26	Amamioshima (Kagoshima)	2015	6.4	3.1	0.4	NA
27	Miyakojima (Okinawa)	2015	0.9	2.8	0.5	NA
28	Ishigakishima (Okinawa)	2015	2.7	5.4	3.3	NA

Source: Ministry of the Environment, FY 2015 Comprehensive investigation on measures to tackle beach debris [in Japanese], 2015b; Ministry of the Environment, FY 2016 Comprehensive investigation on measures to tackle beach debris [in Japanese], 2016c.

Note: NA, not analyzed. ND, not detected; According to the method of International Pellet Watch (IPW, www.pelletwatch.org), median values of the concentrations in five samples at each point are given.

been studied over the last few decades, to my knowledge, there has been only one study of the distribution of particulate plastics in fish. To observe the distribution of particulate plastics in fish bodies, Kashiwada (2006) used 40-nm-size fluorescent polystyrene particles and see-through medaka (*Oryzias latipes*). He succeeded in visualizing the distribution of polystyrene particles in these fish. The particles were distributed mainly in the intestine, but also in the gills, blood, liver, kidney, gonads, and brain. A particle 40 nm in size was thus small enough to penetrate through the intestinal or gill biological membranes and reach each organ via the blood. Although we have limited information on the size range of materials that are able to penetrate biological membranes, generally, such penetration should be more difficult for larger materials. Recently, Tanaka and Takada (2016) reported that Japanese anchovies (a planktivorous fish) caught in Tokyo Bay had microplastics and microbeads in their digestive tracts. Plastics were detected in 49 out of 64 fish (52.0% polyethylene and 43.3% polypropylene), with an average of 2.3 pieces (up to 15 pieces) per individual. The size range of most (80%) detected plastics was 150–1000 μm. The reported concentration of PCBs (ng/g) in pellets on a beach in Tokyo was <28 to 2300 ng/g (Endo et al. 2005). Polyethylene resin pellets weigh about 0.1 g/40 pieces (Teuten et al. 2009), which means Japanese anchovies could contain an average of 0.00575 g of plastic and <0.161 to 13.225 ng PCBs per individual. To date, there is limited information of the toxicity of PCBs for humans. The lowest oral LD50 in the rat was 1.0 g/kg body weight for Aroclor 1254 (World Health Organization 1993). Hence, at this point, anchovies, and even humans as consumers of anchovies, may not be suffering from any toxic effects from PCBs in plastics. Even if resin pellets were to have high concentrations of PCBs, it would not always mean that they would have ecotoxicological effects on the organisms in which they were found. The concentration of contaminated chemicals in the pellets means that they have passed through a contaminated environmental area. Because fish take up xenobiotics mainly through their gills from the ambient water, it is well known that the contribution of contaminated ambient water is much higher than those of other environmental/ecological factors. Rather than just considering the toxicological effects of xenobiotics in microplastics, we therefore must also consider the physiological and physical effects of microplastics on fish gills and intestines. Also, if one has found chemically contaminated plastics in an ocean and has information of the country of origin and/or area of the plastics and ocean currents, one could give some political recommendation for environmental protection to the country and/or area. Furthermore, people are unlikely to be willing to eat fish that contain plastics, even if there were no problem with toxicity. This is another issue that could affect public perceptions.

19.4 PARTICULATE PLASTICS AND HUMAN HEALTH

In 2009, a new environmental law (Act on Promoting the Treatment of Marine Debris Affecting the Conservation of Good Coastal Landscapes and Environments to Protect Natural Beauty and Variety) came into force to protect seas and seashores in Japan and manage and minimize marine debris. In 2018, this law was revised to consider further the marine environment (Act on Promotion of Disposal of Articles Washed Ashore for Good Coastal Views and Environment for Conserving Beautiful Rich Nature (Act No. 82 of 2009; Washed-Ashore Articles Disposal Promotion Act/2009 Legislation Introduced by Diet Members). The law is based on concepts of conservation and restoration of comprehensive coastal environments; clarification of the responsibility for, and promotion of, efficient processing of marine debris; effective suppression of marine debris by promotion of the 3Rs (reuse, reduce, and recycle); conservation of marine environments (including addressing the microplastics issue); diverse and appropriate assignment of roles and coordination for achievement of conservation of marine environments; and promotion of international cooperation. In regard to microplastics, the law indicates clearly that the industry should strive to suppress the use of microplastics in products, as well as the discharge of waste plastics, which will run off into rivers; the government is to act promptly to consider how to manage marine microplastics on the basis of scientific knowledge and international trends, and to take necessary actions.

Marine plastic debris is a global environmental problem. However, so far, we have no idea of the nature of the major ecological impact of marine plastics and particulate plastics. Although the impact is still unclear, to protect human health from indeterminate environmental hazards and risks, there is an obvious need to avoid mismanagement of plastic waste and leakage of plastics from land to ocean. In Asia, contamination by plastics is more severe than elsewhere because of the input from China and other developing countries. Without better political environmental regulation and education, plastic contamination is likely to continue to worsen. It is very important to manage plastic waste, but Asia is composed of 48 countries and more than 4 billion people. There are different civilizations, cultures, and lifestyles. Hence, it is impossible to control waste management by using a united conceptual method. Instead, to understand the ecological impacts of environmental particulate plastics, we need to understand and regulate the life cycles of plastics, simulate the fate of global plastics, understand their biological effects, and perform ecological risk assessments. For sustainable development by humans, we need comprehensive investigations of plastics.

REFERENCES

Andrady, A.L. (2011). Microplastics in the marine environment. *Marine Pollution Bulletin* 62(8): 1596–1605.

Bonefeld-Jørgensen Eva, C., Long, M., Hofmeister Marlene, V., and Vinggaard Anne, M. (2007). Endocrine-disrupting potential of bisphenol A, bisphenol A dimethacrylate, 4-n-nonylphenol, and 4-n-octylphenol in vitro: New data and a brief review. *Environmental Health Perspectives* 115(1): 69–76.

Cole, M., Lindeque, P., Halsband, C., and Galloway, T.S. (2011). Microplastics as contaminants in the marine environment: A review. *Marine Pollution Bulletin* 62(12): 2588–2597.

Council for PET Bottle Recycling. PET BOTTLE RECYCLING 2018 (in Japanese).

Endo, S., Takizawa, R., Okuda, K., Takada, H., Chiba, K., Kanehiro, H., Ogi, H., Yamashita, R., and Date, T. (2005). Concentration of polychlorinated biphenyls (PCBs) in beached resin pellets: Variability among individual particles and regional differences. *Marine Pollution Bulletin* 50(10): 1103–1114.

House of Councilors proceedings record information. (1970). The 064th National Diet of Pollution Control Special Committee No. 3 (in Japanese).

IARC: IARC Monograph on the Evaluation of Carcinogenic Risk to Humans, 1987, Supplement 7.

International Pellet Watch. www.pelletwatch.org (accessed April 4, 2019).

Isobe, A., Uchiyama-Matsumoto, K., Uchida, K., and Tokai, T. (2017). Microplastics in the Southern Ocean. *Marine Pollution Bulletin* 114(1): 623–626.

Jambeck, J.R., Geyer, R., Wilcox, C., Siegler, T.R., Perryman, M., Andrady, A., Narayan, R., and Law, K.L. (2015). Plastic waste inputs from land into the ocean. *Science* 347(6223): 768–771.

Japan Environmental Agency (1998). Strategic Problem on Environmental Endocrine Disruptors '98 (SPEED) '98 (in Japanese). www.env.go.jp/chemi/end/speed98.html (accessed May 18, 2019).

Japan Coast Guard (2006). Annual report of current marine pollution in 2006 (in Japanese).

Japan Expanded Polystyrene Association (2017) (in Japanese) www.jepsa.jp/recycle/results.html (accessed April 4, 2019).

Japan Meteorological Agency (2009). Annual Report on Atmospheric and Marine Environment Monitoring. Observation Results for 2007. Chapter 6, Marine Pollution, pp. 197–206 (in Japanese).

Japan Petrochemical Industry Association (2018). Current Petrochemical Industry (annual statistical data book) for 2018 (in Japanese).

Kasai, E. (2008). Recent research trends on the dioxin emissions in high temperature process. *Journal of High Temperature Society* 34(3): 104–110 (in Japanese).

Kashiwada, S. (2006). Distribution of nanoparticles in the see-through medaka (*Oryzias latipes*). *Environmental Health Perspectives* 114(11): 1697–1702.

Lebreton, L.C.M., van der Zwet, J., Damsteeg, J.-W., Slat, B., Andrady, A., and Reisser, J. (2017). River plastic emissions to the world's oceans. *Nature Communications* 8: 15611.

Lee, H.J. (2003). Comparison of RIAMOM and MOM in modeling the East Sea/Japan Sea circulation. Comparison of RIAMOM and MOM in modeling the East Sea/Japan Sea circulation. *Ocean Polar Research* 25(3): 287–302.

Matsumura, S., and Nasu, K. (1997). *Distribution of Floating Debris in the North Pacific Ocean: Sighting Surveys 1986–1991.* Berlin, Germany, Springer.

Ministry of the Environment (2003). A report of the first Meeting for Endocrine Disrupting Chemicals Study Group (in Japanese). www.env.go.jp/press/press.php?serial=4159 (accessed May 18, 2019).

Ministry of the Environment (2006a). FY 2006 Basic research report on drifted driftwood (in Japanese).

Ministry of the Environment (2006b). FY 2006 International reduction measures survey report on floating marine litter and beach litter (in Japanese).

Ministry of the Environment (2007). FY 2007 International reduction measures survey report on floating marine litter and beach litter (in Japanese).

Ministry of the Environment (2010a). FY 2010 Research report on measures to tackle beach debris (in Japanese).
Ministry of the Environment (2010b). FY 2010 Investigation of causes, foreign outflow investigation report on floating marine litter and beach litter (in Japanese).
Ministry of the Environment (2011). FY 2011 Research report on measures to tackle beach debris (in Japanese).
Ministry of the Environment (2012). FY 2012 Research report on measures to tackle beach debris (in Japanese).
Ministry of the Environment (2013). FY 2013 Research report on floating marine litter and seabed marine litter (in Japanese).
Ministry of the Environment (2014a). FY 2014 Research report on marine debris floating on the ocean surface and settled on the sea bed in offshore areas around Japan (in Japanese).
Ministry of the Environment (2014b). FY 2014 Comprehensive investigation on measures to tackle beach debris (in Japanese).
Ministry of the Environment (2015a). FY 2015 Research report on floating marine litter and seabed marine litter in coastal areas around Japan (in Japanese).
Ministry of the Environment (2015b). FY 2015 Comprehensive investigation on measures to tackle beach debris (in Japanese).
Ministry of the Environment (2016a). Environmental Statistic Data Set 2016 (in Japanese) www.env.go.jp/doc/toukei/h28.html (accessed April 4, 2019)
Ministry of the Environment (2016b). FY 2016 Research report on marine debris floating on the ocean surface and settled on the sea bed in offshore areas around Japan (in Japanese).
Ministry of the Environment (2016c). FY 2016 Comprehensive investigation on measures to tackle beach debris (in Japanese).
Nakashima, E., Isobe, A., Kato, S., Itai, T., and Takahashi, S. (2012). Quantification of toxic metals derived from macroplastic litter on Ookushi Beach, Japan. *Environmental Science & Technology* 46(18): 10099–10105.
Plastics Waste Management Institute JAPAN. Annual Report of Materials Flow of Plastic in Japan 2017 (in Japanese).
PlasticsEurope (2018). Plastics—The Facts. www.plasticseurope.org/en/resources/publications/619-plastics-facts-2018 (accessed April 4, 2019)
Schmidt, C., Krauth, T., and Wagner, S. (2017). Export of plastic debris by rivers into the sea. *Environmental Science & Technology* 51(21): 12246–12253.
Tanaka, K., and Takada, H. (2016). Microplastic fragments and microbeads in digestive tracts of planktivorous fish from urban coastal waters. *Scientific Reports* 6: 34351.
Teuten Emma, L., Saquing Jovita, M., Knappe Detlef, R.U., Barlaz Morton, A., et al. (2009). Transport and release of chemicals from plastics to the environment and to wildlife. *Philosophical Transactions of the Royal Society B: Biological Sciences* 364(1526): 2027–2045.
Tokyo University of Science. Press release of Tokyo University of Science in December 12, 2018 (in Japanese) www.tus.ac.jp/ura/pressrelease/pdf/181221.pdf (accessed April 4, 2019).
Vinyl Environmental Council, Japan (in Japanese) www.vec.gr.jp/lib/lib2_4.html (accessed May 10, 2019).
World Health Organization. (1993). *Environmental Health Criteria 140, Polychlorinated Biphenyls and Terphenyls*, 2nd ed. Geneva, Switzerland, WHO.

20 Particulate Plastic Distribution and Ecotoxicity in Marine Ecosystems and a Case Study in Thailand

Suchana Chavanich, Voranop Viyakarn, Somkiat Khokiattiwong, and Wenxi Zhu

CONTENTS

20.1 Introduction 355
20.2 Pathways and Distribution of Particulate Plastics in Marine Ecosystems 356
20.2.1 Pathways 356
20.2.2 Distribution in Waters, Sediments, and Beaches 357
20.2.3 Distribution in Marine Organisms 357
20.3 Effects of Particulate Plastics on Marine Ecosystems 358
20.4 Particulate Plastic Studies in Marine Ecosystems in Thailand 359
Acknowledgments 360
References 360

20.1 INTRODUCTION

Marine ecosystems, including oceans, deep sea, salt marshes, estuaries, mangroves, and coral reefs, are the largest aquatic ecosystems on earth. They provide multiple ecosystem services to the world's populations. Unfortunately, marine ecosystems are rapidly changing due to several threats, such as global warming, overfishing, pollution, sedimentation, land development, and reclamation of land (Townsend et al. 2018). Among the threats, marine debris, and, in particular, plastic pollution, has become of global concern recently, and it has grown out of control since the introduction of packaging in the middle of twentieth century (STAP 2011).

Plastics are part of the everyday life of billions of people. More than 400 million tons of plastic are produced globally every year, and packaging accounts for more than one-third of all plastics produced (UNEP 2016; Plastic Atlas 2019). In 2025, the production of plastic is expected to be as high as 600 million tons per year (Plastic Atlas 2019). The problem with plastic pollution is that it is not only found on land; more than 10 to 20 million tons of plastics are finding their way into the oceans (UNEP 2016). Only a small percentage is recycled (UNEP 2016). Some evidence also suggests that plastics do not float for a long time. Because of degradation and biological interactions, they move to shallower waters, sink to the sea floor, or are washed onto the shores (Plastic Atlas 2019).

When marine debris and marine plastic pollution enter marine ecosystems, they impact fisheries, aquaculture, human health, and food safety. Plastics usually persist in the marine environment over a long period of time without decomposing (Plastic Atlas 2019). Incidents concerning entanglement and ingestion have been widely reported for a variety of marine mammals, reptiles, and birds, which lead to chronic injury and death (Allen et al. 2012; Campani et al. 2013; Thevenon et al. 2014).

Marine debris and plastic pollution can also cause extensive damage to reefs (Valderrama Ballesteros et al. 2018). Some types of coastal debris found in oceans are related to economic- and land-based activities in the area (Thushari et al. 2017a). The majority of coastal debris includes plastic bottles, caps, lids, food and beverage containers, rubber bands, and cigarette buds (UNEP 2016; Thushari et al. 2017a). Jambeck et al. (2015) estimated that top contributors to plastic marine litter were from middle-income countries.

20.2 PATHWAYS AND DISTRIBUTION OF PARTICULATE PLASTICS IN MARINE ECOSYSTEMS

Based on the UNEP (2016), plastic refers to a group of synthetic polymers. There are two main types of plastics: thermoplastic [such as polyethylene (PE), polyethylene terephthalate, polypropylene (PP), polyvinyl chloride, and polystyrene (PS, including expanded EPS)] and thermoset [such as polyacrylic (PA), polyurethane, and epoxy resins or coatings].

Particulate plastics are particles of plastics less than <5 mm, and their size goes down to the nano-meter range (GESAMP 2015). Two types of particulate plastics can be found in marine ecosystems: primary and secondary particulate plastics. Primary particulate plastics are originally manufactured and used as raw materials or components of industrial products, while secondary particulate plastics are from the breakdown of larger items through physical and chemical processes, such as ultraviolet radiation and mechanical abrasion (GESAMP 2015).

20.2.1 PATHWAYS

Both primary and secondary particulate plastics in marine ecosystems can come from land-based sectors and sea-based sectors (UNEP 2016). Large quantities of particulate plastics in marine environments have originated from land-based sources (Thompson et al. 2009). However, at present, reliable estimations of total quantities of particulate plastics from land, and then entering the oceans are not available. Potential land-based sources of particulate plastics include cosmetics and personal care products, textiles and clothing, the tourism industry, food and drink, plastics producers, household, terrestrial transportation, cleaning ships and buildings, manufacturing, plastics recyclers, construction, and agriculture (GESAMP 2015; UNEP 2016). The tourism industry and food and drink are considered to be important sources on land, and they are likely to be regionally dependent (UNEP 2016). Thus, effectiveness of solid waste and wastewater collection and treatment in the area where particulate plastics are generated can play a significant role in preventing them from entering marine ecosystems (UNEP 2016). Recently, plastic dusts and car tires have been recognized as potential major sources of particulate plastics that contaminate the sea (NEA 2014; Verschoor et al. 2014). The emission of plastic dust is from tire wear. Some plastic dust flies into the air, and the rest lands on soil around roads. Because of rain and runoff, this dust from land can end up in rivers and oceans (UNEP 2016).

Sea-based sectors, such as many maritime activities, lead to the release of both primary and secondary particulate plastics directly to the sea. Primary particulate plastics from sea-based sectors can come from the loss of cargos, while secondary particulate plastics result from routine wear and tear of fishing gear (UNEP 2016; Thushari et al. 2017b).

The abundance of particulate plastics in marine environments depends on several environmental factors, such as wind currents, coastline geology, and anthropogenic activities (Sharma and Chatterjee 2017). Industrial areas tend to have high amounts of particulate plastics that land in the sea (Sharma and Chatterjee 2017). One of the dominant pathways of transportation of particulate plastics in the ocean is ocean circulation (Isobe et al. 2015). This has a significant influence on the distribution of any floating objects (Isobe et al. 2015). Long-distance transport of floating objects also can occur by a combination of ocean circulation and winds (Isobe et al. 2015, 2017). At present, modeling techniques are being attempted to simulate the distribution of particulate plastics. Models allow one to investigate the relative importance of different sources, when accurate data are absent; there is a need to fill in gaps in the absence data (Isobe et al. 2019b).

20.2.2 Distribution in Waters, Sediments, and Beaches

At present, particulate plastics can be found not only in water bodies, but also in sediments and beaches around the globe (Cózar et al. 2014; Eriksen et al. 2014; Isobe 2016; GESAMP 2019). Because of currents and ocean circulation, they can be found as far south as in the southern oceans, where no humans live permanently (Isobe et al. 2017). Contamination by particulate plastics in sediments and beaches can account for more than 90% of the contamination in an area (Fok and Cheung 2015; Sharma and Chatterjee 2017). Types of particulate plastics found in different areas depend on proximity to residential areas, commercial fishing, and the activities on land and shorelines (Ng and Obbard 2006; Thushari et al. 2017b). For example, in Singapore, types of particulate plastics found in the marine environment were PE, PP, and PS (Ng and Obbard 2006), while, in Thailand, PS and PA were more prevalent (Thushari et al. 2017b). In addition, the abundance and spatial patterns of particulate plastics can depend on whether or not it is the rainy season and other natural phenomenon (Ng and Obbard 2006; Thushari et al. 2017b).

During the last few years, studies have been done related to the surveying and monitoring of particulate plastics in waters, sediments, and beaches (Cózar et al. 2014; Eriksen et al. 2014; GESAMP 2019; Isobe et al. 2019a). However, the data generated and published lack standardization of sampling methods. Thus, this makes the comparison of data impossible or problematic (GESAMP 2019). To solve this problem, standardized monitoring methods and sampling protocols have been developed by several international and national organizations (GESAMP 2019; Isobe et al. 2019a). It is important to establish standard methodologies to consider the behavior of particulate plastics in marine environments.

For example, in the Western Pacific Region, under the Intergovernmental Oceanographic Commission of the UNESCO Sub-Commission for the Western Pacific, a marine microplastic program has been established. This program aims to establish a marine microplastic monitoring and research network and to develop joint monitoring and research activities in order to understand the sources, distribution, fate, and effects of marine microplastics in the Western Pacific and its adjacent regions. In addition, standard operating procedures to analyze marine microplastics in beaches, water columns, and selected marine organisms in the region have been developed under the program.

20.2.3 Distribution in Marine Organisms

Because of their pervasive contamination, particulate plastics can enter into the food chain and cause substantial risk to marine biota. Particulate plastics are passive particles that float on water surfaces, are suspended in water columns, or sink and are deposited on the seafloor. They can enter the food chain through ingestion, entanglement, and inhalation by marine organisms. Particulate plastics can be either accidentally ingested or selectively ingested by animals (Schuyler et al. 2012). The rates of ingestion vary depending on animals' feeding behaviors, plastic sizes, shapes, and chemical characters (GESAMP 2015, 2019). When marine organisms ingest non-degradable particulate plastics, they bioaccumulate in the food chain. They are passed from lower trophic levels to higher ones.

Suspended feeders or filter feeders and other marine benthic organisms ingest particulate plastics from seawater (Thushari et al. 2017b). Examples of such organisms are oysters, barnacles, mussels, and lobsters (Goldstein and Goodwin 2013; Nerland et al. 2014). When these organisms serve as food sources for omnivores, particulate plastics are then consumed by them via feeding processes (Murray and Cowie 2011). The sizes of particles that can be ingested depend on feeding modes and structural apparatus (Riisgård and Larsen 2010). In addition to physical contacts, chemo- and mechanoreceptors can influence the capture rates (Riisgård and Larsen 2010). Some environmental factors, such as temperature and salinity, also affect the feeding and excretion rates of organisms (Garrido et al. 2013). Marine organisms, like whales and mussels, with long lifespans and a high capacity to filter large water quantities, can encounter more particles than others (Kautsky and Kautsky 2000).

When marine organisms ingest particulate plastics, they can either be egested or retained by the organisms. If a predator consumes organisms with particles, those particles can be transferred in the trophic cascade (Murray and Cowie 2011). Particles can be observed in the stomachs or intestinal tracts of organisms (Murray and Cowie 2011; Lusher et al. 2013). However, Cheryl et al. (2018) demonstrated that some marine organisms, such as corals, can ingest and egest particulate plastics. Thus, the numbers of particles found during sampling may not provide true numbers (Cheryl et al. 2018).

More than 30% of fish and seabirds have particulate plastics in their stomachs (Lusher et al. 2013). Benthic and bottom-feeding fish tend to be more contaminated with particulate plastics in their stomachs than other species (Ramos et al. 2012). Even though particulate plastics can be found in seabirds' stomachs, they can be removed through feces and regurgitation (Lindborg et al. 2012).

20.3 EFFECTS OF PARTICULATE PLASTICS ON MARINE ECOSYSTEMS

Particulate plastics become available to both benthic and pelagic marine organisms. Many organisms, such as bivalves, fish, echinoderms, and polychaetes, have shown evidence of ingestion of particulate plastics (Sharma and Chatterjee 2017). Once the particles are inside the animals, they may have health effects that are both physical and biological (Figure 20.1). Physical effects include physical obstruction, damage of feeding appendages, damage of feeding activities, damage of the digestive tract, and harm to mobility. Several organisms also show biological effects when consuming particulate plastics. Examples of biological effects include impaired health, abnormal heartbeats, growth reduction, delay in reproduction, damage to the nervous system, cancers, and other fatal diseases. Particulate plastics ingested by plankton can block their digestive track and disturb feeding and digestion (Cole et al. 2013). In benthic worms, high particulate plastic consumption and chronic exposure cause reduction in feeding and reduction in weight (Besseling et al. 2013). A study of mussels, *Mytilus edulis*, showed that small particulate plastics were translocated into the circulatory fluid (von Moos et al. 2012). If particulate plastics are as small as the nano size (<100 nm diameter),

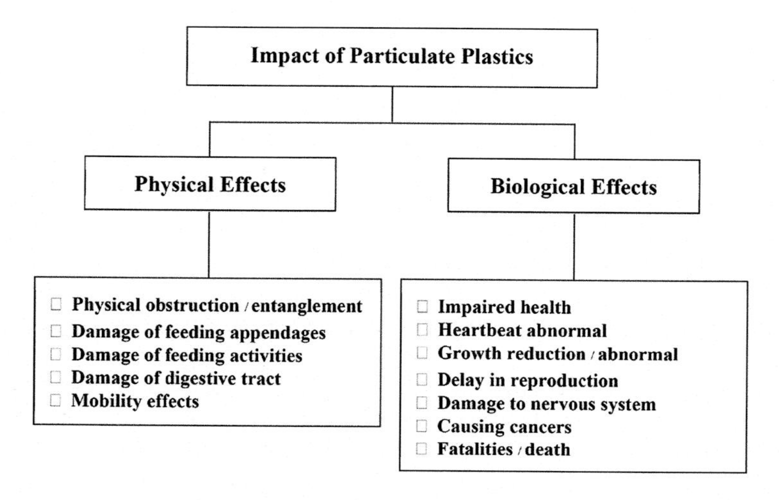

FIGURE 20.1 Negative impacts of particulate plastics on the health of marine organisms.

they can enter epithelia tissue and circulatory fluids (Rescigno et al. 2001). Accumulation of particles in the guts of fish can lead to starvation and malnourishment and eventual fatalities (Boerger et al. 2010). Particulate plastics can be found not only in marine organisms, but also in seabirds. Even though seabirds have regurgitation processes to remove particulate plastics from their bodies, still, more particles are left inside them, and they cause a negative impact on their feeding habits leading to starvation and a loss of fitness (Tanaka et al. 2013). Particulate plastics have also been reported in large marine animals, such as sea turtles, whales, and dolphins; however, hazardous effects on these animals tend to be more from large plastics than particulate plastics (UNEP 2016).

In addition to accumulation in marine organisms, the capacity of particulate plastics in absorbing pollutants from water and transferring them along the food chains is a serious concern. Persistent organic pollutants are found to be taken up by particulate plastics (Browne et al. 2007). The accumulation of these toxic compounds in particulate plastics leads to shifting the chemicals into the body tissues of organisms causing chronic health risks, such as carcinogenic effects and hormonal imbalances (Thevenon et al. 2014). Exposure of particulate plastics to harmful phycotoxins from algae also can cause adverse effects on human health (Campbell et al. 2005). The toxins can accumulate in particulate plastics and pass from one trophic level to the next by feeding (Campbell et al. 2005). However, so far there is no quantitative measure of the quantity of absorbed pollutants on particulate plastics that are passed from one trophic level to another and how much of an effect the particulate plastics have on animal health. Therefore, more studies related to animal health effects are needed.

Particulate plastics have been demonstrated to be ingested by many commercial marine species. Thus, consequences of humans exposed to particulate plastics through seafood and other diets are of concern (van Cauwenberghe and Janssen 2014). They pose a threat to food safety (van Cauwenberghe and Janssen 2014). Although particulate plastics are found in commercial and edible fish and mussels, they are usually found in the stomach or gastrointestinal tracts of animals, which are often removed in the depuration process before consumption by humans (Lusher et al. 2013). Therefore, the risk of exposing humans to particulate plastics through seafood can be minimized (Lusher et al. 2013). The effect of particulate plastics on human health may include disruption of the body's balanced endocrine system, asthma, cancers, infertility, premature puberty, and obesity (Plastic Atlas 2019). In addition, because of their large surface area, particulate plastics can absorb a high quantity of chemical pollutants from marine environments, leading to an increase in their toxicity (Browne et al. 2007). Particulate plastics can also harbor pathogenic microorganisms, such as bacteria and viruses, which can cause harm to humans and marine organisms (UNEP 2016).

20.4 PARTICULATE PLASTIC STUDIES IN MARINE ECOSYSTEMS IN THAILAND

At present, unfortunately, Thailand has been ranked as the sixth worst contributor of marine debris in the world (Jambeck et al. 2015). Both governmental and non-governmental sectors are focused on advancing awareness and implementation of the best practices to address the causes and solutions for marine debris and microplastics. At the Special Association of Southeast Asian Nations Ministerial Meeting on Marine Debris in March 2019 in Bangkok, the government of Thailand made a commitment to reduce marine debris by at least 50% by 2027, and reduction of marine debris has been added to the 20-year plan (2017–2036) for marine ecosystem management in Thailand. In addition, the government has also established a plastic material flow database throughout the country.

Even though particulate plastics are an emerging threat to marine ecosystems in Thailand, only a few research studies have been done and published related to particulate plastic accumulation in marine ecosystems (Thushari et al. 2017b; Azad et al. 2018a,b; Akkajit et al. 2019). Studies done by Thushari et al. (2017b) and Azad et al. (2018a) showed that more than 50% of fish and bivalves in Thailand were contaminated with particulate plastics. Thushari et al. (2017b) found that the most abundant and common snails, oysters, and barnacles were contaminated with particulate plastics. Because they are consumable, the contamination can pose potential health risks for seafood consumers (Thushari et al. 2017b). Azad et al. (2018a,b) investigated the accumulation of particulate plastics in different types of fish (demersal fish, pelagic fish, and reef-associated fish) in the lower Gulf of Thailand. The results showed that 66.67% of fish samples were contaminated with particulate plastics. The distribution of particulate plastics on beaches at Phuket Province in the southern part of Thailand was also investigated (Akkajit et al. 2019). The surveys showed that particulate plastics can be found on beaches along the coast of Thailand (Akkajit et al. 2019). Even though studies related to particulate plastic accumulation in marine environments in Thailand have been conducted, more studies are needed to investigate the impact on both marine organisms and the health of the Thai people.

Because of the threats to marine organisms and human health, it is important to control both macroplastics and particulate plastics that land in the ocean off Thailand. Policies to regulate the sources of plastic debris and various recycling processes are also needed. In addition, programs using various media that increase awareness of plastic pollution are a must. Beginning January 1, 2020, the Ministry of Natural Resources and the Environment of Thailand has launched a campaign to promote the ban of single-use plastics. Forty-six major shopping malls and convenience stores in Thailand have joined this program. No single-use plastic is given out when customers purchase items, and the general public is encouraged to carry tote or cloth bags. The goal is that, by 2021, Thailand will eliminate single-use plastics. At present, according to the Pollution Control Department, 45 billion plastic bags are used each year; 40% of the usage comes from fresh markets, 30% from grocery stores, and 30% from department stores and convenience stores. According to the Department of Environmental Quality Promotion, Thai people generate 1.14 kg of waste per person per day.

The government has formulated many rules and implemented fines for use of plastic bags. These steps can help to reduce the use of plastic bags. To tackle the plastic waste problem and to reduce particulate plastics in the environment, the Thai cabinet also approved a roadmap for tackling plastic waste in the years 2018–2030. By the end of 2019, microbeads, cap seals, and oxo-degradable plastics were eliminated and banned in Thailand. In addition, plastic bags, styrofoam food containers, single-use plastic cups, and plastic straws will be out by 2022, except for those who still need to use them, such as the elderly, patients, and children. This strategy will be another way to reduce both macro- and particulate plastic pollution, both on land and in the ocean.

In conclusion, plastic pollution in marine ecosystems is an alarming situation. It has harmful effects on marine organisms and human health. To reduce effectively plastic wastes with both macro- and particulate sizes, cooperation among the public, private, and governmental sections is a key for sustainability of plastic and waste management.

ACKNOWLEDGMENTS

This work is partially funded by Thailand Research Fund (RSA6080087), National Research Council of Thailand–Thai Ocean Waste Free, National Research Council of Thailand–Japan Society for the Promotion of Science-Core-to-Core program (NRCT-JSPS Core-to-Core Program), and EU-Horizon 2020 TASCMAR (Tools and strategies to access original bioactive compounds by cultivating marine invertebrates and associated symbionts).

REFERENCES

Akkajit, P., S. Thongnonghin, S. Sriraksa, and S. Pumsri. 2019. Preliminary study of distribution and quantity of plastic-debris on beaches along the coast at Phuket Province. *Applied Environmental Research* 41:54–62.

Allen, R., D. Jarvis, S. Sayer, and C. Mills. 2012. Entanglement of grey seals *Halichoerus grypus* at a haul out site in Cornwall, UK. *Marine Pollution Bulletin* 64:2815–2819.

Azad, S. M. O., P. Towatana, S. Pradit, B. G. Patricia, H. T. T. Hue, and S. Jualaong. 2018a. First evidence of existence of microplastics in stomach of some commercial fishes in the lower Gulf of Thailand. *Applied Ecology and Environmental Research* 16:7345–7360.

Azad, S. M. O., P. Towatana, S. Pradit, B. G. Patricia, and H. T. Hue. 2018b. Ingestion of microplastics by some commercial fishes in the lower Gulf of Thailand: A preliminary approach to ocean conservation International. *Journal of Agricultural Technology* 14:1017–1032.

Besseling, E., A. Wegner, E. M. Foekema, M. J. van den Heuvel-Greve, and A. A. Koelman. 2013. Effects of microplastic on fitness and PCB bioaccumulation by the lugworm *Arenicola marina* (L.). *Environmental Science and Technology* 47:593–600.

Boerger, C. M., G. L. Lattin, S. L. Moore, and C. J. Moore. 2010. Plastic ingestion by planktivorous fishes in the North Pacific Central Gyre. *Marine Pollution Bulletin* 60:2275–2778.

Browne, M. A., T. Galloway, and R. Thompson. 2007. Microplastic—An emerging contaminant of potential concern? *Integrated Environmental Assessment and Management* 3:559–561.

Campani, T., M. Baini, M. Giannetti, F. Cancelli, C. Mancusi, F. Serena, L. Marsili, S. Casini, and M. C. Fossi. 2013. Presence of plastic debris in loggerhead turtle stranded along the Tuscany coasts of the Pelagos Sanctuary for Mediterranean Marine Mammals (Italy). *Marine Pollution Bulletin* 74:225–230.

Campbell R. G., G. J. Teegarden, A. D. Cembella, and E. G. Durbin. 2005. Zooplankton grazing impacts on *Alexandrium* spp. in the nearshore environment of the Gulf of Maine. *Deep-Sea Research II* 52:2817–2833.

Cheryl, H., D. Allyn, and D. Kathryn. 2018. Scleractinian coral microplastic ingestion: Potential calcification effects, size limits, and retention. *Marine Pollution Bulletin* 135:587–593.

Cole, M., P. Lindeque, E. Fileman, C. Halsband, R. Goodhead, J. Moger, and T. S. Galloway. 2013. Microplastic ingestion by zooplankton. *Environmental Science and Technology* 47:6646–6655.

Cózar, A., F. Echevarría, J. I. González-Gordillo, X. Irigoien, B. Úbeda, S. Hernández-León, Á. T. Palma, et al. 2014. Plastic debris in the open ocean. *Proceeding of the National Academy of Sciences of the United States of America* 111:10239–10244.

Eriksen, M., L. C. M. Lebreton, H. S. Carson, M. Thiel, C. J. Moore, J. C. Borerro, F. Galgani, P. G. Ryan, and J. Reisser. 2014. Plastic pollution in the world's oceans: More than 5 trillion plastic pieces weighing over 250,000 tons afloat at Sea. *PLoS One* 9 (12):e111913. doi:10.1371/journal.pone.0111913.

Fok, L., and P. K. Cheung. 2015. Hong Kong at the Pearl River Estuary: A hotspot of microplastic pollution. *Marine Pollution Bulletin* 99:112–118.

Garrido, S., J. Cruz, A. M. P. Santos, P. Ré, and E. Sauz. 2013. Effects of temperature, food type and food concentration on the grazing of the calanoid copepod *Centropages chierchiae*. *Journal of Plankton Research* 35:843–854.

GESAMP. 2015. Sources, fate and effects of microplastics in the marine environment: A global assessment. In: Kershaw, P. J., (ed.), IMO/FAO/UNESCO-IOC/UNIDO/WMO/IAEA/UN/UNEP/UNDP Joint Group of Experts on the Scientific Aspects of Marine Environmental Protection. *Report and Studies GESAMP* No. 90, 96 pp.

GESAMP. 2019. Guidelines or the monitoring and assessment of plastic litter and microplastics in the ocean. In: Kershaw P. J., A. Turra, and F. Galgani (editors), IMO/FAO/UNESCO-IOC/ UNIDO/WMO/IAEA/ UN/UNEP/UNDP/ISA Joint Group of Experts on the Scientific Aspects of Marine Environmental Protection. *Report and Studies GESAMP* No. 99, 130 pp.

Goldstein, M., and D. S. Goodwin. 2013. Gooseneck barnacles (*Lepas* spp.) ingest microplastic debris in the North Pacific Subtropical Gyre. *Peer J* 184. doi:10.7717/peerj.184.

Isobe, A., K. Uchida, T. Tokai, and S. Iwasaki. 2015. East Asian seas: A hot spot of Pelagic microplastics. *Marine Pollution Bulletin* 101:618–623.

Isobe, A. 2016. Percentage of microbeads in pelagic microplastics within Japanese coastal waters. *Marine Pollution Bulletin* 110:432–437.

Isobe, A., K. Uchiyama-Matsumoto, K. Uchida, and T. Tokai. 2017. Microplastics in the Southern Ocean. *Marine Pollution Bulletin* 114:623–626.

Isobe, A., N. T. Buenaventura, S. Chastain, S. Chavanich, A. Cózar, M. DeLorenzo, P. Hagmann, et al. 2019a. An interlaboratory comparison exercise for the determination of microplastics in standard sample bottles. *Marine Pollution Bulletin* 146:831–837.

Isobe, A., S. Iwasaki, K. Uchida, and T. Tokai. 2019b. Abundance of non-conservative microplastics in the upper ocean from 1957 to 2066. *Nature Communications* 10:417.

Jambeck, J. R., R. Geyer, C. Wilcox, T. R. Siegler, M. Perryman, A. Andrady, R. Narayan, and K. L. Law. 2015. Plastic waste inputs from land into the ocean. *Science* 347:768–771.

Kautsky, L., and N. Kautsky. 2000. The Baltic Sea, including Bothnian Sea and Bothnian Bay (Chapter 8). In: Sheppard, C. R. C. (Ed.), *Seas at the Millennium: An Environmental Evaluation*. Elsevier Science Ltd., Amsterdam, the Netherlands. pp. 1–14.

Lindborg, V. A., J. F. Ledbetter, J. M. Walat, and C. Moffett. 2012. Plastic consumption and diet of Glaucous-winged Gulls (*Larus glaucescens*). *Marine Pollution Bulletin* 64:2351–2356.

Lusher, A. L., M. McHugh, and R. C. Thompson. 2013. Occurrence of microplastics in the gastrointestinal tract of pelagic and demersal fish from the English Channel. *Marine Pollution Bulletin* 67:94–99.

Murray, F., and P. R. Cowie. 2011. Plastic contamination in the decapod crustacean *Nephrops norvegicus*. *Marine Pollution Bulletin* 62:1207–1217.

NEA. 2014. Sources of microplastics to the marine environment. Norwegian Environment Agency (Miljødirektoratet). 108 pp.

Nerland, I. L., C. Halsband, I. Allanm, and K. V. Thomas. 2014. Microplastics in marine environments: Occurrence, distribution and effects. Project No. 14338, Report No. 6754-2014, NIVA Oslo. 71 pp.

Ng, K. L., and J. P. Obbard. 2006. Prevalence of microplastics in Singapore's coastal marine environment. *Marine Pollution Bulletin* 52:761–767.

Plastic Atlas. 2019. *Facts and Figures about the World of Synthetic Polymers*. Lili Fuhr, L., and M. Franklin (editors), *Plastic Atlas*, Berlin, Germany. 50 pp.

Ramos, J. A. A., M. Barletta, and M. F. Costa. 2012. Ingestion of nylon threads by Gerreidae while using a tropical estuary as foraging grounds. *Aquatic Biology* 17:29–34.

Rescigno, M., M. Urbano, B. Valzasina, M. Francolini, G. Rotta, R. Bonasio, F. Granucci, J. P. Kraehenbuhl, and P. Ricciardi-Castagnoli. 2001. Dendritic cells express tight junction proteins and penetrate gut epithelial monolayers to sample bacteria. *Nature Immunology* 2:361–367.

Riisgård, H. U., and P. S. Larsen. 2010. Particle capture mechanisms in suspension-feeding invertebrates. *Marine Ecology Progress Series* 418:255–293.

Schuyler, Q., B. D. Hardesty, C. Wilcox, and K. Townsend. 2012. To eat or not to eat? Debris selectivity by marine turtles. *PLoS One* 7:e40884. doi:10.1371/journal.pone.0040884.

Sharma, S., and S. Chatterjee. 2017. Microplastic pollution, a threat to marine ecosystem and human health: A short review. *Environmental Science and Pollution Research* 24:21530–21547.

STAP (Scientific and Technical Advisory Panel). 2011. Marine debris as a global environmental problem: Introducing as a solutions based framework focuses on plastic. A STAP Information Document. Global Environment Facility, Washington, DC. 32 pp.

Tanaka, K., H. Takada, R. Yamashita, K. Mizukawa, M. Fukuwaka, and Y. Watanuki. 2013. Accumulation of plastic-derived chemicals in tissues of seabirds ingesting marine plastics. *Marine Pollution Bulletin* 69:219–222.

Thevenon, F., C. Carroll, and J. Sousa. (editors). 2014. Plastic debris in the ocean: The Characterization of marine plastics and their environmental impacts, situation analysis report. *IUCN Global Marine and Polar Programme*, Gland, Switzerland. 52 pp.

Thompson, R. C., S. H. Swan, C. J. Moore, and F. S. vom Saal. 2009. Our plastic age. *Philosophical Transactions of the Royal Society B* 364:1973–1976.

Thushari, G. G. N., S. Chavanich, and A. Yakupitiyage. 2017a. Coastal debris analysis in beaches of Chonburi Province, eastern of Thailand as implications for coastal conservation. *Marine Pollution Bulletin* 116:121–129.

Thushari, G. G. N., J. D. M. Senevirathna, A. Yakupitiyage, and S. Chavanich. 2017b. Effects of microplastics on sessile invertebrates in the eastern coast of Thailand: An approach to coastal zone conservation. *Marine Pollution Bulletin* 124:349–355.

Townsend, M., K. Davies, N. Hanley, J. E. Hewitt, C. J. Lundquist, and A. M. Lohrer. 2018. The challenge of implementing the marine ecosystem service concept. *Frontiers in Marine Science* 5:359. doi:10.3389/fmars.2018.00359.

UNEP. 2016. Marine plastic debris and microplastics—Global lessons and research to inspire action and guide policy change. *United Nations Environment Programme*, Nairobi. 252 pp.

Valderrama Ballesteros, L., J. L. Matthews, and B. W. Hoeksema. 2018. Pollution and coral damage caused by derelict fishing gear on coral reefs around Koh Tao, Gulf of Thailand. *Marine Pollution Bulletin* 135:1107–1116.

van Cauwenberghe, L., and C. R. Janssen 2014. Microplastics in bivalves cultured for human consumption. *Environmental Pollution* 193:65–70.

Verschoor, A., L. de Poorter, E. Roex, and B. Bellert. 2014. Quick scan and prioritization of microplastic sources and emissions. *RIVM Advisory Letter* 250012001. 41 pp.

von Moos, N., P. Burkhardt-Holm, and A. Köler. 2012. Uptake and effects of microplastics on cells and tissue of the blue mussel *Mytilus edulis* L. after an experimental exposure. *Environmental Science and Technology* 46:11327–11335.

21 The Current Status of Plastics
A New Zealand Perspective

Louis A. Tremblay, Xavier Pochon, Olivier Champeau, Virginia Baker, and Grant L. Northcott

CONTENTS

21.1 Introduction 363
21.2 Characterization of Microplastics 364
21.2.1 Definition 364
21.2.2 Sources 364
21.2.2.1 Land 365
21.2.2.2 Freshwater 365
21.2.2.3 Marine Environment 366
21.2.2.4 Microplastics in New Zealand 366
21.3 Risk of Microplastics 366
21.3.1 Environmental Health Risk 367
21.3.1.1 Biosecurity Risk 369
21.3.2 Human Health Risk 369
21.4 Managing the Risk of Microplastics 370
21.4.1 International Strategies to Manage Microplastics 371
21.4.2 New Zealand Initiatives 372
21.5 Conclusions 373
References 374

21.1 INTRODUCTION

An estimated 5–13 million metric tons of plastics enter the oceans annually, and that amount is projected to continue to increase if no mitigation measures are put in place (Burgess and Ho 2017). There is increasing scientific and public concern over the presence and persistence of this plastic pollution in the environment. There are similar trends in New Zealand, where such environmental issues must be addressed and managed by regional councils that are the local territorial authorities. Research on marine plastic pollution has been the focal point of scientists, the public, and policy-makers. However, recent literature on the presence of microplastics in air, soil, sediments, fresh-waters, oceans, plants, animals, and parts of the human diet raises broad concerns of the impacts of plastics and microplastics in complex global ecosystems. It is established that microplastics are a ubiquitous contaminant, and their impacts on the environment pose the highest risk from plastic pollution. Plastic debris are generally categorized according to two main size range categories: macroplastics >5 mm and microplastics, which are greater than 0.3 mm and less than 5 mm (Moore 2008). The presence of microplastics in all environmental compartments (water, soil, air, and biota) has gained increasing public and political awareness, along with the desire to identify sources and reduction/remediation options (Vollertsen and Hansen 2017). The limited investigations completed to date in New Zealand and the scarce data produced indicate that the types and concentration of microplastics entering and persisting in our environment are likely to be similar to those reported in many other countries.

21.2 CHARACTERIZATION OF MICROPLASTICS

21.2.1 Definition

Microplastics come from the partial degradation of plastic material. Plastic has been defined as a synthetic organic water-insoluble polymer, generally of petrochemical origin, that can be molded on heating and manipulated into various shapes designed to be maintained during use (Burns and Boxall 2018). This includes both thermoplastics, such as polyethylene and polypropylene, and thermoset plastics (i.e., cannot be remolded after successive heating), for example, polyurethane foams and epoxy resins (Burns and Boxall 2018). The physical characteristics of most plastics show high resistance to aging and minimal biological degradation.

The United States National Oceanographic and Atmospheric Administration (NOAA) defines microplastics as any plastic particle <5 mm in dimension (Rochman et al. 2019). However, there is some debate about the lower size limit for microplastics, and nano-sized plastics (<0.1 mm) are often included in this definition. To address this issue over current sampling and processing practices, a recommendation has been made to report microplastics data in three size classes: $1 \leq 100$ μm; $100 \leq 350$ μm; and from 350 μm to ≤5 mm (Frias and Nash 2019). Overall, the size definition of microplastics remains a source of debate among the scientific community (Rochman et al. 2019). Relatively few studies have directly assessed microplastics in nature at or below the 10–50 μm size range because this size range typically falls below the limit of resolution for most of the readily available equipment for analysis. However, researchers are continuously expanding their analytical techniques to detect and identify ever smaller microplastics (Rochman et al. 2019) and nanoplastics (Lambert and Wagner 2016; Mattsson et al. 2018). The presence and risks of (nano) plastic have been difficult to ascertain, as there are technical challenges for isolating and quantifying them. However, there is a consensus that they can be ingested by organisms at the base of the food-chain and pose a risk to the environment and human health (da Costa et al. 2016).

Microplastics come in many shapes and colors and can be a source of contaminants from the chemical additives incorporated into plastic materials during manufacturing processes. They also act as a substrate that can directly accumulate pollutants from the environment. The shape of a microplastic particle is often used to assign it to a common category, which helps identify the source. Generally, researchers use between four and seven different categories defined by shape or morphology including fiber, fiber bundle, fragment, sphere (or bead), pellet, film, and foam (Rochman et al. 2019). For example, fibers and fiber bundles tend to shed from clothing, upholstery, or carpet; pellets, or plastic knurdles, are generally associated with pre-plastic feedstock; spheres may be microbeads from personal care products or industrial scrubbers; and foam often comes from expanded polystyrene foam products, such as insulation, construction materials, or food packaging (Rochman et al. 2019).

21.2.2 Sources

The exceptional properties of plastics (versatility, durability, strength, lightness, and transparency) make them "unique material" for applications in industry, construction, medicine, and food safety (Guzzetti et al. 2018). Primary microplastics, including pellets, granules, and microbeads, are produced for specific purposes, while secondary microplastics arise from the fragmentation of larger plastic items during use or once released into the environment (Triebskorn et al. 2019). Plastics become brittle and fragment into smaller pieces through exposure to sunlight, ultraviolet B (UVB) radiation, and degradation in the atmosphere and seawater to the point of becoming bioavailable and posing potential risks to exposed organisms (Moore 2008).

Human behavioral patterns are responsible for plastic production and pollution through discarding plastic and from the use of plastic-enabled products that over time break down and release microplastics. This is a global problem rooted in systems of production and consumption, including human behavior fuelled by convenience and compounded by often absent or inadequate waste management practices and infrastructure (Burgess and Ho 2017). Sources of microplastic pollution include single-use plastic items (bags, bottles, straws, and food packaging), textiles, abrasion of vehicle tires, general waste, products containing microplastics, and equipment/products used in fisheries, agriculture, and industry (Rochman et al. 2019). There is a need to distinguish between primary and secondary microplastics and their respective sources.

Microplastics can be found worldwide in the water and sediment phases of marine and freshwater ecosystems even in the most remote areas of the world, including the deep sea, the Arctic, mountain lakes, and in atmospheric deposition (Laskar and Kumar 2019; Law and Thompson 2014; Lima et al. 2015; Miller et al. 2017). Continental plastic litter enters the ocean largely through stormwater runoff into riverine systems, and it is dumped on shorelines during recreational activities or directly discharged at sea from ships (Walker et al. 2019). While a significant proportion of microplastics entering wastewater treatment plants (WWTPs) are concentrated and retained within sewage sludge and biosolids, most WWTP technology does not fully remove microplastics. Over 5% of microplastics, in the form of microfibers from washing clothing, remain in the effluent process stream to be subsequently released into the environment via the discharge of treated effluent into water, by effluent irrigation to land, or land application of sewage sludge/biosolids (Keswani et al. 2016; Mahon et al. 2017; Mason et al. 2016; Nizzetto et al. 2016).

21.2.2.1 Land

The amount of microplastics in terrestrial environments is currently estimated to be equal or greater than it is in the world's oceans, and it is continuing to increase (Rillig 2012). The amount of plastic residues within soil in an industrial area represented 0.03%–6.7% of the mass of soil (Bläsing and Amelung 2018). Several sources of plastic pollution are associated with a range of land use practices within terrestrial environments (Windsor et al. 2019). Agricultural runoff may incorporate microplastics produced from the degradation of greenhouse films, plastic mulch, irrigation systems, and planters (Koelmans et al. 2017). Urban land use and associated activities also provide several sources of plastic pollution. In particular, loss during waste collection and disposal, industrial spillage, and release from landfills provide significant sources of plastic to land (Windsor et al. 2019). For instance, plastic comprised at least 10% of the mass of municipal solid waste in 58% (61 out of 105) of countries contributing data in 2005 (Jambeck et al. 2015).

21.2.2.2 Freshwater

The major microplastic sources of freshwater ecosystems include land-based plastic litter carried by wind, deliberate dumping, stormwater discharges from urban areas, roadway drainage systems, agricultural runoff, and WWTP effluent discharges (Ziajahromi et al. 2016). Macroplastic material entering freshwater ecosystems by these same means are eroded and fragmented to microplastics by exposure to sunlight and wind and abrasion by sediments and water flow (Triebskorn et al. 2019). Current evidence strongly suggests that rivers in both rural and urban areas contain some of the highest concentrations of plastic and are hotspots of plastic pollution (Siegfried et al. 2017). River systems are pivotal conduits for plastic transport within terrestrial, floodplain, riparian, benthic, and transitional ecosystems to which they connect (Windsor et al. 2019). Evidence suggests that freshwater systems share similarities to marine systems in the types of processes that transport microplastics (e.g., surface currents); the prevalence of microplastics (e.g., numerically abundant and ubiquitous); the approaches used for sampling, detection, identification, and quantification (e.g., density separation, filtration, sieving, and infrared spectroscopy); and the potential impacts (e.g., physical damage to organisms that ingest them and chemical transfer of toxicants) (Eerkes-Medrano et al. 2015). In comparison to the marine environment, relatively few studies have investigated the risk presented by microplastics in freshwater ecosystems (Wagner et al. 2014). The laboratory assessments completed to date have typically used high concentrations of microplastics that are not representative of environmentally realistic concentrations, and only two studies observed adverse effects (Triebskorn et al. 2019).

21.2.2.3 Marine Environment

In seawater, the plastic polymers commonly present as microplastics are polypropylene, polyethylene, polystyrene, polyvinylchloride, and polyethylene terephthalate (Guzzetti et al. 2018). It is well documented that microplastics are abundant and widespread in the marine environment (Botterell et al. 2019). Worldwide data on solid waste, population density, and economic status have been combined to estimate that 275 million metric tons of plastic waste were generated in 192 coastal countries in 2010, with 4.8–12.7 million metric tons entering the ocean (Burgess and Ho 2017; Jambeck et al. 2015). The main human activities linked to the release of microplastics into the marine environment are aquaculture, fishing, tourism, the food and consumer goods packaging industries, industrial and domestic wastewater systems, and plastic litter (Guzzetti et al. 2018). Key receptor species negatively impacted by microplastics include several zooplankton taxa that readily ingest microplastics due to their small size, particularly benthic organisms exposed to high densities of microplastics concentrated in sediment (Botterell et al. 2019). Most organism-impact studies have been conducted under controlled laboratory conditions using concentrations of microplastics that are not representative of those prevalent in the marine environment (Botterell et al. 2019), with the consequence that it is very difficult to assess the risk that they may present to biota in natural marine ecosystems.

21.2.2.4 Microplastics in New Zealand

There is limited information on the quantity and fate of microplastics in the New Zealand environment. However, the use of plastic, recycling, and management of waste, together with the quantity of plastic waste and litter produced per capita in New Zealand, present similar trends to those anywhere else in the world. We can therefore expect that the quantity, distribution, and impacts of microplastics in New Zealand's environment are likely to be comparable to those observed in other developed countries.

The limited studies of microplastics completed to date in New Zealand have demonstrated the concentration of microplastics in exposed coastal beaches is significantly greater than those in harbor and estuarine environments, suggesting coastal beaches are exposed to microplastics from coastal transport (Clunies-Ross et al. 2016).

The concentration of microplastics in urban streams in Auckland is similar to that within large rivers in Europe and the United States (Dikareva and Simon 2019), suggesting that local-scale factors may be more important than catchment-scale processes in determining microplastic pollution in small urban streams (Dikareva and Simon 2019).

Currently there have been no studies investigating the prevalence and types of microplastics in larger rivers, lakes or ponds, soil, or groundwater in New Zealand.

21.3 RISK OF MICROPLASTICS

In recent years, plastic pollution has become an issue of environmental concern that has intensively been discussed in the scientific literature and public media (Triebskorn et al. 2019). A similar trend has occurred in New Zealand, and the issue of plastic pollution is increasingly recognized within our communities. Research on the complex topic of microplastics and their potential risk to ecosystems and humans is incomplete and many knowledge gaps remain. Some scientists are concerned that the environmental risks from some types of microplastics are exaggerated. For example, the ban on microbeads introduced into many countries including New Zealand may lead to negligible risk reduction, as the dominant type of microplastic debris discharged to the environment from wastewater treatment plant effluents are polyester fibers and fragments (Burton 2017). In view of the limited data currently available on the distribution and fate of microplastics in the New Zealand environment, we recommend that a precautionary approach is adopted until the significant knowledge gaps are addressed and more robust risk assessments can be made.

It has been suggested that the process of characterizing microplastic risk must be based on an analysis of "true" risk, incorporating realistic laboratory exposure scenarios based on environmentally relevant exposure and concentrations, followed by field-based evaluations (Burton 2017). Additive or even synergistic effects of microplastics may occur in already stressed ecosystems impacted by excess nutrients, low dissolved oxygen, solids from erosion, pathogens, altered flows, degraded habitats, temperature, and loss of shading. It has been suggested that microplastics may represent a relatively minor risk to ecosystems compared to other major stressors (Burton 2017). There is a call for the science on microplastics to move away from talking in terms of "potential" risks, to encompass "a more rigorous and mature risk assessment of plastic debris" (Koelmans et al. 2017).

The level of uncertainty, knowledge gaps, and the nature of the risks as latent and cumulative mean that any analysis of "realistic," "true," or "actual risk" of microplastics being called for in the literature should be truly interdisciplinary and include robust input from different academic, disciplinary, and community knowledge perspectives. It is long established in the social studies of science, risk, and decision-making that robust risk assessment and risk management approaches ought to include social and political deliberation in addition to biophysical assessments (Kasperson and Kasperson 1996; Nowotny et al. 2003).

Importantly, risk assessments and acceptable levels of risk need to be debated from diverse perspectives, as risk can mean different things in different contexts. For example, many expert weightings of risk are determined by exposure guidelines and toxicological standards that are based on Western rather than indigenous customary practices around food gathering and consumption. This is particularly relevant for the New Zealand context. Considering different disciplines and sets of expertise is important in formulating assessments and judgments about risk. Where uncertainties exist, it is important for science and policy to have a normative and ethical orientation to acknowledge that risk is subjective rather than absolute. The issue of microplastics has an inherent high level of complexity and uncertainty and is also characterized as a "wicked problem" that is symptomatic of a number of intersecting issues (Brown et al. 2010). As such, wider engagement in a risk assessment process enables different contextual connections to be made, leading to more robust, scalable, and multipronged interventions to achieve meaningful change to help reduce the risks and potential impacts to human and environmental health.

There are also potential financial and economic risks to consider. The presence of microplastic contamination in the environment could threaten both ecosystems and the economic services that they support. This is particularly relevant to New Zealand, where primary export industries rely upon terrestrial and aquatic environments that need to be largely free of contaminants in order to produce high quality food.

21.3.1 Environmental Health Risk

The growing amount of interest among the general public, researchers, and media has caused plastic debris to be perceived as a major threat to environment and human health (Galgani et al. 2019). However, many knowledge gaps need to be addressed before the environmental and human health risks of microplastics can be fully assessed and ranked against other emerging environmental issues (Koelmans et al. 2017). It is suggested that microplastics can induce physical and chemical toxicity to exposed organisms. These can result in physical injuries, inducing inflammation and stress, or they can result in a blockage of the gastrointestinal tract and a subsequent reduced energy intake or respiration (Guzzetti et al. 2018). The microplastic concentrations detected in the environment are typically orders of magnitude lower than those reported to affect endpoints, such as biochemistry, feeding, reproduction, growth, tissue inflammation, and mortality in organisms (Burns and Boxall 2018). Consequently, there is currently limited evidence to suggest that microplastics are causing significant adverse impacts, or, that they increase the uptake of hydrophobic organic compounds into organisms (Burns and Boxall 2018).

Although plastic is often described as an inert material because of its chemically stable polymeric structure, every piece of plastic contains a complex chemical cocktail of monomers, oligomers, and other chemical additives. Plastic materials come from a multitude of sources and comprise different sizes, shapes, thickness, density, colors, and types of material. Chemical additives are incorporated into the polymers during production, sometimes accounting for a large proportion of the overall weight (e.g., phthalates, which are used to alter the properties of plastics, can comprise up to 50% of a polyvinylchloride product's total weight). There are several categories of additives, including antioxidants, plasticizers, colorants, reinforcements or fillers, flame retardants, and UV stabilizers (Rochman et al. 2019). The potential bioavailability of compounds added to plastics at the time of manufacture, as well as those adsorbed from the environment, is a complex issue that merits more widespread investigation (Moore 2008). The hypothesis that microplastics are vectors for hydrophobic organic chemicals that increase their bioavailability to aquatic organisms and ultimately humans is a topic that has been both supported and challenged in research and review papers.

It is undeniable that plastic residue and microplastics adsorb hydrophobic organic chemicals and metals from the environment and can accumulate concentrations many times higher than that of natural organic particulate matter. Although the relative role of microplastics as vectors of hydrophobic organic chemicals to organisms is generally considered minor in comparison to that of natural exposure pathways (such as water, food, and natural particulate matter), it is important to emphasize that microplastic concentrations and environmental conditions change over time, and spatio-temporal hotspots of microplastics do (and will) occur (Fok and Cheung 2015). However, there is little evidence that microplastics play a major role in the bioaccumulation of persistent organic chemicals by biota when compared to their total dietary and environmental exposure (Lohmann 2017). Regardless, the transfer of hydrophobic organic chemicals from microplastics into biota needs to be comprehensively investigated to better understand the effects of weathering, sorption, and desorption processes among different polymers under varying conditions (Hartmann et al. 2017).

Microplastics could present a pathway for organisms to be exposed to chemical additives that otherwise would not be easily transferred into the environment, and proportions of these additives according to their functions are illustrated in Figure 21.1. Many chemicals associated with plastic

FIGURE 21.1 Word cloud to illustrate the proportions of plastic additives that include plasticizers, flame retardants, antioxidants, stabilizers and UV stabilizers, heat stabilizers, slip agents, curing agents and cross-linkers, dyes and pigments, and biocides.

packaging are persistent, bioaccumulative, and toxic, and some have endocrine disruption activity (Groh et al. 2019). It has been suggested that research should focus on the release of phenolic additive-derived chemicals (i.e., alkylphenols, bisphenol A, UV stabilizers, and antioxidants) from microplastics to the food web, but there are no data to date demonstrating that the ingestion of microplastics presents a pathway for the uptake of these compounds by biota (Lohmann 2017).

Some caution is warranted when interpreting the significance of the outcome of earlier studies assessing the risk of chemical contaminants in microplastics. The wide range of experimental conditions, test concentrations, and test organisms used in these investigations is likely to either over- or underestimate the resulting exposure risk. Significantly, concentrations of microplastics and contaminants used in many of these studies are orders of magnitude higher than those typically found in natural ecosystems, where the concentrations of natural particles, algae, and invertebrates that organisms are feeding on are greater than those of microplastics and therefore preferentially ingested by biota (Burton 2017). Importantly, there is growing demand from scientists to use standardized methods for collecting, quantifying, and characterizing microplastics, combined with common species of biota and test conditions so the results obtained from different studies can be compared (Burton 2017).

21.3.1.1 Biosecurity Risk

Microplastics are a suspected biosecurity risk, acting as mobile substrates or microrafts for the spread of pathogens and invasive species within environments in which they would otherwise be absent from (Eckert et al. 2018; Gregory 2009; Lamb et al. 2018). There are several known sources for the accumulation and spread of plastic litter and associated organisms. It has also been hypothesized that microplastics may provide a vector for rafting sewage-associated pathogenic microbes that survive wastewater treatment in WWTPs, thereby providing a vector for the pathogens and/or antibiotic resistant microbes into the environment via the discharge of treated effluent (Eckert et al. 2018; Keswani et al. 2016). Finally, plastic at sea may transport alien species over long distances or act as substratum for mobile and fixed organisms, providing a support for colonization (Casabianca et al. 2019).

21.3.2 Human Health Risk

For ecosystems and biota, the evidence so far is that the ecological risks of microplastics are low, apart from locations where microplastics are likely to be concentrated. For plastics of sizes below 5 mm, there are some locations in coastal waters and sediments where ecological risks might occur (SAPEA 2019).

Likewise, the potential human health effects of microplastics are unknown (Wright and Kelly 2017). The risk of microplastic exposure to humans could occur via diet (food and water) or inhalation, as evidenced by the observations of plastic microfibers in lung tissue biopsy samples (Wright and Kelly 2017). For example, a recent Science Advice for Policy by European Academies report, based on an interdisciplinary analysis by independent scientists, highlights that occupational exposure of workers to microplastics can lead to granulomatous lesions, causing respiratory irritation, functional abnormalities, and other conditions, such as flock worker's lung in humans (SAPEA 2019). The chemical additives in microplastics can have additional (and difficult to assess) human health effects, such as reproductive toxicity and carcinogenicity, but the risk is probably small at present (SAPEA 2019). Overall, while there is strong evidence of the impacts of microplastics in animal models under laboratory conditions, it is not known if this translates to a potential risk to humans. There is limited information on the potential transfer of microplastics and associated contaminants from seafood to humans and the implications for human health, although there are concerns that indigenous communities may bear a greater burden of risk, due to, for example, greater frequency of consumption of shellfish than assumed in standard exposure and risk assessment models (Gismondi and Sherman 1996). A significant knowledge gap in this respect is the absence of bioaccumulation factors for microplastics in commonly consumed types of seafood, which is a prerequisite to establishing the potential human health impacts of microplastics in seafood (Carbery et al. 2018).

21.4 MANAGING THE RISK OF MICROPLASTICS

There is no single solution to a global problem that is deeply rooted in human systems of production and consumptive behavior, fueled by convenience, and compounded by often absent or inadequate waste management practices and infrastructures (Burgess and Ho 2017). One potential solution is the development of alternatives to petrochemical-based plastics, e.g., using plant-based materials to produce truly biodegradable plastics. However, the misconception that these so-called biodegradable plastics are viable alternatives to conventional plastics needs to be acknowledged (Burgess and Ho 2017).

While some of the risks associated with microplastics have been investigated, many remain unaddressed, as scientists, politicians, and the wider society continue to debate the magnitude of the problem that plastic waste and microplastics may represent (Burgess and Ho 2017). Burgess and Ho (2017) facilitated a specialists' debate to continue the dialogue regarding the risks of microplastics in aquatic environments by bringing together seven representatives and viewpoints from industry, government, academia, and a non-governmental organization. These specialists agreed a coordinated approach was a prerequisite to resolving this complex issue. There is a strong consensus that some caution is warranted when interpreting the results and conclusions from many of the published studies on the impact and risks of microplastics in the environment. It is critical that studies of effects and monitoring use robust and validated experimental designs in order to identify the materials, activities, and practices representing the highest contribution to the problem (Burns and Boxall 2018). As such, the experimental design and analytical methods that have been employed in studies assessing the impact and risk of microplastics in the environment must be carefully reviewed before considering the significance of the stated risk.

Although the evidence of harm from microplastics remains to be confirmed, efforts to reduce the release of plastic material into the environment should remain a priority. The view of non-government organizations is that "we know enough to act," illustrating the rising public sentiment and concern that the issue of plastic pollution needs to be more proactively managed (World Economic Forum 2016).

Technical solutions are emphasized. For example, microfibers are among the most common types of microplastics found in environmental samples, and the production, use, and washing of synthetic textiles, including clothing, are recognized to be significant sources (Hartline et al. 2016). As such, filters on washing machines may be a simple solution to prevent the release of microfibers into WWTPs and subsequently into the environment. The increasing adoption of plastic-based materials used in the construction industry represents another potentially significant source for microplastics, particularly as these materials weather and age. In addition, tire wear particles are known to be the source of a large fraction of microplastics within stormwater, urban, and roading network runoff entering the environment. Interception and capture methods including bioretention cells, rain gardens, and sand filters have the potential to reduce the amount of microplastics entering urban catchments from these sources (Rochman et al. 2019).

It is recognized that community awareness and behavioral change are part of the suite of required solutions. The plastic issue is similar to other environmental issues, based on the scientific literature, in that it requires increased emphasis on consumer behavior, behavioral change, and consumer choice to better manage human impacts on the environment (Pahl and Wyles 2017).

One way to reduce plastic pollution in the environment is to minimize the amount of single-use plastic used in our daily lives (Xanthos and Walker 2017). However, relying on consumer behavioral change is problematic for many reasons, including that plastics are ubiquitous and embedded in all aspects of our daily life – in food production, transport, communications, hygiene, medical, and personal use. While consumer choice can contribute to sustainable change, avoiding plastics requires a significant change in regulation, infrastructure, technologies, and social practices in order to influence household consumption and patterns of use. Consumer pressure has been effective in encouraging businesses to consider more sustainable alternatives to disposable plastic products including reducing the use of plastic bags, straws, and packaging, but far more attention, research, and effort is needed in the redesign of viable socio-technical alternatives to plastics if we are to achieve meaningful sustainable change.

While small choices and individual household actions can add up, we lack the tools to measure, consolidate, and prove this. There are also significant socio-economic factors – many households are time poor, cash poor, transport poor, and geographically isolated and therefore have very limited options available to participate in exercising a "choice" to reduce plastics in everyday life. The most effective way to reduce plastics pollution is upstream design/redesign to stop producing plastics in the first place, and to invest in finding viable alternatives in manufacturing, production, and distribution. This requires structural change, including regulation, industry investment in change, and investment in science, innovation, and disruptive technologies.

Consumer lobbying of industry and politicians is arguably a far more effective pathway for building sustainable change, especially given increased access and participation in social media technologies, and the speed, influence, and power of social networking (Pearson et al. 2016). This suggests that the value of consumer action lies in social networking for organized and tactical lobbying for political, regulative, and industry change. Consumer choice and voluntary consumer action offer an ineffective downstream response, unless combined with regulatory and policy levers.

It is important to reduce the entry of additional plastic waste into the environment through better litter collection and recycling capacity. These initiatives could be effective through the combined actions of the public, industry, scientists, policymakers, and possibly increased funding for cleaning our oceans (Rochman et al. 2019). There is recognition that voluntary consumer initiatives and citizen science-oriented beach cleanups are downstream interventions that are not very effective in terms of reducing risks and impacts (Burgess and Ho 2017). It is important to note that consumer behavioral change is only a very small aspect of the multipronged solutions required to address this wicked problem. Some authors have stated that more respect for the environment and its ecosystems by industry and the general public is needed (Guzzetti et al. 2018).

The Intergovernmental Science-Policy Platform on Biodiversity and Ecosystem Services (IPBES) 2019 global assessment synthesis of 15,000 publications related to biodiversity decline notes that changes in consumption make limited contribution to actual reductions in waste (United Nations 2019). This comprehensive global study notes the importance of multiactor governance interventions, leverage points, strategic policy mixes, scaling, and coordination of effort. Other studies emphasize that effective risk management strategies ought to focus on regulation, politics, industry practice, circular economies, green chemistry, and "de-materialization" to get plastics out of the economy (World Economic Forum 2016).

A risk assessment framework for plastic debris of all sizes and in all habitats has been proposed (Koelmans et al. 2017). This framework aligns with other global environmental studies (United Nations 2019; World Economic Forum 2016), suggesting that widening the boundaries of inquiry from microplastics to encompass addressing the impacts of macroplastics and waste in the environment would be prudent for risk management frameworks.

21.4.1 International Strategies to Manage Microplastics

Numerous policy and regulatory developments have been implemented around the globe to reduce the use and emissions of microplastics (Xanthos and Walker 2017). Perhaps the most publicized are the ban of microbeads in all wash-off cosmetic products including the United States Microbead Free Water Act of 2015 and the Environmental Protection (Microbeads; England) Regulations 2017 (Burns and Boxall 2018). The ban was followed up by other countries including Canada and New Zealand (Rochman et al. 2019).

A ban on single-use plastics by 2021 in Canada will consider a wider range of plastic products including not only plastic bags, but straws, cutlery, plates, and stir sticks[1] The objective of these bans is to remove a major plastic pollution source by reducing litter.

[1] https://www.nationalgeographic.com/environment/2019/06/canada-single-use-plastics-ban-2021/.

The development of governance and mitigation strategies to manage the issue of microplastics is very challenging due to the high level of complexity of this issue (Rochman et al. 2019). Despite these limitations, there are policy initiatives for reducing marine litter aiming at: (1) understanding presence and impacts and (2) preventing further inputs or reducing total amounts in the environment (Eerkes-Medrano et al. 2015). Examples of efforts to manage marine litter include the U.S. Interagency Marine Debris Coordinating Committee, which supports the U.S. national/international marine debris activities, and "recommends research priorities, monitoring techniques, educational programs, and regulatory action." The European Commission's Marine Strategy Framework Directive (the Directive) has designated a Technical Subgroup on Marine Litter to provide "scientific and technical background for the implementation of Directive requirements," which include identification of research needs, development of monitoring protocols, preventing litter inputs, and reducing litter in the marine environment. The Directive's "litter" designation includes microplastics and acknowledges a limitation in "knowledge of the accumulation, sources, sinks ... environmental impacts ... temporal and spatial patterns and potential physical and chemical impacts" of microplastics (Eerkes-Medrano et al. 2015).

Although the risk of microplastics is still unclear, there is a need to focus on solutions, such as proper regulations on plastic production, waste management practices, plastic recycling schemes, and politicians encouraging a change of attitude by society (Burgess and Ho 2017). There is more public, policy, and management interest for marine than freshwater ecosystems due to greater knowledge and publicity of the extent and impacts of microplastics in the marine environment, but overall, more effort is needed to address and manage the issues of plastic pollution (Eerkes-Medrano et al. 2015).

There is an increasing number of initiatives lead by non-governmental organizations (NGOs) to address the issue of microplastics in the environment. For instance, a group of international scientists wanting to prevent plastic pollution has formed the Plastic Pollution Emissions Working Group (www.plasticpeg.org). International NGOs raising awareness of the global impact of plastic and microplastics include Algalita and 5 Gyres. In New Zealand, Sustainable Coastlines are actively raising awareness about plastic pollution, encouraging debate of this issue, and identifying mitigation and reduction solutions. Improving waste management infrastructure in developing countries is paramount and will require substantial resources and time. While such infrastructure is being developed, industrialized countries can take immediate action by reducing waste and curbing the growth of single-use plastics (Jambeck et al. 2015).

21.4.2 New Zealand Initiatives

The reason for the microbead ban by the New Zealand government is to *prevent plastic microbeads, which are non-biodegradable, entering our marine environment. They can harm both marine life and life higher on the food chain including humans* (NZ Ministry for the Environment 2019). Furthermore, New Zealand will ban single-use plastic bags, although many retailers have already phased them out (New Zealand Government 2019). In addition to the bans on microbeads and single-use plastic bags, the government is also involved in a range of initiatives to reduce the amount of plastic entering the environment in New Zealand. The Office of the Prime Minister's Chief Science Advisor established a panel to investigate options to reduce the impact of plastic: Rethinking Plastics in Aotearoa, New Zealand (Prime Ministers Chief Science Advisor 2019). New Zealand is a signatory to the United Nations-led CleanSeas campaign to rid our oceans of plastic. New Zealand also signed the New Plastics Economy Global Commitment, an initiative led by the Ellen MacArthur Foundation, in collaboration with the United Nations Environment Programme.

Research capability and expertise in the field of microplastics are growing in New Zealand. One important requirement is the development of science capability to measure and characterize microplastic particles based on their structure and chemical composition and assess their impact and risk. Table 21.1 summarizes the current organizations and equipment available to characterize plastic and microplastics in New Zealand.

TABLE 21.1
New Zealand-Based Capability to Measure and Characterize Microplastics

Organizations	Capability
Institute of Environmental Science and Research (ESR), Scion Research, and the University of Canterbury	Attenuated total reflection–Fourier-transform infrared spectroscopy (ATR-FTIR) and FTIR-microscope instruments for the identification and quantitation of microplastics
	Scanning electron microscope (SEM) and transmission electron microscopy (TEM) instruments with confocal capabilities for physical and chemical analysis of microplastics
University of Auckland	Stereomicroscopes with digital camera systems for visual identification, counting, and size measurement of microplastics
Scion Research	Solid-state nuclear magnetic resonance (NMR) spectroscopy and pyrolysis-gas chromatography (GC) instruments for polymer characterization, laboratory facilities to accelerate polymer weathering, and purpose-built state-of-the-art biodegradation facilities to assess the fate of different types of polymer, and polymer-microbe interactions under environmental conditions

21.5 CONCLUSIONS

There has been an increase in microplastic-related research in New Zealand. The microplastics Aotearoa Impacts & Mitigation of Microplastics project is the single largest project. Summaries of the research program objectives and critical steps are provided in Figure 21.2. The risks that microplastics pose to ecosystems and human health are not fully characterized and many research and knowledge gaps remain. Research on microplastics is a very new topic, and there is limited information about their risks, particularly in New Zealand. The main risks are likely due to the multiple chemical additives used in plastics. Therefore, microplastics may represent a major source of these chemical additives into the environment and to exposed biota. Findings so far suggest

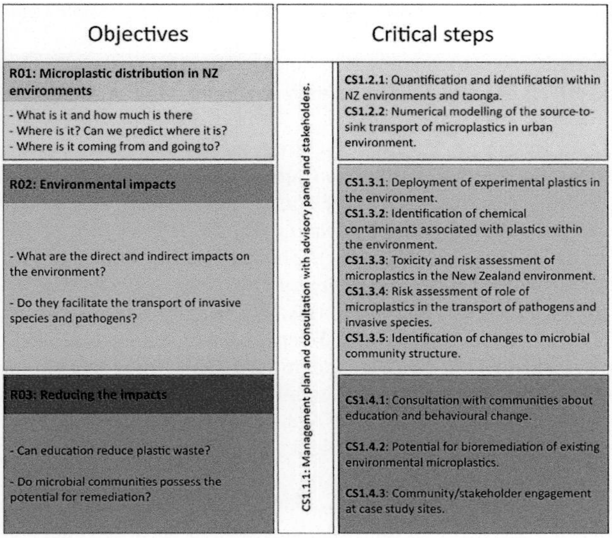

Objectives
Critical steps
R01: Microplastic distribution in NZ environments
- What is it and how much is there
- Where is it? Can we predict where it is?
- Where is it coming from and going to?
CS1.2.1: Quantification and identification within NZ environments and taonga.
CS1.2.2: Numerical modelling of the source-to-sink transport of microplastics in urban environment.
R02: Environmental impacts
- What are the direct and indirect impacts on the environment?
- Do they facilitate the transport of invasive species and pathogens?
CS1.3.1: Deployment of experimental plastics in the environment.
CS1.3.2: Identification of chemical contaminants associated with plastics within the environment.
CS1.3.3: Toxicity and risk assessment of microplastics in the New Zealand environment.
CS1.3.4: Risk assessment of role of microplastics in the transport of pathogens and invasive species.
CS1.3.5: Identification of changes to microbial community structure.
R03: Reducing the impacts
- Can education reduce plastic waste?
- Do microbial communities possess the potential for remediation?
CS1.4.1: Consultation with communities about education and behavioural change.
CS1.4.2: Potential for bioremediation of existing environmental microplastics.
CS1.4.3: Community/stakeholder engagement at case study sites.
CS1.1.1: Management plan and consultation with advisory panel and stakeholders.

FIGURE 21.2 Research and engagement program: A schematic representation of the ESR (Institute of Environmental Science and Research)-led MBIE (Ministry of Business, Innovation and Employment) Endeavour Aotearoa Impacts & Mitigation of Microplastics project showing the main objectives and critical steps and the scientists involved.

that microplastics are likely to pose risks at hot spots, where concentrations will be highest and in combination with other stressors. In view of the significant knowledge gaps, we recommend a precautionary approach until the risks are better characterized. As such, regional councils need to keep abreast of the latest developments through close communication with the main research groups in New Zealand. Some of the research gaps include:

- Assessment of the prevalence and types of microplastics in soil and larger freshwater catchments including rivers and lakes impacted by human activity
- The same applies for groundwater, particularly as there is a trend in New Zealand to remove WWTP discharges from water and instead dispose on land
- As previous research has focused on marine environments, there is a need to better understand the risk that microplastics represent to estuarine, coastal, freshwater, and terrestrial (soil) environments
- Assessment of the potential impact and risk of microplastics on taonga (key native) species in New Zealand
- Characterization of human exposure to microplastics via recreational and customary harvest to assess whether consumers of wild foods (e.g., mana whenua, local Maori tribes, the indigenous people of New Zealand) may particularly be exposed to an increased dietary loading compared to the general population.

It is important to align with the Tiriti o Waitangi (the treaty of Waitangi between the crown and Maori) process and mana whenua (people of the land) as key partners in environmental and resource management. Regional councils should continue to work closely with mana whenua in risk assessment to reflect actual risks in the context of cultural values and practices, and in the design of co-management strategies and coordinated national policy and industry initiatives to reduce the impacts of human waste and production practices on the receiving environment.

REFERENCES

Bläsing, M., Amelung, W., 2018. Plastics in soil: Analytical methods and possible sources. *Science of the Total Environment*. 612, 422–435.

Botterell, Z. L. R., Beaumont, N., Dorrington, T., Steinke, M., Thompson, R. C., Lindeque, P. K., 2019. Bioavailability and effects of microplastics on marine zooplankton: A review. *Environmental Pollution*. 245, 98–110.

Brown, V. A., Harris, J. A., Russell, J. Y., 2010. Tackling wicked problems: Through the transdisciplinary imagination. *International Journal of Climate Change Strategies and Management*. 2, 14–16.

Burgess, R. M., Ho, K. T., 2017. Microplastics in the aquatic environment—Perspectives on the scope of the problem. *Environmental Toxicology and Chemistry*. 36, 2259–2265.

Burns, E. E., Boxall, A. B. A., 2018. Microplastics in the aquatic environment: Evidence for or against adverse impacts and major knowledge gaps. *Environmental Toxicology and Chemistry*. 37, 2776–2796.

Burton, G. A., 2017. Stressor exposures determine risk: So, why do fellow scientists continue to focus on superficial microplastics risk? *Environmental Science and Technology*. 51, 13515–13516.

Carbery, M., O'Connor, W., Palanisami, T., 2018. Trophic transfer of microplastics and mixed contaminants in the marine food web and implications for human health. *Environment International*. 115, 400–409.

Casabianca, S., Capellacci, S., Giacobbe, M. G., Dell'Aversano, C., Tartaglione, L., Varriale, F., Narizzano, R., et al., 2019. Plastic-associated harmful microalgal assemblages in marine environment. *Environmental Pollution*. 244, 617–626.

Clunies-Ross, P. J., Smith, G. P. S., Gordon, K. C., Gaw, S., 2016. Synthetic shorelines in New Zealand? Quantification and characterisation of microplastic pollution on Canterbury's coastlines. *New Zealand Journal of Marine and Freshwater Research*. 50, 317–325.

da Costa, J. P., Santos, P. S. M., Duarte, A. C., Rocha-Santos, T., 2016. (Nano)plastics in the environment—Sources, fates and effects. *Science of the Total Environment*. 566–567, 15–26.

Dikareva, N., Simon, K. S., 2019. Microplastic pollution in streams spanning an urbanisation gradient. *Environmental Pollution*. 250, 292–299.

Eckert, E. M., Di Cesare, A., Kettner, M. T., Arias-Andres, M., Fontaneto, D., Grossart, H. P., Corno, G., 2018. Microplastics increase impact of treated wastewater on freshwater microbial community. *Environmental Pollution*. 234, 495–502.

Eerkes-Medrano, D., Thompson, R. C., Aldridge, D. C., 2015. Microplastics in freshwater systems: A review of the emerging threats, identification of knowledge gaps and prioritisation of research needs. *Water Research*. 75, 63–82.

Fok, L., Cheung, P. K., 2015. Hong Kong at the Pearl River Estuary: A hotspot of microplastic pollution. *Marine Pollution Bulletin*. 99, 112–118.

Frias, J. P. G. L., Nash, R., 2019. Microplastics: Finding a consensus on the definition. *Marine Pollution Bulletin*. 145–147.

Galgani, L., Beiras, R., Galgani, F., Panti, C., Borja, A., 2019. Editorial: "Impacts of marine litter". *Frontiers in Marine Science*. 6, 208.

Gismondi, M., Sherman, J., 1996. Pulp mills, fish contamination, and fish eaters: A participatory workshop on the politics of expert knowledge. *Capitalism Nature Socialism*. 7, 127–137.

Gregory, M. R., 2009. Environmental implications of plastic debris in marine settings- entanglement, ingestion, smothering, hangers-on, hitch-hiking and alien invasions. *Philosophical Transactions of the Royal Society B: Biological Sciences*. 364, 2013–2025.

Groh, K. J., Backhaus, T., Carney-Almroth, B., Geueke, B., Inostroza, P. A., Lennquist, A., Leslie, H. A., et al., 2019. Overview of known plastic packaging-associated chemicals and their hazards. *Science of the Total Environment*. 651, 3253–3268.

Guzzetti, E., Sureda, A., Tejada, S., Faggio, C., 2018. Microplastic in marine organism: Environmental and toxicological effects. *Environmental Toxicology and Pharmacology*. 64, 164–171.

Hartline, N. L., Bruce, N. J., Karba, S. N., Ruff, E. O., Sonar, S. U., Holden, P. A., 2016. Microfiber masses recovered from conventional machine washing of new or aged garments. *Environmental Science and Technology*. 50, 11532–11538.

Hartmann, N. B., Rist, S., Bodin, J., Jensen, L. H. S., Schmidt, S. N., Mayer, P., Meibom, A., Baun, A., 2017. Microplastics as vectors for environmental contaminants: Exploring sorption, desorption, and transfer to biota. 13, 488–493.

Jambeck, J. R., Geyer, R., Wilcox, C., Siegler, T. R., Perryman, M., Andrady, A., Narayan, R., Law, K. L., 2015. Plastic waste inputs from land into the ocean. *Science*. 347, 768–771.

Kasperson, R. E., Kasperson, J. X., 1996. The social amplification and attenuation of risk. *The ANNALS of the American Academy of Political and Social Science*. 545, 95–105.

Keswani, A., Oliver, D. M., Gutierrez, T., Quilliam, R. S., 2016. Microbial hitchhikers on marine plastic debris: Human exposure risks at bathing waters and beach environments. *Marine Environmental Research*. 118, 10–19.

Koelmans, A. A., Besseling, E., Foekema, E., Kooi, M., Mintenig, S., Ossendorp, B. C., Redondo-Hasselerharm, P. E., Verschoor, A., Van Wezel, A. P., Scheffer, M., 2017. Risks of plastic debris: Unravelling fact, opinion, perception, and belief. *Environmental Science and Technology*. 51, 11513–11519.

Lamb, J. B., Willis, B. L., Fiorenza, E. A., Couch, C. S., Howard, R., Rader, D. N., True, J. D., et al. 2018. Plastic waste associated with disease on coral reefs. *Science*. 359, 460–462.

Lambert, S., Wagner, M., 2016. Characterisation of nanoplastics during the degradation of polystyrene. *Chemosphere*. 145, 265–268.

Laskar, N., Kumar, U., 2019. Plastics and microplastics: A threat to environment. *Environmental Technology and Innovation*. 14.

Law, K. L., Thompson, R. C., 2014. Microplastics in the seas. *Science*. 345, 144–145.

Lima, A. R. A., Barletta, M., Costa, M. F., 2015. Seasonal distribution and interactions between plankton and microplastics in a tropical estuary. *Estuarine, Coastal and Shelf Science*. 165, 213–225.

Lohmann, R., 2017. Microplastics are not important for the cycling and bioaccumulation of organic pollutants in the oceans—but should microplastics be considered POPs themselves? *Integrated Environmental Assessment and Management*. 13, 460–465.

Mahon, A. M., O'Connell, B., Healy, M. G., O'Connor, I., Officer, R., Nash, R., Morrison, L., 2017. Microplastics in sewage sludge: Effects of treatment. *Environmental Science and Technology*. 51, 810–818.

Mason, S. A., Garneau, D., Sutton, R., Chu, Y., Ehmann, K., Barnes, J., Fink, P., Papazissimos, D., Rogers, D. L., 2016. Microplastic pollution is widely detected in US municipal wastewater treatment plant effluent. *Environmental Pollution*. 218, 1045–1054.

Mattsson, K., Jocic, S., Doverbratt, I., Hansson, L.-A., Chapter 13—Nanoplastics in the aquatic environment. In: E. Y. Zeng, (Ed.), *Microplastic Contamination in Aquatic Environments*. Elsevier, San Diego, CA, 2018, pp. 379–399.

Miller, R. Z., Watts, A. J. R., Winslow, B. O., Galloway, T. S., Barrows, A. P. W., 2017. Mountains to the sea: River study of plastic and non-plastic microfiber pollution in the northeast USA. *Marine Pollution Bulletin.* 124, 245–251.

Moore, C. J., 2008. Synthetic polymers in the marine environment: A rapidly increasing, long-term threat. *Environmental Research.* 108, 131–139.

New Zealand Government, 2019. "Mandatory phase out of single-use plastic bags confirmed." Accessed December 6, 2019, from www.beehive.govt.nz/release/mandatory-phase-out-single-use-plastic-bags-confirmed.

Nizzetto, L., Futter, M., Langaas, S., 2016. Are agricultural soils dumps for microplastics of urban origin? *Environmental Science and Technology.* 50, 10777–10779.

Nowotny, H., Scott, P., Gibbons, M., 2003. Introduction: "Mode 2" revisited: The new production of knowledge. *Minerva.* 41, 179–194.

NZ Ministry for the Environment, 2019. "Plastic microbeads ban." Accessed December 6, 2019, from www.mfe.govt.nz/waste/waste-strategy-and-legislation/plastic-microbeads-ban.

Pahl, S., Wyles, K. J., 2017. The human dimension: How social and behavioural research methods can help address microplastics in the environment. *Analytical Methods.* 9, 1404–1411.

Pearson, E., Tindle, H., Ferguson, M., Ryan, J., Litchfield, C., 2016. Can we tweet, post, and share our way to a more sustainable society? A review of the current contributions and future potential of #socialmediaforsustainability. *Annual Review of Environment and Resources.* 41, 363–397.

Prime Ministers Chief Science Advisor, 2019. "Rethinking plastics in Aotearoa New Zealand." Accessed December 6, 2019, from www.pmcsa.ac.nz/our-projects/plastics/rethinking-plastics-in-aotearoa-new-zealand/.

Rillig, M. C., 2012. Microplastic in terrestrial ecosystems and the soil? *Environmental Science and Technology.* 46, 6453–6454.

Rochman, C. M., Brookson, C., Bikker, J., Djuric, N., Earn, A., Bucci, K., Athey, S., et al., 2019. Rethinking microplastics as a diverse contaminant suite. *Environmental Toxicology and Chemistry.* 38, 703–711.

SAPEA, A scientific perspective on microplastics in nature and society. In: SAPEA, (Ed.). *Science Advice for Policy by European Academies*, Berlin, Germany, 2019.

Siegfried, M., Koelmans, A. A., Besseling, E., Kroeze, C., 2017. Export of microplastics from land to sea. A modelling approach. *Water Research.* 127, 249–257.

Triebskorn, R., Braunbeck, T., Grummt, T., Hanslik, L., Huppertsberg, S., Jekel, M., Knepper, T. P., et al., 2019. Relevance of nano- and microplastics for freshwater ecosystems: A critical review. *TrAC Trends in Analytical Chemistry.* 110, 375–392.

United Nations, 2019. "Intergovernmental science-policy platform on biodiversity and ecosystem services—Global assessment report on biodiversity and ecosystem services. Summary for Policy Makers of the global assessment report on biodiversity and ecosystem services. Advanced unedited version, 6 May."

Vollertsen, J., Hansen, A., Microplastic in Danish wastewater: Sources, occurrences and fate. 2017, The Danish Environmental Protection Agency, Environmental Project No. 1906, pp. 54.

Wagner, M., Scherer, C., Alvarez-Muñoz, D., Brennholt, N., Bourrain, X., Buchinger, S., Fries, E., et al. 2014. Microplastics in freshwater ecosystems: What we know and what we need to know. *Environmental Sciences Europe.* 26, 1–9.

Walker, T. R., Adebambo, O., Del Aguila Feijoo, M. C., Elhaimer, E., Hossain, T., Edwards, S. J., Morrison, C. E., et al., Chapter 27—Environmental effects of marine transportation. In: C. Sheppard, (Ed.), *World Seas: An Environmental Evaluation* (2nd ed.). Academic Press, London, UK, 2019, pp. 505–530.

Windsor, F. M., Durance, I., Horton, A. A., Thompson, R. C., Tyler, C. R., Ormerod, S. J., 2019. A catchment-scale perspective of plastic pollution. *Global Change Biology.* 25(4), 1207–1221.

World Economic Forum, 2016. "The new plastics economy: Rethinking the future of plastics." Ellen MacArthur Foundation, and McKinsey & Company. Accessed December 6, 2019, from www.newplasticseconomy.org.

Wright, S. L., Kelly, F. J., 2017. Plastic and human health: A micro issue? *Environmental Science and Technology.* 51, 6634–6647.

Xanthos, D., Walker, T. R., 2017. International policies to reduce plastic marine pollution from single-use plastics (plastic bags and microbeads): A review. *Marine Pollution Bulletin.* 118, 17–26.

Ziajahromi, S., Neale, P. A., Leusch, F. D. L., 2016. Wastewater treatment plant effluent as a source of microplastics: Review of the fate, chemical interactions and potential risks to aquatic organisms. *Water Science and Technology.* 74, 2253–2269.

22 Plastic Food for Fledgling Short-Tailed Shearwaters (*Ardenna tenuirostris*)

A Case Study

Jacinta Colvin, Peter Dann, and Dayanthi Nugegoda

CONTENTS

22.1 Introduction ... 378
- 22.1.1 Plastic Ingestion by Short-Tailed Shearwaters ... 378
- 22.1.2 Why Do Short-Tailed Shearwaters Ingest So Much Plastic? ... 378
- 22.1.3 Location and Amount of Plastic within the Gastrointestinal Tract ... 379
- 22.1.4 Adverse Effects Relating to Plastic Ingestion ... 379
- 22.1.5 Identification of Plastics ... 380
- 22.1.6 Research Aims ... 380

22.2 Materials and Methods ... 380
- 22.2.1 Collection of Samples ... 380
- 22.2.2 Dissection Procedure ... 380
- 22.2.3 Initial Analysis of the Plastic ... 380
- 22.2.4 FTIR Analysis ... 381
- 22.2.5 Statistical Analysis ... 381

22.3 Results ... 382
- 22.3.1 Prevalence of Plastic Ingestion by Beachcast Short-Tailed Shearwater Fledglings on Phillip Island ... 382
- 22.3.2 Location of Plastic within the Gastrointestinal Tract ... 382
- 22.3.3 Condition of the Gastrointestinal Tract and Intestinal Fat Stores ... 383
- 22.3.4 Origin and Physical Characteristics of the Ingested Plastic ... 383
- 22.3.5 FTIR Analysis ... 386

22.4 Discussion and Conclusions ... 386
- 22.4.1 Prevalence of Plastic Ingestion by Short-Tailed Shearwaters ... 386
- 22.4.2 Pathology, Visceral Fat Stores, and Amount of Ingesta in the Proventriculus ... 387
- 22.4.3 Physical Characteristics and Origin of the Recovered Plastic ... 388
- 22.4.4 Size and Location of Ingested Plastic ... 390
- 22.4.5 FTIR Analysis ... 390
- 22.4.6 Future Research ... 391

References ... 391

22.1 INTRODUCTION

22.1.1 Plastic Ingestion by Short-Tailed Shearwaters

Plastic pollution is of significant concern in the marine environment given the increasing production and use of plastics, coupled with the generally poor biodegradability of the material (Derraik 2002). Seabirds are of particular concern, being prone to entanglements and ingestion of plastic. Short-tailed shearwaters (*Ardenna tenuirostris*) are an abundant species of seabird over-represented for plastic ingestion (Carey 2011; Hardesty et al. 2014) and have a migratory pattern that can extend as far south as the oceans around Antarctica, to north of the Bering Strait, establishing breeding colonies in southern mainland Australia and Tasmania. Their abundance, distribution, and predilection for ingesting plastic makes them a suitable species for monitoring plastic in the oceans, although comparative studies following specific colonies in the Southern Hemisphere appear to be uncommon (Cousin et al. 2015; Department of the Environment and Energy 2018). Monitoring populations of short-tailed shearwaters can be useful not only from a wildlife health standpoint, but also for detection of changes in plastic pollution in the marine environment in a way that may be less costly than ongoing ocean trawl studies (Lavers et al. 2018). Although currently abundant, there is evidence that the short-tailed shearwater population may be in decline (BirdLife International 2017).

In the Southern Hemisphere, studies conducted during the 1950s and 1960s were finding virtually no plastic in shearwaters; however, in recent decades, high levels of plastic ingestion are now being recorded (Cousin et al. 2015; Vlietstra and Parga 2002). A study involving birds stranded on North Stradbroke Island in 2010 and 2012 found that 67% of birds had ingested anthropogenic debris, with juveniles being significantly more likely to have plastic in their gut than adults (Acampora et al. 2014). A study where necropsies were performed on illegally harvested pre-fledgling chicks from the relatively remote Clifton Beach Colony in Tasmania during 2012 found 96% of birds contained plastic (Cousin et al. 2015), while similar findings were seen in beachcast fledglings on Phillip Island in 2010, with 100% of the birds were found to have ingested plastic. It is worth noting in the case of the latter two studies, all plastic ingested by these chicks and fledglings is likely to have been fed to them by their parents since they are unable to unable to forage for themselves at this age (Carey 2011).

Fledglings unable to make the migratory journey due to circumstances including poor condition, exhaustion, immaturity, unfavorable weather conditions, illness, injury, or predation, may be unable to take off once on the ocean's surface, and are frequently found washed onto the beaches (beachcast birds), with many dying (Rodríguez et al. 2014; 2018; Skira et al. 1986). The mortality rate for immature individuals can be higher than 50%, dropping significantly to an estimated 5% per year once breeding age has been reached (Skira et al. 1986).

22.1.2 Why Do Short-Tailed Shearwaters Ingest So Much Plastic?

It is thought that most seabirds are ingesting plastics as a consequence of mistaking them for food items (Acampora et al. 2014). There are two main theories as to why this may occur. The first relates to the fact that shearwaters belong to the Order Procellariiformes (otherwise known as tube-nosed seabirds), which includes a number of bird types known to frequently have high levels of plastic ingestion including petrels and albatrosses. These birds have evolved to be wide ranging, pelagic, and highly olfactory; using their sense of smell to assist with finding food. They respond to dimethyl sulfide (DMS) as a cue to locate prey. This is produced naturally in pelagic ecosystems by the enzymatic breakdown of dimethylsulfoniopropionate in marine phytoplankton, especially when being grazed by zooplanktons, which are a food source for shearwaters. Plastic in the ocean can become biofouled with organisms including phytoplankton that can produce DMS, which in turn appears to attract the birds, which mistake the plastic for food. It seems probable that DMS plays a role in the shearwater's predisposition to ingest plastic, with predictions based on patterns obtained in the study by Savoca et al. (2016), suggesting that DMS-responsive species will ingest plastic five times as frequently compared to those that are not responsive to DMS (Ryan 2015; Savoca et al. 2016).

A second theory relates to visual cues. Wedge-tailed shearwaters (*Ardenna pacifica*) have been demonstrated to have enhanced color vision, a trait that the closely related short-tailed shearwaters are considered likely to share and could potentially utilize for hunting via sight. Studies on plastic ingestion by short-tailed shearwaters have generally shown an apparent preference for lighter or more brightly colored plastic, which is presumably either more visible or selected due to the colors matching that of the prey local to the area. Shearwaters forage by pursuit diving, which has been shown to predispose seabird species to increased rates of plastic ingestion, and this species often feeds around gyres and upwellings where plastic pollution can accumulate (Acampora et al. 2014).

There is also an anatomical aspect as to why shearwaters may have a high incidence of plastic in their gastrointestinal tracts, as their proventriculus is connected to the ventriculus by a narrow passage, which can potentially trap indigestible items including plastics, making it difficult for birds to regurgitate them even if they wished to. It appears to be a particular problem for chicks fed plastic items by their parents, such as the individuals in this study, as they have a reduced ability to regurgitate compared to adults. This could allow pieces of plastic to accumulate in the gut during the nesting period (Acampora et al. 2014; Carey 2011).

22.1.3 Location and Amount of Plastic within the Gastrointestinal Tract

The majority of ingested plastic is generally recovered from the ventriculus (gizzard) on postmortem examination of shearwaters, with smaller amounts in the proventriculus (located just above the ventriculus in the gastrointestinal tract). There can be some variation between the mean mass and number of plastic pieces per bird between different studies, which could be due to variations in location, time of year, age, number of birds assessed, selection technique for inclusion in the study, and techniques used to recover the plastic (Acampora et al. 2014; Carey 2011; Vlietstra and Parga 2002).

Age does appear to be an important factor for the plastic load per bird, with juveniles often having higher plastic loads than adults. This could relate to younger birds still carrying plastics in their gut that were fed to them before fledging (retention times are thought to be in excess of 6 months in some cases), adults off-loading plastics to their young during the breeding season, or due to juveniles being less experienced hunters leading to increased plastic ingestion (Acampora et al. 2014; Cousin et al. 2015).

22.1.4 Adverse Effects Relating to Plastic Ingestion

Ingestion of plastics by birds has the potential to cause a number of lethal and sub-lethal effects, including reduced body condition or growth rates via ingestion of material with no nutritional value or by reducing the capacity of the stomach to contain food before satiation occurs. Ingested plastic is also capable of causing mechanical damage within the gut of animals, including obstructions, perforations, and/or ulcerations, which can result in lowered survival rates (Acampora et al. 2014; Carey 2011; Rodríguez et al. 2018). Another potential hazard comes in the form of toxicants leaching from ingested plastic and being absorbed, or via the ingestion of prey where this has occurred. Plastics have been known to sorb a range of metals and hydrophobic chemicals that are potentially hazardous to biota, including the environmentally persistent pollutant polychlorinated biphenyls (PCBs). Positive correlations have been discovered between plastic loads and adipose tissue PCB concentrations in seabird species including short-tailed shearwaters. This is of concern because this toxicant is known to have deleterious effects in avian species that can include altered hormone levels, increased risk of disease, reduced breeding success, neurological abnormalities (such as trembling and limb paralysis), and mortalities (Derraik 2002; Prestt et al. 1970; Ryan et al. 1988; Yamashita et al. 2011).

22.1.5 Identification of Plastics

Plastics recovered from the gastrointestinal tracts of birds are frequently sorted by their visual characteristics (i.e., size, color, shape, origin, and mass), as this method is typically fast, inexpensive, and easy. This kind of analysis can be limited by the lack of polymer composition data, and false positives can occur, as there is no chemical confirmation of the suspected plastic particles. For this reason, it can be advantageous to add an additional identification method, such as Fourier-transform infrared (FTIR) spectroscopy (Shim et al. 2017). This type of spectroscopy works on the principle that when infrared radiation interacts with the sample, it will excite molecular vibrations in a way that is specific to the molecular structure and composition of the substance. The infrared spectra produced via the FTIR spectroscopy can, therefore, be used in identification of specific plastic polymers (Löder and Gerdts 2015).

22.1.6 Research Aims

This research sought to provide a comparative analysis of plastic ingestion by fledgling short-tailed shearwaters on Phillip Island in the Southern Hemisphere over a period of 8 years. Aspects that could be contributing to selectivity by the birds, including color, shape, size, and buoyancy, were also investigated and changes in the proportions of plastic type assessed. Additionally, this appears to be the first study to evaluate plastic ingested by short-tailed shearwaters via the FTIR analysis.

22.2 MATERIALS AND METHODS

22.2.1 Collection of Samples

Fifty-two deceased beachcast short-tailed shearwater fledglings were collected from the shores of Phillip Island, Victoria, between the April 30 and May 2, 2018 to coincide with the peak departure period for the fledglings. Only the removed gastrointestinal tracts were available for this study.

22.2.2 Dissection Procedure

Each section of the gastrointestinal tract was dissected separately using standard techniques. Utilizing a scale detailed by van Franeker (2004), the level of visceral fat was assessed (where "0" indicates no fat, "1" indicates a low amount of fat was present, "2" indicates good fat deposits, and "3" indicates very large fat deposits were present); and tissues were examined for the presence of pre-mortem damage or abnormalities if the level of autolysis allowed it. The amount of food present in the proventriculus was assessed using a scale of 1–3 (where "1" indicates little or no material present, "2" indicates a moderate amount of material, and "3" indicates the proventriculus was distended with a large amount of ingesta). The contents from each part of the gastrointestinal tract were examined in a petri dish under a dissecting microscope and any plastic was removed for further analysis.

22.2.3 Initial Analysis of the Plastic

Recovered plastic was assessed for buoyancy in seawater in a measuring cylinder, then cleaned with distilled water, and air dried. Plastic particles were counted and weighed using a digital balance (0.001 g), and the maximum diameter assessed using the Leica Application Suite (Version 4.9.0) from images obtained from the connected Leica MZ9.5 microscope. Only plastics with a minimum length of 0.5 mm in at least one dimension were included in the study. The minimum size to include in the analysis was chosen as studies appear to indicate that particles larger than this do not readily pass through the gut wall without pre-existing damage (Lusher et al. 2017).

Any objects of uncertain type were checked for deformation when a needle was heated in flame until red hot and applied to the surface of the object (Mossman 2008; Semenova 2017). If any doubt remained as to the identity of a sample, it was earmarked to be checked with FTIR to confirm the nature of the material.

Plastics were classified according to their shapes into fragments, beads, pellets, foams, and fibers, as described by Lusher et al. (2017), and into one of three origin types: industrial pellets (beads used in molding into products, as well as in packaging of cargo during transport), user plastic (often fragmented pieces from household items, such as plastic bottles), and other (including polystyrene, string, rubber, plastic bags, or sheets) following the methods used by Vlietstra and Parga (2002) and Ogi (1990).

If the dimensions of the fragment allowed, following the methodology of Acampora et al. (2014), the plastic pieces were scraped using a scalpel to determine the original color (Figure 22.1), then classified as described by Vlietstra and Parga (2002) into light (white, yellow, and yellow-brown), medium (green, blue, red, and brown), and dark shades (dark green, dark blue, dark red, and gray-black.)

22.2.4 FTIR Analysis

The plastic ingested by 31 shearwaters (163 particles) underwent attenuated total reflection FTIR (ATR-FTIR) analysis using a PerkinElmer Frontier FT-IR/FIR Spectrometer, with a PIKE GladiATR technologies attachment. Samples were prepared for analysis by washing in 70% ethanol, then air drying. Where the dimensions of the plastic allowed, the sample was incised with a scalpel blade to present a flat, "clean" surface in contact with the crystal (Jung et al. 2018). Analysis of the FTIR spectra was performed using Spectrum software (PerkinElmer Inc. 2015). Plastics were identified down to their general types (including polyethylene, polypropylene, polystyrene, and polyvinyl chloride) using the search score and best fit of the spectra to the reference libraries. Only spectra with a search score of at least 0.70 when compared to the reference spectra were considered to be a "high match" and included in the study (Caron et al. 2018).

22.2.5 Statistical Analysis

Statistical analysis was performed using Excel (Microsoft 2019) and SPSS Statistics (IBM 2017). Data for comparison of fledgling short-tailed shearwaters at the Phillip Island site in 2010 were obtained from Carey (2011). The significance criteria for the statistical testing was set at $P < 0.05$. All data are presented as the mean ± standard error.

FIGURE 22.1 Particles recovered from the ventriculus with the outer layer of plastic removed to reveal their original color. The two fragments on the left appeared grey-green from the outside due to discoloration; however, once scraped, they were seen to actually be light blue in color.

22.3 RESULTS

22.3.1 Prevalence of Plastic Ingestion by Beachcast Short-Tailed Shearwater Fledglings on Phillip Island

The gastrointestinal tracts of 52 beachcast short-tailed shearwater fledglings gathered in 2018 from Phillip Island were examined for this study and, where possible, compared to the paper by Carey (2011), which included 67 individuals from the same location in 2010. The prevalence of plastic ingestion among the fledglings has remained very high, with all but one (98.1%) of the individuals found to have ingested plastic. This is only a slight decrease from 2010, where 100% of the birds contained plastic. In total, 310 pieces of plastic were recovered, which equated to a mean of six particles (±0.6) (min 0, max 22) or 107.3 mg (±1.9) (min 0, max 640.1 mg) of plastic per bird. There were no statistically significant changes in the number of particles or mass of plastic in birds necropsied in 2018, when compared to 2010 (where the mean was found to be 7.6 (±0.7) particles ($t = -1.7$, $P = 0.08$), or 113 mg (±1.0) of plastic per bird ($t = -0.4$, $P = 0.7$). There was also no statistically significant changes observed in mass or number of particles when only birds containing plastic were included in the statistical analysis.

Although there was variation in the amount of plastic being ingested among birds, the data appeared to be skewed toward individuals containing four or less pieces within their gastrointestinal tracts (Figure 22.2). Some birds did ingest quite a bit more than this though, with up to 22 particles and 640.1 mg of plastic seen in some birds.

22.3.2 Location of Plastic within the Gastrointestinal Tract

Plastic was found in the ventriculus of almost all the necropsied birds, with 96.2% of the individuals containing at least one piece of plastic in this location (Figure 22.3). Plastic was present in the proventriculus of 48.1% of birds. Although the entire gastrointestinal tract was dissected, plastics with a minimum diameter of at least 0.5 mm were not identified elsewhere in these birds. As well as being the most common location for plastic to be found, particles appeared to accumulate more readily in the ventriculus with a mean of 4.7 (±0.6) particles per bird (mean mass: 85.3 ± 13.9 mg). This was significantly higher than that occurring in the proventriculus (mean number of particles: 1.3 ± 0.3, Mann-Whitney U test: $Z = -6.0$, $P < 0.001$ and mean mass: 22.0 ± 5.9 mg per bird, $Z = -5.8$, $P < 0.001$). On average, plastic in the proventriculus (mean maximum diameter: 4.78 mm ± 0.26; min 0.53 mm, max 9.78 mm) tended to be larger than that found in the ventriculus (mean maximum diameter: 4.54 mm ± 0.15; min: 1.85 mm, max: 27.05 mm); however, this difference was not statistically significant (Mann-Whitney U test, $Z = -1.39$, $P = 0.16$).

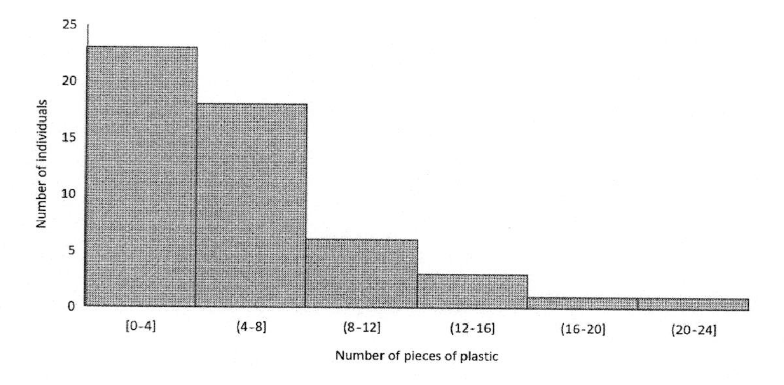

FIGURE 22.2 Histogram displaying the number of pieces of plastic per bird recovered from the gastrointestinal tracts of short-tailed shearwater fledglings in this study. The data appear to be skewed toward birds containing four pieces or less of plastic; however, larger amounts could be ingested with a maximum of 22 pieces recovered from one individual (n = 52).

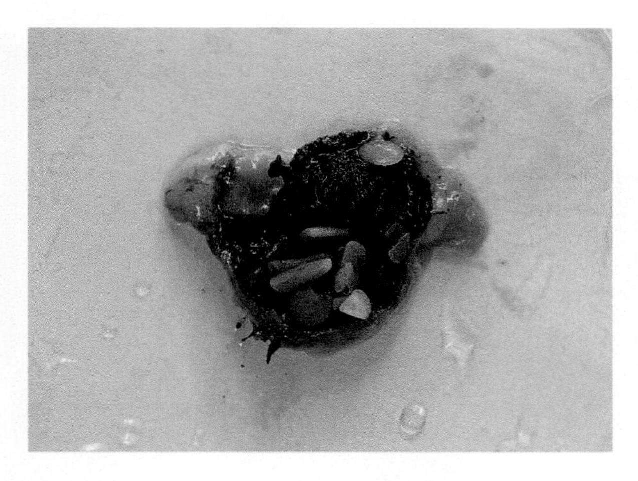

FIGURE 22.3 The ventriculus of a short-tailed shearwater incised during postmortem examination to reveal stomach contents containing multiple pieces of plastic.

22.3.3 Condition of the Gastrointestinal Tract and Intestinal Fat Stores

Despite the high prevalence of plastic ingestion, no birds were found to have suffered perforations of the gastrointestinal tract by foreign bodies, and only a low proportion (two cases) were seen by Carey (2011) in 2010. Additionally, most birds had good visceral fat stores with 59.6% of the individuals having an intestinal fat score of 3 (very large fat deposits), 32.7% of birds having a score of 2, and only 3.8% of the individuals in each of the lower fat score groups of 1 and 0.

The majority (46.2%) of shearwaters had little to no material in their proventriculus (score 1), with the remaining proportion having moderate (score 2) to large (score 3) amounts of ingesta in their proventriculus (25.0% and 28.8%, respectively). There was a visual trend between the mass of food and plastic present in the proventriculus, with 100% of the individuals with the least amount of food in their proventriculus (score 1) also having proventriculus plastic loads of 50 mg or less. By contrast, birds with the highest amount of food in their proventriculus (score 3) also contained the highest proportion of the individuals with proventriculus plastic loads of 100 mg or greater (20.0% of birds with proventriculus food score 3 had plastic loads of ≥100 mg, compared to 15.4% of birds with a proventriculus food score of 2). Although a trend appears to be present, this has not been statistically verified at this stage.

22.3.4 Origin and Physical Characteristics of the Ingested Plastic

The majority of the plastic ingested by the shearwaters in this sample was buoyant in seawater, with all but one piece (99.7%) floating on the water's surface when tested. When classified by shape, irregular-shaped particles made up the majority of the plastic recovered from the gastrointestinal tracts of the birds (78.4%). This was followed by pellets (14.9%), spherical beads (5.5%), fibers (1.6%), and foams (polystyrene: 0.3%). The mean maximum diameter was 4.59 mm (±0.13), with particles ranging in size from 0.53 to 27.05 mm. If fibers that were often long, but thin and flexible, were excluded from the analysis, the largest recorded diameter of an individual plastic piece was 10.38 mm (reducing the mean maximum diameter slightly to 4.43 mm (±0.10)).

The greatest proportion of plastics observed within the gastrointestinal tract was light in color, with white being the most commonly recovered, and dark-hued plastic the least (Figure 22.4). Although this trend was also observed by Carey (2011), there appeared to be statistically significant differences between the proportions of colors in the 2 years, with more light-colored plastics, and less medium- and dark-hued plastics, recovered in 2018 when compared to 2010 ($\chi^2 = 56.4$, $P < 0.001$).

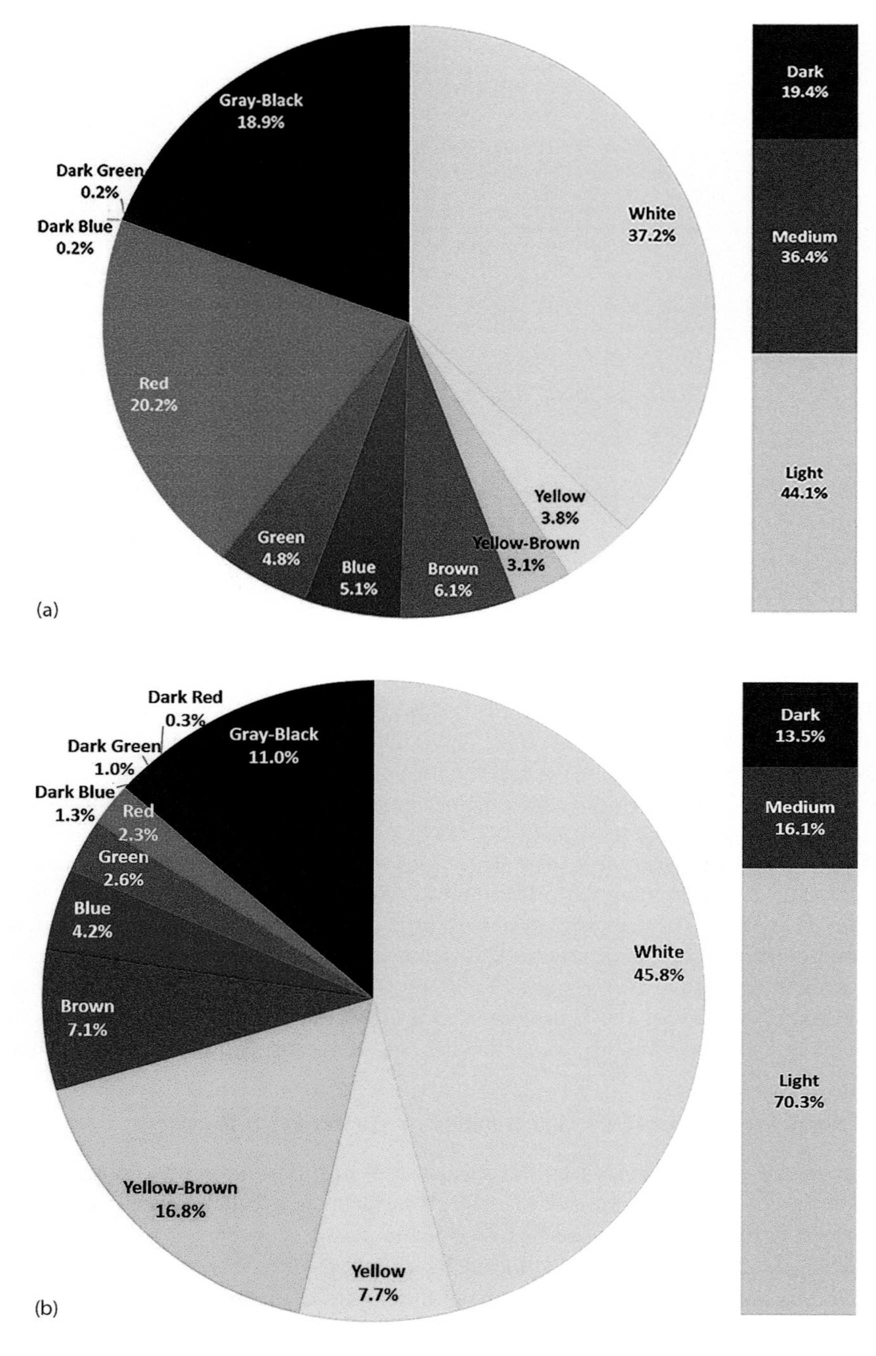

FIGURE 22.4 (a) Percentage of plastic particles by color collected from the ventriculus and proventriculus of short-tailed shearwater fledglings located on Phillip Island during 2010 (Carey 2011) (n = 513) and (b) 2018 (n = 310).

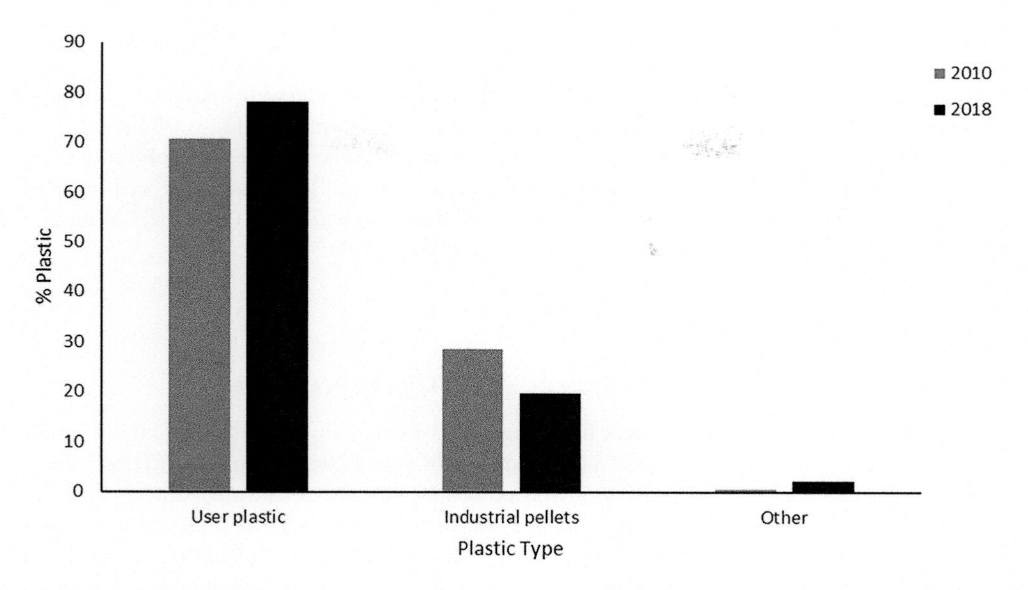

FIGURE 22.5 Percentage of plastic types found in short-tailed shearwater fledglings collected on Phillip Island in 2010 (n = 513) (Carey 2011) and 2018 (n = 310).

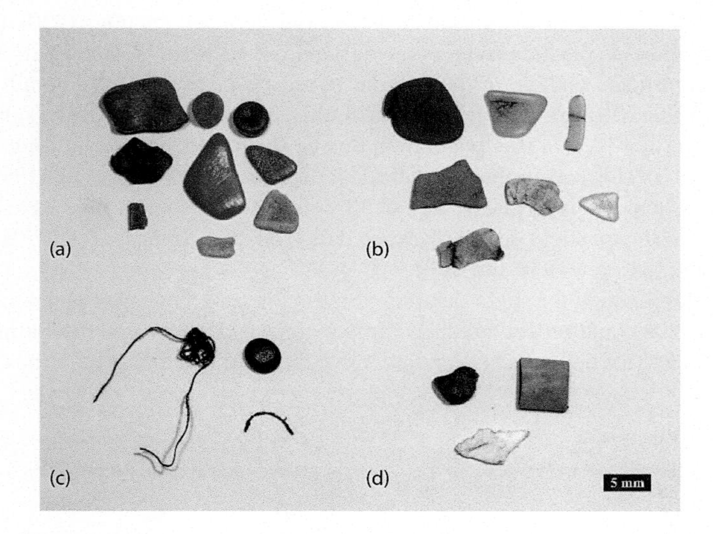

FIGURE 22.6 Examples of debris recovered from the gastrointestinal tracts of four short-tailed shearwater fledglings. (a) User plastic and industrial pellets, (b) user plastic fragments, (c) fibers and a pellet, and (d) user plastic fragments (including a piece of drinking straw on the upper right) and foam (polystyrene fragment present on the upper left).

The most common type of plastic found in the birds was user plastic (78.1%), followed by industrial pellets (19.7%), with a relatively small proportion of other types of plastic (2.3%). Although this trend was also seen in 2010 by Carey (2011), there was a statistically significant difference in the 2 years, with an increase in user and other plastics, and a decrease in industrial pellets (χ^2 = 12.0, P = 0.002) (Figures 22.5 and 22.6).

22.3.5 FTIR Analysis

When 163 pieces of plastic ingested by 31 short-tailed shearwaters underwent analysis by FTIR, polyethylene was found to be most prevalent (66.9%), followed by polypropylene (31.9%) (Figure 22.7). Only a small proportion of samples were identified as polystyrene (0.6%) and polyvinyl chloride (0.6%). When individual birds were considered, most contained at least one piece of polyethylene (93.5%), with ingestion of polypropylene also being relatively high (77.4%). Only 3.2% of birds were found to have ingested either polystyrene or polyvinyl chloride.

22.4 DISCUSSION AND CONCLUSIONS

22.4.1 Prevalence of Plastic Ingestion by Short-Tailed Shearwaters

The proportion of short-tailed shearwater fledglings ingesting plastic on Phillip Island has remained very high, with only a slight decline when the results of this study were compared to previous years (e.g., 98% of birds in 2018 compared to 100% of beachcast birds examined during 2015–2016; Rodríguez et al. 2018, and in 2010 Carey 2011). This continues to place them as one of the seabird species most likely to ingest plastic and may make them useful for monitoring plastic pollution in the ocean (Carey 2011; Vlietstra and Parga 2002).

Although plastic was rarely ingested by short-tailed shearwaters present in Australia during the 1950s and 1960s (Day 1980), by the 1970s, significant increases were being recorded, with almost half the shearwaters examined from Bass Strait during 1978–1979 found to contain plastic (Day 1980; Vlietstra and Parga 2002). This rising trend has been thought to correspond to increases in plastic production and subsequent pollution of the marine environment (Ogi 1990). Despite continuing increases in worldwide plastic production (PlasticsEurope 2018; Wilcox et al. 2015), this study failed to show a significant change in the amount of plastic ingested in the fledglings sampled on Phillip Island between 2010 (Carey 2011) and 2018 (this study). This appears to suggest that plastic ingestion may have stabilized in this population during this period. These findings have also been observed in research performed in the Northern Hemisphere by Robards et al. (1995) and Vlietstra and Parga (2002), where levels of plastic ingested by short-tailed shearwaters have either remained relatively constant or shown slight decreases when data from the 1960s and/or 1970s were compared with periods of time between 1988 and 2001.

This apparent stabilization (or slight decline) of plastic ingested by shearwaters observed in studies since the late 1980s could reflect a reduced amount of plastic entering or remaining in the ocean and becoming available for them to eat. Regulations, such as the enactment of Annex V of MARPOL

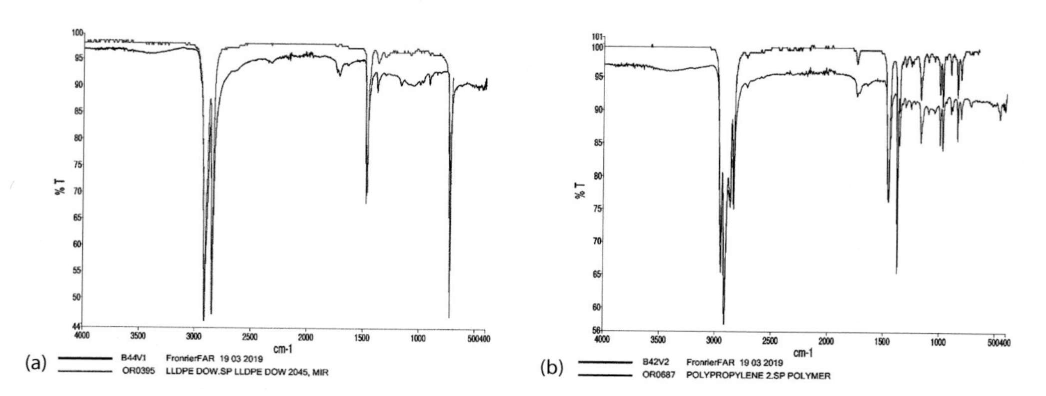

FIGURE 22.7 Examples of spectra produced during FTIR analysis of plastics recovered from the gastrointestinal tracts of short-tailed shearwaters for this study. These samples were identified as being most likely to be (a) polyethylene and (b) polypropylene.

(the International Convention for the Prevention of Pollution from Ships) in 1988 (which prohibits the disposal of rubbish from ships into the ocean) (International Maritime Organization 2018; Vlietstra and Parga 2002), increased recycling (PlasticsEurope 2018), and programs designed to reduce plastic pollution, such as Operation Clean Sweep (American Chemistry Council 2019) may be assisting with decreasing the level of plastic pollution in the marine environment.

While stabilization of plastic levels in the marine environment available to the shearwaters has been hypothesized to be a possible reason for the lack of significant change in its ingestion, the exact amount of plastic present in the world's oceans is unknown and variability is present in proposed estimates that have been put forward (Eriksen et al. 2014). Even if a decrease in the rate of new plastics entering the oceans were found to be present, the biodegradability of most plastics is poor, meaning they could potentially remain available to marine life for extended amounts of time (Derraik 2002; Moore et al. 2001).

Plastic is known to be removed from the ocean's surface (where it is more available to surface feeding birds) by multiple processes including degradation, biofouling decreasing buoyancy, beaching on shorelines, or ingestion by biota, which may allow for a decrease in floating plastic density with time (Eriksen et al. 2014). Further to this, the amount and proportions of plastic recovered can vary depending on location and changes in oceanic conditions and may cause variations in the level of plastic exposure to the birds over time. These factors might account for at least some of the variability in results observed between locations and years (Hardesty et al. 2014; Law et al. 2010; Robards et al. 1995). Trends in the amount of plastics located close to the ocean's surface have been inconsistent; with some studies recording an increase in plastic density (Moore et al. 2001), while others have found no significant trends (Law et al. 2010). Trawl studies to assess plastic present at the shearwater's feeding grounds in the Southern Hemisphere in concert with postmortem examinations of the birds could be useful in helping to assess plastic densities in the sea in relation to the amounts being ingested by the birds. It is also important to consider that any differences between the methodology for recovering or preparing the plastics from these birds for analysis could potentially impact the results. A standardized technique would be beneficial to allow the results from different researchers to be compared with less potential for error (Provencher et al. 2018).

The data were skewed toward birds having lower amounts of plastic in their gut, with 44.2% containing four or fewer pieces of plastic; however, some shearwaters were ingesting a lot more (up to 22 pieces), which could indicate a difference in feeding habits or frequented locations between individual birds (Lavers and Bond 2016; Law et al. 2010).

22.4.2 Pathology, Visceral Fat Stores, and Amount of Ingesta in the Proventriculus

Ingestion of plastic can potentially impair the health of seabirds via direct damage (e.g., ulceration and perforations), obstruction of normal rates of food passage, leaching of toxicants, or by a reduction in functional volume simulating satiation and, therefore, reducing food intake (Pierce et al. 2004; Vlietstra and Parga 2002; Yamashita et al. 2011). While cases in the literature which definitively attribute ingestion of plastic by short-tailed shearwaters as being the primary cause of death appear to be relatively uncommon, obstructions, perforations, and ulcerations have been recorded in a number of seabird species (Carey 2011; Pierce et al. 2004). No gastrointestinal perforations were recorded in the birds assessed for this study, while only two birds (3.0%) showed evidence of this occurring in 2010. The reason for the relatively low rates of gastrointestinal perforations by plastic is uncertain, particularly as many of the fragments did not have rounded edges. One possibility is that their stomach walls are relatively durable since they have to contend with other sharp or abrasive items, which are ingested naturally, such as rocks and squid beaks (Acampora et al. 2014).

As only the gastrointestinal tracts of the birds were obtained for this research, visceral (intestinal) fat stores were used as the measurement of body condition. Changes in visceral fat stores should be a relatively sensitive indicator of decreasing body condition, as birds tend to deplete their fat reserves first before utilizing proteins from muscle (where changes in pectoral muscle condition would become more evident) (van Franeker 2004). The majority of the birds assessed appeared to have good to very good levels of intestinal fat, with 59.6% of the birds scoring 3 (out of a 0–3 scale), and a further 32.7% scoring 2. This indicates that relatively few birds were in poor body condition despite the prevalence of plastic ingestion in this population, and mortalities due to severe malnutrition appear to be uncommon in this sample. This finding has been seen in other studies, such as that by Cousin et al. (2015) in Tasmania (Australia), who found that, although 96% of sampled chicks ingested plastic in their study, the mean intestinal fat score remained high. Although no correlation was noted by Cousin et al. (2015) between the mass of ingested plastic with regards to pectoral muscle or subcutaneous fat scores, a weak (non-statistically significant) correlation was observed between plastic loads and intestinal fat levels.

The chicks are last fed by their parents 2 to 3 weeks before they are fully fledged, leaving them to survive on their own body reserves during this time (Parks and Wildlife Services Tasmania 2014). A variation in the amount of food present in the proventriculus was noticed in the fledglings on postmortem examination, ranging from empty to distended with ingesta. Since plastic foreign bodies are known to have the potential to interfere with normal rates of food passage due to obstruction (Pierce et al. 2004; Vlietstra and Parga 2002), it was hypothesized that higher plastic loads might predispose birds to slower emptying of the proventriculus.

There is a possible visual trend indicating birds with higher amounts of food in their proventriculus also have higher proventriculus plastic loads (by mass). This may indicate that fledglings with higher plastic loads in their proventriculus may be more prone to a degree of gastrointestinal obstruction by plastic; however, a larger sample size is recommended with statistical analysis to confirm this correlation. A reason for the possible correlation involving proventriculus plastic amounts and delayed emptying of the proventriculus could relate to the proventriculus being separated from the ventriculus by a relatively narrow passageway. As this area is narrowed compared to the main body of the ventriculus and proventriculus, and is also known for restricting their ability to regurgitate indigestible items (Pierce et al. 2004), it may be a region where obstructions could be more likely to occur as plastic pieces attempt to pass into the ventriculus.

22.4.3 Physical Characteristics and Origin of the Recovered Plastic

Seabirds belonging to the family Procellariidae (which includes shearwaters) generally have a high frequency of plastic ingestion. Although this could relate at least in part to the use of smell as a cue to identify prey (resulting in ingestion of biofouled plastics by mistake) (Savoca et al. 2016), there appears to be evidence that short-tailed shearwaters exhibit selectivity based on the appearance of the particles and are, therefore, also hunting by sight. This would indicate that the appearance of plastics polluting the oceans is likely an important factor, as well as the density, when it comes to plastic loads in these birds (Acampora et al. 2014; Vlietstra and Parga 2002).

Although short-tailed shearwaters have been recorded to ingest a wide range of colored plastics (Day 1980), many studies have shown lighter-hued plastics to be taken at a proportionally high rate compared to medium- and dark-colored plastics (Carey 2011; Vlietstra and Parga 2002). Acampora et al. (2014) conducted an ocean trawl, finding significant differences in the proportion of colors from the trawl when compared to necropsied birds, which also supports the hypothesis that selectivity is occurring based on the visibility of the particles in the water or similarity to prey items (e.g., Euphausiids are light in color and a favored prey of short-tailed shearwaters). Without direct sampling of the Phillip Island colony's feeding grounds, however, it is difficult to say with certainty if the results of this research reflect color selectivity or if it is an indication of availability (Acampora et al. 2014; Carey 2011; Vlietstra and Parga 2002).

Although the findings of both this study and Carey (2011) supported a preference of light-hued plastic over medium- and dark-colored pieces, the proportions between the two studies were significantly different with greater amounts of light-colored particles found to be ingested during 2018. One explanation could be that there was an increased amount of light-hued plastics available to the birds in 2018 compared to 2010. As no ocean trawls for plastic were performed for either study, this is difficult to determine (Acampora et al. 2014). Another explanation could relate to a difference in methodology. In this study, the outside layer of the plastic was scraped away to determine its original color since many pieces were discolored (e.g., staining from time spent within the gastrointestinal tract, etc.) (Cousin et al. 2015). As not all publications (including the paper by Carey 2011) specify in detail whether certain steps, such as these were done, failing to remove the outside layer of the plastic could result in a lower proportion of lighter colors observed. This highlights the need for a standardized protocol to allow research to be compared accurately, especially where shearwaters are to be used for biomonitoring purposes (Lavers et al. 2018).

The most commonly recovered particle shape were fragments, which also corresponded closely with the proportion of user plastic. Although this could indicate selectivity for irregular-shaped particles, it seems likely that it is simply reflecting a higher availability of user plastics, which were predominantly fragments. This is also supported by a decrease in the proportion of pellets ingested by seabirds seen in other studies, as the density of industrial pellets has decreased in the oceans (Law et al. 2010; Ryan 2008).

Although the amount of plastic ingested appears to have stabilized in this population between 2010 and 2018, there has been a shift in the types of plastic being recovered, with a significant decrease in the proportion of ingestion of "industrial" plastic compared to "user" and "other" plastics. The predominance of user plastic in this sample is in agreement with many of the more recent articles examining plastic ingestion by short-tailed shearwaters in both the Northern and Southern Hemispheres (Carey 2011; Vlietstra and Parga 2002). This has been a shift from earlier research performed by Ogi (1969–1977) and Day (1970–1979), who found industrial pellets to be the most frequently recovered plastic type (Day 1980; Ogi 1990; Vlietstra and Parga 2002). This pattern could be linked to reduced availability of industrial plastic at the feeding grounds, especially as decreases in industrial plastic proportions have also been observed in other seabird species and ocean trawls. This apparent decline has been attributed mainly to industry efforts focusing on reducing the loss of pellets (Lavers et al. 2018).

A factor that has the ability to negatively impact the health of short-tailed shearwaters, which was not investigated by this study, is the potential for any toxicants present in the plastics to leach out while they remain in the gut, exposing the animals to harmful substances. Adverse effects due to toxicants leaching from plastics may become an increasing concern due to the apparent rise in user plastic ingestion by these birds. While industrial plastics are often relatively biologically inert, user plastics frequently have had additional compounds added (including softeners, colorants, etc.), some of which have the potential to be toxic if assimilated (Carey 2011; van Franeker 1985; Vlietstra and Parga 2002). Additional research for monitoring seabirds for toxicant uptake is, therefore, recommended (Ryan et al. 1988).

Short-tailed shearwaters feed primarily on krill, squid, and fish, capturing prey by hydroplaning (propelling themselves across the surface of the water with their beak submerged) or pursuit diving (plunging from height to capture food) (Acampora et al. 2014; Furness and Monaghan 1987; Ogi 1990). Since this species tends to feed close to the surface, a preference for buoyant plastic was hypothesized; however, an alternative source of plastic ingestion could be via eating prey containing plastic (Acampora et al. 2014). The result of this research appeared to indicate a strong preference for buoyant plastics, with 99.7% floating at the surface in seawater, and suggests that the majority of the plastic is being picked up from the water's surface. A preference for buoyant plastics agrees with the paper by Acampora et al. (2014), although the proportion of their buoyant plastics was lower (70%–72%). The reason for this is uncertain, but may relate to different feeding habits or plastic compositions due to a different location being studied, or the inclusion of adult birds as well as juveniles.

22.4.4 Size and Location of Ingested Plastic

Most plastic particles larger than 0.5 mm were recovered from the ventriculus, with the remainder from the proventriculus. The mean maximum diameter was slightly larger for particles in the proventriculus. This distribution is likely occurring because, if not regurgitated after ingestion, plastic will most likely be retained in the ventriculus until broken down into small enough fragments to pass through into the small intestine. The length of time this process takes can depend on the size, shape, and type of plastic, but may require months, allowing a build of plastic to occur in the ventriculus (Provencher et al. 2018).

The shearwaters in this sample were observed to ingest both macro- and microplastics, with the mean maximum diameter being 4.59 mm. Although fibers of up to 27.05 mm were ingested, these were thin and flexible, likely making them easier to swallow. If fibers were excluded, short-tailed shearwaters appeared to select for particles up to about 10 mm in diameter (max 10.4 mm). The maximum size of plastics ingested has been found to correlate positively with increasing head size in many seabird species, and a study by Roman et al. (2019), which included short-tailed shearwaters, also found that debris larger than 10 mm across were unlikely to be swallowed. The most likely explanation for the rarity of larger plastics is either that the birds are selecting against ingesting particles greater than 10 mm; or perhaps if these larger particles are taken, they are less likely to pass through the narrowed area of the gut between the proventriculus and ventriculus and could be more likely to be regurgitated (Roman et al. 2019).

22.4.5 FTIR Analysis

When analyzed with FTIR, polyethylene was found to be the most frequently recovered plastic type, followed by polypropylene, which reflects the trends in world demand for plastics (Figure 22.8) (Geyer et al. 2017; PlasticsEurope 2008). The percentages recovered from the birds were not exactly the same as what is being generated; however, which may indicate factors other than the production amount is at play. These factors could include the following: unequal rates of each plastic type entering the oceans or other localized differences in plastic pollution types (Lacerda et al. 2019); buoyancy (e.g., the proportion of polyvinyl chloride was lower than expected

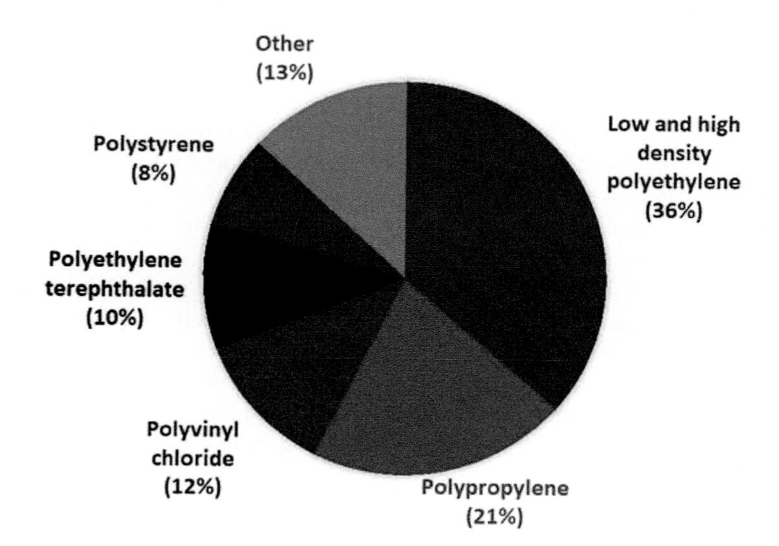

FIGURE 22.8 Plastic demand by polymer type between 2002 and 2014 in China, Europe, India, and the United States. (Adapted from Geyer, R. et al., *Sci. Adv.*, 3, 1–5, 2017.)

and may be due to its tending to have a higher density than polyethylene and polypropylene and, therefore, be more likely to sink where it is less accessible to the birds) (Law et al. 2010); or fragility (e.g., the polystyrene in this sample was comparatively fragile and may be breaking up into microscopic fragments faster).

The method ATR-FTIR appeared to be successful for identifying major plastic types ingested by shearwaters, although contamination and degradation can limit the specificity of more specific identification in some samples. Jung et al. (2018) described a method, whereby differences in specific areas of the spectra (for plastic obtained from the gastrointestinal tracts of sea turtles) could be used to help identify plastics that are difficult to differentiate by FTIR. An example of this involves examining the intensity at 1377 cm^{-1} to help determine the polyethylene density type. In practice, however, some spectra for samples in this study were still found to be quite challenging to identify using this method. Additionally as noted by other authors, ATR-FTIR analysis can be damaging to more fragile particles, making it a less suitable technique if samples must remain undamaged (Shim et al. 2017).

22.4.6 Future Research

Due to the current high rates of plastic ingestion, ongoing monitoring of short-tailed shearwater colonies in Australia is recommended and may provide a way to help monitor changes occurring in marine plastic pollution. Additionally, with the increasing incidence of user plastic ingestion, the potential for increasing deleterious effects from toxicants present in ingested plastic may require further research to determine their impact on marine life (Carey 2011; Ryan 2008).

REFERENCES

Acampora, H., Schuyler, Q. A., Townsend, K. A. and Hardesty, B. D. 2014. "Comparing plastic ingestion in juvenile and adult stranded short-tailed shearwaters (*Puffinus tenuirostris*) in eastern Australia." *Marine Pollution Bulletin* 78 (1):63–68. https://doi.org/10.1016/j.marpolbul.2013.11.009.

American Chemistry Council. 2019. "Value of operation clean sweep." American Chemistry Council, Accessed 5th June. www.opcleansweep.org/about/value-of-ocs.

BirdLife International. 2017. "*Ardenna tenuirostris* (amended version of 2016 assessment). The IUCN Red List of Threatened Species 2017: e. T22698216A110676140." International Union for Conservation of Nature and Natural Resources, Accessed 1st November. http://dx.doi.org/10.2305/IUCN.UK.2017-1.RLTS.T22698216A110676140.en.

Carey, M. J. 2011. "Intergenerational transfer of plastic debris by short-tailed shearwaters (*Ardenna tenuirostris*)." *Emu* 111 (3):229–234. doi:10.1071/MU10085.

Caron, A. G. M., Thomas, C. R., Berry, K. L. E., Motti, C. A., Ariel, E. and Brodie, J. E. 2018. "Ingestion of microplastic debris by green sea turtles (*Chelonia mydas*) in the Great Barrier Reef: Validation of a sequential extraction protocol." *Marine Pollution Bulletin* 127:743–751. https://doi.org/10.1016/j.marpolbul.2017.12.062.

Cousin, H. R., Auman, H. J., Alderman, R. and Virtue, P. 2015. "The frequency of ingested plastic debris and its effects on body condition of Short-tailed Shearwater (*Puffinus tenuirostris*) pre-fledging chicks in Tasmania, Australia." *Emu* 115 (1):6–11. doi:10.1071/MU13086.

Day, R. H. 1980. "The occurrence and characteristics of plastic pollution in Alaska's marine birds." Master of Science, Master of Science thesis, University of Alaska.

Department of the Environment and Energy. 2018. "Short-tailed shearwater." Australian Government, Accessed 2nd October. www.antarctica.gov.au/about-antarctica/wildlife/animals/flying-birds/petrels-and-shearwaters/short-tailed-shearwater.

Derraik, J. G. B. 2002. "The pollution of the marine environment by plastic debris: A review." *Marine Pollution Bulletin* 44 (9):842–852. https://doi.org/10.1016/S0025–326X(02)00220–5.

Eriksen, M., Lebreton, L. C. M., Carson, H. S., Thiel, M., Moore, C. J., Borerro, J. C., Galgani, F., Peter, G., Ryan, P. G. and Reisser, J. 2014. "Plastic pollution in the world's oceans: More than 5 trillion plastic pieces weighing over 250,000 tons afloat at sea." *PLOS ONE* 9 (12):e111913. doi:10.1371/journal.pone.0111913.

Furness, R. W. and Monaghan, P. 1987. *Seabird Ecology*. Glasgow, Scotland: Blackie.
Geyer, R., Jambeck, J. R. and Law, K. L. 2017. "Production, use, and fate of all plastics ever made." *Science Advances* 3 (7):1–5. doi:10.1126/sciadv.1700782.
Hardesty, B. D., Wilcox, C., Lawson, T. J., Lansdell, M. and van der Velde, T. 2014. *Understanding the Effects of Marine Debris on Wildlife*. A final report to Earthwatch Australia. Australia: CSIRO.
IBM. 2017, SPSS Statistics, Version 25, IBM.
International Maritime Organization. 2018. "Prevention of pollution by garbage from ships." International Maritime Organization, Accessed 4th November. http://www.imo.org/en/OurWork/Environment/PollutionPrevention/Garbage/Pages/Default.aspx.
Jung, M. R., Horgen, F. D., Orski, S. V., Rodriguez, V., Beers, K. L., Balazs, G. H., Jones, T. T., Work, T. M., Brignac, K. C. and Royer, S. 2018. "Validation of ATR FT-IR to identify polymers of plastic marine debris, including those ingested by marine organisms." *Marine Pollution Bulletin* 127:704–716.
Lacerda, A. L. F., Rodrigues, L. S., van Sebille, E., Rodrigues, F. L., Ribeiro, L., Secchi, E. R., Kessler, F. and Proietti, M. C. 2019. "Plastics in sea surface waters around the Antarctic Peninsula." *Scientific Reports* 9 (1). doi:10.1038/s41598-019-40311-4.
Lavers, J. L. and Bond, A. L. 2016. "Selectivity of flesh-footed shearwaters for plastic colour: Evidence for differential provisioning in adults and fledglings." *Marine Environmental Research* 113:1–6. doi:10.1016/j.marenvres.2015.10.011.
Lavers, J. L., Hutton, I. and Bond, A. L. 2018. "Ingestion of marine debris by wedge-tailed shearwaters *(Ardenna pacifica)* on Lord Howe Island, Australia during 2005–2018." *Marine Pollution Bulletin* 133:616–621. doi:10.1016/j.marpolbul.2018.06.023.
Law, K. L., Morét-Ferguson, S., Maximenko, N. A., Proskurowski, G., Peacock, E. E., Hafner, J. and Reddy, C. M. 2010. "Plastic accumulation in the North Atlantic subtropical gyre." *Science* 329 (5996):1185–1188. doi:10.1126/science.1192321.
Löder, M. G. J. and Gerdts, G. 2015. "Methodology used for the detection and identification of microplastics—A critical appraisal." In *Marine Anthropogenic Litter*, edited by Melanie Bergmann, Lars Gutow and Michael Klages, 201–227. Cham, Switzerland: Springer International Publishing.
Lusher, A. L., Welden, N. A., Sobral, P. and Cole, M. 2017. "Sampling, isolating and identifying microplastics ingested by fish and invertebrates." *Analytical Methods* 9 (9):1346–1360. doi:10.1039/c6ay02415g.
Microsoft. 2019. Excel, Version 1903, Microsoft.
Moore, C. J., Moore, S. L., Leecaster, M. K. and Weisberg, S. B. 2001. "A comparison of plastic and plankton in the North Pacific central gyre." *Marine Pollution Bulletin* 42 (12):1297–1300. https://doi.org/10.1016/S0025-326X(01)00114-X.
Mossman, S. 2008. "Saving plastics for posterity." *Nature* 455:288–289. doi:10.1038/455288b.
Ogi, H. 1990. "Ingestion of plastic particles by sooty and short-tailed shearwaters in the North Pacific." In *Proceedings of the second international conference on marine debris', 2–7 April 1989, Honolulu, Hawaii. National oceanic and atmospheric administration technical memorandum*, edited by R.S. Shomura and M.L. Godfrey, 635–652. Honolulu, Hawaii: US Department of Commerce.
Parks and Wildlife Services Tasmania. 2014. "Short-tailed shearwater, *Puffinus tenuirostris*." Tasmanian Government, Accessed 30th October www.parks.tas.gov.au/?base = 5100.
PerkinElmer Inc. 2015, Spectrum, Version 10.5.2.636, PerkinElmer Inc.
Pierce, K. E., Harris, R. J., Larned, L. S. and Pokras, M. A. 2004. "Obstruction and starvation associated with plastic ingestion in a Northern Gannet *Morus bassanus* and a greater shearwater *Puffinus gravis*." *Marine Ornithology* 32:187–189.
PlasticsEurope. 2008. *The Compelling Facts About Plastics, Analysis of Plastics Production, Demand and Recovery for 2006 in Europe*. Belgium: PlasticsEurope.
PlasticsEurope. 2018. *Plastics – The Facts 2018. An Analysis of European Plastics Production, Demand and Waste Data*. Belgium: PlasticsEurope.
Prestt, I., Jefferies, D. J. and Moore, N. W. 1970. "Polychlorinated biphenyls in wild birds in Britain and their avian toxicity." *Environmental Pollution* 1 (1):3–26. https://doi.org/10.1016/0013-9327(70)90003-0.
Provencher, J. F., Vermaire, J. C., Avery-Gomm, S., Braune, B. M. and Mallory, M. L. 2018. "Garbage in guano? Microplastic debris found in faecal precursors of seabirds known to ingest plastics." *Science of the Total Environment* 644:1477–1484. https://doi.org/10.1016/j.scitotenv.2018.07.101.
Robards, M. D., Piatt, J. F. and Wohl, K. D. 1995. "Increasing frequency of plastic particles ingested by seabirds in the subarctic North Pacific." *Marine Pollution Bulletin* 30 (2):151–157. doi:10.1016/0025-326X(94)00121-O.
Rodríguez, A., Burgan, G., Dann, P., Jessop, R., Negro, J. J. and Chiaradia, A. 2014. "Fatal attraction of short-tailed shearwaters to artificial lights." *PLOS ONE* 9 (10):e110114. doi:10.1371/journal.pone.0110114.

Rodríguez, A., Ramírez, F., Carrasco, M. N. and Chiaradia, A. 2018. "Seabird plastic ingestion differs among collection methods: Examples from the short-tailed shearwater." *Environmental Pollution* 243:1750–1757. https://doi.org/10.1016/j.envpol.2018.09.007.

Roman, L., Paterson, H., Townsend, K. A., Wilcox, C., Hardesty, B. D. and Hindell, M. A. 2019. "Size of marine debris items ingested and retained by petrels." *Marine Pollution Bulletin* 142:569–575. https://doi.org/10.1016/j.marpolbul.2019.04.021.

Ryan, P. G., Connell, A. D. and Gardner, B. D. 1988. "Plastic ingestion and PCBs in seabirds: Is there a relationship?" *Marine Pollution Bulletin* 19 (4):174–176. doi:10.1016/0025-326X(88)90674-1.

Ryan, P. G. 2008. "Seabirds indicate changes in the composition of plastic litter in the Atlantic and south-western Indian Oceans." *Marine Pollution Bulletin* 56 (8):1406–1409. https://doi.org/10.1016/j.marpolbul.2008.05.004.

Ryan, P. G. 2015. "How quickly do albatrosses and petrels digest plastic particles?" *Environmental Pollution* 207:438–440. https://doi.org/10.1016/j.envpol.2015.08.005.

Savoca, M. S., Wohlfeil, M. E., Ebeler, S. E. and Nevitt, G. A. 2016. "Marine plastic debris emits a keystone infochemical for olfactory foraging seabirds." *Science Advances* 2 (11):1–8.

Semenova, P. 2017. "The study of Tapio Wirkkala's plywood sculptures technique, and typical defect mechanisms." Bachelor in Culture and Art, Bachelor in Culture and Art thesis, Metropolia University of Applied Sciences.

Shim, W. J., Hong, S. H. and Eo, S. E. 2017. "Identification methods in microplastic analysis: A review." *Analytical Methods* 9 (9):1384–1391. doi:10.1039/c6ay02558g.

Skira, I. J., Wapstra, J. E., Towney, G. N. and Naarding, J. A. 1986. "Conservation of the short-tailed shearwater *Puffinus tenuirostris* in Tasmania, Australia." *Biological Conservation* 37 (3):225–236. https://doi.org/10.1016/0006-3207(86)90083-2.

van Franeker, J. A. 1985. "Plastic ingestion in the North Atlantic fulmar." *Marine Pollution Bulletin* 16 (9):367–369. https://doi.org/10.1016/0025-326X(85)90090-6.

van Franeker, J. A. 2004. *Save the North Sea Fulmar-Litter-ecoQO Manual Part 1: Collection and Dissection Procedures*. Wageningen, the Netherlands: Alterra.

Vlietstra, L. S. and Parga, J. A. 2002. "Long-term changes in the type, but not amount, of ingested plastic particles in short-tailed shearwaters in the southeastern Bering Sea." *Marine Pollution Bulletin* 44 (9):945–955. https://doi.org/10.1016/S0025-326X(02)00130–3.

Wilcox, C., Van Sebille, E. and Hardesty, B. D. 2015. "Threat of plastic pollution to seabirds is global, pervasive, and increasing." *Proceedings of the National Academy of Sciences of the United States of America* 112 (38):11899–11904.

Yamashita, R., Takada, H., Fukuwaka, M. and Yutaka Watanuki, Y. 2011. "Physical and chemical effects of ingested plastic debris on short-tailed shearwaters, *Puffinus tenuirostris*, in the North Pacific Ocean." *Marine Pollution Bulletin* 62 (12):2845–2849. https://doi.org/10.1016/j.marpolbul.2011.10.008.

Section V

Management of Particulate Plastics

23 Management of Particulate Plastic Waste Input to Terrestrial and Aquatic Environments

Binoy Sarkar, Nanthi S. Bolan, Raj Mukhopadhyay, Shiv Shankar Bolan, M.B. Kirkham, and Jörg Rinklebe

CONTENTS

23.1 Sources and Volume of Particulate Plastic Wastes Produced 397
23.2 Need for Sustainable Management of Particulate Plastic Wastes 400
23.3 Particulate Plastic Waste Management Practices 400
23.3.1 Source Reduction of Particulate Plastic Wastes 401
23.3.2 Improvements to Infrastructure 403
23.3.3 Regulatory and Legislative Changes 403
23.3.4 Value Addition and Beneficial Utilization of Particulate Plastic Wastes 405
23.4 Innovations in Particulate Plastic Waste Management 407
23.5 Summary and Conclusions 408
References 408

23.1 SOURCES AND VOLUME OF PARTICULATE PLASTIC WASTES PRODUCED

In 2017, globally, 348 million tons of plastics were produced, of which, 29.4% was contributed by China (Statista 2019). In the United States, 3.14 million tons of plastic wastes were recovered in 2015 in municipal solid wastes, which constituted 9.1% of the total municipal waste stream (34.5 million tons) in the country (Statista 2019). In that year, the container and packaging industry contributed to the highest generation of 14.68 million tons of plastics in municipal solid waste in the United States, followed by durable goods (12.5 million tons), and non-durable goods (7.32 million tons) (Statista 2019). According to the PlasticsEurope 2017–2018 annual report (PlasticsEurope 2018), more than 8.4 million tons of plastic wastes were collected for recycling in 2016 inside or outside the European Union. Among all the plastics used in the European Union in 2016, the packaging industry topped the list (39.9%) in plastic usage, followed by building and construction (19.7%), automotive (10%), electrical and electronic (6.2%), household, leisure, and sports (4.2%), agriculture (3.3%), and other industries (16.7%), including manufacturing of furniture and various appliances used in mechanical engineering or medicine.

In 2016–2017, Australia produced about 67 million tons of solid waste (Pickin et al. 2018). Solid waste is grouped into various categories based on the source and characteristics. The major sources of solid waste include masonry materials, municipal organic wastes derived from households (e.g., green waste) and wastewater treatment plants (e.g., biosolids), food wastes, and hazardous chemical and biowastes derived from industries and hospitals. Among the 54 million tons of "core waste," municipal solid waste contributed 13.8 million tons from household and local government activities, followed by the commercial and industrial sector, which contributed 20.4 million tons of wastes (Pickin et al. 2018). The construction and demolition sector contributed another 20.4 million tons of wastes during this time (Pickin et al. 2018). Municipal solid waste recycling in Australia increased by 31% during the last decade, and the country currently sends about 40% of the generated "core waste" to the landfill (Pickin et al. 2018). About 46% of the core organic wastes (14 million tons) were comprised of municipal solid waste including food wastes, commercial and industrial wastes, and construction and demolition wastes (Pickin et al. 2018). The food waste exclusively comprised 14% of the total organic wastes in 2016–2017 (Pickin et al. 2018). It was estimated that Australia loses more than $20 billion per year as the total economic cost due to food wastage (Pickin et al. 2018; Bolan and Tsang 2018b). The non-core organic wastes (16 million tons) were mainly generated from the agriculture and fishery industries (Pickin et al. 2018).

In Australia, over 3.5 million tons of plastic were consumed in 2016–2017, which equates to approximately 144 kg of plastic per person (O'Farrell 2018). About 2.5 million tons (103 kg per capita) of plastic waste were generated in 2016–2017, of which, only about 12% was recycled, with 87% sent to landfills, and 1% sent to an energy-from-waste facility (Pickin et al. 2018). In 2016–2017, 5.66 billion single-use plastic bags were used in Australia (O'Farrell 2018).

Terrestrial and aquatic pollution by particulate plastics originates from urban and industrial waste sites, sewage outlets, stormwater, litter transported by systems, and litter discarded by the general public (Reisser et al. 2013). Usage of plastic is increasing day by day due to pressure from a rising population, and the management of plastic wastes is a serious concern in the twenty-first century (Xanthos and Walker 2017). PlasticsEurope (2018) reported that the massive amount of plastics generated from various sources created an alarming ecological imbalance in the marine and other aquatic environments. In addition, urban litter includes lost or abandoned plastic items, i.e., items that fall out of garbage cans due to overfilling or windy weather and plastic debris that is inadequately secured during transportation.

Plastic debris found in the terrestrial and aquatic environments is either large debris or small particles. Based on the particle size, plastics can be classified in different groups, *viz.* nanoplastics (<100 nm), microplastics (1–<5 mm), mesoplastics (5–<25 mm), and macroplastics (≥25 mm) (Lee et al. 2013; Ng et al. 2018). Macroplastics are composed of a wide variety of industrial, commercial, and consumer items. Of particular concern are beverage containers and single-use plastic bags. For example, Clean Up Australia estimated that over 5 billion plastic bags are consumed each year, with less than 4% of these bags being recycled and over 7000 bags entering into landfills each minute (Clean Up Australia 2019).

Nanoplastics and microplastics are tiny plastic particles in different forms like foams, fragments, and fibers. There are a number of major sources of microplastics in the environment. The sources of microplastics and nanoplastics can be categorized broadly as primary and secondary sources including: (i) intentionally produced items; (ii) inherent by-products of other products or activities; (iii) emitted through accidental or unintentional spills; and (iv) degradation (thermal, UV, and biodegradation). The important primary sources of plastics are the spillage during plastic production and use of plastic-based personal care products that have particle sizes ranging up to 2 mm (Browne 2015). About 93% of the microplastics and nanoplastics in cosmetic products are prepared with polyethylene, and some are made up of nylon and polypropylene (Eriksen et al. 2013). The secondary sources include broken fragments of larger plastics, industrial or agricultural wastes, discharge from landfills, and particles from bio- and thermo-degradation (Zhao et al. 2015). The colored plastics (micro and nano-sized) indicate the synthetic source and enrichment with organic substances, which may be more detrimental to the aquatic environment than plastics without organic substances. Low-density polyethylene films, which are used in large amounts as plastic mulching in agriculture to improve crop growth and yields, to control weeds, and to maintain the soil temperature and moisture, are also a significant source of both micro and nano-sized plastics in the environment

(Lambert et al. 2014). Polyurethane and styromull are used as soil conditioners to improve soil quality and grow horticultural crops, such as fruits and vegetables. Therefore, plastic wastes can enter into the environment through various commodities including personal care products, agricultural consumables, and many other industrial consumables (Do and Scherer 2012; Duis and Coors 2016).

In many countries including Australia, sewage and other domestic waste is often added to soils to improve plant nutrient availability and reduce water loss. This practice can lead to contaminant inputs including microplastics and nanoplastics, which eventually enter the marine environment through sediment movement. Microbeads are commercially produced in particle sizes from 10–1000 μm (1 mm). Microbeads are used in products such as abrasives, including exfoliating personal care products for face and body wash and toothpaste, while other personal care products use microbeads for bulking or to provide a slipping effect, and they include shaving foam, lipstick, mascara, or sunscreen (Guerranti et al. 2019). For example, a single tube of deep facial cleanser can contain 350,000 microbeads. Industrial products intentionally utilizing microplastics include plastic blasting grit, speciality products used in oil and gas exploration and printing, and medical products, such as dentistry polish. Because these products can end up in wastewater treatment works, wastewater treatment plants are identified as potential sources of microplastics. Microbeads used in personal care products are readily disposed into raw effluents, and, because of their very small size, they may pass the wastewater treatment process easily (Figure 23.1) (Murphy et al. 2016; Wijesekara et al. 2018).

Microplastic by-products include dust from cutting and polishing plastic items and by-products from maintenance of painted metal constructions, such as bridges and buildings, and high pressure washing of painted items. They also include household and commercial building dust created through weathering and abrasion of plastic items and carpets, building maintenance, and clothing fibers loosened during laundering. Road dust contains microplastic by-products from tire friction, road paint, and polymer-modified bitumen. In addition, waste handling by-products often include plastic particles from the shredding and fragmenting of plastic waste, such as mattresses, bottles,

FIGURE 23.1 Stereomicroscopic images of various shapes and colors of plastic microbeads found in biosolids. (Adapted from Wijesekara, H. et al., *Chemosphere*, 199, 331–339, 2018.)

and plastic bags. Microplastics are also formed through the degradation of macroplastic items both within the marine environment and on land. Plastic degrades through oxidation, UV exposure, wave action, and animal and insect digestion and nesting; the degradation thus produces micro and nanoplastic by-products (Ceccarini et al. 2018). Macroplastics also are shredded by boat and ship propellers and released when plastic-contaminated sediment is dredged (Ng et al. 2018).

23.2 NEED FOR SUSTAINABLE MANAGEMENT OF PARTICULATE PLASTIC WASTES

China's recent "ban" on import of recyclables has sent shockwaves throughout the global recycling and waste management sector. The ban is a set of import restrictions imposed by China under its Blue Sky/National Sword program, which tightens inspection efforts to reduce the amount of contaminated materials entering the country by restricting the importation of 24 streams of recyclable materials (Bolan and Tsang 2018a; Downes and Dominish 2018). This ban is imposed by setting stringent "maximum contamination thresholds" and limiting the number of import permits provided to Chinese businesses.

Many countries including Australia and those in the European Union and North America have limited local markets for recyclable wastes, such as paper, plastics, and glass, and they rely heavily on overseas markets to buy and reprocess these wastes. China used to provide the major market for solid waste recycling. For example, around 29% (920,000 tons) of all paper and 36% (125,000 tons) of all plastics collected in Australia were exported to China in 2017 (Downes and Dominish 2018; Pickin et al. 2018). Papers, plastics, and glasses occur as a mixture (average mixing level 6%–10%) in curbside wastes collected in Australia mainly because of the lack of sufficient sorting/segregation at a household level (Downes and Dominish 2018). Even after an additional sorting is undertaken on this waste mixture at a recycling facility following its collection, lowering the mixing level to 0.5% from 6%–10% becomes challenging. As a result, Australian curbside wastes fail to qualify to be sent to China for recycling because according to China's new guidelines, the wastes should contain ≤0.5% mixing level as an acceptable threshold (Downes and Dominish 2018). Losing such a major market for solid wastes is a major challenge to Australia's recycling industry, and, hence, the government and research organizations need to "STEP-In" to provide **In**novative solutions in **S**oil-waste **T**reatment for **E**nvironmental **P**rotection in Australia. For example, the *Waste Management Association of Australia* has been lobbying for a A$150 million action plan to invest in infrastructure and innovations in recycling, and for the promotion of beneficial utilization of recycled products (Downes and Dominish 2018; Bolan and Tsang 2018a).

During January–June 2018, the United States exported most of its plastic wastes to Malaysia (157,300 tons), followed by Thailand (91,510 tons), Vietnam (71,220 tons), India (69,710 tons), Canada (66,100 tons), Hong Kong (60,450 tons), and China (30,250 tons) (Statista 2019). This export amount to China drastically was reduced from the previous year due to China's import restriction as part of the National Sword program. During January–June 2017, the United States exported 379,380 tons to China (Statista 2019). It is alarming that many of the developing countries to which the United States, Europe, and other countries currently send plastic wastes for recycling do not have adequate solid waste management plans and facilities (Silpa et al. 2018). Therefore, the need of the hour is to develop "in-situ" sustainable best management practices of particulate plastic wastes across the world, especially focusing on the availability of regional solid waste treatment facilities.

23.3 PARTICULATE PLASTIC WASTE MANAGEMENT PRACTICES

Usage pattern and plastic production are two key factors that control plastic pollution in the terrestrial and marine environments. To restrict the entry of plastics into the environment, the segregation, identification, and quantification of plastic sources are necessary. In order to establish regional-level best management practices and to develop infrastructure to treat plastics, precise annual data on plastic waste production and information on the types of plastics produced are urgently needed (Singh and Sharma 2016). Currently, many developing countries lack information concerning estimates of production of plastic wastes. The few statistics dealing with plastic waste production are often inconsistent and lack standard terminologies for their classification even at a national level (Singh and Sharma 2016). Management processes should include energy-efficient methods and preferably carbon neutral practices. An integrated approach, therefore, is required that includes: (1) thorough examination of the mixture contaminated with plastics, (2) critical life cycle analysis of plastic materials (starting from the selection of raw materials used to manufacture plastics to manufacturing methods and modification and engineering of the final consumer products), (3) recycling and reuse of materials and resources wherever possible, and (4) appropriate disposal of unwanted wastes at each step (Subramanian 2000). Sustainable solid waste management includes source reduction, value addition, and beneficial utilization (Figure 23.2). Examples of these management strategies for various solid waste streams including particulate plastics are given in Table 23.1.

23.3.1 Source Reduction of Particulate Plastic Wastes

Source reduction, which refers to consuming less, is the most successful and sustainable method of reducing solid waste generation. Common examples of source reduction in solid waste management include household backyard composting, information digitation instead of printing, and utilizing durable, long-lasting environmentally friendly goods. However, manufacturers should be more responsible

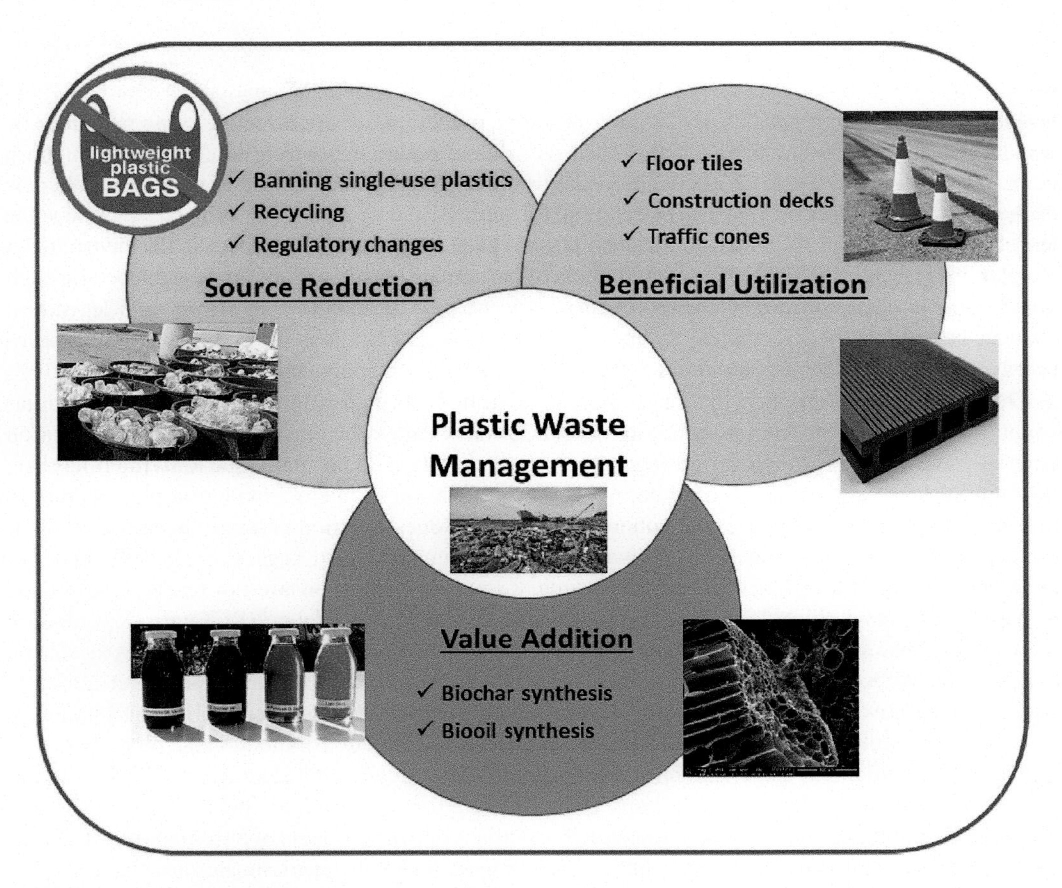

FIGURE 23.2 Sustainable plastic waste management through source reduction, value addition, and beneficial utilization.

TABLE 23.1
Examples of Solid Waste Treatment and Management

Solid Waste Streams	Examples	Source Reduction	Value Addition	Beneficial Utilization
Municipal waste	Green wastes, biosolids	House composting	Nutrient extraction	Soil amendment – nutrient source
Agricultural wastes	Straw, fruit peel	Soil incorporation	Biochar production	Soil amendment – nutrient source
Animal farming waste	Manure, blood, and bone	Open grazing	Nutrient extraction, bioplastics	Soil amendment – nutrient source
Construction industry	Steel, timber, bricks	Long-lasting construction materials	Biochar production	Soil amendment – nutrient source
Heavy industry	Slag and fly ash	–	Nutrient enrichment	Soil amendment – liming material
Food chain waste	Food scraps	In-situ composting	Extraction of chemicals	Niche chemicals (e.g., aconitic acid)
Particulate plastics	Plastic bags, microbeads	Recycling	Extraction of biofuel	Floor tiles, construction decks

Source: Adapted from Bolan, N., and Tsang, D. C. W., STEP In: Innovations in solid-waste treatment for environmental protection, *Waste + Water Management Australia*, 30–32, 2018a.

with their production of plastics. By promoting source reduction, we are directly saving resources by avoiding waste collection and disposal costs. Costs due to pollution control, liability, and regulatory compliance also are reduced. Without analyzing where our waste streams need to be reduced, there is no way of knowing what actions will be essential for successful outcomes. A sustainable solution is the use of green raw materials, which could allow plastics to be consumed by microbes in the environment (Engler 2012; Peng et al. 2017). Source reduction of particulate plastic wastes can be achieved through improvements in the infrastructure to collect plastic wastes. In addition, regulatory and legislative changes can reduce plastic wastes. Source reduction prevents emissions of greenhouse gases, saves energy, conserves resources, and reduces wastes for landfills and combustors (Subramanian 2000).

Degradation is the transformation of polymers from high molecular weight to low molecular weight compounds (i.e., from polymer to monomer) that weaken the plastic materials. Degradation generally takes place through various agents: (a) biodegradation — normally due to living microbes, (b) hydrolysis — due to action of water, (c) thermo-oxidation — a slow oxidation process due to action of medium temperatures, (d) photo-oxidation — due to action of sunlight and UV light, (e) thermal degradation — due to the prevalence of high temperatures (Ragaert et al. 2017; Andrady 2011). Synthetic plastics derived from the primary source of crude oil are not readily degradable in nature. Polyvinyl chloride, polystyrene, and polyethylene terephthalate (PET) are degraded by microorganisms, such as *Pseudomonas putida*, *Aspergillus niger*, and *Brevibacillus borstelensis* that secrete extracellular polymer enzymes to degrade the plastic materials (Asmita et al. 2015; Nishida and Tokiwa 1993; Yamano et al. 2008). Degradation of plastics under sunlight and solar UV radiation is very effective when exposed to the air in the marine environment. Gear-related plastics like nylon monofilament and twine have been reported to be degraded successfully by sunlight and solar UV radiation (Andrady 2011; Sil and Chakrabarti 2010; Gewert et al. 2015).

Biodegradable and chemically degradable particulate plastic products are frequently offered as better alternatives to traditional plastic items. For example, polyhydroxyalkanoates are polymers that can be used as bio-derived and biodegradable plastics. The polyhydroxyalkanoates are synthesized by microorganisms. They are compatible with living organisms and biodegradable thermoplastics that are insoluble in water, with high tensile strength, and non-toxic. These compounds have characteristics similar to those of synthetic plastic sources, such as polypropylene and polyethylene. Hence, they serve as an excellent replacement for conventional plastics with the additional benefit of being completely and relatively rapidly degradable by microorganisms under both aerobic and anaerobic conditions (Jendrossek 1998, 2001). Similarly, thermoplastic polyesters, such as poly[(R)-3-hydroxybutyric acid] and chemosynthetic polyesters can be derived biologically. These polymers are prone to degradation under water- and CO_2-enriched environments. Microorganisms such as *Alcaligenes faecalis, Pseudomonas jhorescens GK 13*, and *Amycolutopsis* strains have been found suitable to degrade such polyesters (Jendrossek 1998). These microorganisms decompose the polymers by secreting extracellular polyester depolymerase enzymes; they then utilize the low molecular weight reaction by-products as their food source (Jendrossek 1998). However, degradable plastics may not always offer an alternative to traditional plastics, because the former may disintegrate into microplastics and nanoplastics quicker than the expected lifetime of the plastic products.

23.3.2 Improvements to Infrastructure

Components of the infrastructure, such as stormwater drainage systems and garbage cans, are both contributors to the problem of plastic pollution in the marine environment and important source reduction measures. Stormwater drainage systems, in particular, are known to facilitate the transport of plastics from the urban environment into the marine environment (Kataoka et al. 2019). However, installation into the infrastructure of structures, such as gross pollutant traps provides an opportunity for urban litter to be collected and removed before it reaches the marine environment. For example, gross pollutant traps are designed to trap and isolate pollutants, only allowing filtered stormwater to continue on to the marine environment. There are a variety of gross pollution traps available, and they can remove contaminants, such as plastic, litter, oil, grit, and sediment. In addition, by trapping the suspended microplastics, gross pollution traps can reduce significantly the plastic load in water treatment plants, which improves the efficiency of the treatment plant. Membrane bioreactors have proved to be 99.9% efficient in removing microplastics in suspensions (Talvitie et al. 2017). However, this method can be expensive and energy intensive for its implementation on a large scale.

Litter, which is overflow from public garbage cans, has been found to contribute to marine plastic pollution. However, the provision of public garbage cans in many countries has been found to change consumer behavior and reduce levels of littering. Like stormwater systems, public garbage cans can be a source of marine plastic pollution, and they require important mitigation measures. The single largest source of marine plastic pollution is waste from the beverage sector, which includes plastic bottles, lids, straws, and cups (Geyer et al. 2017). Return of the bottles after use is an eco-friendly way of managing the plastic wastes. For example, in Australia, container deposit schemes refer to programs for the collection of used beverage containers in exchange for a small amount of cash. Containers can be returned to manufacturers via retailers, collected at designated depots, returned though reverse vending machines, or recovered as part of existing waste or recycling collection systems (Figure 23.3). Market-based financial incentives could promote a reduction for plastic entering the waste stream. The manufacturers and distributors of plastic products should be expected to develop strategies to recover them after use. Financial benefit, such as tax incentives, could be provided when the strategies are successfully implemented.

23.3.3 Regulatory and Legislative Changes

The main objective of reducing plastics usage is directly related to controlling the mishandling and loss of plastic parent materials in the supply usage chain. A recent ban on the use of plastic carrier bags already has been implemented in about 50 countries all over the world, and some of the countries follow the rule of minimum film thickness of plastic bags so that they can be reused (Kish 2018). The banning of lightweight single-use plastic bags and products containing microbeads

FIGURE 23.3 Plastic collection measures for recycling in Australia (a: yellow top bin – recycles; green top bin – green waste for composting; and red top bin: landfill; b: recyclables only sign; and c: plastics and cardboards in yellow top bin).

could result in a reduction of over 70% in marine plastic pollution within 3–5 years (Kish 2018). A range of marine fauna including turtles often ingests single-use plastic bags. They also break down to form microplastics in the marine environment. In Australia, single-use plastic bags have been banned or levied a tax in a number of jurisdictions. Municipalities in some countries, including Australia, also have banned the usage of plastic water bottles. Bans on the sale of some personal care products containing microplastic beads have occurred in countries like the United States and United Kingdom (Welden 2019). In some instances, levies have been used as transitional measures to change consumer behavior. The banning of single-use plastic bags was seen as an effective and easy way of reducing the amount of plastic entering the marine environment, but the consumer alternatives to plastic bags need urgent research attention (Kish 2018).

Creating awareness in the general population must be implemented by educating people at private and governmental levels concerning the adverse effects of plastics on biota. The United Nations Environmental Programme started an awareness program covering over 40 million people from 120 countries; it developed educational measures to promote less plastic use and to encourage recycling, reuse, and disposal facilities (GESAMP 2015). Several non-governmental organizations also started various awareness programs to create safe marine and human life at regional and state levels. The Joint Group of Experts on the Scientific Aspects of Marine Environmental Protection started working on minimizing the amount of plastics entering into the ocean by adopting the reduce-reuse-recycle circular economy, which proposed an inexpensive path of reducing the quantity of plastic objects and microplastic particles entering into the ocean (GESAMP 2015). In 2011, the plastic industry put forth a Joint Declaration of the Global Plastics Associations for management of marine litter, which reduced litter entry into the marine environment (Auta et al. 2017). The "four-in-one programme" in Taiwan got huge popularity across the country, and it generated awareness in the public. The program involved collaboration of different communities to collect recyclable plastic wastes in the whole country. This program involved separation of waste items into four categories such as metal containers, glass containers, plastic containers, and waste electrical appliances (Chow et al. 2017). Reports about the program suggested that education could change the attitude, behavior, and knowledge of plastic users.

The Australian National Environmental Law Association recommended that the Australian Government legislate for the substitution and phasing-out of microbeads, where manufactured locally, and the restriction on the import of products containing such content. Similarly, the Total Environment Centre in Australia described microbeads as "problematic" and called for a ban on microbeads in "cosmetics, personal care products, laundry detergents and cleaning products and paints." In 2012, Unilever announced that all its products worldwide would be microplastic-free by 2015, and, subsequently, a number of other multinationals such as Oral B (Procter and Gamble Australia), L'Oreal, and Johnson & Johnson made similar announcements.

The U.S. House of Representatives passed the Microbead Free Waters Act of 2015, which commenced the phase-out of microbeads in cosmetic and personal care items from July 1, 2017. In Australia, cosmetics are defined broadly in the Industrial Chemicals Notification and Assessment Act (ICNA 2016) and include a range of personal care products, including those in which microbeads might be found, such as facial cleansers, shampoos, and toothpaste. However, other products containing microbeads, such as cleaning and laundry products, are not covered by the Industrial Chemicals Notification and Assessment Act, and therefore, microbeads still can be added into the environment through those products. The Australian National Environmental Law Association noted that classification of plastics as a hazardous substance would be a potential avenue to manage them, particularly in relation to nurdles and microbeads.

23.3.4 Value Addition and Beneficial Utilization of Particulate Plastic Wastes

Value addition refers to increasing the utilization and commercial value of waste resources through creating new products. For example, organic solid wastes including food wastes are emerging as a resource with a significant potential to be utilized as a feedstock material for the production of fuels, chemicals, and biomaterials, given the presence in them of diverse functionalized chemical components (i.e., carbohydrates, proteins, triglycerides, fatty acids, and phenolics) (Lin et al. 2013). An array of valuable chemicals can be synthesized from these wastes (Pfaltzgraff et al. 2013).

The biomass from wastes provides a major source of building block chemicals, which are molecules with multiple functional groups that possess the potential to be transformed into new secondary chemicals (Rahimi and García 2017). For example, 12 major building block chemicals have been identified that can be produced from sugars (one of the dominant components of food chain wastes) via biological or chemical conversions (Bolan and Tsang 2018b). The 12 sugar-based building blocks are 1,4-diacids (succinic, fumaric, and malic), 2,5-furan dicarboxylic acid, 3-hydroxy propionic acid, aspartic acid, glucaric acid, glutamic acid, itaconic acid, levulinic acid, 3-hydroxybutyrolactone, glycerol, sorbitol, and xylitol/arabinitol (Lin et al. 2013). A second-tier group of building block chemicals includes gluconic acid, lactic acid, malonic acid, propionic acid, the triacids, citric and aconitic, xylonic acid, acetoin, furfural, levoglucosan, lysine, serine, and threonine. These building block chemicals can be converted subsequently to a number of high-value bio-based chemicals or materials.

An ideal solution would economically incentivize plastic collection to produce chemicals and materials that have a higher intrinsic value than the polymer itself. Such a solution could encourage recycling and even provide an economic justification for the recovery of the plastic from the environment. Innovations in this area will, in turn, deliver new solutions to high-value problems in energy, pharmaceuticals, composites, and healthcare that could further transform and revolutionize society. State-of-the-art polymer sorting techniques (e.g., magnetic density separation, triboelectric separation, hyperspectral imaging, and laser-induced breakdown spectroscopy) can be used with some limitations to delineate differences among various polymers in order to decide their subsequent chemical treatments for value added chemical synthesis (Rahimi and García 2017). The "melt-and-remold" method with or without using a catalyst can be implemented as an initial step of plastic recycling. Finding novel, environmentally benign catalytic methods might overcome the existing bottlenecks in chemical transformation of used plastics into value added new chemicals.

Concrete is one of the most commonly used materials for the construction of roads, and utilization of plastic wastes in pavement or road preparation is another innovative idea for recycling the plastic wastes (Siddique et al. 2008; Islam et al. 2016). Recycled plastic waste was used as a partial replacement of fine aggregates, and modulus of rupture, tensile strength, and compressive strength were decreased when the level of replacement was increased (Jibrael and Peter 2016). Similarly, about 15% and 20% replacement of fine aggregates with plastic waste reduced the compressive strength by 9% and 10%, respectively (Jaivignesh and Sofi 2017). PET is one of the frequently used plastics for soft-drink containers and packaging. Recycled PET was used as a binder in high performance polymer concrete (Marzouk et al. 2007). The use of plastic wastes as a replacement of fine aggregates improved the mechanical properties and workability of concrete mixtures that showed no adverse impact on the quality of the concrete (Marzouk et al. 2007).

Thermal degradation of plastic wastes is one of the prospective ways to solve the problem. This way the plastic wastes can be converted into fuel for the petrochemical industry (Panda 2018; Budsaereechai et al. 2019). Polyethylene, polypropylene, and polystyrene are the targeted polymers. Plastic polymers can be depolymerized to reproduce monomers, and they can be purified by distillation and polymerized again to form new polymers (Rahimi and García 2017). Polymers, like other high molecular weight organic compounds, such as the alkanes in oil, crack at high temperatures to form small molecules. Mixtures of polymers can be converted into useful compounds by pyrolysis or oxidation.

Pyrolysis is a process by which plastics are heated in the absence of oxygen until the waste plastic material decomposes into gases and oils (Budsaereechai et al. 2019; Al-Salem et al. 2017). During pyrolysis, plastic polymers are broken down into small molecules. Pyrolysis at high temperatures (>600°C) favors the production of small gas molecules, while low temperatures (<400°C) produce more viscous liquids. This process is a viable route for the recycling of waste plastics, and it converts them into fuels and gases. It solves an environmental problem, because most of plastics commonly contain toxic and halogen flame retardants. Char produced from pyrolysis can be activated under standard conditions and used in wastewater treatment, heavy-metal removal, and smoke and odor removal (Jamradloedluk and Lertsatitthanakorn 2014; Miandad et al. 2018). The gases produced from pyrolysis are hydrogen (H_2), carbon monoxide (CO), and carbon dioxide (CO_2), and they can be used as energy carriers (Lee et al. 2017; Kannan et al. 2017).

Beneficial utilization of solid wastes is achieved through recycling and direct reutilization in various industries. Through the recycling process, materials like glass, metal, plastics, and paper are collected and separated and sent to processing centers, where they are made into new products. A number of waste products, including compost and biosolids derived from organic wastes and slags derived from steel and coal-fired power plants, can be reutilized as soil amendments. Compost provides a valuable resource of nutrient and carbon inputs to improve soil health and productivity. Similarly, some of the slag materials can be used as a liming material to manage soil acidity and for the remediation of metal-contaminated soils including mine soils. Beneficial utilization through recycling generates a host of environmental, financial, and social benefits. It prevents the emission of many greenhouse gases that affect global climate, saves energy, supplies valuable raw materials to industry, creates jobs, stimulates the development of green technologies, conserves resources for the future, and reduces the need for new landfills and combustors. The use of waste commodity plastics in binder modification carries the advantage of a cheap and effective means of enhancing conventional bitumen binder performance and is an alternative way to utilize plastic wastes.

23.4 INNOVATIONS IN PARTICULATE PLASTIC WASTE MANAGEMENT

Particulate plastic solid wastes represent a major environmental problem, but innovations in the treatment and sustainable management of these waste resources can provide economic and social benefits. Innovation in solid waste treatment and management is critical not only for environmental protection, but also for the beneficial utilization of these wastes for sustainable production. Figure 23.4 shows various ways for beneficial utilization of plastic wastes for innovative sustainable management. Several strategies to utilize solid wastes have been implemented, including recycling by composting, which, however, cannot process all these waste resources. In this regard, advanced waste utilization practices aimed at achieving sustainable development should focus on innovative technologies able to convert these waste resources into value-added products. These include integrated green chemistry technologies involving biochemical processes to synthesize high-value chemicals, advanced materials, and biofuel precursors, and extractive processes for the recovery of valuable compounds, such as antioxidants (Dhawan et al. 2019; Rahimi and García 2017; Veksha et al. 2017). For example, organic wastes will play a key role in the near future around the "biorefinery and bio-economy" concept that will contribute to a greener and more sustainable society. The diversity of functionalized chemical components found in organic wastes, including food wastes, reflects the range of chemical industry sectors that could benefit from using particulate plastics as a renewable feedstock, thereby improving their green credentials. Through the increased utilization of wastes for non-food applications, such as the synthesis of chemicals, the public perception about wastes may change, thereby helping to establish the new supply chains needed for a future sustainable society. However, in order to achieve a true bioeconomy for the utilization of solid waste, we also need governments to take this opportunity to go beyond recycling and invest in waste reduction and reuse through various incentive and support schemes, such as a carbon credit scheme. Recently, through the pyrolysis method, diesel fuel has been prepared using used grocery plastic bags for sustainable utilization of waste plastic bags (Sharma et al. 2014a; Al-Salem 2019). In addition, granulate plastic bags have been mixed with concrete to form concrete bricks for roads (Safinia and Alkalbani 2016; Marzouk et al. 2007). Hi-tech carbon nanotubes (Bazargan and McKay 2012; Wu et al. 2014; Zhuo and Levendis 2014) have been synthesized using the plastic wastes. For example, using a solvent-free catalytic thermal dissociation process, successful conversion of PET bottle waste into carbon nanostructures including fullerenes and carbon nanotubes was reported (El Essawy et al. 2017). Similarly, single-crystal graphene was synthesized successfully

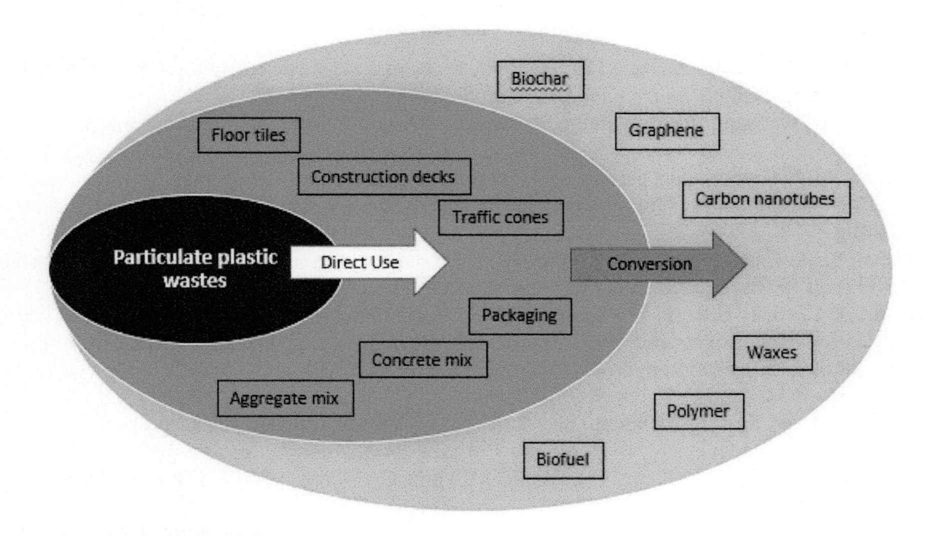

FIGURE 23.4 Innovations in sustainable plastic waste management and beneficial utilization.

on polycrystalline Cu foil from polyethylene- and polystyrene-based plastic wastes under ambient pressure (Sharma et al. 2014b). Thus, plastic wastes could find numerous high value applications in various industries (Figure 23.4) contributing to sustainable recycling of these materials, which otherwise are a menace to the environment.

23.5 SUMMARY AND CONCLUSIONS

Sustainable plastic waste management includes source reduction, value addition, and beneficial utilization. Source reduction, which refers to consuming less, is the most successful and sustainable method of reducing plastic waste generation. Source reduction of particulate plastic wastes can be achieved through improvements in infrastructure to collect plastic wastes and via the regulatory and legislative changes to reduce the wastes. The installation of infrastructure, such as gross pollutant traps provides an opportunity for urban litter to be collected and removed before it reaches the marine environment. Litter overflowing from public garbage cans contributes to marine plastic pollution; hence, there is a need for appropriate collection and segregation before the litter reaches oceans. The provision of public garbage cans has been found to change consumer behavior and reduce levels of littering. Value addition and beneficial utilization of plastic wastes include synthesis of secondary chemicals and utilization of plastic wastes in landscaping, concrete mixtures, nanomaterial preparations, and road construction.

REFERENCES

Al-Salem, S. M. 2019. Thermal pyrolysis of high density polyethylene (HDPE) in a novel fixed bed reactor system for the production of high value gasoline range hydrocarbons (HC). *Process Safety and Environmental Protection* 127:171–179.

Al-Salem, S. M., A. Antelava, A. Constantinou, G. Manos, and A. Dutta. 2017. A review on thermal and catalytic pyrolysis of plastic solid waste (PSW). *Journal of Environmental Management* 197:177–198.

Andrady, A. L. 2011. Microplastics in the marine environment. *Marine Pollution Bulletin* 62(8):1596–1605.

Asmita, K., T. Shubhamsingh, and S. Tejashree. 2015. Isolation of plastic degrading micro-organisms from soil samples collected at various locations in Mumbai, India. *International Research Journal of Environment Sciences* 4(3):77–85.

Auta, H. S., C. U. Emenike, and S. H. Fauziah. 2017. Distribution and importance of microplastics in the marine environment: A review of the sources, fate, effects, and potential solutions. *Environment International* 102:165–176.

Bazargan, A., and G. McKay. 2012. A review – Synthesis of carbon nanotubes from plastic wastes. *Chemical Engineering Journal* 195–196:377–391.

Bolan, N., and D. C. W. Tsang. 2018a. STEP In: Innovations in solid-waste treatment for environmental protection. *Waste + Water Management Australia*, September 2018:30–32.

Bolan, N., and D. C. W. Tsang. 2018b. Value-added chemicals from food wastes. *Chemistry in Australia*, April 2018:16–19.

Browne, M. A. 2015. Sources and pathways of microplastics to habitats. In *Marine Anthropogenic Litter*, edited by M. Bergmann, L. Gutow and M. Klages, 229–244. Cham, Switzerland: Springer International Publishing.

Budsaereechai, S., A. J. Hunt, and Y. Ngernyen. 2019. Catalytic pyrolysis of plastic waste for the production of liquid fuels for engines. *RSC Advances* 9(10):5844–5857.

Ceccarini, A., A. Corti, F. Erba, F. Modugno, J. La Nasa, S. Bianchi, and V. Castelvetro. 2018. The hidden microplastics: New insights and figures from the thorough separation and characterization of microplastics and of their degradation byproducts in coastal sediments. *Environmental Science and Technology* 52(10):5634–5643.

Chow, C. F., W. M. W. So, T. Y. Cheung, and S. K. D. Yeung. 2017. Plastic waste problem and education for plastic waste management. In *Emerging Practices in Scholarship of Learning and Teaching in a Digital Era*, edited by S. C. Kong, T. L. Wong, M. Yang, C. F. Chow and K. H. Tse, 125–140. Singapore: Springer Singapore.

CleanUpAustralia. 2019. Clean Up Australia; https://www.cleanup.org.au/plastic-bags; Accessed on May 24, 2019.

Dhawan, R., B. M. S. Bisht, R. Kumar, S. Kumari, and S. K. Dhawan. 2019. Recycling of plastic waste into tiles with reduced flammability and improved tensile strength. *Process Safety and Environmental Protection* 124:299–307.

Do, T. C. V., and H. W. Scherer. 2012. Compost and biogas residues as basic materials for potting substrates. *Plant Soil Environment* 58(10):459–464.

Downes, J., and E. Dominish. 2018. China's recycling 'ban' throws Australia into a very messy waste crisis. *The Conversation, April 2018*, 26 May.

Duis, K., and A. Coors. 2016. Microplastics in the aquatic and terrestrial environment: Sources (with a specific focus on personal care products), fate and effects. *Environmental Sciences Europe* 28(1):2.

El Essawy, N. A., A. H. Konsowa, M. Elnouby, and H. A. Farag. 2017. A novel one-step synthesis for carbon-based nanomaterials from polyethylene terephthalate (PET) bottles waste. *Journal of the Air and Waste Management Association* 67(3):358–370.

Engler, R. E. 2012. The complex interaction between marine debris and toxic chemicals in the ocean. *Environmental Science and Technology* 46(22):12302–12315.

Eriksen, M., S. Mason, S. Wilson, C. Box, A. Zellers, W. Edwards, H. Farley, and S. Amato. 2013. Microplastic pollution in the surface waters of the Laurentian Great Lakes. *Marine Pollution Bulletin* 77(1):177–182.

GESAMP. 2015. Sources, fate and effects of microplastics in the marine environment: A global assessment. In *Joint Group of Experts on the Scientific Aspects of Marine Environmental Protection (GESAMP) Reports and Studies*, edited by P.J. Kershaw: IMO/FAO/UNESCO-IOC/UNIDO/-WMO/IAEA/UN/UNEP/UNDP.

Gewert, B., M. M. Plassmann, and M. MacLeod. 2015. Pathways for degradation of plastic polymers floating in the marine environment. *Environmental Science: Processes and Impacts* 17(9):1513–1521.

Geyer, R., J. R. Jambeck, and K. L. Law. 2017. Production, use, and fate of all plastics ever made. *Science Advances* 3:e1700782.

Guerranti, C., T. Martellini, G. Perra, C. Scopetani, and A. Cincinelli. 2019. Microplastics in cosmetics: Environmental issues and needs for global bans. *Environmental Toxicology and Pharmacology* 68:75–79.

ICNA, 2016. Industrial Chemicals (Notification and Assessment) Act 1989 Compilation No. 38, Federal Register of Legislation, Australian Government.

Islam, M. J., M. S. Meherier, and A. K. M. R. Islam. 2016. Effects of waste PET as coarse aggregate on the fresh and harden properties of concrete. *Construction and Building Materials* 125:946–951.

Jaivignesh, B., and A. Sofi. 2017. Study on mechanical properties of concrete using plastic waste as an aggregate. *IOP Conference Series: Earth and Environmental Science* 80:012016.

Jamradloedluk, J., and C. Lertsatitthanakorn. 2014. Characterization and utilization of char derived from fast pyrolysis of plastic wastes. *Procedia Engineering* 69:1437–1442.

Jendrossek, D. 2001. Microbial degradation of polyesters. *Advances in Biochemical Engineering/Biotechnology* 71:293–325.

Jendrossek, D. 1998. Microbial degradation of polyesters: A review on extracellular poly(hydroxyalkanoic acid) depolymerases. *Polymer Degradation and Stability* 59:317–325.

Jibrael, M. A., and F. Peter. 2016. Strength and behavior of concrete contains waste plastic. *Journal of Ecosystem and Ecography* 6:186.

Kannan, P., G. Lakshmanan, A. Al Shoaibi, and C. Srinivasakannan. 2017. Equilibrium model analysis of waste plastics gasification using CO_2 and steam. *Waste Management and Research* 35(12):1247–1253.

Kataoka, T., Y. Nihei, K. Kudou, and H. Hinata. 2019. Assessment of the sources and inflow processes of microplastics in the river environments of Japan. *Environmental Pollution* 244:958–965.

Kish, R. J. 2018. Using legislation to reduce one-time plastic bag usage. *Economic Affairs* 38(2):224–239.

Lambert, S., C. Sinclair, and A. Boxall. 2014. Occurrence, degradation, and effect of polymer-based materials in the environment. In *Reviews of Environmental Contamination and Toxicology*, Volume 227, edited by D. M. Whitacre, 1–53. Cham, Switzerland: Springer International Publishing.

Lee, J., T. Lee, Y. F. Tsang, J. I. Oh, and E. E. Kwon. 2017. Enhanced energy recovery from polyethylene terephthalate via pyrolysis in CO_2 atmosphere while suppressing acidic chemical species. *Energy Conversion and Management* 148:456–460.

Lee, J., S. Hong, Y. K. Song, S. H. Hong, Y. C. Jang, M. Jang, N. W. Heo, et al. 2013. Relationships among the abundances of plastic debris in different size classes on beaches in South Korea. *Marine Pollution Bulletin* 77(1):349–354.

Lin, C. S. K., L. A. Pfaltzgraff, L. Herrero-Davila, E. B. Mubofu, S. Abderrahim, J. H. Clark, A. A. Koutinas, et al. 2013. Food waste as a valuable resource for the production of chemicals, materials and fuels. Current situation and global perspective. *Energy and Environmental Science* 6(2):426–464.

Marzouk, O. Y., R. M. Dheilly, and M. Queneudec. 2007. Valorization of post-consumer waste plastic in cementitious concrete composites. *Waste Management* 27(2):310–318.

Miandad, R., R. Kumar, M. A. Barakat, C. Basheer, A. S. Aburiazaiza, A. S. Nizami, and M. Rehan. 2018. Untapped conversion of plastic waste char into carbon-metal LDOs for the adsorption of Congo red. *Journal of Colloid and Interface Science* 511:402–410.

Murphy, F., C. Ewins, F. Carbonnier, and B. Quinn. 2016. Wastewater treatment works (WwTW) as a source of microplastics in the aquatic environment. *Environmental Science and Technology* 50(11):5800–5808.

Ng, E.-L., E. H. Lwanga, S. M. Eldridge, P. Johnston, H. W. Hu, V. Geissen, and D. Chen. 2018. An overview of microplastic and nanoplastic pollution in agroecosystems. *Science of The Total Environment* 627:1377–1388.

Nishida, H., and Y. Tokiwa. 1993. Distribution of poly(β-hydroxybutyrate) and poly(ε-caprolactone) aerobic degrading microorganisms in different environments. *Journal of Environmental Polymer Degradation* 1(3):227–233.

O'Farrell, K. 2018. 2016–17 Australian Plastics Recycling Survey—National Report. Melbourne: Department of the Environment and Energy, Australian Government.

Panda, A. K. 2018. Thermo-catalytic degradation of different plastics to drop in liquid fuel using calcium bentonite catalyst. *International Journal of Industrial Chemistry* 9(2):167–176.

Peng, J., J. Wang, and L. Cai. 2017. Current understanding of microplastics in the environment: Occurrence, fate, risks, and what we should do. *Integrated Environmental Assessment and Management* 13(3):476–482.

Pfaltzgraff, L. A., M. De Bruyn, E. C. Cooper, V. Budarin, and J. H. Clark. 2013. Food waste biomass: A resource for high-value chemicals. *Green Chemistry* 15(2):307–314.

Pickin, J., P. Randell, J. Trinh, and B. Grant. 2018. *National Waste Report 2018*. Melbourne, Australia: Department of the Environment and Energy, Australian Government.

PlasticsEurope. 2018. *PlasticsEurope Annual Review 2017–2018*. Brussels: Association of Plastics Manufacturers.

Ragaert, K., L. Delva, and K. Van Geem. 2017. Mechanical and chemical recycling of solid plastic waste. *Waste Management* 69:24–58.

Rahimi, A., and J. M. García. 2017. Chemical recycling of waste plastics for new materials production. *Nature Reviews Chemistry* 1:0046.

Reisser, J., J. Shaw, C. Wilcox, B. D. Hardesty, M. Proietti, M. Thums, and C. Pattiaratchi. 2013. Marine plastic pollution in waters around Australia: Characteristics, concentrations, and pathways. *PLOS ONE* 8(11):e80466.

Safinia, S., and A. Alkalbani. 2016. Use of recycled plastic water bottles in concrete blocks. *Procedia Engineering* 164:214–221.

Sharma, B. K., B. R. Moser, K. E. Vermillion, K. M. Doll, and N. Rajagopalan. 2014a. Production, characterization and fuel properties of alternative diesel fuel from pyrolysis of waste plastic grocery bags. *Fuel Processing Technology* 122:79–90.

Sharma, S., G. Kalita, R. Hirano, S. M. Shinde, R. Papon, H. Ohtani, and M. Tanemura. 2014b. Synthesis of graphene crystals from solid waste plastic by chemical vapor deposition. *Carbon* 72:66–73.

Siddique, R., J. Khatib, and I. Kaur. 2008. Use of recycled plastic in concrete: A review. *Waste Management* 28(10):1835–1852.

Sil, D., and S. Chakrabarti. 2010. Photocatalytic degradation of PVC–ZnO composite film under tropical sunlight and artificial UV radiation: A comparative study. *Solar Energy* 84(3):476–485.

Silpa, K., L. Yao, P. Bhada-Tata, and F. Van Woerden. 2018. What a Waste 2.0: A Global Snapshot of Solid Waste Management to 2050 Overview booklet. Washington, DC: World Bank.

Singh, P., and V. P. Sharma. 2016. Integrated plastic waste management: Environmental and improved health approaches. *Procedia Environmental Sciences* 35:692–700.

Statista. 2019. Dossier about plastic waste in the United States. https://www.statista.com/study/60094/plastic-waste-in-the-us/

Subramanian, P. M. 2000. Plastics recycling and waste management in the US. *Resources, Conservation and Recycling* 28(3):253–263.

Talvitie, J., A. Mikola, A. Koistinen, and O. Setälä. 2017. Solutions to microplastic pollution – Removal of microplastics from wastewater effluent with advanced wastewater treatment technologies. *Water Research* 123:401–407.

Veksha, A., A. Giannis, and V. W. C. Chang. 2017. Conversion of non-condensable pyrolysis gases from plastics into carbon nanomaterials: Effects of feedstock and temperature. *Journal of Analytical and Applied Pyrolysis* 124:16–24.

Welden, N. 2019. Microplastics: Emerging contaminants requiring multilevel management. In *Waste: A Handbook for Management*, 2nd ed., edited by T.M. Letcher and D.A. Vallero, 405–421. Amsterdam, the Netherlands, Academic Press.

Wijesekara, H., N. S. Bolan, L. Bradney, N. Obadamudalige, B. Seshadri, A. Kunhikrishnan, R. Dharmarajan, et al. 2018. Trace element dynamics of biosolids-derived microbeads. *Chemosphere* 199:331–339.

Wu, C., M. A. Nahil, N. Miskolczi, J. Huang, and P. T. Williams. 2014. Processing real-world waste plastics by pyrolysis-reforming for hydrogen and high-value carbon nanotubes. *Environmental Science and Technology* 48(1):819–826.

Xanthos, D., and T. R. Walker. 2017. International policies to reduce plastic marine pollution from single-use plastics (plastic bags and microbeads): A review. *Marine Pollution Bulletin* 118(1):17–26.

Yamano, N., A. Nakayama, N. Kawasaki, N. Yamamoto, and S. Aiba. 2008. Mechanism and characterization of polyamide 4 degradation by *Pseudomonas* sp. *Journal of Polymers and the Environment* 16(2):141–146.

Zhao, S., L. Zhu, and D. Li. 2015. Microplastic in three urban estuaries, China. *Environmental Pollution* 206:597–604.

Zhuo, C., and Y. A. Levendis. 2014. Upcycling waste plastics into carbon nanomaterials: A review. *Journal of Applied Polymer Science* 131(4). doi:10.1002/app.39931.

24 Evaluation and Mitigation of the Environmental Impact of Synthetic Microfibers

Francesca De Falco, Mariacristina Cocca, Emilia Di Pace, Maria Emanuela Errico, Gennaro Gentile, and Maurizio Avella

CONTENTS

24.1 Introduction 413
24.2 Evaluation of Microplastic Release from Synthetic Clothes during Washing 415
24.3 Mitigation Actions 419
References 421

24.1 INTRODUCTION

The washing of synthetic fabrics has been identified recently as one of the major contributors to the global release of primary microplastics to the oceans (Boucher and Friot 2017). First records of the presence of synthetic fibers coming from clothes washing machines, which end in sludge and sewage treatment plant effluents, were reported two decades ago (Habib, Locke, and Cannone 1998; Zubris and Richards 2005). Such fibers were shown to be persistent in both treated wastewater effluents and sludge produced by wastewater treatment plants (WWTPs) (Zubris and Richards 2005). Microplastics with fibrous shape were also found by Thompson et al. (2004) during a campaign sampling different beaches, estuaries, and coasts in the United Kingdom. The identified synthetic polymers were acrylic, polyamide, polyester, and polypropylene (Thompson et al. 2004). Despite these first accounts, it was only in 2011 that the presence of microplastics with a fibrous shape found in marine sediments was directly correlated to the washing of garments; this was done by comparing the proportions of polyester and acrylic fibers used in apparel with those found in habitats receiving sewage discharges and effluents (Browne et al. 2011). Synthetic fibers like polyester and polyamides represent a large cut (60%) of the global consumption of fibers in the apparel industry. The same global amount of fibers for apparel has been constantly increasing in the last two decades, exactly due to a 300% growth of the market of synthetic fibers in the same period (Boucher and Friot 2017). The mechanical and chemical stresses of a washing process can cause the detachment of "microfibers" from the yarns composing a fabric, which cannot be completely removed by WWTPs. Several studies have investigated the capability of WWTPs to remove microplastics. A WWTP on the west coast of Sweden was able to block >99% of microplastics, but, nevertheless, the amount of microplastics with dimensions $\geq$300 μm was consistent (Magnusson and Wahlberg 2014). A great abundance of microfibers was recorded in the wastewater effluents from eight wastewater treatment plants that discharge into the San Francisco Bay (California) (Sutton et al. 2016). Also investigations of three WWTPs in Sydney (Australia) revealed an average number of 0.28, 0.48, and 1.54 microplastics per liter of final effluent in tertiary-, secondary-, and primary-treated effluent, respectively (Ziajahromi et al. 2017).

Another study on the stepwise removal of microlitter in a tertiary level WWTP in Finland found that 1.7×10^6 to 1.4×10^8 microplastics per day were discharged into the sea through wastewater effluents (Talvitie, Mikola, and Setälä et al. 2017). Such results cannot be scaled up to obtain global estimations of the amount of microplastics that can pass through WWTPs, since the type of plant and its efficiency strongly depend on the country, and, moreover, some countries with poor infrastructure do not collect and treat most of their wastewater (UNEP 2016).

Even if the concentrations of microplastics escaping from WWTPs are low, the volumes of effluents discharged daily to aquatic ecosystems are large, which means that WWTPs are potentially an important entrance route for microplastics, particularly for microfibers. Cesa, Turra, and Baruque-Ramos (2017) reviewed all the studies which found textile fibers dominant in field samples; their review showed that microfibers are present in beaches all over the world, in the North Sea, in the Atlantic and Pacific Oceans, in the Arctic, and even in the deep sea as illustrated in Figure 24.1 (Cesa, Turra, and Baruque-Ramos 2017; Taylor et al. 2016). Of course, such a ubiquitous presence poses some questions on the

FIGURE 24.1 Images of specimens in situ (a, b) and close-up images (c, d) of microplastic fibers exhibiting their interference colors (used to aide classification) under cross-polarized illumination (c) and under plain polarized light (d); (a) sea pen, JC066-3717; (b) hermit crab with zoanthid symbionts, JC066-702; (c) polyester microfiber, JC066-702-09; (d) acrylic microfiber, JC066-702-10. (Reprinted from Taylor, M.L. et al., *Sci. Rep.*, 6, 33997, 2016. Open access article with unrestricted use and distribution.)

possible interaction with marine fauna. For instance, polyethylene terephthalate microfibers ingested by the zooplankton crustacean *Daphnia magna* were found responsible for an increased mortality of the species (Jemec et al. 2016). Textile fibers were also found in fish and shellfish on sale for human consumption in some markets in Makassar, Indonesia, and in California (Rochman et al. 2015).

In conclusion, considering all information currently available on microplastic pollution caused by the washing process of synthetic clothes, it is clear that more uniform and consistent investigations are needed to better understand the mechanisms that lead to such release of microfibers and possible parameters that influence their release. There is also a need to estimate the actual amount of microfibers that are released from washing. Only starting from this point, will it be possible to assess which mitigation actions could be applied to reduce this source of pollution.

24.2 EVALUATION OF MICROPLASTIC RELEASE FROM SYNTHETIC CLOTHES DURING WASHING

The procedures used worldwide to wash fabrics are difficult to estimate due to the evident differences in uses and consumption among countries. There is also a lack of information from continents like South Americas and Africa. However, it has been assessed that the number of washing machines globally used is more than 840 million, along with an annual consumption of about 100 TWh of energy and 20 km^3 of water (Cesa, Turra, and Baruque-Ramos 2017). The European Life+ Mermaids project performed a survey in 2014 of household washing habits; the majority of respondents came from Italy, Spain, Belgium, the Netherlands, Germany, France, and Portugal (Life+ Mermaids project 2014). The results of the survey showed that the most common brands of washing machines purchased are Whirlpool, Bosch, LG, AEG, and Indesit, of which, 90% have a capacity between 5 and 8 kg. The most used washing program, temperature, and time are the cotton program, 30°C–40°C, and 1–1 hour 30 minutes, respectively. The consumption of water for each washing is around 30–50 L, and the average number of washes per household each year is 352.54, that is 6.7 washes/household each week. Taking into account such data, and the fact that this survey represents an estimation of washing habits only in Europe, it is clear that the amount of wastewater coming from washing machines is substantial and is a pathway for the transport of microplastics with a fibrous shape. With such a scenario, it is a high priority to evaluate the real environmental impact of washing processes of synthetic clothes, starting from quantifying the microplastics that can be released during a wash and identifying possible parameters that influence their release. The most relevant studies on this topic are summarized below.

Browne et al. (2011) made the first attempt to evaluate the number of fibers released by a washing process. They used for their tests three different front-loading washing machines (Siemens Extra Lasse XL 1000, John Lewis JLWM1203, and Bosch WAE24468GB). They washed three types of polyester clothes (fleeces, shirts, and blankets) at 40°C and 600 rpm, without using any detergents since they clogged the filter-papers used to filter the wastewater. The outcome was that all garments released more than 100 fibers per liter of effluent. They concluded that each garment can shed more than 1900 fibers per wash (Browne et al. 2011).

Such estimation was the most cited and the only one available for some time, but doubts arose about its reliability since the study provided no detailed information about the procedure used for the quantification, in particular, regarding the filter pore size and the counting procedure of the fibers.

After this first investigation, several studies were published between 2016 and 2018 that evaluated microplastic release from the washing of synthetics.

Hartline et al. (2016) used a gravimetric method to estimate the release of microplastics during the washing of five types of commercial jackets, which were characterized by different compositions of synthetic fibers. They performed two types of washing tests: using new garments and garments that were mechanically aged. The tests were performed without any detergents using a front-load (Samsung model WF42H5000AW/A2) or top-load (Whirlpool model WET3300XQ) washing machine. For the trials in the front-load machine, they applied the following conditions: 29°C–41°C warm cycle for 24 minutes, which was made up of an 8-minute wash, 10-minute rinse, and 6-minute spin (1200 rpm); for the top-load washing machine, they used the following program: 29.6°C warm cycle for 30 minutes, which was divided into a 12-minute wash, 14-minute rinse, and 4-minute spin. They sampled only an aliquot of the wastewater of each washing test, which was filtered through two in-line hand-cut Nitex nylon filters, the first with a 333 μm pore size and the second with a 20 μm pore size. Results showed that the microfiber mass per garment collected across all experiments ranged from 0 to 2 g, and that the quantities recovered during the tests with the top-load machine were about 7 times greater than those collected using the front-load machine. In addition, the garments mechanically aged released more microfibers than the new ones (Hartline et al. 2016).

A front-load washing machine (Bosch model Maxx7 VarioPerfect) was also used by Pirc et al. (2016), who tested six identical fleece blankets. Using up to ten consecutive washing cycles, they washed them with only water, water with detergent, or water with detergent plus softener. Their washing program was characterized by a temperature of 30°C, duration of 15 minutes, and spinning at 600 rpm. The wastewater was filtered through a stainless steel filter with 200 × 200 μm pores, and the relative microplastic release was calculated as a percentage of the initial blanket mass. This study concluded that the usage of detergent and softener did not significantly influence emission, and that after eight washing cycles, the emission decreased and stabilized at around 0.0012 wt% (Pirc et al. 2016).

An approach similar to Pirc et al. (2016) was applied by Napper and Thompson (2016). They tested three synthetic jumpers with different chemical compositions: 100% acrylic, 100% polyester, and 65% polyester/35% cotton blend. Different from the previous studies, they did not wash the whole garment, but cut a 20 × 20 cm square from the back of each garment, sewing the edges with cotton thread. The square samples were washed one per time in a Whirlpool WWDC6400 washing machine for 1 hour and 15 minutes at 1400 rpm and with varying temperatures (30°C and 40°C), detergents (absent – i.e., with only water, biodetergent, and non-biodetergent), and conditioners (absent or present). Data were recorded after the fifth wash, because they observed a stabilization of the amount of released microfibers after the first five washes. The evaluation procedure consisted of filtering the wastewater effluent through a nylon sieve with a 25 μm pore size, which was attached to the end of the drain pipe. The quantity of microfibers released was calculated through a conversion formula. They obtained the following numbers of microplastics released per 6 kg of washing load: 728,789 microfibers for the acrylic fabric, 496,030 microfibers for the polyester fabric, and 137,951 microfibers for the polyester-cotton blend. No clear trends were observed regarding the use of detergents and conditioners (Napper and Thompson 2016).

Another study by Sillanpää et al. (2017) tested four different types of polyester textiles and two garments of cotton. They washed them with liquid detergent for up to five washing cycles. The tests were performed by using a front-load washing machine (Bosch WAE28477SN) at 40°C, with a duration of 75 minutes, and spin-dry rate of 1200 rpm. From each washing test, only a sample of the total wastewater was collected, and it was filtered under vacuum through filters of 0.7 μm pore size. Then, a portion of the filter surface, around 10%, was observed under an optical stereomicroscope to count the fibers released. The number and mass of microfibers released from cotton and polyester textiles during the first washing cycle were in the range of 2.1×10^5–1.3×10^7 and 0.12–0.33% w/w, respectively, and these amounts decreased during sequential washes (Sillanpää and Sainio 2017).

Textile parameters were the focus of the work of Carney Almroth et al. (2018), who studied polyamide, polyacrylic (polyacrylonitrile), and polyester (polyethylene terephthalate) fabrics characterized by different knitting factors. The wash trials were carried out not using a washing machine, but using a laboratory simulator of real washing processes, a Gyrowash one bath 815. Fabric samples were washed at 60°C for 30 minutes with a liquid detergent, and the washing water was filtered through a glass filter with 1.2 μm pore size. Only a part of the filter surface was observed by light microscopy to evaluate the number of microfibers released. Their results highlighted that polyester fleece fabrics released more microfibers (an average of 7360 fibers m^{-2} L^{-1} in one wash) than polyester fabrics (87 fibers m^{-2} L^{-1}). Moreover, they concluded that loose textile structures can release more microfibers, as did worn fabrics, and high twist yarns can decrease the release of microfibers (Carney Almroth et al. 2018).

In several studies, De Falco and co-workers (De Falco et al. 2018a,b, 2019a) developed two effective analytical procedures, at the laboratory scale and at the real scale, to assess the release of microplastics from the washing of fabrics, and to investigate textile and washing parameters that may have an influence on their release. In their first study (De Falco et al. 2018b), they analyzed the release of microplastics from three types of standard fabrics (woven and knitted polyester and woven polypropylene) and evaluated the effect of washing detergents (liquid and powdered ones), additives (i.e., softener, bleach, and oxidizing agents), washing parameters (i.e., time, temperature, mechanical action, and water hardness), and washing conditions (domestic and industrial). Washing tests were performed in a laboratory simulator of real washing processes (Linitest apparatus, URAI S.p.A.). Fabric samples were washed at 40°C for 45 minutes and the effluents were filtered through filters with 5 μm pore size. The filter surfaces were observed by scanning electron microscopy, applying a counting method (statistically validated) to evaluate the number and dimensions of microfibers released. The outcomes of this study revealed that the number of microfibers released per gram of washed fabric ranged from 60 to 200 (for washes with only distilled water) and 600–1300 (for washes with liquid detergent) to 1700–3500 (for washes with powdered detergent). It was clear that, in general, the use of the detergent increases the release and the powdered one has the worst effect, probably due to the friction with the fabric caused by the presence of inorganic compounds insoluble in water, like zeolite. Moreover, independent from the detergent used, the greatest release of microplastics was recorded from woven polyester, probably due to the short staple fibers composing the yarns, which could more easily slip away during the washing process. Further tests carried out on woven polyester by using different additives showed that the release of the microplastics could be reduced by more than 35% if a softener is used, due to its property of reducing the friction among the fibers. Moreover, tests with changing the washing parameters indicated that higher temperature, water hardness, and mechanical action can increase the microplastic release. The length of microfibers released ranged from 50 to 900 μm, with a mean diameter of 14–20 μm, and these dimensions are compatible with those found in marine environments and organisms.

To evaluate the impact of domestic washings on the environment, another study (De Falco et al. 2019a) on the release of microfibers during washings of real commercial clothes was performed. Four different garments were tested, all made with 100% polyester, but one was composed of a blend of 50% polyester and 50% cotton. This time, washing tests were performed in a household washing machine (Bosch Series 4 VarioPerfect WLG24225i) at 40°C, for 107 minutes, and 1200 rpm, and a commercial detergent was used. A real washing load was considered by washing garments together, and loads of 2–2.5 kg were reached. The entire volume of effluents was filtered to obtain reliable data not affected by errors due to the sampling of only an aliquot. The whole wastewater was filtered through filters with decreasing pore size (400, 60, and 20 μm), and, due to clogging problems, only 300 mL were filtered through a 5 μm pore size filter. The main aims of this investigation were to obtain reliable, quantitative data about the contribution of real washings of commercial synthetic clothes to microfiber pollution, and to identify possible effects of textile parameters on their release. Compared to other studies on the same topic, this work introduced three simultaneous novelties: (1) the whole wastewater coming from the washing machine was filtered; (2) real washing conditions were tested using real washing loads, a commercial detergent, and a real washing program for synthetics; and (3) four different filter pore sizes were used in a multistep filtration procedure, allowing the evaluation of the dimensions of released microfibers. Regarding this last part, the multistep filtration procedure was effective in the quantification of the microfiber release, showing amounts that ranged from 124 to 308 mg per kg of washed fabrics (Figure 24.2). The highest release was recorded

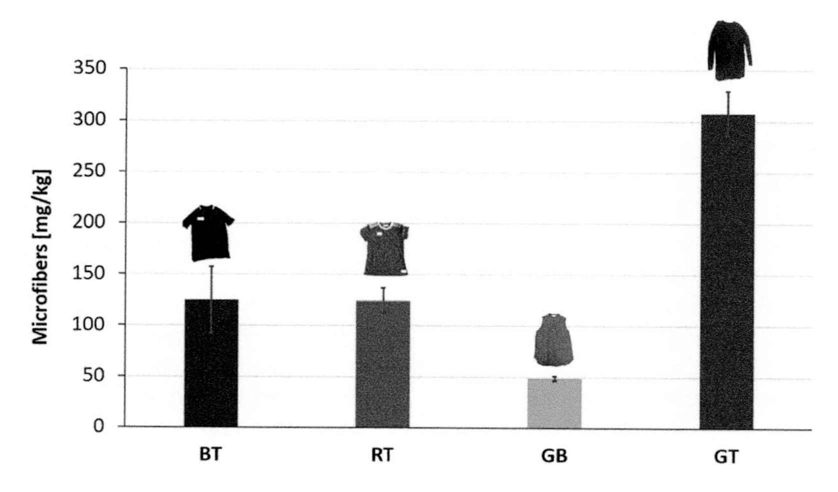

FIGURE 24.2 Microfibers released (expressed in mg/kg, mean ± SD, $n = 2$) from a blue T-shirt (100% polyester, code BT), a red T-shirt (100% polyester, code RT), a green sleeveless blouse (100% polyester of which 65% was recycled polyester, code GB), a green long-sleeved top (the front was made of 100% polyester and the back was made of a blend of 50% cotton and 50% rayon, code GT). (Reprinted from De Falco, F. et al., *Sci. Rep.*, 9, 6633, 2019. Open access article with unrestricted use and distribution.)

for the garment made of the polyester/cotton blend, but further analysis on the chemical composition of the microfibers released revealed that 80% of them were of cellulosic nature. It was possible to single out some parameters that may have a beneficial effect on the release of microfibers, i.e., yarns made of continuous filaments, high twist, and low hairiness. Additional trials were performed in which up to 10 washes were done, and they showed that the release of microfibers decreases and reaches a plateau after 4–5 cycles in the case of 100% polyester garments. The multistep filtration procedure pointed out that the greatest fraction of microfiber released was collected on 60 μm pore size filters, indicating the dimensions that could pass through WWTPs and pose a threat for marine organisms.

The two analytical procedures developed by De Falco and co-workers were compared in another systematic study (De Falco et al. 2018a). The study aimed to compare the amount of microfibers released from synthetic fabrics during washings in a household washing machine and in a laboratory simulator (Gyrowash, James H. Heal & Co, UK). The fibers collected in both trials were analyzed to determine their dimensions and to quantify the overall microfiber release. The same 100% polyester fabric was washed in both tests, and it was possible to compare the amounts of microfibers released (mg/kg in real tests, number of microfibers/kg in laboratory tests) by using a conversion formula based on the mean dimensions and density of the fibers. In the case of real scale tests, 125 ± 32 mg of microfibers per kg of washed fabric were released, with dimensions of 645 ± 408 μm in length and 18 ± 1 μm in diameter. Applying the conversion formula, this quantity corresponded to 549,913 microfibers per kg of washed fabric. In contrast, the amount released in laboratory scale washing tests was 1,733,000 microfibers per kg of washed fabric, with microfiber dimensions of 376 ± 82 μm in length and 18 ± 4 μm in diameter, corresponding to 219 mg of microfibers per kg of washed fabrics. Comparing the amount of microfibers released from the two approaches, the laboratory scale tests seem to produce more microfibers. This result is due not only to the difference between the two types of tests, but also to the different filtration procedures, because the laboratory scale test includes the smallest microfibers, which the real scale test does not collect. Furthermore, the test at the laboratory scale probably simulates more than one washing cycle in real conditions, and the more aggressive conditions are reflected also by the smaller length of the microfibers shed. Taking into account that currently a standard method for the simulation of the release of microfibers is still missing, it is possible to conclude that the laboratory scale procedure can represent a useful tool to perform low time/low cost evaluations of the microplastic release from synthetic fabrics, because the amount of water and fabric used in this procedure is very low. In conclusion, this approach could be adopted as an effective analytical procedure to perform comparative experiments on microplastic release as well as mitigation solutions.

24.3 MITIGATION ACTIONS

In 1974, a plankton survey in the North Atlantic found plastic particles, different in shape and polymer type, but most of them with dimensions less than 5 mm (Colton, Knapp, and Burns 1974). Even if it was one of the first studies that revealed the presence of microplastic particles in marine ecosystems, it already called for prevention measures to avoid this type of pollution. In fact, the study suggested the following were needed: increased efforts in plastic recycling; the development of efficient, non-atmospheric polluting incinerators to replace open dumping and sanitary landfills; increased effort in the technological development of plastic reclamation systems; and the development of water-soluble and photodegradable polymers. From this first work, microplastic pollution has reached worldwide media attention with many studies that have recorded their occurrence in different environments and interaction with aquatic fauna. The current scenario calls for the urgent development of remediation, mitigation, and prevention actions. In this respect, recently, Wu et al. (2017) have summarized possible solutions to tackle the problem of microplastic pollution, highlighting the following issues:

- Improved reuse, recycling, and recovery of plastics
- Development of cleanup and bioremediation technologies
- Improved separation efficiency at WWTPs
- Use of biodegradable materials
- Removing plastic microbeads from personal care products

Regarding this last point, legislation at the national level already has banned the use of microbeads in cosmetics and other products ("Beat the microbead" 2018). The European Commission has decided to apply strong actions against microplastic pollution. In January 2018, the "European Strategy for Plastics in a Circular Economy" was launched with the objective to change how plastic products are designed, produced, used, and recycled (European Commission 2018a). Furthermore, in May 2018, the European Commission proposed new rules to tackle ten single-use plastic products, which are the most prevalent on European beaches and in seas (European Commission 2018b). Targeted policies on the use and consumption of plastic materials, and on the industries that produce them, could contribute to speeding up the process of finding and developing solutions to plastic pollution.

For instance, a solution could be the replacement of traditional plastics with biodegradable polymers in many applications, especially for single-use plastic items. This could have beneficial effects for marine pollution, because the environmental risk depends on the concentration of the "stressor" (in this case, plastics) and on its residence time in the environment, and biodegradability reduces the residence time of the potential "stressor" (Degli Innocenti 2016). Biodegradable polymers are designed to degrade during their disposal, thanks to the enzymatic action of microorganisms (i.e., bacteria, algae, and fungi) or to non-enzymatic processes (i.e., chemical hydrolysis). Their market ranges from packaging, hygiene products, disposable non-woven materials, and agricultural tools to consumer goods (Gross and Kalra 2002). However, up to now, biodegradable polymers have been designed to degrade in soil, an environment completely different from seawater, where salt acts as a preservation agent, and with different bacteria compared to soil. More studies are needed to assess the degradation rate of biodegradable polymers in water and to develop water-degradable polymers (Miller 2013).

Also, WWTPs should have an important role as barriers for the entrance of microplastics into aquatic environments, particularly for microfibers from the washing of synthetic clothes. Talvitie, Mikola, Koistinen, and Setälä (2017) showed the efficiency in removing microplastics at WWTPs, and those that use different advanced, final-stage treatment technologies, such as rapid sand filtration, dissolved air flotation, disc filters, and membrane bioreactors, are effective. However, besides the action at the final stage of the entrance route of microplastics into the environment, it is also important to act at the very beginning of the source of pollution, for example, during the washing of synthetic garments.

Different filters for washing machines have been proposed, but they are not easy to design because they must be able to retain fibers of micro dimensions without blocking the flux of water. The Canadian company Environmental Enhancements is selling the Lint LUV-R Washing Machine Discharge Filter as a device capable of screening out synthetic microplastic particulates, but no studies on its actual efficiencies are available (Environmental Enhancements website 2019). Another filtration system for washing machines was patented by the Slovenian company Planet Care, which developed an external filtration system with a layered filter structure designed to distribute fiber capture through the entire depth of the filter, preventing clogging and prolonging the lifetime of the filter. The company is actively working with research centers to assess the effectiveness of their product, with promising preliminary results (Planet Care website 2019). Other solutions involve the use of microfiber-catching devices that are inserted into the washing machine: the Cora Ball (Cora Ball website 2019) and the Guppyfriend washing bag (Guppyfriend website 2019). The first is a ball whose design is inspired by the structure of corals, which should collect entangled fibers catching about a third of the microfibers per load. The latter is a polyamide bag that encloses the clothes to be washed, retaining the microfibers released. No information on efficiency tests on the Cora Ball are available, whereas the Guppyfriend bag has been tested by research institutes, which claim that it is able to reduce by 86% the amount of shed fibers, but no more data on the tests carried out have been disclosed.

All these type of solutions, particularly advanced WWTPs and filters for washing machines, could contribute to the overall reduction of microplastic release from the washing of synthetic clothes. But another important factor needs to be considered: the synthetic textile itself. Microfibers detach from the yarns that constitute the textile structure, because they are damaged or they just slip away from the yarn. By treatment of the textile surface, it could be possible to prevent the shedding of microfibers during washing. A first approach, based on the development of a protective coating on the surface of the fabric, was investigated by the Life+ Mermaids project. In this work, conventional textile auxiliaries of synthetic nature (i.e., acrylic resins and silicone emulsions) were applied (LIFE13 ENV/IT/001069 Mermaids 2016). Starting from the results of the Life+ Mermaids project, two alternative finishing treatments were developed at the laboratory scale (De Falco et al. 2018c; De Falco et al. 2019c): protective coatings on fabric surfaces using natural or biodegradable polymers, preserving the eco-sustainability of the process, and avoiding the introduction of other polluting agents that could jeopardize the final mitigation purpose.

The first treatment (De Falco et al. 2018c) is based on the chemical grafting of a natural polysaccharide, pectin, on the surface of polyamide 6.6 fabrics. Polyamide 6.6 is a type of polyamide made by polycondensation of two monomers each containing six carbon atoms, hexamethylenediamine, and adipic acid. Pectin is an interesting product because it is cheap and abundantly available and a waste product of fruit juice, sunflower oil, and sugar manufacturing. The treatment involved two reaction steps: chemical modification of pectin by reaction with the monomer glycidyl methacrylate and grafting of the product pectin-glycidyl methacrylate on the surface of the polyamide fabric. Different ratios and concentrations involved in the reactions were tested to find the best compromise between the formation of a homogeneous and continuous thin coating on the fibers and the preservation of the hand of the fabric. The effectiveness of the treatment was tested through washing tests at the laboratory scale, and they revealed that the pectin treatment was able to reduce the release of microplastics by more than 80%. Moreover, the treatment did

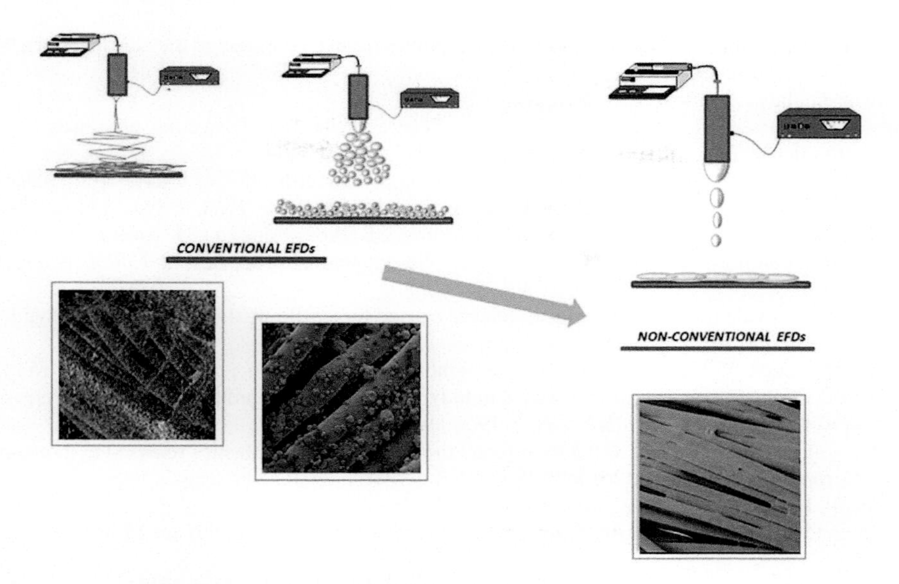

FIGURE 24.3 Example of coating deposition by EFD process. (Reprinted from *J. Colloid Interf. Sci.*, 541, De Falco, F. et al., Design of functional textile coatings via non-conventional electrofluidodynamic processes, 367–375, Copyright 2019, with permission from Elsevier.)

not alter textile properties like surface roughness and tearing strength. Such finishing treatment is compatible with padding processes commonly used in the textile industry, so its scaling up at the industrial scale is promising.

The second finishing treatment developed by De Falco et al. (2019c) was based on a non-conventional electrofluidodynamic (EFD) process, and involved the usage of biodegradable polymers, such as polylactic acid and polybutylene succinate adipate, always on polyamide 6.6 fabrics The EFD process parameters were optimized in order to deposit a uniform and continuous thin layer on the polyamide surface (De Falco et al. 2019b), without altering the key textile properties of the fabric (Figure 24.3). Washing tests on the treated fabrics revealed a reduction of microfiber release of more than 80% compared to untreated fabrics. Up to five consecutive washing cycles were performed on the treated fabrics, showing a good resistance to the washing stresses of the polylactic acid-based coating, whereas improvements must be undertaken for the polybutylene succinate adipate-based one.

In conclusion, an effective prevention of microplastic release from the washing processes of synthetic clothes can be achieved only by applying mitigation actions at different stages including textile design, finishing treatments, washing method, and the treatment at WWTPs.

REFERENCES

"Beat the microbead." Campaign. www.beatthemicrobead.org/impact/global-impact/. 2018.

Boucher, J., and D. Friot. *Primary Microplastics in the Oceans: A Global Evaluation of Sources.* Gland, Switzerland: IUCN, 2017.

Browne, M.A., et al. "Accumulations of microplastic on shorelines worldwide: Sources and sinks." *Environ. Sci. Technol.*, 45 (2011): 9175–9179.

Carney Almroth, B., L. Aström, S. Roslund, H. Petersson, M. Johansson, and N. Persson. "Quantifying shedding of synthetic fibers from textiles; a source of microplastics released into the environment." *Environ. Sci. Pollut. Res.*, 25 (2018): 1191.

Cesa, F.S., A. Turra, and J. Baruque-Ramos. "Synthetic fibers as microplastics in the marine environment: A review from textile perspective with a focus on domestic washings." *Sci. Total Environ.*, 598 (2017): 1116–1129.

Colton, J.B. Jr., F.D. Knapp, and B.R. Burns. "Plastic particles in surface waters of the Northwestern Atlantic." *Science*, 185 (1974): 491–497.

Cora Ball website. https://coraball.com/pages/faqs. 2019.

De Falco, F., E. Di Pace, M. Cocca, and M. Avella. "The contribution of washing processes of synthetic clothes to microplastic pollution." *Sci. Rep.*, 9 (2019a): 6633.

De Falco, F., G. Gentile, E. Di Pace, M. Avella, and M. Cocca. "Quantification of microfibres released during washing of synthetic clothes in real conditions and at lab scale." *Eur. Phys. J. Plus*, 133 (2018a): 257.

De Falco, F., V. Guarino, G. Gentile, M. Cocca, V. Ambrogi, L. Ambrosio, and M. Avella. "Design of functional textile coatings via non-conventional electrofluidodynamic processes." *J. Colloid Interf. Sci.*, 541 (2019b): 367–375.

De Falco, F., et al. "Evaluation of microplastic release caused by textile washing processes of synthetic fabrics." *Environ. Pollut.*, 236 (2018b): 916–925.

De Falco, F., M. Cocca, V. Guarino, G. Gentile, V. Ambrogi, L. Ambrosio, and M. Avella. "Novel finishing treatments of polyamide fabrics by electrofluidodynamic process to reduce microplastic release during washings." *Polym. Degrad. Stabil.*, 165 (2019c): 110–116.

De Falco, F., et al. "Pectin based finishing to mitigate the impact of microplastics released by polyamide fabrics." *Carbohyd. Polym.*, 198 (2018c): 175–180.

Degli Innocenti, F. *Bioplastics Magazine* 2, no. 11 (2016): 16–17.

Environmental Enhancements website. www.environmentalenhancements.com/Lint-LUV-R-solutions.html, 2019.

European Commission. http://ec.europa.eu/environment/waste/plastic_waste.htm, 2018a.

European Commission. http://europa.eu/rapid/press-release_IP-18-3927_en.htm, 2018b.

Gross, R.A., and B. Kalra. "Biodegradable polymers for the environment." *Science*, 297 (2002): 803–806.

Guppyfriend website. http://guppyfriend.com/en/faq_guppyfriend, 2019.

Habib, D., D.C. Locke, and L.J. Cannone. "Synthetic fibers as indicators of municipal sewage sludge, sludge products, and sewage treatment plant effluents." *Water Air Soil Pollut.*, 103 (1998): 1–8.

Hartline, N.L., N.J. Bruce, S.N. Karba, E.O. Ruff, S.U. Sonarand, and P.A. Holden. "Microfibre masses recovered from conventional machine washing of new or aged garments." *Environ. Sci. Technol.*, 50 (2016): 11532–11538.

Jemec, A., P. Horvat, U. Kunej, M. Bele, and A. Kržan. "Uptake and effects of microplastic textile fibers on freshwater crustacean *Daphnia magna*." *Environ. Pollut.*, 219 (2016): 201–209.

Life+ Mermaids project, LIFE13 ENV/IT/001069. "Report on localization and estimation of laundry microplastics sources and on micro and nanoplastics present in washing wastewater effluents." Deliverable A1: http://life-mermaids.eu/en/deliverables-mermaids-life-2/, 2014.

LIFE13 ENV/IT/001069 Mermaids, project. "Report of the reduction of fibres loss by the use of textiles auxiliaries." Deliverable B1: http://life-mermaids.eu/en/deliverables-mermaids-life-2/, 2016.

Magnusson, K., and C. Wahlberg. *Screening of Microplastic Particles in and Down-Stream of a Wastewater Treatment Plant.* Technical Report, Stockholm, Sweden: Swedish Environmental Research Institute, 2014.

Miller, S.A. "Sustainable polymers: Opportunities for the next decade." *ACS Macro. Lett.*, 2 (2013): 550–554.

Napper, I.E., and R.C. Thompson. "Release of synthetic microplastic plastic fibres from domestic washing machines: Effects of fabric type and washing conditions." *Mar. Pollu Bull.*, 2016: 39–45.

Pirc, U., M. Vidmar, A. Mozer, and A. Kržan. "Emissions of microplastic fibers from microfibre fleece during domestic washing." *Environ. Sci. Pollut. Res.*, 23 (2016): 22206–22211.

Planet Care website. https://planetcare.org/en/, 2019.

Rochman, C.M., et al. "Anthropogenic debris in seafood: Plastic debris and fibres from textiles in fish and bivalves sold for human consumption." *Sci. Rep.*, 5 (2015): 14340.

Sillanpää, M., and P. Sainio. "Release of polyester and cotton fibers from textiles in machine washings." *Environ. Sci. Pollut. Res.*, 24 (2017): 19313–19321.

Sutton, R., S.A. Mason, S.K. Stanek, E. Willis-Norton, I.F. Wren, and C. Box. "Microplastic contamination in San Francisco Bay, California, USA." *Mar. Pollut. Bull.*, 109 (2016): 230–235.

Talvitie, J., A. Mikola, A. Koistinen, and O. Setälä. "Solutions to microplastic pollution - Removal of microplastics from wastewater effluent with advanced wastewater treatment technologies." *Water Res.*, 123 (2017): 401–407.

Talvitie, J., A. Mikola, O. Setälä, M. Heinonen, and A. Koistinen. "How well is microlitter purified from wastewater? A detailed study on the stepwise removal of microlitter in a tertiary level wastewater treatment plant." *Water Res.*, 109 (2017): 164–172.

Taylor, M.L., C. Gwinnett, L.F. Robinson, and L.C. Woodall. "Plastic microfibre ingestion by deep-sea organisms." *Sci. Rep.*, 6 (2016): 33997.

Thompson, R.C., et al. "Lost at sea: Where is all the plastic?" *Science*, 304 (2004): 838.

UNEP. *Marine Plastic Debris and Microplastics: Global Lessons and Research to Inspire Action and Guide Policy Change.* Nairobi, Kenya: United Nations Environment Programme (UNEP), 2016.

Wu, W., J. Yang, and C.S. Criddle. "Microplastics pollution and reduction strategies." *Front. Environ. Sci. Eng.*, 11 (2017): 6.

Ziajahromi, S., P.A. Neale, L. Rintoul, and F.D.L. Leusch. "Wastewater treatment plants as a pathway for microplastics: Development of a new approach to sample wastewater-based microplastics." *Water Res.*, 112 (2017): 93–99.

Zubris, K.A.V., and B.K. Richards. "Synthetic fibers as an indicator of land application of sludge." *Environ. Pollut.*, 138 (2005): 201–211.

25 Biodegradable Bioplastics
A Silver Bullet to Plastic Pollution?

Steven Pratt, Nanthi S. Bolan, Bronwyn Laycock, Paul Lant, Emily Bryson, and Leela Dilkes-Hoffman

CONTENTS

25.1 Plastic Pollution ... 425
25.2 What Are Bioplastics? ... 426
25.3 The Bioplastics Conundrum: Volume, Compatibility, and Fate ... 427
25.3.1 Volume ... 428
25.3.2 Compatibility ... 428
25.3.3 Degradation ... 429
25.4 Where Bioplastics Make Sense ... 430
25.4.1 Biodegradable Bioplastics for Agricultural Applications ... 430
25.4.2 Biodegradable Bioplastics for Food Packaging ... 431
25.5 Conclusions ... 432
References ... 432

25.1 PLASTIC POLLUTION

Over 300 million tons per annum of plastic are produced globally, and this rate of production is increasing exponentially (PlasticsEurope 2015). Due to its durability, almost all of the plastic that has ever been produced still persists, whether it be in use, present as litter, or in landfills (Geyer et al. 2017; Thompson et al. 2009). Our understanding of the current flows of plastic, from production to final destination, is shown in Figure 25.1 (Dilkes-Hoffman et al. 2019d).

The majority of plastic ends up in landfills, but a considerable proportion also will escape into the environment, including the oceans. Leakage – and therefore plastic pollution – is both inevitable and concerning. A recent survey of the Australian public indicated that they view plastic pollution as a serious environmental problem, ranking plastic in the ocean top among environmental issues – even higher than climate change (Dilkes-Hoffman et al. 2019b); at the same time, the public are very positive about bioplastics (Dilkes-Hoffman et al. 2019c).

The question that could be posed is: are bioplastics a "silver bullet" for plastic pollution? The short answer is no. But plastic pollution is a wicked problem with no single obvious solution, and bioplastics do have a role in mitigating some of the issues with plastic pollution, as outlined in this chapter.

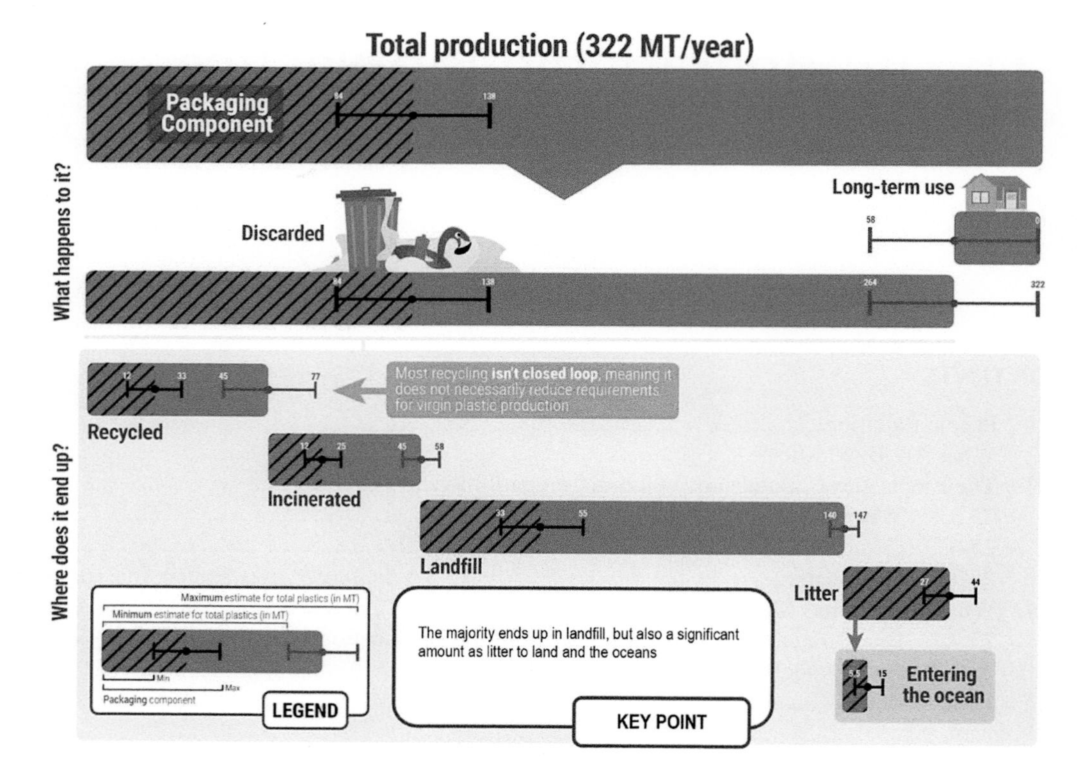

FIGURE 25.1 Predicted destination (by 2035) of all the plastics produced in 2015 – as presented in Dilkes-Hoffman et al. (2019d), and derived from Geyer at al. (2017), Jambeck et al. (2015), Hoornweg and Bhada-Tata (2012), and data from PlasticsEurope and World Economic Forum; Ellen MacArthur Foundation; McKinsey & Company. Litter includes all directly littered items as well as mismanagement and illegal dumping of waste.

25.2 WHAT ARE BIOPLASTICS?

To assess the role bioplastics could play in addressing plastic pollution, it is necessary to understand what bioplastics actually are. There is considerable ambiguity, and consequently confusion, regarding what is meant by "bioplastic." The term bioplastic refers to a broad group of plastics that are bio-based and/or biodegradable,[1] but not necessarily both (see Table 25.1). For example, it encompasses bioderived polyethylene (BioPE), which is bio-based, but not biodegradable; polycaprolactone (PCL), which is biodegradable, but not bio-based; polybutylene succinate (PBS), which is biodegradable and can be bio-based, but is currently mainly petroleum derived; and thermoplastic starch (TPS), polyhydroxyalkanoates (PHAs), and polylactic acid (PLA), which are all bio-based and biodegradable (Dilkes-Hoffman 2019d; Shen et al. 2009). This chapter focuses on the biodegradable subset of bioplastic alternatives, including TPS, PLA, and PHA, which offer significant potential for tackling long-term plastic pollution.

The certification of biodegradability is governed by standards, which stipulate that biodegradable plastics must fully break down into CO_2, water, and biomass or CH_4 and CO_2, as a result of the action of microorganisms (Rudnik 2019). While there is no universal standard for determining bioplastic biodegradation, there are only minor differences in requirements and test methods

[1] BioPE: Bio-Polyethylene (bio-based, non-biodegradable). PCL: Polycaprolactone (petroleum-based, biodegradable). PBS: Polybutylene succinate (generally petroleum-based, biodegradable). TPS: Thermoplastic starch (bio-based, biodegradable). PLA: Polylactic acid (bio-based, biodegradable). PHA: Polyhydroxyalkanoate (bio-based, biodegradable).

TABLE 25.1
Example Subsets of Bioplastics

	Fossil-Derived	Bio-Derived
Biodegradable under ambient conditions	**Polybutylene succinate (PBS)** **Polycaprolactone (PCL)** **Polybutylene adipate-co-terephthalate (PBAT)**	**Polyhydroxyalkanoate (PHA)** **Thermoplastic Starch (TPS)** **Cellophane**
Biodegradable at elevated temp. (compostable)		**Polylactic Acid (PLA)**
Non-biodegradable	**Polyethylene (PE)** **Polypropylene (PP)** **Poly(ethylene terephthalate) (PET)** **Polystyrene (PS)** **Polyvinyl chloride (PVC)**	**BioPE** **BioPP** **BioPET** **Polyethylene furanoate**

Note: Sections that fall into the bioplastics category are colored green (light gray in print version).

between regions. The most common standards used for certifying bioplastics are ASTMD6400 (United States), EN13432 (Europe), ISO17088 (International), and AS4736 (Australia) (Briassoulis et al. 2010; Gutiérrez 2018). Industry compliance with biodegradability standards is voluntary. To attain product packaging certification, assessments of bioplastic materials must be performed by independent laboratories (Narancic et al. 2018).

Some bioplastic polymers, especially PLA, require high temperatures and controlled aeration and humidity to degrade completely as intended. In practice, this means the vast majority of available bioplastics require industrial composting facilities to meet biodegradability standards (Briassoulis et al. 2010). By contrast, polymers such as PHA and starch biodegrade under ambient conditions (see Table 25.1). Bioplastics shown to degrade effectively in ambient composting environments can be labeled as "home compostable" (Narancic et al. 2018). While the test method is the same for both industrial and home compostable bioplastics, standards for home compostable products take into account lower temperatures and longer degradation times (Rudnik 2019).

25.3 THE BIOPLASTICS CONUNDRUM: VOLUME, COMPATIBILITY, AND FATE

Theoretically, bioplastic alternatives have the potential to replace most conventional plastic polymers. Shen et al. (2009) estimated that bio-based plastics could substitute for 94% of conventional plastics, with biodegradable plastics potentially replacing 31%, and bio-based, but non-biodegradable plastics, standing in for 63% of petroleum-based polymers. Biodegradable bioplastics are already used in disposable items like packaging, containers, straws, bags, and bottles. However, the widespread adoption of biodegradable bioplastics is complicated by a number of factors:

- The volume of biodegradable bioplastics currently produced represents only a very small fraction of global plastic production. This relative volume is not forecast to change any time soon because of existing investment in infrastructure for conventional plastic materials and concurrent challenges in securing feedstocks for the alternative materials.
- Biodegradable bioplastics are not readily compatible with some circular economy objectives, as they threaten existing recycling systems, and potentially even encourage leakage.
- The fate of biodegradable bioplastics in natural and engineered environments is poorly understood and potentially problematic. Methane, a potent greenhouse gas, is a product of

biodegradation in anaerobic environments (landfills). Further, some materials can retain their mechanical integrity for many years under ambient conditions, meaning little reduction in the physical and ecotoxicological risk for ecosystems.

25.3.1 Volume

The global bioplastic market is projected to grow from 2.11 Mt in 2018 to 2.62 Mt in 2023 (European Bioplastics 2018). While this seems like rapid growth, bioplastics represent only a very small portion of the plastics market (about 1%), and the subset of materials that are biodegradable is even smaller (<0.5%) (Figure 25.2; European Bioplastics 2018). Transitioning to bioderived and biodegradable bioplastics would require enormous capital investment and the sourcing of hundreds of megatons of biomass feedstocks. Considering that many bio-based, biodegradable plastics are currently produced from plants, such as sugar cane, wheat, and corn, rapid development in this space would present food-versus-biomaterials-versus-biofuel issues. The Institute for Bioplastics and Biocomposites (2017) notes that bioplastic production already accounts for 1% of arable land. Switching fully to biodegradable bioplastics would not be possible without serious disruption to land use.

25.3.2 Compatibility

Enthusiasm for a "circular economy" is increasing, which the Ellen MacArthur Foundation defines as "decoupling economic activity from the consumption of finite resources, and designing waste out of the system" (World Economic Forum et al. 2016). This concept suggests we should aspire

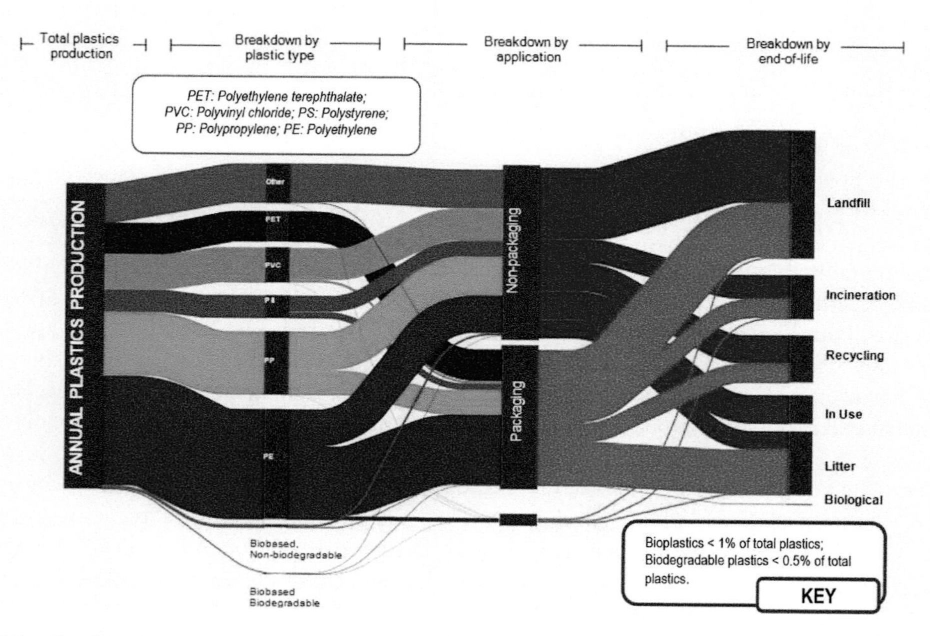

FIGURE 25.2 Predicted destination of 2015 plastics by 2035. (Modified from Dilkes-Hoffman, L.S. et al., *Resour. Conserv. Recycl.*, 151, 104479, 2019c.) Polymer type breakdown based on PlasticsEurope (2017) and European Bioplastics (2018); End-of-life destinations based on references presented in Figure 25.1.

to design products to be reusable as opposed to disposable. Sheehan (2017) argues that in almost all cases, only a small number of reuses are required before the environmental "break even" point is reached. When considering longer timeframes in this way, reusable products will generally have reduced environmental impacts when compared to disposable products. Biodegradable bioplastics, as a "silver bullet" for plastic pollution, are not compatible with this agenda, as they are normally being designed from the outset to be disposed of after one use.

There are further concerns that biodegradable bioplastics disrupt mechanical recycling systems. Thermoplastic starch, PHA, and PLA polymers are all recyclable, and recycling of these materials has been demonstrated with material properties being retained for up to 5–10 cycles (La Mantia et al. 2002 for starch; Shah et al. 2012 for PHA; Żenkiewiczl et al. 2009 for PLA). This is competitive with the recycling lifetimes of conventional plastics. However, the concern is that conventional recycling systems are designed specifically for the commodity plastics. Introducing small volumes of new materials into waste streams would require additional costs for sorting, which could otherwise lead to contamination issues (Cornell 2007).

25.3.3 Degradation

A critical question when considering the use of biodegradable bioplastics is what happens to polymers in the environment. Understanding the fate of these materials starts with understanding the mechanisms through which biodegradation can occur. There are many abiotic and biotic processes that contribute to biodegradation through a complex interplay (as reviewed by Laycock et al. 2017). Ultimately, biopolymer degradation occurs by chain cleavage through hydrolysis (both enzyme-promoted and abiotic), with biological activity then converting the carbon in the lower molecular weight fragments to CO_2 in aerobic environments and CH_4 and CO_2 in anaerobic environments. The rate of water penetration is a controlling factor. When the rate of water penetration is high and exceeds the rate of surface hydrolysis (either biotic or abiotic), then degradation is defined as bulk erosion; when the rate of water penetration is relatively low relative to the rate of chain scission, then degradation is defined as surface degradation (Laycock et al. 2017).

TPS is water soluble, and readily (bio)degradable (Du et al. 2008). But the stories for PLA and PHA are not so clear. PLA is marketed as a biodegradable polymer (e.g., Corbion 2019), but hydrolysis, and consequently biodegradation, of the synthetic material is severely restricted at room temperature, at least initially before autocatalytic depolymerization accelerates the rate of polymer chain hydrolysis. Rudnik and Briassoulis (2011) showed that PLA buried under ambient conditions showed no sign of degradation even after a year. Mild thermal treatment is necessary to stimulate chemical hydrolysis. For example, Ho et al. (1999) analyzed molecular weight reduction of thin films at 25°C, 40°C, and 55°C and showed that the rate of decline at 55°C is 20 times higher than at room temperature. Thus, composting at 55°C has been shown to be effective for PLA biodegradation. Significant for this analysis is that PLA's glass transition temperature is 50°C–60°C.

The effect of biological activity in instigating degradation of PLA is still not clear. It has been reported that microorganisms do not enhance PLA degradation and that polymer cleavage proceeds solely through abiotic hydrolysis of ester linkages in the presence or absence of microorganisms (Agarwal et al. 1998). Other studies provide evidence that degradation of PLA is accelerated in non-sterile conditions (Karamanlioglu and Robson 2013; Karamanlioglu et al. 2017). Either way, PLA products are not readily (bio)degradable under ambient conditions, meaning that they must be collected and composted in well-managed facilities. Any PLA that leaks from the system will persist for a long time in the environment (Karamanlioglu and Robson 2013).

PHA is known to be biodegradable in natural environments, both on land and at sea. It is marketed as "marine degradable" (e.g., Full Cycle Bioplastics 2019). It is biodegradable in natural (ambient) conditions, as microorganisms facilitate enzymatic attack of the PHA polymer. But the material is hydrophobic, resisting water penetration, meaning that the biodegradation of PHA primarily occurs, at least initially, through surface erosion, where microbes and their enzymes can

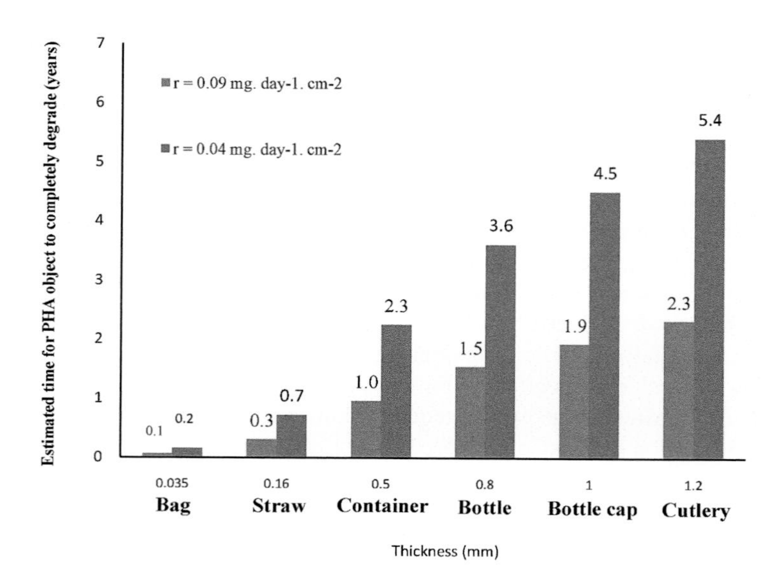

FIGURE 25.3 How long will it take a PHA item to degrade? Lifetime values estimated using the 95% confidence interval for the mean of the rate of biodegradation of PHA in the marine environment. (Adapted from Dilkes-Hoffman, L. et al., *Mar. Pollut. Bull.*, 142, 15–24, 2019a.)

access the material. However, adsorption of dissolved organic carbon by PHA leads to biofilm formation, thereby facilitating subsequent biodegradation (Morohoshi et al. 2018). Therefore, the rate of mass loss of PHA is related to the accessible surface area. This means that the lifetime of PHA products is strongly a function of their morphology; bottles, caps, and cutlery (which rank high on lists of marine debris, Dilkes-Hoffman et al. 2019d and Andrady 2015) would persist in the environment for years before being completely mineralized (Figure 25.3).

The threats associated with all these biodegradable plastics to natural ecosystems should not be ignored. PLA products, and even some bulky PHA products, will retain their mechanical integrity for years under natural conditions, potentially causing physical harm if they are then ingested by marine or terrestrial animals. Their hydrophobic nature means that they could concentrate organotoxins over their lifetime, similar to polyethylene (Rios et al. 2007), posing a risk to local ecosystems and global food chains. Still, we can be confident that the half-life PHA, at least in the environment, is orders of magnitude less than that of fossil fuel-based plastics. The products of degradation are also benign (i.e., CO_2 and H_2O) rather than conventional plastic microfragments, particularly residual cross-linked microfragments.

25.4 WHERE BIOPLASTICS MAKE SENSE

The majority of plastic ends up in landfills, but also a considerable proportion will escape into the environment, including the oceans. Even in countries with well organized waste management, leakage to the environment is currently inevitable. This section will present two areas where moving to biodegradable bioplastics makes sense: agricultural applications and disposable food packaging.

25.4.1 Biodegradable Bioplastics for Agricultural Applications

The agricultural use of plastic is increasing, with approximately 5 Mt/year of plastic being applied primarily as mulch film, but also as sheets, rods, tubing, and transplanting pots (Malinconico 2017;

Razza and Innocenti 2012). In China alone, the use of plastic mulch film increased from about 0.3 Mt in 1991 to over 1.2 Mt in 2011 (Liu et al. 2014). Mulching offers significant agronomic benefits, including increased yield, water retention, and weed control. Despite the many advantages of using plastic mulch for crop production, appropriate disposal after use can be a complicated challenge. In places like China where ultra-thin films are used (Liu et al. 2014), the rate of collection and recovery is especially low. The mechanical cultivation of crops makes collecting and recycling the thin films non-viable, and they become lost in the environment (Malinconico 2017).

Bioplastic agrichemical products (i.e., plastic matrices) can also be used on farms to provide controlled release of fertilizers, such as urea to soils (Timilsena et al. 2014). Again, the agronomic benefits can be significant. But the distribution across farmland and the microplastic nature of the residue make it impossible to collect and properly manage the plastic waste. As with mulch films, residues will leak into terrestrial and, potentially, marine environments.

Bioplastic alternatives could have significant merit in agricultural applications. Biodegradable mulch films and encapsulants or coatings for fertilizers and other agrichemicals would not have to be removed from the soil and disposed of, thereby simplifying operations. However, for bioplastic agricultural mulch films and agrochemical formulations to be superior to conventional products, biopolymers must be capable of degrading in on-farm soil environments. Products made from polymers incompatible with on-farm disposal conditions (e.g., non-industrial composting or in-soil biodegradation) and/or exhibiting ecotoxic effects on crops are not necessarily preferable to conventional agricultural plastics (Hottle et al. 2013).

Starch-based mulch films have been shown to perform similarly to conventional plastic in terms of agronomic benefit (Razza and Innocenti 2012), although control of lifetimes and mechanical properties can be a challenge. By contrast, Rudnik and Briassoulis (2011) found that PLA agricultural bioplastic films did not sufficiently degrade in on-farm soil or composting conditions. In comparison, PHA films performed well under these ambient environments and, therefore, the PHA polymer is more appropriate for application in agriculture (Rudnik and Briassoulis 2011). Fortunately, the development of PHA biodegradable encapsulants for agrichemicals is well underway (Levett et al. 2019).

25.4.2 Biodegradable Bioplastics for Food Packaging

Over 40% of plastic production is directed toward packaging materials, with the largest application being food packaging (Halley and Avérous 2014; PlasticsEurope 2017) (Figure 25.2). In fact, food packaging represents the largest and most rapidly growing sector of the plastics industry due to an increased desire for packed fresh goods and ready-made meals (World Packaging Organisation 2008). It is also an important area to focus on when considering plastic pollution reduction. Five of the top ten categories of items found in coastal cleanups are plastic food packaging or single-use plastic items associated with food (Andrady 2015).

It is estimated that up to 30% of plastic packaging will never be eligible for recycling without fundamental redesign (World Economic Forum et al. 2016). Materials that are included within this 30% are organically soiled packaging, multilayer packaging, items such as small sachets that are physically or impractically recyclable, and other food packaging-related materials (Barlow and Morgan 2013; Gross and Kalra 2002; World Economic Forum et al. 2016). Redesigning and replacing plastic food packaging materials with bioplastic alternatives is a complex process due to specific packaging requirements, but doing so can contribute to waste reduction (Williams and Wikstrom 2011).

Multilayered films play an important role in reducing food wastage, because they have high gas and water barrier properties; but they are hard to recycle (Barlow and Morgan 2013). There are three key ways in which redesigning of multilayered films could be achieved in an attempt to retain their desired properties, but improve their waste management options. The first is through the invention of monolayer materials that have comparable barrier properties, but would be fully recyclable; the second is through recycling innovation to develop chemical recycling technologies that can separate out the multilayers that then can be mechanically recycled; and the third is to shift focus from recycling toward organic waste processing and produce biodegradable, multilayered materials. The idea of producing biodegradable, multilayered food packaging has been considered both in the literature (Dilkes-Hoffman et al. 2018), as well as by several companies (Elk Packaging 2018; Tipa Compostable Packaging 2019).

Another example where biodegradable plastic materials are being considered is for sachet alternatives. In particular, relatively new seaweed-based polymers are being trialed for single-serve sachets of sauces and powders (Evoware 2019; Notpla 2019).

25.5 CONCLUSIONS

Overall, biodegradable and bioderived plastics could have significant merit as we move toward a more sustainable plastics economy, particularly in specialty applications, such as agricultural applications and in disposable food packaging. While bioplastic products are generally not designed for reuse, they can be tailored for such applications; even in biodegradable applications, they can contribute to a circular economy by continually recycling organic materials through production, use, and waste streams.

However, they do not, of their own accord, represent a silver bullet for solving the plastics pollution problem. There are significant challenges in introducing bioplastics into broader commodity use, particularly with the need to consider the source and separation of this new stream of materials. It is also important to understand the full life cycle implications of the manufacture and use of such materials, when considering their use as replacements for fossil-derived plastics. Further, despite the fact that materials such as starch and PHA do biodegrade under ambient environmental conditions, the rate of such biodegradation is still unclear and dependent on their form and fate – for example, the potential for such materials to have a physical impact if ingested. This impact and others, such as through sorption of organo-toxins from the environment while still intact, need to be assessed. In other words, plastics pollution is a wicked problem and bioderived, biodegradable plastics will likely play a role, potentially a significant one, in solving this problem. But the path forward needs to be well considered with good underpinning techno-economic, social, and environmental cost-benefit analyses.

REFERENCES

Agarwal, M., K. W. Koelling, and J. J. Chalmers. 1998. "Characterization of the degradation of polylactic acid polymer in a solid substrate environment." *Biotechnology Progress* 14 (3):517–526.

Andrady, A. L. 2015. "Persistence of plastic litter in the oceans." In *Marine Anthropogenic Litter*, edited by M. Bergmann, L. Gutow, and M. Klages, 57–72. Cham, Switzerland: Springer International Publishing.

Barlow, C., and D. Morgan 2013. "Polymer film packaging for food: An environmental assessment." *Resources, Conservation and Recycling* 78:74–80.

Briassoulis, D., C. Dejean, and P. Picuno. 2010. "Critical review of norms and standards for biodegradable agricultural plastics part II: Composting." *Journal of Polymers and the Environment* 18 (3):364–383.

Corbion. 2019. PLA bioplastics: A significant contributor to global sustainability. https://www.corbion.com/media/75646/corbion_bioplastics_brochure.pdf. (Accessed June 26, 2019.)

Cornell, D. D. 2007. "Biopolymers in the existing postconsumer plastics recycling stream." *Journal of Polymers and the Environment* 15 (4):295–299.

Dilkes-Hoffman, L. S., P. A. Lant, B. Laycock, and S. Pratt. 2019a. "The rate of biodegradation of PHA bioplastics in the marine environment: A meta-study." *Marine Pollution Bulletin* 142:15–24.

Dilkes-Hoffman, L. S., S. Pratt, B. Laycock, P. Ashworth, and P. A. Lant. 2019b. "Public attitudes towards plastics." *Resources, Conservation and Recycling* 147:227–235.

Dilkes-Hoffman, L. S., S. Pratt, B. Laycock, P. Ashworth, and P. A. Lant. 2019c. "Public attitudes towards bioplastics—Knowledge, perception and end-of-life management." *Resources, Conservation and Recycling* 151:104479.